FROM ANIMALS TO ANIMATS 7

Complex Adaptive Systems
John H. Holland, Christopher G. Langton, and Stewart W. Wilson, advisors

Adaptation in Natural and Artificial Systems: An Introductory Analysis with Applications to Biology, Control, and Artificial Intelligence, John H. Holland

Genetic Programming: On the Programming of Computers by Means of Natural Selection, John R. Koza

Intelligent Behavior in Animals and Robots, David McFarland and Thomas Bösser

Genetic Programming II: Automatic Discovery of Reusable Programs, John R. Koza

Turtles, Termites, and Traffic Jams: Explorations in Massively Parallel Microworlds, Mitchel Resnick

Comparative Approaches to Cognitive Science, edited by Herbert L. Roitblat and Jean-Arcady Meyer

Artificial Life: An Overview, edited by Christopher G. Langton

An Introduction to Genetic Algorithms, Melanie Mitchell

Catching Ourselves in the Act: Situated Activity, Interactive Emergence, and Human Thought, Horst Hendriks-Jansen

Elements of Artificial Neural Networks, Kishan Mehrotra, Chilukuri K. Mohan, and Sanjay Ranka

Growing Artificial Societies: Social Science from the Bottom Up, Joshua M. Epstein and Robert Axtell

An Introduction to Natural Computation, Dana H. Ballard

An Introduction to Fuzzy Sets: Analysis and Design, Witold Pedrycz and Fernando Gomide

From Animals to Animats 5: Proceedings of the Fifth International Conference on Simulation of Adaptive Behavior, edited by Rolf Pfeifer, Bruce Blumberg, Jean-Arcady Meyer, and Stewart W. Wilson

Artificial Life VI: Proceedings of the Sixth International Conference, edited by Christoph Adami, Richard K. Belew, Hiroaki Kitano, and Charles E. Taylor

The Simple Genetic Algorithm: Foundations and Theory, Michael D. Vose

Advances in Genetic Programming: Volume 3, edited by Lee Spector, William B. Langdon, Una-May O'Reilly, and Peter J. Angeline

Toward a Science of Consciousness III: The Third Tucson Discussions and Debates, edited by Stuart R. Hameroff, Alfred W. Kasniak, and David J. Chalmers

Truth from Trash: How Learning Makes Sense, Chris Thornton

From Animals to Animats 6: Proceedings of the Sixth International Conference on Simulation of Adaptive Behavior, edited by Jean-Arcady Meyer, Alain Berthoz, Dario Floreano, Herbert Roitblat, and Stewart W. Wilson

Learning and Soft Computing: With Support Vector Machines, Neural Networks, and Fuzzy Logic Models, Vojislav Kecman

From Animals to Animats 7: Proceedings of the Seventh International Conference on Simulation of Adaptive Behavior, edited by Bridget Hallam, Dario Floreano, Gillian Hayes, Jean-Arcady Meyer, and John Hallam

FROM ANIMALS TO ANIMATS 7

Proceedings of the Seventh International Conference
on Simulation of Adaptive Behavior

edited by
Bridget Hallam, Dario Floreano, Gillian Hayes, Jean-Arcady Meyer, and John Hallam

A Bradford Book

The MIT Press
Cambridge, Massachusetts
London, England

Printed and bound in the United States of America.

This publication may be ordered from The MIT Press, 5 Cambridge Center, Cambridge, Massachusetts 02142 USA; by telephone: 1-800-356-0343 (toll-free) or 617-625-8569; by fax: 617-625-6660; by e-mail: mitpress-orders@mit.edu or http:/ / mitpress.mit.edu.

ISSN 1089-4365
ISBN 0-262-58217-1

CONTENTS

EVOLUTION

COLLECTIVE AND SOCIAL BEHAVIOR

PREFACE

The papers in this book were presented at the seventh international Conference on the Simulation of Adaptive Behavior (SAB'02), held at the University of Edinburgh from August 5-9 2002. The objective of the biennial SAB conference is to bring together researchers from a wide range of backgrounds including ethology, psychology, ecology, artificial intelligence, artificial life, robotics, computer science, engineering, and related fields in order to further our understanding of the behaviors and underlying mechanisms that allow adaptation and survival in uncertain environments.

Adaptive behavior research is distinguished by its focus on the modeling and creation of WORKING animal-like systems, which we believe to be one of the best routes to understanding intelligence in natural and artificial systems. The field received initial recognition on the occasion of the first SAB conference, which was held in Paris in September 1990. Subsequent SAB conferences in Honolulu (1992), Brighton (1994), Cape Cod (1996), Zürich (1998) and Paris (2000) drew increasing numbers of papers and participants. In 1992, the MIT Press introduced the quarterly journal Adaptive Behavior. The establishment of the International Society for Adaptive Behavior (ISAB) in 1995 further underlined the emergence of adaptive behavior as a fully-fledged scientific discipline. The present proceedings are a comprehensive and up-to-date resource of the latest progress in this exciting field.

The 33 papers and 36 poster summaries published here were selected from 137 submissions after a two-pass review process designed to ensure high and consistent overall quality. The authors focus on robotic and computational experimentation with well-defined models that help to characterize and compare various organizational principles or architectures underlying adaptive behavior in both natural animals and synthetic animats. The papers are ordered according to the scale at which adaptive behavior takes place, ranging from immediate adaptation in sensorimotor control, to learning within an animat's lifetime, to adaptive behavior exhibited by successive generations of animats, and finally to adaptive collective behavior of animats in groups.

The conference and its proceedings could not have happened without the substantial help of a wide range of people. First and foremost, we would like to thank members of the Program Committee who thoughtfully reviewed all the submissions we received.

We are indebted to all our sponsors. The ones committed before these Proceedings went to press were:

International Society for Adaptive Behavior
Syddansk University
Laboratoire d'Informatique de Paris 6
Ecole Polytechnique Fédérale de Lausanne
Division of Informatics, University of Edinburgh
Mobile Robot Research Group, University of Edinburgh
International Society for Artificial Life
Society for the Study of Artificial Intelligence and the Simulation of Behaviour
Applied AI Systems, Inc.
Cyberbotics
Hewlett Packard Research Labs
K-team

The enthusiasm and hard work of numerous individuals was essential to the conference's success. Above all, we would like to acknowledge the significant contributions of Deirdre Burke, Pete Ottery, and members of the Mobile Robot Research Group to ensuring local arrangements went smoothly.

Finally, we are greatly indebted to Jean Solé for the artistic conception of the SAB'02 poster and the proceedings cover.

We invite readers to enjoy and profit from the papers in this book, and look forward to the next SAB conference in 2004.

Bridget Hallam, Dario Floreano, John Hallam, Gillian Hayes, and Jean-Arcady Meyer.

Programme Committee

Ronald Arkin, Georgia Tech, College of Computing, USA
Christian Balkenius, Lund University, Lund, Sweden
Mark Bedau, Reed College, Portland, Oregon, USA
Aude Billard, CS Dept, University of Southern California, USA
Andrea Bonarini, Politecnico di MIlano - DEI, Milan, Italy
Lashon Booker, The MITRE Corporation, USA
Rodney Brooks, MIT Artificial Intelligence Lab, USA
Joanna Bryson, Harvard Primate Cognitive Neuroscience, USA
Larry Bull, University of the West of England, UK
Seth Bullock, University of Leeds, Leeds, UK
Jose Carmena, IPAB, University of Edinburgh, UK
Philippe Codognet, Université Paris VI - Inria, Paris, France
Thomas Collett, University of Sussex, UK
Holk Cruse, Biology, University of Bielefeld, Germany
Kerstin Dautenhahn, University of Hertfordshire, UK
Marco Dorigo, IRIDIA - Université Libre de Bruxelles, Belgium
Alexis Drogoul, Université Paris 6 / LIP6, France
Yiannis Demiris, Imperial College, University of London, UK
Ezequiel Di Paolo, University of Sussex/COGS, UK
Michael Dyer, University of California, Los Angeles, USA
Dario Floreano, Institut de Systemes Robotiques, EPFL, Switzerland
Philippe Gaussier, Neurocybernetics, Cergy-Pontoise University, France
Stephen Grossberg, Boston University, Boston, USA
Agnès Guillot, Université Paris X, Paris, France
Bridget Hallam, Biology Institute, Syddansk University, Denmark
John Hallam, University of Edinburgh, UK
Inman Harvey, University of Sussex, Brighton, UK
Gillian Hayes, IPAB, University of Edinburgh, UK
Phil Husbands, University of Sussex, Brighton, UK
Auke Jan Ijspeert, University of Southern California, Los Angeles, USA
Yasuo Kuniyoshi, University of Tokyo, Japan
Pier Luca Lanzi, Politecnico di Milano, Milano, Italy
Henrik Hautop Lund, University of Southern Denmark, Odense, Denmark
Hanspeter Mallot, Max-Planck Institute for Biological Cybernetics, Tübingen, Germany
Chris Melhuish, University of the West of England, Bristol, UK
Jean-Arcady Meyer, Université Paris VI, Paris, France
Francesco Mondada, Swiss Federal Inst. of Technology, EPFL, Lausanne
Takayuki Nakamura, Robotics Lab, Nara Institute of Science and Technology, Japan
Ryohei Nakano, Nagoya Institute of Technology, Japan
Ulrich Nehmzow, University of Essex, Colchester, UK
Jason Noble, University of Leeds, Leeds, UK
Stefano Nolfi, Institute of Psychology, Rome, Italy
Domenico Parisi, Institute of Psychology, Rome, Italy
Rolf Pfeifer, AI lab, University of Zürich, Switzerland
Tony Prescott, University of Sheffield, UK
Herbert L. Roitblat, Dept of Psychology, University of Hawaii, USA
Olivier Sigaud, Université Paris VI, Paris, France

Olaf Sporns, Indiana University, Bloomington, USA
Luc Steels, Sony Computer Science Laboratory, Paris, France
Wolfgang Stolzmann, Daimler-Chrysler, Research & Technology, Germany
Jun Tani, Brain Science Institute, RIKEN, Tokyo, Japan
Tim Taylor, IPAB, University of Edinburgh, UK
Peter Todd, Max Planck Institute for Human Development, Berlin, Germany
Barbara Webb, University of Stirling, Stirling, UK
Myra S. Wilson, University of Wales, Aberystwyth, UK
Stewart W. Wilson, Prediction Dynamics, USA
Brian Yamauchi, iRobot Corporation, USA
Tom Ziemke, University of Skovde, Sweden

THE ANIMAT APPROACH TO ADAPTIVE BEHAVIOR

Constructing complex minds through multiple authors

Mark Humphrys[1,3]
[1]School of Computer Applications,
Dublin City University,
Glasnevin, Dublin 9, Ireland
www.compapp.dcu.ie/~humphrys

Ciarán O'Leary[2,3]
[2]School of Computing,
Dublin Institute of Technology,
Kevin St, Dublin 8, Ireland
comp.dit.ie/coleary

[3]The World-Wide-Mind project
w2mind.org

Abstract

The World-Wide-Mind (WWM) was introduced in [Humphrys, 2001]. For a short introduction see [Humphrys, 2001a]. Briefly, this is a scheme for putting animat "minds" online (as WWM "servers") so that large complex minds may be constructed from many remote components. The aim is to address the *scaling up* of animat research, or how to construct minds more complex than could be written by one author (or one research group).

The first part of this paper describes how a number of existing animat architectures could be implemented as WWM servers. Any *unified* mind can easily map to a *single* WWM server. So most of the discussion here is on *action selection* (or behavior or goal selection), where each module could be a different WWM server (written by a different author).

The second part of this paper describes the first implementation of WWM servers and clients, and explains in particular how to write a WWM server. Most animats researchers are programmers but not network programmers. Almost all protocols for remote services (CORBA, SOAP, etc.) assume the programmer is a networks specialist. This paper rejects these solutions, and shows how any animats researcher can put their animat "mind" or "world" online as a server by simply converting it into a command-line program that reads standard input and writes to standard output.

1 Introduction

1.1 The AI problem

"AI" refers here to all artificial modelling of life, animals and humans. In the sense in which we use it, classic symbolic AI, sub-symbolic AI, Animats, Agents and ALife are all subfields of "AI".

It is generally agreed that the AI problem is much harder than researchers used to think (though it is not clearly understood *why*). Early optimism has given way to a sober respect for the magnitude of the problem, and a number of approaches have evolved:

1. The standard AI approach has been to work on *subsections* of the postulated mind, such as computer vision, or language processing. The criticism of this approach [Brooks, 1986, Brooks, 1991] is that the whole mind never actually gets built [Nilsson, 1995].
2. The *Animats* approach [Wilson, 1990] is to start with *simple whole creatures* [Dennett, 1978] and work up gradually to more complex whole creatures.
3. The *evolutionary* approach is to say that control systems are too hard to design and must be evolved [Harvey et al., 1992]. In practice this has usually taken the animat approach of starting with simple whole creatures.

It may be time to ask questions about how the animats and evolutionary approaches scale up. Both seem to share an implicit assumption that *one lab can do it all.* As a result, the complexity of the minds produced is limited to the complexity that can be grasped by a single research team (or even a single individual). Perhaps the Cog project [Brooks, 1997, Brooks et al., 1998] is beginning to hit the limits of what a single coherent team can understand.

1.2 Constructing complex minds through multiple authors

What is the alternative? The alternative is to *link the work of multiple laboratories* - to construct minds out of sub-minds written by many diverse authors. This is done of course already within research groups, but we propose a public system, where perhaps hundreds of diverse authors may contribute parts to a large, complex mind. Other researchers may specialise *entirely* on just finding different ways of combining other people's components. No one individual need understand the entire mind.

This is, of course, the problem of *building whole minds out of specialist components* that standard AI never solved, and we suggest why. We argue that only by using a public, open system on the Internet as the infrastructure on which to build the mind can this problem be solved. We do not suggest the abolition of the animats approach, but rather modifying it to building *simple whole creatures* out of components written by multiple authors, and scale up to building *complex whole creatures* out of components written by multiple authors.

1.3 "Not enough stuff" in AI

AI has a history of people saying that the systems simply need *more* of some property and there will be a breakthrough:

1. **Not enough speed.** - "If we just get faster machines we will have full AI". This thinking is common when AI is viewed primarily as a *search* problem - and there is obviously a lot of truth in it (see AI's recent triumph at chess [McDermott, 1997]). This thinking has also been revived with the discussion of hypothetical super-powerful quantum computers that might be able to solve NP-complete problems in polynomial time. Some writing on this (e.g. [Berger, 1998]) suggests that if such super-powerful computers can be made, then AI will be solved.
2. **Not enough neurons/memory.** - "Brain-building" [de Garis, 1996] will, it is claimed, lead to breakthroughs once enough neurons are put together.
3. **Not enough data.** - The CYC project [cyc.com], and the online projects MindPixel [mindpixel.com] and Open Mind [openmind.org], take the approach that what is needed is to build up vast rule sets.

Here we suggest another one:

1. **Not enough authors.** - Not enough diversity in the mind.

The criticism of "not enough stuff" arguments is that *theoretical* breakthroughs are needed. However, if you look at the brain, it *is* true that it has more "stuff" - i.e. neurons, connections, memories, experience, multiple learning algorithms, multiple specialised structures - than artificial models. The brain is much bigger and richer than any machine we have built so far (perhaps even bigger and richer than *all* of our machines plus the Internet). So the "not enough stuff" argument must be true on one level (while not disagreeing with the idea that it is *also* true that we will need theoretical breakthroughs).

The "not enough authors" approach is the one that has not yet been *tried*. Many researchers have emphasised the vast and *heterogenous* nature of the mind, notably Minsky [Minsky, 1986, Minsky, 1991]. In the Animats world it is at least accepted that complex minds will have "Action Selection" among competing sub-minds. So far, the case has been made for heterogenous minds, but no one has shown how to build *really* heterogenous minds.

2 The World-Wide-Mind (WWM)

The World-Wide-Mind (WWM) was introduced in [Humphrys, 2001]. For a short introduction see [Humphrys, 2001a]. Briefly, it is proposed that researchers construct their animat "minds" and "worlds" as *servers* on the Internet. Each WWM "server" is a program residing on a normal Web server. There are two basic types of server:

1. A **World** and Body server. This server can be queried for the current state of the world x as detected by the body, and can be sent actions a for the body to perform in the world. For example, one could put Tyrrell's action-selection world [Tyrrell, 1993] online as a server, and then people could write minds to drive the agent in it.
2. A **Mind** server, which is a behavior-producing system, capable of suggesting an action a given a particular input state x. This does not mean it is stimulus-response. It may remember *all* previous states. It may take actions independent of the current state. It may work by any AI methodology or algorithm and may contain within itself any degree of complexity.

A *user* "runs" a Mind server in a World server, using some dedicated *client software*. The user is typically remote from both Mind and World, and starts the client by giving it the (remote) World URL and Mind URL. The client then (repeatedly) queries the World for state, passes this to the Mind to get a suggested action, sends this to the World for execution, and so on.

2.1 How to construct a Society of Mind

Even in its basic form above, the scheme would allow remote re-use of other people's minds and worlds, something for which there is no easy scheme at present. But we may also consider using someone else's mind as merely a *component* in a larger mind. At the top level, there is always exactly one Mind server running in exactly one World server. But that Mind server may *itself* be calling many other Mind servers in order to select its action. We define the following types of Mind servers:

1. $Mind_M$ - a Mind server that calls other Mind servers.

2. $Mind_{AS}$ (or Action Selection or AS server) - a $Mind_M$ server that resolves competition among multiple Mind servers. Each Mind server i suggests an action a_i to execute. The AS server queries them and somehow produces a winning action a_k. To the outside world, the AS server looks like just another Mind server, producing action a given state x.
3. $Mind_i$ - a Mind server that accepts it may not be the only mind in the body (and may support additional queries to *cooperate* with the Action Selection).
4. $Mind_{Feu}$ - a Mind server that accepts *Feudal* commands of the form: "Take me to state c", rather than just commands of the form "What do you want to do now?".
5. $Mind_L$ - a Mind server that learns (and may support additional queries turning on and off learning).

Societies of mind may be incrementally built up using the servers that are online, for example:

1. 1st party makes World.
2. 2nd party makes Mind for World.
3. 3rd party makes $Mind_M$ which in state x does something, otherwise does what 2nd party Mind does.
4. 4th party makes different Mind for World.
5. 5th party makes $Mind_M$ which in state y does what 4th party Mind does otherwise does what 3rd party $Mind_M$ does. And so on, with people modifying and cautiously overriding what already works.

2.2 What is the definition of x and a?

Clearly, we *cannot* define a universal format for state x or action a that works across *all* worlds. We have no choice but to allow each World server define its own format of what the agent sees [e.g. gridworld state, 3D world description, neural network input vector, robotic sensory inputs] and what actions it can execute. Each Mind server that wants to run in that World will have to understand the format used.

What may prevent complete chaos, however, is first, that popular worlds will serve as benchmarks for testing. "Islands" of compatible worlds and minds may develop around each popular basic problem. Also, we *can* define a *client* that will work with all World servers and Mind servers. And finally, it *may* be the case that Minds can be written that will run in all worlds, or at least in a lot of quite different worlds. For example, one could write a *generic* Q-learning Mind server [Watkins, 1989]. When set to run in a new World server, it queries the world to learn that it has a finite number of states, numbered state 1 to state n, and a finite number of actions, action 1 to action m, and the World server will occasionally generate a numeric reward after an action has been taken. The Q-learning Mind server can then attempt to learn a policy without knowing anything more about what the world represents or what the problem is.

2.3 Testing and methodological issues in AI

Once systems are brought online in an open, public way, we can see that this could address some general methodological issues in the animats field (and other AI fields):

1. **Lack of re-use** - Sharing work has been so difficult that researchers tend to build their own animat minds and worlds from scratch, often duplicating work that has been done elsewhere. There have been a number of attempts to re-use animat or agent minds [Sloman and Logan, 1999, Sutton and Santamaria] and worlds [Daniels, 1999], but the model of re-use often requires installation, or even a particular programming language.
2. **Taking results on trust** - Often, the only person who *ever* does experiments with an animat or agent is its author. In this field it has become acceptable not to have direct access to many of the major systems under discussion. How many action selection researchers have ever *seen* Tyrrell's world running [Tyrrell, 1993] for example? This lack of direct experience is even greater when it comes to other researchers' *robotic* projects. We accept that we will never experiment with many of the systems under discussion ourselves, but only read papers on them.
3. **Lack of re-testing** - Lack of re-use has serious consequences for *scientific progress* in the sense of being able to *repeat* experiments and being able to *prove* that one system is better than another. [Bryson, 2000] points out that, essentially, no one uses each other's agent architectures, because they are not convinced by each other's tests. [Guillot and Meyer, 2000] make the same point about the animats field - that the number of architectures has grown faster than the number of comparisons. Having systems publicly available for indefinite re-testing by 3rd parties is, we argue, the only solution to this.

3 How to express animat architectures as networks of WWM servers

We now discuss how a number of existing animat architectures might be expressed as networks of WWM servers using simple remote queries. An attempt to define the full set of WWM server queries is in [Humphrys, 2001].

3.1 A single (sub-symbolic or symbolic) Mind server

A hand-coded mind program can clearly be implemented as a single Mind server, receiving x and returning a.

There are a vast number of models of animat or agent mind, whether hand-coded, learnt or evolved, symbolic or non-symbolic, that could be implemented as a single WWM server without raising any particular issues.

3.2 Multiple sub-symbolic Mind servers

The difficulty arises when we consider *competition* between multiple Minds. Any *unified* mind (such as a single learner trained to solve a task) can easily map to a *single* WWM server. So most of the discussion here is on *action selection* (or behavior or goal selection) among competing modules, where each module could be a different remote WWM server (written by a different remote author). Many of the Action Selection methods discussed below are surveyed in detail in [Humphrys, 1997], or, for a brief introduction, see [Humphrys, 1996].

3.3 This is easier to do at sub-symbolic level than at symbolic level

We concentrate initially on defining *sub-symbolic* level queries, where, for example, competition is resolved using numeric weights rather than by symbolic-level negotiation. This avoids the problem of symbolic *knowledge representation schemes* [Ginsberg, 1991] and *agent communication languages* [Martin et al., 2000] that has been the graveyard of so many previous attempts. Obviously the symbolic level will have to be addressed at *some* point, but we show below how much can be done before we get to that level, and how the WWM is currently more suitable for the *sub-symbolic minds* popular in the Animats, ALife and Neural Networks fields.

3.4 The Subsumption Architecture

A Subsumption Architecture model [Brooks, 1986, Brooks, 1991] could be implemented as a hierarchy of $Mind_M$ servers, each one building on the ones below it. Each one sends the current state `x` to the server below it, and then either uses their output or overrides it. As in Brooks' model, a set of lower layers will still work if the higher layers are removed. On the WWM, there may be many choices for (remote, 3rd party) higher layers to add to a given collection of lower layers.

3.5 Reinforcement Learning

An ordinary Reinforcement Learning (RL) agent, which receives rewards and punishments as it acts [Kaelbling et al., 1996], can clearly be implemented as a single Mind server. For example a Q-learning agent [Watkins, 1989] builds up Q-values ("Quality"-values) of how good each action is in each state: `Q(x,a)`. When learning, it can calculate a reward based on `x`, `a` and the new state `y`. So, so long as the client informs this Mind server what state `y` resulted from the previous action `a`, it can calculate rewards, and learn.

3.6 Hierarchical Q-Learning

Hierarchical Q-Learning [Lin, 1993] is a way of driving multiple Q-learners with a master Q-learner. It can be implemented on the WWM as follows. The client talks to a single $Mind_{AS}$ server, sending it `x` and receiving `a`. The $Mind_{AS}$ server talks to a number of Mind servers. The $Mind_{AS}$ server maintains a table of values `Q(x,i)` where `i` is which Mind server to pick in state `x`. Initially its choices are random, but by its own reward function, the $Mind_{AS}$ server fills in values for `Q(x,i)`. Having chosen `i`, it passes on the action suggested by Mind server `i` to the client. To save on the number of server queries (which is a more serious issue on the WWM than in a self-contained system), the $Mind_{AS}$ server does not query *any* of the Mind servers until it has picked an action `i`, and then it only queries a *single* Mind server `i`. There are a number of interesting possibilities:

1. The $Mind_{AS}$ server *need not know its list of Mind servers in advance.* It can be passed this list by the client at startup.
2. The subsidiary Mind servers *need not be Q-learners.* They could be any type of Mind server (including *symbolic* Mind servers), and the $Mind_{AS}$ server simply learns which one to let through.
3. A further possibility [thanks to Dave O'Connor] is that the Hierarchical Q-Learner could build `Q(x,i)` values for every Mind server on the Net. It acts as a spider, finding *new*, random Mind servers, and trying their actions out in pursuit of its goal. The result of all of this learning will be the construction of a huge map telling it which server `i` to pick in each state `x`. It may use hundreds of servers to implement its goal.

Because the number of queries made is a more important issue on a network system than on a *self-contained* system, we distinguish different types of $Mind_{AS}$ server:

1. AS_s **server** - makes a single query of each Mind server before making its decision.
2. AS_o **server** - does not even query *all* Mind servers *once.* It just makes one query of one Mind server.
3. AS_m **server** - makes multiple queries of each Mind server before making its decision.

Hierarchical Q-Learning is an AS_o server.

3.7 W-learning

We consider a number of schemes where Mind servers promote their actions with a weight W, or "W-value" [Humphrys, 1997]. Ideally the W-value will depend on the state `x` and will be higher or lower depending on how much the Mind server "cares" about winning the competition for this state. A *static* measure of the W-value is one in which the Mind server promotes its action with a value of W independent of the competition (e.g. W=Q). Any such method can clearly be implemented as a $Mind_i$ server. A *dynamic* measure of W is one in which the value of W changes depending on whether the Mind server was obeyed, and on what happened if it was not obeyed. Clearly this is an AS_s server that queries once, lets through the highest W, and then *reports back* afterwards to each Mind server whether or not it was obeyed. The server may then *modify* its W-value next time round in this state.

W-learning [Humphrys, 1996] is a form of dynamic W where W is modified based on (i) whether we were obeyed or not, and (ii) what the new state y is as a result. This can clearly be implemented as an AS_s server. In the pure form of W-learning the Minds do not even share the same suite of actions, and so, for example, cannot simply get together and *negotiate* to find the optimum action. The inspiration was simply to see if competition could be resolved between Minds that had as little in common as possible. That work was unable to give convincing examples where this might arise. Now with the WWM, we see this is the *kind* of model we need when parts of the mind have great difficulty understanding each other (e.g. are written by different remote authors).

3.8 Global Action Selection decisions

If Minds *do* share the same suite of actions, then we can make various global decisions. Say we have n Mind servers. Mind server i's preferred action is a_i. It can quantify "how good" action a is in state x by returning: $Q_i(x,a)$, and can quantify "how bad" action a is in state x by returning: $Q_i(x,a_i) - Q_i(x,a)$. Then we have 4 basic approaches [Humphrys, 1997]:

1. **Maximize the Best Happiness:**

$$\max_a \max_i Q_i(x,a)$$

which is equal to static W=Q above, and can be implemented as an AS_s server, with just one query to each Mind server to get its best action and Q-value.

2. **Minimize the Worst Unhappiness:**

$$\min_a \max_i (Q_i(x,a_i) - Q_i(x,a))$$

which is an AS_m server, requiring multiple queries of each Mind server.

3. **Minimize Collective Unhappiness:**

$$\min_a \left[\sum_i (Q_i(x,a_i) - Q_i(x,a)) \right]$$

which is an AS_m server.

4. **Maximize Collective Happiness:**

$$\max_a \left[\sum_i Q_i(x,a) \right]$$

which is an AS_m server.

A number of authors [Aylett, 1995, Tyrrell, 1993, Whitehead et al., 1993, Karlsson, 1997, Ono et al., 1996] implement, using a variety of notations, one of the 4 basic AS methods defined above.

3.9 Nested Mind servers

Digney [Digney, 1996, Digney, 1998] defines Nested Q-learning, where *each* Mind in a collection is able to call on any of the others. Each Mind server has its own set of actions $Q_i(x,a)$ *and* a set of actions $Q_i(x,k)$ where action k means "do whatever server k wants to do" (as in Hierarchical Q-learning). In a WWM implementation, each Nested server has a list of Mind URLs, either hard-coded or passed to it at startup. So the Nested server *looks* like a $Mind_{AS}$ server co-ordinating many Mind servers to make its decision. But of course it is *not* making the final decision. It is merely *suggesting* an action to the master $Mind_{AS}$ server that coordinates the competition between the Nested servers themselves.

Some of the Nested servers might actually be outside the Action Selection competition, and simply *wait* to be called by a server that *is* in the competition. [Humphrys, 1997] calls these "passive" servers. We have the same with hand-coded $Mind_M$ servers, where some Mind servers may have to wait to be called by others. A server may be "passive" in one Society and at the same time "active" (i.e. the server is in the Action Selection loop) in a different Society.

3.10 Feudal Mind servers

Watkins [Watkins, 1989] defines a *Feudal* (or "slave") Q-learner as one that accepts commands of the form "Take me to state c". In Watkins' system, the command is part of the current state. Using the notation `(x,c),a -> (y,c)` the slave will receive rewards for transitions of the form: `(*,c),a -> (c,c)` So the master server drives the slave server by *explicitly* altering the state for it. We do not have to change our definition of the server. It is just that the server driving it is constructing the state x rather than simply passing it on from above.

3.11 The sub-symbolic Society of Mind

The Nested and Feudal models are combined in [Humphrys, 1997, Fig. 18.4] showing the general form of a Society of Mind based on Reinforcement Learning. Indeed, the whole model of a complex, overlapping, competing, duplicated, sub-symbolic Society of Mind here is based on the generalised form of a Society of Mind based on Reinforcement Learning.

3.12 Multiple symbolic Mind servers

So far we have only defined a protocol for conflict resolution using numeric weights. Higher-bandwidth communication leads us into the field of *Agents* and its problems with defining *agent communication languages* (formerly symbolic AI knowledge-sharing protocols) that we discussed above. We imagine that numeric weights will be more easily generated by *sub-symbolic* Minds, and are harder to generate in symbolic Minds. This is because symbolic Minds often know *what* they want to do but not "how much" they want to do it. Sub-symbolic Minds, who prefer certain actions precisely because numbers for that action have risen higher than numbers for other actions, may be able to say precisely "how much" they want to do something, and quantify how bad alternative actions would be.

It may be that in the symbolic domain we will make a lot more use of hand-coded $Mind_M$ servers instead of having Minds generating weights to resolve competition. The drawback, of course, is that the $Mind_M$ server needs a lot of intelligence. It needs to understand the goals of all the Mind servers. This relates to the "homunculus" problem, or the need for an intelligent headquarters. Another possibility is the subsidiary Mind servers can be symbolic, while the master $Mind_{AS}$ server is sub-symbolic - e.g. a Hierarchical Q-learner.

3.13 How about robots?

Any scheme of remote re-use is clearly more adapted to virtual agents than real (embodied) ones. But, as discussed above, it is interesting to consider the current lack of direct access to other researchers' robotic experiments. If software animat experiments are hard to replicate, robotic ones are doubly so.

"Telerobotics" is the ability to control a robot remotely. Telerobotic systems have in fact been used in animats [Wilson and Neal, 2000], though not on the network. Outside of the animats field there are in fact a number of "Internet telerobotics" robots that can be controlled remotely over the Internet. [Taylor and Dalton, 1997] discuss some of the issues:

1. We may want a scheme where only one client can control the robot at a time. Whereas with a software-only world one can always allow multiple clients (e.g. by creating a new *instance* of the world for each).
2. The robot owner may want to restrict *who* is able to run a mind on his machine, since some control programs may cause damage.
3. A virtual world server requires little or no maintenance. The author can put the virtual world up on the server and then forget about it. A robotic world server, however, demands much more of a commitment. As a result most of the Internet robots so far have run for a limited time only.

For example, [Stein, 1998] allows remote control of the robot until the client gives it up, or until a timeout has passed. [Paulos and Canny, 1996] operate in a special type of problem space where each action represents the completion of an entire goal, and so actions of different clients can be interleaved. In the robotic tele-garden [Goldberg et al., 1996] users could submit discrete requests at any time, which were executed later by the robot according to its own scheduling algorithm. The robotic Ouija board [Goldberg et al., 2000] is a special type of problem where the actions of multiple clients can be *aggregated*. It seems that all of these schemes could be implemented under the model discussed here. The focus so far in Internet telerobotics has been on remote human-driven control rather than remote *program*-driven control, but this may change.

3.14 How about speed? (robotic worlds and real-time virtual worlds)

Obviously there are inherent performance problems in any system of *remote* re-use. These problems may not be so acute in virtual worlds, where the world can freeze and *wait* for the agent to make a decision. We can *speed up* and *slow down* time in a virtual world to allow for delays caused by the network, without changing the nature of the problem.

This is not, however, an option in *real-time* virtual worlds, such as ones where other users or agents are changing the environment. Here the system may share some of the features of real-time *multi-player online games* (see survey in [Smed et al., 2001]). A large, nested Society of Mind may resemble a peer-to-peer game with low-bandwidth communication, which should scale well. A possible bottleneck is the top-level Mind server, depending how it is designed. [Abdelkhalek et al., 2001] considers performance issues with centralised game servers.

A top-level Mind server is unavoidable because the diversity of suggested actions must be reduced at *some* point, and a decision made. This point is the potential bottleneck. In many of the Action Selection schemes above, the top-level mind is reduced more or less to a *router* rather than a processor in its own right, in an effort to decentralise the intelligence. We now see that such an approach may also be useful in distributing the network *load*.

In the real physical world, a *robotic* animat also needs to make decisions quickly. It may be that a system such as this will be used for *prototyping* - experimenting with different Mind server combinations out of the choices online. Once a combination is chosen, one attempts to get a local installation of all the Mind servers involved. *Why* we are trying to avoid local installation is considered below. If we reject local installation, we cannot avoid network delays.

3.15 What if the remote server is down?

One problem with scaling up AI is that researchers do not want to be dependent on other people's work. What if the remote server is down? Or removed permanently?

Part of the problem is, we argue, models of mind in which the loss of a single server *would* be a serious issue. Instead of models of mind where hundreds of similar servers compete to do the same job, researchers have been assuming the use of *parsimonious* minds where each component does a particular task that is not done by others. A better strategy is to keep adding "unnecessary" duplicated minds to your society. The master $Mind_{AS}$ server asks all Mind servers to suggest actions, and times-out if it does not receive an answer in a short time. So in a highly-duplicated model, if the action does not arrive from one Mind server, it will have arrived from another similar one. In a mind with enough duplication, the temporary network failure (or even permanent deletion) of servers may never even be noticed. Obviously, *some* servers will be essential - like the World server, for instance. The basic answer for how to cope with essential servers is that if it is important to us, we will copy it (if it is free) or buy it or rent it.

[Humphrys, 1997] describes a multiple-minds model of AI that can survive brain damage by re-organising. The reader might have wondered what is the point of that. After all, if the AI is damaged, you just fix it or reinstall it surely? Here is the point - a model of AI that can survive *broken links*.

3.16 How about multi-agent systems?

In distributed intelligence, there are two major camps:

1. Multiple minds in one body, competing for expression (the fields of action selection, motivation, goal conflict, emotion, Society of Mind). We will refer to this as the "AS" camp.
2. Multiple bodies, which can act independently (multi-agent systems, collective behaviour). We will refer to this as the "MAS" camp.

The field of Animats focuses on both AS and MAS *offline*. The field of Internet Agents focuses on MAS online. The WWM focuses on AS online - on addressing the issue of how to construct really complex agent minds, and implementing action selection across servers. This, we argue, is the neglected area.

There is perhaps one other neglected area, which is *sub-symbolic* MAS online. Most work on MAS online is at the symbolic level (see *agent communication languages*, as discussed previously). One interesting issue is whether the Animats work on MAS, which involves what might be called *sub-symbolic* communication or signalling, can be brought online. Ongoing research by Walshe [Walshe, 2001] will attempt to interface sub-symbolic AS and MAS online.

4 Implementation of the WWM

We now describe the first implementation of the WWM, and the decisions made in reaching that implementation.

4.1 Rejecting local installation

First note that we reject the solution that would have been imagined for most of the history of AI - local installation. Given the huge diversity of, and incompatibility of, operating systems, platforms, files, libraries, versions, environments, programming methodologies and programming languages in use in AI (a diversity perhaps actually greater in AI than in any other field of computing), we view it as highly unlikely that local installation could lead to widespread re-use. We clearly reject the idea of asking all animats researchers to use a certain programming language (e.g. Java) or platform.

How can one avoid these compatibility problems and allow researchers use whatever platform they want? By *server-side* programs rather than client-side programs. The Web demonstrates this highly successful model of re-use - *leaving* the program on the remote Web server, and running it from there. One strange aspect of adopting this model for the WWM is that the mind may consist of components which are physically at different remote sites, and which stay there, and just communicate with each other remotely. Hence the mind is *literally* decentralised across the world - something which has never existed in nature. Hence the name, the "World-Wide-Mind".

4.2 Rejecting models designed for network programmers

Given that we propose a remote solution, we might look at the emerging *web services* architectures for program-to-program transactions online. These are called "web services" because they run over the existing HTTP network (rather than demand a new network be set up). The emerging standard is to send messages to remote objects or programs using the SOAP message system [w3.org/TR/SOAP], which runs over normal HTTP. SOAP messages are based on XML [w3.org/XML], a standard meta language used to describe data in tagged plaintext (i.e. it looks similar to HTML).

While we agree with this general scheme (run on HTTP, the data format should be tagged and extensible), we reject using the full complexity of the web services protocols. Why? Because of the unique nature of our audience. Most animats researchers are programmers, but not network programmers. These protocols - and indeed almost *all* protocols in computer networks, web services, distributed objects or Internet Agents - assume the programmer is a networks specialist (or is willing to become one). SOAP messages are complex, and you require an API to parse them. Doing it yourself is difficult.

4.3 Rejecting unforgiving data formats

We also reject strict XML. XML has moved away from the *forgiving* nature of HTML. In XML, opening and closing tags must both be present, the tags in a document must form a *tree*, and so on. Any failure results in the document being rejected by the parser. XML parsers are also extremely complex to use.

We still agree with the idea of *tagged plain text* for our data. Plain text is important so humans can read the data, and programmers can parse it themselves. With tagged plain text (each piece of data is delimited by tags, whitespace is ignored) it is much easier to create a *tolerant* parser than with untagged plain text (where, say, *precise* column number or line number defines which piece of data is which). Tagging allows *extensible* systems - we can ignore new tags that we don't recognise.

4.4 Lightweight Web Services

We describe our approach as *Lightweight* Web Services [O'Leary, 2002a]. These are web services that anyone can create without having to learn a whole family of new protocols. It should be (almost) as easy to create a Lightweight Web Service as it is to create a Web Page. Under the system we have now developed, any animats researcher can put their animat "mind" or "world" online as a WWM server by converting it into a command-line program that reads standard input and writes to standard output. The program can be written in *any* language and runs as a "CGI script". The input and output is in a stripped-down, forgiving, XML-like language called AIML. We now explain this.

4.5 CGI

All Web servers support a system of server-side programs called CGI. CGI is *not* a difficult technology - indeed, there is almost nothing to it except placing a command-line program in the CGI directory of your Web server. Any programming language may be used. Programs read plaintext input (text, HTML, XML, or any XML-like format) on standard input and write plain text output to standard output. All browsers (and other clients) can run remote CGI programs.

CGI is the command-line of the Internet. Network enthusiasts often neglect CGI and describe *much* more complex technologies (JavaScript, VBScript, Java applets, Java Servlets, ASP, JSP, etc.) we suspect precisely *because* CGI is so simple and was worked out long ago (1993, before the Web took off). But CGI is not obsolete. CGI is still a *far* simpler technology than these or *any* other technology for programs online either at client-side or server-side.

4.6 AIML

The plaintext input to and output from the WWM servers is in a simple, loosely-defined XML-like language we call AI Markup Language (AIML). Strictly speaking, AIML is not XML since we reject strict XML formatting. AIML is closer to HTML in that we try never to reject messages because of their being badly formed. A best effort is always attempted. We will *have* to be tolerant of loosely-defined servers on the WWM because often there will be *no alternative* to the AI author. If Bloggs does not put the Bloggs learning algorithm online, it will often be the case that *no one else will.* So we *can't* just refuse to use his server if it generates sloppy AIML.

This tolerance does not mean we cannot issue *recommendations.* The situation will be like the Web. The portal site `w3.org` defines the official HTML spec. (e.g. "tables should end with an end-table tag"). But the browser can't just choke on bad HTML, not if there is scope to make a guess and display it (e.g. if end of file comes with no end-table tag, then insert end-table tag). The browser *must* tolerate bad HTML, or users will switch to browsers that do. And the pool of authors would never have grown so big if authors had to write strict HTML. It is often forgotten that the Web *does not* run on strict HTML, and never could have.

Similarly, the portal site `w2mind.org` will define an official AIML spec. (e.g. "WWM query responses should end with an end-response tag"). But no matter how we define it, there will always be room for the client to make some guesses with bad AIML. Clients must try to tolerate AIML "close to" the spec. - though obviously there can be no guarantees once one deviates from the spec.

4.7 How to write a WWM server

Now we bring this all together. To write a Mind server that can suggest an action given that the world is in some state `x`, one writes a command-line program that can parse something like this on standard input:

```
<request type="GetAction" runid="RUNID">
    <data name="x"> x </data>
</request>
```

where the format of x is decided by the World server. Clearly, this plaintext input can easily be parsed by any programmer using simple string searching mechanisms in any language (the first author's parser is just 5 lines of UNIX Shell, and is tolerant of many different variations in the input AIML). The Mind server then outputs to standard output something like:

```
<response type="GetAction" runid="RUNID">
    <data name="a"> a </data>
</response>
```

where the format of a is decided by the World server. We say "something like" because AIML is still in a state of revision. An agreed standard will be released in mid-2002. For the current draft, AIML v1.1, see [O'Leary, 2002].

That is all one needs - to agree on the format of AIML - and even full agreement is not necessary if one writes a tolerant parser. The program can be written in any language. Input and output can be debugged using an ordinary web browser (though for *repeated* queries one would want to use one of our dedicated clients).

4.8 Existing work

To date, we have put online:

1. World servers representing simple toroidal "grid worlds" with "food" and mobile "predators", written in C++ and Java by multiple authors.
2. Mind servers to drive the animat in these worlds, written in C++ and Java by multiple authors.

The servers are hosted on *separate* remote Web (HTTP) servers. Both GUI and command-line clients have been written for Windows and UNIX. Using the clients, Mind A (Java) was able to explore World B (C++), and Mind B (C++) was able to explore World A (Java).

4.9 Further issues

There are many further implementation issues, such as:

1. How to save the *state* of the world (or mind, if it is learning) in between requests.
2. How to write CGI programs that are *persistent* in memory between requests.
3. How to handle multiple users.
4. How to display graphically what is happening in the world (or inside the mind) and where to display this.
5. How does a server call another server.

We deal with all of these issues in [Humphrys, 2001, O'Leary, 2002a]. But it remains that a basic WWM server can still be got running with just a program that parses plaintext and outputs plaintext as above.

5 Future work

From the WWM viewpoint the next immediate thing to do is finalise a standard for AIML and then publicly release this standard, plus servers and clients that use it, and make these available from the portal site `w2mind.org`. Then other researchers can start building their own servers.

From the *animats* viewpoint the next things to do are: (a) Put existing well-known minds and worlds online as servers (we already have Tyrrell's world running [Tyrrell, 1993] but not yet as a server) and: (b) Construct *network action selection* mechanisms for Action Selection across multiple *remote* minds by different authors.

6 Conclusion

This paper has argued for the need to decentralise the work in AI so that researchers may specialise on different parts, *and* a mind may be constructed from these multiple specialist parts. Such a future (of specialists coming together) has been imagined (at least implicitly) in many branches of AI, but no practical scheme for implementing it has yet emerged. We believe that now, with server-side programming ubiquitous on the Internet, such a scheme is finally possible.

We have a new vision of a mind: no *single* author could write a high-level artificial mind, but perhaps the *entire* scientific community could. Each *piece* will be understood by someone, but the whole may be understood by no-one. Perhaps we need a new respect for the *magnitude* of the AI problem - that building a high-level artificial mind may be on the same scale as constructing something like a national *economy*, or the city of London. No single individual or company built London or New York. But humanity as a whole did.

Acknowledgements

The software for this system (servers, clients and server support software) and the design of AIML is the joint work of the authors of this paper and Dave O'Connor and Ray Walshe. We are grateful to two anonymous referees for their comments.

References

Abdelkhalek, A.; Bilas, A. and Moshovos, A. (2001), Behavior and Performance of Interactive Multiplayer Game Servers, *Proc. Int. IEEE Symposium on the Performance Analysis of Systems and Software.*

Aylett, R. (1995), Multi-Agent Planning: Modelling Execution Agents, *14th UK Planning and Scheduling SIG.*

Berger, H.W. (1998), *Is The NP Problem Solved? (Quantum and DNA Computers)*, www.pcs.cnu.edu/~hberger/Quantum_Computing.html

Brooks, R.A. (1986), A robust layered control system for a mobile robot, *IEEE Journal of Robotics and Automation* 2:14-23.

Brooks, R.A. (1991), Intelligence without Representation, *Artificial Intelligence* 47:139-160.

Brooks, R.A. (1997), From Earwigs to Humans, *Robotics and Autonomous Systems*, Vol. 20, pp. 291-304.

Brooks, R.A. et al. (1998), The Cog Project, *Computation for Metaphors, Analogy and Agents*, Springer-Verlag.

Bryson, J. (2000), Cross-Paradigm Analysis of Autonomous Agent Architecture, *JETAI* 12(2):165-89.

Daniels, M. (1999), Integrating Simulation Technologies With Swarm, *Workshop on Agent Simulation*, Univ. Chicago, Oct 1999.

de Garis, H. (1996), CAM-BRAIN: The Evolutionary Engineering of a Billion Neuron Artificial Brain, *Towards Evolvable Hardware*, Springer.

Dennett, D.C. (1978), Why not the whole iguana?, *Behavioral and Brain Sciences* 1:103-104.

Digney, B.L. (1996), Emergent Hierarchical Control Structures, *SAB-96.*

Digney, B.L. (1998), Learning Hierarchical Control Structures for Multiple Tasks and Changing Environments, *SAB-98.*

Ginsberg, M.L. (1991), Knowledge Interchange Format: The KIF of Death, *AI Magazine*, Vol.5, No.63, 1991.

Goldberg, K. et al. (1996), A Tele-Robotic Garden on the World Wide Web, *SPIE Robotics and Machine Perception Newsletter*, 5(1), March 1996.

Goldberg, K. et al. (2000), Collaborative Teleoperation via the Internet, *IEEE Int. Conf. on Robotics and Automation (ICRA-00).*

Guillot, A. and Meyer, J.-A. (2000), From SAB94 to SAB2000: What's New, Animat?, *SAB-00.*

Harvey, I.; Husbands, P. and Cliff, D. (1992), Issues in Evolutionary Robotics, *SAB-92.*

Humphrys, M. (1996), Action Selection methods using Reinforcement Learning, *SAB-96.*

Humphrys, M. (1997), *Action Selection methods using Reinforcement Learning*, PhD thesis, University of Cambridge, Computer Laboratory. `www.compapp.dcu.ie/~humphrys/PhD`

Humphrys, M. (2001), *The World-Wide-Mind: Draft Proposal*, Dublin City University, School of Computer Applications, Technical Report CA-0301, February 2001. `www.compapp.dcu.ie/~humphrys/WWM`

Humphrys, M. (2001a), Distributing a Mind on the Internet: The World-Wide-Mind, *ECAL-01*, Springer-Verlag LNCS/LNAI 2159, September 2001.

Kaelbling, L.P.; Littman, M.L. and Moore, A.W. (1996), Reinforcement Learning: A Survey, *JAIR* 4:237-285.

Karlsson, J. (1997), *Learning to Solve Multiple Goals*, PhD thesis, University of Rochester, Department of Computer Science.

Lin, L-J (1993), Scaling up Reinforcement Learning for robot control, *10th Int. Conf. on Machine Learning.*

Martin, F.J.; Plaza, E. and Rodriguez-Aguilar, J.A. (2000), An Infrastructure for Agent-Based Systems: an Interagent Approach, *Int. Journal of Intelligent Systems* 15(3):217-240.

McDermott, D. (1997), "How Intelligent is Deep Blue?", *New York Times*, May 14, 1997.

Minsky, M. (1986), *The Society of Mind.*

Minsky, M. (1991), Society of Mind: a response to four reviews, *Artificial Intelligence* 48:371-96.

Nilsson, N.J. (1995), Eye on the Prize, *AI Magazine* 16(2):9-17, Summer 1995.

O'Leary, C. (2002), *AIML v1.1 - Artificial Intelligence Markup Language*, `comp.dit.ie/coleary/research/phd/wwm/`

O'Leary, C. (2002a), *Lightweight Web Services for AI Researchers*, `comp.dit.ie/coleary/research/phd/wwm/`

Ono, N.; Fukumoto, K. and Ikeda, O. (1996), Collective Behavior by Modular Reinforcement-Learning Animats, *SAB-96.*

Paulos, E. and Canny, J. (1996), Delivering Real Reality to the World Wide Web via Telerobotics, *IEEE Int. Conf. on Robotics and Automation (ICRA-96).*

Sloman, A. and Logan, B. (1999), Building cognitively rich agents using the SIM_AGENT toolkit, *Communications of the ACM*, 43(2):71-7, March 1999.

Smed, J.; Kaukoranta, T. and Hakonen, H. (2001), Aspects of Networking in Multiplayer Computer Games, *Proc. Int. Conf. on Application and Development of Computer Games in the 21st Century.*

Stein, M.R. (1998), Painting on the World Wide Web, *IEEE / RSJ Int. Conf. on Intelligent Robotic Systems.*

Sutton, R.S. and Santamaria, J.C., A Standard Interface for Reinforcement Learning Software, www-anw.cs.umass.edu/~rich/RLinterface/RLinterface.html

Taylor, K. and Dalton, B. (1997), Issues in Internet Telerobotics, *Int. Conf. on Field and Service Robotics.*

Tyrrell, T. (1993), *Computational Mechanisms for Action Selection*, PhD thesis, University of Edinburgh.

Walshe, R. (2001), The Origin of the Speeches: language evolution through collaborative reinforcement learning, *Proc. 3rd Int. Workshop on Intelligent Virtual Agents (IVA-2001).*

Watkins, C.J.C.H. (1989), *Learning from delayed rewards*, PhD thesis, University of Cambridge.

Whitehead, S.; Karlsson, J. and Tenenberg, J. (1993), Learning Multiple Goal Behavior via Task Decomposition and Dynamic Policy Merging, *Robot Learning*, Kluwer.

Wilson, M. and Neal, M. (2000), Telerobotic Sheepdogs: How useful is autonomous behavior?, *SAB-00.*

Wilson, S.W. (1990), The animat path to AI, *SAB-90.*

Agent-Based Modelling and the Environmental Complexity Thesis

Anil K Seth
The Neurosciences Institute, 10640 John Jay Hopkins Drive, San Diego, CA 92121, USA
seth@nsi.edu

Abstract

A variation of Godfrey-Smith's 'environmental complexity thesis' is described which draws together two broad themes; the relation of functional properties of behaviour to environmental structure, and the distinction between behavioural and mechanistic levels of description. The specific idea defended here is that *behavioural and/or mechanistic complexity can be understood in terms of mediating well-adapted responses to environmental variability.* Particular attention is paid to the value of agent-based modelling within this framework.

1. Introduction

"The function of cognition is to enable the agent to deal with environmental complexity". So states, with admirable brevity, the 'environmental complexity thesis' (ECT) of Peter Godfrey-Smith, a philosophical project which first appeared in his 1996 book, *Complexity and the Function of Mind in Nature,* and which has subsequently been the focus of considerable and largely positive attention (for example Hardcastle, 1999; McShea, 1996; Oyama, 1996; Bedau, 1996; Belew, 1996).

My aim in this paper is to use a discussion of the ECT as a way of drawing together two broad themes: the relation of functional properties of behaviour to environmental structure (central to the ECT as described by Godfrey-Smith), and the distinction between behavioural and mechanistic levels of description (*not* so central). The motivation for this is to describe a framework within which the relations between behaviour, mechanism, and environment, as they appear in agent-environment systems in general, can be usefully articulated. I shall also give examples throughout of how agent-based modelling techniques complement this framework in advancing our understanding of adaptive behaviour.

The upshot of all this will be a new interpretation of the ECT, which is that *behavioural and/or mechanistic complexity can be understood in terms of mediating well-adapted responses to environmental variability.* Undoubtedly this is similar to Godfrey-Smith's own formulation, but there are differences, the significance of which will be illustrated by critiquing the original ECT from three directions: (i) the use, and meaning, of the term 'cognition', (ii) the relation of the ECT to W. Ross Ashby's 'law of requisite variety', and (iii) the role played by agents in the 'construction' of their environments. Whilst distinct in substance, these three elements share a common interest, unlike Godfrey-Smith's original thesis, in distinguishing between behavioural and mechanistic levels of description. It is nevertheless important to emphasise that the purpose of this paper is not to try to undermine the ECT *per se,* but rather to exploit some of its many riches.

2. Setting the stage

We begin with some definitions, which may be regarded as assumptions of this paper, and with a discussion of the distinction between behaviour and mechanism.

Behaviour is defined in this paper as *observed ongoing agent-environment interactivity,* and *mechanism* as the *agent-side structure subserving this interactivity.* Defined this way, all behaviours (eating, swimming, fleeing, building-a-house) depend on continuous patterns of interaction between agent and environment; there can be no eating without food, no building-a-house without bricks, no swimming without water. In addition, it is ultimately up to the external observer to decide which segments of agent-environment interactivity warrant which behavioural labels. In principle, different observers may both privilege different junctures in observed activity and label the same segments differently, either way causing problems for any explanation of mechanism framed in the language of behaviour. Behaviour is a product of the joint activity of agent, environment, and observer, therefore - and this is the crucial point - the (agent-side) mechanisms underlying the generation of any behaviour should not be assumed to be identical to the behaviour itself. This is what is meant by distinguishing between behavioural and mechanistic levels of description.

An instructive example of the importance of this distinction comes from Lorenz's (1937) observations of the 'parenting behaviour' of ducks (see also Hendriks-

Jansen, 1996). Mother ducks engage in a number of different patterns of interaction with their offspring, and Lorenz subsumed all such observed patterns under the label of 'parenting behaviour'. This is clearly valuable for descriptive classification, but it should *not* be taken as reason to believe in the existence of any 'behavioural icon' for parenting behaviour, internal to the mother duck, as a mechanistic explanatory locus. And indeed, as Lorenz subsequently discovered, from the perspective of the mother duck the various interaction patterns are all triggered by quite different stimuli. The only point at which they 'intersect' in any sense is on the duckling itself, as an object in the eyes of the external observer.

Moreover, if a behaviour appears *complex* (to an external observer), this does *not* imply that the underlying internal mechanisms are also complex. The classic illustration of this is Herbert Simon's description of an ant on a beach (Simon, 1988). The internal mechanism of this (hypothetical) ant consists of a simple obstacle-avoidance rule - if there is a clump of sand to the left, go right, and vice-versa. Thus the ant responds to every tiny clump of sand, veering first left then right as it negotiates its terrain. Simon's point is that from the perspective of the external observer - who cannot perceive the small-scale heterogeneity of the beach surface, and who is not aware of the simplicity of the ant's 'algorithm' - the trajectory traced by the ant is strikingly and perhaps irretrievably complex.

So how can one begin to trace the relationship between behaviour and mechanism? A first step is to be clear about the meaning of environment, which we may think of in two ways: the environment as it appears to us, as external observers, and the environment as it appears to the agent that we (as external observers) are observing. The former (the 'external' environment, following Brandon, 1990, or 'distal' environment, following Brunswik, 1952, and Nolfi, 1998) is that which features in the definition of behaviour as agent-environment interactivity, and is perhaps the most intuitive way to understand the term. The latter is perhaps best labelled by the term *Umwelt* ('proximal' will also do), coined by Jakob von Uexküll (1934) to refer to the space of sensorimotor cues relevant to an organism, containing those features which constitute stimuli for the organism, to which the organism can potentially muster a response.[1] Consider, then, that behaviours, being elements in the vocabulary of an observer, are located in the external environment. In the case of Lorenz's duck, the descriptions of the various interaction patterns that constitute parenting behaviour are framed in terms of the external environment, and intersect only on the duckling *as an entity in the external environment.* The relation of these behavioural descriptions to mechanism may then require that the relations between the external environment and the *Umwelt* be traced - Lorenz had to identify what constituted stimuli for the mother-duck in order to understand the nature of its parenting behaviour.[2]

[1]See Ziemke and Sharkey (2001) for an interesting commentary on von Uexküll and recent developments in adaptive robotics.

3. The environmental complexity thesis

We may now move on to the second of our primary themes, the relation between behaviour and environmental structure. This is the province of the ECT.

"The function of cognition is to enable the agent to deal with environmental complexity". A strong claim, and one which Godfrey-Smith is keen to place in the context of what he labels 'externalism', the explanation of internal organism properties in terms of their relations to the external. In fact Godfrey-Smith describes the ECT as expressing 'c-externalism', which he defines as the attempt to understand internal *complexity* in terms of external *complexity.* The intuition is that 'cognition' is more complex than 'no cognition', thus 'cognition' is argued to be a complex response to a complex environment. Importantly the ECT is also a claim about necessity, not sufficiency. The idea is that environmental complexity is necessary for there to be cognition, but *not* that environmental complexity will always result in there being cognition.

A wider aim that Godfrey-Smith holds for his book is the exploration of externalist explanation in general, and it will repay us to understand something of this aim. In this broad context the ECT stands along with adaptationist explanations in biology, empiricism in philosophy, and associationism and behaviourist learning theory in psychology. The obvious (and equally influential) complement of externalism is internalism, the attempt to explain internal organism properties in terms of (their relation to) other internal properties. Biological internalism is manifest in the structuralist accounts of Goodwin (1994), and, earlier, d'Arcy Thompson (1917). Psychological internalism is strident in Chomsky, and in philosophy the rationalist tradition is internalist insofar as it argues for the necessity of mental 'pre-structure' in the formation of beliefs and judgements.

Of course any distinction between internalism and externalism naturally presupposes a distinction between the external and the internal, yet as many have argued this can be difficult to justify, and theories which take as their explanandum the *existence* (whether apparent or 'real') of a distinction between external and internal naturally resist classification as either externalist or internalist (see, for example, Oyama, 1996, 1985). In view of this it can be argued that the ECT ought to ex-

[2]To take another example, the catching behaviour of cricketers has been greatly elucidated by appreciating as a stimulus the acceleration of the tangent of elevation of gaze from player to ball; if this acceleration is kept at zero, the cricketer *will* meet the ball before the ball meets the ground (McLeod & Dienes, 1996).

press a 'pragmatic externalism', retaining the emphasis of accounting for internal properties in terms of external properties, but at the same time displaying consistency with arbitrary boundaries between external and internal. Importantly, not all versions of the ECT are pragmatically externalist in this sense.

Getting back to the ECT itself, Godfrey-Smith is quick to point out that its plausibility rests entirely on the meanings ascribed to the terms 'function', 'cognition', and 'environmental complexity'. The vision of *environmental complexity* held by Godfrey-Smith is one of *heterogeneity*:

> Complexity is changeability, variability; having a lot of different states or modes, or doing a lot of different things. Something is simple when it is all the same. (1996a., p.24)

This definition of complexity is not without its difficulties, and in particular bears a controversial relationship with unpredictability (McShea, 1996), but for present purposes we may let it stand. 'Function' is interpreted 'teleonomically'; the teleonomic function of something is the effect it has which explains why it is there, usually in view of some selective process, so from this vantage cognition is understood to be an *adaptation* to the problem of environmental complexity, the ECT an adaptationist thesis. Finally, what Godfrey-Smith means by 'cognition' is, unfortunately, less clear, despite being that which the ECT is ostensibly directed towards accounting for. We will have more to say about this in section 4.1.

Godfrey-Smith also identifies an informative historical context for the ECT which takes the form of a contrast between the positions of the philosophers Herbert Spencer and John Dewey. Spencer believed in a general 'law of evolution' which prescribed a universal dynamic from states of "indefinite, incoherent homogeneity" to states of "definite, coherent heterogeneity". Increases in environmental heterogeneity were supposed to lead to corresponding increases in internal heterogeneity by "the continuous adjustment of internal relations to external relations". Godfrey-Smith interprets this as suggesting that "organic properties are not seen so much as solutions to environmental problems but rather as bearing the imprint of the environment's pattern" (ibid., p.89). There is therefore no sense, for Godfrey-Smith, in which a Spencerian ECT implies that cognition 'deals with' environmental complexity by way of judiciously deployed responses, rather, organism 'accommodates' environment in the sense of the external somehow being captured 'inside' the internal. Interpreted this way, the Spencerian ECT is endlessly problematic. Most seriously, the idea of organic properties 'accommodating' environmental properties demands a reification of the distinction between external and internal, one cannot bring the external within the confines of the internal without asserting a pre-existing distinction between the two. Because of this the Spencerian ECT is inconsistent with pragmatic externalism.

Dewey receives greater favour from Godfrey-Smith, and it is in his later work - most particularly in *Experience and Nature* (1929) - that a version of the ECT is identified. For example:

> The world must actually be such as to generate ignorance and inquiry; doubt and hypothesis, trial and temporal conclusions; [...] The ultimate evidence of genuine hazard, contingency, irregularity and indeterminedness in nature is thus found in the occurrence of thinking. (1929, p.69, quoted in Godfrey-Smith, 1996a, p.100)

It is Dewey's pragmatism that makes him more acceptable. As Godfrey-Smith explains, "Dewey was opposed to the idea that in solving problems organic systems merely accommodate environmental demands. Rather, they intervene in the environmental processes which generated the problem, and alter the environment's intrinsic course" (ibid., p.139).

Godfrey-Smith's reading of this version of the ECT improves on Spencer's version in at least two ways. First, consistency with pragmatic externalism is ensured because the boundary between external and internal can be retained as arbitrary (to the extent that the response is *not* framed in terms of the environment itself, otherwise one is once again forced to 'internalise' the environment). Second, Dewey helps sharpen our ideas about the kinds of environment that might provide adaptive advantage for cognition, focusing on a mixture of predictability and unpredictability:

> The incomplete and the uncertain give point and application to the ascertainment of regular relations and orders. (1929, p.160, quoted in Godfrey-Smith, 1996a, p.130)

Godfrey-Smith restates this idea in the form of two conditions to be satisfied if a role for cognition is to be mandated. First, that there exists *variability* "with respect to distal conditions that make a difference to the organisms's well-being". Second, that there be *stability* "with respect to the relations between these distal conditions and proximal and observable conditions" (ibid., p.118). We shall return to these conditions in section 4.2.

In summary, we may say that Godfrey-Smith favours an adaptationist, teleonomic version of the ECT in which complexity is interpreted as heterogeneity, in which the environment is seen to comprise of a mix of stability and variability, and in which cognition is interpreted as response rather than accommodation. As such, the ECT provides a clear way to think about how the functional properties of behaviour relate to environmental structure: they cope with environmental complexity.

4. Critiquing the ECT

Whilst both the ECT and the behaviour/mechanism distinction make good sense when considered individually, the picture they paint when considered together is hardly an integrated whole. Our efforts are now in this direction in the form of a critique of Godfrey-Smith's version of the ECT. The first element of this critique concerns 'cognition'.

4.1 Cognition

Without a clear understanding of what Godfrey-Smith means by cognition it is not possible to understand what the ECT is trying to explain. Is it just 'response' and nothing more? A good place to begin is the Precis of his book:

> A core set of cognitive capacities, including the capacities for perception, internal representation of the world, memory, and decision making, have the function of making possible complex patterns of behavior that enable organisms to deal effectively with complex patterns and conditions in their environments. (1996b, p. 453)

Elsewhere throughout the book itself, however, ambiguity returns, with cognition discussed at a variety of levels of abstraction and in several different senses (a worry partly shared by Belew, 1996). In one place it is interpreted "as a means to the production of behavioural complexity" (1996a, p.26), in another, the possibility of cognition being an advanced kind of homeostatic device is advanced (ibid., pp.76-79), and there is also reference to a "basic mental tool-kit" (ibid., p.127). This variety makes it clear that although the ECT may provide a detailed discussion of how the functional properties of behaviour may relate to environmental structure, there will be less to say about how such behaviour is to be related to underlying mechanisms. For Godfrey-Smith, both levels of description are subsumed under the term 'cognition', a term which he loads with many of its customary connotations such as internal representation and decision making (cf. Neisser, 1967).

This dalliance with what one may call 'classical' cognitive terminology leads his version of the ECT into some unfortunate territory. Shapiro (1999), for example, gets things very much the wrong way around, suggesting that the ECT may be used to answer the question of whether the observation of a particular behaviour warrants the ascription of *psychological* mechanism, rather than just *physiological* mechanism. If there is sufficient environmental complexity, his argument goes, then psychological (read 'cognitive') mechanism must be involved, the implication being that physiological (read 'non-cognitive') mechanism just could not cope in such circumstances). Unfortunately, in proposing this argument, Shapiro succeeds only in reviving a prejudice effectively dispelled by Pavlov in the 1920s; that any process analysable in physiological terms *ipso facto* cannot be a psychological process (see Lorenz, 1948, p.204).[3]

This, then, is the first element of my critique of Godfrey-Smith's formulation of the ECT. Slippery usage of the term 'cognition' enables him to gloss over the distinction between behaviour and mechanism. The functional properties of behaviour are the proper targets of the adaptationist claims of the ECT, mechanisms are certainly required to subserve behaviour, and so mechanisms will also be indirectly subject to the same forces of selection that operate directly on the functional entities of behaviour, but there is no necessary, direct link between complex behaviour and complex mechanism, cognitive or otherwise (recall Simon's ant). Therefore, there is no way the ECT should be used to justify claims for the existence of 'psychological mechanism' over 'physiological mechanism'. In other words, part of the appeal of the ECT comes from the apparent link between environmental structure and internal mechanism, yet this misses out the essential link provided by behaviour. Part of the difficulty with Godfrey-Smith's formulation is that the term 'cognition' obscures any clear distinction between behaviour and mechanism, rendering any direct implications of the ECT for internal mechanism unclear.

There are some straightforward implications for agent based modelling here, quite simply that such models allow distinctions between behaviour and mechanism to be fully realised and empirically interrogated, obviating the need for problematic catch-all terms such as cognition. In any but the most trivial concrete model there will be both observable behaviour and analysable behaviour-generating mechanisms. Causal networks linking behaviour, mechanism, and environment can in principle be fully unravelled, and the explanatory leverage of the ECT in accounting for *mechanistic* complexity empirically assessed.

To take a concrete example, consider Beer's work on the evolution of 'minimally cognitive' behaviour (Beer, 1996, 2000). Beer analysed an agent equipped with seven sensory 'rays', set at various angles relative to the midline, reporting the distance to the edges of a falling object, either a circle or a diamond. The agent could move along a horizontal line and was faced with the task of avoiding diamonds and centering circles. Beer used a genetic algorithm to evolve successful agents, con-

[3] Dennett (1987) makes a similar point, arguing that behaviour of a sufficient apparent complexity drives observers to take the 'intentional stance' with respect to the behaving agent, such that in order to understand and/or predict the behaviour of the agent it is necessary to (or at least it *helps* to) ascribe 'classical' cognitive states to the agent. However, as Dennett is careful to emphasise, the act of taking the intentional stance does *not* imply that the actual agent-side behaviour-generating mechanisms are cognitive in this sense; the intentional stance is an *observational* strategy.

trolled by recurrent neural networks, and then asked how they worked. Without going into details, he observed that "subtle interplay between sensory input and internal state is crucial to accurate discrimination" (Beer, 2000, p.94), thus accounting for an arguably 'cognitive' phenomenon in terms of dynamic interrelations between behaviour, mechanism, and environment. A second example which more directly probes the relationship between environmental structure and mechanism complexity is provided by Biró and Ziemke (1998), who analysed simple evolved neural networks controlling agents able to locate objects within a walled arena. They noticed that the dynamics of successful networks could often be analysed in terms of clusters of activity that correlated well with distinct behavioural segments (search, tracking, avoidance), in this case arguing for a simple relationship between behaviour and mechanism (see also Ziemke, 2000). One can easily imagine a productive extension of this work, challenging this simple relationship by varying both the structure of the environment and the sensory mapping from the environment to the controlling network.

4.2 The law of requisite variety

The second element of the critique analyses the ECT in terms of Ashby's (1956) 'law of requisite variety' (henceforth LRV). A product of the mid-twentieth century cybernetic school, the LRV may be considered intuitively as asserting that only agent-side variation can cope with - or 'force down' - environmental variation (as opposed to, for example, agent-side *stasis*) . This is clearly similar to the basic premise of the ECT, however although the LRV is mentioned by Godfrey-Smith it is never for him a focus of attention. This I argue is a missed opportunity for the reason that the LRV can help the ECT bring more clearly into focus the important concepts of variability and homeostasis.

Let's start with a more formal account of the LRV. Consider that for an agent to maintain relative stability in certain (internal) essential variables (for example heart rate, body core temperature), it must *prevent* the transmission of environmental variability through to these essential variables. In the same way that a good thermostat prevents the transmission of environmental variations in temperature through to a particular object (for example, the interior of a refrigerator should remain at a constant cool temperature despite the fluctuating temperature of a kitchen on a midsummer day), a well adapted agent prevents the transmission of certain environmental variables (for example the prevalence or scarcity of food, the proximity, or otherwise, of predators) through to such essential internal variables as blood sugar level or heart rate. With this in mind, the LRV can be easily formulated: Consider a set of possible environmental disturbances D, a set of possible responses on the part of the agent, R, and a set of possible outcomes, O. Consider also that for each D_i, there is distinct outcome O_i, and a particular response R_i. Stability in the essential variables requires minimising the variation in O, and this then requires that the variety in D is matched by the variety in R. Ashby himself provides a more concise summary:

> If R's move is unvarying, then the variety in outcomes will be as large as the variety in D's moves; only variety in R's moves can force down variety in the outcomes.[4] (1956, p.206)

So how can the LRV express a version of this? Let's rehearse once again: "the function of cognition is to enable the agent to deal with environmental complexity".

Most obviously, environmental complexity may be interpreted as the set of environmental disturbances D. 'Function' can be interpreted, teleonomically, as the maintenance of stability in essential variables (the minimisation of variation in O). This leaves 'cognition', which instead of being associated with the variety of interpretations favoured by Godfrey-Smith, is here related directly to the regulation effected by the various responses (the set R) deployed by the agent in 'forcing down the variety' in O.

In this way the LRV expresses a quite specific version of the ECT, one with several interesting properties. First, there is a clear emphasis on environmental (distal) variability, which for Ashby is significant in two distinct manifestations:

> There is that which threatens the survival of the gene-pattern. This part must be blocked at all costs. And there is that which, while it may threaten the gene-pattern, can be transformed (or re-coded) through the regulator R and used to block the effect of the remainder (1956, p.212).

This may be usefully compared with Godfrey-Smith's dual condition for the adaptive significance of cognition. For Godfrey-Smith, without distal variability, cognition is not necessary, and without stability in distal-proximal relations, cognition is not possible. Ashby offers a third

[4] A small qualification is worthwhile. It is not usually necessary for essential variables to be maintained at a precise value; variation within a certain range is usually permissible and often inevitable. In mammals, for example, changes in heart rate and blood sugar concentration are in fact essential contributions to adaptive behaviour, for example in ensuring physiological preparedness for 'fight' or 'flight' responses. McFarland and Sibly (1975) speak of a physiological state-space, with essential variables defining axes in this space such that its dimensionality is determined by the number of essential variables. It would be more accurate to interpret the LRV in terms of maintaining the state of the organism *not* at a single point in physiological state-space, but rather in a region of this space circumscribed by a set of 'lethal boundaries'. However, for the present purposes this distinction may be overlooked; I shall be concerned with a broadly construed stability in essential variables only.

alternative, identifying the importance of 'potentially beneficial' variability, the suggestion being that without this kind of variability, the agent would not be able to act - or respond - at all. One example of this kind of variability would be a change in odour intensity that correlates with proximity to a food source, another would be some kind of environmental stochasticity that enables protean behaviour in a predator-prey situation. The perspectives of Ashby and Godfrey-Smith are therefore complementary, and an interpretation of the ECT emerges in which all three characteristics combine: stability in distal-proximal relations, and distal variability that both threatens the agent and facilitates its activity.

Secondly, by framing the components of the ECT in terms of potentially *measurable* variety - in environment, response, and outcome, - the LRV emphasises the value of concrete agent-based modelling. Fletcher, Zwick, and Bedau (1998, 1996), for example, investigate how the manipulation of environmental texture in a toroidal grid-world environment (interpolating between flat, random, and sinusoidal resource distributions) relates to the variety of responses deployed by well-adapted agents. They employ Shannon entropy to track three distinct measures of variety: the information content of the environment (from the perspective of the agent), the information content of look-up tables of sensorimotor rules that constitute agent mechanism (which represents the variety of response), and the *between-agent* variety in look-up table structure. Although they do not mention the ECT, they take two results to exemplify the LRV. First, that response variety approaches environment information content; thus variety in R is matching variety in D. Second, that between-agent variety falls to zero, indicating that the variety in R is not random, such that for every D_i there is indeed a particular response R_i.

In a second example, Seth (in press) explores an evolutionary iterated prisoner's dilemma (IPD) model, focusing on the relationship between the introduction of noise at various loci in the model and the evolution of strategy memory (which can informally be thought of as strategy complexity). Analysis in terms of the LRV predicts which loci promote the evolution of memory when noise is applied, and which do not, enabling a distinction to be drawn between *adaptive* and *non-adaptive* evolution of memory, without reference to fitness statistics.

It is worth considering why the LRV was given such short shrift by Godfrey-Smith; one possible reason is the suspicion that the LRV implies that cognition be understood as *homeostasis,* and this is an implication that Godfrey-Smith tries hard to avoid (pp.76-79). Ever since Cannon (1932), the term 'homeostasis' has been used to describe mechanisms in which stasis in certain properties of a mechanism is maintained by variation in others. Cannon originally discussed homeostasis in a physiological context, but it is clear that the notion generalises, and the LRV certainly has a homeostatic interpretation in terms of the maintenance of stability in essential variables (in whatever context) through the deployment of appropriate responses. So let us briefly consider the question: can 'cognition' be construed as homeostasis?

For Godfrey-Smith, the answer is 'sometimes'. In some cases, cognition will lead to actions that are genuinely homeostatic, for example in the intelligent use of fire to maintain bodily warmth. 'Genuine' homeostasis, for Godfrey-Smith, obtains when "there is some intermediate organic property [body temperature] such that complex activity contributes to the maintenance of stasis in this intermediate property, where this property makes a real contribution to survival" (1996a, p.79). However, in other cases there will be no non-trivial homeostatic interpretation of the action of cognition, for example in the adept use of perception and coordination to evade a sudden rock slide. In these cases, cognition "is like hibernation - it is adaptive, but the explanation for why it is adaptive goes *directly* from organic variation to survival" (ibid., p.79).

Godfrey-Smith is therefore right to be cautious of *equating* cognition with *non-trivial* homeostasis, even if some instances of cognition - or indeed many - can be understood in this way. However, reading the LRV as a version of the ECT does *not* necessitate this commitment. Admittedly, there is a temptation to equate Ashby's essential variables with Godfrey-Smith's 'intermediate organic properties' - body temperature, for instance, has been used in examples of both - and such an equivalence would indeed encourage an interpretation of the LRV as non-trivially homeostatic. But at the abstract level of the LRV, it does not appear possible to uphold any distinction between an 'intermediate organic property', and a property that is *constitutive* of survival. After all, for Ashby, the essential variables are *defined* as precisely those variables for which their stability is a condition of survival. So to the extent that essential variables are constitutive of survival, rather than mediatory of it, the LRV asserts only a 'trivial' homeostasis: as Godfrey-Smith says, the explanation for why a response is adaptive can go *directly* from variation to survival (survival being, by definition, stability in essential variables). The mechanisms underlying the generation of behaviour may be genuinely homeostatic in some cases, not so in others. An interpretation of the ECT drawn from the LRV is consistent with this position, and provides a useful conceptual tool for exploring those situations in which genuine homeostasis *does* apply. (For a recent agent-based exploration of non-trivial homeostasis see Di Paolo, 2000, in which neuronal homeostasis underlies adaptation to visual field inversion in a model of phototaxis.)

It may be suggested, in summary, that the ECT fails to avail itself of the extended understanding offered by the LRV in several ways: (1) a particular emphasis on distal variability, together with the distinction between that which threatens the organism and that which facilitates its activity, (2) a potential for quantitative expression in terms of measurements of variety, and for empirical exploration in terms of instantiation of disturbances, responses, and essential variables, and lastly (3) an engagement with the idea of cognition as homeostasis, but *not* a blanket commitment to this proposition.

4.3 Construction

Externalism and internalism together define a third position, which we have not so far discussed, that of *construction;* the explanation of properties of the external in terms of properties of the internal. The third and final element of our critique concerns the role of construction in the ECT.

The most obvious interpretation of construction is when the actions of an agent alter structures of the external environment, for example when a beaver builds a dam. Something about the external environment has changed, and, because this change also figures in the *Umwelt* of the beaver, it can elicit subsequent responses from the animal. For Godfrey-Smith, the key feature of this 'narrow', or 'causal' sense of construction is that "some change is made to an intrinsic property of something external to the organic system" (ibid., p.146). For reasons explored below, this is the only sense that Godfrey-Smith entertains in his formulation of the ECT.

However, construction can also be interpreted in a 'constitutive' or 'ontological' sense. For Godfrey-Smith this is the sense in which "[f]eatures of the environment which were not physically put there by the organism are nonetheless dependent on the organisms's faculties for their existence, individual identity or structure" (ibid., p.145). The key features of this sense of construction are that properties of an organism entail the existence of features in the *Umwelt* of that organism, that the relations between these features and features of the external environment need not be straightforward, and that no change to intrinsic properties of things external to the agent need be involved.

Constitutive construction can be manifest in several ways. Simply by moving around, an agent can influence what features of the external environment can influence its activity, without necessarily altering these features as they appear in the external environment (although relational properties of the external environment may have changed, intrinsic properties of things external to the agent will have remained the same). Even the fact that an agent is a particular *size* can influence the statistical structure of its *Umwelt*, for example by determining whether or not a field is homogeneous with respect to temperature, light intensity *etc.* A third manifestation of constitutive construction attaches to the way in which organisms "transduce the physical signals that reach them from the outside world" (Lewontin, 1983, p.100). For example, an (external) environmental change - an approaching rattlesnake - that may entail a change in the rate of vibration of air molecules, which is then transduced by the organism into some feature of the agent's *Umwelt*, perhaps associated with - or identifiable with - changes in the concentration of particular chemicals, which are themselves "transformed by the neuro-secretory system into the chemical signals of fear" (ibid., p.100).

Constitutive construction therefore describes the process by which *Umwelt* is generated from the external environment, and as we have argued previously, an understanding of this process may facilitate - or perhaps be essential - in effectively relating behavioural and mechanistic levels of description. Any interpretation of the ECT that respects the distinction between behaviour and mechanism must therefore be considered weak unless it entertains a role for constitutive construction (similar concerns are entertained in Oyama, 1996; Bedau, 1996).

Why, then, does Godfrey-Smith explicitly limit his formulation of the ECT to entertain only narrow, causal construction? One reason seems to be that he fears an admission of constitutive construction would necessitate an interpretation of the mechanisms underlying behaviour as *accommodatory.* This is evident from his choice of an example of constitutive construction:

> Suppose an organism develops a way to detoxify some chemical in its environment which was formerly highly poisonous to it. This organism has made an internal change to its chemistry, and it has also made a change to the relational properties of the external chemical. There is now one less thing the chemical can poison. But if the organism has not, in doing this, made any change to an intrinsic property of any external feature then this is a paradigm case of an internal accommodation of the environment. It is the type of thing to be *contrasted* with [narrow] constructive actions such as physically removing the chemical from the environment or spraying something on it to change its intrinsic nature. (1996a, p.147)

Two points arise here. First, this example is *not* a paradigmatic example of internal accommodation in the sense in which accommodation is understood in this paper (the external somehow becoming 'internalised', a transgression of pragmatic externalism). The 'internal change to its chemistry' could perhaps be interpreted as a response, and if so, then the kind of constitutive construction alluded to in this example would *not* necessitate accommodation, would be consistent with pragmatic externalism, and as such should not pose a problem for Godfrey-Smith.

Secondly, and more importantly, it is certainly not the case that this example is representative of constitutive construction in general. Consider another hypothetical example. Suppose an organism has a property such that some types of fluctuation in light intensity constitute relevant stimuli for it, whereas others do not. The *Umwelt* of the agent thereby contains these relevant stimuli (or transformations of them). The organism responds to these stimuli (or their transformations) in order to maintain some internal variable within a certain range (perhaps by moving away from certain kinds of bright lights in order to maintain body temperature). In this example, as in the examples described in the previous section, it is clear that constitutive construction plays a role - the generation of *Umwelt* - in determining how the organism should respond to (not accommodate) the environment. Godfrey-Smith is, however, reluctant to discuss this potentially useful interpretation of constitutive construction:

> What do we say about the role which organisms play in determining which properties of the environment are relevant to them? [...] We should say that and nothing more: *relevance* is a good concept to capture these phenomena [...] The organism plays a role in making it the case that its environment contains *relevant complexity* or not. (ibid., pp.148-154)

"But what then is left of the ECT?" Godfrey-Smith asks this very question (ibid., pp.154), and he answers with a 'concession' and a 'bet'. The concession - which he describes as a concession to internalism - is that "the organic system in question does play a role in determining whether or not a given environmental pattern is relevant to it" (ibid, p.155). The bet is that once this role has been played, "there will be other organic properties that can be explained in terms of this environmental pattern" (ibid., p.155). This 'externalist bet' is to be contrasted with its internalist counterpart, that once the first role has been played (of generating *Umwelt*), there would be little or nothing left to explain about the organic system. This way of putting things seems to suggest that constitutive construction be understood as a *precondition* of application of the ECT, the generation of *Umwelt*, not as something to be explained *by* it. But this, as we have seen, is a step Godfrey-Smith appears unwilling to take, for reasons that seem to stem from an attachment to the idea that construction must feature, if at all, as a consequence of the ECT, and that as such, must involve accommodation.

But that's not all. It may also be possible to interpret construction as a consequence of the ECT, *without* this entailing an identification of construction with accommodation, by considering the generation of an appropriate *Umwelt*, in some cases, as constituting a *response* to external environmental variability. Before placing much faith in this idea, however, we must face some difficulties to do with mechanism. As remarked above, to the extent that the generation of *Umwelt* is understood as a precondition of application of the ECT, those aspects of mechanism involved in this generation can no longer be explained by the ECT; only those aspects of mechanism mediating responses, given the composition of the *Umwelt*, would fall within its explanatory domain. However, if the generation of *Umwelt* is construed as response, then those mechanism structures mediating the generation of *Umwelt do* fall within the explanatory domain of the ECT after all. The problem is that there may be no *a priori* way of disambiguating these two interpretations for any given agent-environment system. Furthermore, there may be no way of unambiguously identifying those aspects of mechanism involved in the generation of *Umwelt* and those involved in responding to stimuli in the *Umwelt* (indeed to expect this to be possible at all would be to place questionable faith being able to meaningfully distinguish between perception and action).

Here the value of concrete agent-based models is perhaps clearest of all. If there is no escaping the need for pre-existing internal structure in subserving the translation from external environment to *Umwelt* - and there isn't - then this must simply be accepted, and as we have said before (section 4.1), any concrete model of agents and environments in continuous interaction and mutual specification must *necessarily* endow agents with some initial structure, if the model is not to be infinitely trivial. In this way, once one starts building concrete models of this sort, issues of constitutive construction attach to operational details of the model and cease to be philosophical obstacles.

To give some examples, models that endow agents with stable sensors (but other potentially modifiable aspects of mechanism) can treat the constitutive construction mediated by these sensors as a precondition of application of the ECT. This then clears the way for understanding - in terms of the ECT - the behaviour patterns that respond to features of the external environment, and for understanding - also in terms of the ECT - the mechanistic structures (other than the sensors) that respond to features of the *Umwelt*. The work of Beer (1996, 2000) again comes to mind (see section 4.1), as does Seth (1998) in which reliable and noisy sensors were compared in an evolutionary robotics task involving homing navigation; it was found that evolved mechanisms integrated multi-modal sensory data in the latter case, but utilised only a single modality in the former, and the qualitative complexity of behaviour also differed across the two conditions.

By extension, models that allow the sensors themselves to adapt can admit the constitutive construction associated with this adaptation into the explanatory domain of the ECT. Harvey, Husbands, and Cliff (1994), for example, explore the artificial evolution of sensor morphologies for robots faced with the task of discriminating between triangles and squares, as they appear in the external environment. This, on the face of it a difficult problem, becomes trivial for the robot once it has adapted its sensors to generate a very simple *Umwelt*. Going even further, models that allow the entire agent morphology to adapt can be interpreted in the same way (Pfeifer, 2000; Pfeifer & Scheier, 1999).

In either case, constitutive construction associated with agent *movement* can fall within the explanatory domain of the ECT to the extent that this movement can itself be interpreted a response to the external environment. Todd and Yanco (1996), for example, explore several ways in which the adaptive significance of (externally apparent) resource 'clumps' depends upon the movement of simulated agents in a simple concrete model, and Nolfi and Parisi (1993) assess the performance of systems which have the ability to expose themselves only to sub-classes of stimuli to which they can effectively respond.

A last comment on this issue is that I am nowhere denying that 'narrow' or 'causal' construction is also important (and the distinction between causal and constitutive varieties is certainly worth keeping). By intervening in formerly autonomous external environmental processes, an agent may well bring about modifications in that environment - which may entail changes in the *Umwelt* - which can then elicit distinctive responses. The point is simple that causal construction on its own *is not always enough.*

5. Agent-based modelling

We have discussed the utility of agent-based models at several stages in this paper. To summarise: they explicitly realise the distinction between behaviour and mechanism, getting around the slipperiness of 'cognition'. They allow the quantitative exploration of the ECT from the perspective of Ashby's LRV, and, by opening *Umwelt* generation to empirical analysis, they find a place for constitutive construction in the ECT, as both precondition *and* explanandum.

We have not said much about the particular kinds of agent-based models best suited for these purposes beyond the general requirements that they separate intuitively into agent and environment components and display activity at both behavioural and mechanistic levels of description. Whilst it is beyond the scope of this paper to comprehensively cover all relevant paradigms (for example the dynamical systems approach well articulated in Beer 1995, or the various agent-based methodologies described in Pfeifer & Scheier, 1999), it is worth mentioning one broad class of model that may be particularly useful, indeed many of the examples previously discussed have been of this kind. Artificial evolution models not only satisfy the above general requirements, but also they are *not* required to prefigure the relations between behaviour and mechanism (see Nolfi, 1998). The designer can specify a (teleo-functional) fitness function, and leave the evolutionary process to work out the details of the underlying mechanism (it remains incumbent on the designer, of course, to specify in advance *some* aspects of mechanism structure). They are also (pragmatically) externalist in character, at least superficially, in that properties of the internal are moulded, by selection, to engage with properties of the external. Indeed a useful case can be made for understanding such models as extensions of biological 'optimal foraging' models in biology, which operate on exactly this premise but which are overly constrained with regard to the relations between behaviour and mechanism (Seth, 2000b).

Agent-based modelling (in general) also allows the issue of pragmatic externalism to be tackled head on. Whilst it has been stated as a requirement that such models separate intuitively into agent-environment components, and should initially be analysed as such, further analysis could examine how the explanatory story of a model would differ if *different* boundaries were demarcated between internal and external, an enterprise which could shed light on the nature of pragmatically externalist explanation, and possibly also on our intuitions about how agents and environments should normally be distinguished.

Lastly, it would be very remiss not to note that Godfrey-Smith does what few philosophers dare (as Belew, 1996, remarks) in presenting some concrete models of his own in the second part of his book. There are certainly points of contact between these models and those advocated here, one examines the costs and benefits of phenotypic plasticity, for example, and another asks with signal detection theory how much sensory acuity an agent should 'invest in' in different environments, but in general the focus is on mathematically quantifying agent and environmental variability, and less on agent-based elucidations of the relations between behaviour and mechanism. Nevertheless, they are to be highly recommended, not only for their specific insights, but also for their alternative perspectives on modelling the ECT, and not least for reassurance that effective modelling of a philosophical project is indeed possible.

6. Summary

The organising feature of this paper has been a courtship between two broad themes: distinguishing behaviour from mechanism, and relating behavioural function to environmental structure. In its simplest form, the resulting message is this: the ECT articulates the pragmatically externalist hypothesis that environmental complexity can incur behavioural complexity, but to understand how such behavioural complexity relates to underlying mechanism it is of enormous importance to understand how the agent perceives its environment, how the external environment is translated into *Umwelt.* This kind of understanding can figure both as a precondition, and as a consequence of application of the ECT.

More specifically, we have considered three related critiques of Godfrey-Smith's ECT which, in combination, lead us to a new interpretation of his idea which is that *behavioural and/or mechanistic complexity can be understood in terms of mediating well-adapted responses to environmental variability.* This interpretation differs from Godfrey-Smith's in three important ways, corresponding to the three elements of the critique offered above. The first is that there is no commitment to an interpretation of 'cognition' which obscures the essential distinction between behaviour and mechanism. This is important because the adaptationist claims of the ECT attach to behaviours, yet any behaviour can be subserved by a variety of mechanisms. This is why there an obvious ambiguity in my interpretation of the ECT; the relations between behaviour and mechanism cannot be prespecified in advance of consideration of any particular agent-environment system.

The second difference is that the interpretation of the ECT favoured here accepts - and exploits - the parallel with Ashby's LRV. This extends the understanding of the role of homeostasis, lights the way to quantitative modelling, and highlights the distinction between environmental variability threatening to the agent and environmental variability which facilitates its activity.

The third difference is the attempt to explicitly incorporate constitutive construction, on the one hand as a precondition of application of the ECT (fixed sensor structure, for example), and on the other as a response explicable by the ECT (adaptive sensors, for example, or *Umwelt* generation through movement). This third difference is intimately related to the first insofar as understanding the relations between behaviour and mechanism can be facilitated - or may even require - tracing of the relations between the external environment and the *Umwelt.*

Finally, two other distinguishing features are apparent from the wording. The term 'variability' is preferred over (environmental) complexity, largely in view of the specific ideas about variability in the perspectives of Dewey and Ashby, and the greater handle afforded for empirical analysis. Also, I speak of behavioural/mechanistic *complexity;* this is perhaps the most easily falsifiable element of the present formulation (recall, one final time, Simon's ant), but this only goes to emphasise that the ECT is, above all, a hypothesis, hypotheses are after all to be tested, and a wonderful thing about the ECT is that falsifying evidence is likely to be just as interesting as evidence adduced in its support.

Acknowledgements

I am grateful to the CCNR, the Neurosciences Research Foundation, my anonymous reviewers, and also to Peter Todd and Andy Clark for a combination of financial support and constructive discussion. The material in this paper is drawn in part from my D.Phil. thesis (Seth, 2000a).

References

Ashby, W. (1956). *An introduction to cybernetics.* Chapman Hall, London.

Bedau, M. (1996). The extent to which organisms construct their environments. *Adaptive Behaviour, 4*(3/4), 476–483.

Beer, R. (1995). A dynamical systems perspective on agent-environment interaction. *Artificial Intelligence, 72,* 173–215.

Beer, R. (1996). Toward the evolution of dynamical neural networks for minimally cognitive behaviour. In Maes, P., Mataric, M., Meyer, J., Pollack, J., & Wilson, S. (Eds.), *From animals to animats 4: Proceedings of the Fourth International Conference on the Simulation of Adaptive Behavior,* pp. 421–429 Cambridge, MA. MIT Press.

Beer, R. (2000). Dynamical approaches to cognitive science. *Trends in Cognitive Sciences, 4,* 91–99.

Belew, R. (1996). Developments across the internalist/externalist dichotomy. *Adaptive Behaviour, 4*(3/4), 483–486.

Biró, Z., & Ziemke, T. (1998). Evolution of visually-guided approach behaviour in recurrent artificial neural network robot controllers. In Pfeifer, R., Blumberg, B., Meyer, J., & Wilson, S. (Eds.), *From animals to animats 5: Proceedings of the Fifth International Conference on the Simulation of Adaptive Behavior,* pp. 73–77 Cambridge, MA. MIT Press.

Brandon, R. (1990). *Adaptation and environment.* Princeton University Press, Princeton.

Brunswik, E. (1952). *The conceptual framework of psychology.* University of Chicago Press, Chicago.

Cannon, W. (1932). *The wisdom of the body.* Norton, New York.

Dennett, D. (1987). *The intentional stance.* MIT Press, Cambridge, MA.

Dewey, J. (1929). *Experience and nature.* Dover, New York. revised edition.

Di Paolo, E. (2000). Homeostatic adaptation to inversion of the visual field and other sensorimotor disruptions. In Meyer, J., Berthoz, A., Floreano, D., Roitblat, H., & Wilson, S. (Eds.), *From animals to animats 6: Proceedings of the Sixth International Conference on the Simulation of Adaptive Behavior,* pp. 440–449 Cambridge, MA. MIT Press.

Fletcher, J., Zwick, M., & Bedau, M. (1996). Dependence of adaptability on environmental structure. *Adaptive Behaviour, 4*(3-4), 275–307.

Fletcher, J., Zwick, M., & Bedau, M. (1998). Effect of environmental texture on evolutionary adaptation. In Adami, C., Belew, R., Kitano, H., & Taylor, C. (Eds.), *Proceedings of the Sixth International Conference on Artificial Life*, pp. 189–199 Cambridge. MA. MIT Press.

Godfrey-Smith, P. (1996a). *Complexity and the function of mind in nature.* Cambridge University Press, Cambridge.

Godfrey-Smith, P. (1996b). Precis of *Complexity and the function of mind in nature*. *Adaptive Behaviour*, *4*(3/4), 453–466.

Goodwin, B. (1994). *How the leopard changed its spots: The evolution of complexity.* Phoenix, London.

Hardcastle, V. (Ed.). (1999). *Where biology meets psychology: Philosophical essays.* MIT Press, Cambridge, MA.

Harvey, I., Husbands, P., & Cliff, D. (1994). Seeing the light: Artificial evolution, real vision. In Cliff, D., Husbands, P., Meyer, J., & Wilson, S. (Eds.), *From animals to animats 3: Proceedings of the Third International Conference on the Simulation of Adaptive Behavior*, pp. 392–401 Cambridge, MA. MIT Press.

Hendriks-Jansen, H. (1996). *Catching ourselves in the act: Situated activity, interactive emergence, and human thought.* MIT Press, Cambridge, MA.

Lewontin, R. (1983). The organism as subject and object of evolution. In Levins, R., & Lewontin, R. (Eds.), *The Dialectical Biologist*, pp. 85–106. Harvard University Press, Cambridge, MA.

Lorenz, K. (1937). The nature of instinct: The conception of instinctive behaviour. In Schiller and Lashley (1957), pp. 129-175.

Lorenz, K. (1948). *The natural science of the human species.* MIT Press, Cambridge, MA. edited from author's posthumous works (1944-48) by A. von Cranach, trans. R.D. Martin, 1996.

McFarland, D., & Sibly, R. (1975). The behavioural final common path. *Philosophical Transactions of the Royal Society of London: Series B*, *270*(907), 265–293.

McLeod, P., & Dienes, Z. (1996). Do fielders know where to go to catch the ball, or only how to get there?. *Journal of Experimental Psychology: Human Perception and Performance*, *22*, 531–543.

McShea, D. (1996). Unpredictability! and the function of mind in nature. *Adaptive Behaviour*, *4*(3/4), 466–471.

Neisser, U. (1967). *Cognitive psychology.* Appleton, New York.

Nolfi, S. (1998). Evolutionary robotics: Exploiting the full power of self-organisation. *Connection Science*, *10*(3/4), 167–185.

Nolfi, S., & Parisi, D. (1993). Self-selection of input stimuli for improving performance. In Bekey, G. (Ed.), *Neural networks and robotics*, pp. 403–418. Kluwer Academic, Boston, MA.

Oyama, S. (1985). *The ontogeny of information.* Cambridge University Press, Cambridge.

Oyama, S. (1996). The ins and outs of nature and mind. *Adaptive Behaviour*, *4*(3/4), 471–476.

Pfeifer, R. (2000). On the role of morphology and materials in adaptive behavior. In Meyer, J., Berthoz, A., Floreano, D., Roitblat, H., & Wilson, S. (Eds.), *From animals to animats 6: Proceedings of the Sixth International Conference on the Simulation of Adaptive Behavior*, pp. 23–32 Cambridge, MA. MIT Press.

Pfeifer, R., & Scheier, C. (1999). *Understanding intelligence.* MIT Press, Cambridge, MA.

Seth, A. (1998). The evolution of complexity and the value of variability. In Adami, C., Belew, R., Kitano, H., & Taylor, C. (Eds.), *Artificial Life VI: Proceedings of the Sixth International Conference on the Simulation and Synthesis of Living Systems*, pp. 209–221 Cambridge, MA. MIT Press.

Seth, A. (2000a). *On the relations between behaviour, mechanism, and environment: Explorations in artificial evolution.* Ph.D. thesis, University of Sussex.

Seth, A. (2000b). Unorthodox optimal foraging theory. In Meyer, J., Berthoz, A., Floreano, D., Roitblat, H., & Wilson, S. (Eds.), *From animals to animats 6: Proceedings of the Sixth International Conference on the Simulation of Adaptive Behavior*, pp. 471–481 Cambridge, MA. MIT Press.

Seth, A. (in press). Distinguishing adaptive from nonadaptive evolution using Ashby's law of requisite variety. *To appear in Proc. 2002 IEEE Congress on Evolutionary Computation (CEC2002).*

Shapiro, L. (1999). Presence of mind. In Hardcastle, V. (Ed.), *Where biology meets psychology: Philosophical essays*, pp. 83–99. MIT Press, Cambridge, MA.

Simon, H. (1988). *The sciences of the artificial.* MIT Press, Cambridge, MA. 3rd edition.

Thompson, W. (1917). *On growth and form.* Cambridge University Press, Cambridge.

Todd, P., & Yanco, H. (1996). Environmental effects on minimal behaviours in the minimat world. *Adaptive Behavior*, *4*(3/4), 365–413.

von Uexküll, J. (1934). A stroll through the worlds of animals and men. In Lashley, K. (Ed.), *Instinctive behavior.* International Universities Press, New York.

Ziemke, T. (2000). On the parts and wholes of adaptive behavior: Functional modularity and diachronic structure in recurrent neural network robot controllers. In Meyer, J., Berthoz, A., Floreano, D., Roitblat, H., & Wilson, S. (Eds.), *From animals to animats 6: Proceedings of the Sixth International Conference on the Simulation of Adaptive Behavior*, pp. 115–124 Cambridge, MA. MIT Press.

Ziemke, T., & Sharkey, N. (2001). A stroll through the worlds of robots and animals: Applying Jakob von Uexküll's theory of meaning to adaptive robotics and artificial life. *Semiotica*, *134*, 701–746.

PERCEPTION AND MOTOR CONTROL

The origin of "feel"

J. Kevin O'Regan
Laboratoire de Psychologie Expérimentale
Centre National de Recherche Scientifique,
Université René Descartes,
92774 Boulogne Billancourt, France
http://nivea.psycho.univ-paris5.fr

Alva Noë
Department of Philosophy
University of California, Santa Cruz
Santa Cruz, CA 95064
http://www2.ucsc.edu/people/anoe/

Abstract

This paper proposes a way of bridging the explanatory gap between physical processes in the brain and the "felt" aspect of sensory experience. The approach is based on the idea that experience is not generated by brain processes themselves, but rather is constituted by the way these brain processes enable a particular form of "give-and-take" between the perceiver and the environment. From this starting-point we are able to characterize the phenomenological differences between the different sensory modalities in a more principled way than has been done in the past. We are also able to approach the issue of the experiential quality of sensory awareness and consciousness in a satisfactory way. Finally we describe one of the empirical consequences of the theory, namely the phenomenon of "change blindness".

1 Introduction

1.1 Sensory experience as a way of doing things

What is the "feel" of driving a Porsche as compared to other cars? To answer this question you would probably say: the feel of Porsche-driving comes from the particular way the Porsche handles, that is, the way it responds to your actions. When you press on the gas, the car accelerates particularly fast. When you turn the wheel, the car responds in a way typical of Porsches. Note that at a given instant you might actually not be doing anything at all (as when, for example, you cruise along at high speed without moving the steering wheel or pressing the accelerator). Despite the fact that you perform no physical movements, you continue to undergo the experience of driving the Porsche.

The experience of driving a Porsche would seem to consist not so much in the occurrence, as it were all at once, of a certain sensation (something like a twinge or a wave of dizziness), but rather in the characteristic pattern of integrated activity in which the actual driving of the car consists. The particular experience of Porsche-driving comes from the typically Porsche-like give-and-take between you and the car when you drive the Porsche.

This analysis of the experience of driving a Porsche in terms of a temporally extended pattern of activity has two consequences. First, it suggests the possibility that to investigate the nature of experience we must direct our investigations not to some ineffable inner event, but rather to the temporally extended activity itself, to the laws that govern this activity. We shall see that on this approach, the problem of understanding experience may become more tractable.

Second, it comes to seem doubtful that the Porsche-driving experience is the sort of thing that can be generated by activity in the brain. It seems clear that there is no special circuit that itself underlies the Porsche-driving experience. There are of course brain mechanisms that participate in all the various Porsche-driving behaviors, but none that can be thought of as sufficient to produce the Porsche-driving experience.

In the present paper we attempt to show how an analogous approach can be applied to sensory experience in general. Take seeing for example. Like Porsche-driving, seeing is a temporally extended pattern of activity. To see is to be skilled in this activity. Like Porsche-driving, we shall claim that visual experience does not consist in the occurrence of "qualia" or such like. Rather, it is a kind of give-and-take between you and the environment.

Moreover, we claim, there are no states or processes in the brain that *generate* the experience of seeing. Brain processes *participate* in seeing, but none deserves to be thought of as "the locus of seeing in the brain". Seeing is something we do, not something that happens in our brains (even though, of course, a lot goes on in the brain when we see).

A first advantange of this activity-based approach is that it enables us to overcome the problem of what has been called the "explanatory gap" (c.f. Levine [1983]; also e.g. Chalmers [1996]) -- that is, the problem of understanding how something physical like the brain, can generate something non-physical, namely experience. We have solved the problem by noting that experience is not generated in the brain at all.

Second, the approach allows us to actually characterize the quality of sensory phenomenology, and to explain why nervous influx coming, for example, into the visual cortex appears to give rise to qualitatively different experiences than those arising from nervous influx coming into the auditory, or any other sensory cortex. Just as driving is different from fishing or cooking, seeing is different from hearing because it involves *doing different things*. This explanation of the difference in phenomenology between the different senses has the advantage of not requiring invocation of any kind of "specific nerve energy" (Müller [1838]) or other essence or mechanism that somehow endows each sensory pathway with its specific "feel".

A third advantage of taking this approach to seeing, is that it sheds a new light on a number of classic problems in visual science, sometimes causing them simply to evaporate. A problem which evaporates is the problem of why the world seems stable despite eye movements and the problem of why we do not notice the blind spot in our visual field. Another problem which becomes a non-problem is the so-called binding problem (that is, the problem of how information about different visual attributes like color, texture, depth, form, movement, etc., are combined into a single spatially and temporally unified percept). The theory also makes empirical predictions, among them the phenomenon of change blindness (in which large portions of a visual scene can be modified without this being noticed), and the phenomenon of sensory subsitution (in which blind persons can obtain a form of vision through tactile stimulation). The details of these applications of the approach can be found in O'Regan & Noë [2001].

In the next sections we shall concentrate on the visual sense modality. We shall distinguish one aspect of the activity of seeing which can be related to what psychologists call "sensation", and one aspect which can be related to what they call "perception". We shall then define a specific sense of the notion of visual "awareness" and go on to consider the problem of accounting for the "feel" associated with seeing.

2 Sensation

2.1 Exercising mastery of apparatus-related sensorimotor contingencies

Consider a missile guidance system. When the missile is following an airplane, there are a number of things the missile can do which will result in predictable changes in the information that the system is receiving. For example, the missile can go faster, which will make the image of the airplane in its camera become bigger. Or it can turn, which will make the image of the airplane in its camera shift. We shall say that the missile guidance system has *mastery of the sensorimotor contingencies* of airplane tracking, if it "knows" the laws that govern what happens when it does all the things it can do when it is tracking airplanes.

Note that the sensorimotor contingency laws that the missile guidance system has mastered are laws which depend on the nature of three-dimensional space and how the missile's apparatus senses things in that space: the type of optics and the type of projection used, the way the image is formed in the camera, the way it is sampled, etc., will all play a role in determining the laws that link the missile's possible behaviors to the resulting sensory inputs. If the system were using radar or sonar instead of visual information to track airplanes, the laws would be quite different.

Having mastery of the sensorimotor contingency laws governing airplane tracking does not mean that the missile guidance system is *currently* tracking an airplane -- it might currently be cruising with no airplane in view, or it might not be flying at all. Only when the laws of airplane tracking are currently being *put to use*, that is, only if the missile guidance system is *exercising* its mastery of airplane tracking is it true to say that the system is visually tracking the airplane.

But note also that visually tracking the airplane does not actually require any input to be coming into the missile guidance system. For example, suppose that the airplane has momentarily slipped out of the missile's view. Then the system's airplane detector circuits will not currently be active. Yet so long as the missile guidance system is, for example, tuned to the fact that it can turn to bring the airplane back into the camera's sights, we still would say that the missile guidance system is currently visually tracking the airplane. To say that it is "tuned to the fact that it can turn, etc.", is to say something about what the tracking system could do if it were presented with particular types of input.

We wish to make the parallel between airplane tracking by the missile guidance system, and sensation in humans. We shall say that perceivers have sensations in a particular sense modality, when they *exercise their mastery of the sensorimotor laws* that govern the relation between possible actions and the resulting changes in incoming information in that sense modality. Note that this use of the word sensation is somewhat unconventional, because it treats sensations as something that might be construed as unconscious. After all, on this view, the missile guidance system has sensations of some sort. We shall take this very limited definition of sensation (perhaps we should call it "proto-sensation") at this stage, and show later how it can be used within a more complete framework to deal with the "experiential" aspects of sensation.

2.2 The sensation of red

As a concrete example of what it means to have a visual sensation under this approach, consider what it means to have the sensation of a red patch of color.

Under the classical view, the sensation of red occurs when a particular ratio of excitation occurs in the long, medium and short wavelength sensitive mechanisms in the visual system. There are additionally some color constancy mechanisms which compensate for the effect of ambient lighting conditions. The output of these processes then is read by some other cortical process which then presumably generates the sensation of red. But how does this come about? What mechanism in the brain could take neural excitation from some cortical process, and convert it to experience? This is an outstanding and hotly-debated topic in recent philosophy and cognitive science.

Under the view espoused here, the sensation of seeing a red patch of color does not arise from activation of some cortical mechanism (even if that activity is necessary for the experience). On the contrary, the sensation of red occurs when the brain is tuned to certain very particular things that will happen to the neural influx if we do certain things with respect to the red patch of color. For example, one thing we can do is close our eyes, at which point there should be a very dramatic change in the neural influx. This particular sensorimotor contingency is an indication that we are *seeing* the red patch of color, and not hearing or touching it, since eye-closing has no effect in the auditory or tactile modalities. Another thing we can do is move our eyes off the patch and back onto it. As the red patch moves from fovea into periphery, retinal sampling becomes more sparse and color information changes its nature, with influx from monochromatically tuned rod photoreceptors taking over as compared to influx from the three different cone types present in central vision. Another difference between sampling the red patch with central and peripheral vision derives from the fact that the central retina is covered with a yellow macular pigment, which absorbs light in the short wavelengths.

As a result of such differences, lawful changes in the neural influx occur as a function of the eyes' position. The laws underlying these changes, that is, the sensorimotor contingencies, are indicative of the fact that the patch is being sampled by the visual apparatus, and not via, say, the olfactory or tactile modalities.

Another kind of sensorimotor contingency is related not to the effect of eye movements, but to the effect of moving the red patch itself, or moving ourselves with respect to it. Yellowish sunlight or reddish lamplight have different spectra. Depending on how we move our bodies or how we move the patch, the proportion of sunlight or lamplight reflected off the patch will change, and the spectrum of the light coming into our eyes will change in a way that is typical of red, and not of other colors. Broackes [1992], in a theory of color which could be considered a special case of our theory, says that "color is the way surfaces change the light". D'Zmura & Iverson [1994] have developed a mathematical approach to color constancy related to this idea.

Exercising our mastery of all these possible kinds of sensorimotor contingency constitutes the sensation of red. The sensation of red is the exercise of our mastery of the way red behaves as we do things. The sensation of red is a way of doing things, not something that emerges from neural excitation. (Reminder: we are using sensation in the limited, proto-sensation sense defined above.)

But note that though we identify the sensation with a pattern of skillful activity, one can have the sensation even when one is, at the moment, inactive. (Just as the the missile guidance system can be tracking the airplane even though no information about the airplane is currently being received.) We can exercise our mastery of the sensorimotor contingencies that signal the presence of redness without actually blinking, or moving our eyes or our bodies with respect to the red patch. This exercise consists in our practical understanding that if we *were* to move our eyes or bodies or blink, the resulting changes would be those that are typical of red, and not of green patches of light. Indeed, psychologists over the last century have extensively studied vision under tachistoscopic presentation conditions, where presentation durations are so short that observers cannot properly exercise their mastery of visual contingencies. Because visual stimuli are relatively redundant it is nevertheless possible to deduce from such a momentary stimulation what the consequences of one's actions would be if one were allowed to move. Under the view taken here, visual sensation that occurs in tachistoscopic experiments is constituted by this kind of practical knowledge, and does not derive in a direct fashion from the neural influx that activates the visual pathways.

Note also that to have the sensation of red, there need actually be no excitation currently coming into the brain which by itself might correspond to red: this is the situation when we blink. Thus, unless we pay attention to the fact that we are blinking, the sensation of redness (and for that matter the sensation of seeing) does not go away during the blink. This is also the situation when we see a red object first in central vision and then in peripheral vision: for a given object, the information about color available in peripheral vision is dramatically reduced as compared to what is available in central vision -- and yet the perception of redness persists. This is because what provides the sensation of redness is not the neural influx, but the knowledge of how the influx would change if you were to move your eyes.

2.3 *"Knowledge," "mastery," "attuned"*

On the theory of sensation presented here, to have a sensation is to exercise one's mastery of the relevant sensorimotor contingencies and in this sense to be "attuned" to the ways in which one's movements will affect the character of input. We characterize this attunement as a form of practical knowledge. These terms are vague and may be somewhat confusing. Perhaps it would clarify our point to consider the example of a phototactic device such as one of Braitenberg's simple "vehicles" (Braitenberg [1984]). The imagined vehicle is equipped with two light sensors on the front of the wheeled vehicle. The left sensor is linked to the right rear wheel driving mechanism and the right sensor is linked to the left rear wheel driving mechanism. As a result of this wiring, the vehicle will orient itself toward light sources and move towards them. Such a simple mechanism can track and hunt light sources.

Note that the ability of this extremely simple mechanism to "sense light" does *not* consist just in the fact of activity in the sensor. Light causally affects the sensors, to be sure. But the vehicle's ability to sense the light consists in the ways in which it makes use of the raw stimulation of the light on the sensor. If there were no linkage between the activity of the sensor and the driving mechanism, then activation in the sensor would have no behavioral or ecological implications for the vehicle and it would be as good as blind. Having sensations, as we are using the term, consists in the use the system makes of input (whether it is a simple Braitenberg vehicle or a complex organism). Any system capable of making use of input is a system that is attuned to patterns of sensorimotor contingencies, and so is one that can be regarded as having practical knowledge or mastery of these contingencies.

Note one final consequence of this account of sensation. The sensation does not occur in the brain any more than it occurs in the sensors of the vehicle. The occurrence of sensation is a system or creature-wide phenomenon. Mastery of sensorimotor contingencies may be neurally encoded, but this mastery does not itself reside in the brain. To the extent that it makes sense to speak of mastery residing anywhere, then it resides in the creature as a whole, in the whole neurally enlivened body.

2.4 Sensation concerns apparatus-related sensorimotor contingencies

Let us examine more closely the different kinds of sensorimotor contingencies that are typical of sensation in the visual, as opposed to nonvisual, sensory modality.

We have seen from the example of the sensation of red that one aspect of the sensorimotor contingencies is related to the structure of the visual apparatus: blinks cause a drastic change in the sensory input; moving the eye causes lawful changes in the color information available from a stimulus. Other laws concern the spatial sampling of the image: foveal resolution is much finer than peripheral resolution, so as the eye moves, the resolution of the sampled information changes in a lawful way which is typical of the visual modality. Another lawful relationship is determined by the optics of the eye. Owing to the spherical shape of the eyeball, there are certain distortions that occur in the image when it shifts -- for example the curvature of the arc formed on the inside of the ocular sphere by looking at a straight line changes as the eye moves on and off the line (this point has been discussed by Platt [1960], and in a more general fashion by Koenderink [1984]). These changes are also affected by the eyes' state of accomodation, and by spherical and chromatic aberrations in the eyes' optics.

While the sensorimotor contingencies we have just discussed are predominantly a consequence of the structure of the visual apparatus, another somewhat different class of sensorimotor contingency is related to the 3D structure of space. Because the visual apparatus is sampling a two-dimensional projection of three-dimensional space, certain laws apply. For example, when you move forwards or backwards, there is an expanding or contracting flow-field on the retina; the amount of light impinging on the eye is governed by the inverse square law. There are also more complicated laws which, as pointed out by Piaget [1937/1977] in his studies on how the notion of space is acquired, are related to the (mathematical) group structure of three-dimensional space (see also Poincaré [1905] and Husserl [1907/1991] for related ideas). Thus for example, if you move one step forward and one step sideways, there is a diagonal direction you can move back along such that the nervous influx comes to be the same as it was to begin with.

We have distinguished two classes of sensorimotor contingencies that are typical of vision: one that is a consequence of the structure of the visual apparatus, and one that is a consequence of the nature of 3D space. What should be noted is that both kinds of sensorimotor contingencies are independent of the nature of the objects which are being seen. The laws that govern the changes that are created by blinks, eye movements and body movements are global laws that apply to whatever objects are in the visual scene. Thus, these particular types of sensorimotor contingency, because they are independent of objects but particular to the visual modality, could be considered to constitute the defining characteristics of visual sensation, and they are what distinguish vision from sensation in other modalities.

Take the tactile modality as a further example. Here the sensorimotor contingencies are quite different from the visual modality. Moving one's hand over a surface causes the texture elements to move in a way analogous to how the image shifts when one moves one's eye. But there are a number of important differences. For example, the way tactile acuity varies over the surface of the hand is nothing like the way spatial acuity varies over the retina. Furthermore, bringing one's hand nearer and farther from an object does not create an expanding or contracting tactile flow-field like that created on the retina. On the contrary there are laws of contact and pressure which have no equivalent among the sensorimotor laws of vision.

It is a very important aspect of our approach to sensation that we claim that what determines the particular, visual, tactile, auditory, olfactory, etc. nature of a stimulation is nothing directly to do with the sensory pathways or brain areas which carry the nervous influx. Rather, what determines the experienced sensory modality of a stimulation are the sensorimotor laws governing that stimulation. This important concept opens the possibility for sensory substitution. If we can, through tactile stimulation for example, create sensorimotor contingencies similar to those usually obtained via visual pathways, we should be able to create sensations through touch which will be perceived of as having a visual nature. There is indeed evidence that this is possible.

Thus, Bach-y-Rita [1972] (see also Bach-y-Rita [1967]; Lechelt [1986]) mounted a 20 x 20 array of tactile vibrators on the back or abdomen of a blind or blindfolded person, and found that with surprisingly little practise, the subjects report that they no longer attend to the tickling stimulation on the skin, but "see" objects in front of them. Furthermore, suddenly increasing the magnification of the video camera by jogging its zoom adjustment can cause users to involuntarily jump back, as though they saw that objects were going to collide with them (Bach-y-Rita [1972]) -- and this is true irrespective of whether the stimulator array is mounted on the observer's back, abdomen, or other cutaneous area.

The discussion remains lively of course as to what extent we should say people really "see" with such a device (Morgan [1977]), and it is interesting to note that blind people seem not as interested in using the devices as the engineers who construct them had hoped (Easton [1992]; Bach-y-Rita [1983]). Nonetheless, first-person reports by people who have learnt to use the devices certainly favor the notion that some kind of vision is involved (Guarniero [1974]; Guarniero [1977]; Apkarian [1983]).

3 Perception

3.1 *Exercising mastery of object-related sensorimotor contingencies*

Consider again the missile guidance system. When it is tracking an airplane, it is exercising its mastery of the sensorimotor contingencies imposed by the structure of its sensing apparatus and by the nature of 3D space. But note that it may be following different kinds of airplanes. Perhaps there are ones that turn slowly, and others which make fast, sophisticated evasion maneuvers. The missile guidance system may adapt to one or other type of airplane, and, to do this, may invoke different strategies.

Contrary to the sensorimotor contingencies which we considered in the previous section, therefore, there are other sensorimotor contingencies which are related, not to the nature of the visual apparatus or to the 3D space in which it is embedded, but rather to the nature of objects themselves. A pitcher, for example, is a thing that has the property that, depending on how you turn it, a protuberance (the handle) appears and disappears. A glass, in contrast, is a thing that does not have this property. We suggest that *perception* could be considered to be the exercise of mastery of this kind of object-related sensorimotor contingency.

Note that, just as was the case for our definition of sensation, there is nothing about our definition of perception which prevents a machine from having perception. We shall come to the specifically human aspects of perception later.

As a concrete illustration of what perception is under this approach, consider first an example from the domain of tactile perception: the children's game in which a household object like a cork, a potato, or a pencil sharpener, for example, is put in an opaque bag, and the child must attempt to identify the object by feeling it with his hand in the bag. The striking aspect of this game is that at first, when you have not yet identified the object, you are aware of local bits of texture, protuberances, edges, etc., but not of holding a particular object. Suddenly however, the "veil falls", and the previously unrelated parts come together into a whole. You no longer have the impression of a collection of incomprehensible protuberances, smoothnesses, edges, but of holding, say, a swiss army knife. It is worth playing this game in order to understand this sudden feeling of recognition, like an illumination. Once the illumination has occurred, you no longer feel the local sensations that you were feeling before, but you feel the object as a whole object. Even parts of the object that your fingers are not currently in contact with, somehow are perceived as being present.

We suggest that this feeling of presence derives from the fact that once the object has been recognized, you "have tabs" on it, you can exercise your mastery of the way it "behaves" under your grasp. You know that if you move your fingers upwards, you will encounter the ring attached to one end of the knife, and if you move the other way, you will encounter the smoothness of the plastic surface, and the roughness of the corkscrew, etc. It is this knowledge which *constitutes* the haptic perception of the object. In fact you need not do anything at all, and yet you have the acute feeling of holding a swiss army knife.

Notice that the fact that the sensitivity of different parts of your hand (fingertips, palm, finger nails, etc.) is quite different, and the fact that your hand is composed of fingers of different shapes and lengths and having spaces between them, has no impact on your ability to recognize the object. You know the difference that will occur if you use your fingernails rather than your fingertips to touch the surfaces, and if you lodge the object in the palm of your hand or put it inbetween finger and thumb. In fact your fingernails and the spaces between your fingers, far from being useless non-sensitive zones of the hand, can be used actively as a way of assessing the texture or the size of parts of an object.

The same analysis can be applied to visual perception -- indeed the idea that vision might consist in "palpation" by the eyes was already mentioned by Merleau-Ponty [1968] and developed by MacKay [1962; 1967; 1973], who suggested considering vision like a giant hand that samples the environment. In O'Regan [1992] and O'Regan & Noë [2001] we present an illustration of how this approach can explain why we do not see the blind spots in our eyes, and why eye movements do not cause the world to appear to shift. The explanation we propose is simpler and more natural than the explanations in terms of special compensatory mechanisms that are usually invoked, like "filling-in" or "saccadic suppression".

4 Perceptual Consciousness

This account of sensation and perception in terms of the mastery of sensorimotor contingencies goes some way towards characterizing what it is like to see. But to the extent that we have stressed that everything we have said could be applied to a machine, we still need to explain what is additionally needed to provide "feel." This is crucial if we are to give an account of full-fledged human perceptual awareness.

4.1 *Awareness: Integrating sensorimotor mastery into planning behavior*

Let us restrict use of the word "awareness" to "awareness of something in particular". Let us say that a person (or system or machine) is perceptually aware of something if the system makes use of perceptual information about the thing for the purpose of planning, rational thought or linguistic behavior. This definition of awareness has also been used by Chalmers [1996] -- it corresponds to what Block [1995] calls "access-consciousness".

A driver, for example, would be said to be *aware* of a red traffic light if, in addition to the mastery of sensorimotor contingencies associated with the red light, his attunement to these sensorimotor contingencies is integrated into his planning, rational thought or linguistic behavior.

Depending on the extent to which the seeing of the red light is incorporated into his planning or thought, the driver would be said to be aware of the red light to varying degrees. For example if he is driving while at the same time engaged in animated conversation with a friend, he may not be aware of the red light at all. Or he may be indirectly aware of it, because he realizes that stopping is going to make him late for an appointment. If the driver is an artist interested in shades of red, on the other hand, he may notice that this traffic light has a different hue than usual, and therefore be particularly aware of the color of the light, though he may not notice its shape.

Notice that there is nothing about this characterization of awareness that prevents its attribution to a machine -- it is conceivable that a sufficiently complex machine, with a sufficiently wide range of possible behaviors, in a sufficiently rich environment, could reasonably be said to be aware in this sense. Certainly in order for the notion to be applicable however, the machine would have to have quite a sophisticated reasoning ability and would have to reside in quite a complicated environment. A medical diagnosis system or chess-playing machine could be said to evaluate possible plans before deciding on what actions to undertake. But one would not want to say that the machine is "aware" of the decisions it is making. Awareness seems to require broader social capabilities than simply making a diagnostic or advancing pieces on a chessboard. If the machine could additionally evaluate the impact that they were going to have on the patients or chess opponents (say, by not telling a patient he has cancer, or purposefully losing a game in order not to upset a child), then one would be more willing to use the word "aware" in describing the machine's behavior.

Nevertheless, the important point to note about this discussion is that there is nothing about our notion of awareness which logically precludes it from being described in scientific terms, and from being implemented in a machine.

4.2 *Visual experience: awareness of vision*

The question is now, given this definition of awareness, whether what we generally call the perceptual experience of red, say, corresponds precisely to nothing more or less than: awareness of having mastery of the sensorimotor contingencies associated with red.

Certainly by adopting this definition we have *partially* accounted for what it is like to see something red. After all, we have presented, in admittedly rough outline, the factors that differentiate the seeing of something red from the haptic feel of something red, or the seeing of something blue. What more is there to be accounted for?

It is sometimes suggested that when you see something red, you have a feeling -- a "raw feel" -- that is distinct from your mere awareness of what you see. A mere robot, so it is reasoned, could have mastery of the relevant sensorimotor contingencies (and so have the corresponding sensation and perception), and it could also make use of its mastery for the purpose of appropriately guiding its actions and making plans (thus exercising awareness), without undergoing the *feel* of red. It is the presence or absence of this distinctive qualitative state that makes it the case that there is, (in the phrase of Nagel [1974]) something that it is like to experience red. Unless we can account for this, we have failed to explain perceptual consciousness.What then is missing from our account?

One might respond: one further not-yet-explained element is that distinctive feeling of the *ongoing presence* of redness. But the explanation of this feeling of the continual presence of redness is ready to hand. Suppose you look at a red colored wall. The redness is on the wall, there to be appreciated. Because we have continuous access to the present redness, it is as if you are continuously in contact with it. This would explain the fact that the redness would seem to be continuously present in experience. This point can be sharpened. The "feeling of ongoing presence of redness" that would seem to accompany the seeing of something red is to be explained by the fact that we understand (in a practical sense) that at any moment we can direct our attention to the redness of the wall. That is to say, our feeling of present redness consists in our *awareness* of our immediate perceptual *access* to environmental redness.

This account gains in credibility when we recall the feeling of "illumination" that occurs when a stimulus is recognized in the child's hand-in-the-bag game: one suddenly really "feels" the presence of a whole object. This feeling is clearly nothing more than the coming together of the knowledge associated with the different parts of the object: a kind of "being at home" with the various things that one can do in regard to the object. This example suggests how it might be possible that a state of knowledge about existing contingencies could actually *constitute* an ongoing, continuous feeling of presence.

Another illustration which may help in rendering our approach more plausible is to consider cases where the word "feeling" is used outside what is normally taken as the domain of sensory experience. Consider what is meant when a person says he "feels rich". What he means is that he can, if he wishes, buy a yacht or an expensive car, take cash out of the bank or go on a cruise around the world, etc. His feeling of richness actually consists in knowledge of things he can do, without actually having to do them (other examples might be feeling British, feeling lonely, virtuous, etc.).

Let us here note another important point. Obviously the "feeling" of richness does not have the same acute ongoing quality as the experience of a red light or a bell ringing or a pain. This difference could reside in what we might call "bodiliness": the fact that, unlike the case of richness, the knowledge associated with perceptual experience is intimately linked to motor activity: the minutest eye-muscle twitch creates an associated change in the neural influx from the retina, but no muscle twitch changes the availability of cash at the bank. Another fact that gives perceptual feelings a more "real" quality could be that the associated sensorimotor contingencies are highly ingrained, learnt during maturation, and known perhaps only tacitly. Finally, another important difference with other mental activities is what we call "grabbiness": the fact that low-level perceptual systems possess genetically determined orienting responses which cause the perceptual apparatus to exogenously orient to sudden changes in stimulation. Thus, when a light turns on, the eyes are attracted to the change, but when your bank account goes empty, there is no immediate, automatic reaction of your body. Sensory stimulation thus provides sensorimotor contingencies which are profoundly driven by local external events, whereas feelings like those of "being rich" or "being virtuous" are not so tightly sensorimotor.

In sum, we claim that sensory experience is nothing more or less than: awareness of exercising mastery of sensorimotor contingencies. The differences in experienced quality of the different sensory modalities derives from the different laws of sensorimotor contingency involved. The particular, "felt" quality of sensory consciousness derives from the "bodiliness" and "grabbiness". of sensory input channels. This is what gives sensory experience its continuous, ongoing, "present" quality, and differentiates it from memory, knowledge, and states like richness or virtuousness, for example, which do not really provide any kind of "raw feel" at all.

4.3 *"Change blindness" and the world as an outside memory*

In the course of the arguments presented here, we have already alluded to examples of implications of our theory for empirical research. One implication concerned the possibility of sensory substitution. Another implication concerned the filling in of the blind spot and the apparent stability of the visual world despite eye movements. These and other implications are discussed in more detail elsewhere (O'Regan [1992]; O'Regan & Noë [2001]). Here we wish to mention an implication of the theory which concerns the nature of the internal representation of the world, and show how this was tested using experiments on "change blindness".

The subjective impression people have of their visual world is one of great richness, with the feeling of almost infinite detail and color spread out before their eyes. The most natural explanation for this would appear to be the idea that in the brain, there is an internal representation of the world whose activation provides this experience. In order to provide the perceived richness of the experience, the internal representation must presumably also be very rich.

The theory we have developed here rejects the notion that experience is caused by activation of any brain mechanism -- for if it were, then we would have to explain how activation of some brain mechanism (something material) provides experience (something immaterial). On the contrary, our theory states that the character of the visual experience of the external world depends on the *fact that the visual world is immediately accessible to our exploration.* We see something, not because its image is impinging on our retinas, but because we know we can, by moving our eyes, our bodies, or the object itself, change the retinal stimulation that the object creates in certain known ways.

A way of thinking about this view of what seeing consists of, is to adopt the analogy of what might be called the "the world as an outside memory" (O'Regan [1992]; Minsky [1988]). Instead of supposing that we have visual experience because an internal representation of the world is activated, we claim that the outside world functions in some sense as an external memory store. Consider what happens with normal memory: The slightest flick of your thoughts to something you want to remember -- say the color of your grandmother's eyes -- suffices to bring that thing to your attention, and you can immediately recall it. Having memory of your grandmother's eyes thus consists in the fact that you know you can, by a mental effort, bring that bit of information into your mind. Similarly, with vision, the slightest flick of your eyes or of attention can bring up details about something before you. Thus, seeing something consists in the fact that you know you can, by the appropriate eye, body, or object movement, cause changes that provide information about that thing.

But the view we are taking is rather curious, because it says that to have the ongoing, occurrent, perception of richness of the visual world, the richness does not actually have to be continually impinging on the retina or activating some internal representation. It just has to be *potentially* obtainable by the appropriate movement.

From the point of view of brain storage and efficiency of calculation, it is of course a great advantage not to have to store the contents of the entire visual field, but rather to use the world itself as its own storage buffer.

On the other hand, this point of view makes a rather counter-intuitive prediction. It predicts that if something in the visual world were to change, unless at the moment of the change you happened to be attending to it, you would not notice the change. Since attentional capacity is extremely limited -- we can only consciously think about one thing at a time -- the question then arises of why we do not normally miss all sorts of things that change in our visual field?

The answer undoubtedly lies in "grabbiness": the fact that, as we have pointed out before, the low-level visual system possesses automatic transient-detection mechanisms which cause attention to be exogenously oriented towards any change in luminance, color or position in the visual field. The best known example of such mechanisms might be motion detectors. When anything changes in the visual field, the eye, or attention, is immediately directed to the location of the change, and the changing element, because it is being attended to, is seen. This low-level "watchdog" mechanism therefore provides a way of accounting for perceptual awareness of details in the environment without supposing that everything is actually stored in the brain.

These considerations lead to the possibility of some interesting empirical verifications. If somehow we could render the normal attention-grabbing transient-detection mechanisms inoperative, then changes in a scene should not be noticed unless they were being directly attended to. We tested this prediction in a series of experiments in which the attention-grabbing transient was swamped in a variety of ways by the occurrence of other, irrelevant transients.

In one set of experiments we used a brief flicker that covered the whole image at the moment that the "true" scene change occurred (Rensink, O'Regan, & Clark [1997]; Rensink, O'Regan, & Clark [2000]). In another experiment the scene change was synchronized with the occurrence of a blink (O'Regan, Deubel, Clark, & Rensink [2000]). Other workers have synchronized scene changes with the occurrence of eye saccades (Grimes [1996]; McConkie & Currie [1996]; Irwin & Gordon [1998]). The principle of these experiments is that in all cases, the "true" change occurs simultaneously with a brief global change in the retinal image. This global change produces a large transient that overloads the local transient detectors which would normally signal the change location. Attention is therefore prevented from orienting to the location of the change. Another experimental paradigm that we have used involved what we called "mudsplashes": like mud splattered briefly on a car windshield, these created brief diversions which, when synchronized with the occurrence of a change in the visual field, acted as decoys, and attracted attention away from the location of the "true" change (O'Regan, Rensink, & Clark [1999]).

As predicted from the theory, in all these experiments very large changes could be made without observers noticing them. The changes could be so large that they occupied a significant portion of the picture, and were flagrantly obvious when not synchronized with the flicker, blink, saccade, or mudsplash (examples can be seen on the website http://nivea.psycho.univ-paris5.fr). The phenomenon was particularly striking in the "blink" experiment, where, it was shown by measuring eye movements, that even when people were looking directly at the change, in almost 50% of the cases they did not notice it (O'Regan et al. [2000]). Of course the fact that a person can be directly looking at something and yet not see it is quite compatible with our theory, since, as we have said, what determines whether you are aware of seeing something is not the fact that it impinges on your retina, but rather the fact that you are making use of it in your planning, thought, decisional or linguistic behavior. Even when fixating something directly, there may be some aspects of what you are fixating which you are, in this sense, not making use of, and so which you will not notice if it changes. For example, fixating in the middle of a word, you may be recognizing the word and not notice that the shape of the very letter you are fixating has changed, or that its color or the background color visible behind the letter has changed.

5 Conclusion

We have presented a new framework within which to study sensory experience. Our most basic idea is that attempts to explain perceptual consciousness (experiences of hearing a bell, seeing a light, etc) by postulating a brain mechanism which is thought of as sufficient to produce the experience must fail. Such theories, we believe, must always fall afoul of the explanatory gap: that is, they must always fail to explain how it is that neural activation in the brain can give rise to the experience.

On our view, experience is not something that happens in us but is something we do. Perceptual experience consists in ways of exploring the environment. We have decomposed these ways of exploring the environment into several parts.

The first two parts correspond to what is usually called sensation and perception and occur when a person exercises mastery of the laws of sensorimotor contingency that govern how actions affect sensory inputs. Sensation occurs when a person exercises mastery of those sensorimotor contingencies which are typical of a sensory modality (in general those contingencies that are related to the way the modality samples the space in which it is embedded).

Perception occurs when a person has mastery of those sensorimotor contingencies which are typical of the way the attributes of objects are sampled by the particular sensory modality.

When one's sensation and perception are integrated into current planning, rational thought and speech behavior, we say that the perceiver is perceptually aware of that which is perceived.

We have also shown that what could be called the "raw feel" of sensory experience can be accounted for under our theory. This is because what people mean by "raw feel" can be understood in terms of differences in the patterns of sensorimotor contingencies governing exploration of the world in different modalities. The apparent "ongoingness", "presence", and the peculiar sensory, "felt" nature of perceptual experience as compared to other mental states can be understood in terms of what we have called the "bodiliness" and "grabbiness" of sensory input channels.

Acknowledgments

We thank Ken Knoblauch for discussion on the draft of this paper. A.N. gratefully acknowledges the support of faculty research funds granted by the University of California, Santa Cruz. (This paper is a considerably shortened and modified version of a paper that appeared previously, with a different title, in *Synthese, [2001], 129,1,79-103*)

References

Apkarian, P. A. (1983). Visual training after long term deprivation: a case report. *Int J Neurosci, 19*(1-4), 65-83.

Bach-y-Rita, P. (1967). Sensory plasticity. Applications to a vision substitution system. *Acta Neurol Scand, 43*(4), 417-426.

Bach-y-Rita, P. (1972). *Brain mechanisms in sensory substitution*. New York: Academic Press.

Bach-y-Rita, P. (1983). Tactile vision substitution: past and future. *Int J Neurosci, 19*(1-4), 29-36.

Block, N. (1995). On a confusion about a function of consciousness. *Behavioral and Brain Sciences, 18*(2), 227-247.

Braitenberg, V. (1984). *Vehicles. Experiments in synthetic psychology*. Cambridge, Mass.: Bradford, MIT Press.

Broackes, J. (1992). The autonomy of colour. In K. Lennon & D. Charles (Eds.), *Reduction, Explanation, and Realism* (pp. 421-465). Oxford: Oxford University Press.

Chalmers, D. J. (1996). *The conscious mind: In search of a fundamental theory*. New York, NY, USA: Oxford University Press.

D'Zmura, M., & Iverson, G. (1994). Color constancy. III. General linear recovery of spectral descriptions for lights and surfaces. *J Opt Soc Am A, 11*(9), 2398-2400.

Easton, R. D. (1992). Inherent problems of attempts to apply sonar and vibrotactile sensory aid technology to the perceptual needs of the blind. *Optom Vis Sci, 69*(1), 3-14.

Grimes, J. (1996). On the failure to detect changes in scenes across saccades. In K. Akins (Ed.), *Perception* (pp. 89-110). New York & Oxford: Oxford University Press.

Guarniero, G. (1974). Experience of tactile vision. *Perception, 3*(1), 101-104.

Guarniero, G. (1977). Tactile Vision: a personal view. *Visual Impairment and Blindness*, 125-130.

Husserl, E. (1907/1991). *Ding und Raum. Vorlesungen 1907*. Hamburg: Meiner.

Irwin, D. E., & Gordon, R. D. (1998). Eye movements, attention, and transsaccadic memory. *Visual Cognition, 5*(1-2), 127-155.

Koenderink, J. J. (1984). The concept of local sign. In A. J. van Doorn, W. A. van de Grind, & J. J. Koenderink (Eds.), *Limits in perception* (pp. 495-547). Zeist, Netherlands: VNU Science Press.

Lechelt, E. C. (1986). Sensory-substitution systems for the sensorily impaired: the case for the use of tactile-vibratory stimulation. *Percept Mot Skills, 62*(2), 356-358.

Levine, J. (1983). Materialism and qualia: The explanatory gap. *Pacific Philosophical Quarterly, 64*(354-361).

MacKay, D. M. (1962). Theoretical models of space perception. In C. A. Muses (Ed.), *Aspects of the theory of artificial intelligence* (pp. 83-104). New York: Plenum Press.

MacKay, D. M. (1967). Ways of looking at perception. In W. Wathen-Dunn (Ed.), *Mopdels for the perception of speech and visual form* (pp. 25-43). Cambridge, Mass.: MIT Press.

MacKay, D. M. (1973). Visual stability and voluntary eye movements. In R. Jung (Ed.), *Handbook of sensory physiology, Vol. VII/3A* (pp. 307-331). Berlin: Springer.

McConkie, G. W., & Currie, C. B. (1996). Visual stability across saccades while viewing complex pictures. *Journal of Experimental Psychology: Human Perception & Performance, 22*(3), 563-581.

Merleau-Ponty, M. (1968). *Résumés de cours au Collège de France*. Paris: Gallimard.

Minsky, M. (1988). *The society of mind*: Simon & Schuster.

Morgan, M. J. (1977). *Molyneux's question. Vision, touch and the philosophy of perception*. Cambridge: Cambridge University Press.

Müller, J. (1838). *Handbuch der Physiologie des Menschen*. (Vol. V). Coblenz: Hölscher.

Nagel, T. (1974). What is it like to be a bat? *Philosophical Review, 83*(435-456).

O'Regan, J. K. (1992). Solving the "real" mysteries of visual perception: the world as an outside memory. *Canadian Journal of Psychology, 46*, 461-488.

O'Regan, J. K., Deubel, H., Clark, J. J., & Rensink, R. A. (2000). Picture changes during blinks: Looking without seeing and seeing without looking. *Visual Cognition, 7*(1-3), 191-211.

O'Regan, J. K., Rensink, R. A., & Clark, J. J. (1999). Change-blindness as a result of 'mudsplashes'. *Nature, 398*, 34.

Piaget, J. (1937/1977). *La construction du réel chez l'enfant*. (6th ed.). Neuchatel: Delachaux & Niestlé.

Platt, J. R. (1960). How we see straight lines. *Scientific American, 202*(6), 121-129.

Poincaré, H. (1905). *La valeur de la science*. Paris: Flammarion.

Rensink, R. A., O'Regan, J. K., & Clark, J. J. (1997). To see or not to see: The need for attention to perceive changes in scenes. *Psychological Science, 8*(5), 368-373.

Rensink, R. A., O'Regan, J. K., & Clark, J. J. (2000). On the failure to detect changes in scenes across brief interruptions. *Visual Cognition, 7*(1-3),127-146..

Environment-Specific Novelty Detection

Stephen Marsland* **Ulrich Nehmzow**** **Jonathan Shapiro***

*Department of Computer Science
University of Manchester
Oxford Road
Manchester M13 9PL
UK
{smarsland, jls}@cs.man.ac.uk

**Department of Computer Science
The University of Essex
Wivenhoe Park
Colchester CO4 3SQ
UK
udfn@essex.ac.uk

Abstract

Novelty detection, recognising features that differ from those that are normally seen, is a potentially useful ability for a mobile robot. Once a robot can detect those features that are novel the amount of learning that has to be done can be reduced (as only new things need to be learnt), the attention of the robot can be focused onto the new features, and the robot can be used as an inspection agent.

However, features that are novel in one place could be completely normal elsewhere – for example, tables and chairs are usually seen in offices, but very rarely seen in corridors. This paper suggests a method by which a set of novelty filters can be trained for different environments and the correct filter autonomously selected for the environment that the robot is currently travelling in. The method can also extend itself, so that further environments that are seen by the robot can be added without any retraining.

1 Introduction

Recognising stimuli that differ from the usual inputs in some way is a very useful ability for both natural and artificial learning agents. This capability is known as novelty detection. For animals it can be a crucial survival instinct, enabling them to avoid potential predators, while for robots and other learning agents it can help to select particular inputs of interest, reducing the computational cost of dealing with the world.

Neural networks that can detect novelty have been used to highlight potential problems in fields such as medical diagnosis and machine fault detection. In these cases there is a lot of data where the result of the test is negative (no disease diagnosed or no machine fault), but relatively few of the important class that the network should detect. This means that normal neural network training is not suitable, as many instances of the disease may be missed. The novelty detection approach operates by having the neural network learn a model of the 'normal' data that does not show any examples of the class that should be detected, and having the novelty detector highlight inputs that do not fit into the pattern of the training set.

There are two important properties that the training set for the novelty detector should have. It should contain no examples of the inputs that should be detected, otherwise this will be learnt and so these inputs will not be found to be novel, and it should contain examples of every possible kind of 'normal' inputs, otherwise these inputs will be found to be novel.

In previous work (Marsland et al., 2000) the novelty detection approach has been used to enable a mobile robot to be used as an inspection agent. A robot equipped with a novelty filter can learn a model of that environment, perceiving its environment through whatever sensors it is equipped with, and then explore other environments, highlighting features that were not found in the original training environment.

However, there are a few problems with this basic approach. When the robot is training there is no guarantee that the training set of inputs will satisfy the two required properties of the input set. This problem is exacerbated by the fact that most novelty detection algorithms are not capable of learning on-line, so that further data cannot be added at a later date. It would also be nice if the filters could quantify the amount of novelty, so that inputs that are only fairly novel – a few similar inputs have been seen during training, but not that many – can be marked, but with less emphasis than completely novel features. Finally, in some cases features that are perfectly normal in one part of an environment should be found to be novel in other places. One example of this is that when exploring an office environment, chairs are perfectly usual within offices, but rarely found in corridors.

The novelty filter that is described in this paper is a proposed solution to these problems. The filter is based on a neural network that is capable of growing during use, so that it can be used for continuous learning. A part of the filter is a set of habituating synapses, so that the filter can quantify the amount of novelty in the current input with respect to the inputs that were seen during training. This also means that the filter has some robustness to incorrect information in the training set – if there are a small number of inputs in the training set that should not be there, they will still be found to be novel after training because they have been seen only infrequently. Finally, an extension to the algorithm is described that allows multiple novelty filters to be trained in different environments and the correct filter for the current inputs to be selected from those available. If none of the filters is suitable then a new filter can be created and trained, meaning that the system is capable of boot-strapping during the training phase.

2 Related Work

There have been a number of novelty detection techniques proposed in the literature. A more complete review is given in (Marsland, 2001). The first was Kohonen's Novelty Filter (Kohonen and Oja, 1976, Kohonen, 1993). This is an autoencoder network that is trained using back-propagation of error, so that the network extracts the principal components of the input. The network is trained on a dataset and, after training, any input presented to the network produces one of the learnt outputs, and the bitwise difference between input and output highlights novel components of the input.

(Ypma and Duin, 1997) proposed a novelty detection mechanism based on the self-organising map. They describe a number of measures by which the goodness of a SOM with respect to a particular dataset can be evaluated. In particular, they measure the average quantisation error over the dataset, and also measure how far away from each other map units that respond to similar inputs are. By training the SOM on data that are known to be normal, and then evaluating the measures on a new dataset, it can be seen whether or not the new dataset fits the same distribution as the data that generated the SOM.

Another approach using the SOM is to calculate the distance of the winning neuron from neighbourhoods that fired when training data known to be normal was introduced, and counting as novel those inputs where the distance is beyond a certain threshold. This method was used by (Taylor and MacIntyre, 1998) to detect faults when monitoring machines. The network was trained on data taken from machines operating normally, and data deviating from this pattern was taken as novel. This is a common technique when faced with a problem for which there is very little data in one class, relative to others. Examples include machine breakdowns (Nairac et al., 1999, Worden et al., 2000) and mammogram scans (Tarassenko et al., 1995). Often, supervised techniques such as Gaussian Mixture Models or Parzen Windows are used, and the problem reduces to attempting to recognise when inputs do not belong to the distribution which generates the normal data (Bishop, 1994). This is the problem of kernel density estimation. The method proposed by (Taylor and MacIntyre, 1998) relies very strongly on the choice of threshold and on the properties of the data presented to the network, which must form strictly segmented neighbourhood clusters without much spread.

Growing networks such as Adaptive Resonance Theory (ART) (Carpenter and Grossberg, 1988) can be used to define as novel those things that have never been seen before, by using a new, uncommitted node to represent them.

3 The On-line Novelty Filter

3.1 Habituation

Habituation is a reversible reduction in the behavioural response to a stimulus when it is presented repeatedly. It enables the animal to ignore stimuli that are seen often, so that it can concentrate on other, potentially more important, stimuli. Habituation is thought to be one of the fundamental mechanisms of adaptive behaviour, and can be seen in animals as diverse as the sea slug *Aplysia* (Bailey and Chen, 1983), cats (Thompson and Spencer, 1966), toads (Wang and Arbib, 1992) and humans (O'Keefe and Nadel, 1978). In contrast to other forms of behavioural decrement, such as fatigue, the response can be restored to its original level without the organism resting by introducing a change in the stimulus. An overview of the effects and causes of habituation can be found in (Thompson and Spencer, 1966).

There have been several attempts to model the effects of habituation computationally and to explain the interaction between habituation and dishabituation, the process whereby an habituated response returns to its original strength. (Groves and Thompson, 1970) suggested that dishabituation was an instance of sensitisation and therefore an independent construct that interacts with habituation to produce a net response. Their model was used by (Stanley, 1976) to simulate habituation data from experiments of the spinal cord of a cat. He described the decrease of synaptic efficacy y by the first-order differential equation

$$\tau \frac{dy(t)}{dt} = \alpha \left[y_0 - y(t) \right] - S(t), \tag{1}$$

where y_0 is the initial value of y, $S(t)$ is the external stimulation and τ is a time constant governing the rate of habituation, while α controls the recovery rate. A graph showing the effects of this equation is given on the left of figure 1. The values of the variables given in figure 1 are the ones that were used.

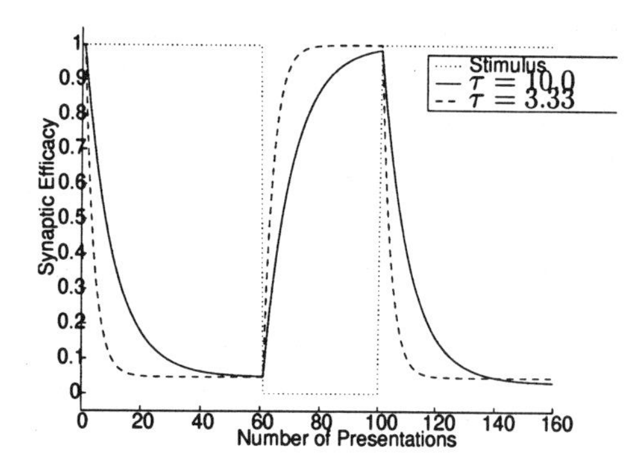

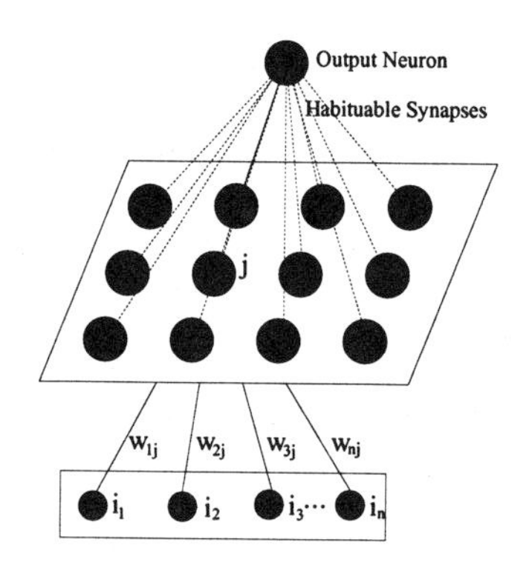

Figure 1: *Left:* The dynamics of habituation using equation 1 ($\alpha = 1.05$, $y_0 = 1$). *Right:* A diagram of the on-line novelty filter.

3.2 Using Habituation in a Novelty Filter

The effect of habituation is to filter out those perceptions that have been seen before, meaning that only novel stimuli are noticed. This is exactly the behaviour that a novelty filter should demonstrate. Habituation allows novelty to be defined more specifically as those things which have not been seen in the current context. The novelty filter described here operates by learning an on-line, adaptive representation of the current environment, testing whether the neural network has already habituated to each new perception. Habituation also allows the novelty of a stimulus to be evaluated, so that the novelty reduces with perception over time.

Before it is known whether a particular input is novel, it needs to be recognised. A schematic of the novelty filter that is described in this paper is shown on the right of figure 1. The filter uses a clustering network to identify the current perception. Each node of the clustering network is attached to an output neuron by a synapse that habituates with use (controlled by equation 1). This means that the first time a node fires the output is strong (the current input is novel), but as that node of the network fires frequently its synapse habituates and so the input is found to be normal clustering network classifies the input, selecting the node whose weights best match the current input vector. The output of this winning node is propagated along the habituable synapse, which modifies the strength of the signal. If that node has not fired before, or fired only rarely, then the signal is passed on without attenuation, and the activity at the output neuron is high. If that node has fired often, meaning that the perception is not novel, then the efficacy of the synapse is low and so the output neuron receives a very weak signal. As well as the synapse connecting the winning node to the output habituating, the synapses of nodes that are neighbours of the winning node in the map space also habituate, although to a lesser extent.

This abstract representation of the novelty filter does not require that any particular clustering network be used. The network needs to cluster similar inputs together, be able to deal with unlabelled data and should be capable of operating on-line. Previous work (Marsland et al., 2000) has shown that networks such as the Self-Organising Map are ideal except that they are not capable of on-line operation. A new neural network was therefore devised that can add new nodes according to the data that is presented, so that it can be used on-line. This network, termed the Grow When Required (GWR) network, is the subject of the next section.

3.3 The Grow When Required (GWR) Algorithm

This section first describes the GWR network and then gives details of the algorithm (section 3.3.1). The network decides when to add nodes according to the activity of the best-matching node. If a node matches an input well, then the activity of that node is close to 1. There are two reasons why the activity of the node could be low – either the node is still being positioned by the learning rule, having been added recently, or there is a mismatch, and that node is already representing another feature. If the node is a new one then it will not have fired often, and so the habituation counter that is attached to the node for the novelty detection part of the filter will be high. In that case the node should be trained to be a better match to the input.

So, if the habituation value is close to 0, then the node should be well placed in the map field and hence the activity should be high. If this is not the case then we need to introduce an extra node to match the current input better. This node is added between the winning node, which caused the problem, and the input, with the weights of the new node being initialised to be the mean average of the weights for the best matching node and the input. If the node is well placed then a neighbourhood connection is put between the winning node and the second best matching node (if it does not already exist), and the weight vectors of all nodes in the neighbourhood of the winning node (that is, nodes that have a direct neighbourhood connection to the winning node) are updated.

Thus, two thresholds are needed to decide whether or not to insert a node on the current iteration: a minimum activity, a_T, below which the current node is not considered to be a sufficiently good match, and a maximum habituation value, above which the current node is not considered to have learnt sufficiently well. In practice, the value of the habituation threshold does not seem to matter very much, and is usually set to 5 presentations. The value of a_T does make a considerable difference. If the value is set very close to 1 then more nodes will be produced to make a better match with the inputs.

3.3.1 The Algorithm

Let A be the set of map nodes, and $C \subset A \times A$ be the set of connections between nodes in the map field. Let the input distribution be $p(\boldsymbol{\xi})$, for inputs $\boldsymbol{\xi}$. Define $\boldsymbol{w_c}$ as the weight vector of node c.

3.3.2 Initialisation

The network is initialised using:

- Create two nodes for the set A,

$$A = \{c_1, c_2\} \tag{2}$$

  with their weights $\boldsymbol{w_{c_1}}$, $\boldsymbol{w_{c_2}}$ initialised randomly from $p(\boldsymbol{\xi})$

- Define C, the connection set, to be the empty set,

$$C = \emptyset \tag{3}$$

- Let $y(0) = y_0$

3.3.3 Iteration

Each iteration of the algorithm follows the following steps:

1. Generate a data sample $\boldsymbol{\xi}$ for input to the network

2. For each node in the network, calculate the distance from the input $\|\boldsymbol{\xi} - \boldsymbol{w}_i\|$

3. Select the best matching node, and the second best, that is, the nodes $s, t \in A$ such that

$$s = \arg\min_{c \in A} \|\boldsymbol{\xi} - \boldsymbol{w}_c\| \tag{4}$$

   and

$$t = \arg\min_{c \in A/\{s\}} \|\boldsymbol{\xi} - \boldsymbol{w}_c\|, \tag{5}$$

   where $\boldsymbol{w}_c$ is the weight vector of node c.

4. If there is not a connection between s and t, create it with age 0

$$C = C \cup \{(s,t)\}, \tag{6}$$

   otherwise, set the age of the connection to 0.

5. Calculate the activity a of the best matching unit (s),

$$a = \exp(-\|\boldsymbol{\xi} - \boldsymbol{w}_s\|) \tag{7}$$

6. If we should add a node, i.e., if activity $a_s <$ activity threshold a_T and habituation $y_s(t) <$ habituation threshold h_T

   - Add the new node, r

$$A = A \cup \{r\}. \tag{8}$$

   - Create the new weight vector, setting the weights to be the average of the weights for the best matching node and the input vector

$$\boldsymbol{w}_r = (\boldsymbol{w}_s + \boldsymbol{\xi})/2. \tag{9}$$

   - Insert edges between r and s and between r and t

$$C = C \cup \{(r,s), (r,t)\}, \tag{10}$$

   - Remove the link between s and t

$$C = C/\{(s,t)\} \tag{11}$$

7. Adapt the positions of the winning node and its neighbours, i, that is the nodes to which it is connected.

$$\Delta \boldsymbol{w}_s = \epsilon_b(\boldsymbol{\xi} - \boldsymbol{w}_s) \tag{12}$$

$$\Delta \boldsymbol{w}_i = \epsilon_n(\boldsymbol{\xi} - \boldsymbol{w}_i) \tag{13}$$

   where the learning rates are such that $0 < \epsilon_n < \epsilon_b < 1$

8. Age edges with an end at s.

$$age_{(s,i)} = age_{(s,i)} + 1. \tag{14}$$

9. Habituate the winning node and its neighbours using

$$\tau \frac{dy_i(t)}{dt} = \alpha[y_0 - y_i(t)] - S(t), \tag{15}$$

   where $y_i(t)$ is the strength of synapse i, y_0 is the initial strength, and $S(t)$ is the stimulus strength, usually 1. α and τ are constants controlling the behaviour of the curve. The winner habituates faster than its neighbours. Values of the parameters used in the experiments are $\alpha = 1.05$, $y_0 = 1$ and $\tau = 3.33$ for the winning node, $\tau = 10.0$ for the neighbours.

10. Check if there are any nodes or edges to delete, i.e., if there are any nodes that no longer have any neighbours, or edges whose age is greater than some constant $a_{\max}$.

3.4 Selecting Different Novelty Filters

The second part of the system that is described in this paper is the part that enables the robot to choose which, if any, of a set of previously trained filters should be used, or whether a new filter should be trained. As the robot travels through an environment it monitors how well each perception fits into the model of each of the novelty filters that has been trained. At the end of a run through that environment the robots makes a choice about which environment it is in. At this stage a set of different behaviours could be used to decide how the robot should react.

A vector of 'familiarity indices' is used to keep a record of how familiar each of the different trained novelty filters finds the current perceptions of the robot. This familiarity vector (with one element for each of the m trained novelty filters) is updated after each perception has been presented to all of the novelty filters, each of which has produced a novelty value n.

All of the elements of the familiarity vector are initialised to be $1/m$. For each input to the novelty filters the following steps are taken:

- compute the novelty value n_i (i.e., the output of the novelty filter) for the current filter, i
- update the element of the familiarity vector $\mathbf{f}$ for that network:

$$\mathbf{f}_i = \mathbf{f}_i - c \times n_i, \tag{16}$$

  where c is a scaling constant
- update all the other elements so that the sum of the elements remains normalised:

$$\mathbf{f}_j = \mathbf{f}_j + \frac{c \times n_i}{n_i - 1}, \ \forall j \neq i \tag{17}$$

- repeat for all the other novelty filters

In the experiments reported in the next section a value of $c = 0.1$ was used. Investigations showed that the value was not critical, although obviously it does affect how quickly the familiarity vector responds to inputs that the filters find to be novel.

Once the robot had travelled through the environment a decision was made about which environment it was. If one element of the familiarity index was significantly larger than the others (i.e., one familiarity index was above 0.7), then the corresponding environment was taken as the one being explored. However, the algorithm also stores the accumulated novelty of each novelty filter as the robot explores it. If the best-matching novelty filter has very high accumulated novelty then it could be that it is not actually a good match to the inputs, merely a better match than the others. In this case a new novelty filter should be made and trained.

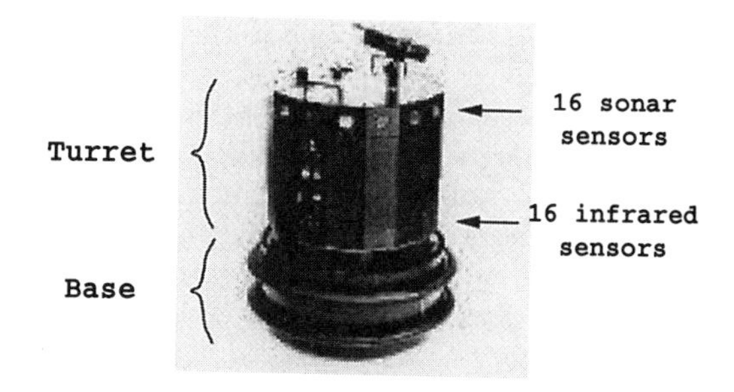

Figure 2: The Nomad 200 robot used in the experiments.

In the case where no environment is obviously more familiar than any of the others it is assumed that this is because the robot has just explored a novel environment. The accumulated novelty should also be high in this case. If the environment was novel then it would be suitable to generate a new filter and further explore the environment, training the filter. In this way the algorithm could extend itself as required.

4 Experimental Results

4.1 Training a Novelty Filter

This section describes how a novelty filter can be trained. The robot (shown in figure 2) explores two small environments, 10 m sections of corridor. In each case the robot used a pre-trained wall-following behaviour based on infra-red sensors to travel through an environment, taking sonar scans as it travelled and produced an input vector by taking the average of these readings over the last 10 cm of travel. This input vector was then presented to the novelty filter, which categorised it and produced an output of how novel that perception was according to the strength of the habituation synapse for the best-matching node.

In the first experiment the robot, initially equipped with an uninitialised novelty filter, travelled along a 10 m section of corridor. Figure 3 shows the filter learning about this environment (labelled environment A) during three learning runs. Spikes show the amount of novelty found in each input, with high spikes denoting novel features and very small spikes completely normal inputs. Initially the filter finds all perceptions novel, shown by the burst of spikes. However, it rapidly learns to recognise the wall that is seen on either side, whereupon the only novelties are around the area of the doorway on the right-hand side of the robot. By the second run these have mostly been learnt, and in the third run nothing is found to be novel.

This trained network was then used in a modified version of the environment, environment A* in figure 4. The

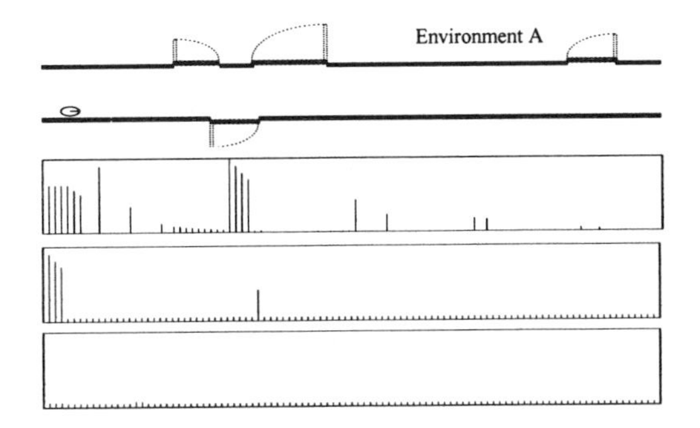

Figure 3: The novelty filter learning about environment A with no initial training. Spikes show the output of the novelty filter, with a high spike denoting a novel input and a low spike a normal input. During the first run many features are found to be novel, but these are learnt about and are not found novel by the third run.

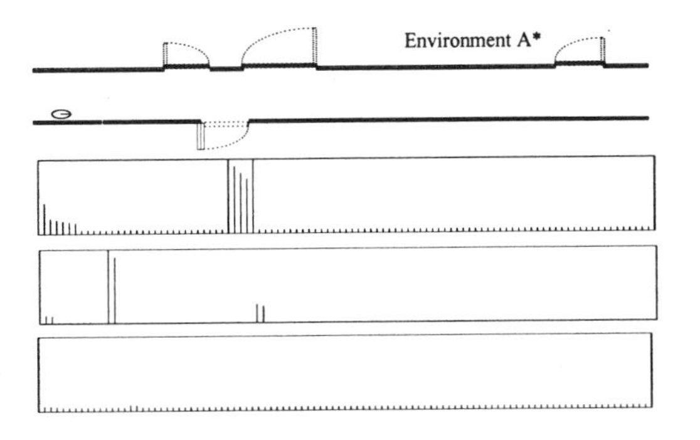

Figure 4: The novelty filter learning about environment A* after training in environment A. It can be seen that the only place where novelty is found (i.e., there are high spikes in the graph) on the first run is the area around the now-open doorway. In the second run, the spikes are caused by a crack in the wall, which is only detected occasionally, but that, when detected, dominates the sonar readings.

door on the right of the robot was opened, which meant that the sonar signals reflected off a wooden barricade amount 1.5 m further away. A cardboard box was placed in the doorway so that the infra-red sensors that controlled the wall-following motion could see it, but the sonar sensors could not. Figure 4 shows the results of this experiment. Only the area around the now-open door is found to be novel, and this is learnt after the second run through the environment.

4.2 *Selecting a Suitable Filter*

Using the technique described in the previous section four separate GWR-based novelty filters were trained. The first was trained in environment A, a schematic of which is shown at the top of figure 3, the second in environment A* (figure 4), and the third in a very similar area of corridor, labelled environment B. Finally, a different type of corridor in another part of the building was used as environment C. This corridor is wider and is built of different materials and is wider than the others. The appearance of the three different corridors used can be seen in figure 5.

The training in each environment was as described in section 4.1. The robot made three training runs in each of the environments, sampling the environment with its sonar sensors and presenting an average sonar reading over the last 10 cm of travel every 10 cm. Each network was initially completely untrained, and after the three training runs the filter had stopped finding any features in the environment novel. An insertion threshold of $a_T = 0.9$ was used. Since four novelty filters were trained, one in each of the environments A, A*, B and C, so the elements of the familiarity vector were initialised to $\frac{1}{4}$.

After training each of these filters the robot was exposed to an unknown environment. Five different environments were used for this purpose, each of the four used for training (A, A*, B and C) and a control environment that was completely different. This novel environment was part of the robot laboratory, which is wider than the corridors and has obstacles placed in the path of the robot, so that the perceptions were very different to those in the corridor environments.

Once the robot was placed in an environment it explored that environment using the wall-following behaviour to follow the wall to the right of the robot. As the robot travelled in this test environment the algorithm had to decide which environment the robot was exploring – one of the known environments or the control environment. As in the training, the robot moved 10 cm taking sonar scans as it moved, and after moving computed the average sonar scan over the 10 cm. It presented this average sonar scan as input to each of the novelty filters, which produced a novelty value n for the perception. The algorithm described in section 3.4 was then used to update the familiarity indices of each of the environments and the robot moved on another 10 cm.

4.2.1 *Testing the Complete System*

Five testing runs were performed in each of the five environments, the environments A, A*, B and C and the control environment. Table 1 shows the results of the experiments averaged over five testing runs. It can be seen that in each case the algorithm picks the correct network (shown in bold) and that the familiarity indices for the other environments are all small. Where the algorithm is tested in the control environment, which does not have a trained filter, all four of the filters have similar scores of about $\frac{1}{4}$.

Figure 6 shows a sample output when the algorithm is run in each of the environments. It can be seen that for some of the environments it takes a long time before any one of the networks is clearly the winner, while for others

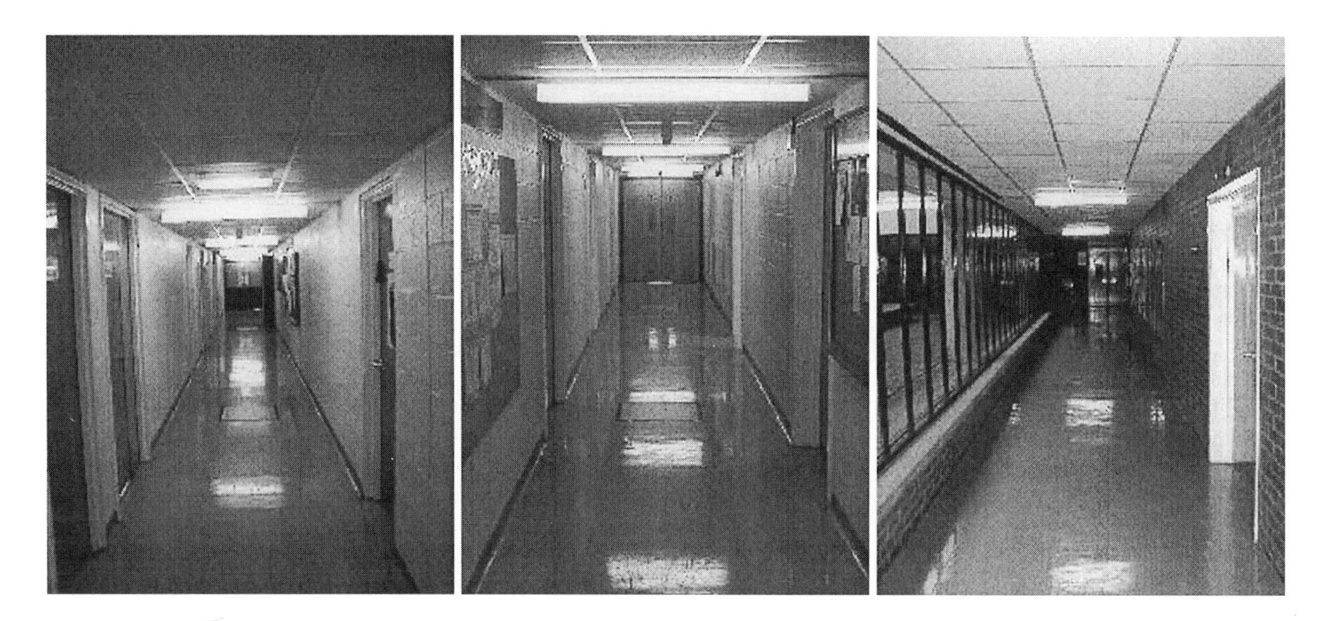

Figure 5: Photographs of environments A (*left*), B (*centre*) and C (*right*). Environments A and B are similar sections of corridor, while environment C is in a different part of the building and has brick walls instead of breezeblock.

Training Environment	Testing Environment				
	A	A*	B	C	Control
A	**0.809** ± 0.055	0.042 ± 0.031	0.115 ± 0.038	0.026 ± 0.029	0.255 ± 0.067
A*	0.157 ± 0.150	**0.723** ± 0.191	0.014 ± 0.0128	0.106 ± 0.103	0.242 ± 0.073
B	0.012 ± 0.009	0.073 ± 0.078	**0.899** ± 0.093	0.016 ± 0.032	0.245 ± 0.045
C	0.068 ± 0.060	0.102 ± 0.082	0.034 ± 0.024	**0.796** ± 0.099	0.263 ± 0.102

Table 1: The familiarity index for each of the environments for each of the trained networks, averaged over five runs. Each table entry gives mean ± standard deviation.

the winner is apparent very quickly. It is particularly interesting that when the robot explores environment C, which is very different to the others, the filter for environment C is initially very low in familiarity, and only late in the run does the correct hypothesis overtake the others. This is probably because the start of this environment contains a door very similar to those seen in the other environments and only later is the different brick, etc. apparent.

In environment B, initially all of environments A, A* and B (which do look similar) are equally likely, and it takes quite a while to settle for environment B. It does this towards the end, where the perceptions of the boxes on the wall appear. These boxes do not appear in any of the other environments. To differentiate between environments A and A*, the algorithm has to wait until the perceptions of the doorway that can be open or closed are seen. However, for the novel environment, none of the possible environments were ever seen to be similar to the perceptions. This is very encouraging.

As a further test of the system, a different type of testing was performed. Here, one of the four environments was missed out of the bank of trained filters, so that there were only three trained filters. The results of this are shown for environments B and C in figure 7. It can be seen that in the case where the robot explores environment B, environments A and C are considered equally likely, but environment A*, which contains the open door, is unlikely. In this kind of case the algorithm also looks at the total amount of novelty that has been found in the environment. This is computed by summing the novelty found at each timestep. For testing in environment B this was at least 14.4, as compared to under 5.2 for non-novel environments.

In the case where environment C is explored, environment A is initially favoured, presumably because it also has a door on the right-hand side of the robot, as does environment C, but environment B becomes more likely as the run progresses. The minimum amount of novelty found during testing in environment C in any of the five runs was 19.8. In both of these cases it is very unclear which of the environments was more likely, and the total novelty found during the testing run would be sufficiently high to suggest that this was a novel environment and

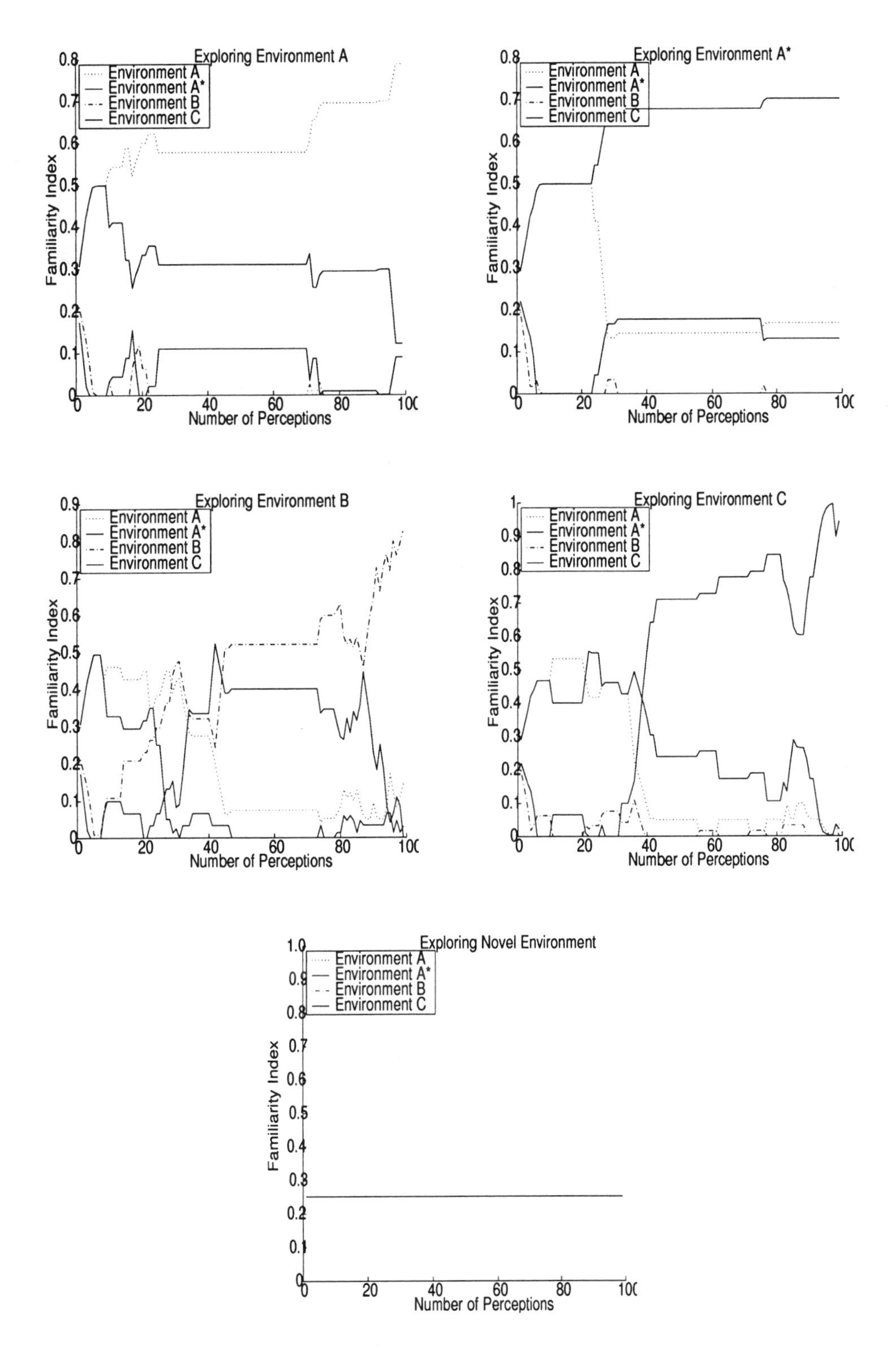

Figure 6: Plots of the familiarity indices for the trained novelty filters during sample runs in each of the five test environments. The x−axis shows the number of perceptions of the environment. *Top left:* Environment A. *Top right:* Environment A*. *Middle left:* Environment B. *Middle right:* Environment C. *Bottom:* Novel (control) environment. In all cases the correct environment is selected.

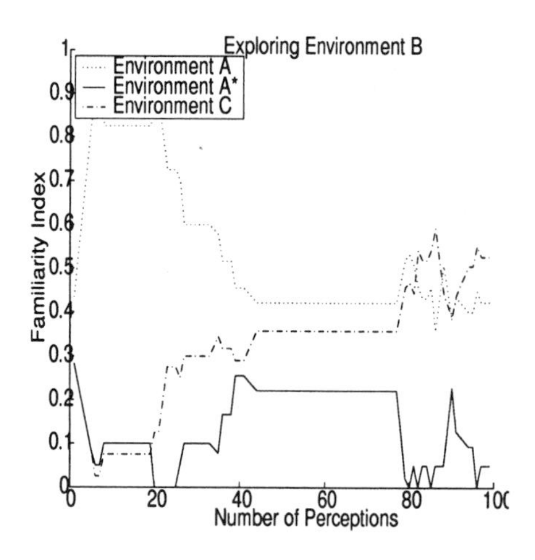

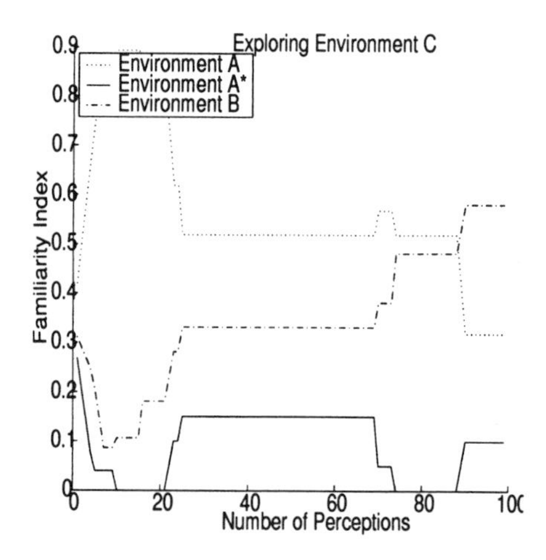

Figure 7: *Left:* Familiarity indices for the novelty filters trained in A, A* and C, for tests in environment B. *Right:* Testing on environment C after training on A, A* and B. All of the trained filters are found to be unlikely in both cases.

so a new filter should be trained.

5 Discussion and Conclusions

The novelty filter that is described in this paper is capable of learning on-line about a series of environments as the robot travels through them. This means that the robot can be used as an inspection agent, exploring a set of environments that are known to be normal (that is, to contain no unusual features or features that are known to be possible problems). Once this training set has been learned, the robot can be used to inspect further environments, highlighting any inputs that do not fit into the training set, and so are potential faults. One benefit of the novelty filter proposed here is that it learns on-line, which means that if any important feature is missing from the training set it can be added in at a later date without any requirement for retraining on the rest of the data.

The extension to the novelty filter system that has been described here means that a bank of novelty filters can be trained for the different places that the robot can visit. In this way the fact that each of these different places has different properties, and that different things may be found to be novel in each is encoded. The robot can then decide autonomously which of its bank of filters should be used to evaluate the current inputs, or whether none of them is sufficient and a new filter should be trained. It has been demonstrated that in the experiments presented here the correct filter was selected 100% of the time.

In this paper the only inputs that have been used have been the sonar sensors of the robot. It would be necessary to use more sensors, especially cameras, in order to enable the robot to act properly as an inspection agent in any sort of real world environment. While this has been considered for simple camera images (Marsland et al., 2001), it has not been covered for more complex images, nor has the question of sensor fusion yet been addressed.

Acknowledgements

This work was supported by an EPSRC studentship.

References

Bailey, C. and Chen, M. (1983). Morphological basis of long-term habituation and sensitization in *aplysia*. *Science*, 220:91–93.

Bishop, C. M. (1994). Novelty detection and neural network validation. *IEEE Proceedings on Vision, Image and Signal Processing*, 141(4):217–222.

Carpenter, G. A. and Grossberg, S. (1988). The ART of adaptive pattern recognition by a self–organising neural network. *IEEE Computer*, 21:77 – 88.

Groves, P. and Thompson, R. (1970). Habituation: A dual-process theory. *Psychological Review*, 77(5):419–450.

Kohonen, T. (1993). *Self-Organization and Associative Memory, 3rd ed.* Springer, Berlin.

Kohonen, T. and Oja, E. (1976). Fast adaptive formation of orthogonalizing filters and associative memory in recurrent networks of neuron-like elements. *Biological Cybernetics*, 25:85–95.

Marsland, S. (2001). *On-line Novelty Detection Through Self-Organisation, With Application to Inspection Robotics.* PhD thesis, Department of Computer Science, University of Manchester.

Marsland, S., Nehmzow, U., and Shapiro, J. (2000). Novelty detection on a mobile robot using habituation. In *From Animals to Animats: Proceedings of the 6th International Conference on Simulation of Adaptive Behaviour (SAB'00)*, pages 189 – 198. MIT Press.

Marsland, S., Nehmzow, U., and Shapiro, J. (2001). Vision-based environmental novelty detection on a mobile robot. In *Proceedings of International Conference on Neural Information Processing (ICONIP'01)*.

Nairac, A., Townsend, N., Carr, R., King, S., Cowley, P., and Tarassenko, L. (1999). A system for the analysis of jet system vibration data. *Integrated Computer-Aided Engineering*, 6(1):53 – 65.

O'Keefe, J. and Nadel, L. (1978). *The Hippocampus as a Cognitive Map*. Oxford University Press, Oxford, England.

Stanley, J. C. (1976). Computer simulation of a model of habituation. *Nature*, 261:146–148.

Tarassenko, L., Hayton, P., Cerneaz, N., and Brady, M. (1995). Novelty detection for the identification of masses in mammograms. In *Proceedings of the 4th IEE International Conference on Artificial Neural Networks (ICANN'95)*, pages 442 – 447.

Taylor, O. and MacIntyre, J. (1998). Adaptive local fusion systems for novelty detection and diagnostics in condition monitoring. In *SPIE International Symposium on Aerospace/Defense Sensing*.

Thompson, R. and Spencer, W. (1966). Habituation: A model phenomenon for the study of neuronal substrates of behaviour. *Psychological Review*, 73(1):16–43.

Wang, D. and Arbib, M. A. (1992). Modelling the dishabituation hierarchy: The role of the primordial hippocampus. *Biological Cybernetics*, 76:535–544.

Worden, K., Pierce, S., Manson, G., Philp, W., Staszewski, W., and Culshaw, B. (2000). Detection of defects in composite plates using lamp waves and novelty detection. *International Journal of Systems Science*, 31(11):1397 – 1409.

Ypma, A. and Duin, R. P. (1997). Novelty detection using self-organizing maps. In *Proceedings of International Conference on Neural Information Processing and Intelligent Information Systems (ICONIP'97)*, pages 1322 – 1325.

Navigation in Unforeseeable and Unstable Environments: A Taxonomy of Environments

Laurent Signac*
*IRCOM-SIC – Université de Poitiers
signac@sic.sp2mi.univ-poitiers.fr

Jean-Denis Fouks**
**Université de Poitiers
fouks@esip.univ-poitiers.fr

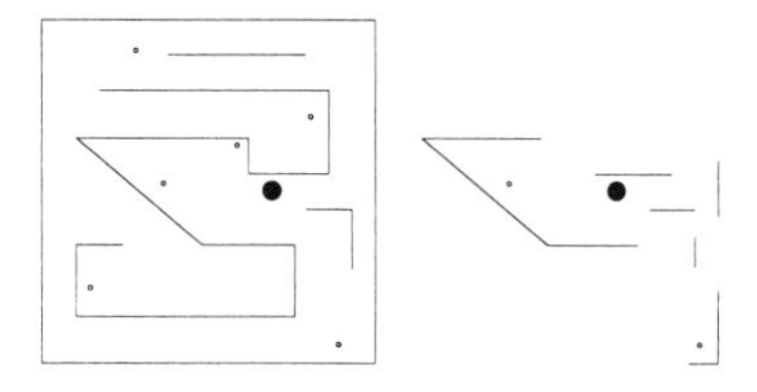

Figure 1: An example of world (left) with walls, beacons and the mobile (large circle), and the set of objects that can be seen by the mobile (right).

1 Introduction

In their everyday life, most living beings adapt themselves to configuration changes in their environment. Even in the case of unforeseeable disruptions, they are able to find new paths. We propose a formal model of environments and show how unforeseeability of natural world is characterized in our model. We give the outline of our algorithms able to drive a virtual mobile in such an environment, whose main characteristic is to be time-varying and whose variations are unforeseeable.

2 A Simple Environment

Our environment is a two dimensional plane made of line segments (walls) and beacons. The goal, for the mobile, is to find several beacons each day. Figure 1 shows an example world, and the set of objects that can be seen by the mobile. The vision (the set of pieces of objects the mobile can see) is the only information available to the mobile.

Numerous works deal with this problem if the mobile itself does not influence its environment (Tsuboushi and Rude, 1996). On the contrary, if an external user changes the position of beacons, destroy walls, or add new ones, *in order to change the paths the mobile uses*, the problem has not yet been studied. It must be clear that several animals (such as dogs) are able to solve such a problem. They are able to explore their environment, to find new paths and to adapt themselves if the environment changes. In the next section we will try to formalize this kind of problems on simpler worlds

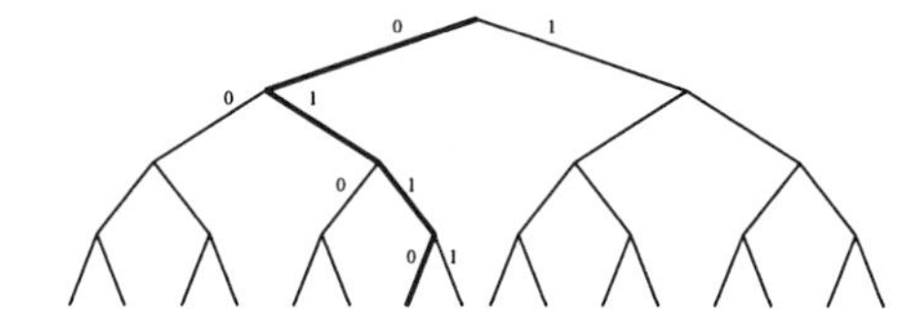

Figure 2: A binary tree of depth 4 and the path from the root to a leaf encoded by the binary word 0110.

3 Taxonomy of Environments

In this section, we will biefly show (due to lack of place) that our study about the recognition of binary stimuli (Fouks and Signac, 2001) can be applied to the navigation problem.

Let's consider a complete binary tree of depth n. A path from the root to a leaf can be encoded by a binary word of length n (Figure 2). Our mobile takes place at the root and must reach a cup of coffee at one of the leaves and return to the root. At each move, its energy level E (initialized to E_0) is decreased by γM ($\gamma > 0$, M is the amount of memory used by the mobile). When the mobile reaches the cup, E is increased by α ($\alpha > 0$). The mobile survives if it moves and if E remains positive. A human experimenter may or may not move the cup from one leaf to another, each time the mobile reaches the cup (but not while it is looking for it). We are now going to explain how the way the human moves the cup urges the mobile to use a particular kind of algorithm.

If the human sets the cup before the mobile is designed (and does not move it anymore), then no learning is required, and the mobile will survive provided $\alpha > 2n\gamma M$ and $E_0 > n\gamma M$.

If the human sets the cup after the mobile is designed, bounded-time learning is required. Provided $E_0 > (2^{n+2} - 4)\gamma M$ and $\alpha > 2n\gamma M$, the mobile will survive. Even if a simple algorithm (using a small amount of memory compared to n) is used to move the cup, the mobile could discover it (provided E_0 is large enough) and survive with small values of α (Of course, to discover such an algorithm may be very hard). We call this type of environements: ruled worlds. If the rules of the world are hard to find, on can use a probabilistic approach (Markovian fields in (LaValle and Sharma, 1996)).

If the human moves the cup each time it has been found by choosing a new random leaf, no strategy exists but exhaustive exploration. It requires that α is greater than $(2^{n+2} - 4)\gamma M$ (prohibitive exponential value).

The most interesting case occurs between random and ruled worlds. If the human moves the cup each time it has been found by randomly choosing the bit number to be changed in the path to the cup, then learning is required (n tries to find the cup instead of 2^n), as well as forgetting (as old paths consume memory and are useless to find the cup). This type of environment has not yet been studied. Although it is difficult to establish a comparison with biological issues, we said that a dog is able to solve this kind of problem, learning new paths, and adapting to changes.

4 Back to Our Maze

We give a short description of how the mobile can find beacons in our maze. Its behavior may be divided into two parallel tasks : exploration and adaptation.

During exploration, the mobile tries to discover new "important" places (beacons location and points for which one has a broad viewpoint). Then it adds these places to a graph, that stands for the representation of the environment. It also links the nodes of the graph each time it passes through two nodes per a straight free path. Broad wiewpoints are approximated by intersections of the generalized Voronoï graph of the walls of the maze, as these points can be computed using only a local point of view (using the biggest empty circles property (Aurenhammer, 1991)). Very important is the fact that each time a node is added to the graph, the corresponding viewpoint is also recorded. The mobile "takes a picture" of important locations to be able to detect changes in the environment. Another important point is the way the mobile tries to connect known nodes: If two nodes can be reached by a straight free path that is very shorter than the path in the graph, it will try to link them. This is of great importance for the mobile to adapt its graph and is called "shortcuts search".

Each times the mobile reaches a known node, it compares the viewpoint with the recorded one. The only thing to do if the environment has changed is to destroy the links from the current node. Because the mobile tries to reach new unknown nodes and tries to find shortcuts, the graph will be updated "automatically". Furthermore, parts of the graph that are now inaccessible are still "known" by the mobile and subsequent changes may reconnect disconnected parts of the graph. This algorithm ensures that known nodes will be corrected and that new important nodes will be added. However, no node is deleted and the graph could grow drastically. A solution is to delete nodes that are not used, after a certain amount of time.

Last is the high level algorithm of the mobile. It tries to collect five beacons a day. It first tries to collect the known beacons, by finding shortest paths in its graph.

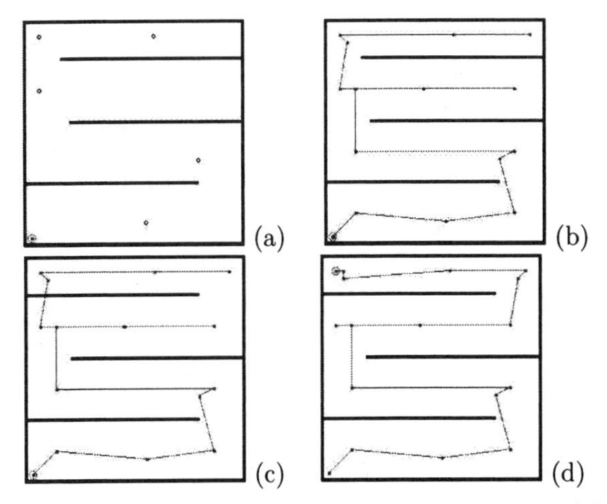

Figure 3: The mobile must collect the five beacons of figure (a). Figure (b) shows the graph constructed by the mobile. On fig. (c) a wall has been translated. Figure (d) shows how the graph has been modified by the mobile to get the five beacons.

If there is time left, or if there is no more beacons (and the mobile has not discovered five of them), it explores the environment, adapting itself to changes (Figure 3).

5 Conclusion

In this paper we adapted to navigation environments the taxonomy we introduced for binary images, based on knowledge management. More particularly, we focused on the problem of environments that urge the mobile to learn and to forget, what best fit biological issues. We proposed a method to face this type of world, and illustrated how it works on an implementation of our model. In future works, we plan to generalize our implementation to more realistic environments.

References

Aurenhammer, F. (1991). Voronoï diagrams – a survey of a fundamental geometric data structure. *ACM Computing Surveys*, 23:345–405.

Fouks, J.-D. and Signac, L. (2001). The problem of survival from an algorithmic point of view. *Artificial Intelligence Journal*, 133(1–2):87–116.

LaValle, S. M. and Sharma, R. (1996). On motion planning in changing, partially-predictable environments. citeseer.nj.nec.com/302769.html.

Tsuboushi, T. and Rude, M. (1996). Motion planning for mobile robots in a time-varying environment: A survey. *Journal of Robotics and Mechatronics*, 8(1).

On the Use of Sensors in Self-Reconfigurable Robots

K. Støy* **W.-M. Shen**** **P. Will****

*The Maersk Institute,
University of Southern Denmark,
Campusvej 55,
DK-5230 Odense M,
Denmark
kaspers@mip.sdu.dk

**Information Sciences Institute
University of Southern California
4676 Admiralty way,
Marina del Rey, CA 90292,
USA
{shen,will}@isi.edu

Abstract

In this paper we investigate the use of sensors in self-reconfigurable robots. We review several physically realized self-reconfigurable robots and conclude that little attention has been paid to the use of sensors. This is unfortunate since sensors can provide essential feedback that can be used to guide self-reconfiguration and control. In the systems that do use sensor feedback, the feedback is used locally on each module. However we identify a need in some situations to use sensor feedback globally. We therefore propose an approach where raw sensor values are abstracted and propagated to all modules. The sensor values are abstracted differently depending on the position of the producing sensor on the robot. We combine this approach with role based control, a control method for self-reconfigurable robots that we have developed earlier. We demonstrate that by combing these two approaches it is possible to make a self-reconfigurable robot consisting of six modules walk and avoid obstacles. However the reaction time of the robot is slow and therefore we discus possible ways of reducing the reaction time.

1 Introduction

In this paper we focus on self-reconfigurable robots built from a possibly large number of physically independent modules. The modules of these robots are able to connect and disconnect autonomously and can have sensors, actuators, a processor, a power source, and a communication system.

Self-reconfigurable robots have several advantages over traditional fixed shaped robots. 1) The modules can connect in many ways making it possible for the same robotic system to solve a range of tasks. This is useful in scenarios where it is undesirable to build a special purpose robot for each task. 2) Self-reconfigurable robots can adapt to the environment and change shape as needed. This could for instance be useful in a retrieval scenario where the robot has to snake its way through the rubble of a collapsed building and at some point change shape to recover an object from the rubble. 3) Since the robot is build from many independent modules it can be robust to module failures. If one module is defect it can be ejected from the system and the robot can still perform its task. 4) Self-reconfigurable robots are built from many identical modules. These modules can be mass produced and therefore the cost can be kept low despite their complexity. The advantages of self-reconfigurable robots can be summarized as: versatility, adaptability, robustness, and cheap production compared to the complexity and versatility of the resulting robot.

2 Research Challenges

In order to realize the potential of self-reconfigurable robots a number of research challenges have to met. There are interesting challenges to be met both in hardware and software.

2.1 Hardware Issues

In hardware some of the fundamental questions are: Does each module need computation power on-board? Where from do the modules get their energy? What kind of sensors are needed? What kind of communication system does the modules need? Several systems have been build to try to answer these questions and their properties are summarized in Table 1.

In Table 1 the systems are approximately sorted in order of increasing autonomy. It can be can seen from the table that very few of these systems are fully autonomous. This is not encouraging, because in order to realize the potential of self-reconfigurable robots it is important that they are autonomous. One important step toward achieving autonomy is to understand how to use sensors to make the robot able to sense and react to its environment. In this paper we take one small step toward understanding this.

Robot	CPU on-board	power on-board	sensors used for	Communication	Reference
JHU Hexagonal	yes[1]	no	n/a	n/a	(Pamecha et al., 1996)
JHU Rectangular	yes	no	n/a	n/a	(Pamecha et al., 1996)
MEL 3d unit	no	no	joint position	serial w. host	(Murata et al., 1998)
RIKEN vertical	no	yes	none	radio w. host	(Hosokawa et al., 1998)
PolyPod	yes	no	force/torque joint position	bus	(Yim, 1994)
Xerox PARC PolyBot	yes	no	joint position docking aid	CANbus	(Yim et al., 2000)
MEL 2000	yes[1]	no	none	serial bus with IDs[2]	(Murata et al., 2000)
MEL fractum	yes	no	none	inter-unit optical	(Murata et al., 1994)
Dartmouth molecule	yes[1]	no	not reported	serial w. host	(Kotay et al., 1998)
CMU ICES-cube	yes[1]	yes	joint position	serial w. host	(Ünsal and Khosla, 2000)
Dartmouth crystalline	yes	yes	joint position	sync. signal f. host	(Rus and Vona, 2000)
USC CONRO	yes	yes	docking aid	inter-unit optical	(Shen et al., 2000a)
CEBOT	yes	yes	docking & obst. avoid.	on-board	(Fukuda and Nakagawa, 1990)

Table 1: Overview of physically realized self-configurable robots.

2.2 Software Issues

There are two main categories of approaches to the control of self-reconfigurable robot: centralized control and distributed control.

In centralized control a central host dictates the actions of each module (Castano et al., 2000b). The advantage of this approach is that it is potentially easier for a controller to handle the complexity of the system when global knowledge is available. The disadvantage is that when the number of modules increase the host becomes the bottleneck. This scalability problem can be reduced by having modules do low-level control and having the central host coordinate the actions of the modules (Yim, 1994). The problem can also be side stepped by having a reconfiguration sequence computed off-line and afterward downloaded into the modules (Rus and Vona, 2001). The central host in this situation works as a conductor telling the modules how far they are in their action sequences. Calculating the reconfiguration sequence off-line unfortunately has the disadvantage that no adaptation can be made when the system is online. Furthermore systems based on centralized control are not robust since they all rely on a single host to function.

In order to address these problems distributed control has been used. Two classes of distributed control systems exist: synchronous and asynchronous distributed control systems. In synchronous control the strict control of the system is maintained, but the problem of robustness is removed. In synchronous control a distributed synchronization algorithm can be used to synchronize the actions of the modules connected in a low bandwidth network. Basically the synchronization signal that before came from a central host now is produced using a distributed synchronization algorithm. The hormone based algorithms developed by Shen, Salemi, and others are examples of this approach (Shen et al., 2000a, Shen et al., 2000b, Salemi et al., 2001). This solves the problem of robustness: there is no central host and the system still works even though a module is taken out. However insisting that all the modules should be strictly synchronized has a cost in terms of efficiency.

Another approach to distributed control is asynchronous distributed control. In synchronous control the modules were considered part of the whole. In asynchronous control each module is considered the whole. A module can be combined with more modules to make a bigger whole, but it in itself represents the whole. In this approach the asynchronous nature of the system is embraced and the idea of having a strictly controlled system is abandoned. The autonomy of each module is increased and the focus is on the local interaction between modules (Murata et al., 1994). The problem in asynchronous control is how to get coherent global behavior to emerge out of local interactions between many modules. The advantage of these systems

[1] but controlled off-board.

[2] local communication under development.

Figure 1: A CONRO module.

is that since all information is handled locally the systems scale. These methods take their inspiration from multi-agent systems (Bojinov et al., 2000b), and cellular automata (Hosokawa et al., 1998, Butler et al., 2001). Their similarities to minimalist collective robotics are also apparent see for instance (Beckers et al., 1994, Beckers et al., 2000, Støy, 2001).

In *role based control* which we will present in detail in Section 5 we combine some of the ideas from synchronous and asynchronous control. We acknowledge the need for each module to only use local information to insure scalability and robustness. However in situations where the modules are cooperating tightly, for instance to produce a locomotion gait, there is also a need to keep the modules synchronized. In role based control synchronization is achieved by having modules synchronize with neighbors from time to time. Over time this leads the entire system to be synchronized. This way the synchronization mechanism is decoupled from the control of the individual module and a robust, scalable, and synchronized system is the result.

3 Sensors and Self-Reconfigurable Robots

In research on sensor fusion there has been some work on how to combine and abstract sensor values into logical (Henderson and Shilcrat, 1984) and virtual sensors (Rowland and Nicholls, 1989). This work has been further extended with a commanding sensor type (Dekhil et al., 1996). The focus was on improving fault-tolerance of sensor systems and aiding development by making the sensor systems modular (Hardy and Ahmad, 1999). These ideas are relevant to the use of sensors in self-reconfigurable robots. However using sensors in self-reconfigurable robots is different because of the unique features of self-reconfigurable robots.

In a self-reconfigurable robot it can not be assumed that the position of the sensor is fixed. It can be moved through reconfiguration or maybe just by movements of the modules. This means that we need to understand how to extract meaningful sensor data from a network of sensors connected in time-varying ways. The previously proposed approaches also mainly deal with one consumer of the sensor date. If distributed control is employed there are many controllers that act on the sensor data. This means that system should be able to deal with inconsistent sensor data.

In distributed systems the problem of many decision makers can be dealt with in two ways. One way is to consider the modules independent and always handle sensor information locally. That is, on the module that receives the sensor input. Butler et al (Butler et al., 2001) have in simulation made a system where each module is a cellular automaton that reacts to its local configuration and surrounding obstacles. Using seven rules the modules are able to role over and across each other to produce "water-flow" like locomotion through an environment with obstacles. A similar idea was explored earlier on a real robot by Hosokawa et al (Hosokawa et al., 1998). Another approach explored by Bojinov et al (Bojinov et al., 2000a, Bojinov et al., 2000b) is to have the structure of the robot grow from seed modules. The growth is accomplished by having the seed module attract spare modules to a specific position with respect to the seed by using a virtual scent. When a spare module reaches that position the old seed module stops being a seed and the newly arrived module becomes the seed. The behavior of the seed module is controlled based on events it can sense in the environment. In these approaches the modules are decoupled in the sense that the modules only interact through stigmergy (Beckers et al., 1994).

In some systems the modules are highly coupled and sensor information can not always be handled locally: a sensor input might have effects in other modules than in the one in which it originated. This raises a fundamental questions which is the main focus of this paper: how do we distribute sensor information in order for it to arrive at the modules that need it? In this paper we present a system where sensor information is abstracted and propagated to all modules in the systems. Each module in the system then independently decides what action to take based on these propagated sensor values. Our use of sensors is inspired by the use of sensors in behavior based robotics (Arkin, 1998, Matarić, 1997) where sensors are used directly to control motors and not to build a geometrical model. We combine this communication system with role based control which we have developed earlier for the control of self-reconfigurable robots (Støy et al., 2002a, Støy et al., 2002b).

4 The CONRO module

Before describing this approach in more detail we will describe the CONRO self-reconfigurable robot. The CONRO modules were developed at University of Southern California's Information Sciences Institute (Castano et al., 2000a, Khoshnevis et al., 2001) (see figure 1). The modules are roughly shaped as rectangular boxes measuring 10cm x 4.5cm x 4.5cm and weigh 100grams. The modules have a female connector located at one end facing south and three male connectors located at the other end facing east, west, and north. Each connector has an infra-red transmitter and receiver used for local communication and sensing. The modules have two controllable degrees of freedom: pitch (up and down) and yaw (side to side). Processing is taken care of by an onboard Basic Stamp 2 processor. The modules have onboard batteries, but these do not supply enough power for the experiments reported here and therefore the modules are powered through cables. For the experiments reported here we also equipped the modules with flex sensors. The flex sensors are mounted on small circuit boards and Velcro is used to attach them to the modules. This mounting strategy is not very robust, but it is very flexible making it easy to experiment with different sensor morphologies. The flex sensors are 11cm long. Their resistance increase as they are bent and can therefore be used for rich tactile sensing. Refer to http://www.isi.edu/conro for more details and for videos of the experiments reported later in this paper.

5 Short Introduction to Role Based Control

In previous work we have introduced role based control. Role based control is a simple minimalist approach to the control of self-reconfigurable robots. We have shown earlier how this control method can be applied to chain and tree configurations to implement caterpillar like locomotion, locomotion similar to that of a sidewinding snake, and rolling track locomotion (Støy et al., 2002a). We have also used the method to make the CONRO robot configured as a hexapod and a quadruped robot walk (Støy et al., 2002b). Here we summarize role based control.

5.1 A Role

A role r consists of three components. The first component is a function $A(t)$ that specifies the joint angles of a module given an integer $t \epsilon [0:T]$. Where T is the period of the motion and the second component that needs to be specified. The third component is a set of delays D. A delay $d_i \epsilon D$ specifies the delay between the child connected to connector i and the parent. That is, if the parent is at step $t_{parent} = t_1$ the child is at $t_{child} = (T + t_1 - d_i)\, modulus\, T$. Below is some examples of roles used in the quadruped robot shown in Figure 4. The spine modules play the spine role:

$$A(t, spine) = \begin{cases} pitch(t) = 0° \\ yaw(t) = 25° \cos(\frac{2\pi}{T}t + \pi) \end{cases} \quad (1)$$

$$d_{east} = \frac{T}{4} \quad (2)$$

$$d_{south} = \frac{2T}{4} \quad (3)$$

$$d_{west} = \frac{3T}{4} \quad (4)$$

$$T = 180 \quad (5)$$

The legs play the forward role below or the backward role where t is replaced by $2\pi - t$ giving the same motion, but in the opposite direction.

$$A(t, forward) = \begin{cases} pitch(t) = 35° \cos(\frac{2\pi}{T}t) - 55° \\ yaw(t) = 40° \sin(\frac{2\pi}{T}t) \end{cases} \quad (6)$$

$$T = 180 \quad (7)$$

5.2 Playing a Role

The algorithm that we now will describe is used to make a module play a role. However first some assumptions need to be made: a parent connector is specified and the remaining connectors are considered child connectors, connections can only be made between a parent connector and a child connector. Furthermore we assume that there are no loops in the configuration. These assumptions limit the configurations the algorithm can handle to tree configurations.

The algorithm has two components. One component makes sure that the actions are executed as specified in the role definition. This component also is responsible for synchronization with neighboring modules. The second component is discovering what role the module should play in case it can play more than one role. What role to play is discovered based on information propagated down from the parent and the local configuration.

The role playing component is visualized in Figure 2. The algorithm starts by setting $t = 0$ and continues to the main loop. Here the algorithm first checks if t is equal to the delay specified for each connector. In case t equals one of these delays d_i a signal is send through the corresponding child connector i. If the module has received a signal from its parent, t is reset. After that the joints are moved to the position described by $A(t)$. Finally t is incremented unless a period has been completed in which case t is reset and another iteration of the loop is initiated.

In some situations it is desirable for a module to be able to play different roles depending on its location in

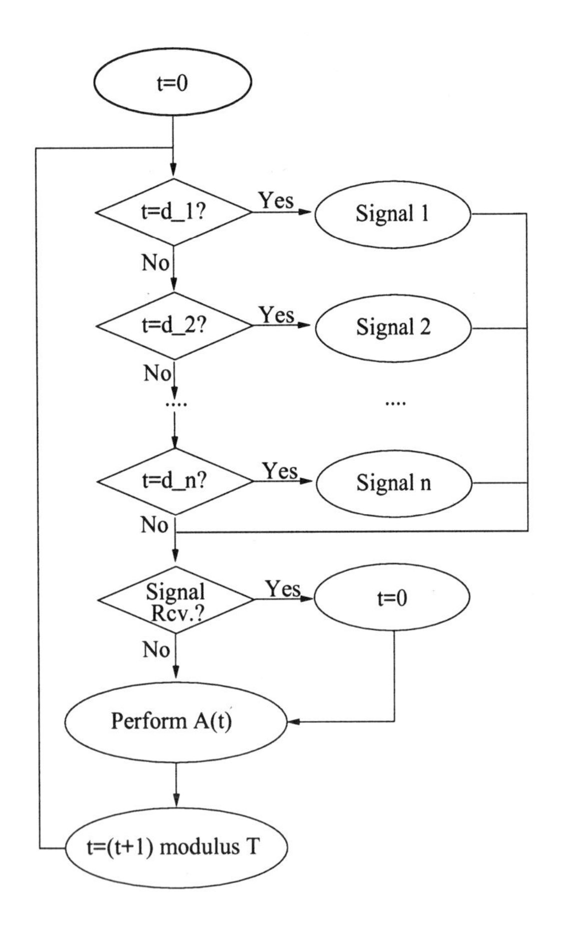

Figure 2: Visualization of the role playing part of the algorithm. See section 5.2 for an explanation.

```
r = <start role>
t = 0
while(1)
  if (t=d(r)_1) then
    <send message M(r,1) to child connector 1>
    <update r>
  endif
  ...
  if (t=d(r)_n) then
    <send message M(r,n) to child connector n>
    <update r>
  endif

  if <message m received from parent connector> then
    t=0
    <update r based on m>
  endif

  <perform action A(r,t)>
  t = (t+1) modulus T(r)
endwhile
```

Figure 3: The algorithm used to play multiple roles. Refer to section 5.2 for further explanation.

the configuration tree. This is taken care of by the role selection component of the algorithm. The role can be selected based on the local configuration. Therefore every time a signal has successfully been sent through a child connector meaning that a child module is connected to that connector there is a check to see if the role should be changed because of that. In the basic algorithm the parent signaled the children to keep them synchronized the parent now sends a message which is a function of the parents role and the connector to which the child is connected. These two algorithmic components combined are shown in pseudo code in figure 3.

We have used this algorithm to make the CONRO self-reconfigurable robot walk. In the walker two modules are connected to form a spine. One module is connected on each side of the spine modules (see Figure 4). In this configuration the modules can play three different roles: east leg, west leg, and spine. A modules decide which role to play using the following rules: if communication to the sides (to the legs) is successful the module plays the spine role. It plays the role of a west leg if its parent is a spine module and it received the synchronization message through the west connector of the parent module. A module plays east leg if the synchronization message was send through the east connector.

Role based control is an example of an synchronous control method that does not insist on all the modules being synchronized at each step, but achieves this over time. This makes a role based system like the walker efficient because the modules work independently most of the time and only occasionally share information and synchronization information with neighboring modules. In this system all modules run identical programs and therefore modules can be interchanged and switch roles accordingly resulting in a very robust system. For more information on this system and role based control refer to (Støy et al., 2002b).

We have now summarized how role based control can be used to make the CONRO self-reconfigurable robot walk. However the system is open-looped in the sense that no sensor input from the environment is used in the control. Therefore we want to extend role based control to include sensor feedback. This is the subject of the following sections.

6 Role Based Control using Propagated Sensor Information

The CONRO self-reconfigurable robot is now configured into a quadruped robot as shown in Figure 4. Two flex sensors are attached to the front spine module and one is attached to each of the front legs. We now want to use feedback from these sensors to make the robot steer away from obstacles. In general the direction of loco-

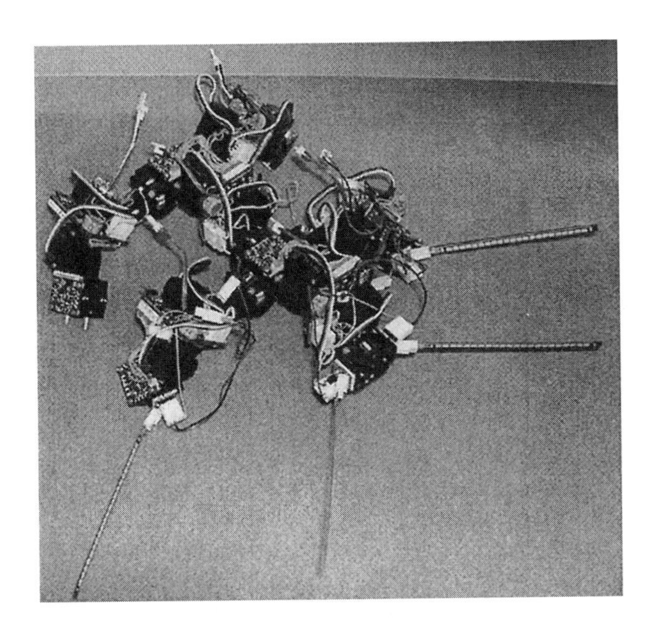

Figure 4: The CONRO robot in a walker configuration. The spine is made from two modules and the legs are made from one module each.

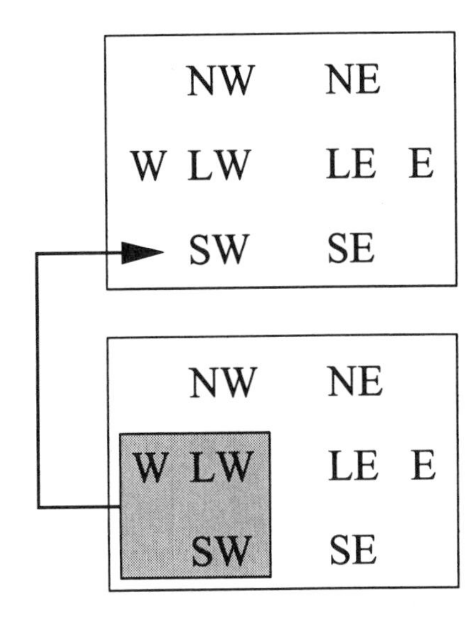

Figure 5: This figure shows how sensor values are propagated north from one spine module to the next. The south module (bottom) sums up the variables in the gray box and sends the sum to the north module (top). The north module receives this sum and write it in the variable indicated by the arrow.

motion can be changed in two ways: the motion of the legs can be biased so legs on the side pointing away from the obstacle take shorter steps and those on the other take longer steps. This approach will enable the robot to make soft turns away from obstacles. An alternative is to have legs on the side pointing away from the obstacle to move backwards and thus producing a turning on the spot motion. We found in initial experiments that the sensor based bias of the locomotion pattern does not produce a sharp enough turn to avoid obstacles. Therefore we decided to implement roles that make it possible for the robot to turn on the spot.

The goal is to make the quadruped robot turn on the spot away from an obstacle detected using one of the four flex sensors. Below we describe how this is achieved. A module has up to two flex sensors mounted: the front legs have one each, the front spine module has two, and the rest zero. The modules continuously sample these sensors and write the analog value into a local variable. What variable dependents on the position of the sensor. If the flex sensor is pointing toward the east the sensor value is written in a variable named local east (LE) and if it points west in local west (LW). If there are no sensors attached these variables contain zero. Each module has an additional six variables: northeast (NE), northwest (NW), east (E), west (W), southeast (SE), southwest (SW). These variables represent the sensor activity in the direction indicated by the names. For instance if all the west variables including local west are added up it will give the sum of the sensor activity on the west side of the robot. The same is true for the east values.

We will now describe how the sensor values are propagated in the system to produce the contents of the variables as it is described informally above. When a spine module sends sensor information to a module connected to its north connector it works as follows. The south module adds up the variables west, southwest, and local west and sends the sum to the north module. The sum is received by the north module and is written in the southwest variable. Note that this satisfy the invariant that the southwest variable of the north module now contains the sum of the sensor activity to the southwest. This mechanism is summarized in figure 5. At the same time the sum of the east variables is propagated and written in the southeast variable of the north module.

This mechanism will sum up the sensor inputs all along the spine and the northern most module will have information about the sensor activity to the southeast and southwest. However we want all the modules to have information of sensor activity in all directions therefore a similar, but independent mechanism is also propagating sensor activity southward. These two propagation mechanisms are summarized in figure 6.

All the spine modules now have information about sensor activity along the spine. For instance the modules can sum the northwest, local west, and southwest variables to find the sensor activity on the west side of the robot. This is not enough in our situation, because there are also two legs attached to the spine modules. Therefore sensor information should also be propagated from and to them. In our setup an east leg can only have one piece of sensor information that the spine does not have:

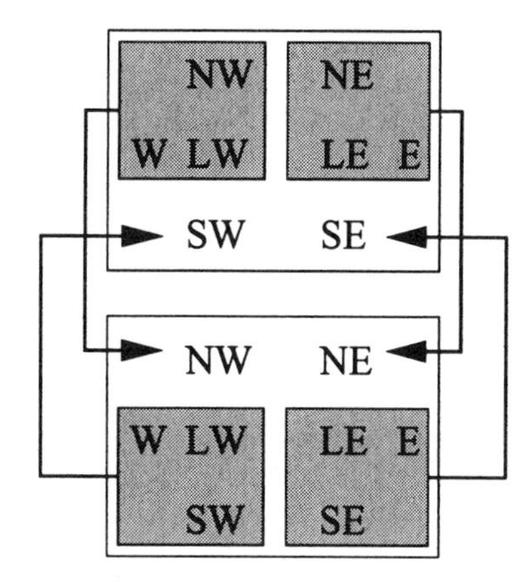

Figure 6: This figure shows how sensor values are exchanged between two spine modules. The modules sum the variables in the gray boxes and send these two values to the other module. The receiving module then writes these values in the variables as indicated by the arrows.

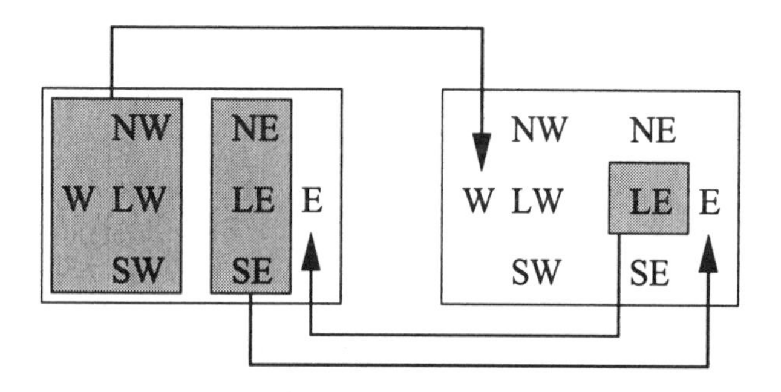

Figure 7: This figure show how sensor values are exchanged between a spine module (left) and an east leg module (right). The modules adds up the numbers in the gray boxes and send these values to the other module. The receiving module then write these values in the variables as as indicated by the arrows.

the value of the sensor connected to that leg. Therefore the local east value is propagated from the leg to the spine. The leg on the other hand receives the sum of all the sensor activity on the west side of the robot and writes that in the west variable. The sum of the variables northeast, local east, and southeast of the spine is written in the east variable of the leg. This is summarized in figure 7.

In role based control synchronization information from the parent module is sent to child modules each period of locomotion. The sensor information is also exchanged at this time as described above. How the sensor information is exchanged depends on what role the module plays. This is decided by the role playing algorithm. Therefore modules can still be exchanged and the system will continue to function if the sensors are placed correctly.

All the modules now have access to global sensor information and can make their decisions based on this information. In order to make a decision we sum the variables for the west and east side of the robot to have a measure of the activity on each side of the robot. A leg then decides to move backward if the sensor activity on the other side of the robot is above a small threshold and higher than the sensor activity on the leg's side. Otherwise it will move forward.

7 Results

First we will note some general properties of the system. One step of the robot corresponding to one period of locomotion takes two seconds. The step length is 15cm. Note that a step is quite long compared to the length of a module (10cm). The long steps are achieved by actively using the spine to make the steps longer. The robot achieves a speed of 7.5cm/second.

In four separate experiments the robot was placed so it approached an obstacle from four different angles. These experiments were videotaped using an overhead camera. We then manually analyzed the tape and for every two seconds recorded the position of the front end of the robot, the rear end, and whether a flex sensor was touching the obstacle. The results of this analysis can be seen in Figure 8.

The sensor values are exchanged when modules synchronize. We know the spine synchronizes with the east leg at $T/4$, the spine module to the south at $2T/4$, and the west leg at $3T/4$. We can use this information to calculate upper and lower bounds on communication delays. In the worst case where the sensor change happens just after synchronization it takes 2 periods to get sensor information from a front leg to the rear leg on the other side. In the best case where the sensor change happens just before synchronization it takes 1 period. This means that the whole system has a reaction time between two and four seconds or a reaction distance of 15cm to 30cm. Note that the reaction time is much better for the front legs. We can see these slow reaction times in Figure 8. The robot can only successfully avoid the obstacle when it approaches at an angle. In trial number four where the robot does not have time to react it bumps into the obstacle. This also explains why we decided to implement the turning on the spot behavior.

8 Discussion

If we look at how our approach can be used in general. We note that there are two things that make out system work. 1) The sensor data is abstracted based on the sensors position in a way that is useful for the receiving modules. 2) The abstracted sensor values are propagated at a constant slow rate to all modules.

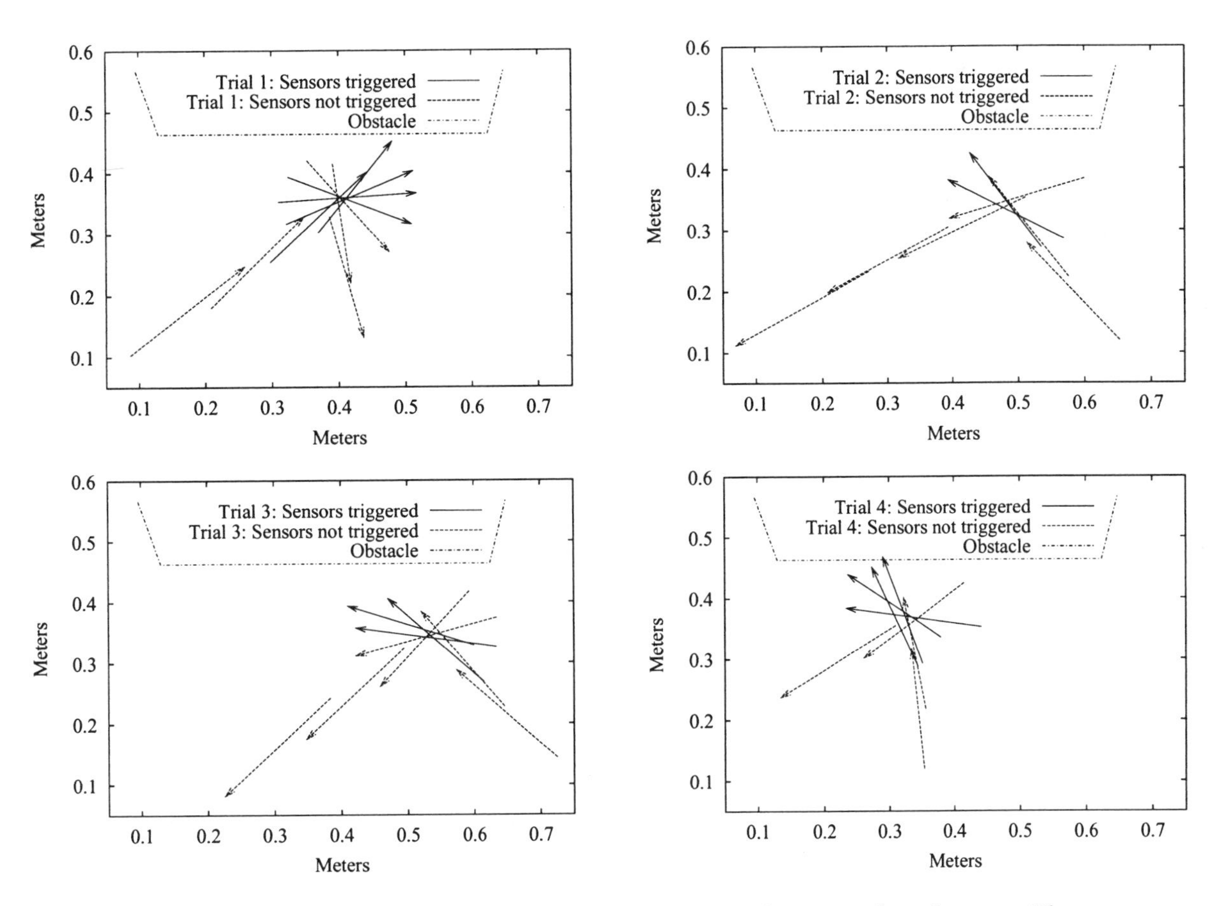

Figure 8: These figures show the robot approaching an obstacle, turning on the spot, and moving away. The arrows represent the positions of the front end and rear end of the robot recorded every 2seconds. The direction of the arrow shows the direction of movement. A solid arrow indicate that a flex sensor was triggered in that time step. A dashed that none were triggered.

In order to keep the amount of communication manageable we abstract sensor values to maintain an estimate in each module of the sensor activity to the left and right of the robot. It is possible for all the modules to agree on left and right, because of the properties of the CONRO hardware. The modules can only be connected in a tree structure (with one loop) and be connected in four ways and therefore the transformation of direction from module to module is easy. In general it seems to be important to have some relative position information about the sensor with respect to the acting module. This means that in systems where the relative position of two connected modules can be found it is possible to abstract the sensor information in a useful way. This is also what makes this approach different from previous work on sensor fusion. The position of the sensor is variable and this is in our case handled by the abstraction mechanism.

Sensor values flow around in our system at a constant slow rate. This rate could be increased significantly to reduce the reaction time. The problem of doing this is that our modules have limited resources and therefore if time is spent on communication less time can be spent on control of the motors resulting in a decrease in speed. Therefore in order to decrease the reaction time of the system without sacrificing speed we need to use less communication and achieve a shorter reaction time at the same time. One solution would be to have the module monitor a sensor and if it goes above a certain threshold it can be propagated. When the sensor later drops below the threshold another message can be propagated. This might improve the response time of the system, because when communication takes place only when it is needed it can be made efficient.

Another orthogonal way to decrease the amount of communication would be to only propagate sensor information to the modules that need it. For instance in the walker sensor information from one side could be propagated to the other side. In this way the sensors on the left control the legs on the right and the other way around. In general directed diffusion (Intanagonwiwat et al., 2000) could be used for this. In directed diffusion information is propagated from the producer to the consumer through networks with a time varying configuration. In this framework it is possible for the consumer to show his interest in a specific kind of data and have that routed to it from the producer.

9 Summary

In a self-reconfigurable robot sensors can be used in two ways. One way is to use the sensor information locally: sensors can be used to bias the motions of modules to produce the desired global behavior. The advantage of this is efficiency since the sensor values do not need to be communicated anywhere.

In some situations it is not possible to handle all sensor information locally. Therefore another way to handle sensor information has been presented. We have experimented with an approach where sensor information is abstracted based on the sensors position on the robot. This abstracted sensor information then flows from the modules that produce the information to all the modules of the system. An important aspect of this system is that the acting modules have abstract information about the position where the sensor value originated. We have shown that by combining this communication mechanism with role based control we were able to make a six module self-reconfigurable robot walk and avoid an obstacle.

Acknowledgments

This research is funded under the DARPA contract DAAN02-98-C-4032, the EU contract IST-20001-33060, and the Danish Technical Research Council contract 26-01-0088.

References

Arkin, R. (1998). *Behaviour-Based Robotics.* MIT Press.

Beckers, R., Holland, O., and Deneubourg, J. (1994). From action to global task: Stigmergy and collective robotics. In *Proceedings of Artificial Life 4*, pages 181–189, Cambridge, Massachusetts, USA.

Beckers, R., Holland, O., and Deneubourg, J.-L. (2000). From local actions to global tasks: Stigmergy and collective robotics. *Prerational intelligence: Adaptive behavior and intelligent systems without symbols and logic*, 2:549–563.

Bojinov, H., Casal, A., and Hogg, T. (2000a). Emergent structures in modular self-reconfigurable robots. In *Proceedings of the IEEE int. conf. on Robotics & Automation*, volume 2, pages 1734 –1741, San Francisco, California, USA.

Bojinov, H., Casal, A., and Hogg, T. (2000b). Multiagent control of self-reconfigurable robots. In *Proceedings of the Fourth int. conf. on MultiAgent Systems*, pages 143 –150, Boston, Massachusetts, USA.

Butler, Z., Kotay, K., Rus, D., and Tomita, K. (2001). Cellular automata for decentralized control of self-reconfigurable robots. In *ICRA 2001 Workshop on Modular Self-Reconfigurable Robots*, Seoul, Korea.

Castano, A., Chokkalingam, R., and Will, P. (2000a). Autonomous and self-sufficient conro modules for reconfigurable robots. In *Proceedings of the 5th int. Symposion on Distributed Autonomous Robotic Systems*, pages 155–164, Knoxville, Texas, USA.

Castano, A., Shen, W.-M., and Will, P. (2000b). Conro: Towards deployable robots with inter-robot metamorphic capabilities. *Autonomous Robots*, 8(3):309–324.

Dekhil, M., Sobh, T., and Efros, A. (1996). Commanding sensors and controlling indoor autonomous mobile robots. In *Proceedings of the 1996 IEEE International Conference on Control Applications*, pages 199 –204, Dearborn, Michigan, USA.

Fukuda, T. and Nakagawa, S. (1990). Method of autonomous approach, docking and detaching between cells for dynamically reconfigurable robotic system cebot. *JSME int. Journal*, 33(2):263–268.

Hardy, N. and Ahmad, A. (1999). De-coupling for reuse in design and implementation using virtual sensors. *Autonomous Robots*, 6:265–280.

Henderson, T. and Shilcrat, E. (1984). Logical sensor systems. *Robotic Systems*, 1(2):169–193.

Hosokawa, K., Tsujimori, T., Fujii, T., Kaetsu, H., Asama, H., Kuroda, Y., and Endo, I. (1998). Self-organizing collective robots with morphogenesis in a vertical plane. In *Proceedings of the IEEE int. conf. on Robotics & Automation*, pages 2858–2863, Leuven, Belgium.

Intanagonwiwat, C., Govindan, R., and Estrin, D. (2000). Directed diffusion: A scalable and robust communication paradigm for sensor networks. In *Proceedings of the Sixth Annual International Conference on Mobile Computing and Networking (MobiCOM '00)*, pages 56–67, Boston, Massachussetts, USA.

Khoshnevis, B., Kovac, B., Shen, W.-M., and Will, P. (2001). Reconnectable joints for self-reconfigurable robots. In *Proceedings of the IEEE/RSJ int. conf. on Intelligent Robots and Systems*, Maui, Hawaii, USA.

Kotay, K., Rus, D., Vona, M., and McGray, C. (1998). The self-reconfiguring robotic molecule. In *Proceedings of the IEEE int. conf. on Robotics & Automation*, pages 424–431, Leuven, Belgium.

Matarić, M. J. (1997). Behavior-based control: Examples from navigation, learning, and group behavior. *Journal of Experimental and Theoretical Artificial Intelligence*, 9(2–3):323–336.

Murata, S., Kurokawa, H., and Kokaji, S. (1994). Self-assembling machine. In *Proceedings of the IEEE int. conf. on Robotics & Automation*, pages 441–448, San Diego, USA.

Murata, S., Kurokawa, H., Yoshida, E., Tomita, K., and Kokaji, S. (1998). A 3-d self-reconfigurable structure. In *Proceedings of the IEEE int. conf. on Robotics & Automation*, pages 432–439, Leuven, Belgium.

Murata, S., Yoshida, E., Tomita, K., Kurokawa, H., Kamimura, A., and Kokaji, S. (2000). Hardware design of modular robotic system. In *Proceedings of the IEEE/RSJ int. conf. on Intelligent Robots and Systems*, pages 2210–2217, Takamatsu, Japan.

Pamecha, A., Chiang, C., Stein, D., and Chirikjian, G. (1996). Design and implementation of metamorphic robots. In *Proceedings of the ASME Design Engineering Technical conf. and Computers in Engineering conf.*, pages 1–10, Irvine, USA.

Rowland, J. and Nicholls, H. (1989). A modular approach to sensor integration in robotic assembly. In Puente, E. and Nemes, L., (Eds.), *Information control problems in Manufacturing Technology*, pages 371–376. IFAC, Pergamon Press, Oxford, UK.

Rus, D. and Vona, M. (2000). A physical implementation of the crystalline robot. In *Proceedings of the IEEE int. conf. on Robotics & Automation*, pages 1726–1733, San Francisco, USA.

Rus, D. and Vona, M. (2001). Crystalline robots: Self-reconfiguration with compressible unit modules. *Autonomous Robots*, 10(1):107–124.

Salemi, B., Shen, W., and Will, P. (2001). Hormone controlled metamorphic robots. In *Proceedings of the IEEE int. conf. on Robotics & Automation*, pages 4194–4199, Seoul, Korea.

Shen, W.-M., Salemi, B., and Will, P. (2000a). Hormone-based control for self-reconfigurable robots. In *Proceedings of the int. conf. on Autonomous Agents*, pages 1–8, Barcelona, Spain.

Shen, W.-M., Salemi, B., and Will, P. (2000b). Hormones for self-reconfigurable robots. In *Proceedings of the int. conf. on Intelligent Autonomous Systems*, pages 918–925, Venice, Italy.

Støy, K. (2001). Using situated communication in distributed autonomous mobile robots. In *Proceedings of the 7th Scandinavian conf. on Artificial Intelligence*, Odense, Denmark.

Støy, K., Shen, W.-M., and Will, P. (2002a). Global locomotion from local interaction in self-reconfigurable robots. In *Proceedings of the 7th int. conf. on Intelligent Autonomous Systems IAS-7 (to appear)*, Marina del Rey, California, USA.

Støy, K., Shen, W.-M., and Will, P. (2002b). How to make a self-reconfigurable robot run. In *Proceedings of the First International Joint Conference on Autonomous Agents & Multiagent Systems (to appear)*, Bologna, Italy.

Ünsal, C. and Khosla, P. (2000). Mechatronic design of a modular self-reconfiguring robotic system. In *Proceedings of the IEEE int. conf. on Robotics & Automation*, pages 1742–1747, San Francisco, USA.

Yim, M. (1994). New locomotion gaits. In *Proceedings of int. conf. on Robotics & Automation*, pages 2508 –2514, San Diego, California, USA.

Yim, M., Duff, D., and Roufas, K. (2000). Polybot: A modular reconfigurable robot. In *Proceedings of the IEEE int. conf. on Robotics & Automation*, pages 514–520, San Francisco, USA.

Whisking: An Unexplored Sensory Modality

Max Lungarella* **Verena V. Hafner*** **Rolf Pfeifer*** **Hiroshi Yokoi****

*Artificial Intelligence Laboratory
Dept. of Inf. Technology, University of Zurich
Winterthurerstr. 190, 8057 Zurich, Switzerland
{lunga,vhafner,pfeifer}@ifi.unizh.ch

**Research Group of Complex Systems Eng.
Graduate School of Engineering
Hokkaido Univ., Sapporo 060-8628, Japan
yokoi@complex.eng.hokudai.ac.jp

Abstract

Whiskers are widespread in the animal kingdom because they play an essential role in adaptive behavior. In spite of the enormous potential of whiskers, they have yet to be systematically investigated and exploited by roboticists. In this work, we present a first series of experiments with prototype artificial whiskers that have been developed in our laboratory. These experiments have been inspired by neuroscience research on real rats. The experiments provide the foundation for future work including active sensing, whisker arrays, and cross-modal integration.

1 Introduction

Whisker signals can be exploited to deliver a wide range of information about the environment such as texture, distance, shape, and orientation. Thus, there is a potential overlap with other sensory modalities, for example vision, which yields precise spatial information. In spite of the enormous potential of whiskers, they have been up to now almost completely neglected by the robotics community: For the better part, the research has investigated sensors based mostly on binary touch devices, called whiskers, tactile whiskers (Russell, 1992), or active antennae used for distance measurements (Kaneko et al., 1998). However, natural whisker systems - according to our hypothesis - yield much richer information.

The rodent somatosensory system is characterized by a prominent representation of their mystacial vibrissae (called whiskers). Though rats and mice strongly rely on visual and olfactory cues, their somatosensory system (here: their whiskers) is an important and accurate sensory modality. Rats are able to distinguish sandpaper surfaces purely on the basis of cues from the whiskers (Guic-Robles et al., 1989). It is even suggested that the accuracy of rat whiskers is comparable to that of primates finger tips (Carvell and Simons, 1990).

In this work, we present a prototype of a realistic artificial whisker sensor, which returns signals with properties, that can potentially be processed in a way similar to the rat somatosensory system. The artificial whisker system consists of a whisker-shaped material probe, which was chosen among a small set of materials such as plastic, human hair, and real rat whiskers.

2 Artificial Whisker and Experimental Apparatus

The core element of our sensor is a standard off-the-shelf *electret* microphone capsule (see figure 1a), which is a particular type of electrostatic sound sensor. In order to build the actual *whisker sensor*, we glue a small piece of a whisker-shaped (straight or curved thin and circular) rod of whatever material we wish to employ on the diaphragm of the microphone with a relatively hard glue (vanilla *cyanoacrilic super-glue*). Upon contact with an object, the whisker-shaped rod transmits the resulting contact force to the plate capacitor, where a small, but detectable change of its plate distance is induced.

The experimental setup consists of a plastic cylinder on which different types of material samples can be attached. It is possible to have samples with smoothly or roughly textured surfaces (e.g., plastic, paper or sandpaper). The cylinder is placed on a $12V$ DC-motor, whose turning speed is controlled by means of a microcontroller unit, connected via a serial cable to a host computer. The distance between the plastic cylinder and the base of the whisker sensor can be adjusted (see figure 1b).

3 Analysis Methods

The analysis of the time series data collected for this paper have been performed in the frequency domain. We are interested in basic oscillations, or fundamental modes, that are the result of the interaction of sensing device and environment. The discrete Fourier transform (DFT) of N data points of a sequence of samples x_k is defined as $X_n = A_n + iB_n = \sum_{k=0}^{N-1} x_k e^{-2\pi ikn/N}$. The power spectral density (PSD) describes how the power (or variance) of a time series is distributed with frequency. In our case, we make use of a popular nonparametric PSD scheme developed by Welch (Welch, 1967).

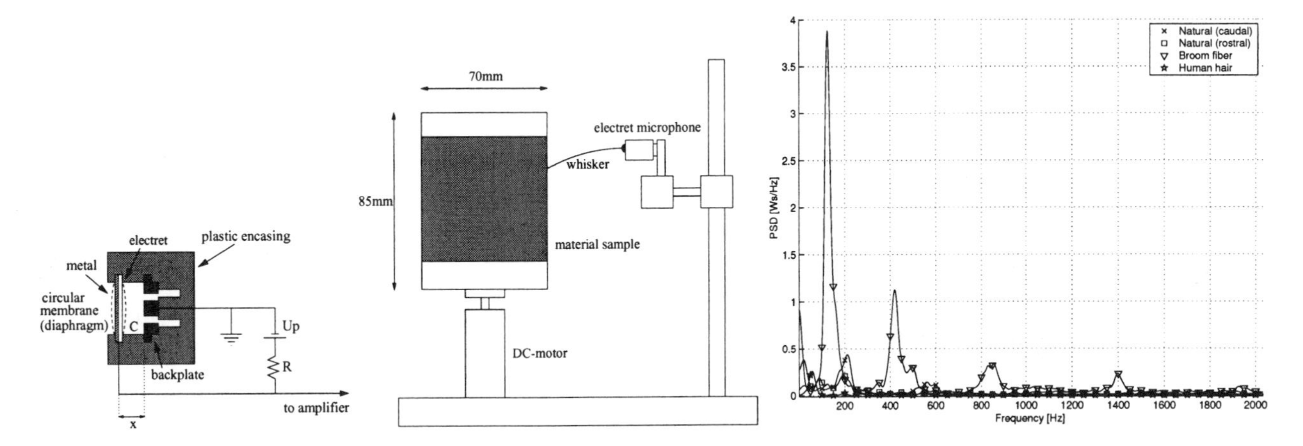

Figure 1: a) Basic schematic of an electret microphone. The deflection of the circular membrane, in response to a change of pressure, is measured by the change of capacitance. b) Schematic drawing of the experimental apparatus used to perform the experiments. c) Power spectral density (PSD) of four different type of whiskers.

4 Results of First Experiments

For a complete and thorough understanding of this sensing device, many parameters have to be taken into account. These include length and mechanical properties of the whisker (elasticity, stiffness, moment of inertia), material characteristics of the glue, location where the whisker is glued on the diaphragm, amplifier gain, radius of curvature of the whisker, and so forth. The goal of the initial set of experiments was to get a feel for the extent to which the material properties of the whisker are responsible for the characteristics of the sensory signal. Four different whiskers have been built and tested, one made of a small hair-shaped piece of polyvinyl (plastic) of length $l = 38mm$, a human hair ($l = 37mm$), and two types of rat whiskers (*caudal*, $l = 51mm$, and *rostral*, $l = 35mm$). Figure 1c displays the PSD of the time series of the different whiskers over a $1s$ time interval. Stiffness seems to play a major role. For the polyvinyl whisker, the ratio of the amplitude of the 1st peak and the amplitude of the 2nd peak and the one of the 2nd and the 3rd peak are bigger than the same ratios for the other three whiskers. Further experiments have been performed to categorize different cylinder surfaces made of sandpaper of different roughness (not shown here).

5 Conclusions

We have built an artificial whisker sensor consisting of an electret microphone on which we have glued a small whisker made of different materials. By performing palpation experiments with a rotating cylinder, we have shown that the artificial whisker sensor is producing complex data, which depend on various parameters. Comparing different materials for the whisker, we found out that a real rat whisker attached to our sensory system produces more clear and stable data. We assume that the natural whiskers have good damping properties compared with the artificial materials. Another interesting aspect of our prototype is, that by acoustically listening to the amplified output of the raw whisker sensor data, a human observer was able to distinguish different surfaces explored by the whisker. It will need further investigation to take all possible parameters into account.

6 Acknowledgements

This research has been supported in part by the grants #11-57267.99 and #20-61372.00 of the Swiss National Science Foundation, and by the IST-2000-28127 European project (AMOUSE). The natural rat whiskers were kindly provided by SISSA, Cog. Neuro. sector, Trieste.

References

Carvell, G. and Simons, J. (1990). Biometric analyses of vibrissal tactile discrimination in the rat. *Journal of Neuroscience*, 10(8):2638–2648.

Guic-Robles, E., Valdivesco, C., and Guajardo, G. (1989). Rats can learn a roughness discrimination using only their vibrissal system. *Behavioral Brain Research*, 31:285–289.

Kaneko, M., Kanayama, K., and Tsuji, T. (1998). Active antenna for contact sensing. *IEEE Trans. on Robotics and Automation*, 14(2):278–291.

Russell, R. (1992). Using tactile whiskers to measure surface contours. In *Proc. IEEE Int. Conf. Robot. Automat.*, pages 1295–1300.

Welch, P. (1967). The use of fast fourier transform for the estimation of power spectra: A method based on time averaging over short, modified periodograms. *IEEE Trans. Audio Electroacoust.*, AU-15:70–73.

Using IIDs to Estimate Sound Source Direction

Leslie S. Smith
Department of Computing Science and Mathematics
University of Stirling, Stirling FK9 4LA, UK
lss@cs.stir.ac.uk

Abstract

Poster. Sounds recorded using a binaural head are analysed to find the azimuthal direction of a sound source. Two techniques for inter-aural intensity difference (IID) estimation are compared. In both, signals were filtered into a number of wideband logarithmically spaced frequency bands. In method 1, an estimate of the IID in each frequency band was made every 20ms, and in method 2, estimates were made only when a cluster of onsets had been found. Onsets were detected using a biologically plausible spike-based technique. IID vectors were converted to directions using estimates of the impulse response of the binaural head. The onset-based technique provides better results, particularly in reverberant environments.

1 Introduction

Locating the source of a sound is an important task for an animal. It is normally achieved using a mixture of IIDs (resulting from ears having a non-uniform omindirectional response) and inter-aural temporal differences (ITDs) (resulting from differing path lengths from the sound source to each ear) (Blauert, 1996). However, sound generally reaches the ears not only from the direct path, but also from paths which include reflections. The onset of a sound always comes from the shortest, direct path. Human IIDs tend to be small below about 1000Hz, due to diffraction round the head. At high frequencies ITDs tend to become ambiguous because each ITD estimate e is actually $e \pm np$ (p is the signal period). Using the same model head (Smith, 2001), we showed that ITDs at low frequencies can be used to estimate the azimuth of sound source. Here we use IIDs measured at frequencies above 1000Hz.

In method 1, we compute the IID in each frequency band every 20ms. One problem in using IIDs for azimuthal source determination is that the instantaneous IID depends on all the paths sound has taken to reach the ears. In method 2, we examine IIDs only at onsets. Below, we describe how we measure IID vectors, how these are turned into directions, and we compare the effectiveness of methods 1 and 2. The most similar approach is in (Macpherson, 1991), but he uses brief pulses, whereas we are using normal utterances.

2 Methods

Sounds are played using a small amplifier and loudspeaker to the binaural recording system. This consists of two matched omni-directional microphones, placed at the outer ends of the auditory canals of a model head. This head consists of a realistic model skull with a latex covering modelling real flesh and skin with latex pinnae. The head was mounted on a simple model torso. Signals were played to the binaural recording system at the same height as the pinnae, at a distance of 1.5m. They were played at 10° intervals. After digitisation at 96000 samples/second, 16 bit, the signal was filtered into 32 overlapping bands using a gammatone filterbank with centre frequencies and bandwidths based on (Moore and Glasberg, 1983).

For both methods 1 and 2, an instantaneous IID value for each filtered band was calculated as follows. First the sum of the squares of the sample values inside an interval of length 5ms was computed for both the left and the right channels. Then the sum from the left channel was divided by the sum from the right channel. For method 1, IID vectors were calculated every 20 ms throughout the sound. For method 2, these values were calculated only at onsets, using only the channels in which onsets had occurred. Onset times need to be estimated accurately, with low latency, and across a wide range of signal levels. Firstly each channel signal was coded as a set of 20 spike trains, with spike probability in each spike train depending on signal strength but saturating, and with spikes generated in phase with the signal. These spike trains were fed into a set of depressing synapses on an integrate-and-fire neuron with a high leakage. This results in a low latency phase locked onset detector similar to that in (Smith, 2001). We considered as onsets only those occasions in which at least a certain number of onset spikes had occurred within a brief period of time, in both left and right channels. One problem with this approach is that at high source angles, the signal strength in the contralateral ear is low, and some onsets went undetected.

Both methods produce IID vectors. Computing azimuths requires estimation of the difference in response between the ears for the frequencies of the filterbank at different angles. The impulse response of each ear was computed using the MLS technique (Rife and Vanderkooy, 1989) at a range of azimuthal angles. This was truncated to 2 ms, as we were interested only in the effect of the head and torso, not environmental reflections. The difference between the left and right ear response was computed by taking the ratio of the power spectra of the Fourier transforms of the truncated impulse responses. The IID vector was compared with these, and the closest used to estimate the source angle.

3 Results

The figure shows the results of processing one second of male speech in a slightly reverberant environment. Method 1 underestimates the angle, and has a larger standard deviation than method 2. Method 2 overestimates the angle and discriminates poorly at large angles, partly due to the low energy of the contralateral signal. The instantaneous results at the bottom of the figure show that even single onsets can provide reasonable estimates of source azimuth. At large angles, some onsets are undetected due to the low energy of the contralateral signal. Other results (not shown) demonstrate that in highly reverberant environments, estimating IIDs only at onsets is even more crucial.

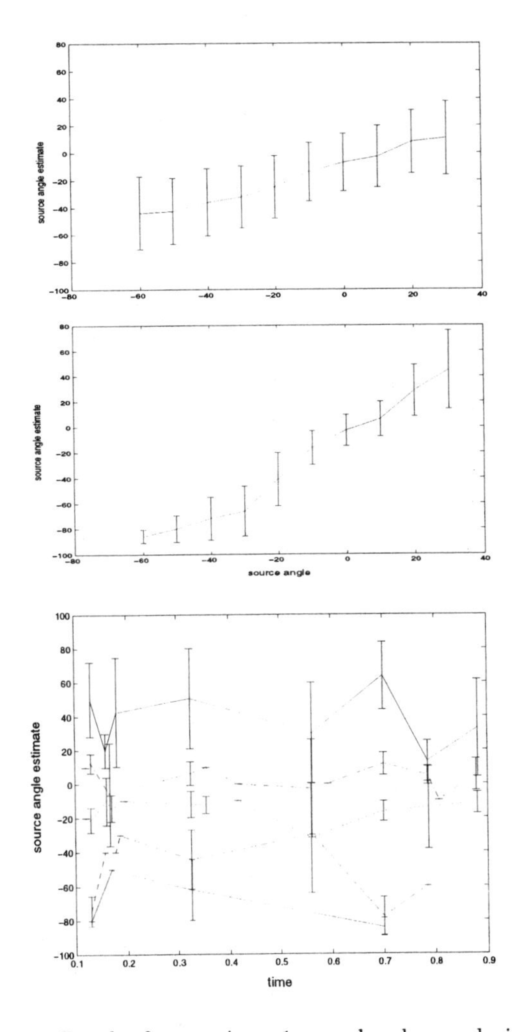

Figure 1: Result of processing a 1 second male speech signal. Top, middle: IID angles computed for varying source angles using method 1 (top) and method 2 (middle). Bottom: IID angle computed at each onset, for angles from -50 to + 30 degrees for 1 second of speech. line styles: line for -50, +30 degrees, dashed for -30 degrees, dotted line for -10 degrees and dash-dot for +10 degrees.

4 Discussion and further work

We have shown that using IIDs at onsets can permit sound source azimuthal estimation for sounds of short duration, even in a reverberant environment. This IID based technique should be combined with the ITD based technique in (Smith, 2001) which works at lower frequencies. Together, they should be able to substantially improve sound direction estimation (including elevation, given impulse responses at different elevations) for wideband sounds such as speech. The use of onsets makes the system emulate the precedence effect (Blauert, 1996) since a secondary onset after the first one (but without any intermediate offset) will be ignored because the depressing synapses will not have time to recover. In addition, clustering onsets should allow the location of more than one simultaneous (but not simultaneously onsetting) sound source.

References

Blauert, J. (1996). *Spatial Hearing.* MIT Press, revised edition.

Macpherson, E. (1991). A computer model of binaural localization for stereo imaging measurement. *Journal of the Audio Engineering Society,* 39:604–622.

Moore, B. and Glasberg, B. (1983). Suggested formulae for calculating auditory-filter bandwidths and excitation patterns. *Journal of the Acoustical Society of America,* 74(3):750–753.

Rife, D. and Vanderkooy, J. (1989). Transfer-function measurement with maximum-length sequences. *Journal of the Audio Engineering Society,* 37:419–444.

Smith, L. (2001). Using depressing synapses for phase locked auditory onset detection. In Dorffner, G., Bischof, H., and Hornik, K., (Eds.), *Artificial Neural Networks: ICANN 2001,* volume 2130 of *LNCS,* pages 1103–1108. Springer.

Visual Orientation and Motion Control of MAKRO – Adaptation to the Sewer Environment

Marina Kolesnik
Fraunhofer Institute for Media Communication,
Schloss Birlinghoven, D-53754 Sankt Augustin, Germany.
marina.kolesnik@imk.fraunhofer.de

Hermann Streich
Fraunhofer Institute for Autonomous Intelligent Systems,
Schloss Birlinghoven, D-53754 Sankt Augustin, Germany.
hermann.streich@ais.fraunhofer.de

Abstract

Adaptation has become an important aspect of robot design. The work here describes the perception and motion control of MAKRO - an autonomous robot for sewer inspection - from the point of view of MAKRO's adaptation to specific features of the sewer environment. Two features are crucial for MAKRO's adaptation. First, narrow sewer pipes connected into a unified system via junctions, compose a graph-like structure with rather constraint surface geometry. Second, a sewer interior is absolutely dark. The visual sensing of MAKRO is not only well adapted to these specific conditions, in fact it benefits from them. Visual orientation by a hybrid vision system gives rise to a rather simple vision model, which is capable of supporting real time orientation in the sewer. This instantiates an important principle of embodied cognition, which states that adaptation of an agent to an environment allows the use of simple principles of "cheap vision" for navigation purposes. Moreover, a fast visual processing enables MAKRO to react rapidly to events in its surroundings. This in turn, changes our approach to movement control: MAKRO does not act in the "plan – move" fashion; instead, it explores the environment, updates its heading and finds the right direction for the next move in real time. This leads to a second principle of the current work: if visual orientation of an agent operates in real time, all that is required for its successful navigation is to continuously update the right direction of motion. Navigation of MAKRO gives a powerful demonstration of how adaptation to an ecological niche and the exploitation of environmental constraints can lead to extraordinarily robust performance in a mobile robot.

1. Introduction

All systems, whether biological or artificial must fit their environments if they are to survive. Specifically, the development of an artificial agent must be guided by the environment where it has to operate. Conversely, in order for agent's movement to be regulated by the environment, the agent must be able to detect structures and events in its surroundings. Moreover, its reaction to events must be rapid, i.e. in real time.

Over the last decade, great advances have been made in the field of Embodied Intelligence and Situated Cognition (see, e.g. Brooks, 1999; Clark, 1997; Hendriks-Jansen, 1997; Lakoff and Johnson, 1999; Pfeifer and Scheier, 1999; Thelen and Smith, 1994), which have presented us with examples of artificial agents that display extraordinarily robust performance (see e.g. Horswill, 1992; Chahl and Srinivasan, 1996; Lambrinos, et. al, 2000; Möller, 1999). Typically, these agents are navigated by rather simple computation models applied for analysis of a sensory input which is itself quite limited. Still, their performance is impressive. Why and how is this possible? One common feature is careful design ensuring that the agent fits the particular ecological niche where it is to operate. The agent's sensory system, the way sensory data is processed and the way movements are triggered and controlled are all carefully adapted to the given task. As a result of this adaptation, a navigation strategy emerges, which is only functional under given conditions, but under these correct conditions the strategy becomes reliable, fast and highly precise.

Efficient visual processing is a crucial component of any autonomous agent not only because vision provides the most comprehensive information about the outside world but also, because of the sheer intensity of the flow of video data which has to be processed. Extensive computations that are required for solving such classical tasks as landmark identification or stereo reconstruction may easily block an agent's ability to react to events in real time. Yet again, if the agent is well adapted to the particular environment, i.e. it uses *a priori* knowledge about the environment in order to constrain the task of visual analysis, simple principles of "cheap vision" (Horswill, 1992) may support the navigation. Fast visual analysis opens up a new kind of navigation strategy: the agent no longer needs to perform long-range planning or to operate in a "sense-think-move" fashion. In the new navigation paradigm the two previously separated steps "think" and "move" can be effectively fused into a single "reactive motion" step, in which the agent moves and reacts to events simultaneously. Several works modelling vision-based navigation by insects illustrate this point [see e.g. Weber et. al, 1997; Srinivasan et. al, 1998; Iida and Lambrinos, 2000;]

Navigation scenarios are commonly divided into *global* and *local navigation*. The global navigation guides the agent's motion towards destination point that is not visible most of the way. By contrast, the task of local navigation is to guide the agent in such a way so as to dodge obstacles in the "visible" area immediately ahead of the agent. It follows that fast visual sensing, gained by the agent in the course of adaptation to the environment, has the power to transform the local navigation into reactive motion. Given a fast visual feedback on events, which are happening just ahead of the agent, the need for local path planning is eliminated altogether. Instead, the agent follows its global trajectory while continuously updating its heading in order to dodge close obstacles.

The motivation behind this work is to present a convincing example of the role of agent adaptation to task and environment and to demonstrate how this adaptation may boost the agent's performance. Our agent is an autonomous robot for sewer inspection called MAKRO (Figure 1). The task of MAKRO is to navigate through a system of sewer pipes (Figure 2) and collect a video record of sewer conditions. All aspects of MAKRO design are considered in the context of a complete agent, which is an agent that is autonomous, self-sufficient, embodied and situated (Pfeifer and Scheier, 1999). In fact, the shape of the MAKRO platform, the way MAKRO perceives its surroundings as well as the motion control it uses, are all specifically adapted to the sewer environment. We especially focus on two aspects of this adaptation: sensing and motion control. Our approach is to support reactive motion based on fast visual analysis.

In light of these ideas we, first, describe the sewer environment and the task of MAKRO navigation. Next, we elaborate the design of the MAKRO platform and its sensory system (Section 2). After that, we describe vision-based orientation aimed at recovering MAKRO's heading in real time. We will see that this capability has direct implications for motion control (Section 3). Our experiments demonstrate MAKRO navigating using this algorithm (Section 4). Finally, we discuss the principles of MAKRO's extraordinary performance, which is, of course, due to its adaptation to the specific ecological niche called sewer (Section 5).

2 Sewer environment and the robot

If MAKRO is to fit the sewer environment, its design must be regulated by the task and conditions of its operation. Therefore we start by looking at geometry of the sewer environment, the task of MAKRO navigation and the implications of these factors on the design of the MAKRO's sensory system.

2.1 Sewer geometry and the navigation task

Modern concrete sewers consist of cylindrical pipe segments having a 30 or 60cm diameter. These are joined together into the longer straight pipe portions (Figure 2). The straight portions intersect each other in T-, L- or X-shaped junctions sometimes called *manholes*. The latter are regions where humans can access the sewer from outside. Manholes are constructed out of preformed standard-shaped blocks, which are portions of vertical cylinders of about 2m diameter with the pipe entrances in perpendicular directions inside them. There are also four little stairs between the pipe entrances.

These elements of sewer construction define a dominating constraint of the sewer environment – the restricted geometry of its inner surfaces. The second constraint is obvious: the underground sewer world is absolutely dark.

MAKRO moves in this graph-structured environment, characterized by: 1) straight portions of the pipe separated by 2) junctions. The task of MAKRO's navigation is to reach a point of final destination, which is usually a given manhole. To do this, MAKRO must be able to move safely along straight pipes, identify moments when entering into junctions, choose the right direction to turn in, and execute the turn.

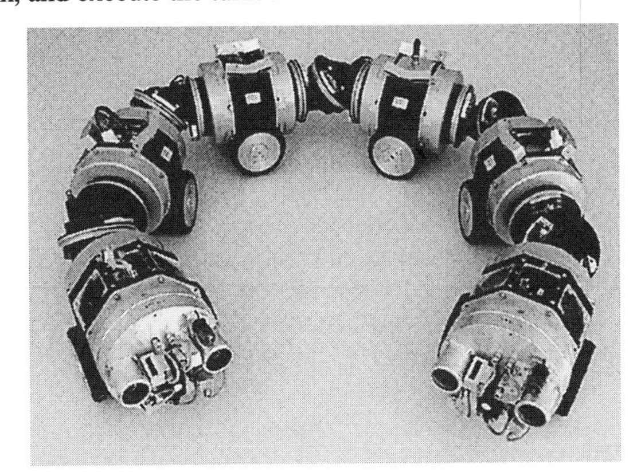

Figure 1: MAKRO, an autonomous robot for sewer inspection. The design of the MAKRO platform makes it well adapted for motion within narrow sewer pipes. The size of the cross-section of MAKRO segments is about 20 cm, which is well below the diameter of smallest sewer pipes which are typically 30cm.

Figure 2: The full-scale on-ground model of a typical modern sewer. The sewer model, with a total length of 80 meters, consists of pipes with a diameter of 60cm and of 7 manholes (GMD-AiS). The main difference between the model and a real sewer is that the model does not contain running water and is dry inside.

2.2 The navigation strategy

Any navigation scenario is ultimately driven by two mechanisms. The first one is *global navigation* which guides the robot's motion towards its final destination. The objective of the second, *local navigation,* is mainly to find a safe, collision-free path in the close vicinity of the robot. It is concerned with conditions in the robot's proximity and requires a capability to react rapidly to events as they occur on the move. Similarly, MAKRO's navigation strategy has two components.

A graph-like structure of the sewer environment reduces the global navigation of MAKRO to orientation on a 2-D map, which is rather simple. The robot's path is defined by a list of T-, X-, and L-junctions, which the robot has to pass, complemented with a sequence of corresponding directions indicating whether the robot has to turn or go straight. The robot uses its proximity sensors to check whether it is currently within a pipe or has entered a junction. The robot records the number of pipes and junctions that it has passed and can check its current position against the list. This mechanism reliably supports global navigation.

Local navigation of the robot is characterised by motion 1) along straight pipe portions and 2) within junctions. While moving along the straight portion of a pipe, it is important to maintain the robot's heading along the pipe. This implies the robot's ability to update its direction along the pipe axis continuously. When at a junction, the robot has to solve a different kind of problem. It has to get oriented among surfaces inside the junction, find the desired exit pipe, orient towards the pipe entrance and move in. Yet again, the local navigation can be treated as the continuous update of the robot's heading relative to the surfaces inside the junction.

2.3 The design of the platform and sensors

MAKRO's design embodies the constraints of the sewer environment and the task of navigation, whereas architecture of the sensory system is regulated by perceptual abilities which have to support platform movements.

MAKRO has a snakelike 1.5m long platform composed out of 6 equal sized segments enabling it to crawl along narrow pipes. MAKRO's adjacent segments are connected via flexible joints that can bend by 90 degrees in any direction. This facilitates MAKRO's ability to turn inside junctions while entering into emanating pipes and makes it possible to climb over a step or an obstacle of up to 30 cm height. Because MAKRO cannot turn around within a pipe, it is equipped with two entirely similar end segments: the configuration of sensors on the head and the tail segment is identical. This enables MAKRO to move forward as well as backward.

The task of MAKRO's visual perception is to support its local navigation. In agreement with our strategy, an analysis of video data must be sufficiently fast in order to find the right direction of motion in real time. In other words, the visual processing must be able of updating robot's heading continuously.

What kind of visual perception is proper for MAKRO survival in the sewer? A natural way to decide on the locally right direction for the robot's heading is just to check it visually. But visual analysis is a complex task, because camera images carry an enormous amount of information. In fact, only a tiny fraction of that information is needed for orientation. In the sewer, where the shape of typical surfaces is limited, a small set of characteristic points on the sewer wall may hint at robot's orientation. Because the sewer environment is absolutely dark, these characteristic points have to be highlighted in order to make them clearly visible in the camera image.

Figure 3: The LASIRIS laser crosshair projector.

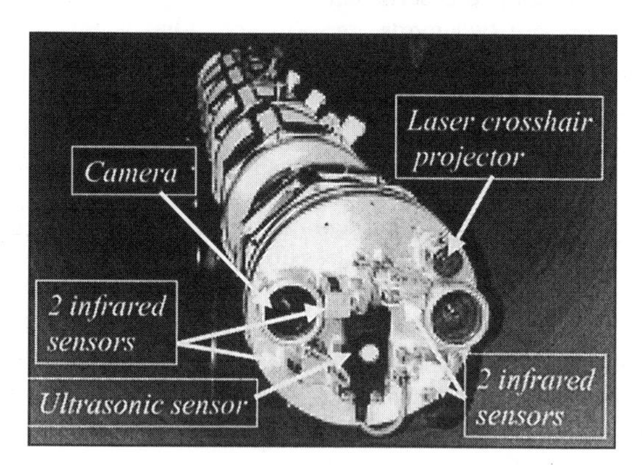

Figure 4: Configuration of sensors on the head (tail) segment of MAKRO. Even though the hybrid system does not require any calibration, its geometry must comply with certain requirements. The camera and the laser crosshair are mounted on the opposite sides of the segment to ensure the largest distance possible between them. The laser is elevated (or lowered) with respect to the camera thus avoiding their horizontal configuration. The laser is rotated around its axis so that the cross hair image appears to be roughly parallel to the image horizontal and vertical axis.

These considerations have lead to the idea of using a hybrid vision system that consists of two components: 1) a pen-size laser crosshair projector and 2) an optical camera (Figure 4). The laser is equipped with a special optical head which generates a high quality crosshair from the laser beam (Figure 3). The crosshair consists of two perpendicular planar sheets spanned within an accurately known fan angle. By projecting the ideal laser generated pattern onto the sewer surface whose geometrical features are roughly known, we extract a small number of surface points that, like a condensed print, carry information about the surface geometry. This information may be enough to recover orientation of the robot relative to the sewer surface. An important point of this approach is in its exploitation of the major environmental constraint – the restricted geometry of sewer surfaces. Clearly, dark conditions within the sewer will enhance the image of the laser footprint cast on the sewer surface. Thus, the visual system directly exploits the environmental constraints and, in fact, benefits from them.

The visual system is complemented by a set of proximity sensors such as infrared and ultrasonic sensors. There are four infrared sensors situated on the head segment. These monitor distances in four directions to the left, right, up and down from the robot (Figure 4). In addition, there are four infrared sensors situated on the left and right side of each body segment. These sensors monitor the distance between the robot and the sewer walls. Each infrared sensor is capable of measuring distances within range of about 8cm – 80cm. A single ultrasonic sensor on the head segment monitors obstacles in the immediate proximity ahead of the robot.

Actual position of each MAKRO segment is provided by the internal angle sensors, which are built into the joints connecting pairs of adjacent segments. Angular data recorded by these sensors during robot's motion carry full information about the current position of robot segments. These are used to describe the robot's trajectory.

There are two gyros that measure robot's inclination and give the robot a sense of vertical direction. Odometry sensors situated on each driving axle count wheel revolutions and provide the information about the speed and the distance covered.

3 Orientation and motion

The question to ask now is what kind of orientation is crucial for MAKRO navigation. These are of two different kinds for motion along a pipe and inside a junction.

While moving along the straight portion of a pipe, it is only important to "sense" the direction along the pipe, or, in case of a circular pipe, the direction along its axis. To illustrate this, let us consider motion within a dark pipe with the only source of light coming from the open end of the pipe far ahead. Motion towards the pipe's end can be successfully completed even though nearby pipe surroundings are not visible. It is enough to keep the right instantaneous orientation in the direction along the pipe. This direction is given by the point of the light source at the pipe's end. Therefore, in order to move along the pipe efficiently, MAKRO must only find the direction along the pipe axis.

The situation is different at a junction, where the robot must be able to orient itself properly for entering into a next emanating pipe.

3.1 Real time orientation

Real time orientation of MAKRO in the sewer is based on the visual analysis of a shape of the laser crosshair footprint. An optical head splits the laser beam onto two mutually orthogonal planar sheets. The two sheets originate at the beginning of the laser beam and are spanned within a fan angle which is accurately known. When the two laser-generated planar sheets are projected on a plane, the laser footprint is a cross (see Figure 3). When these are projected on the cylindrical pipe wall they give rise to a pair of conic sections whose shape encodes the orientation of the robot head within the pipe. However, the image of the laser footprint acquired by the camera depends on the distance between the two. This is easy to see in a thought experiment, in which the center of the laser coincides with the camera optical center. Because the camera center lies on the both planar sheets, the footprint observed by the camera in this case would always be an ideal cross regardless of its actual 3-D shape. The larger the distance separating the camera from the laser, the bigger the variations in the shape of the footprint captured in the image. Simulations (Kolesnik 1999) have indicated that, the distance, which is about one third that of the radius of the cylindrical pipe, is enough to record pronounced variations in the shape of the laser footprint in images acquired for different pointing directions of the laser. Test images acquired by the robot's camera within the typical sewer pipe, have confirmed that the sufficient distance is feasible.

If the above distance constraint is satisfied, the shape of the crosshair footprint in the image is uniquely related to the instantaneous orientation of the laser and, consequently, to the robot heading inside the pipe (see Appendix for a detailed geometrical clarification). The horizontal and the vertical parts of the crosshair footprint in the cylindrical pipe give rise to a pair of quadratic curves in the image (Figure 5). Each curve has a single point of maximum curvature. The distance between the two points of maximum curvature on the horizontal and vertical curves is directly related to the deviation of the laser pointing direction from the central axis of the cylindrical pipe. If these two points coincide the laser is oriented along the pipe. Because a field of view for the camera is known, a discrepancy between the two points of maximum curvature measured in pixels can be transformed into degrees. With that the instantaneous deviation of the robot's heading from the central pipe axis is expressed as an angle, which is used to correct the heading along the pipe.

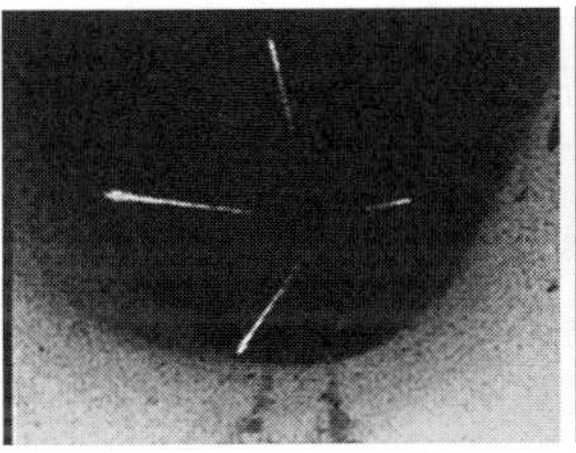
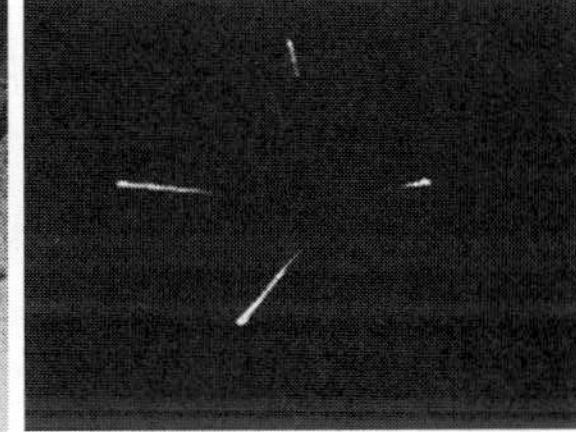

Figure 5: Left: The image of the straight pipe segment with the overlaid laser footprint. Right: The footprint image used for recovery of instantaneous orientation. The shape of the footprint indicates that the laser is well oriented towards the pipe axis.

Finally, the process of the visual orientation of the robot head within a cylindrical pipe, whose technical details can be gathered in (Kolesnik 2000), is straightforward. When MAKRO moves along the straight pipe, it continuously projects the crosshair pattern, and the camera acquires the image of the laser footprint. The footprint is then extracted from the image and its shape is analyzed so as to define an instantaneous orientation of the robot heading. Because the image of the laser footprint is acquired when the flashlight is off, the camera records a predominantly dark image with two bright intersecting curves (Figure 5, right). The analysis of such an image is extremely fast and can the update robot's heading in real time.

When arriving at a junction, MAKRO has to identify an entrance into an emanating pipe. The robot, facilitated by *a priori* knowledge of the possible configuration of surfaces at the junction, uses the images of the laser footprints to identify roughly the type of the surface it looks at. There are 4 different surface types considered: 1) pipe; 2) stair; 3) left wall; 4) right wall. Each surface type is characterized by the specifically shaped footprint in the image (Figure.5, Figure 6). The robot looks around by bending its head segment and finds the next adjacent pipe by analyzing the shape of the laser footprint. When the pipe is located, the robot finds the right direction leading into the pipe, orients its head segment in this direction and moves in. The heading is continuously corrected along the pipe axis.

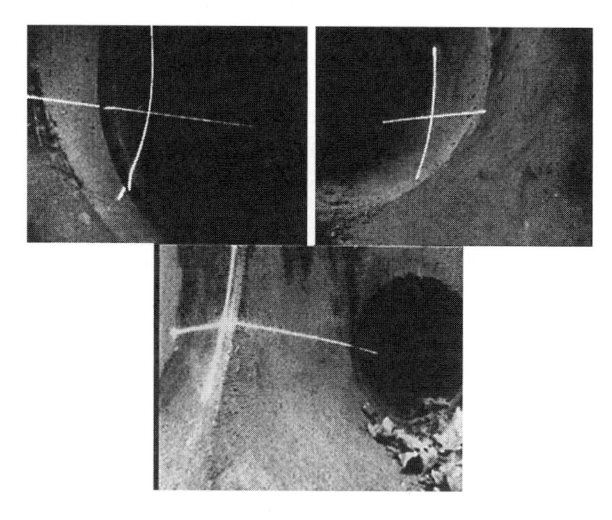

Figure 6: Three surface types with overlaid laser footprints. Left upper image illustrates the footprint typical for the surface type left wall. Its characteristic feature is indicated by the horizontal stripe which goes downwards when followed from the left to the right. This stands in contrast to the horizontal stripe of the footprint typical for the surface type right wall, which is shown in the upper row, right. Here the horizontal stripe goes upward when followed in the same left to right direction. The image in second row illustrates the footprint, which is typical for the surface type ridge. Its characteristic feature is indicated by the horizontal stripe which is fractured upwards.

3.2 Motion control

Two different movement patterns guide the motion of MAKRO within a pipe and at a junction. As in the case of the visual analysis, these patterns reflect geometrical differences between the two types of the sewer environment, i.e. the cylindrical pipes and the junctions.

First movement pattern supports MAKRO's motion along a straight pipe. In this case the movement pattern is defined by a task, which is to follow along the pipe up to its end. After this task is accomplished and the robot arrives at the end of the pipe (the robot detects the pipe end using its infrared sensors on the head segment), the movement pattern switches to the second one, which is motion within a junction.

At the junction the robot turns in the direction defined by the global trajectory. Let say the robot has to turn into a pipe to its left. The robot performs a sequence of standard movements that takes the robot head segment about 90 degrees to its left. During the turn the visual sensing (1) continuously updates the heading in the direction along the pipe the robot is entering into and (2) checks orientation of the tail segment. As soon as the robot has bent its head segment strong enough to identify the adjacent pipe, its heading is oriented into the right direction along the pipe axis. From this point the robot moves into the pipe. The other segments follow motion of the robot head.

Even though MAKRO is able to update the heading in real time, it must also coordinate movements of its all segments. MAKRO segments move along a certain local trajectory defined by its heading, which is continuously updated using the laser footprint analysis. Initially, the trajectory coincides with the position of all segments and has the length of the MAKRO's platform. MAKRO extends the trajectory for its next move by about 3-5 cm in the heading direction. Next, MAKRO executes the extended trajectory while continuously updating its heading and checking position of all segments. This motion strategy is common for both movement patterns, i.e. during the motion within pipes and at junctions.

A major coordinating principle of MAKRO movements is that the head segment simply causes the other segments to follow its own local trajectory: motion of the head segment is repeated by other segments when they reach same position on the trajectory. Angular values for rotation of all MAKRO joints and the speed of its all wheels is computed in real time. Movements of MAKRO segments are then triggered according to these values so as to keep the MAKRO platform on the trajectory.

The local trajectory is given by 3-D coordinate of virtual points, which model positions of MAKRO joints. Movements executed along the local trajectory are stored may be used again in those emergency situations when the robot has to move backward. In this case the robot can reproduce its motion backward along the same trajectory even without the analysis its sensor data. The stored trajectory insures a safe backward motion of MAKRO.

However, it may happen that robot's actual position deviates from the ideal local trajectory which it has to follow. Slippery surface, different traction affecting the segments, etc., may cause the robot to slip off the ideal pre-computed trajectory in an uncontrolled manner. In all these cases MAKRO has to check whether its actual position is really safe. One source of information for this check is provided by the proximity sensors that guard the robot from moving into a dangerous proximity towards sewer walls. Other information comes from the laser based visual sensing, which updates the heading of both the head and the tail segments by orienting them properly within pipes. This gives the correct reference position for all intermediate body segments.

Finally, the motion control of MAKRO is structured onto two layers: 1) the micro-controller and 2) the main processor layer. Each of MAKRO's six segments contains a C167 micro-controller to which sensors and actuators are connected. Sensor data collected by the micro-controllers are sent to the onboard main processor (166 MHz Pentium II). The micro-controller layer consists of low-level loops for moving joints and controlling their velocities. The main processor layer is hierarchically structured into three sub-layers (from low to high):

- activity-layer;
- context-layer;
- mission layer.

The activity layer contains a number of concurrently executed tasks, *(activities)*, for collision detection, trajectory generation, movements along the trajectory, etc. The context layer contains control structures for execution of various activities (*contexts*). Contexts themselves are the building blocks to execute the mission-layer. *Missions* define the global navigation of MAKRO and describe its global path through the sewer. Missions are prescribed by a human operator and are executed autonomously.

4. Experiments

Navigation of MAKRO has been tested during many trials, in which the robot has moved along straight pipes, approached junctions, executed turns and entered into emanating pipes. MAKRO has displayed highly stable performance in all our trials. Here we outline major technical aspects of MAKRO navigation.

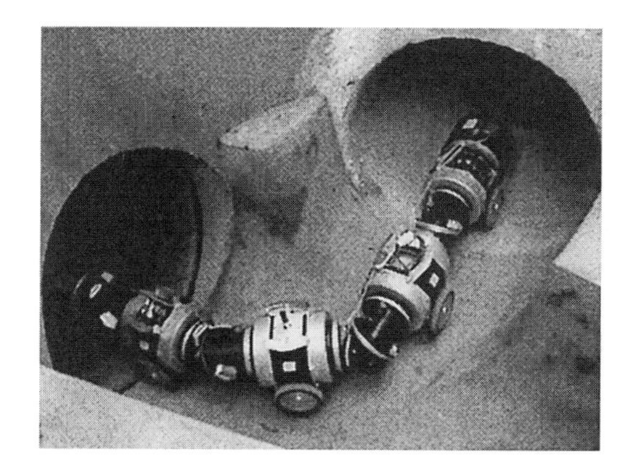

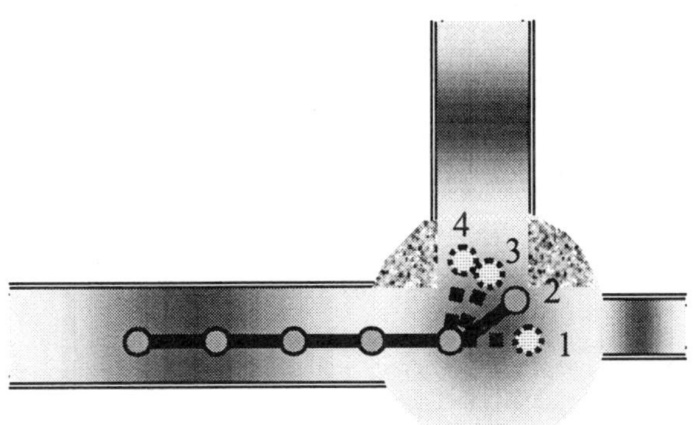

Figure 7: MAKRO turning in the T-shaped junction (upper image) and the schematic illustration of MAKRO executing this turn (lower image). Numbers 1-4 depict 4 optional positions of the MAKRO head segment while looking for the direction into the emanating pipe. In positions 1 and 4 the visual system detects a pipe. In position 2 the visual analysis detects a ridge, which is a small stair used by a human operator when getting into the manhole. The Right wall is identified in position 3. In all four cases, the tail segment detects the pipe.

While navigating along a straight pipe, MAKRO always moves in the direction prescribed by its head visual system. It takes 200msec for acquisition and the visual analysis of a single footprint image processed by the main processor. Because the visual analysis is that fast and does not require high quality images, the robot updates its heading continuously. When MAKRO approaches a junction, its orientation along the relatively short pipe portion remaining ahead of the robot may become imprecise. Likely deviations from the correct heading are compensated by visual orientation of the rear segment. The correct direction obtained by the vision system looking backward provides indispensable reference for correction of all MAKRO segments.

The moment when MAKRO enters into a junction is reliably detected by the infrared sensors located on the left, right and upper sides of the head segment. After this, the movement pattern of MAKRO changes to motion at the junction and the robot performs a turn in the direction defined by the global navigation. A process of turning is initiated by the head segment which bends slowly in the required direction while exploring the shape of surfaces it is looking at. The expected sequence of surface types, which has to be identified by the head segment while bending, is as follows: pipe – ridge - right wall – pipe (Figure 7). In case of unexpected false detection, the robot performs 2-3 more attempts to identify the surface type while bending its head segment further. If the identified surfaces are in agreement with the expected sequence, the robot proceeds executing the turn. In case if a false or unidentified surface type is persistently detected for all extra attempts (what has never happened to MAKRO during our tests!), the robot stops its operation. Position of the tail segment is always continuously updated during the turn. As long as the tail segment remains looking along the previous pipe (Figure 7), its orientation is corrected according to the direction given by the footprint analysis. This “look-back-check” provides MAKRO with independent control over position of its whole platform.

5. Discussion and conclusions

No vision system, however advanced and sophisticated it is, can provide a robot with a universal tool for its visual orientation. However, adaptation of the vision system to specific environmental conditions permits the robot to move around successfully using rather simple visual processing. We illustrated this point by looking at motion and visual sensing of the sewer robot MAKRO. MAKRO’s hardware, its sensory system and visual orientation are all perfectly adapted for action in the sewer environment. The point to be made here concerns the implications of this adaptation on motion control of MAKRO and, more generally, interconnections between the adaptation and the navigation strategy of an autonomous agent.

There is no doubt that MAKRO’s visual sensing by the hybrid vision system presented in this paper is well adapted to the sewer environment. The vision system exploits the two major constraints of the environment in a very efficient manner. First, the typical geometry of the modern sewer is exploited by projecting the ideal geometrical pattern of the laser crosshair. Second, sewer darkness enhances images of the laser footprint by providing a dark background. The laser footprint is clearly seen in the dark sewer from respectable distances, about 25m. This provides MAKRO with simple visual cues for encoding instantaneous orientation in the sewer. Thus, the visual sensing, emerging as a consequence of the “complete” adaptation to the sewer environment, is extremely simple in terms of computational complexity. The fact that it takes only 50msec for either classifying the type of surface in front of the robot or computing deviation of the robot's heading from the pipe axis, suggests an entirely new approach for motion control.

MAKRO does not stop to perform any thorough visual analysis of all surrounding objects but updates its heading continuously during motion. Note that the only information which has to be detected and updated for the successful navigation of MAKRO in the sewer, is the *right direction*. And this task is solved in real time. This is very different from the traditional navigation methods based on distance measurements which are often fairly difficult to perform accurately. Another advantage that comes for free is that the hybrid vision system does not require calibration. There are only a couple of simple initialization requirements that must hold.

Another advantageous aspect of the visual perception of MAKRO is that it is fairly stable as far as the quality of images is concerned. The laser footprints may be blurred or have discontinuities due to minor disruptions in the smoothness of sewer surface. But the dominating overall geometry of the sewer pipes overrides small deviations in the shape of footprints and makes the visual perception utterly stable. One more advantage of MAKRO vision is that it is economical in terms of power consumption. MAKRO is powered by rechargeable batteries and has rather limited energy resources. The fact that the vision system uses low power consuming laser projector makes it extremely attractive.

How do all these advantages become possible and why can such a simple visual analysis support the efficient navigation of MAKRO? The right answer seems to be this one: vision should never be viewed in isolation, but as part of a complete agent adapted to the task and the environment (Pfeifer, Lambrinos, 2000). If environmental constraints are properly taken into account vision algorithms become much simpler. It is then no problem to perform real time visual analysis. This, in turn, greatly simplifies motion control and the navigation strategy of MAKRO. All these, of course, fits into the principles of "cheap vision" (Horswill, 1993) and navigation of MAKRO beautifully demonstrates this point.

One broader proposition which follows is that if a robot is capable of reacting to events in real time, its navigation does not need any path planning step. All that is required is to find the *right direction* for the robot's next move and the ability to update this direction rapidly. This is what we call the task of real time orientation. In the case of MAKRO the orientation task is solved by the hybrid vision system. An underlying principle of operation of the hybrid system is based on the geometric traits of the sewer environment and the known geometrical pattern generated by the laser crosshair. It does not mean, however, that this principle is only applicable for the use in the sewer. Generalization of this principle leads to a second broader proposition that may prove useful for solving the task of orientation in environments other than the sewer. When a laser pattern is projected onto a scene, the shape of the footprints is like a condensed print, which carries information about the scene geometry. If the geometrical features of an environment are roughly known, information vital for deducing instantaneous orientation of a robot, may be readily extracted using a reference set of specifically shaped footprints.

Finally let us mention further challenging problems for MAKRO navigation. Even though MAKRO navigates successfully in the concrete sewer model, it remains unknown how well the visual system can be adapted to the wet conditions that are common for any real sewer. New kind of reflections caused by wet walls and a surface of water streaming down the sewer pipes may critically affect images of the laser footprint. Adaptation of the existing visual analysis to address this problem is one focus of our future work. Another challenge to the visual system will be posed by detection of the step which sometimes precedes entrances into pipes with a smaller diameter. In our future work we will address these problems also.

Acknowledgement

This work has been partially supported by the German Federal Ministry of Education, Research and Technology (BMBF) in the project MAKRO (02-WK9702/4), project partners being rhenag, FZI, GMD, and Inspector Systems Rainer Hitzel.

This support and cooperation is gratefully acknowledged.

References

Brooks, R.A. (1999). Cambrian intelligence. *The early history of the New AI.* Cambridge, Mass.: MIT Press.

Chahl, J. and Srinivasan, M. V. (1996) Visual computation of egomotion using an image interpolation technique. *Biol. Cybernetics* 74, 405-411

Clark, A. (1997). *Being there. Putting brain, body, and world together again.* Cambridge, Mass.: MIT Press.

Horswill, I. (1992). Characterizing adaptation by constraint. In F. J. Varela and P. Bourgine (Eds.), *Toward a practice of autonomous systems: Proceedings of the First European Conference on Artificial Life,* 58-64. Cambridge, MA:MIT Press.

Horswill, I. (1993). A simple, cheap, and robust visual navigation system. *From animals to animats: Proc. of the second International Conference on Simulation of Adaptive Behaviour.* Cambridge, MA: MIT Press (A Bradford Book).

Hendriks-Jansen, H. (1996). *Catching ourselves in the act: Situated activity, interactive emergence, evolution, and human thought.* Cambridge, MA: MIT Press (A Bradford Book).

Iida, F, and Lambrinos, D. (2000). Navigation in an autonomous flying robot by using a biologically inspired visual odometer. Sensor Fusion and Decentralized Control in Robotic System III, Photonics East, *Proceeding of SPIE*, vol. 4196, pp.86-97.

Kolesnik, M. (1999). View-based method for relative orientation in the pipe. *Proceedings of SPIE, "Sensor Fusion: Architecture, Algorithms, and Applications".* Vol. 3719, (1999).

Kolesnik, M. (2000). On Vision-Based Orientation Method of a Robot Head in a Dark Cylindrical Pipe. *SOFSEM'2000 - Theory and Practice of Informatics. Lecture Notes in Computer Science,* Volume 1963, 2000, pages 364-372, (©Springer-Verlag).

Lakoff, G., and Johnson, M. (1999). *Philosophy in the flesh. The embodied mind and its challenge to Western thought.* New York: Basic Books.

Lambrinos, D., Möller,, R., Labhart,, T., Pfeifer, R., and Wehner, R. (2000). A mobile robot employing insect strategies for navigation. *Robotics and Autonomous Systems, special issue on Biomimetic Robots*, Vol. 30, 39-64, © Elsevier.

Möller, R. (1999). Visual homing in analog hardware. *In Proc. 2nd European Workshop of Neuromorphic systems*

Pfeifer, R., and Scheier, C. (1999). *Understanding intelligence.* Cambridge, Mass.: MIT Press.

Pfeifer, R., and Lambrinos, D. (2000). Cheap Vision – Exploiting Ecological Niche and Morphology. *Lecture Notes in Computer Science* 1963, pp. 202-226.

Srinivasan, M., V., Chahl, J., S., Weber, K., Venkatesh, S., Nagle, M., G., and Zhang, S., W. (1998) Robot navigation inspired by principles of insect vision. In: *Field and Service Robotics,* A. Zelinsky (ed), Springer Verlag, Berlin, New York, 12-16.

Thelen, E., and Smith, L. (1994). *A dynamic systems approach to the development of cognition and action.* Cambridge, Mass.: MIT Press, (A Bradford Book).

K. Weber, S. Venkatesh and M.V. Srinivasan (1997) Insect inspired behaviours for the autonomous control of mobile robots. *In: From Living Eyes to Seeing Machines,* M.V. Srinivasan and S. Venkatesh (eds), Oxford University Press, U.K. pp. 226-248.

Appendix

Let C be an infinite circular cylinder modelling a sewer pipe, as in Figure 2. Let L be the central point of the laser crosshair, which is the origin of the two perpendicular planar sheets π_1 and π_2, generated by the laser. π_1 and π_2 intersect in a line ℓ which is the laser beam. When the crosshair is projected onto the pipe surface, each planar sheet slices through C in a quadratic curve (a segment of an ellipse). Let e_1 and.e_2 be the two ellipse segments, cut by the planar sheets π_1 and π_2, respectively. Let V_1 and V_2 be the vertices of e_1 and.e_2, respectively.

Orientation of the laser within the pipe is related to the direction of its beam, i.e. with ℓ. As the laser orients along the pipe and ℓ becomes more parallel to the axis of C, the vertices V_1 and V_2 move further away from L. At some point, when ℓ is parallel to the axis of C, both V_1 and V_2 will lie at infinity. Because all parallel lines intersect in a single point on the plane at infinity, V_1, V_2 and the axis of C will meet at this point.

Consider a camera located at a point O distinct from L. The camera acquires images of e_1 and e_2, which are two quadratic curves. Given a certain distance between O and L, the vertices V_1 and V_2 are viewed as two points of maximum curvature on the quadratic curves. As the laser orients along the pipe causing both V_1 and V_2 to slip away closer to infinity, the camera records a smaller distance between V_1 and V_2. At some point this distance will become too small and the images of V_1 and V_2 will eventually coincide. This is illustrated in Figure 3. Note that by including infinity in our considerations, we go beyond constructions of Euclidean geometry into an infinite world of Projective geometry. This is justified by the fact that a pinhole camera performs perspective projection mapping of 3-dimensional projective space onto a projective plane.

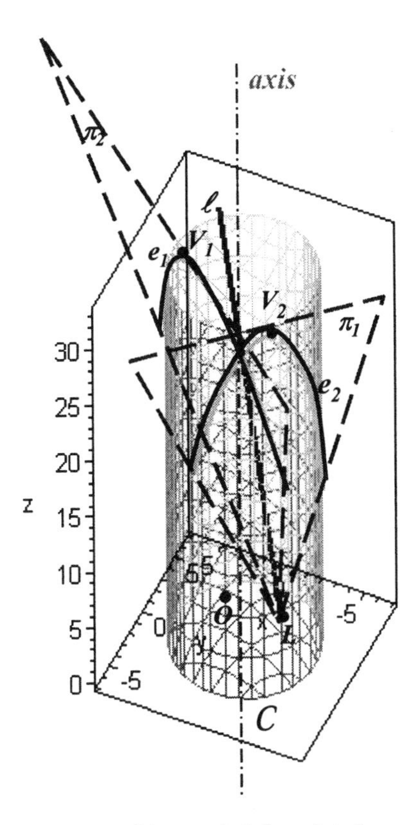

Figure 8: Geometry of the crosshair footprint when projected on the surface of a cylinder.

It follows from these geometrical considerations that the image of the laser footprint in a cylindrical pipe is uniquely connected to the instantaneous orientation of the robot head inside the pipe. The discrepancy between the images of V_1 and V_2 reflects robot's deviation from the direction along the pipe axis.

Adaptive leg placement strategies in the fruit fly set an example for six-legged walking systems

Simon Pick **Roland Strauss**
Lehrstuhl fuer Genetik und Neurobiologie
Biozentrum, Am Hubland,
97074 Wuerzburg, Germany
pick@biozentrum.uni-wuerzburg.de strauss@biozentrum.uni-wuerzburg.de

Abstract

Poster. Leg placement of the fruit fly *Drosophila melanogaster* was studied under various visual conditions while walking on a square-wave like linear array of narrow "stepping stones". Three-dimensional high speed video analysis revealed elaborate and error tolerant trajectories of the different legs which depend only to a minor extent on visual influences. Flies do not endeavor to place their legs always on the top surfaces of the treads but equally often cling to the edges or to the side surfaces of the stepping stones. A successful fly-like walking robot would rely on visual influences only for far-field orientation but would exploit a rich mechanosensation at its lower extremities and the ability to attach its legs to various surfaces. Technical solutions for these requirements are currently investigated using a single-leg test stand and a six-legged robot.

1. Introduction

Since the fruit fly *Drosophila* is a highly mobile and agile runner on all terrains, it lends itself as a source of inspiration for walking animates. Even more so since a wealth of molecular and classical genetic tools and of mutant stocks is available for this species, opening the path to an in-depth analysis of the neuronal underpinning of this highly adaptive behavior.

Previous work of our group has dealt with the walking behavior of *Drosophila* on a smooth horizontal surface (Strauss and Heisenberg 1990). We also revealed behavioral adaptations which enable the fly to walk on rough surfaces (Ernst and Strauss 2000, 2001, Ernst et al. in preparation). When flies walked over a linear array of small stepping stones (Figure 1) the speed and the stepping efficiency (average number of steps needed to proceed one tread on the linear array) were largely independent of variations in the contrast of the tread surfaces with regard to the gaps and of the visibility of the immediate foreground. The latter result was achieved by partially or completely occluding one or both eyes with light-tight paint. For each leg the fly has evolved elaborate trajectories which are under tactile control, and also efficient recovery strategies that follow a step into the void. Just to a minor extent particular eye regions help to further reduce the low rate of particular placement errors of particular legs. Placement information of anterior legs is conveyed to more posterior legs on the same body side.

In contrast, visual far-field orientation does have a strong influence on walking. The possibly motivating presence of an attractive landmark increases the walking speed in the stepping-stone paradigm by about 35%. Because of the associated increase in stride length the average rate of placement errors on a given array of treads can either increase or decrease depending on the spacing of the treads.

However, the high stepping rate (each leg moves up to 16 steps/s) makes *Drosophila* a technical challenge as an experimental animal. Moreover, the previously used overhead analog high-speed camera delivered necessarily incomplete 2-D information. To overcome both limitations we have set up a digital dual high-speed camera system permitting 3-D analysis of the complex walking behavior. Now we investigated error rates, leg trajectories and attachment sites. We are currently testing a single robot leg suspended on linear guides and a six-legged robot.

2. Method

Leg placement of wild type *Drosophila melanogaster* walking on square-wave like linear arrays of narrow "stepping stones" (0.5 mm tread width and 1.0 or 1.1 mm gap width) was recorded. Two digital cameras, each with 200 frames/s were used (Dalsa CA-D1). Steps were analyzed frame-by-frame and the positions of the tarsal tips were digitized. Software was written in C++ using the Common Vision Blox image library. Six different single leg behaviors were discriminated. In some experiments two vertical black stripes were presented as attractive landmarks on a white cylinder surrounding the setup in such a way that they were aligned with the ends of the platform. Wings of the flies were clipped to 1/3rd of their previous length under cold anesthesia at least 4h before an experiment. If eyes had to be occluded black air-brush paint was used (Schminke Aerocolor 28770).

3. Biological Results

Figure 1 shows a side view of a female fly walking on the stepping stone paradigm. A recovering leg reaches to the next stepping stone most of the time but occasionally will also retreat to its starting tread ("same") or will reach to the next but one tread. All events necessitating corrections taken together amount to 15% - 18% of the steps.

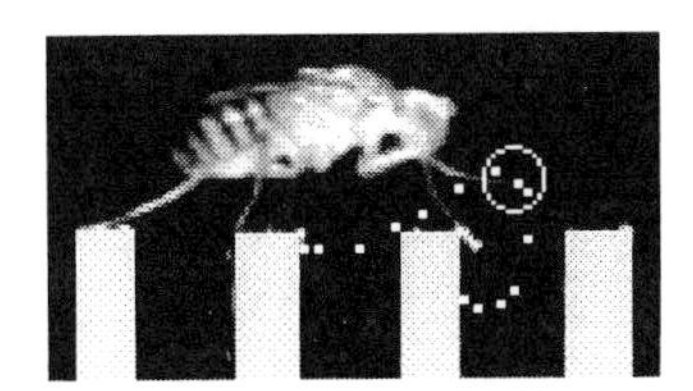

Figure 1: *Drosophila* walking on stepping stones. The trajectory of the left front leg's claw is marked at 5-ms intervals, its current position is circled. Treads are marked in light gray.

In Figure 2 error-free steps of all legs were classified with regard to their landing tread and, in a fourth category, all steps with an initial miss ("errors") were counted. Groups of 14 to 18 flies were tested under three visual conditions: (1) intact eyes and a landmark, (2) intact eyes but no landmark, (3) blind-folded flies. Seeing flies without a landmark perform not better than blind flies. We conclude that visibility of the treads does not improve placement performance.

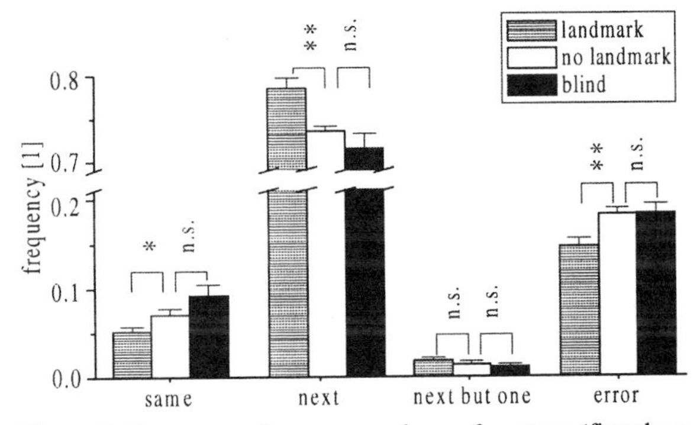

Figure 2: Frequency of ocurrence of error free steps (first three groups) and initially faulty steps under three visual conditions.

A comparison of groups (1) and (2) shows significant differences which can only be explained by the influence of the distant attractive landmark. We conclude that leg placement in *Drosophila* is largely under tactile control.

The tarsal tips of *Drosophila* can cling equally well to horizontal as to vertical surfaces. Figure 3 shows a statistics on the attachment sites of legs for flies walking with intact eyes in the stepping-stone paradigm toward an attractive landmark. In accord with the different orientations of the attachment organs on the different legs the front legs tend to cling more often to the distal vertical surfaces of the treads, whereas the hind legs more frequently make contact to the proximal edges. Claws or pulvilli (adhesive pads) facilitate reliable contact to a large variety of substrates. Unlike for many walking robots it is dispensable for the fly to aim its legs to the top surface of a tread. This minimizes the need for subsequent corrections of leg placement. Special trajectories of the tarsal tips during recovery strokes feature final contact to a support even if a side surface is reached.

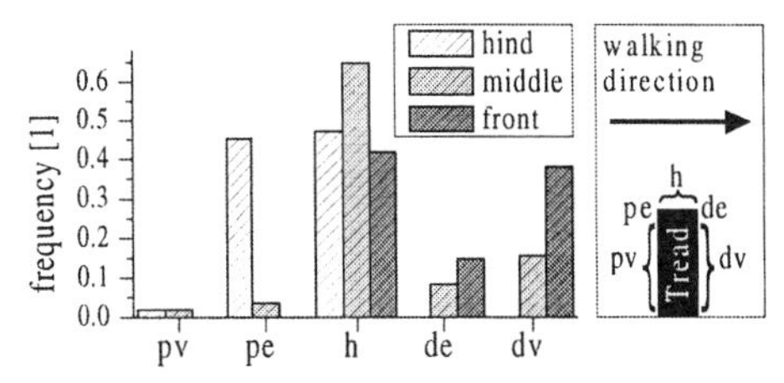

Figure 3: Location of supporting points on a stepping stone; proximal ("p") or distal ("d") vertical surface ("v") or edge ("e"); "h" horizontal surface, see inset. n=108 per leg pair.

4. Application to Robotics

As a proof of concept for our biological findings it is our goal to implement a *Drosophila*-like walking system with equally high stepping efficiency and error tolerance on rugged terrain. Since no visual targeting of legs was found in *Drosophila*, we will not resort to camera guided leg placement. Instead, the robot will rely on special leg trajectories, an effective attachment system and mechanosensory control. We are currently testing a single robot leg which is suspended on linear guides and "walks" on an uneven model landscape. The leg moves under tactile control along trajectories which are derived from the fly. This setup is used to investigate various attachment systems, trajectories, mechanosensors and the interplay of these components. In an initial phase the model attachment system used relies on a special surface of the walking substrate. Particular interest is taken in fast attachment and release, adhesive strength and direction-independent attachment efficiency. Moreover, we are modifying a TARRY II-type 6 legged robot (original layout Frik et al., 1998) by adding attachment systems and multiple contact sensors on the legs. Both systems, robot and single-leg test stand, are driven by servo motors controlled by an industrial PC and customized electronics. (Supported by BMBF grant no. 03118559)

References

Ernst R and Strauss R (2000) Influence of vision on leg placements in Drosophila walking on an uneven surface. *Europ J Neurosc.* **12 (Supp. 11)**, 501.

Ernst R and Strauss R (2001) Leg placement and coordination in *Drosophila* walking on rough terrain. In: *Göttingen Neurobiology report 2001* (ed. N. Elsner and G. W. Kreutzberg), p.346. Stuttgart, New York: Thieme.

Ernst R, Pick S and Strauss R. (in prep.) Leg placement in *Drosophila melanogaster* walking on uneven surfaces.

Frik, M., Guddat, M., Losch, D.C. and Karatas, M. (1998) Terrain Adaptive Control of the Walking Machine Tarry II. Proc. European Mechanics Colloquium, Euromech 375, Munich, pp. 108-115.

Strauss R and Heisenberg M (1990). Coordination of legs during straight walking and turning in *Drosophila melanogaster. J Comp Physiol A* **167**, 403-412.

ACTION SELECTION AND BEHAVIORAL SEQUENCES

Comparing a brain-inspired robot action selection mechanism with 'winner-takes-all'

Benoît Girard * **Vincent Cuzin** * **Agnès Guillot** * **Kevin N. Gurney** **
Tony J. Prescott **

* AnimatLab-LIP6
8, rue du capitaine Scott
75015 Paris, France
{benoit.girard,vincent.cuzin,agnes.guillot}@lip6.fr

** Department of Psychology
University of Sheffield
Western Bank, Sheffield S10 2TP, UK
{k.gurney,t.j.prescott}@sheffield.ac.uk

Abstract

We present a new robotic implementation of a brain-inspired model of action selection described by Gurney et al. (Gurney et al., 2001a, Gurney et al., 2001b) based on neural circuits located in the basal ganglia and thalamus of the vertebrate brain. Compared to an earlier robot implementation (Montes-Gonzalez et al., 2000), the new model demonstrates the capacity of the selection system to produce efficient 'energy' consumption/conversion in a 'feeding/resting' task whilst maintaining essential state variables within a 'zone of viability'. Generating appropriate action selection in this new setting entailed using biologically plausible Sigma-Pi units that can exploit correlated and anti-correlated dependencies between input signals when computing the 'salience' (urgency) of competing actions. A comparison between this brain-inspired selection mechanism and classical 'winner-takes-all' showed that the former can provide better behavioral persistence leading to more efficient energy intake.

1 Introduction

If the behavior of an animal or a robot is viewed as a discrete sequence of actions, then an understanding is needed of the mechanisms underlying the switching of behavior from one action to the next. In ethology several speculative hypotheses have been proposed concerning the action selection mechanisms underlying animal behavior switching. These hypotheses generally suppose that the motivational systems associated with a given act could win because they directly or indirectly activate, inhibit, or disinhibit their competitors. In the 1970's and 1980's, the mechanisms proposed for such interactions tried to explain the transitions between various behaviors in fishes, birds and rodents (Baerends et al., 1970, Ludlow, 1976, McFarland, 1977, Slater, 1978, Houston and Sumida, 1985). Eventually, ethologists lost interest in these models as they were unable to find a relationship between these speculative mechanisms and plausible biological equivalents.

Since the 1990's, with the rise of the animat approach, these models have been rediscovered and, with the improvement of computer methods, more precisely investigated (see Prescott et al. 1999, Guillot and Meyer, 2000, for reviews). However, many of the issues deriving from the earlier animal studies remain to be resolved (see Snaith and Holland 1991, Tyrrell, 1993, for reviews).

In recent years, a growing number of neurobiologists have become interested in a group of centrally-located brain structures known as basal ganglia as a possible neural substrate for action selection (for reviews see Redgrave et al., 1999, Prescott et al. 1999). According to Redgrave et al. (1999) centralised action selection could be important for large brains in order to achieve effective conflict resolution between competing sensorimotor systems whilst maintaining a cap on the connectivity and energy costs of the arbitration mechanisms. Several computational models of these neural structures have been investigated in a variety of simulation tasks (see Houk et al., 1995 for a representative selection). However, only that due to Gurney, Prescott and Redgrave (2001a,b) (henceforth the GPR model) has demonstrated the capacity of the basal ganglia to provide effective action selection in a real robot (Montes-Gonzalez et al., 2000). Based on the connectivity of the rat's basal ganglia, the GPR model (more precisely described below) is composed of two main circuits, one that computes the selection of the action *per se* and another that modulates the function of the first and which controls how this selection is done. The inputs to the model are variables called 'saliences', that are weighted functions computed from sensory, proprioceptive and contextual information, denoting the urgency associated with each act. The outputs of the model are inhibitions assigned to each potential action. At each time-step, the act which is least inhibited is performed. A third circuit provides, via the thalamus and cortex, a feedback loop whereby the output of the basal ganglia can influence its own future input, and in particular, enhance the salience signals of currently selected actions (Humphries and Gurney, 2001).

As noted by the authors, the GPR model exhibits three properties that are important for such mechanisms (Snaith and Holland, 1991, Prescott et al., 1999). The first is *clean switching* between actions: a competitor with a slight edge over its rivals should see the competition resolved rapidly in its favor. The second is *lack of distortion*: the presence of other candidates for the control of an effector should not interfere with the performance of the winning sub-system, once the competition has been resolved. The third is *persistence*: a winning act should remain active with lower input levels than were initially required for it to overcome the competition.

When embedded in a complete 'creature', in this case a Khepera robot, the GPR model displayed effective transitions between five actions (Montes-Gonzalez et al., 2000). The task of this robot was to mimic some of the behaviors of a hungry rat placed in a novel environment. Specifically, the robot was required to avoid open-spaces by moving towards wall and corners when the level of simulated fear was high at the start of the experiment, and to forage (by collecting wooden cylinders) when simulated hunger was relatively high (and fear relatively low) later in the experiment. This work also focussed on the effects of simulated dopamine modulation on the behavioral display. Dopamine is a neuromodulator known to have a critical effect on the function of the basal ganglia and behavioral switching more generally (see Redgrave et al., 1999).

In the current paper, we describe a second robot implementation of this model using a different robot platform, the Lego Mindstorms robot, and a task more typical of the type used in earlier action selection studies. Here the robot is required to select efficiently between four actions –wandering, avoiding obstacles, 'feeding' and 'resting'– in order to 'survive' in an environment where it can find 'food places' and 'rest places'. Its control architecture should be sufficiently adaptive to generate sequences of actions allowing it to remain as long as possible in its so-called *viability zone* (Ashby, 1952). This requires maintaining two essential state variables above minimal levels: *Potential energy* (obtained via 'feeding') and *Energy* (converted from Potential energy via resting). Spier and McFarland (1996) note that a 'two resource' problem of this type is a minimal scenario for evaluating an action selection or decision-making mechanism.

A further objective of this work is to investigate if and how the GPR model implements more than a simple 'winner-takes-all' (WTA) mechanism; a classical selection mechanism proposed long ago by engineers and ethologists (Atkinson and Birch, 1970). The WTA is based on selecting for execution the action that corresponds to the highest 'motivation' (integration of internal and external factors), whilst inhibiting all competitors. Whilst the GPR model has a superficially similar property of selecting (albeit by disinhibition) the most highly motivated action, this is modulated by the effects of the control and feedback circuits, potentially resulting in different pattern of behavior switching compared to simple WTA. For instance, according to Prescott et al. (1999), although a WTA can display both clean switching and lack of distortion, the lack of a mechanism to support appropriate persistence could lead it to generate unadaptive 'dithering' between actions, an issue in action selection previously noted by ethologists (Atkinson and Birch, 1970, Houston and Sumida, 1985). A comparison of the two control architectures, embedded in the same robot in the same environment, should therefore demonstrate precisely what benefits the GPR control circuits can bring to the action selection process.

Following a summary of the GPR model in section 2, we will describe, in section 3, how this model was reimplemented within the control architecture of a Lego Mindstorms robot. In section 4, the results obtained with the model will be presented and compared with those of a WTA, and these will be discussed, in section 5, from the perspective of biological plausibility.

2 The GPR model

The details of the computational model and its correspondence with the neural anatomy are fully described in Gurney et al.(2001a,b). We will only summarize here the main features of the model as shown in Fig. 1.

The terminology used for component structures is based on those comprising the basal ganglia: the striatum, the globus pallidus (with subcomponents GPe and GPi), the sub-thalamic nucleus (STN), and the substantia nigra (SNr). The selection and control sub-circuits of the Basal Ganglia-based model are designated here for conciseness by *BGI* and *BGII* respectively.

In each compoment structure, each action is associated with a discrete channel, which is represented by a single artificial neuron. Each artificial neuron consists of a leaky integrator whose activation is driven by a weighted sum of inputs (in the work presented here, this is modified to include nonlinear contributions). Each neuron is supposed to represent a biological neural population so that the activity in the model of each unit represents the mean activity of the population as a whole. While these model neurons are not as physiologically realistic as those that use conductance based methods with multiple membrane compartments, they are configured in circuits that are anatomically realistic and afford a useful tool for investigating models at the systems level of

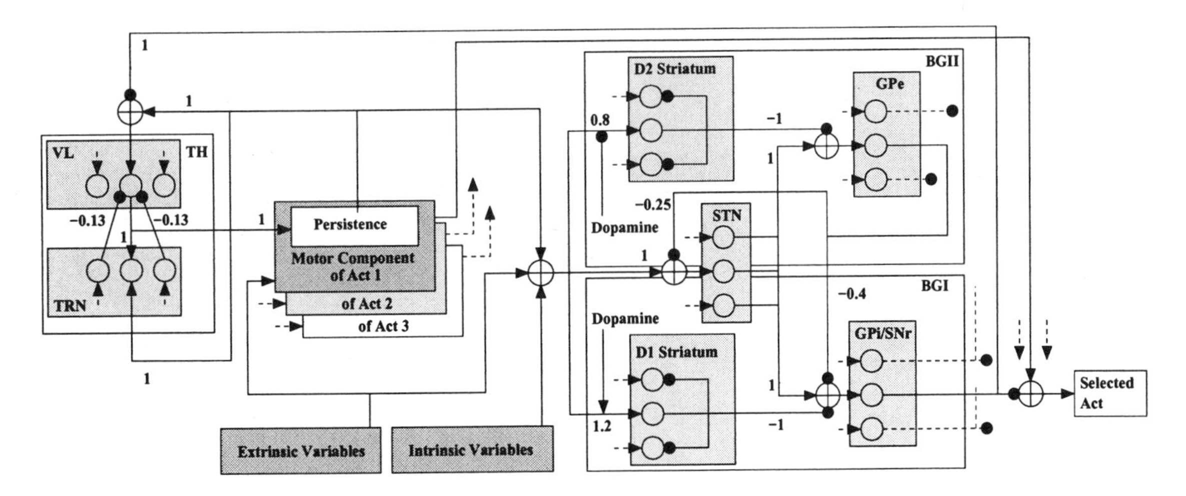

Figure 1: The GPR model. Arrows represent excitatory connections, blobs inhibitory connections. Weights are shown next to their respective pathways. See text for details.

description.

In *BGI*, selection is mediated by two separate mechanisms. First, there are local recurrent inhibitory circuits within the input component *D1 striatum.* [1] The second selection mechanism is comprised of an off-centre on-surround, feedforward network in which the 'on-surround' is supplied by excitation from STN and the 'off-centre' via inhibition from D1 striatum.

A similar arrangement prevails in *BGII*, except the 'output' of this structure (provided by the GPe) sends signals to *BGI*. In particular, it may be shown that the inhibition supplied to STN –the source of excitation for the feedforward selection network– is just sufficient to automatically scale this excitation with the number of channels n in the model, in such a way as to ensure appropriate selection. If this were not the case, the magnitude of the weights from STN and striatum would have to be crafted to be in an approximate ratio of $1 : n$. In the model, these weights have approximately the same magnitude and the scaling is performed by the automatic 'gain control' supplied by outputs from *BGII*.

Humphries and Gurney (2001) embedded the two circuits *BGI, BGII* into a wider anatomical context that included the thalamo-cortical excitatory recurrent loop. The thalamus was decomposed into two constituent structures: the thalamic reticular nucleus (TRN) and the ventro-lateral thalamus (VL). Both thalamic structures have the same segregated channels as *BGI* and *BGII*. This entire circuit is designated by *TH* in Fig. 1. The *TH* circuit not only improves the *clean switching* and *lack of distortion* mechanisms of the basic model, but also reinforces the salience of selected actions thereby fostering persistence of their state of being selected.

[1] the labels D1, D2 refer to types of dopamine synaptic receptor.

3 Implementation

3.1 The robot and its environment

The environment is a 2m x 1.60m flat surface surrounded by walls. It is covered by 40cm x 40cm tiles of three different kinds: 16 uniformly gray tiles (this neutral-gray represents 'barren' locations), 2 tiles with a gray to black gradient ('food' locations), and 2 tiles with gray to white gradient ('nest' locations) (Fig. 2). The robot is equipped with two frontal light sensors pointed to the ground –one behind the other– and with two bumpers, on the front-right and front-left sides (Fig. 2). These sensors provide the four *extrinsic* variables used in the salience calculations (see 3.4 below). Each light sensor produces a raw value corresponding to the color of the ground. The mean of these two values is filtered using a median filter with a 10 time-step window and then used to compute two variables, *Brightness* and *Darkness*, designated L_B, L_D respectively. L_B (resp. L_D) is equal to 0 for all grays darker (resp. brighter) than the neutral-gray, and increases linearly with brighter (resp. darker) grays, reaching 1 for the central white (resp. black) spots. Each of the two bumpers produces a binary value, B_L, B_R set to 1 when the robot hits an obstacle on the left and right respectively.

The 'metabolism' of the robot is based on two *intrinsic* variables: *Potential Energy*, E_{Pot} and *Energy*, E, that initially take on values between 0 and 255. Any action sub-system consumes *Energy* at a rate of 0.5 units per second (except for the variable rate of the resting behavior, see below). Then, these variables are normalised to lie between 0 and 1 for the salience computation.

When E reaches zero, the robot 'dies'. The procedure to reload *Energy* is:

1. to 'eat' on a black place, in order to get *Potential*

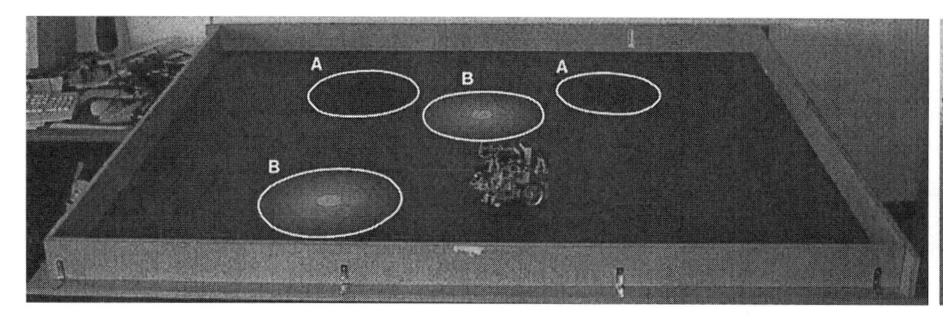

Figure 2: Left: The environment showing 'food' (A) and 'nest' (B) locations. Right: the Lego Mindstorm robot. (A): the light sensors; (B): the bumpers. See text for further details.

Energy, E_{Pot}. The gain $\Delta_E Pot$ in E_{Pot} during this time is proportional to the duration T_{eat} (in seconds) of the eating behavior and to the *Darkness*:

$$\Delta E_{Pot} = 7T_{eat}L_D$$

2. to 'rest' on a white place, in order to 'assimilate' *Potential Energy* and convert it into *Energy*. When there is no *Potential Energy* to assimilate, *Energy* is decreased with the standard 0.5 units/sec rate, otherwise the changes in *Energy* and *Potential Energy* are proportional to the resting duration T_{rest}

$$\Delta E = T_{rest}(7L_B - 0.5)$$

$$\Delta E_{Pot} = -7T_{rest}L_B$$

These relations imply that, when the robot activates these action sub-systems at an inappropriate location (eating on a neutral-gray or bright place or resting on a neutral-gray or dark place), it consumes *Energy* without any benefit.

3.2 Robot: hardware details

The controller (the RCX) for the Lego Mindstorms robot has only 32 KB of memory, some of which is used by the operating system (LegOS). This limited the computation available on-board the robot to the sensory, metabolism and action sub-systems. A Linux-based PC performed all the GPR model-specific computations, calculating and returning inhibitory output signals based on the sensory inputs received from the RCX.

The RCX-PC communication occurred through the Lego MindStorms standard IR transceivers at roughly 10 Hz. This low communication rate required that the GPR model be allowed to compute up to four cycles with the same sensory data in order to have the GPR model working at equilibrium.

3.3 The action sub-systems

In all experiments, the robot has to select efficiently between four action sub-systems. Note that each of these sub-systems corresponds to one channel in the GPR model. When activated, each action sub-system generates a predefined, but interruptible, sequence of elementary acts chosen among the following four available commands for the wheel actuators: *move forward, move backward, turn on the spot, stop*.

The action sub-systems are:

1. *Wander*: a random walk in the environment, programmed as a succession of forward and turning acts of random duration. This action provides the only means for the robot to move around and find the black or white areas; it should, for instance, be activated when the robot is on neutral-gray places, when the current level of either *Energy* or *Potential Energy* is low.

2. *AvoidObstacles*: a short backward movement followed by a rotation triggered when one or both bumpers are activated. Note that there is no movement if the behavior is selected while no bumper is active, therefore it should only be activated when the robot detects it hit an obstacle.

3. *ReloadOnDark*: the robot stops, and, as previously stated, it 'eats' on a dark place, that is, it reloads the *Potential Energy*. This action should therefore only be activated when the robot is on a dark place while *Potential Energy* is low.

4. *ReloadOnBright*: the robot stops and 'rests', that is, it reloads *Energy* and consumes *Potential Energy* when activated on a white place. This action should therefore be activated only when the robot is on a white place while *Energy* is low and *Potential Energy* is high enough for assimilation to be productive.

3.4 The GPR model implementation

The configuration and parameters of the GPR model used in these experiments are the same as in the 'full' embodied model (with normal dopamine modulation) described in Montes-Gonzalez (2001) (see Fig. 1), but there are also several key differences. These are concerned with modifications to processing of inputs and basal ganglia outputs which have been modified to take into account our different embodiment, environment, and tasks.

One important difference concerns the calculation of input saliences. In Montes-Gonzalez (2001), these were always computed as a linear, weighted sum of sensory, proprioceptive, and contextual variables. However, using any simple weighted sum does not allow salience to depend on a *coupling* of two variables. For instance, in our setting, the activation of *ReloadOnDark* should be correlated to the extrinsic variable *Darkness* and anti-correlated to the intrinsic variable *Potential Energy* (i.e. activated when the one is high and the other low). Activating it on a neutral-gray (or bright) place or while there is no need for *Potential Energy* just wastes *Energy* without any benefit. This situation can eventually lead to 'death', because the salience corresponding to this channel is reinforced by its feedback persistence and prevents other behavior from taking control of the robot. A similar problem also arises with *ReloadOnBright.* We therefore modified the salience computation to use Sigma-Pi units. These are artificial neurons that allow non-linear (multiplicative) combinations of inputs that can convey interdependencies between variables (Feldman and Ballard, 1982).

For the GPR and the WTA architectures, the weights of salience calculations were 'hand-crafted' over a series of pilot experiments in an attempt to find setting that were close to optimal. The following equations[2] show how the salience for each sub-system was computed as a function of the extrinsic sensory variables (*Brightness* L_B, *Darkness* L_D, *Bump left* B_L, *Bump right* B_L), the intrinsic sensory variables (*Potential Energy* E_{Pot}, *Energy* E) and the *Persistence* signal P for the given channel.

GPR salience calculations:

- *Wander*:
  $-B_L - B_R + 0.8(1 - E_{Pot}) + 0.9(1 - E)$
- *AvoidObstacles*:
  $3B_L + 3B_R + 0.5P$
- *ReloadOnDark*:
  $-2L_B - B_L - B_R + 3L_D(1 - E_{Pot}) + 0.4P$
- *ReloadOnBright*:
  $-2L_D - B_L - B_R + 3L_B(1-E)[1-(1-E_{Pot})^2]^{\frac{1}{2}} + 0.5P$

WTA salience calculations:

- *Wander*:
  $-B_L - B_R + 0.5(1 - E_{Pot}) + 0.7(1 - E)$
- *AvoidObstacles*:
  $3B_L + 3B_R$
- *ReloadOnDark*:
  $-2L_B - B_L - B_R + 3L_D(1 - E_{Pot})$
- *ReloadOnBright*:
  $-2L_D - B_L - B_R + 3L_B(1 - E)[1 - (1 - E_{Pot})^2]^{\frac{1}{2}}$

A second difference is in our use of the inhibitory output signal of the GPR model. A characteristic of the GPR model is that, in some cases where there is more than one channel with high salience, there can be partial disinhibition of the motor output of more than one channel. In the earlier robot implementation (Montes-Gonzalez, 2001) the motor outputs of all action sub-systems were therefore combined by weighting each one according to its degree of disinhibition, and Gurney et al (2001a) use the term 'soft switching' to describe an action selection mechanism that can generate a mixed/combined motor output of this kind. Clearly, when conflicting action sub-systems are involved, a merging of motor signals may result in distortion of the selected action(s). On the other hand, however, there are circumstances in which 'soft switching' may be desirable, for instance, where the outputs of two action sub-systems are fully compatible. For the current experiments, we were interested in making comparisons with the WTA mechanism which allows for only one winner (all losers are fully inhibited), a situation that can be termed 'hard switching'. In order to make comparisons between the two models the 'soft switching' characteristic of the GPR model was therefore disabled, in other words, the motor output of the most fully disinhibited action system was always enacted, and that of any partially disinhibited competitors ignored.

A final difference concerns the use in that model of an additional intrinsic variable termed the 'busy signal' whereby an active action sub-system could provide an additional signal to the selection mechanism that would give a temporary and short-term boost to its own salience. In the current robot task setting, the required behavior switching has so far been effectively implemented without including this feature of the original model.

Both architectures -GPR and WTA- were tested with the same robot, the same task, and in the same environment. As shown before, the saliences of the WTA and GPR were computed alike with the exception of the persistence signal P, which is included only in the GPR model. In the GPR architecture, the action sub-system with the least inhibition at each time-step is selected; in the WTA architecture, the action sub-system with the highest salience at each time-step is selected. In either architecture, where there were multiple winning outputs, the sub-system previously selected remained active.

[2]The term containing E_{Pot} in the *ReloadOnBright* salience is not a simple product. However, it may be reduced to such a form if we assume an intermediate variable $[1 - (1 - E_{Pot})^2]^{\frac{1}{2}}$ has been pre-computed first.

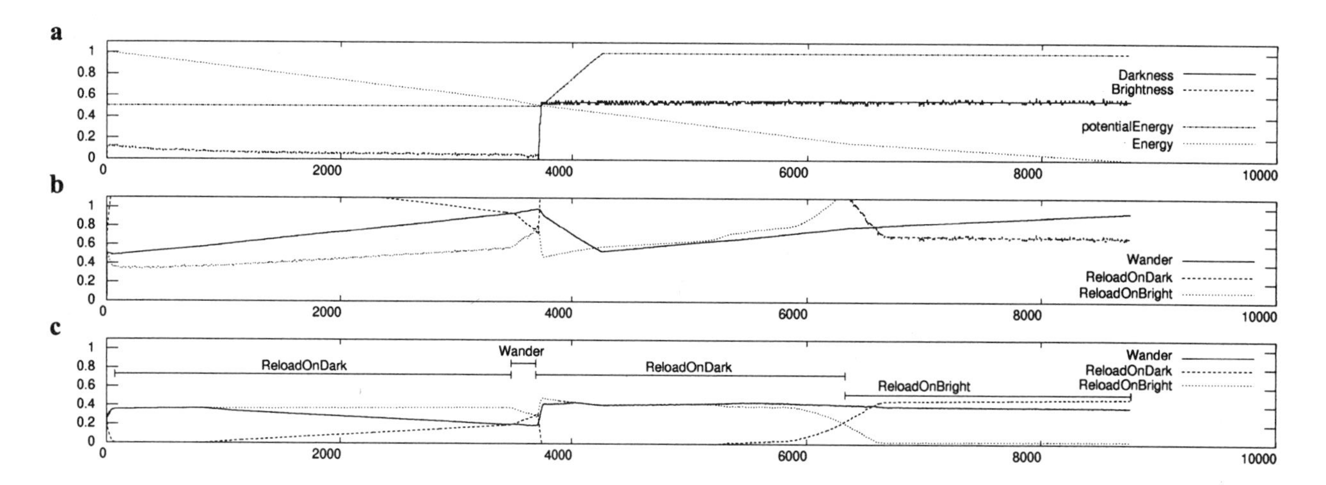

Figure 3: A typical GPR run without use of Sigma-Pi units, a) intrinsic and extrinsic sensory variables, b) corresponding saliences c) output GPi/SNr signals and behavioral sequence. The abscissa on all plots is the number of cycles where 1450 cycles correspond to 100 sec.

4 Results

4.1 Salience computation

Initial experiments with both architectures used simple linear weightings to compute action saliences. However, during the total 12hrs of experiments with the best hand-crafted weightings obtainable, the lifespan of the GPR and WTA robots never exceeded 1.5 time its minimum (8 minutes). Such a situation is depicted in Fig. 3. Here, the simple sum of *Potential Energy* and *Darkness* (3a, b) leads, with the same set of weights, to two inappropriate selections of *ReloadOnDark* (each for different reasons), and a final, and fatal inappropriate selection of *ReloadOnBright*. The first bout of *ReloadOnDark* occurs away from a dark square because the robot lacks *Potential Energy*; the second bout, on the other hand, is the result of a very dark sensor-reading, even though *Potential Energy* is no longer needed. The final ineffective *ReloadOnBright* occurs away from a bright tile because of a profound lack of *Energy*, and this act seals the fate of the animat.

Barring technical problems (such as communication glitches), the use of Sigma-Pi units enormously enhanced the life expectancy of both robots architectures (GPR and WTA), the longest uninterrupted experiment lasting 4 hrs and 20 minutes. Note that the robots can however still die, due to the intrinsic randomness in the *Wander* behavior. In the remainder of the paper we are exclusively concerned with experiments using the Sigma-Pi salience calculations.

4.2 GPR/WTA comparison

During experiments totalling more than 10hrs duration, we did not find any substantial difference between the GPR and WTA architectures with respect to life expectancy, simply because both robot architectures can outlive the time available for a single experiment. This first result led us to further analyze the structure of the behavior generated in the two conditions. In Fig. 4, graphs (a-c) shows the saliences, outputs, and behavior sequences of a typical run with the GPR model, while graphs (d) and (e) show the salience and behavior of the WTA architecture. Note first the substantial difference between the input saliences in the two runs Fig. 4 (a) and (d) which are primarily due to effects of persistence (positive feedback) in the GPR model. The output signals in Fig. 4 (b) show that the control circuit (*BGII*) and feedback loop (*TH*) have also increased the contrast between the action saliences (recall that, with GPR, the action sub-system with lowest inhibition is selected). Finally, in both behavioral sequences, we can observe that similar clean switching is displayed.

	activations per hour		avg. duration	
	GPR	WTA	GPR	WTA
W	302.3	488.0	4.0	3.8
ROD	41.7	62.5	16.0	8.8
ROB	65.1	81.6	15.1	8.0
AO	137.8	363.7	3.3	1.6

Table 1: Activation of each action sub-system showing average bouts per hour and average bout duration in seconds (W : Wander ; ROD : ReloadOnDark ; ROB : ReloadOnBright ; AO : Avoid Obstacle).

Table 1 shows that, with the exception of *Wander*, bouts of individual acts generally last longer with the GPR architecture than with WTA. This can be explained by effects of the persistence mechanism: posi-

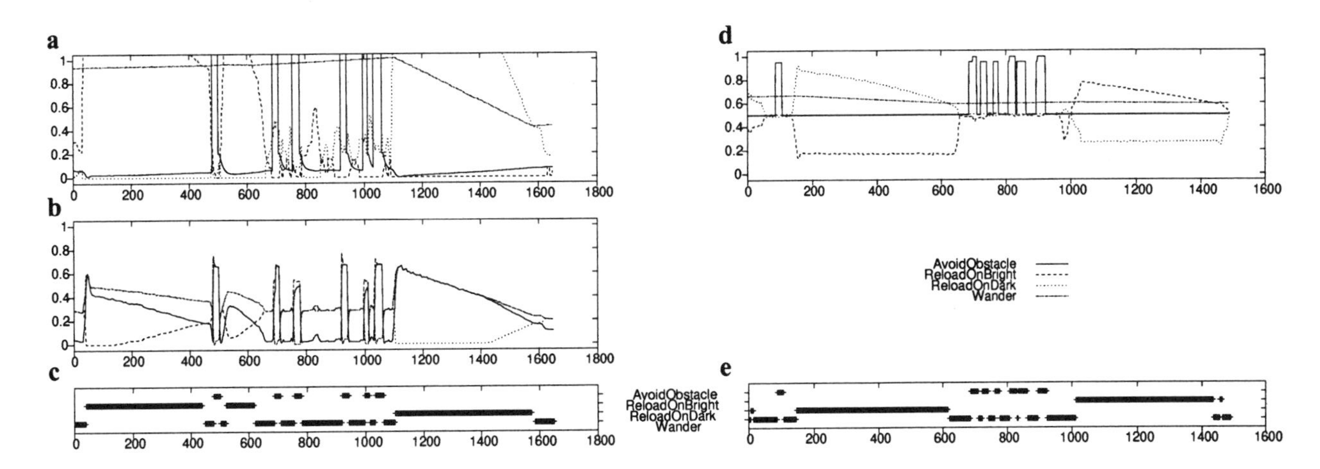

Figure 4: Left: a) Input saliences, b) output GPi/SNr signals and c) corresponding behavioral sequence generated by the GPR model. Right: d) Input and output signals (saliences) and e) the corresponding behavioral sequence generated by a WTA. The abscissa shows the number of cycles where 1450 cycles correspond to 100 sec.

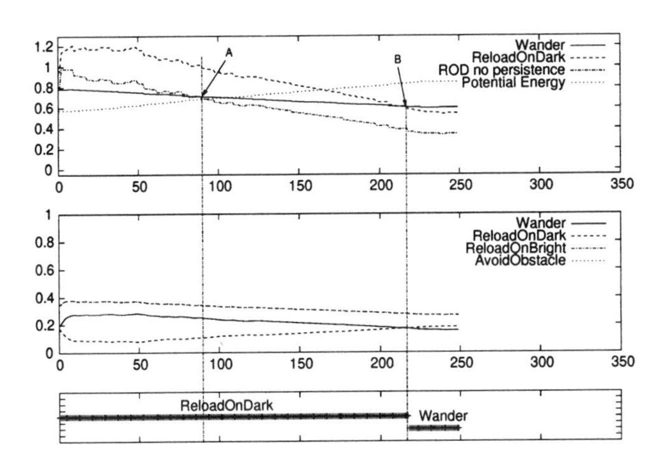

Figure 5: Effect of persistence in GPR. From top to bottom: 'raw' salience (i.e. without persistence) of ReloadOnDark ; output GPi/SNr signals ; the corresponding behavioral sequence generated by the GPR model where (A) points to the time where the switch would happen without persistence, and (B) points to where the switch actually takes place. The abscissa on all plots is the the number of computation cycles, where 1450 cycles correspond to 100 sec.

tive feedback allows a behavior to remain active for some time after its 'raw' salience has fallen below that of other behaviors (see Fig. 5 for an illustration). *Wander* is an exception to this pattern because there is zero weighting on the persistence input.

Although bouts of 'feeding' and 'resting' behavior are shorter in the WTA condition, their frequencies are correspondingly higher. This serves to substantially compensate for their shorter durations, to the point that the average *Potential Energy* and *Energy* end up having similar values ($E = 0.748, E_{Pot} = 0.711$ for WTA and $E = 0.76, E_{Pot} = 0.76$ for GPR). We suspect that this was helped by the relatively short distance between the *Energy* and *Potential Energy* sources in our environment. One may then ask whether the behavioral differences exhibited in Table 1 are reflected in the way the energies are collected.

As can be seen in Fig. 6, there is indeed a major difference between the temporal distribution of E_{Pot} in WTA and GPR. Specifically, the GPR manages to maintain its *Potential Energy* at over 95% of the maximum charge for 25% of the time, while the WTA does so for less than 10%. Unsurprisingly, since the energy sources are inexhaustible, the fact that a reloading action is allowed to last longer allows it to eventually reach the maximum charge most of the time.

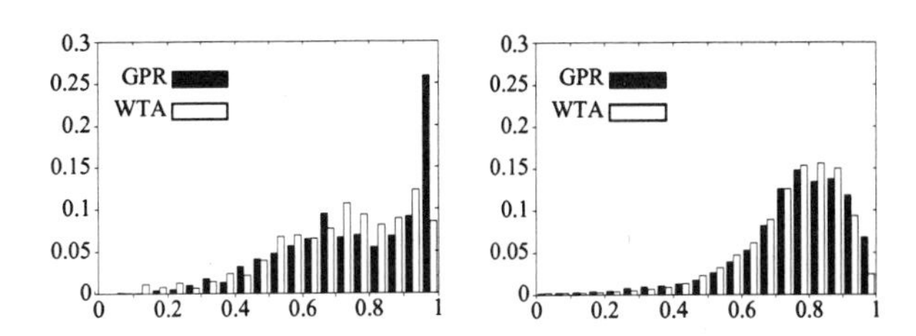

Figure 6: Histograms of the percentages of overall time during which Potential Energy (left) and Energy (right) are reloaded at the values shown on the abscissa.

The same occurs, though to a lesser extent, with the *Energy* (6.8% of maximal charge with GPR, 2.4% with WTA). The effect is less pronounced, because while *Potential Energy* only diminishes when assimilated by *ReloadOnBright*, all actions consume *Energy* therefore the constant decay levels the difference. Whilst persistence can be increased still further, a point is soon reached where the robot continues to recharge beyond the point where further energy can be usefully consumed.

The preceding results showed that both models can display clean and efficient switching between actions but, due to the effects of *Persistence*, the GPR robot performs fewer transitions than the WTA robot, making it possible for it to load more energy.

5 Discussion

In the following, we discuss the biological plausibility of the modifications we have made to the computation of input and output signals and consider the role of persistence in our model in relation to the notion of positive feedback in animal behavior switching.

5.1 Merging sensory information

In our implementation, the action sub-systems are assumed to depend on saliences, which correspond to the causal motivational factors depicted by ethologists. The general issue of how the display of an action is related to internal and external stimuli is not yet resolved. It seems to depend highly on environmental context and on the animal's previous experiences. In our work, salience is calculated using nonlinear relationships processed by Sigma-Pi units. Such units have been already used for solving similar problems before, while dealing with learning in neural networks (Rumelhart and McClelland, 1986, Gurney, 1992) or context processing in animats (Balkenius and Moren, 2000), but the question arises as to whether such a computation is anything more than an engineering solution. Mel (1993) argues that the dendritic trees of neocortical pyramidal cells can compute complex functions of this type, thus it is at least plausible to assume that second-order functions of the relevant contextual variables could be extracted by the neurons in either the cortex or the striatum that compute action saliences.

5.2 Merging motor signals

In the earlier robot implementation of GPR (Montes-Gonzalez et al., 2000), the motor components of all action sub-systems that were not fully inhibited could influence the displayed output behavior. In the current work, in order to facilitate comparison with a widely-used engineering solution to the action selection problem, we have not merged the output motor vectors. The literature on animal behavioral switching seems to indicate a wide-range of possible outcomes in situations where there is more than one highly salient action. Possibilities include the merging of the motor outputs (with potential positive or negative consequences), rapid switching between alternative actions (dithering), or the substitution of the salient actions by a third, non-salient 'displacement activity' that is unrelated to the current context (for example eating or grooming in a situation where both fight/flight are similarly primed (Hinde, 1970). The neurobiological substrates that support these various alternatives remain to be understood. However, it is worth noting that the behavior of animals in these (generally) unusual situations may reveal some of the processing characteristics and limitations of the underlying neural mechanisms.

The merging of multiple motor commands is also an issue with respect to the problem of generating appropriate behavior in animats with multiple actuators. In this case, action sub-systems with non-conflicting requirement to use different actuators (like walking and chewing-gum) can be selected at the same time. Neurobiological evidence of a somatotopic organization in the basal ganglia (Redgrave et al., 1999) suggests that there may, indeed, be distinct selection circuits subserving conflict resolution in relation to different limbs or body parts. Some preliminary work, derived from this idea, has been performed by replicating the GPR model as many times as there are actuators, with each GPR copy granting access to only one actuator.

5.3 Persistence and positive feedback in animats and animals

The main quality of the GPR model demonstrated in this study is that it provides a mechanism for ensuring appropriate persistence of a selected action. Though it is possible to add persistence to a WTA via a simple feedback loop, a control circuit (like the BGII in GPR) is then mandatory to avoid overload. The choice of such a circuit would divert the WTA from the zeroth-level action-selection mechanism we need to compare our system with.

Persistence has real adaptive effects. As stated before, it can maintain the animat's internal variables more effectively within their limits, helping it to survive any temporary upset in the availability of resources. It also serves to avoid dithering, which may be particular deleterious where there are significant costs associated with unnecessary switching between one action and another.

Another less intuitive effect of positive feedback is that it can 'prime' the animat to anticipate forthcoming opportunities for action. For instance, in our experiments we noticed that, due to the low communication frequency between RCX and PC, the WTA-robot often stops only after it has driven past the central brightest (or darkest) patch on the gradient tiles whereas the GPR version generally manages to stop closer to the center-most patch. What appears to be happening here is that the corresponding salience increases slightly as the robot enters the brighter (or darker) area. Although this is not, in itself, enough to prompt a change in the selected action, the positive feedback begins to build up the salience so that, when the robot eventually reaches the center, it is able to select the appropriate action more rapidly. This increased responsiveness is possible because the lightness gradient serves to prime the appropriate behavior.

The importance of persistence as an adaptive process for animals has already been pointed out by ethologists (e.g. McFarland, 1971). They wished to explain how an activity could continue, in spite of a rapid decrease in its drive. To do so, they supposed that mechanisms of positive feedback or hysteresis are initiated at the start of a bout, enabling the animal to maintain its activity until sufficiently satisfied (Wiepkema, 1971). For instance, in the model of Houston and Sumida (1985), persistence was induced in a competition between two independent motivational systems by a positive feedback pathway similar to a simplified version of the current model. The current experiments provide a useful embodied demonstration of this principle, and of the hypothesis of Redgrave et al. (1999) that the basal ganglia thalamo-cortical loop may serve as the neural substrate that carries this feedback path.

Ethologists have also noted that persistence is more than simply the consequence of closed-loop positive feedback, and can emerge in a variety of different ways. For instance, in the model of Ludlow (1976), hysteresis emerges as the consequence of reciprocal inhibition between multiple motivational systems. Such a configuration can provide for a form of persistence in the selected action. However, the advantages and disadvantages of this solution compared with explicit positive feedback control remain to be fully explored. There is also a sense in which an action may show a 'hidden' persistence, even after its execution has been interrupted. For example, in the 'time-sharing' model of McFarland and Lloyd (1973), a 'dominant' act may be temporarily suspended to allow an alternative behavior to be expressed, only later resuming its performance. In this case, the 'salience' of the dominant act persists even though the behavior itself is deselected. The neural substrate that might underlie a time-sharing mechanism in the vertebrate brain has yet to be investigated. Finally, the duration of any observed behavioral persistence varies according to contextual factors. For example, McFarland (1971) pointed out that the duration of feeding bouts in rats could be diversely triggered by the stimulation of oral and of gut receptors. In Le Magnen (1985) and Guillot (1988), the persistence effect on feeding and drinking bouts in rats and mice was also shown to depend on learning, diurnal and nocturnal conditions. In the current model, the duration of behavioral persistence will also be sensitive to contextual variables since salience is a function of many factors of which positive feedback is only one. The weight on the persistence pathway could also, itself, be subjected to contextual modulation.

These considerations confirm the importance for biology of investigating biomimetic models of action selection such as the GPR model. Compared to the earlier ethological hypotheses, this model is fully computationally specified, is identified with specific neural circuits, and has now been tested in two different embodied implementations. This model is also significantly more complex than earlier proposals and further work is needed to determine both the consequences of this additional complexity for observable behavioral switching, and to consider what potential advantages these may bring to the animal.

6 Conclusion and Perspective

Building on the work of Gurney et al. (2001a,b) and Montes-Gonzalez et al. (2000), our objective was to demonstrate the robustness of the brain-inspired GPR model of action selection. We have shown that the model is able to generate adaptive behavioral sequences when embedded in a different robot, performing different actions, and situated in a different environment. The new implementation also revealed that more flexible mechanisms (Sigma-Pi) can be useful for salience computations than a simple weighted sum. Finally, the comparison with WTA served to highlight several adaptive properties specific to the GPR model, and in particular, its capacity to generate appropriate behavioral persistence.

Further research is planned in three principle directions. First, we will investigate the 'soft switching' capability of the GPR model (merging of output signals coming from multiple sub-systems), in order to explore the capacity of the model to generate compromise behaviors, and also to replicate some of the consequences of mixed motor output observed in animals. Second, we will submit the salience and persistence parameters of the model to learning processes, which will automate the process of tuning the system to new tasks and may also enhance switching efficiency. Third, we will utilise this model within an ongoing, multi-partner project which aims to synthesizing an 'artificial rat' in which biomimetic mechanisms for action selection are combined with a biomimetic mechanism for navigation, both inspired by existing structures in the rat brain.

Acknowledgements

This work was supported by Robea, an interdisciplinary program of the French Centre National de la Recherche Scientifique.

References

Ashby, W. (1952). *Design for a brain.* Chapman and Hall.

Atkinson, J. and Birch, D. (1970). *The dynamics of action.* John Wiley & Sons.

Baerends, G., Drent, R., Glas, P., and Groenewold, H. (1970). An ethological analysis of incubation behaviour in the herring gull. *Behaviour (Supplement)*, 17:135–235.

Balkenius, C. and Moren, J. (2000). A computational model of context processing. In Meyer, J.-A., Berthoz, A., Floreano, D., Roitblat, H., and Wilson, S., (Eds.), *From animals to animats 6*, pages 256–265. Cambridge, MA: The MIT Press.

Feldman, J. and Ballard, D. (1982). Connectionist models and their properties. *Cognitive Science*, 6:205–254.

Guillot, A. (1988). *Contribution à l'étude des séquences comportementales de la souris: approches descriptive, causale et fonctionnelle.* PhD thesis, University of Paris 7.

Guillot, A. and Meyer, J.-A. (2000). From sab94 to sab2000: What's new animat? In Meyer, J.-A., Berthoz, A., Floreano, D., Roitblat, H., and Wilson, S., (Eds.), *From animals to animats 6*, pages 3–12. Cambridge, MA: The MIT Press.

Gurney, K. (1992). Training nets of hardware realizable sigma-pi units. *Neural Networks*, 5(2):289–303.

Gurney, K. N., Prescott, T. J., and Redgrave, P. (2001a). A computational model of action selection in the basal ganglia i. a new functional anatomy. *Biological Cybernetics*, 84:401–410.

Gurney, K. N., Prescott, T. J., and Redgrave, P. (2001b). A computational model of action selection in the basal ganglia ii. analysis and simulation of behaviour. *Biological Cybernetics*, 84:411–423.

Hinde, R. (1970). *Animal behaviour: a synthesis of ethology and comparative psychology.* Mc Graw Hill.

Houk, J., Davis, J., and Beiser, D., (Eds.) (1995). *Models of information processing in the basal ganglia.* Cambridge, MA: The MIT Press.

Houston, A. and Sumida, B. (1985). A positive feedback model for switching between two activities. *Animal Behaviour*, 33:315–325.

Humphries, M. D. and Gurney, K. N. (2001). The role of intra-thalamic and thalamocortical circuits in action selection. Submitted to : Network: Computation in Neural Systems.

LeMagnen, J. (1985). *Hunger.* Cambridge, UK: Cambridge University Press.

Ludlow, A. (1976). The behaviour of a model animal. *Behaviour*, 58:131–172.

McFarland, D. (1971). *Feedback mechanisms in animal behaviour.* London: Academic Press.

McFarland, D. (1977). Decision making in animals. *Nature*, 269:15–21.

McFarland, D. and Lloyd, I. (1973). Time-shared feeding and drinking. *Quaterly Journal of Experimental Psychology*, 25:48–61.

Mel, B. W. (1993). Synaptic integration in an excitable dendritic tree. *Journal of Neurophysiology*, 70(3):1086–1101.

Montes-Gonzalez, F. (2001). *A robot model of action selection in the vertebrate brain.* PhD thesis, University of Sheffield, UK.

Montes-Gonzalez, F., Prescott, T. J., Gurney, K. N., Humphries, M., and Redgrave, P. (2000). An embodied model of action selection mechanisms in the vertebrate brain. In Meyer, J.-A., Berthoz, A., Floreano, D., Roitblat, H., and Wilson, S. W., (Eds.), *From animals to animats 6*, volume 1, pages 157–166. Cambridge, MA: The MIT Press.

Prescott, T. J., Redgrave, P., and Gurney, K. N. (1999). Layered control architectures in robots and vertebrates. *Adaptive Behavior*, 7(1):99–127.

Redgrave, P., Prescott, T. J., and Gurney, K. (1999). The basal ganglia: a vertebrate solution to the selection problem? *Neuroscience*, 89:1009–1023.

Rumelhart, D. and McClelland, J. (1986). *Parallel Distributed Processing*, volume 1. Cambridge, MA: The MIT Press.

Slater, P. (1978). A simple model for competition between behaviour patterns. *Behaviour*, 57(3):236–257.

Snaith, S. and Holland, O. (1991). An investigation of two mediation strategies suitable for behavioural control in animals and animats. In Meyer, J.-A. and Wilson, S. W., (Eds.), *From animals to animats 1*, pages 255–262. Cambridge, MA: The MIT Press.

Spier, E. and McFarland, D. (1996). A fine-grained motivational model of behaviour sequencing. In Maes, P., Mataric, M. J., Meyer, J.-A., Pollack, J., and Wilson, S. W., (Eds.), *From Animals to Animats 4*, pages 255–263. Cambridge, MA: The MIT Press.

Tyrrell, T. (1993). The use of hierarchies for action selection. *Adaptive Behavior*, 1(4):387–420.

Wiepkema, P. (1971). Positive feedback at work during feeding. *Behaviour*, 39:266–273.

Simulations of Learning and Behaviour in the Hawkmoth *Deilephila elpenor*

Anna Balkenius
Vision Group
Cell and Organism Biology
Lund University
Helgonavägen 3
S-223 62 LUND
anna.balkenius@zool.lu.se

Almut Kelber
Vision Group
Cell and Organism Biology
Lund University
Helgonavägen 3
S-223 62 LUND
almut.kelber@zool.lu.se

Christian Balkenius
Lund University Cognitive Science
Kungshuset, Lundagård
S-222 22 LUND
christian.balkenius@lucs.lu.se

Abstract

We describe a behavioural experiment with the hawkmoth *Deilephila elpenor* and show how its behaviour in the experimental situation can be reproduced by a computational model. The aim of the model is to investigate what learning strategies are necessary to produce the behaviour observed in the experiment. Since very little is known about the nervous system of the animal, the model is mainly based on behavioural data and the sensitivities of its photoreceptors. The model consists of a number of interacting behaviour systems that are triggered by specific stimuli and control specific behaviours. The ability of the moth to learn the colours of different flowers and the adaptive processes involved in the choice between stimulus-approach and place-approach strategies is also modelled. The behavioural choices of the simulated model closely parallel those of the real animal. The model has implications both for the ecology of the animal and for robotic systems.

1. Introduction

Deilephila elpenor is a hawkmoth that feeds nectar from flowers. It is most active at night. While many other insects land on flowers when foraging, *D. elpenor* hovers in front of them while extending its long proboscis to retrieve the nectar. It is extremely good at compensating for drift while hovering under windy conditions. It has small wings and flies fast with a high wing beat frequency. Since this flight behaviour is very energy consuming, it is essential that it can feed effectively. To do this, it must continuously adapt to changes in its environment and learn where to find nectar.

D. elpenor has superposition compound eyes with three different photoreceptor types, an ultraviolet-, a blue- and a green sensitive receptor instead of blue, green and red as in humans.

Like most insects, *D. elpenor* can adapt their behaviour to accommodate changes in the environment. Although the total size of an insect brain is about one cubic millimetre or less, they are still able to show many of the types of learning that have been studied in mammals. However, the situations where each type of learning occurs is much more restricted than in mammals. The stimuli and responses must be selected carefully if any learning is to be shown.

For example, the moth *Spodoptera littoralis* can be classically conditioned to associate an odour with the proboscis extension reflex (PER) when rewarded with sucrose solution (Fan, 2000). *S. littoralis* can also learn discrimination and discrimination reversals as well as feature positive and negative discriminations (Fan & Hansson, 2000). However, classical conditioning was not possible with *S. littoralis* when colour stimuli were used instead of odour (Fan, Kelber & Balkenius, unpublished study).

During classical conditioning, the moths were constrained in a plastic tube. This prevents *D. elpenor* from being used in classical conditioning experiments of this type since they must be hovering to extend the very long proboscis. For free-flying animals, instrumental conditioning is more tractable. A suitable response is the approach of an artificial flower that is rewarded with sucrose solution (Kelber, Warrant & Balkenius, in preparation). Instrumental conditioning has been shown both in the moth *Macroglossum stellatarum* and in *D. elpenor* (Kelber, 1996, Kelber & Pfaff, 1997, Kelber, Warrant & Balkenius, in preparation, Balkenius, 2001).

We have performed experiments with *D. elpenor* where they were trained to search for food at differently coloured artificial flowers. The positions and colours of the flowers were manipulated to investigate how the moths would adapt. By constructing a computational model of *D. elpenor* we hope to generate hypotheses about its behaviour and learning ability that can later be tested in experiments. An additional goal is to find principles that can be used in constructing artificial animals and robots.

Figure 1: *Deilephila elpenor* hovering while foraging in the natural habitat (courtesy of M. Pfaff).

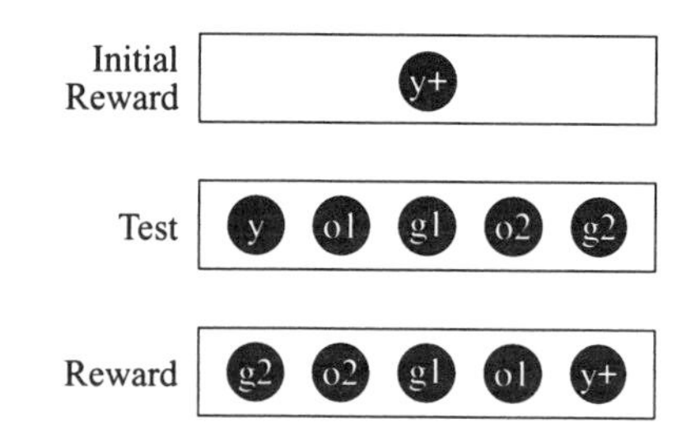

Figure 2: Positions of colours in the stimulus array during initial training, test and reward trials.

2. Experimental Study

This section describes the experimental study with *D. elpenor* and the results that were used to derive the computational model presented in section 3.

2.1 Materials and Methods

Moths were collected in July 1999 and kept at 4°C for hibernation. Three weeks before the experiment they were placed on a 12:12 hour light-dark cycle in a flight cage with a temperature around 20°C. Experiments were performed when the eyes were in the dark adapted state. The room was shielded from daylight and the cage was illuminated from above with a white lamp. The stimuli used were five artificial flowers with different colours. The rewarded colour was yellow (y). The other colours were yellow-orange (o1), orange (o2), light-green (g1), and green (g2). For a moth, these colours look relatively similar (see Fig. 6). The stimuli were presented at different positions in a vertical array on the wall of the flight cage.

Six moths were used in the study. Experiments started on day two after eclosure. The moths were initially trained to associate a single colour with a reward of 20% sucrose solution and later tested when five test stimuli were present. During initial training, each moth was fed the sucrose solution administrated through a 3 mm wide hole in the centre of a yellow artificial flower using a tube connected to a syringe. After one or two days of training, the moths had learned to forage at the artificial flower and the test phase began.

During tests, no food was present. Each animal was tested once every day. Between experiments, animals were released into the flight cage with a day-night cycle. Each test trial consisted of the presentation of five differently coloured flowers. For this paper we chose 6 experimental days when the colours where in the positions shown in the Fig. 2. The animals were allowed to approach the wall of the cage with the artificial flowers four times in a row. Each time the animal touched an artificial flower with its proboscis was counted as one visit. Each trial thus started with the approach of the flowers followed by one or several visits to the different flowers.

After four trials, the positions of the colours were temporarily changed as shown in Fig. 2 and the moths where fed at location 5 at the yellow flower. Without these rewards, the animals would loose interest in the experiment and stop flying.

2.2 Results of Behavioural Tests

Fig. 3 shows a typical example for the behaviour of the moth during the experiment. The first day, the moth is fed at a single yellow flower in the middle of the stimulus array until it has learned to approach and forage at the artificial flower.

When a moth was released in the cage at the test day, it first warmed up before it started to fly and approach the stimuli. It would stop at approximately 3-5 cm distance from the stimulus array and move sideways before choosing one of the stimuli. After a visit, it would either leave or choose a neighbouring stimulus.

In Fig. 3, the visits of flowers within a single trial are connected with a solid line. Dotted lines indicate that the moth left the stimulus array and approached again. This counted as the start of a new trial.

After the first reward, a moth would possibly visit the rewarded colour first but it would more often visit the artificial flower in the position where it received the reward (Fig. 3). This would be even more obvious after the second reward.

The distribution of visits to the different colours is shown together with the simulation results in Fig. 4. Fig. 4a shows the choices made by the moths during the first trial, before they were rewarded. The yellow flower is at position 1, and the generalisation to the other locations depends on the similarity between the colour at each position and the yellow colour. To the moth, the yellow-orange (o1) at position 2 and the orange (o2) at position 4 are more similar to the learned yellow than the light-green (g1) at position 3 or the green (g2) at position 5 (Compare positions of colours in the colour triangle, Fig. 6). The moth clearly uses colour to select which flowers to visit.

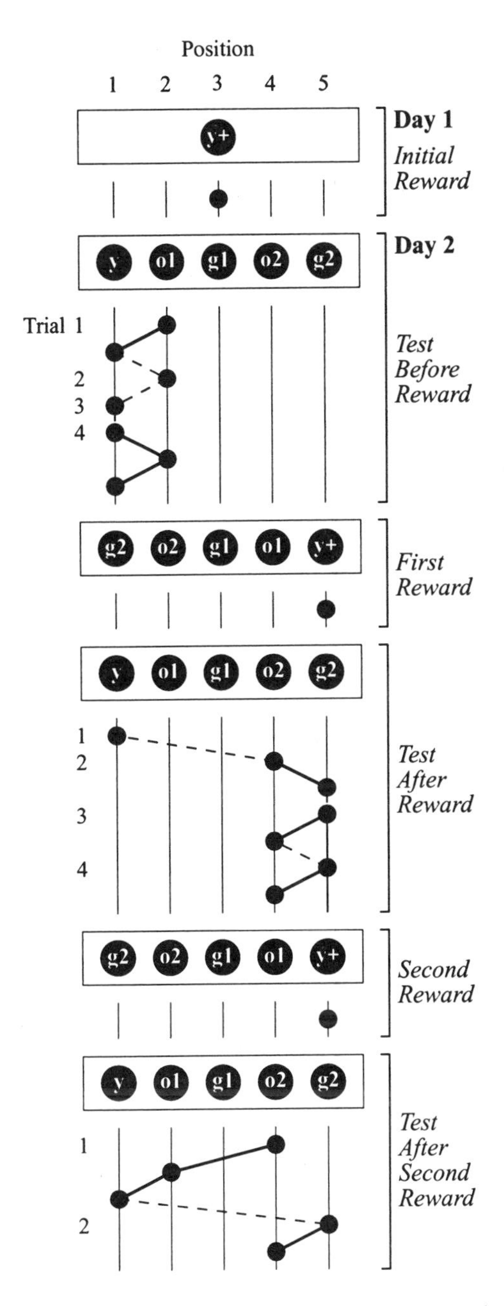

Figure 3: Typical behaviour of a moth during the experiment. See text for explanation.

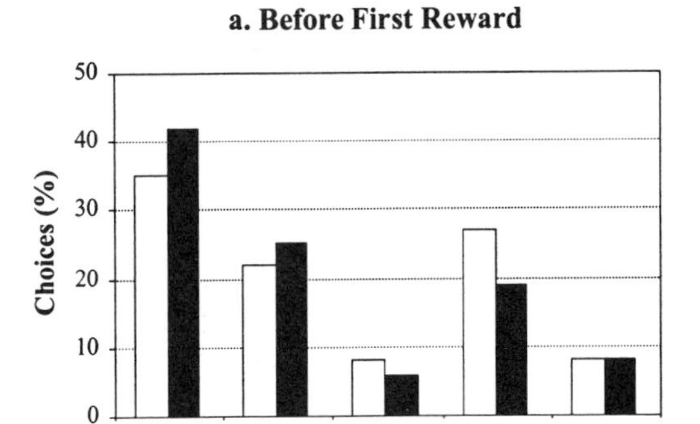

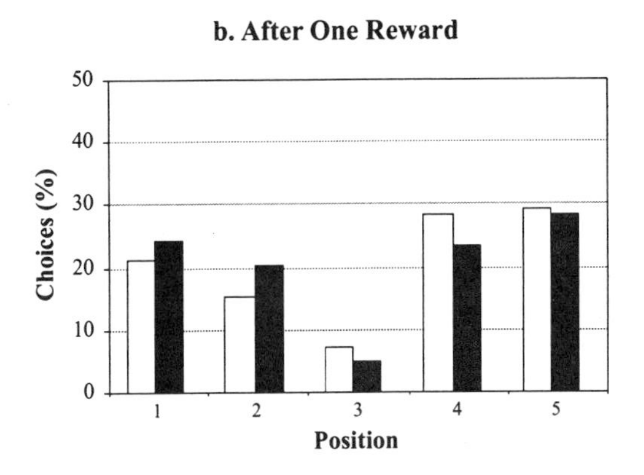

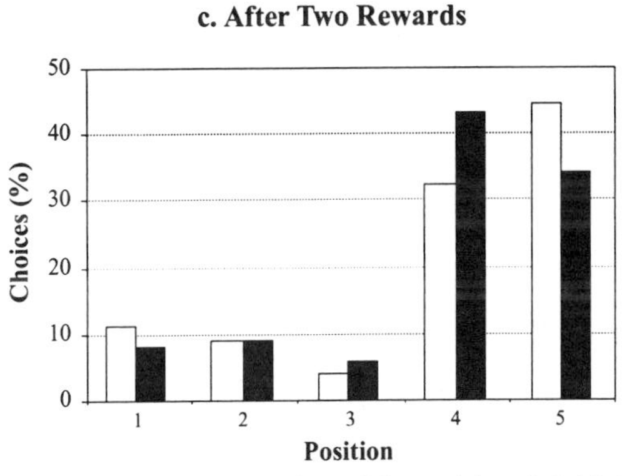

Figure 4: The behavioural choices of the model moth (white) and the real moths (black). The bars correspond to the sums of all visits for all moths. (a). Choices before first reward. 48 choices by 6 animals. (b). Choices after one reward. 79 choices by 6 animals. (c). Choices after two rewards. 35 choices by 6 animals.

The behaviour of the moths after they had been rewarded once is shown in Fig. 4b. The visits shift from the rewarded yellow colour in favour of the position where the animals were rewarded, in this case, position 5. Finally, in Fig. 4c, the distribution of visits after two rewards are shown. The animals now select stimuli to visit according to the rewarded position most of the time. The distribution of choices before reward (Fig. 4a) and after two rewards (Fig. 4c) are significantly different (G-test, $P<0.001$).

The position of the yellow flower during reward trials was the same throughout the experiment and after two rewarded trials the moth had learned that it always received the reward at a specific position and started to ignore colour. This shows that *D. elpenor* can use a place strategy to select flowers.

The difference between the stimulus-approach strategy and the place strategy can easily be seen during the experiment since they are qualitatively very different. The

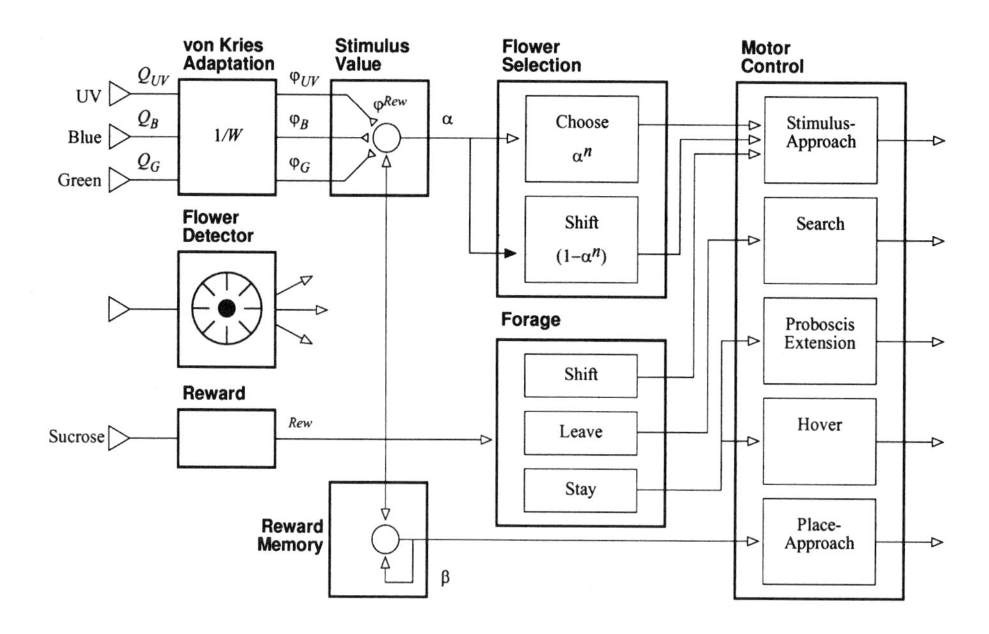

Figure 5: The model of *Deilephila elpenor*. The boxes to the left correspond to different types of stimulus processing. The next column are the learning processes. The boxes in the middle control different behaviours and the boxes to the right correspond to motor control systems.

stimulus-approach behaviour stops well before any flower is reached and is followed by what looks like an evaluation phase where the moth moves sideways in front of the stimulus array. The place-approach behaviour, on the other hand, is much faster and does not stop until the moth is directly ahead of a flower.

Interestingly, when the moths were tested again the next day, they no longer used the place strategy. Instead they returned to using colour to select flowers. The choice to use position rather than colour is temporary. In summary, the experiment shows that moths can use two different strategies when it chooses artificial flowers. They can be instrumentally conditioned to choose a flower according to its colour or position.

3. Simulation Study

The result of the experiment described above has been used to design a computational model of the behaviour in *D. elpenor*. Section 3.1 describes the model, and its performance is described in section 3.2.

3.1 A Computational Model

The behaviour selection of *D. elpenor* depends on both external and internal factors. The external factors used in the model are the colour of the flower in front of it, the location of the moth relative to the stimulus array, and whether it is currently being rewarded. The internal factors are the learned colour and position of the rewarded flower and a memory for how many rewards it has recently received.

Colour

To model the colour vision system of the animal, we calculated the receptor responses corresponding to the different colours used in the experiment. Let $Q = \langle Q_{UV}, Q_B, Q_G \rangle$ be the vector formed by the number of light quanta absorbed by the three photoreceptor types of the animal. The light reflected from the flower in front of it is assumed to excite each receptor type Q_i according to the formula,

$$Q_i = \int_{\lambda=300nm}^{700nm} I(\lambda)S(\lambda)R_i(\lambda)d\lambda,$$

where I is the spectrum of the illumination, S is the reflectance spectrum for a surface, and R_i is the spectral sensitivity of the photoreceptor of type i and λ ranges over the wavelengths visible to the moth.

Fig. 6 shows the location of the colours used in the experiment in the Maxwell colour triangle for the moth. Each corner corresponds to one of the three photoreceptor types of the moth eye. The location of a colour in the triangle represents the relative excitation of the three receptor types for that colour. As can be seen, the five colours are very similar.

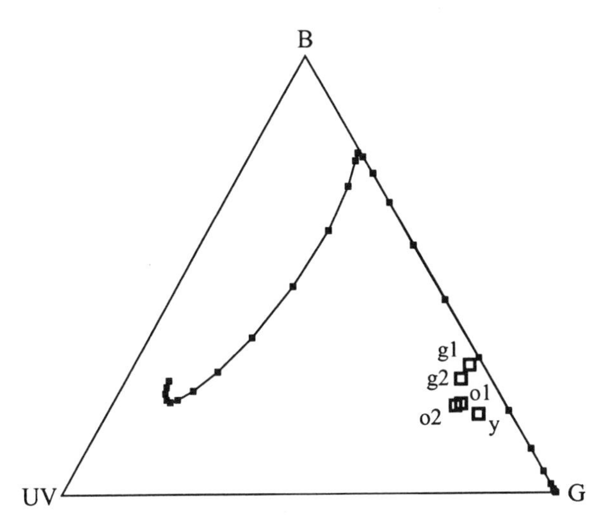

Figure 6. The Maxwell colour triangle for *Deilephila elpenor* with the five different colours used in the experiment. The curve illustrates the location of the monochromatic lights. The corners represent the three different photoreceptor types.

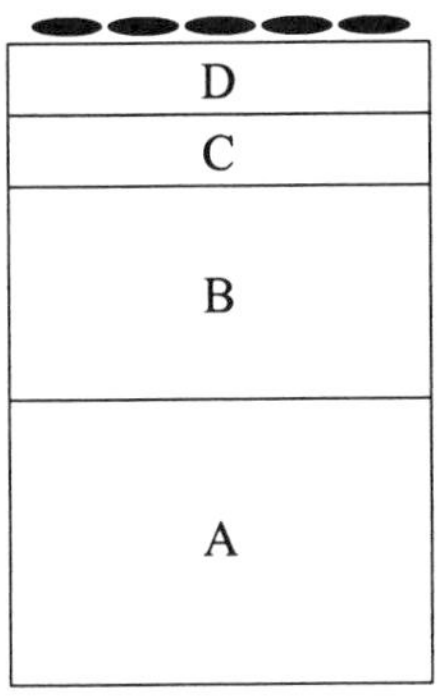

D
C
B
A

Figure 7. The cage seen from above with the stimulus array and the four states corresponding to the different locations.

The loci in the colour triangle were calculated using the spectral sensitivity curves for the photoreceptors of *D. elpenor* (Höglund, 1973) and the spectral reflectance of each test colour measured using a spectrophotometer (S2000, Ocean Optics).

To compensate for fluctuations in the illumination, the animal is assumed to use a von Kries adaptation mechanism working at the receptor level (von Kries, 1902). This mechanism scales the sensitivity of each receptor with the average activation of the receptor in the environment. The von Kries adaptation mechanism leads to an incomplete form of colour constancy (von Kries, 1902). If we assume that the average reflectance in the environment is white (as is the case in the experimental cage) and that the receptor responses for white are given by the constants W_i, the von Kries coefficients ρ_i are given by,

$$\rho_i = \frac{1}{W_i}.$$

We can now calculate the colour coordinates φ_i for a flower as,

$$\varphi_i = \rho_i Q_i$$

Locations and States

We distinguish between four different states of the moth that are not meant to be internal states but roughly correspond to locations in the flight cage: In state A (and location A) the moth is far away from the test stimulus array and does not take notice of it. In location (and state) B the moth is close enough to the array to take notice and react to the stimuli. Location C corresponds to the distance of 3 to 5 cm where the moth usually stops a stimulus approach (see section 2.2) and location D is directly in front of the flowers where the moth can reach them with the proboscis. For a description of the corresponding behaviour see the section on behaviour selection.

Reward

The reward signal that reaches the model is either 0 or 1. A value of 1 indicates that the moth receives sucrose solution. This is only possible when the model has extended its proboscis in front of a simulated flower.

Colour Memory and Matching

When the moth is rewarded at a flower, the colour coordinates for that flower φ are stored in the variable φ^{Rew}. This is the simplest possible learning mechanism that can account for the behavioural data. This variable acts as a colour memory for the rewarded colour and is subsequently used to calculate the similarity between other colours and the rewarded colour. This similarity is used to derive the probability of visiting the flower in front of the moth.

The normalised scalar product is used as a similarity measure for the two vectors corresponding to the rewarded colour φ^{Rew} and the colour of the stimulus in front of the moth,

$$\alpha = \frac{\varphi^{Rew}\varphi}{|\varphi^{Rew}||\varphi|}$$

A perfect match will thus give the value α=1, while two orthogonal colour vectors would give the value α=0. In practice, however, two colour vectors are never completely orthogonal since the spectral sensitivities of the different receptor types overlap. Since the match is always between 0 and 1, it can easily be used as a probability.

Place Memory

When *D. elpenor* is rewarded, it becomes more likely to fly directly to the position where it was rewarded rather than to use the colour of the flowers. This implies that the moth has a memory for the position where it was last rewarded. To model this, we use a position variable p^{Rew} to hold the position where the moth was rewarded. We do not attempt to model how the moth knows where it is or the sensory processing involved in navigation.

Reward Memory

Since the moth becomes more likely to select a flower based on position than colour each time it is rewarded we let a value β indicate the probability that a place strategy will be used and increase this value each time the moth is rewarded. This probability starts out at 0 and is increased by 0.3 each time the moth is rewarded with the restriction that $0< \beta <1$. An increase of 0.3 gives a good fit to the experimental data. To model that the moth returns to a colour based stimulus-approach strategy with time, the value β is assumed to decay slowly with time at a rate that makes sure that it has reached 0 the next day. The variable β is thus essentially a memory for recent rewards.

Behaviour Selection

The behaviour selection of the model depends on the values of α and β together with the current state of the moth. As described above, α represents the similarity between the colour of the stimulus ahead and the rewarded colour and β described the probability of flying directly to the rewarded position instead of evaluating the colours of the flowers. The different behaviours of the moth are summarised in Table 1.

In state A, the model moth is in its search phase where it can either decide to fly directly to the place where it has previously been rewarded or continue to fly around until it finds flowers. If the moth does not start a place-approach behaviour, the model moth will either find flowers or continue searching with equal probability as shown in Table 1.

In state B, the moth has found flowers during its search phase, and approaches them, which will lead it to location C.

In state C, the model moth is flying in front of the flowers and needs to determine whether to try to forage from the flower in front of it or not. This choice depends on the similarity between the colour of that flower and the learned rewarded colour as explained above. If its chooses the flower, it will approach it and enter state D. If it chooses not to approach the flower, it will either shift to the flower to the left or right in the stimulus array.

To derive the probability of choosing the stimulus in front, the similarity α was raised to a power n to sharpen the choice between the different colours. A value of n=4 gave a good fit to the experimental data.

State D represents the situation when the moth is hovering in front of a flower and has extended its proboscis. The behaviour in this situation depends on whether the moth is rewarded or not. If it is not rewarded it will leave the flowers half of the time and start a new search phase. In the other cases, it will either shift to the flower to the left or right. When it has been rewarded, it will leave the flower and start a new search phase.

Table 1: The probability of each action in the simulation.

State	Rew	Probability	Action
A	0	β	*fly to* p^{Rew}
		$0.50\ (1-\beta)$	*approach*
		$0.50\ (1-\beta)$	*stay*
B	0	1.00	*approach*
C	0	α^n	*choose*
		$0.50(1-\alpha^n)$	*left*
		$0.50(1-\alpha^n)$	*right*
D	0	0.50	*leave*
		0.25	*left*
		0.25	*right*
	1	1.00	*leave*

3.2 Simulation Results

Simulation A

We run a simulation of the model presented above in the experiment described in section 2. Data was collected from 500 simulated test sessions. Like the real moth, the model first learned to select the yellow flower before the simulated experiment started. The number of visits before the first reward on the test day is shown in Fig. 4a together with the data from the real moth. In the 4b, the behaviour of the model is shown after a single reward, and finally in 4c, the behaviour after two rewards is presented. As can be seen, the behaviour of the model closely matches that of the real moth. When the behavioural data from the moths were compared to the simulation results, there was no significant difference between the behaviour of the model and the real moths before reward (G-test, $p>0.4$), after one reward (G-test, $p>0.5$) or after two rewards (G-test, $p>0.4$).

Simulation B

In the second simulation, we changed the stimuli in the experiment to much more different colours. Five spectral colours with wavelengths of 350, 400, 450, 500 and 550 nm were used. These were arranged in the stimulus array as shown in Fig. 8.

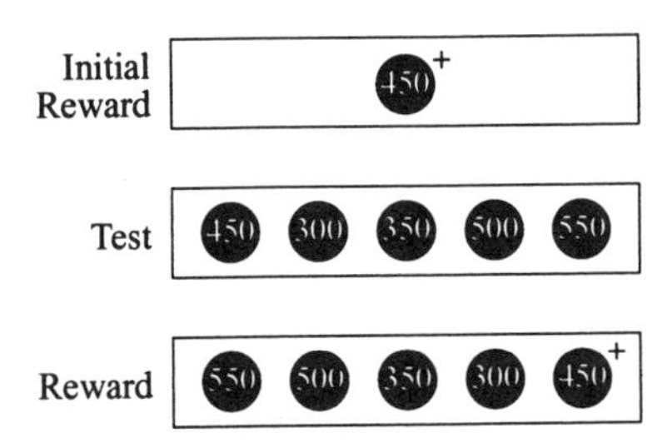

Figure 8: Placement of the five spectral colour of different wavelengths in simulation B.

The simulation result is illustrated in Fig. 9. The graph shows the distribution of visits before the reward and after the first and second reward. As can be seen, the model predicts that the moth will not use a place strategy when the colours can easily be distinguished.

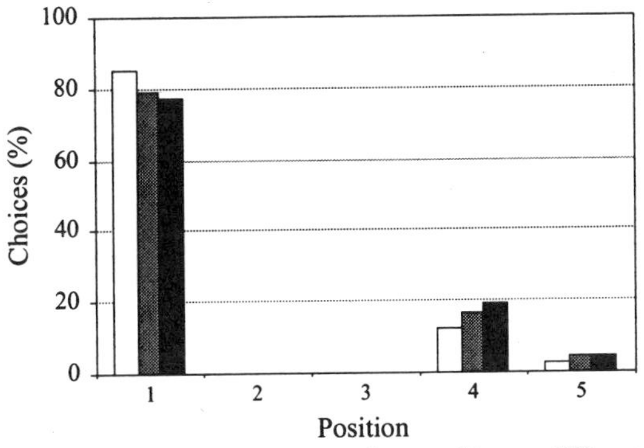

Figure 9: Simulation of the experiment with very different colours. White. Before the first reward. Grey. After one reward. Black. After two rewards.

4. Discussion

The simple model describes the behaviour of the moth very accurately although it makes minimal assumptions about colour matching and learning processes. The moth can not only learn to associate a food reward with a visual stimulus (here colour) but it can also change between two strategies as described in the model.

Other moths, like the diurnal hawkmoth *Macroglossum stellatarum*, visit a large number of artificial stimuli even when they are not rewarded. *D. elpenor* makes only few approaches without reward. This indicates that it needs to keep energy expenses as small as possible. For the same reason, the animal needs to fastly adapt its choice strategy.

When foraging, animals can either use a stimulus-approach or a place-approach strategy to move from one location to another (Balkenius, 1995). In the first case, the movement in space is guided by a single stimulus that the animal moves towards. In the second case, the goal location is given by the relative position of a number of spatial cues, for example, distal landmarks. These types of behaviours have been proposed as alternatives to traditional stimulus-response explanations of spatial behaviour. The adaptive advantage of these strategies is that they can make use of negative feedback from sensory organs to control the behaviour in a goal-directed way.

When both strategies are available, discrimination learning could be used to determine which strategy is most adaptive. The result of the experiment shows that *D. elpenor* can approach flowers using either strategy, but the learning involved is not discrimination learning. Since it is always rewarded at the yellow colour at position 5, but never at position 1, where it appears during test trials, the reward contingency of using colour or position does not differ. This implies that the choice between stimulus-approach and place-approach strategy depends on other factors. In the model, we hypothesised that it was the amount of reward received recently that was used to determine which strategy to use. This may reflect an innate win-stay strategy that is activated by reward (Gallistel, 1990).

In the simulations, the value β was increased by 0.3 each time the animal was rewarded. The value of 0.3 was used to reflect that the animals almost completely switched to using a place-approach strategy when they had been rewarded three times. This is a reasonable strategy when the colours of the flowers are very similar and cannot be distinguished by other means. However, when the different flowers can easily be identified based on their colour, it may be more adaptive to stay with a stimulus-approach strategy. This is the prediction made in simulation B, where the colours were very different and the effect of place learning became very weak. In agreement with this prediction, it has been observed that moths do not appear to use a place strategy in experiments where the colours are easily distinguishable (A. Kelber, unpublished observations). We are currently planning experiments to test the predictions of the model.

In the model, the normalised scalar product is used to measure colour similarity. This measure has the advantage that it stays between 0 and 1, but it does not have the properties of a metric. It would, of course, be possible to use also other forms of matching between the colours (Brandt & Vorobyev, 1996). When a metric is used to calculate colour similarity, a natural choice of probability function would be $G_\sigma(||\varphi-\varphi^{Rew}||_m)$, where G is the Gaussian of the distance between the two colour vectors with variance σ and m indicates the metric used. However, very accurate data would be needed to determine what kind of metric describes the colour space of the moth best.

The model reproduces the behavioural choices of the moth although there are essentially only two parameters: The increase of β and the exponent n used in flower selection. This indicates that a very simple learning mechanism can be used to explain the change in behaviour when the moths are rewarded which is reasonable for an animal with a very small brain.

Acknowledgements

We would like to thank two anonymous reviewers for their insightful comments. This research was supported in part by the Swedish Foundation for Strategic Research (SSF) and the Swedish Research Council (VR).

References

Balkenius, A. (2001). Colour constancy in diurnal and nocturnal hawkmoths. In *International Conference on Invertebrate Vision*, p. 172, Bäckaskog Castle, Swden.

Balkenius, C. (1995). *Natural Intelligence in Artificial Creatures*, Lund University Cognitive Studies, 37.

Brandt, B. & Vorobyev, M. (1996). Metric Analysis of Threshold Spectral Sensitivity in the Honeybee. *Vision Research* 37, 425-439.

Fan, R.-J. & Hansson, B. S. (2000). Olfactory discrimination conditioning in the moth Spodoptera littoralis. *Physiology and Behaviour* (in press).

Fan, R.-J. (2000). *Learning and Memory in Moths*, PhD Thesis. Lund University, Sweden.

Gallistel, C. R. (1990). *The Organization of Learning.* Cambridge, MA: MIT Press.

Höglund, G., Hamdorf, K. & Rosner, G. (1973). Trichromatic visual system in an insect and its sensitivity control by blue light. *Journal of Comparative Physiology* 86, 265-279.

Kelber, A. (1996). Colour learning in the hawkmoth Macroglossum stellatarum. *Journal of Experimental Biology* 38, 1127-1131.

Kelber, A. & Pfaff, M. (1997). Spontaneous and learned preferences for visual flower features in a diurnal Hawkmoth. *Israel Journal of Plant Sciences* 45, 235-245.

von Kries, J. (1902). Chromatic Adaptation. In *Sources of Color Vision* (ed. D. L. MacAdam), pp. 109-119. Cambridge, MA: MIT Press.

Behaviour selection on a mobile robot using W-learning

A.O. Martin Hallerdal* **John C.T. Hallam†**
*Division of Informatics
University of Edinburgh
5 Forrest Hill
Edinburgh EH1 2QL
United Kingdom
marty@home.se

†The Maersk Mc-Kinney Moller Institute for Production Technology
University of Southern Denmark
Campusvej 55
DK-5230 Odense M
Denmark
john@mip.sdu.dk

Abstract

A common approach in complex reinforcement learning tasks is to divide the problem into functional parts, or behaviours, and then to assign a sub-agent to solve each task. The action selection problem then becomes to negotiate between sub-agents with conflicting desires. W-learning is a method whereby agents build up W-values in each state that indicate how important that state is for that agent. These values are then used as basis for selecting agents. In this paper we present the first results, as far as we know, of applying W-learning on a mobile robot in solving a task in the real world. Results from the experiments are presented and the suitability of W-learning for real world robot tasks is discussed.

1. Introduction

In a reinforcement learning task, the main component that relates the agenda of the agent with the one of the designer is the reward function. It is through this channel that the designer communicates his or her norms about what is good and bad. Because of the distal reward problem — the fact that rewards are given only on the completion of a task and that learning therefore might take a long time — it is tempting to try break the task up into sub goals by trying to figure out what constitutes a good performance and giving rewards at these milestones. This makes the rewards less distal and learning will be faster. For example, we might want to reward a chess-playing agent for taking knights in the belief that taking knights is a good way to win a chess game.

The problems with this approach are many. Firstly, we might be wrong in our analysis of the task, leading to the agent performing very poorly on the task as a whole. Second, the milestones might distract the agent from concentrating on what, in our opinion, is the most important goal. For example, our agent might choose to take a knight in a situation where an alternative action would lead to a position with a substantially better chance of winning the game. Third, the agent has no reason not to treat these milestones as ends in themselves, as opposed to just means in achieving what we see as the ultimate goal. This implies that the designer has to set the different rewards very carefully, with rewards proportional to how important and distal they are. Last, because of the fact that we cannot explore every state-action pair an infinite number of times and that agents often get more greedy over time, there is a distinct possibility that the agent will converge in a local maximum (Mahadevan and Connell 1992).

There is a way, however, in which we can inject more reward in the system without falling into any of these pitfalls. The idea is to create many *sub-agents* who each solve smaller, less complex tasks and by combining them, we can achieve the desired behaviour. This concept is not confined to reinforcement learning but has been used in other areas as well (Arkin 1989).

The sub-agents have their own personal reward functions, related to the task they are trained to perform. Since each sub-agent only senses a part of the total sensor space, the total state-action space for the sub-agents is usually much smaller than for a monolithic learner. Consider a situation with four actions and six different sensors, each of which can assume five different values. A monolithic learner will have a state-action space with

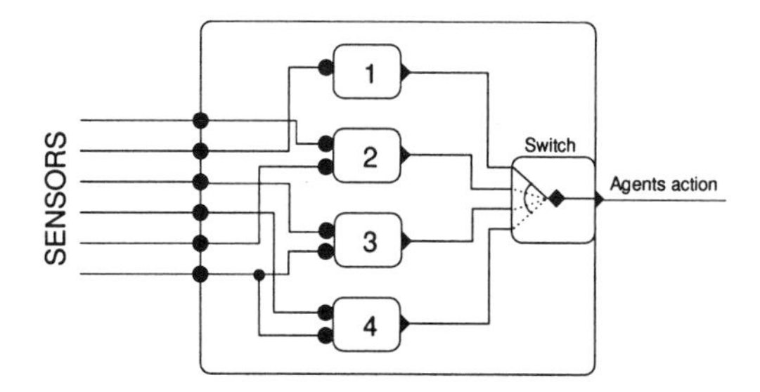

Figure 1: A schematic of a sub-agent architecture. Each of the four sub-agents uses some subset of the total sensors available and proposes some action. A switch then decides which sub-agent gets to decide what action to take in the current state. The grounds on which sub-agents suggest actions can vary but it is often the action with the highest Q-value that is picked. The switch also needs some additional information to choose a sub-agent but this information is different in different architectures and is not shown here.

a size of $4*5^6 = 62500$. If we break the task up into three sub-agents, each of which uses two different sensors, our state-action space is reduced to $3*4*5^2 = 300$.

This may seem like a free lunch and, as we all know, they occur less often than one would like. The catch here is that we have to learn, with respect to the global reward function, which sub-agent to pick in each state. One such switch is known as *hierarchical Q-learning* (Lin 1993) and is just another agent which learns $Q(state, sub-agent)$ and receives its reward from the global reward function. The size of its state-action space is as big as the monolithic learner's, $4*5^6 = 62500$, but since the individual agents already have learned sensible actions, a bad action by the switch is usually better than a bad action by the monolithic learner.

In simple tasks one can also construct the switch manually, as in (Mahadevan and Connell 1992) where a *subsumption architecture* (Brooks 1986) is used. The sub-agents are ranked according to some metric, e.g. how important they are, and in each state each agent is checked for applicability for learning. The highest ranked agent which is applicable is the one which is allowed to select actions and learn.

2. W-learning

W-learning (Humphrys 1995, 1996a,b) is an alternative form of switch in which sub-agents, trained using Q-learning (Watkins 1989), organise themselves with respect to their own personal rewards. Each sub-agent maintains a weight value (W-value) for each state which, simply put, indicates how important that state is for that agent. The switch can then be made very trivial by just selecting the sub-agent with the highest W-value in a given state.

More formally, a W-value for a state and sub-agent represents how much future discounted reward the sub-agent expects to lose in that state if its preferred action isn't selected. To learn the W-values, sub-agent k updates its W-value for state s_t^k in the following way. At time step t, k requests an action a_k to be taken, a request which may or may not be honoured by the switch. After some action has been taken, k will find itself in state s_{t+1}^k. $W^k(s_t^k)$ is then updated towards the difference between what the sub-agent had expected to receive had it been obeyed, $Q^k(s_t^k, a_k)$, and what it actually received, $r_{t+1}^k + \gamma \max_{a'} Q^k(s_{t+1}^k, a')$. The update rule is thus

$$
\begin{aligned}
W_{t+1}^k(s_t^k) \quad \leftarrow \quad & (1-\alpha)W_t^k(s_t^k) + \alpha(Q^k(s_t^k, a_k) - \\
& (r_{t+1}^k + \gamma \max_{a'} Q^k(s_{t+1}^k, a')))
\end{aligned}
$$

The sub-agent whose preferred action was selected, the winner, does not have its W-value updated for the following reason; if the sub-agent has learned its value function correctly and if its preferred action a_k is the optimal action, then the difference between expected reward and received reward should be 0. If updated, its W-value would therefore decrease until another agent would be in the lead (has the highest W-value for that state) after which it would increase again, leading to a never-resolved competition.

From their initial values[1], it is possible that the W-values for a state will converge to stable values without any change in leadership. A more likely scenario though is that, as the W-values converge, some agent will come to have a W-value higher than the current leader's and will become the new leader. With a change of leader comes also a new update target for the W-values and the convergence process will restart. Eventually though the competition will be resolved. This occurs when the sub-agent in the lead causes losses to the rest that are all lower than the loss the previous leader had caused the sub-agent. For a more extensive treatment on the subject of convergence and a more formal approach to W-learning in general, see (Humphrys 1995) or, for a more accessible introduction, (Humphrys 1996a) .

The memory requirements for W-learning *per se* are very modest. In our previous example, you would need to store, in addition to the sub-agent's Q-values, the W-values for each agent and state. This means that the total number of values needed to be stored would be $5^2*4*3+3*5^2 = 15*5^2$.

To get a feel for W-learning, consider this example from (Humphrys 1996a). Table 2 shows the Q-values of two actions for two different sub-agents with the Q-value for the preferred action for each agent in brackets.

[1] The W-values should be initialised to zero because if a sub-agent's W-value is set too high initially, the other sub-agents might not have a chance of catching up with it.

	$a = 1$	$a = 2$
Q_A	[1.1]	1
Q_B	0	[0.9]

Figure 2: Q-values for two sub-agents, A and B, for two actions with the Q-value for the best action for each agent shown in brackets.

From this table it is apparent that if agent A was to pick an action, it would pick action 1 which would be very good for agent A but very bad for agent B. Agent B's preferred action however would only cause a slight loss to agent A. So in a W-learning race agent B would build up a higher W-value since it expects to lose more if it doesn't get to decide.

Clearly, for W-learning to work there has to be some connection between the sub-agent's personal reward functions and the global reward function. The agent's behaviour is a function of the sub-agents that have the highest W-value in the different states and the thing that determines a sub-agent's chances of winning a state is its potential loss, which in turn depends on the rewards of that and the other sub-agents.

From this simple analysis it is easy to see that it is the *relative* strengths of the different agents that determine the behaviour of the agent as a whole and ultimately its success with respect to the global reward function. This means that finding the right *combination* of sub-agent rewards is pivotal and also raises the question about how one goes about determining the size of the individual rewards.

In previous work (Humphrys 1996a), a genetic algorithm has been employed to search through the space of different individual reward functions. The fitness of a genotype — an array of sub-agent rewards — was the total reinforcement from the global reward function while interacting with the environment for a limited amount of time. The genetic algorithm was able to find a good set of sub-agent reward functions within just a couple of generations with satisfactory overall results.

2.1 Negotiated W-learning

Sub-agents may differ not only in what they sense but also in what kind of actions they can perform. This is not a problem for W-learning since sub-agents build up their W-values only by looking at what they themselves would have done and what actually happened at the next time step. They don't need to estimate the value of somebody else's action.

However, if all the sub-agents share the same action set, we can resolve the competition without having to build up W-values by using an algorithm called *negotiated W-learning* (Humphrys 1995). This is of course a good thing since it means we don't have to learn the W-values, a process which can take some time. The algorithm, listed in figure 3, "learns" in one time step what W-learning produces over time and finds a winner within n steps in a collection of n sub-agents. As a result of using this algorithm, memory requirements are further reduced since no W-values need to be stored.

Algorithm *Negotiated W-learning*(s : *state*)
(∗ Finds a winner in the state s. ∗)
(∗ $pref_j$ is the preferred action of agent j ∗)
(∗ i is the current leader ∗)
1. i ←random agent
2. W_i ←0
3. stable ←**true**
4.
5. **repeat**
6.   **for** all agents j except i
7.   **do** $W_j \leftarrow Q_j(s, pref_j) - Q_j(s, pref_i)$
8.   **if** $\max_{j \neq i} W_j > W_i$
9.   **then**
10.   i ←$\arg\max_j W_j$
11.   stable ←**false**
12. **until** stable
13. **return** agent i as winner

Figure 3: Negotiated W-learning

Because of the reduced learning time, negotiated W-learning is more suitable than ordinary W-learning in real world robot learning tasks and was also the method used in our experiments.

3. Experiments

In testing how W-learning would perform on a real world problem, we decided that it would be appropriate to start with a relatively easy task in order to try to gain some useful experiences before moving on to more complex tasks. We settled on a simple foraging task where the agent was supposed to collect food, represented by black wooden cylinders, and thereafter to return the food to its nest, represented by a light bulb.

All the experiments were performed in an arena of approximately one meter square with the walls and floor covered in white paper. The environment was simplified during sub-agent training in that it only contained the things relevant for the sub-agent that was currently being trained. For example, when training the nest agent there would be no food present in the arena. The robot used was the popular Khepera from K-Team (Mondada et al. 1993) with a vision turret and a gripper module added on.

3.1 Sub-agent design

As mentioned previously, the design of the sub-agents and the size of their rewards must always be done with a global reward function in mind. So what is our global reward function?

In (Humphrys 1995), a genetic algorithm searches the space of individual reward functions and the global reward function is implemented as the fitness function used to evaluate the genotypes. However, in robotics this method would take a very long time because of the need to evaluate the individuals in the physical world. So setting the rewards may become something that the designer must do manually, without the assistance of something so effective and fast as a genetic algorithm.

The way we went about doing this was to examine some few key scenarios or states that might arise when performing the task and determining who the winning sub-agent should be in those cases. For example, we studied the case when the agent was not carrying food, decided that the food-seeking agent should win and then used that conclusion and others in setting the rewards for the sub-agents. But when using this method, it turned out to be very difficult to use a global reward function expressed in numerical terms to guide the process of setting the individual rewards. Therefore, our global reward function was a purely qualitative one: when the agent was not carrying food it should try to find food and when the agent had food it should return the food to the nest

We came up with three sub-agents who we thought could be suitably combined to solve the task.

- A *nest agent* who wants to move towards the nest.
- For finding food, a *food agent* who wants to find and approach food.
- Finally, an *explorer agent* who should be a sort of default sub-agent that is active when neither of the other two sub-agents have any preferences.

3.1.1 Action set

To speed up learning, we kept the action set as small as possible and equal for all sub-agents, which enabled the use of negotiated W-learning. The actions were, with the Khepera motor speeds shown in parentheses:

- *Forward* (2,2) — Moves the robot forward.
- *Left* (3,-3) — Turns the robot left on the spot.
- *Right* (-3,3) — Turns the robot right on the spot.

When it comes to applying the Q-learning update rule, one must approach the subject with some things in mind. Consider an agent who updates its Q-values with regular and fixed intervals. Because of the coarseness of the agent's state space, the agent might sense the same state on two consecutive occasions, even though the state of the physical world has changed. Applying the update rule in these cases will often lead to decaying or malformed Q-values. Like Asada et al. (1996), we acknowledge this problem of *state-action deviation*, and the Q-learning update rule was applied whenever the state changed for a sub-agent.

3.2 Nest agent

When the robot has picked up some food, it has to bring it back to its nest for storage. For this purpose, a nest agent was designed whose only wish is to go to the nest. The "nest" in this case was a small light bulb inside a cylinder of transparent acetate-sheet material which acts as support on which the whole thing stands and was located in one of the corners of the arena. The direction to the nest is presented to the sub-agent using the *NestDirection* variable which can take on eight different values (0-7), one for each ambient light sensor, and is at any given moment that ambient light sensor which receives the most light.

Additionally, the sub-agent also knows whether it is at the nest or not through the *AtNest* variable. Its value is determined simply by looking at the value of the lowest sensor and if that value is under a threshold (66), AtNest is set to true.

The reward function for the nest agent is simple; if the sub-agent has arrived at the nest and is facing it, a reward of 1 is given. Since this agent was the first one to be trained, there was no need to consider how big the reward should be since it is difference between the sub-agents' Q-values and not the Q-values *per se* that determine the behaviour of the agent in the end.

The learning process consisted of repeating 10 learning runs followed by 8 evaluation runs until a total of 100 learning runs and subsequent evaluations had been done. The learning runs were random in terms of the initial direction and distance to the nest but care was taken in the first 20-30 runs to keep the starting positions relatively easy. This is known as *Learning from Easy Mission* (LEM) (Asada et al. 1996) and can speed up learning quite a lot since the complexity of the task grows exponentially with the number of state transitions necessary to reach the goal (Asada et al. 1996, Whitehead 1991). All of the sub-agents selected their actions using the Soft-Max procedure.

We wanted training of the sub-agents to be rapid but we also wanted properly converged Q-tables since the Q-values will play a direct role in the action selection procedure. And since the sub-agents tasks only contained one source of reward which meant that the need for exploration was relatively low, we decided to start off with

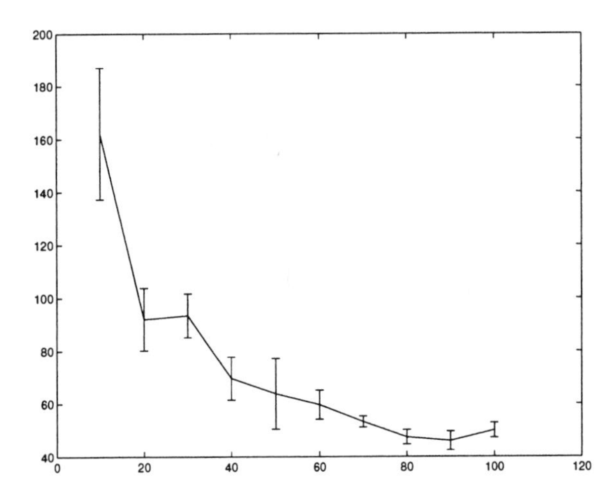

Figure 4: Graph showing the total number of state transitions required for the nest agent to get from eight different starting positions to the nest, averaged over five trials and plotted as function of the number of learning runs.

a generally low and greedy Soft-Max temperature in the experiments and then increase it as learning progressed. This way a lot of reward was injected into the Q-table in the beginning and later, as the temperature was increased, this reward information would propagate out to the different actions in states more distal with respect to any reward.

The actual initial Soft-Max temperature that was used for a sub-agent was calculated using knowledge about its reward structure, a desired probability for picking a given action, and then fine-tuned by testing it on the robot. The Soft-Max temperature was only important in so far that it allowed for efficient training of the sub-agents. When evaluating the overall behaviour, the robot was controlled using W-learning and action selection using Soft-Max was not used.

The learning rate, α, for the nest agent was 0.2 for the first 60 steps, after which it was set to 0.1. The initial Soft-Max temperature was 0.1 and increased by 0.1 every 10 learning runs. The discount factor γ was 0.7.

The evaluation runs were performed using a Soft-Max temperature of 0.05 and with the agent starting 20 cm away from the nest with each evaluation run having a different initial direction to the nest (different values of NestDirection). An evaluation run was stopped when the robot either reached the nest or got stuck at a wall. Each evaluation run was then given a numeric score which was the number of state transitions required to reach the goal. If the robot got stuck, it was given a score of F and the value of F was defined, after all the experiments had been done, to be the maximum number of state transitions for any evaluation run where the agent had eventually reached the nest. However, if the starting position of the evaluation run was Left back or Right back, the value of F was set to be only half of what it otherwise would have been. This is because these states are relatively far away from reward and so the distinction between good and bad actions, in terms of difference in Q-values, is harder to make than for other states for a given Soft-Max temperature. This discounting enables a fairer gauge of progress since, if later on in the learning the agent fails mostly on the hard states, this should also be evident in any statistic that you draw from the evaluation to measure performance.

3.3 Explorer agent

The motivation behind the explorer agent is that it should be a sort of "default agent" that is in control when none of the other agents have a clear preference of action. The explorer agent should wander around exploring the arena, which implies that it always should be moving around and not hitting anything. To achieve this behaviour, we decided to let the agent sense the distance to obstacles around it using the proximity sensors on the Khepera.

Because of the limited action set and the simplicity of the task, we determined to keep the state space coarse and simple as well since the robot probably wouldn't be able to exploit more information about the world to perform better on the task anyway. The sub-agent senses two state variables, *ObstacleLeft* and *ObstacleRight*, each of which can take on two values, true or false. To determine what value ObstacleLeft should assume, the three proximity sensors on the left and front of the Khepera (Left 10, Left 45, Left 90) are examined. If any one of these values is over a threshold, ObstacleLeft is set to true. ObstacleRight is dealt with in an identical way but using the sensors on the right instead. Additionally, because of fluctuations, three consecutive measurements of each proximity sensor were taken and the median value for each proximity sensor was the value that would form the basis for determining the state.

The agent was trained using four different initial positions, each corresponding to a different internal state. The robot was placed at one of the walls of the arena, either facing it directly, having the wall on its left side, having the wall on its right side, or having the wall directly behind it. A learning run was stopped when the robot got stuck or when it had "escaped" the wall in which case an obstacle was placed in front of it, giving it a reward and stopping the run.

A positive reward was given when the Forward action was taken when both ObstacleLeft and ObstacleRight were false. Conversely, a negative reward was given when the Forward action was taken but when both ObstacleLeft and ObstacleRight were true. The need for this negative rewarded originates from the fact that, when

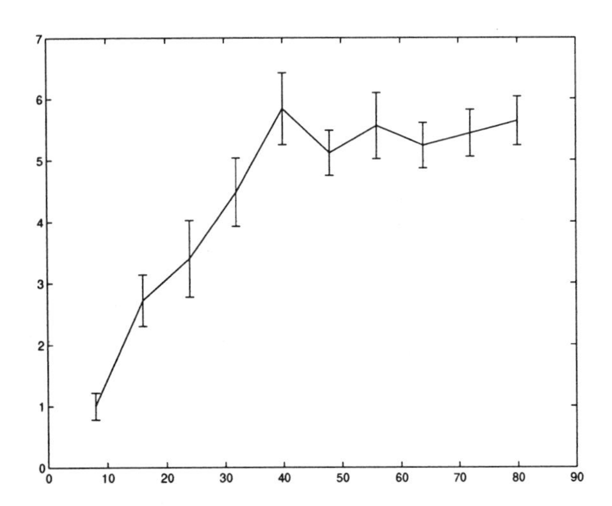

Figure 5: Graph displaying the total amount of reward received by the explorer agent during four evaluation runs averaged over five trials and shown as function of the number of learning runs. The error bars are the standard errors.

the robot performed the Forward action in that state, it would sometimes, due to spurious state detection or because the obstacles would move when hit, end up in a state where one of ObstacleLeft and ObstacleRight were false. This would cause the Forward action to have a very similar Q-value to the Left and Right Q-values and so the sub-agent would be very unfocused when facing an obstacle, a situation where the Forward action is very bad.

The Soft-Max temperature was 0.05 initially and was increased by 0.05 after every 8 learning runs. The learning rate, α, was initially set to 0.2 and was decreased to 0.1 after 48 learning runs. The trial was stopped after 80 learning runs. Because this task is very reactive the discount factor should be set low (Asada et al. 1994) and was set to 0.1 for this agent.

After every 8 learning runs, 4 evaluation runs were performed in a 25 cm meter square arena with a Soft-Max temperature of 0.007. The reason for using a smaller arena is that the robot would face more obstacles in the same amount of time resulting in the evaluation process taking less time. All the evaluation runs started with the Khepera in the middle of the arena but they differed in the direction in which the robot was facing the wall. In two of the runs, it would face a wall head on while in the other two it would face the wall slightly from the left and right respectively. Each evaluation run would last for 180 time steps (about half a minute) and the score for each run was the amount of reward collected. Figure 5 shows the reward received during learning, averaged over five different trials. As can be seen, the agent learns the good actions quite quickly and, after 40 runs, there is no visible progress.

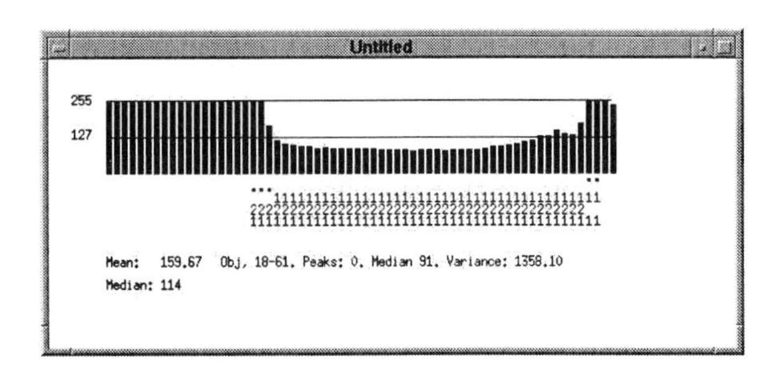

Figure 6: The vision data when a black cylinder was placed 5 cm in front of the Khepera. The cylinder has been detected by the edge detection algorithm, as can be seen by the numbers below the pixels, indicating that they belong to an object.

3.4 Food agent

A separate sub-agent was designed to handle the task of locating food in the arena. The "food" was represented by wooden cylinders of about 1.5 cm in diameter. The information needed to locate the food came from the Khepera vision system and therefore, in order to make the identification process easier, the contrast against the background against the arena walls was increased by covering the food cylinders in black paper.

The vision system on the Khepera consists of a lens with a 36-degree view angle and image data is stored in a linear vision array of 64 grey scale pixels, each of which can assume a value in the range 0-255. Because of the relatively large amount of data, a key issue to resolve in designing this sub-agent was how to present the image data to the sub-agent. Clearly, just giving the sub-agent raw quantisised pixel data would be impractical in terms of learning time — the state space would be too large. Additionally, the sub-agent really doesn't need that much information, it only needs enough information to decide whether to turn left or right or go forward because these are all the actions it has at its disposal. This convinced us to try to reduce the image data to just one of four different states; if food was visible to the left, centre, right, or not at all.

However, this mapping from raw image data was not entirely straightforward partly due to the relatively limited capabilities of the Khepera vision system. In the end, our solution was based on a hand-crafted algorithm which detected so called *edge pixels*. These were pixels that were located in an area of large gradient in the image and so could represent a boundary between two objects, in effect between the background and a food cylinder. Having identified the edge pixels, we used them to form candidate objects, objects which were subsequently classified as either a food cylinder or a spurious detection. This classification process used the variance of the pixel values within the object: if the variance was very low the object was classified as a food cylinder.

After the objects have been classified, determining what state to present to the sub-agent is a relatively trivial task. If more than one food object is present in the image, the one with the biggest apparent size (how many pixels the object occupies in the image) is used. The sub-agent senses the direction to the food through the FoodDirection variable. If the mid point of the object is located to the left of the 25th pixel in the image data, the value of this variable is Left, otherwise if it is located to the left of the 36th pixel the value is Centre and otherwise the value is Right. If no object is present, the value is Not visible. Additionally, the agent senses whether it is at food or not through the *AtFood* variable. If the apparent size of the object is bigger than 37 pixels, AtFood is set to true and false otherwise. Also, because the robot should go to the nest only when it has food to drop off, either the nest agent or the food agent must sense whether the robot is carrying food. This is necessary because, in order for one of the sub-agents to able to yield in W-learning to the other sub-agent in one situation and win in the other situation, it must be able to tell the different states apart. Without any particular rationale in mind, we decided to let the food agent sense if the robot is carrying food.

The agent was trained by putting food in three different initial positions, with the states for the sub-agent in these positions being [Left, False], [Right, False], and [Centre, False] respectively. The initial Soft-Max temperature was 0.1 and was increased by 0.1 every 9 learning runs. α was set initially to 0.2 and then decreased to 0.1 after 54 learning runs.

The reward was 2 when the agent arrived at a food cylinder, facing it (AtFood = True, FoodDirection = Centre). When the Forward action was taken when there was no food visible, a similar strategy as with the explorer agent was used and the agent was given a punishment of -1. The reward was set to be bigger than the reward for the nest agent because if the agent doesn't have food, it should pick up food rather than go to the nest. The exact size of the reward was harder to determine and we decided to do some initial experiments to see exactly how much reward was needed for the food agent to beat the nest agent in negotiated W-learning. γ was set to 0.3.

After every 9 training runs, alternating between the three different initial positions, four evaluation runs were performed with a Soft-Max temperature of 0.2. To receive a reward from the first two evaluation runs the sub-agent would have to make a number of consecutive Left or Right actions. In the last two evaluation runs the sub-agent had to include one or more Forward actions as well in order to receive the reward. The score for a batch of evaluation runs was the total number of states the sub-agent experienced during the four evaluation runs, i.e. the lower the score for an evaluation run the better. Figure 7 shows the result of the evaluation runs.

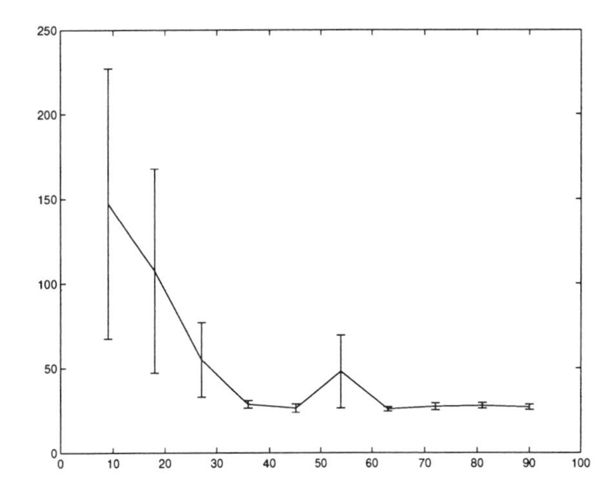

Figure 7: This graph displays the number of state changes required to reach the food summed up over four evaluation runs, averaged over six trials and shown as function of the number of learning runs. The lowest possible number of state changes is 20. The error bars are the standard errors.

4. W-learning

After having trained the sub-agents in isolation it was now to time to integrate them and to control the robot using W-learning. Since no global reward function existed, the evaluation was done entirely on a qualitative basis. This included monitoring behavioural issues, e.g. what the robot did once it had food, as well as implementational ones, such as the robustness of the food cylinder detection routine.

Our analysis in the sub-agent design phase had only dealt with the case when food was visible and in those situations, the experimental results confirmed the validity of our analysis; the food agent would win the W-learning competition and the robot would go towards the food until it was close enough for the food to be picked up. However, an interesting behavioural pattern would emerge when no food was visible. Remember that the nest agent senses the direction to the nest at all times and it will want to turn to face the nest, at which point it will want to go forward. The food agent, if not sensing the presence of any food, will not object to the nest agent's action proposal of turning towards the nest since, according to the food agent, turning actions are good when no food is visible because that will eventually result in food being located. But when the robot eventually faces the nest, the nest agent will propose a forward action, something that the food agent will object to very much. The food agent will thus win in this state and eventually, because the food agent proposes a turning action, the robot will not face the nest any more, at which point the whole process repeats itself.

We solved this by letting the nest agent sense whether the robot is carrying food or not, which, in combination with changing the reward function so that the nest agent only is rewarded when it reaches the nest with food, will result in that the nest agent will only have preferences when the robot is carrying food. Practically, this was accomplished by augmenting the state space of the nest agent with another variable, *HaveFood*, and just copying the values of the old state space to the part of the new state space where *HaveFood* = True, and setting the Q-values of the other states to Zero.

A second problem occurred when the robot had picked up food and had the nest more or less behind it. In this situation, the food agent has no preferences, all its Q-values are zero, and the competition is essentially between the nest agent and the explorer agent. The explorer agent will want to go forward and the nest agent will want to go left or right, depending on which side of the robot the nest is. We want the robot to bring the food back to the nest but, because of the small differences between the different Q-values for the nest agent in states where the nest is behind it, the explorer agent would win and take the robot forward, away from the nest. This predicament was solved by increasing the nest agent's reward sufficiently.

When changing the reward of a sub-agent who will participate in W-learning, one must consider the implications this change will have with regards to other sub-agents. In this case though, the nest agent and food agent will never have any preferences simultaneously — when the robot has food the food agent has no preferences and vice versa — and so we can change the nest agent's Q-values without having to worry about possible side effects with the food agent.

A side effect of increasing the nest agent's reward was that the explorer agent became totally redundant — it would never win a W-learning competition in any state. This became very apparent in some situations when the robot was heading for the nest with food but would get stuck against a wall. The explorer agent would want to turn away from the wall but, since the robot had food in these cases, the nest agent would win and, if the nest was sufficiently in front of the robot, the nest agent would select the Forward action. However, if the robot was very close to a wall and the nest was not exactly in front of the robot, the Forward action could drive the robot into the wall and cause it to get stuck there. A desirable solution would be some sort of weighting of actions, as in (Arkin 1989), but compromising actions or *action averaging* is not within the scope of W-learning.

We therefore decided to re-design the explorer agent with the objective of just transforming it into an obstacle avoidance agent. To achieve this, we rewarded the explorer agent when it reached a state where obstacles were no longer detected from a previous state where there was an obstacle on either side of the agent.

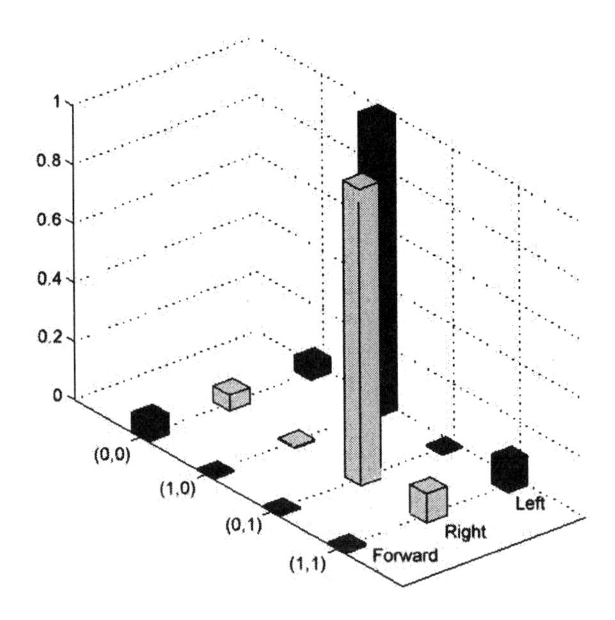

Figure 8: A graph displaying the Q-table of the new simulated explorer agent, averaged over 100 episodes, each lasting 20,000 steps. The state notation (L,R), indicates if there is an obstacle (1) or not (0) on the left and right side respectively.

The training procedure for this sub-agent was slightly different than for the other sub-agents. We trained the sub-agent in a simulated environment, which meant defining a complete Markov model with state-action transition probabilities and reward probabilities. Because the state-action space was quite small and the environment relatively predictable, this wasn't as hard as it sounds but we acknowledge that on more complex tasks, this method is neither practical nor likely to be correct.

The transition probabilities were based on various assumptions, including those that the agent operated in an empty rectangular arena and sensed its current state perfectly. The agent was trained by letting it interact with the environment for a long time, an episode, during which actions were selected using Soft-Max and standard Q-learning was performed after each state transition. See (Hallerdal 2001) for further details on the transition probabilities and parameter values. The resulting Q-values for the sub-agent are shown in figure 8.

Using this new explorer agent in negotiated W-learning was more interesting and solved part of the original problem; when the robot had food the food agent would win in all cases except where the explorer agent detected an obstacle, in which case the explorer agent would win. But what happened when the robot got

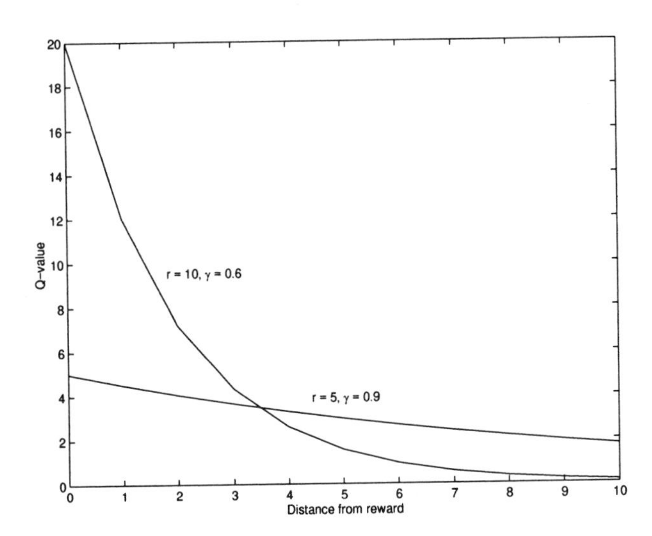

Figure 9: Q-values for two sub-agents plotted as a function of distance from their rewards. The first agent has a reward of 20 and a discount factor of 0.6 while the respective values for the second sub-agent are 5 and 0.9.

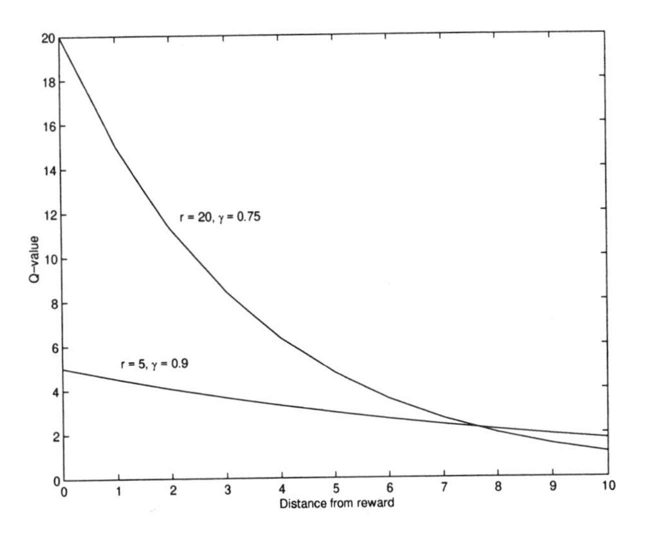

Figure 10: Lowering the discount for the first sub-agent by increasing its discount factor, making its chances of winning negotiated W-learning against the second agent higher.

close to the wall was that the explorer agent would turn away from the wall, and the nest, until no obstacles were sensed. At that point, the nest agent would win and start turning the other way, towards the nest and the wall. When the wall was detected by the explorer agent, this process would repeat itself. For more details on these experiments, see (Hallerdal 2001).

5. Discussion

The task we have implemented in this paper, one of locating food objects in an rectangular arena and returning them to a nest, is not a very difficult one but one which still presented us with some problems. These problems were not really related to the problem of solving the different components of the task, but to the challenge of combining them suitably to solve the task as a whole.

As we have hopefully shown, the process of implementing W-learning on a robot is not a very elegant or effective one. Given a high level specification of the task, expressed either in quantitative terms, for example a reward function, or a description of behaviour, the human designer is forced to map this down to decisions about which sub-agent should win in some key states. But, since executing such a decision involves manipulating the sub-agent's reward, this will change all of the sub-agent's Q-values and have implications not only in the intended state but in many other states. As we have seen, it might not be possible to find rewards to satisfy all of these constraints, given the structure of the reward functions and the way in which the sub-agents are designed, what they sense and what actions are available to them. As the number of sub-agents grows, this problem grows exponentially. The fundamental problem is that the basis on which W-learning operates, Q-values, is a concept which is really too low level for us as humans. We function easier on a higher level and have difficulties seeing how a particular reward strategy will propagate into Q-values and, subsequently, what consequences these values will have in the context of other sub-agents in negotiated W-learning. For example, as illustrated by figures 9 and 10, the discount factor affects a sub-agent's Q-values and can have a great effect on the functioning of the system.

This dependence on Q-values, combined with the inability of the human designer to predict them, implies that a lot of changes may have to be made after all the sub-agents have been trained. These can range from the most straightforward types of changes, such as altering a sub-agent's reward, to more tedious and hard ones, such as re-designing a sub-agent. But, whatever the type of change, each situation needs be analysed carefully, a solution has to be prescribed and implemented, and finally the result has to be evaluated. This procedure takes a lot of time and effort on behalf of the designer and is likely to be very disheartening in the long run as well.

However, the fine results achieved by Humphrys (1996a) in his simulations dissuades us from dismissing W-learning altogether. But how can a good reward set be found without having to perform many lengthy performance evaluations on the robot? There are a few options one can explore:

- The environment can be modelled and the Q-values can be learned in a simulation. Learning is then transferred to the real robot so that the Q-values can be fine tuned to adjust for any discrepancies between the real world and the modelled environment. This approach has the severe disadvantage that only very simple environments can be modelled and even in simple cases it can be hard to guarantee that the model is a sufficiently good representation of the real world. But worse, learning in a simulation is only a shortcut for learning Q-values, it doesn't help us *per se* in finding suitable rewards for the sub-agents.

- One could imagine a reasoning system where the designer can express wishes about what sub-agents should win in different situations. The system could then infer a set of rewards that are consistent with the designer's requirements. But there is no guarantee that the resulting system will perform well. Also, why should you choose W-learning in this case and not just design a simple sub-agent switch manually that will do the same thing?

Additionally, as we have seen, there is also the possibility that one needs to change not only how much reward is given, but also when the rewards are given as well as the design of the sub-agent itself. These features of the system are also likely to be harder to evolve automatically than just numerical rewards.

6. Conclusion

In this paper we have presented the realisation of W-learning on a robot solving a real world task. As implemented using the method presented in this paper, W-learning is currently not a viable option for real world robot learning — the W-learning process is too complex and opaque for human designers. But with a partially automated approach, better results may be possible. However, the exact nature of such an approach is an open future research issue.

Acknowledgements

Facilities for this work were provided by the University of Edinburgh. Martin Hallerdal was funded by an EPSRC studentship, award number 00401518.

References

Arkin, R. C.: 1989, Motor schema-based mobile robot navigation, *International Journal of Robotics Research, August 1989* **8**(4), 92–112.

Asada, M., Noda, S., Tawaratsumida, S. and Hosoda, K.: 1996, Purposive behavior acquisition for a real robot by vision-based reinforcement learning, *Machine Learning* **23**, 279–303.

Asada, M., Uchibe, E., Noda, S., Tawaratsumida, S. and Hosoda, K.: 1994, Coordination of multiple behaviors acquired by a vision-based reinforcement learning, *International Conference on Intelligent Robots and Systems. IEEE.*
URL: *citeseer.nj.nec.com/asada94coordination.html*

Brooks, R.: 1986, Achieving artificial intelligence through building robots, *Technical Report AIM-899*, Massachusetts Institute of Technology.

Hallerdal, M.: 2001, *Investigating w-learning on a khepera robot*, Master's thesis, University of Edinburgh.
URL: *www.users.wineasy.se/martinh/resume/index_eng.html*

Humphrys, M.: 1995, W-learning: Competition among selfish q-learners, *Technical Report 362*, Computer Laboratory, University of Cambridge.

Humphrys, M.: 1996a, Action selection methods using reinforcement learning, *in* M. P. M. M. M. J.-A. P. J. and W. S. W. (eds), *From Animals to Animats 4: Proceedings of the Fourth International Conference on Simulation of Adaptive Behavior*, pp. 135–144.

Humphrys, M.: 1996b, Action selection methods using reinforcement learning, *Phd thesis (first version)*, University of Cambridge, Computer Laboratory.

Lin, L. J.: 1993, Scaling up reinforcement learning for robot control, *International Conference on Machine Learning*, pp. 182–189.

Mahadevan, S. and Connell, J.: 1992, Automatic programming of behavior-based robots using reinforcement learning, *Artificial Intelligence* (55), 311–365.

Mondada, F., Franzi, E. and Ienne, P.: 1993, Mobile robot miniaturisation: A tool for investigation in control algorithms, *Proc. Third Int. Symposium on Experimental Robotics*, Kyoto.
URL: *citeseer.nj.nec.com/mondada94mobile.html*

Watkins, C. J. C. H.: 1989, *Learning from Delayed Rewards*, PhD thesis, Cambridge University.

Whitehead, S. D.: 1991, A complexity analysis of cooperative mechanisms in reinforcement learning, *AAAI, Vol. 2*, pp. 607–613.

Behavior Coordination for a Mobile Visuo-Motor System in an Augmented Real-World Environment

Dimitrij Surmeli **Horst-Michael Gross**
Dept. of Neuroinformatics, Ilmenau Technical University,
PO Box 100565, D-98684 Ilmenau, Germany
E-mail: Dima.Surmeli@informatik.tu-ilmenau.de

Abstract

We utilize HUMPHRYS' W-Learning on a real robot Khepera to coordinate three behaviors in an augmented maze: first, to drive straight and fast while avoiding obstacles, second, to find a location marked by one projected color (e.g., where food can be found), and third, to escape from another color. We describe the experimental setup and compare results of the individual agents to those of a monolithic agent solving all tasks, and of the agents coordinated by different types of W-Learning. We demonstrate the feasibility of W-Learning on a real visuo-motor system and conclude by discussing why the monolith outperforms all forms of coordination investigated.

1 Introduction

Values from some Reinforcement Learning methods may serve to coordinate some agents sharing the same 'body'.

We investigate W-Learning (Humphrys, 1997) and its applicability on a real robot for an adaptive coordination of multiple goals.

The overall task may be solved by a monolithic learner at the cost of an immense state-action space and training time. Scalability to complex tasks, acceptable adaptation times and an easier design of partial reinforcement functions are the driving forces for multi-agent systems.

2 Theoretical background

2.1 W-Learning

In W-Learning (Humphrys, 1997), fig.1 right, a collection of individual agents is coordinated to each solve a task and is assumed to be a fully trained Q-Learners. It uses the Q-values of these agents to determine just how badly they want to execute their own best action in a certain state and select one.

The resulting effect may be described as follows: agent i, insists not on its own suggested action and relinquishes ownership of a state agent j, who executes an action that to agent i holds less reward, but possibly avoids catastrophic losses for j.

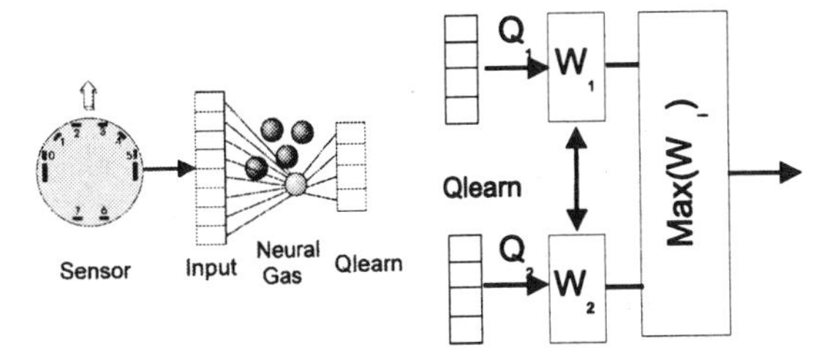

Figure 1: Structure of agents and principle of W-Learning

3 Experimental setup

We use a Khepera miniature robot equipped with 8 infrared sensors and an additional omni-directional color camera, in a special maze with wooden obstacles and projected color patches shown in fig.2 left and actions expressed as *speed* $v \in [-1, 4, 7]$; *angle* $\alpha \in [-30, 0, 30]°$.

3.1 Image preprocessing

The camera produces an omni-directional view by looking straight up at a parabolic mirror. It is transformed into a rectangular image (fig. 2 right top) and reduced to a grid of 6 by 4 regions, where averaged colors (fig. 2 right second) in HSI color space provided the inputs for the agents. The regions preserve color, some heading and distance information.

3.2 Setup of the individual agents

All agents, except the W-Learner, share the architecture shown in figure 1 left, comprised of of a vector quantizer, such as a Neural Gas (Martinetz and Schulten, 1991) and *Sarsa* (Rummery and Niranjan, 1994) in a subsequent supervised layer trained by straight Delta-learning. *Sarsa* was used with ($\gamma = 0.8, \lambda = 0.8, \alpha = 0.5$) with the the learning rate α held constant. Adaptation of the input clustering and the Q-values proceeded concurrently, and a Boltzmann exploration was used.

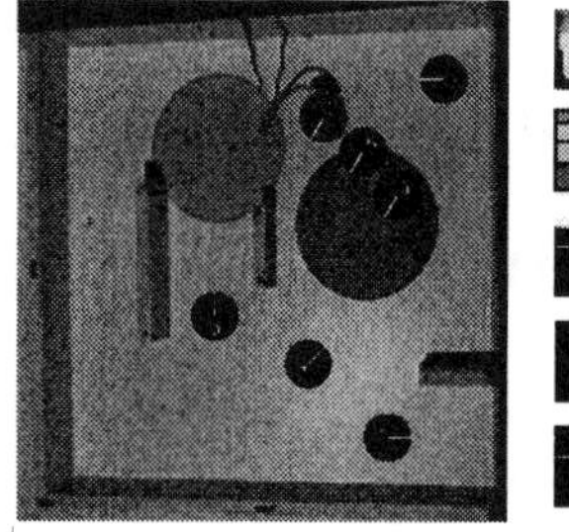
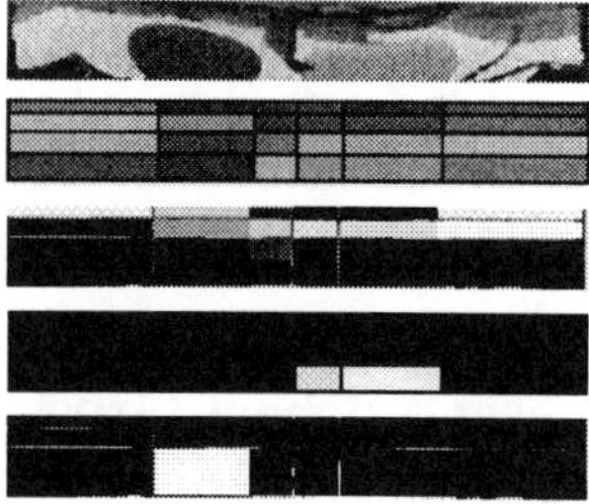

Figure 2: Top-view of the scenario with Khepera and image produced and processed as inputs to Bump (third), Thirst (fourth), and Escape (last). Robot is in Pos.4

Each of the expert agents Bump, Thirst and Escape, used a Neural Gas of 15 nodes, and the monolith 50. We realized the following agents:

- **Bump**, which realizes an obstacle avoidance and otherwise, tries to go as straight and fast as possible,
- **Thirst**, which will try to drive onto a blue patch
- **Escape**, who avoids driving onto the a patch,
- **W-Learner**, who consisted of and coordinated the very same agents above, complemented by a coordination by a form of W-Learning
- **Monolith**, who is charged with solving all three tasks as one overall problem and received as inputs the inputs of all the other agents in one big vector

With IR = normalized infrared readings, v =robot wheel speed, α = steering angle, lRF = lowest regions,

$$\begin{aligned} r(Bump) &= \begin{cases} -8.0, \text{if } \max_{IR} > 0.9 \\ |v| * 0.25 - |(0.025 * \alpha)| & else \end{cases} \\ r(thirst) &= \sum activation(lRF(blue)) \\ r(esc) &= -\sum activation(lRF(red)) \quad (1) \\ r(Mono) = r(WL) &= r(Bump) + r(Thirst) + r(Esc) \end{aligned}$$

4 Results

We measured the performance of the agents from 7 starting points (fig. 2 left) by calculating average total rewards (see eq.2) per step, for each trial individually, and summing those for all starting points, averaged across a number of runs.

The Khepera was set into the respective point in the orientation indicated by the white line in the dark circles in figure 2 left and was allowed to drive until collision or a maximum number of steps of 51, and the reward and trial length were averaged for 10 experiments.

The results are presented numerically for all agents in table 1 and discussed in the following paragraphs.

Agent	cum avg reward (σ)	avg length (σ)
Mono	6.6 (4.3)	238.7 (45.9)
WL(neg)	2.3 (3.4)	193.0 (60.6)
WL(mCH)	1.9 (3.1)	254.8 (40.1)
WL(wl)	1.3 (2.5)	211.4 (27.4)
WL(wlF)	-0.7 (1.9)	134.0 (44.4)
Bump	-3.9 (4.8)	292.3 28.5)
Escape	-7.2 (6.0)	137.2 (29.6)
Thirst	-8.4 (8.0)	183.5 (28.4)
Random	-23.9 (9.1)	100.0 (32.2)

Table 1: Overview of all results. The numbers represent the mean of total reward (as in eq. 2) per step in a trial, and length of trials in steps, summed over all trials and the numbers in parentheses standard deviations. Mono = Monolith, WL = W-Learning, neg = negotiated, wl = learned WL, mCH = maximal collective happiness, wlF = learned WL with full state space

Negotiated WL (Humphrys, 1997) turned out to perform best among the WL-variants, while the monolith outperformed all variants of WL. It was able to use all correlated inputs as landmarks, and so coordinate the different tasks more easily by finding optimal compromises and exploiting mutual support.

It's superior since it solves entire task optimally at the cost of solving subtasks suboptimally. In contrast, in WL the individual agents solve their tasks optimally, but optimality at the global task cannot be achieved.

The monolith uses a total reward signal, which WL never uses, to flexibly weigh the subtasks for an optimal overall task, two disconnected steps for WL.

One reason for WL's suboptimality was the fixed weighting of the agents for the entire state space. Different weight configurations supported better behavior of WL in different states, which is currently unachievable.

The monolith considers costs of transitions between subgoals. With implicit planning through $\gamma > 0$, it optimizes full sequences of coordinated actions, when WL looks at the current situation, plans only for one agent.

References

Humphrys, M. (1997). *Action Selection methods using Reinforcement Learning.* PhD thesis, University of Cambridge, Computer Laboratory.

Martinetz, T. and Schulten, K. (1991). A "Neural Gas" network learns topologies. In Kohonen, T., Mkisara, K., Simula, O., and Kangas, J., (Eds.), *Artificial Neural Networks*, pages 397–402. Elsevier.

Rummery, G. and Niranjan, M. (1994). On-line Q-learning using connectionist systems. Technical Report CUED/F-INFENG/TR 166, Engineering Department, Cambridge University, UK.

Compromise Candidates in Positive Goal Scenarios

Frederick L. Crabbe
U.S. Naval Academy Computer Science Department
572C Holloway Rd, Stop 9F
Annapolis, MD 21402
crabbe@usna.edu

Abstract

Among many properties suggested for action selection mechanisms, one prominent one is the ability to select compromise actions. This paper performs an experimental analysis of compromise behavior in an attempt to determine exactly how much compromise behavior aids an animat. The paper concludes that the ability to select compromise actions in some common cases does provide some additional welfare to the animat, but less than previously postulated.

1. Introduction

In his now classic Ph.D. thesis, Tyrrell (Tyrrell, 1993) introduced a list of fourteen requirements for Action Selection Mechanisms. Of these, number twelve was "Compromise Candidates: the need to be able to choose actions that, while not the best choice for any one sub-problem alone, are best when all sub-problems are considered simultaneously." (p. 174) The ability to consider compromise actions in an uncertain world makes great intuitive sense. When multiple goals interact, solving each optimally is not always optimal for the overall system. Yet, recent work has generated empirical results that seem to contradict the claim that the ability to consider compromise candidates is necessary (e.g., Bryson, 2000). This paper examines a situation where an agent selects between two positive goal and concludes that in this case, compromise actions have a moderate effect the agent's welfare.

2. Problem Formulation

The formulation this paper uses is a spatial navigation task similar to those found in Tyrrell. Imagine an animat in a dynamic environment at some initial location i. The animat can sense one each of two kinds of target objects nearby, T_x and T_y located at x and y respectively. The animat has two independent goals to consume the targets: G_x and G_y. The dynamism in the environment is represented with a probability p, which is the probability that any object in the environment will still exist after each time step. That is, any object will spontaneously disappear from the environment at each time step with probability $1-p$. Notationally, $\overline{ij}$ is the distance from location i to j, such as $\overline{ix}$ or $\overline{xy}$. All distances are measured in the number of time steps it takes the animat to travel that distance. Analysis with Utility Theory (Howard, 1977) determines that the expected utility (EU) of taking a compromise action C when in state S is:

$$\begin{aligned} EU(C|S) = {} & p^{\overline{ic}} p^{\overline{cx}} G_x - p^{2\overline{ic}} p^{\overline{cx}} G_x \\ & + p^{\overline{ic}} p^{\overline{cy}} G_y - p^{2\overline{ic}} p^{\overline{cy}} G_y \\ & + p^{\overline{ic}} \max(p^{\overline{cx}} p^{\overline{xy}} G_y + p^{\overline{cx}} G_x, \\ & \qquad p^{\overline{cy}} p^{\overline{xy}} G_x + p^{\overline{cy}} G_y) \end{aligned} \quad (1)$$

(Crabbe, 2002)

The animat should select action C whenever the expected utility of that action is greater than the EU of all other actions. We perform a series of experiments to determine the properties of this relation.

	1	2	3
% cases better	36.125	99.44	100
% avg improvement	1.487	5.499	8.165
% avg imp. when better	1.493	5.500	8.165
best % improvement	91.36	73.23	76.51

Figure 1: Results of the three experiments.

3. Experimental Set-up

Each experiment consists of 100,000 trials. In each trial a random situation is generated: x, y, p, G_x, G_y From these numbers, values for variables in equation (1) are calculated and plugged in to determine the EU of taking a compromise action. For each trial, the candidate move with the highest EU out of 100,000 randomly generated potential candidates is used. Each potential candidate is generated within the rectangle bounded by x, y and the x-axis.

4. Experimental Results and Discussion

In the first experiment for each compromise candidate the following statistics are gathered: % of cases where

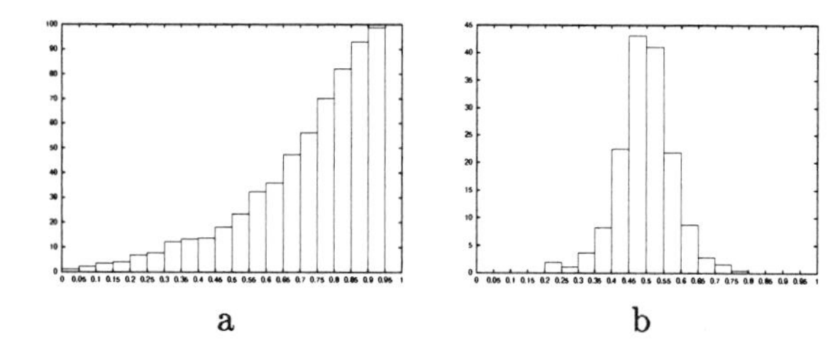

Figure 2: Breakdown of the information gain attributes in the first two experiments. The x-axis is the range of values, and the y-axis is the percentage of situations with a good compromise action.

the compromise candidate C was better than either non-compromise action; the average % improvement of EU when *always* selecting the compromise candidate; the average % improvement in EU when selecting the compromise candidate only when it is better than A or B; and the best % improvement in a single situation offered by the compromise candidate. Figure 1 shows the results of this experiment.

We observe that, assuming that the generated candidate move is a reasonable approximation of the true best compromise candidate, then good compromise candidates are relatively rare (36%), and they offer little EU improvement on average for the animat (1.5%). In order to investigate this further and determine the cases in which compromise candidates are beneficial, we analyze the results with a technique from Information Theory.

In order to analyze the properties of the situations in which there are good compromise candidates, we use the idea of information gain from information theory. This is the same technique used in the decision tree algorithms ID3 and C4.5 (Mitchell, 1997).

By applying information theory, we discover that attribute with the largest information gain (0.4256) is p. Figure 2a shows the the breakdown of the data as classified by the attribute p. We see that the cases where there are good compromise candidates are skewed towards high p. Furthermore while the range of p between 0.8 and 1.0 contains 20.14% of the data, it accounts for 51.26% of the good compromise cases. If we consider the implications of a p less than 0.9, we see that it results in a harsh environment, in that the agent is unlikely to consume any resources. It is more reasonable to assume that p is *at least* 0.9. We will therefore examine the EU space when $p > 0.9$.

In order to investigate EU when p is large, we generate another 100,000 trials, all with $p > 0.9$. Figure 1 shows the results. In this case we see, as expected, that 99% of the generated cases have a good compromise candidate, but the improvement is only 5.5%

When we use an information gain analysis, we find that the most important attribute is the ratio of the distance between i and x to the distance between i and y: $\frac{\overline{ix}}{\overline{ix}+\overline{iy}}$ (figure 2b). Here we see that the most important aspect of the property is in the middle range $(0.4 - 0.6)$, with 52.98% of the examples accounting for 89.96% of the good compromise candidates. When the value of that attribute is in that range, indicating that good compromise candidates are much more likely to exist when the targets are approximately equidistant from the start location. This makes some intuitive sense that when one target is significantly closer than the other, the animat might as well pursue the closer.

Because of the tight range in which good compromises lie, we narrow the range of situations we examine. In the third experiment, we consider only those scenarios where both $0.9 < p < 1.0$ and $0.4 < \frac{\overline{ix}}{\overline{ix}+\overline{iy}} < 0.6$. The results in figure 1 show that the improvement increased to 8.1%. Information theory analysis showed no further clear improvements, so the experiments were stopped.

5. Conclusion

In this paper we have considered whether and how much the ability to select compromise actions is good for animat. We showed that given our Utility Theory based formulation of the problem of two positive goals, the ability to select compromise candidates improves the animat's behavior by only 5.5% in the most reasonable case. We investigated the structure of the expected utility space and determined that good compromise candidates exist in situations where: the probability that that the targets will remain in the environment is large and the distances to the targets are roughly equal.

References

Bryson, J. (2000). Hierarchy and sequence vs. full parallelism in action selection. In *Proceedings of the Sixth International Conference on Simulation of Adaptive Behavior.*

Crabbe, F. (2002). Consideration of compromise candidates in action selection. Technical report, http://www.cs.usna.edu/~crabbe/papers/sab02-long.ps.

Howard. R. (1977). Risk preference. In *Readings in Decision Analysis.* SRI International, Menlo Park, CA.

Mitchell, C. A. (1997). *Machine Learning.* McGraw-Hill, Boston.

Tyrrell, T. (1993). *Computational Mechanism for Action Selection.* PhD thesis, University of Edinburgh.

A context-based architecture for general problem solving

R. C. Peterson
School of Computing, University of Leeds,
Leeds, LS2 9JT, UK.
rickp@comp.leeds.ac.uk

Abstract

Perhaps the most striking feature of the human problem solving ability is its apparent generality. Many animals have well developed path finding and navigational abilities and this suggests a route by which general purpose problem solving skills might have arisen in primates. The idea explored in this paper is that by generalising the notion of 'place' to include contexts defined by factors other than just spatial location a powerful general purpose representational and planning mechanism can be built on the foundations provided by existing models of animal navigation. In the model presented the animat is assumed to have available to it motor primitives that not only allow it to move within and modify its environment but also allow it to change the way in which its sensory system operates, e.g. change its field of view. By selectively directing its attention in this way the animat is able to build up a dynamic multi-scale representation of its environment.

1. Introduction

One of the main long term goals of artificial intelligence is the creation of intelligent systems with which one can converse. In order to have a meaningful conversation it is necessary for both parties to have an understanding of the subject matter being discussed. How one creates a system that actually understands something about its world is still a largely unsolved problem. Classical AI approaches involving the storage of large numbers of facts that are processed using some form of inference engine are unattractive from a philosophical point of view. Furthermore, while considerable effort has been invested over the years in such systems there appears to be a general consensus that we simply don't know how to make this approach work. An alternative strategy that has been proposed is that instead effort be concentrated on developing situated agents that learn causal relationships by direct interaction with their environments and can thus build up sets of symbolic representations that are firmly grounded in their environments (Brooks, 1990).

Since models of such systems are likely to be complex, in their development it makes sense to employ suitably high level descriptions of the architectures involved. The architectures employed in the model presented here have however all been chosen with biological-plausibility in mind; that is to say, physically realisable mechanisms have already been published or are easily constructed using only mechanisms that are understood to have biological counterparts. In particular the assumption is made here that all learning is Hebbian (Kosko, 1992), i.e. that pathways can only be formed between representations that are simultaneously active. While every attempt has been made to ensure that the components of the architecture are biologically plausible no attempt has been made to identify them with specific biological structures.

Numerous successful architectures have been developed for allowing robots to map and then plan paths through their environments (Arkin, 1998, Trullier et al., 1997). One of the most popular approaches involves the construction of what is termed a view-graph, strictly speaking a digraph, the vertices of which correspond to the views seen by the robot at different locations and with different headings, the edges to allowed transitions between views (Scholkopf and Mallot, 1995, Franz and Mallot, 2000).

A variety of techniques are available for finding the shortest path through such a graph, many of the most popular being variants of Dijkstra's algorithm (Payton et al., 1990, Zimmer, 1996, Yamauchi and Beer, 1996). Mataric's (Mataric, 1991) spreading activation model is of particular relevance here in that it suggests a biologically plausible means by which shortest paths may be found.

The model presented here is essentially a generalisation of the view-graph approach to navigation in which the concept of a view is extended by allowing the animat to exercise control over the manner in which its sensory subsystems operate. It is assumed that the animat is able to selectively direct its attention to one of a number of different sensory modes and to direct its attention in one of a limited number of directions associated with that modality. For example when navigating in open spaces distal cues are more likely to form useful landmarks than proximal ones and should thus attract the majority of the animat's attention. Conversely during tasks involving hand-eye coordination proximal cues i.e. objects within reach need to be preferentially attended to. Under normal circumstances, when not performing an action, the animat is assumed to continuously scan the set of locations associated with its current sensory mode. The set of objects the animat detects together with their locations specifies its context.

Associated with each sensory mode there will be a particular notion of saliency. How salient an object is will depend on such factors as an object's apparent size, its distance, its degree of separation from its background, its colour etc. In the present model the animat is assumed to be equipped with an active vision system (Blake and Yuille, 1992, Ullman, 1996) that allows it to recognize a specific object if it is present at the location currently attended to. While the use of an active vision system prevents perceptual aliasing (Franz and Mallot, 2000) occurring during activities such as navigation in which it would be problematic, in many circumstances perceptual aliasing can actually be beneficial in that it allows the animat to generalise its behaviour. Whether or not perceptual aliasing will occur in any given situation will depend upon the animat's current sensory mode. While only visual sensory modalities will be considered here there is nothing to prevent the approach outlined being generalised to other sensory modalities.

2. The model

2.1 Context representation

Figure 1 illustrates the architecture employed here to model context learning and recognition. The design is similar to one proposed by Balkenius (Balkenius and Moren, 2000). Nodes in the binding layer learn to respond specifically whenever a given sensory input (or cue) occurs simultaneously with a particular locational input, such a coincidence being termed a *binding*. Thus each binding encodes the presence of a specific cue at a specific location in the animat's current field of view. The binding layer is modeled using an ART network (Carpenter and Grossberg, 1988), with high vigilance, that allows bindings to be learnt after only a single presentation.

Binding node activities are forwarded to the storage layer, which operates as a competitive storage network (Grossberg, 1986) and is able to store only a small number of the most recent bindings. If the animat is assumed to scan its environment continuously in a fashion

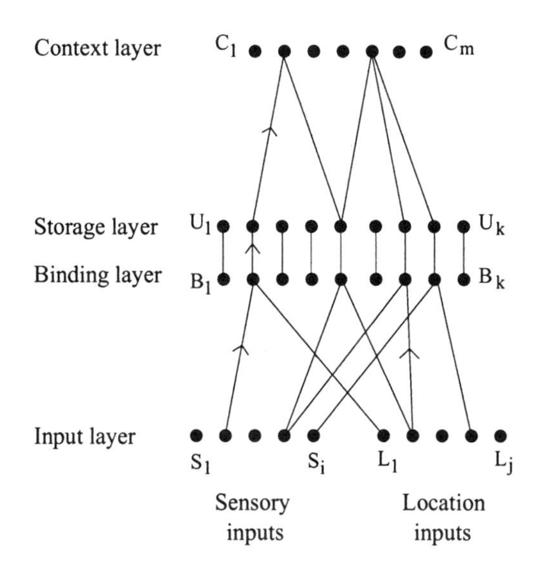

Figure 1: Context learning architecture.

reminiscent of that observed in humans (Yarbus, 1967, Rao et al., 1997, Ryback et al., 1998) then the storage layer will effectively hold a 'picture' of the last few salient cues it has fixated upon. The animat is assumed to be able to fixate upon a single cue every time step. At present a simple sequential scanning mechanism is implemented, the animat viewing each possible location in its field of view in turn before returning to the first.

For simplicity, the storage layer is implemented here in an event-driven fashion using a list of recently activated bindings. As new bindings are detected they are added to the head of the list. Once the list reaches its maximum length the last (oldest) binding in the list is deleted whenever a new binding is added. If a binding is detected while it is already present in the list then it is simply moved to the head of the list. An alternative arrangement would be to allow storage layer representations to decay with time in a continuous fashion, deleting them from the storage layer when their activity levels drop below a given threshold.

The next layer of the architecture is the context layer, the nodes of which learn to respond to specific patterns of activity in the storage layer. The behaviour of the context layer is also modeled using an ART network with high vigilance. Thus whenever the contents of the storage layer change either an existing exactly-matching context node will be activated or a new one will be allocated. In principle there is no reason why the vigilance of the ART network cannot be relaxed to some extent, provided that it can be increased when appropriate. The present model differs from that of Balkenius in that it assumes that a mechanism exists for inhibiting the context classifying subsystem for extended periods of time (here termed *orientation times*) during which the environment is scanned but context classification is not performed.

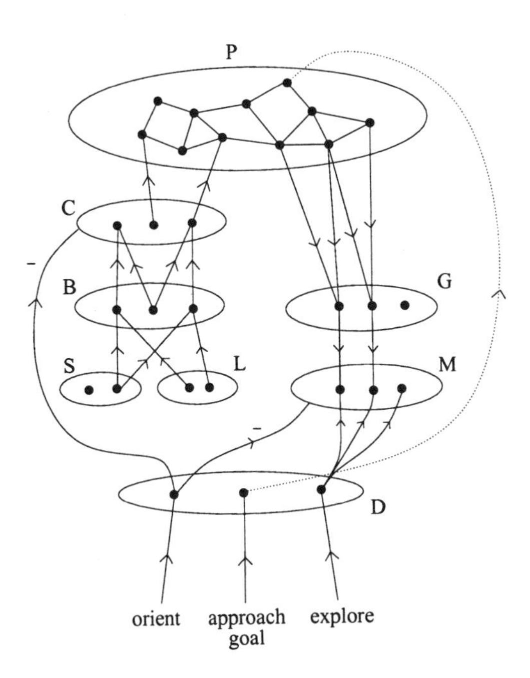

Figure 2: Basic architecture: primary pathways ——, motivational pathways · · · · ·, S sensory field, L location field, C context nodes, P context map, G gradient field, M motor command nodes, D drive representations. All pathways are excitory except for the non-specific inhibitory pathways marked -.

This mechanism is provided by the *orienting subsystem* which is activated whenever the animat moves and remains active until a prescribed number of time steps has elapsed. While active the orienting subsystem also inhibits motor activity by the animat, allowing it time to reorient itself with respect to its new context before selecting its next move.

2.2 Context map formation

The overall architecture of the model is presented schematically in Figure 2 which, for clarity, shows the binding layer and associated storage layer combined as one (B). Each node in the context layer (C) is assumed to have associated with it a node in a context map (P). A directed pathway is formed between a pair of context map nodes whenever the animat experiences a transition between the corresponding contexts.

Figure 3 shows the four conditionable pathways associated with a transition between contexts C_1 and C_2. The context nodes C_1 and C_2 are linked by fixed pathways to their corresponding context map nodes P_1 and P_2. These particular pathways are only employed when a context node is selected as a goal and their primary function is to inject activity into the context map. The context map itself is formed by the conditionable pathways that link the context map nodes. Each context

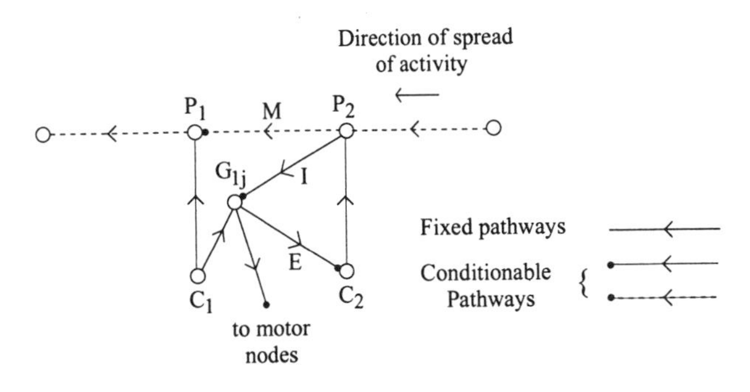

Figure 3: Pathways for context map learning: M context map pathway, I motor incentive pathway, E expectancy pathway.

node is assumed to have associated with it a small number of gradient nodes[1]. Here only a single representative gradient node is shown, labeled $G_{1,j}$. A gradient node is assumed to receive an excitory input from its parent context node whenever it is active. In addition gradient nodes also receive excitory inputs along conditionable pathways from context map nodes that are neighbours of their parent context's context map node in the context map. A conditionable expectancy pathway from the gradient node $G_{1,j}$ to context node C_2 is also shown. The formation of all three of the above classes of conditionable pathway is governed by fast Hebbian learning, allowing the pathways to be formed during a single context transition. The pathway linking the gradient node to the motor command nodes is discussed in Section 2.4.

The expectancy pathways have a variety of roles e.g. the priming of sensory representations, the detection of an unexpected context etc. They also play an important role in controlling the way in which the animat deals with context map pathways that become blocked. Whenever an action fails to lead to the expected context a variable controlling the ability of the corresponding context map node to participate in the spreading of activity is decremented. Once this variable falls below a given threshold the node is inhibited (via the C_i to P_i pathways) and is no longer able to spread activity. Thus all paths through the node will be blocked and the animat will be forced to seek alternative routes to its goal. For normal context map connections the process of forgetting is assumed to be composed of two components: a rapid but reversible short term habituation and a slower but more permanent form of extinction (Witkowski, 2000). This allows the animat to operate in dynamically changing environments in which, for example, a route to a goal might be temporarily blocked but later become available again.

[1]The name is inherited from an earlier formulation of the architecture that employed a path planner based on Laplacian potential fields (Peterson, 2001).

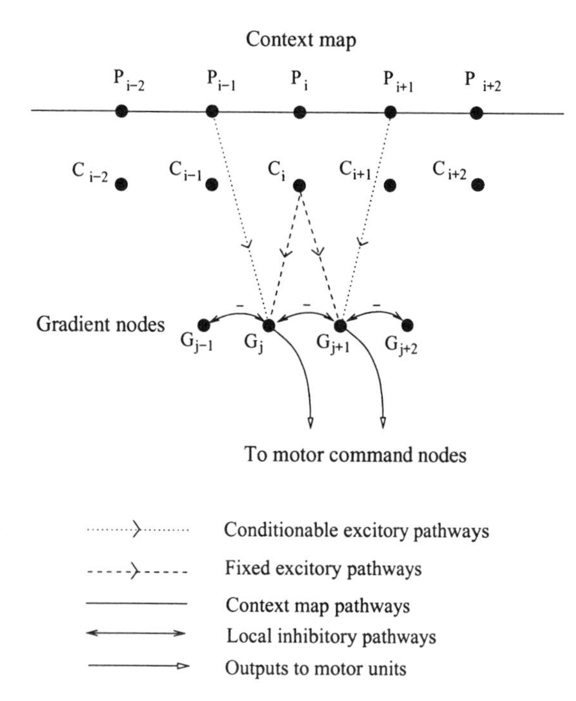

Figure 4: Architecture for motor program selection.

2.3 Drive representations

The animat's current motivational state is governed by a set of drive nodes which form the competitive field (D) shown in Figure 2. At any given time a single drive node is assumed to be active. Drive nodes can influence the behaviour of the animat in two ways. The first is via excitory or inhibitory pathways that link them with motor command nodes (M). While at present only prewired pathways are modeled, in future it is intended that conditionable pathways will be included linking drive nodes and motor nodes.

The other way a drive node can influence the animat's behavior is via motivational pathways linking it to the context map. Whenever a drive node is active any context linked to the drive by a motivational pathway will be selected as a goal and will become a source of spreading activation. Motivational pathways are assumed to be created whenever the currently active drive is positively rewarded as a result of some action performed by the animat.

In contrast to the approach adopted in reinforcement learning schemes (Kaelbling et al., 1996, Sutton and Barto, 1998) in the present architecture knowledge about the environment and knowledge about the locations of goals are acquired and stored independently. Thus skills learnt with one motivation in mind can also be used when a quite different motivation is present.

2.4 Action selection

Action selection is performed using a spreading activation model (Mataric, 1991, Mataric, 1992). Activity is periodically allowed to spread through the context map starting at nodes associated with goals. Activity is passed along a context map edge in the direction corresponding to the reverse of the context transition that created it. When the spreading activity arrives at the node corresponding to the animat's current context the edge along which it arrived will form the first link of the shortest path through the context map to the nearest goal and can thus be used to select the next action that the animat should initiate.

Figure 4 illustrates the basic action selection mechanism. The gradient nodes associated with each context node form a competitive field. Here only the gradient nodes associated with context C_i are shown. In order to become active a gradient node must receive simultaneous signals from both its parent context node and its associated context map node. This ensures that only the gradient node associated with the edge along which activation first arrives at the current context node becomes active. Hebbian learning is also used to form links between gradient nodes and the motor representations that form the competitive field (M) shown in Figure 2. Activation of a motor representation that results in a change in the animat's context will naturally correlate well with the activity of the gradient node corresponding to the context map edge representing the context switch, thus allowing the contingency to be learnt.

In order to make the above mechanism attractive it is necessary to employ not too large a ratio of gradient nodes to context nodes. This can be arranged if it is assumed that gradient nodes are not initially assigned context map nodes from which they receive inputs, such affiliations being formed throughout the life time of the animat in response to the context switches it experiences.

Whenever the animat is not engaging in goal directed activity a default exploratory drive is assumed to be active. This randomly perturbs the activities of nodes in the motor field resulting in exploratory behaviour.

2.5 Chunking

A mechanism for the chunking of common motor sequences has also been successfully implemented (Grossberg, 1986, Moren, 1998). This allows common sequences of motor primitives to become associated with chunked motor command nodes. Once created a chunked motor command node behaves exactly like a primitive motor command node. Grossberg's Outstar Learning Model (Grossberg, 1982, Grossberg, 1986) provides one biologically plausible mechanism by which chunking might be arranged.

Figure 5 illustrates an outstar node schematically. Po-

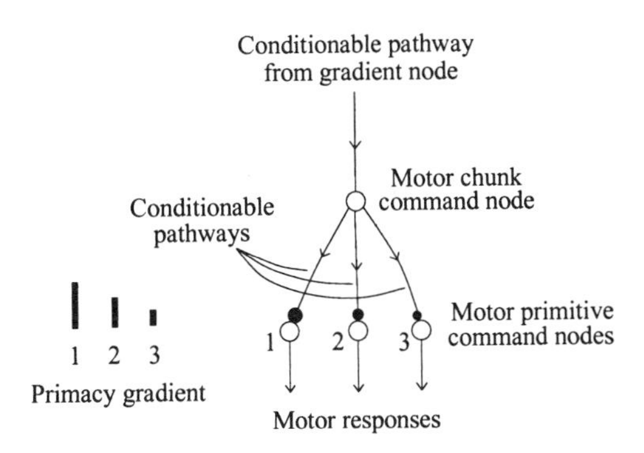

Figure 5: Outstar architecture for motor chunk learning.

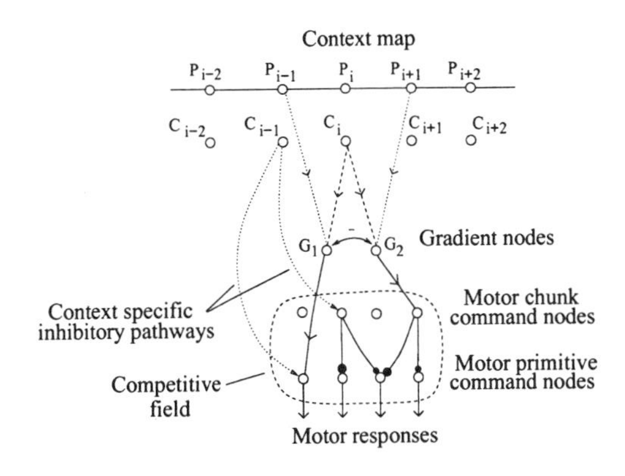

Figure 6: Architecture for motor program selection including chunks.

tentially a pathway exists between each outstar node and each motor primitive command node. Each of these pathways has a weight associated with it, here represented by the size of the filled circle at its end. Sequential activation of the motor primitive command nodes in the sequence 1 2 3 is assumed to result in the setting up of a transient pattern of activity across these nodes, shown on the left, which is known as a primacy gradient (Grossberg, 1982). When such a gradient exists, spontaneous activation of the outstar node will result in the strengthening of the pathways linking it to the motor primitive command nodes, the rate of learning in each pathway being proportional to the activity of the corresponding motor primitive command node, i.e. a function of its position in the primacy gradient. Note that chunks involving multiple instances of the same primitive can also be handled by the mechanism if it is assumed that there are a number of nodes available to represent each primitive's position in the primacy gradient.

Since motor command nodes occurring earliest in the sequence will have the largest weights associated with them, later activation of the outstar or chunking node will result in a pattern of excitatory signals to the motor primitive command nodes that reproduces the original primacy gradient and thus result in the nodes being activated in the original sequence.

One important property of chunked action sequences is that they can be performed automatically without the need for the animat to reorient itself after the execution of each individual motor primitive. Here only a single level of chunking is modeled, the individual chunks being the simplest possible i.e. those involving two motor primitives that are executed in a given order.

Figure 6 illustrates the modifications that might be made to the architecture shown in Figure 4 in order to implement motor chunking. Additional inhibitory pathways linking context map nodes and motor command nodes, that are not part of the basic chunking architecture, are also shown; these are discussed below.

A pool of chunking nodes is assumed to exist, each of which potentially receives an input from any gradient node. Pathways between gradient nodes and motor chunk nodes are assumed to be created only once a motor chunk node has become associated with a given motor sequence or *chunk*. Both primitive and chunking command nodes are assumed to form part of a single competitive winner-takes-all network. Thus, normally, at any one time only a single primitive or chunking command node will be active. In order for a chunking node to be able to activate the motor primitive command nodes that form the chunk it encodes it is necessary to assume that the excitatory signals from the chunking node are capable of overriding the competition between motor command nodes.

Frequent execution of a common motor pattern results in a chunking node becoming associated with the pattern. For simplicity this process is implemented here stochastically, a new chunking node being allocated with a probability of 20% each time a novel pattern occurs.

Once a chunk has been created it is free to compete with the motor primitive command nodes, and any other chunks that exist, for the right to become active. When a chunk is selected for execution in a novel context two possibilities arise. Firstly, execution might fail at any point during the chunk. In such cases it is assumed that no context map learning occurs. Secondly, the chunk might be executed successfully. In this case, once the chunk has been completed and a new context is detected the context map pathway between the original context and the new context is strengthened. The animat is now ready to select a new motor command.

Context map edges that correspond to successfully executed chunks effectively form 'shortcuts' through the context map. Since they are shorter spreading activity taking these paths will arrive at the current context node earlier than that traveling along longer alternative paths. Consequently, when engaged in goal seeking behaviour the animat will have a strong tendency to employ motor chunks wherever possible.

In principle motor chunks may themselves be 'chun-

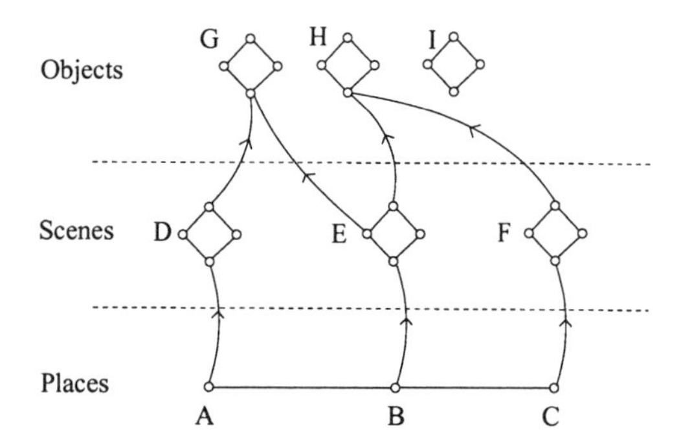

Figure 7: Schematic illustration of the hierarchy of sensory modes showing the disjoint subgraphs of the context map associated with each mode linked by directed arcs represent zoom-in transitions between contexts (circles).

ked', in hierarchical fashion, allowing an arbitrarily complex motor pattern to be controlled by a single motor chunk command node. Where a region has been explored thoroughly, using both chunks and motor primitives, the context map will contain within it a hierarchy of overlapping maps of different resolutions (Balkenius, 1996), each level of the hierarchy corresponding to a different size of chunk.

2.6 *Context specific motor inhibition*

A simple model of aversive conditioning, involving modifiable inhibitory pathways between context nodes and motor command nodes (both primitive and chunking) is also incorporated, the weights associated with these inhibitory pathways being strengthened whenever the execution of the corresponding motor primitive or chunk fails i.e. a collision occurs or no context switch occurs. As a result, after only a small number of failures the probability of an unsuccessful motor program being activated in the given context is greatly reduced.

2.7 *Hierarchical context map formation*

It is assumed that in addition to the primitive motor commands that allow it to control its physical behaviour the animat also has available to it a number of pseudo-motor primitives that allow it to alter the sensory mode in which it operates. Thus for example an animat might have a primitive that allows it to switch its attention from distant landmarks over a large field of view to nearby objects lying directly in front of it. Execution of such a pseudo-motor primitive will naturally result in a change in the animat's context. Context transitions that alter the animat's sensory mode are fundamentally different to transitions between contexts within the same sensory mode. Consequently it is natural to consider the animat's context map as being partitioned into a number of disjoint subgraphs.

Figure 7 illustrates a typical hierarchy of sensory modes. Three sensory modes are shown, separated by dotted lines. In the lower mode contexts correspond to views at the coarsest resolution. When in this mode the animat is assumed to pay attention to distal cues with large angular separations. Thus for example contexts A, B and C might correspond to locations in the environment at which there are tables.

In the intermediate mode contexts correspond to scenes involving groups of objects e.g. a collection of objects on a table. In this mode the attended-to cues are assumed to lie approximately within arm's reach and lie within a field of view of intermediate size. Separate subgraphs of the context map are shown for this mode, each corresponding to a different collection of objects lying on a table. Thus for example context subgraph D might correspond to scenes involving a single block, E and F to scenes involving two and three blocks respectively.

In the upper mode contexts correspond to single objects or contiguous assemblies of objects. In this mode the attended-to field of view is assumed to be so small that only a single object can be perceived at any given time. For example, subgraph G might correspond to a tower of two blocks, viewed from a number of different angles.

The animat is assumed to possess motor primitives that allow it to switch between adjacent modes in the sensory hierarchy. These are referred to as *zoom-in* and *zoom-out* operations. Zooming-in allows the animat to move up a level in the hierarchy while zooming-out causes it to move down a level.

A zoom-in operation essentially involves focusing in on a specific portion of the current field of view. When this happens the new context corresponds in some sense to a part of the old context. Ideally the animat should learn which context to expect when it executes a zoom-in operation in a given context i.e. that 'object' X is at 'location' Y. Providing the animat with a mechanism that allows it to know which context it should expect to return to if it performs a zoom-out operation is however more difficult. The problem is that since it performed the zoom-in operation that caused it to switch to its current sensory mode the animat may have altered its context. Consequently there is no obvious way for it to predict which context a zoom-out operation will return it to given its current context.

The implementation of these two desirable mechanisms must clearly involve both fast learning and fast unlearning of pathways if they are to be useful. Figure 8 illustrates the pathways associated with the zoom-in and zoom-out primitives. Note the asymmetry between the arrangements for the two primitives.

Figure 8(a) shows the pathways created when the animat performs a zoom-in operation that results in a transition from context A to context B. Unlike a normal context transition the expectancy and motor incentive pathways form not to a gradient node associated with context B but to a mode specific ***modal-gradient node.*** A second difference is that the context map pathway (M) has the reverse sense to that which would normally form as a result of a transition between context A and context B. All three of the pathways are assumed to be created anew each time a zoom-in event occurs and replace any preexisting zoom-in pathways between the two sensory modes.

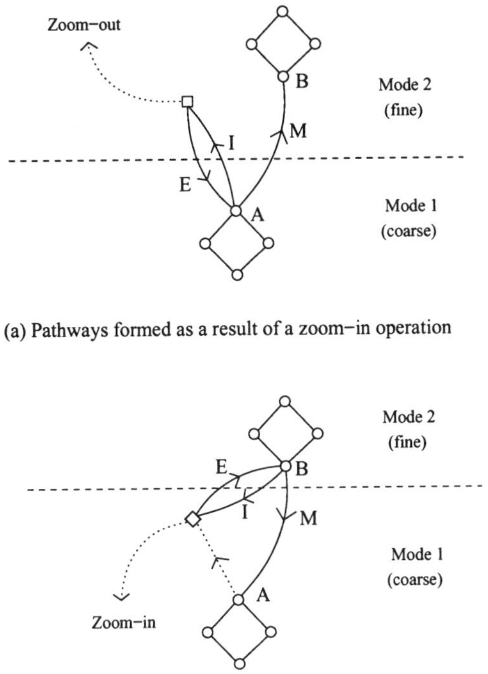

(a) Pathways formed as a result of a zoom–in operation

(b) Pathways formed as a result of a zoom–out operation

Figure 8: Pathways formed as a result of a transition between levels of the sensory mode hierarchy i.e. context A to context B: —— learnt pathways, · · · · · fixed pathways. ◇ gradient node, □ modal-gradient node. For simplicity context nodes and context map nodes are shown combined.

Consider a situation in which the animat has arrived in context B, or some other context in the subgraph containing B. If spreading activation arrives from a goal that is reachable only via A, the arrival of the motor incentive signal (I) associated with the spreading activation causes the modal-gradient node to become active resulting in the execution of a zoom-out operation while at the same time allowing it to provide the appropriate expectancy signal to context A. Such a 'forced' zoom-out is assumed to occur only if the animat is currently in the sensory mode associated with the particular modal-gradient node. If it isn't, i.e. the animat has performed further zoom-in operations, then activity continues to spread along context map pathways until it reaches the level in the mode hierarchy to which the animat's current context belongs, at which point it triggers a forced zoom-out.

Figure 8(b) shows the pathways created when the animat performs a zoom-out operation while in context B, resulting in its return to context A. A comparison with Figure 3 will show that the three pathways created are similar to those that would be formed during a normal transition from context A to context B i.e. the reverse of the context transition actually performed.

To understand why such an arrangement is useful consider the situation in which the animat is in context A and a goal exists somewhere in the subgraph containing context B. When spreading activation arrives at A it activates the gradient node shown and thus triggers a zoom-in operation which results in a transition to context B.

A zoom-out event can be thought of as creating a temporary binding between a 'place' context A and an 'object' context B. The three pathways corresponding to such a binding are assumed to be created anew each time a zoom-out occurs, subject to the constraint that a place can only have one object associated with it at any one time, i.e. any existing binding associated with the place is destroyed when a new one is created.

The arrangements for inter-mode switching described endow the animat with a useful working memory capability, allowing it to remember not only where it is, but also what it saw last time it visited an arbitrary number of locations.

3. Results and discussion

3.1 Navigating in mazes

The first experiment reported here is a simple navigation task in the maze shown in Figure 9. A view-graph navigation scheme is employed, i.e. the motor primitives available to the animat are: move forwards, turn left through 90° and turn right through 90°. At each location the animat can see the contents of the cell it is in together with those of the eight adjacent cells. Note that in the view-graph formulation adopted each grid cell corresponds to four different contexts, the total number of contexts being approximately 850. Cells were labelled with either a digit to indicate an impenetrable barrier (shaded cell) or a randomly chosen letter taken from a subset of the alphabet (unshaded cell) indicating an empty cell. The region was enclosed by an impenetrable barrier (not shown) by placing 1's in the surrounding cells. An orientation time of 12 time steps was employed for this experiment.

After an initial exploratory phase of 60000 time steps, sufficient for approximately 5000 primitive moves, the animat was positioned at the location shown and allowed to navigate towards the goal. The path shown in Figure 9 is typical of that observed after an exploratory period of this duration. The short detour near the end of the approach to the goal is due to incomplete exploration of context space. Note that during the exploratory period of this problem only around 85% of the possible contexts have been found and an even smaller percentage of the possible context transitions have been tried.

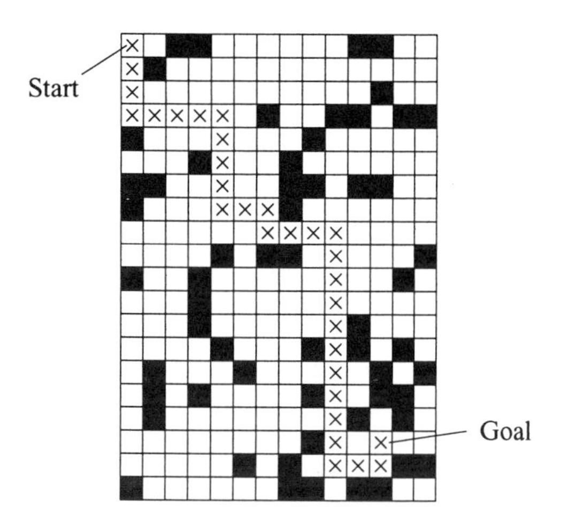

Figure 9: Navigating in a maze.

Table 1 shows statistics relating to a set of runs in which the animat was first allowed to explore the domain for N time steps, and then allowed to navigate to the indicated goal from as many randomly chosen initial positions as were possible within 100000 time steps, holding the goal location fixed. The average numbers of moves (i.e. translations) and turns together with the average number of motor primitives performed are shown in each case. As can be seen from the last column of Table 1 the average number of motor primitives required to reach the goal decreases monotonically as a function of the time spent exploring the maze.

When an animat is engaging in exploratory behaviour the problem arises of how to ensure that the time invested is spent profitably rather than in simply revisiting familiar parts of its environment. To this end an additional drive has been implemented that periodically causes the animat to break-off from purely random exploratory behaviour and purposefully seek out novel stimulae. A 'boredom' variable is maintained which gradually increases whenever the animat is in a familiar context, but decreases more rapidly whenever it encounters a new context. If the value of the boredom variable ever exceeds a specified threshold then novelty seeking behaviour is triggered, the animat selects the most novel context as a goal and sets off towards it. Normal exploratory behaviour recommences once the target context is reached, at which point the boredom variable is reset to zero. This mechanism has some similarities to the counter-based approach described by Thrun (Thrun, 1992). Initial trials suggest that this mechanism is useful when the animat has a large state space to explore.

N	Problems solved	Mean moves	Mean turns	Mean moves+turns
50000	81	20.74	14.38	35.12
100000	83	21.98	12.36	34.34
150000	115	15.17	9.49	24.66
200000	118	15.33	8.96	24.29
250000	130	16.22	6.67	22.89
300000	142	15.84	6.21	22.05

Table 1: Statistics for navigation problems.

3.2 *Problem solving in a blocks world*

The next experiment considered here involves the application of the model to a generalised planning task: the construction of a tower of blocks. Figure 10 illustrates the problem, which involves three blocks lying on a table with six locations. The position of the animat's hand is indicated by the letter H. In this experiment the animat was assumed to have the following degrees of freedom: move hand left, move hand right, grasp block, drop block. The following constraints were imposed: that the hand must always lie over one of the six location; that only one block could be held at any one time; and that blocks could only be stacked at location one.

A series of runs with differing periods of initial exploration were conducted, followed in each case by a test phase of 100000 time steps during which the animat was presented with a series of randomly chosen initial configurations in which each block occupied a different location and the hand was positioned at location three. During the test phase the context map node associated with a tower of three blocks was selected as a goal, and the behaviour of the animat was observed. Note that in this experiment a longer orientation time of 30 time steps was employed. Thus an exploratory period of 100000 time steps corresponds to time for approximately 3300 motor selection episodes.

It was ascertained that from a randomly chosen initial configuration the minimum number of primitives necessary to build a tower is 8, the mean 17.85, the maximum number 30 and the standard deviation approximately 5.1. Figure 11 shows the average number of motor primitives required to build a tower and the standard deviation of the number as functions of the length of the exploratory period. Rapid convergence towards optimal performance is clearly apparent in both the average number of moves required and the standard deviation of the number of moves. Note that during the common part of

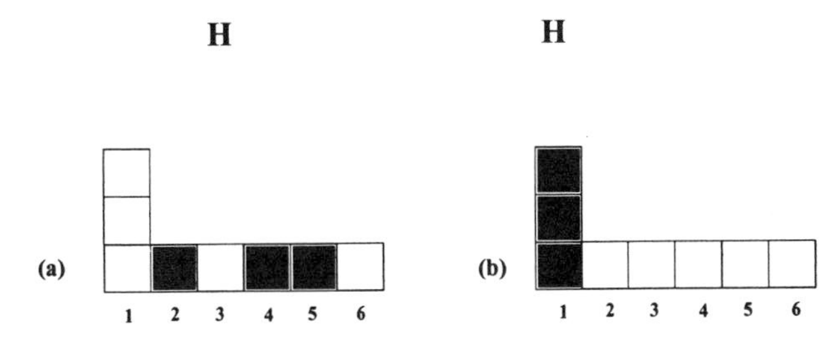

Figure 10: Building a tower: (a) a typical initial configuration; (b) goal configuration.

the exploratory phase the first tower is constructed by chance at time step 17039 by which point only 86 of the 252 possible contexts have been discovered.

3.3 Solving harder problems

While the architecture outlined above has many useful properties it is unclear as to whether it is sufficient to support more sophisticated forms of planning. For example, suppose the animat wishes to build a tower of three blocks but only two are present on the table in front of it. The obvious solution is for the animat to fetch another block from a different table and then build the tower. Unfortunately there is no way that such a plan can be formulated within the current architecture since it involves a path through the context map that intersects itself and thus it would be necessary for spreading activity to pass through the same node in different directions. Such problems arise as a result of the information hiding properties of the context representation employed, which is necessary to allow the animat to generalise its behaviour between similar scenes in different locations. Clearly some sort of additional mechanism is necessary if the animat is to learn to perform tasks of this nature.

One way in which such difficulties could be overcome is through the use of chunks. Since chunks correspond to shortcuts through the context map they can be used to plan paths that would otherwise self-intersect. At present however chunks are only allowed to form from combinations of primitives associated with a single sensory mode i.e. chunks involving zoom-in and zoom-out operations are not permitted.

An alternative approach to the planning of complex tasks involving contexts associated with multiple sensory modes involves the use of subgoals. Thus, for example, an animat wishing to build a tower might first set itself the subgoal of finding an additional block. With this goal achieved, and with the block in hand, it can then return to its original goal, i.e. building a tower.

The advantage of such an approach is that it can be implemented conveniently within the current architecture with a small number of additional modifications. Allowing an animat to set its own goals appears to be psychologically plausible, at least in higher mammals. For example, consider the scenario in which X tells Y to go to London. Most adult humans, given sufficient resources, would have little difficulty in achieving this goal despite the fact that the implicit reward is purely social – to make X happy. We clearly have the ability to internalise and act upon a goal communicated to us by another and thus there appears to be no reason why we cannot assume that humans have the capability to select their own goals, provided, at least, that self-selected goals are not able to override behaviours associated with basic metabolic needs indefinitely.

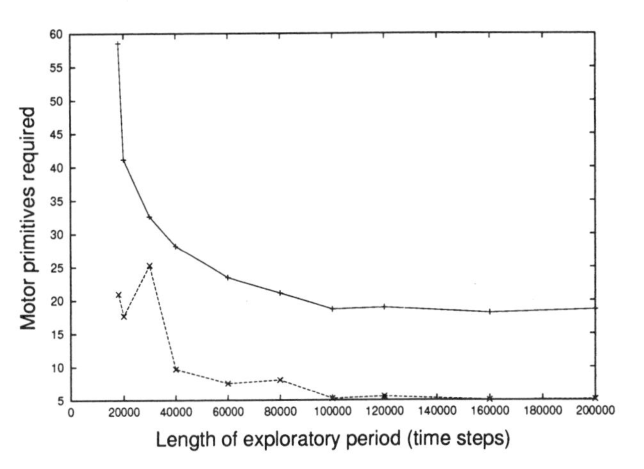

Figure 11: Building a tower: average (upper) and standard deviation (lower) of the number of motor primitives required to build a tower as a function of length of exploratory period.

Currently subgoaling is implemented using the following mechanism. Firstly, a new class of primitives is introduced. These take the form *set_subgoal*(X) where X is a context drawn from a pool of eligible contexts. If the animat executes the primitive *set_subgoal*(A) then context A becomes the new goal. It remains the goal until it is reached or until it expires after a predetermined period of time. If the subgoal is reached before it expires then a set of pathways similar to those associated with a standard intramode context transition is formed between the context in which the goal was set and the subgoal context itself. A subgoal set during exploratory behaviour it is also assumed to expire if the animat performs a zoom-in operation that leads to an unexpected context while pursuing the subgoal. In such circumstances a set of pathways is created linking the context in which the goal was set and the unexpected context. The setting of subgoals is a powerful mechanism since it allows the animat to include potentially open-ended search procedures in its plans.

Figure 12 illustrates the way in which subgoaling can be used to solve the problem of fetching an additional block so that a tower of three blocks can be built on table B. Four disjoint subgraphs, corresponding to con-

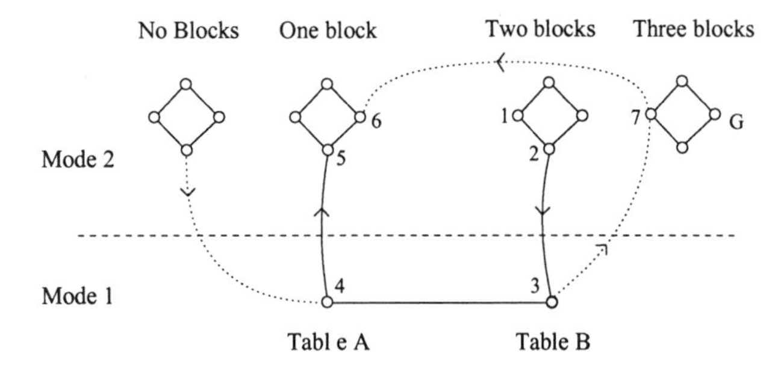

Figure 12: Solving a more complicated blocks-world problem using subgoals: —— pathways existing before the subgoal is set; · · · · · pathways created during the approach to the subgoal.

figurations of none, one, two and three blocks are shown. The animat is assumed to have previously explored its environment and to have ended up at table A in context 6 which corresponds to a configuration in which the animat is currently holding a block. If, continuing its exploration, the animat now selects context 1 as a subgoal (a configuration involving two blocks) and navigates towards it, it will on zooming-in at table B find itself not in context 2, as it expects, but in context 7 which corresponds to a configuration of three blocks, two on the table and one in its hand. At this point the subgoal is extinguished and a new context map pathway (shown as a dotted line) linking contexts 7 and 6 is created. This pathway may be thought of as representing the possibility that a configuration of three blocks can be obtained by starting at context 6 and heading towards a context representing a configuration of two blocks.

If at some later time the animat finds itself in context 1 and with the goal G then, provided it believes table A contains a configuration involving a single block it can immediately plan a course of actions ($1 \to 2 \to 3 \to 4 \to 5 \to 6 \to 7$) that will allow it to retrieve the block from table A.

The model presented here differs from reinforcement learning models (Sutton and Barto, 1998) in that here no attempt is made to assign a 'value' to each state. A value function in effect defines a form of static potential field from which an optimal path may be found using only local information. While many reinforcement learning models guarantee to produce an optimal solution under appropriate conditions convergence is typically slow requiring long exploratory periods. The new architecture, in contrast, is based around the idea that planning should be implemented as a dynamic process, i.e. using the spreading activation model. Experience suggests that near optimal paths to goals can often be found successfully using the new architecture after a far shorter period of exploration that would be required by many reinforcement learning algorithms, the number of exploratory steps required being typically one or two orders of magnitude smaller.

Random exploration of a state space is not what most people would regard as planning. When someone reports that they have planned an operation what they normally mean is that they have visualised the sequence of actions they intend to take and can find no reason to imagine that their plan will fail. It is natural to ask how such reports might be reconciled with the model as it stands. In order for the architecture to be employed for off-line planning the basic architecture requires the following additions. Firstly, mechanisms must exist for disabling sensory input and physical motor activity. Secondly, a mechanism must exist to allow context transitions to occur as if the actual physical motor activity had taken place. This clearly requires some form of expectancy mechanism. With these modifications in place the architecture can be employed for the visualisation of planned sequences of actions. Furthermore, since the sequence of context transitions performed by the model during this off-line planning is identical to that which would actually be carried out in practice, the animat's memories of the locations of objects embedded in its context map can be used to validate the planned course of action and reject it if it conflicts with what the animat thinks it knows about its environment.

4. Conclusions

The architecture outlined has many attractive features that make it suitable for use in controlling autonomous robotic systems in real-time. The model incorporates a powerful general-purpose problem solving capability with a highly flexible representational scheme, and does so within a mechanistic framework that is both biologically plausible and consistent with the psychological evidence.

References

Arkin, R. (1998). *Behavior-Based Robotics*. The MIT Press, Cambridge, MA.

Balkenius, C. (1996). Generalization in Instrumental Learning. In *From Animals to Animats 4: Proceedings of the 4th International Conference on the Simulation of Adaptive Behaviour*, pages 305–314, Cambridge, MA. MIT Press/Bradford Books.

Balkenius, C. and Moren, J. (2000). A Computational Model of Context Processing. In *From Animals to Animats 6: Proceedings of the 6th International Conference on the Simulation of Adaptive Behaviour*, pages 256–265, Cambridge, MA. MIT Press.

Blake, A. and Yuille, A. (1992). *Active Vision*. The MIT Press, Cambridge, MA.

Brooks, R. (1990). Elephants Don't Play Chess. *Robotics and Autonomous Systems*, 6:3–15.

Carpenter, G. and Grossberg, S. (1988). The ART of Adaptive Pattern Recognition by a Self-Organizing Neural Network. *IEEE Computer*, 21:77–88.

Franz, M. and Mallot, H. (2000). Biomimetic Robot Navigation. *Robotics and Autonomous Systems*, 30:133–153. Special issue: Biomimetic Robot Navigation.

Grossberg, S. (1982). Associative and Competitive Principles of Learning and Development: The Temporal Unfolding of STM and LTM Patterns. In *Competition and Cooperation in Neural Networks*, number 45 in Lecture Notes in Biomathematics, pages 295–341. Springer-Verlag Inc. Reproduced in: The Adaptive Brain Ed. S. Grossberg Vol. 1, pp. 449-485, North-Holland, 1987.

Grossberg, S. (1986). The Adaptive Self-Organization of Serial Order in Behaviour: Speech, Language and Motor Control. In *Pattern Recognition by Humans and Machines*, volume 1, pages 187–294. Academic Press Ltd. Reproduced in: The Adaptive Brain Ed. S. Grossberg Vol. 2, pp. 311-391, North-Holland, 1987.

Kaelbling, L., Littman, M., and Moore, A. (1996). Reinforcement Learning: A Survey. *Journal of Artificial Intelligence Research*, 4:237–285.

Kosko, B. (1992). *Neural Networks and Fuzzy Systems.* Prentice-Hall International, Inc.

Mataric, M. (1991). Navigating With a Rat Brain: A Neurobiologically-Inspired Model for Robot Spatial Representation. In *From Animals to Animats: Proceedings of the 1st International Conference on the Simulation of Adaptive Behaviour*, pages 169–175, Cambridge, MA. MIT Press.

Mataric, M. (1992). Integration of Representation Into Goal-Driven Behaviour-Based Robots. *IEEE Transactions on Robotics and Automation*, 8(3):304–312.

Moren, J. (1998). Dynamic Action Sequences in Reinforcement Learning. In *From Animals to Animats 5: Proceedings of the 5th International Conference on the Simulation of Adaptive Behaviour*, pages 366–371, Cambridge, MA. MIT Press.

Payton, D., Rosenblatt, J. K., and Keirsey, D. (1990). Plan Guided Reaction. *IEEE Transactions on Systems, Man, and Cybernetics*, 20:1370–1382.

Peterson, R. (2001). An Integrated Model of Context and Motor Learning for Autonomous Agents. School of Computing Research Report 2001.14 University of Leeds, Leeds, U.K.

Rao, R., G.J.Zelinsky, Hayhoe, M., and Ballard, D. (1997). Eye Movements in Visual Cognition: A Computational Study. Technical Report 97.1, National Resource Laboratory for the Study of Brain and Behavior, Department of Computer Science, University of Rochester.

Ryback, I., V.I.Gusakova, Golovan, A., Podladchikova, L., and Shevtsova, N. (1998). A model of attention-guided visual perception and recognition. *Vision Research*, 38:2387–2400.

Scholkopf, B. and Mallot, H. (1995). View-Based Cognitive Mapping and Path Planning. *Adaptive Behavior*, 3(3):311–348.

Sutton, R. and Barto, A. (1998). *Reinforcement Learning.* Bradford Books, The MIT Press.

Thrun, S. (1992). Efficient Exploration in Reinforcement Learning. Technical Report CMU-CS-92-102, School of Computing, Carnegie-Mellon University, Pittsburgh, Pennsylvania.

Trullier, O., Wiener, S., Berthoz, A., and Meyer, J.-A. (1997). Biologically-based Artificial Navigation Systems: Review and Prospects. *Progress in Neurobiology*, 51:483–544.

Ullman, S. (1996). *High-level Vision.* The MIT Press, Cambridge, MA.

Witkowski, M. (2000). The Role of Behavioral Extinction in Animat Action Selection. In *From Animals to Animats 6: Proceedings of the 6th International Conference on the Simulation of Adaptive Behaviour*, pages 177-186, Cambridge, MA. MIT Press.

Yamauchi, B. and Beer, R. (1996). Spatial Learning for Navigation in Dynamic Environments. *IEEE Transactions on Systems, Man, and Cybernetics - Part B: Cybernetics*, 26:496–505.

Yarbus, A. (1967). *Eye Movement and Vision.* Scientific American Library. Plenum Press, New York.

Zimmer, U. (1996). Robust World-modelling and Navigation in a Real World. *Neurocomputing*, 13:247–260.

An Activation Based Behaviour Control Architecture for Walking Machines *

Jan Albiez **Tobias Luksch** **Karsten Berns** **Rüdiger Dillmann**
Forschungszentrum Informatik an der Universität Karlsruhe (FZI)
Interactive Diagnosis and Servicesystems
Haid-und-Neu-Str. 10-14, 76131 Karlsruhe, Germany
albiez@fzi.de

Abstract

This paper introduces a behaviour network architecture for controlling walking machines. The behavior coordination problem is solved by distributing the activation of the behaviours according to the sensoric information as well as the specified task of the robot. This approach heavily emphasises the loop-back of the behaviour activities and the satisfaction of their goals. The results of initial experiments with this architecture on the four-legged walking machine BISAM are also presented.

1 Introduction

Walking robots have been a field of increasing activity in the last years. Especially the ability to adapt to unstructured terrain and the resulting demands on the control architecture have been in the focus of researchers. These efforts can be separated into two different approaches, one being the classical engineering approach using and refining the known methods of loop-back control structures and dynamic modelling to control the robot, e.g. (Löffler et al., 2001) or (Gienger et al., 2001). The other way is to adopt as much from biological paragons for locomation as possible regarding both mechanical design and control architecture, e.g. (J. Ayers and Massa, 2000) and (Kimura et al., 2001) . The methods proposed in this paper follow the second approach by applying a reflex or behaviour based control architecture to a four-legged walking machine, this way performing sensor-based adaptation to motion on irregular terrain. Difficulties in implementing such a reactive control consist in the handling of the sensoric equipment and the real-time interpretation of its signals as well as in the necessity of adaptation to dynamically changing environment.

Biological research of the last years has identified several key elements being used in nature for adapting locomotion. These range from the geometrical structure of legs (Witte et al., 2001b) and dynamic properties of muscles (Pearson, 1995) to neural networks used for walking by insects (Cruse and Bartling, 1995) and (Cruse et al., 2001). The results of this research suggest a transfer of these principles to legged robots. Due to the high complexity of real walking machines and the impracticality of mimicking especially nature's activators and sensors, up to now only some of the ideas have been transferred to the control architectures of real robots. In (Kimura et al., 1990) and (Kimura and Nakamura, 2000) a neuro-oscillator based pattern generator is introduced. The adaptation to the terrain is solved by directly influencing the activation of the oscillator neurons. (J. Ayers, 2000) also uses neuro-oscillators which are parametrized using the results from the analysis of lobsters. (Hosoda et al., 2000) proposes a reflex based gait generation system, triggered by the input of a camera system mounted on the robot.

In the last years several methods were succefully applied to control the four-legged walking machine BISAM (Berns et al., 1998). These include the usage of coupled neuro-oscillators for gait generation (Ilg et al., 1998a), learning leg trajectories (Ilg and Scholl, 1998) and the application of radial basis function neural networks and reinforcement learning methods for posture control while trotting (Ilg, 2001) and (Albiez et al., 2001a). All these methods were succefull but lacked a certain extensibility when confronted with more demands than they were initially designed for (e.g. both dynamically stable trott and statically stable walking). Thus the necessity arrised to build an architecture being able to handle these demands.

In the following we introduce an activation- and target-rating-based behaviour control and a first implementation of such an architecture on the walking machine BISAM. Afterwards the results of various initial experiments are presented. The paper closes with a conclusion and an outlook on further development.

*This research is funded by the Deutsche Forschungsgemeinschaft (DFG), grants DI-10-1, DI-10-2 and DI-10-3

2 Activation- and Target-Rating-Based Behaviour Control

Behaviour based control has been used for several years in the field of robotics (Endo and Arkin, 2001). Its main area of application has been higher level control like local navigation of mobile robots e.g. (Arkin et al., 2000) or (Mataric, 1997), or the analysis of natural behaviour (Mataric, 1998). These control architectures are mainly tested on robots with only a few degrees of freedom, so the low-level control has not to be implemented unsing behaviours. A lot of research has been done on the critical problem of behaviour coordination. Examples include comfort zones (Likhachev and Arkin, 2000) or case based reasoning (Likhachev and Arkin, 2001); a good overview on behaviour coordination can be found in (Pirajanian, 1999). There have been only a few approaches to use behaviour based control on the lower control levels of robots with many degrees of freedom like walking machines. Noteable are the subsumption architecture (Brooks, 1986), (Ferrell, 1995) or biomimetic methods like the control of the lobster robot in (J. Ayers and Massa, 2000).

In this paper we propose an architecture trying to fulfill the requirements on behaviours and behaviour coordination from the reactive low level control like motor reflexes as well as from the more deliberative higher levels like gait selection. The behaviour coordination problem is solved by distributing the activation of behaviours using feedback not only from the sensors but also from the behaviours themselves.

2.1 Behaviours

A behaviour or reflex [1] $\mathcal{B}$ in the sense of this paper is a functional unit which generates an output vector $\vec{u}$ using an input vector $\vec{e}$ and an activation ι according to a transfer function $b(\vec{e})$. Additionally a target rating criterion $r(\vec{e})$ and a behaviour activity a is calculated. Mathematically this can be combined into a 6-tuple as in (1).

$$\mathcal{B} = (\vec{e}, \vec{u}, \iota, b, r, a) \quad (1)$$

The transfer function b is defined as in (2) for an in-

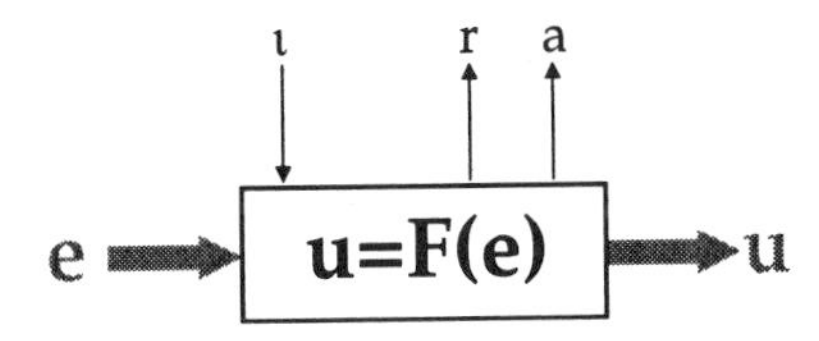

Figure 1: Behaviour design

put dimension of n and an output dimension of m. It implements the fundamental action being performed by the behaviour. The output $\vec{u}$ is generated in two steps. First an unmodified output with respect to the input vector $\vec{e}$ is produced. In the second step this output is scaled with the activity input ι. A ι of 0 means that the behaviour will generate no output at all, one of 1 will lead to the full output. Between 0 and 1 the output is scaled according to the function of the behaviour.

$$b : \Re^n \times [0;1] \rightarrow \Re^m; \quad b(\vec{e}, \iota) = \vec{u} \quad (2)$$

This definition is an abstract view of the behaviour as is. The actual implementation method can vary from simple feed-forward controllers up to more complicated systems like finite state automata.

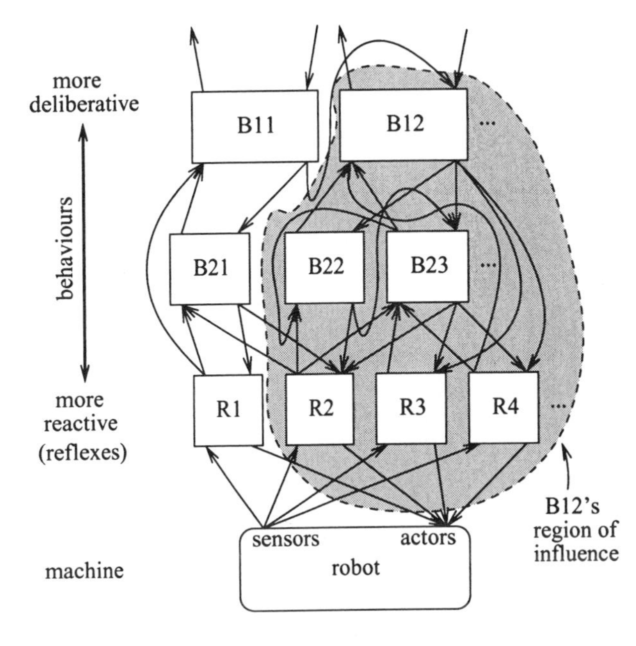

Figure 2: Behaviour coordination network

Furthermore each behaviour generates a target rating r (3) evaluating how far the actual state of the robot matches the aspired goal of the behaviour. For this estimation the input vector $\vec{e}$ consisting of sensor informations from the robot and outputs of other behaviours is used. A value of 0 indicates, that the robot's state matches the beahviours goal, a value of 1, that it doesn't.

$$r : \Re^n \rightarrow [0;1]; r(\vec{e}) = r \quad (3)$$

For monitoring reasons it is desirable to have visualisation of the behaviour's activity. But more important, the activity is used as feedback information for other behaviours. This activity a is defined as in eq. (4).

$$a : \Re^m \rightarrow [0;1] : a(\vec{u}) = ||\vec{u}|| \quad (4)$$

[1] A reflex refers to a simple behaviour close to the hardware thus being more reactive than deliberative

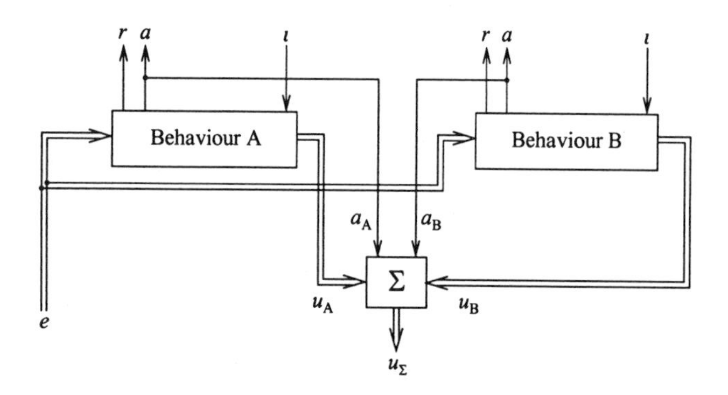

Figure 3: Fusion of different behaviours outputs using their activation as weighting criterion.

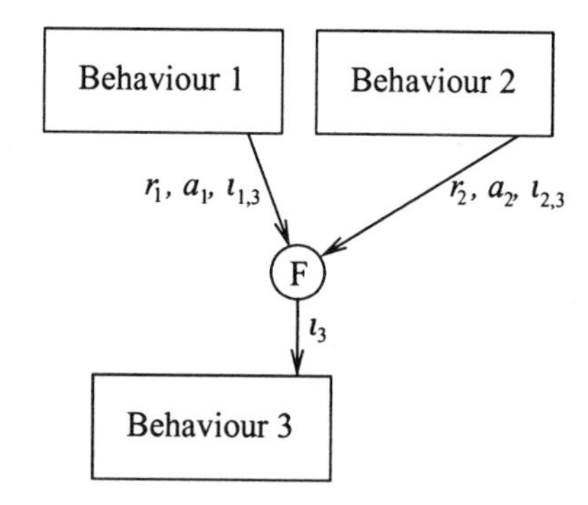

Figure 4: Fusion of different behaviours outputs.

Colloquial, a is the target rating r scaled by the activation input ι. This means that a can be interpreted as indication on the behaviour's effort in transfering the robot to a state achieving its goal.

2.2 Behaviour Coordination

Following the approach of Brook's subsumption architecture, the behaviours are placed on layers depending on how much they actually work on the robot's hardware. The inputs and outputs of each behaviour are connected with each other to form a network structure (see figure 2 for an example). These connections transport control and sensor information as well as loop-back information like r, ι and a of different behaviours. In the case that several outputs of different behaviours are connected to the same input of another behavior an behaviour activity based fusion scheme is used (figure 3). This scheme favors the output of behaviours having an unmet target (i.e. a high r) and a high activity a implying a ι greater than 0. The actual fusion of the outputs is done with the help of a fusion knot. Basically a fusion knot is no more than a fusion function f (eq. (5)) combining the outputs $\vec{u}$, the activities a and the target ratings r of all behaviours involved to generate one output vector $\vec{u}'$.

$$f : \Re^n \times \ldots \times \Re^n \times \Re \times \ldots \Re \rightarrow \Re^n :$$
$$f(\vec{u}_0, \ldots, \vec{u}_n, a_0, \ldots, a_n, r_0, \ldots, r_n) = \vec{u}' \quad (5)$$

The activity in such a network will concentrate inside the influence area or region $\mathcal{R}$, eq. (6), of a higher level behavior.

$$\mathcal{R}(\mathcal{B}_n) = \{\mathcal{B}_a : \mathcal{B}_a \text{is influenced by } \mathcal{B}_n\} \quad (6)$$

A region is an organisational unit of co-operating behaviours. The affiliation of a behaviour to a region is not exclusive. For example a leg's swing behaviour is located in the region of a stride and a trott behaviour. To coordinate the influence of several highlevel behaviours over a lower one, a fusion knot is inserted before the ι input of the one on the lower level. An example for the coordination of the influence of two behaviours on a third is shown in figure 4.

To allow the transition of the activity between different regions, r and a are loop-backed between behaviours on the same level as well as on others. This way it is possible that at a point where one behaviour cannot handle a situation anymore even when it is fully in charge, another behavior which can solve the situation is activated while the other behavior is deactivated. For example if the desired speed of a walking machine cannot be reached by the stride behaviour, the trot behaviour will take over.

3 The Walking Machine BISAM

BISAM (Biologically InSpired wAlking Machine) consists of the main body, four equal legs and a head (see figure 5). The main body is composed of four segments being connected by five rotary joints. Each leg consists of four segments connected by three parallel ro-

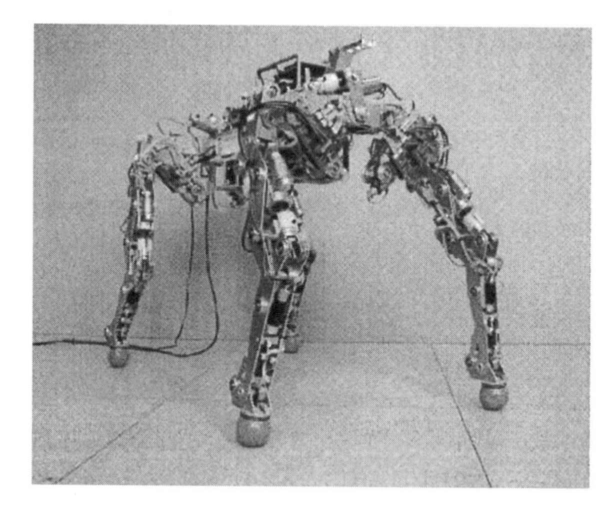

Figure 5: The quadrupedal walking machine BISAM. Due to the five active degrees of freedom in the body and the ability to rotate the shoulder and hip, BISAM implements key elements of mammal-like locomotion.

tary joints and attached to the body by a fourth. The joints are all driven by DC motors and ball screw gears. The height of the robot is 70 cm, its weight is about 23 kg. 21 joint angle encoders, four three dimensional foot sensors and two inclinometers mounted on the central body provide the necessary sensoric input. A more detailed description of the development and specification of BISAM can be found in (Berns et al., 1999) and (Ilg et al., 1998b). Research on BISAM aims at the implementation of mammal-like movement and different gaits like statically stable walking and dynamic trotting with continuous gait transitions. Due to this target, BISAM is developed with joints in the shoulder and in the hip, a mammal-like leg-construction and small foot contact areas. These features have strong impact on the appliable methods for measuring stability and control. For example, caused by BISAM's small feet the ZMP-Criterion (Vukobratovic et al., 1990) is not fully adequate to describe the aspired movements.

The control design has to consider the high number of 21 active joins and especially the five joints in the body. One common way to reduce the model complexity is to combine joins and legs by the approach of the virtual leg, as used in many walking machines (Raibert, 1986), (Kimura et al., 1990), (Yoneda and Hirose, 1992). This approach poses problems when modelling BISAM's body joints and lead to a strong reduction in the flexibility of the walking behaviour (Matsumoto et al., 2000). A second way is to reduce the mechanical complexity of the robot so it is possible to create an exact mathematical model of the robot (Buehler et al., 1999).

Taking the described problems into consideration we used BISAM as the first plattform to implement the proposed behaviour based architecture.

4 Initial Experiments

This section presents initial experiments to study the behaviour based control. First we introduce fuzzy reflexes as the simplest behaviours and present the implemented posture reflexes for the experiments. More complex behaviours will be described in the subsequent part. Experimental results will complete this paper.

4.1 Reflexes as Simple Behaviours

Simple behaviours only built as sensor/action units can be interpreted as reflexes. These small reflex units accept no activation or rating outputs from other behaviours but only sensor data from the machine. In this paper we propose fuzzy reflexes. The use of fuzzy controllers has several benefits, e.g. rapid and easy development making good use of the experience gained by BISAM's operators as well as the possibility to apply machine learning methods. The fuzzy reflexes consist of two main parts: a sensor pre-/postprocessing unit and a fuzzy controller (FC) as shown in figure 6. The pre-/postprocessing unit has the following tasks: scale and filter the sensor inputs $\vec{e}$ (eg. inclinometers) so that they meet the requirements of the FC concerning scaling and range; calculate the criterion r mentioned above of how much the current state of the machine meets the target of the reflex; adapt the output vector $\vec{u}'$ of the FC to the needs of the lower layers of the control architecture.

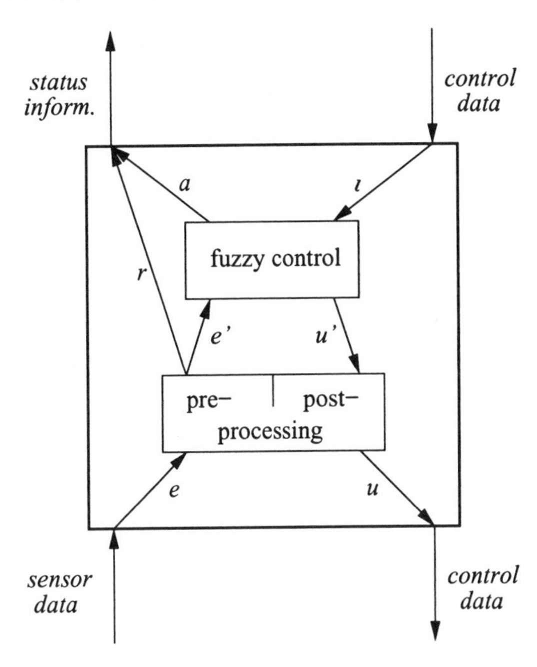

Figure 6: The internal structure of a fuzzy reflex unit.

The fuzzy controller is implemented as a classical Fuzzy-PD-Controller and corresponds to the behaviour function $b(\vec{e})$. Figure 7 illustrates a simple fuzzy rule set where each cell represents a single rule. There are five fuzzy sets (negative large and small, zero, positive small and large) for both input variables and the output variable. The behaviour activity a is calculated by the FC basically as the norm of its output vector $\vec{u}'$. In case of the reflexes the activation input ι scales the output of the FC in a linear way.

These simple behaviours generate outputs directly being used as control data for the robot. The posture reflexes described below hand offset values to the legs and the central body influencing their trajectories.

4.2 Implemented Reflexes

To examine the feasibility of the behaviour based control several experiment were carried out. Below we will present a posture control for stable stand and a statically stable stride. Apart from the more complicated behaviours as descibed in the next section several simple reflexes have to be included. Various experiments (Ilg et al., 2000) and considerations on the in-

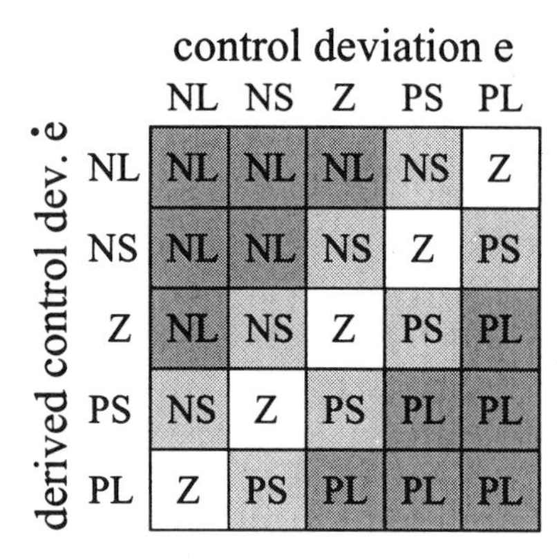

Figure 7: The inference matrix of the Fuzzy-PD controller.

sights from biological research as in (Witte et al., 2001b) and (Cruse and Bartling, 1995) have shown the necessity for the following key reflexes:

- **Even Force X/Y** The purpose of this reflex is to evenly distribute the forces at the feet touching the ground. The SCM_{xy} values (see (Albiez et al., 2001b) for the Sensor Based CoG Model SCM) are used as measurement criterion of the load distribution and as input values e for the FC. $r = \|SCM\|$ is the absolute value of the SCM vector. As reflex output u offset values shifting the central body in the xy-plane are generated by the FC.

- **Even Force Z** Similar to Even Force X/Y this reflex tries to achieve the same goal by individually adapting the length of all four legs in z direction.

- **Body Height** This reflex tries to keep the body on a certain height level. It uses only the internal representation of the machine, and therefore the actual height h of BISAM is calculated from the average length in the z direction of all legs touching the ground

$$h = \frac{\sum_{l_i \in L_g} z(l_i)}{l_g} \tag{7}$$

where L_g is the set of all legs on the ground and $z(l_i)$ is the length in z direction of the i-th leg. This reflex keeps the body at a given height h by influencing the z position of BISAM's foot points.

- **Body Inclination** The inclinometers mounted on BISAM's body are used by this reflex as input e to keep the body parallel to the ground. This results in r to be defined as:

$$r = |\phi - \phi_0| + |\theta - \theta_0| \tag{8}$$

where ϕ and θ are the roll and pitch angles. The rotation of the body (the output u) is implemented by equally stretching the legs on one side while bending the legs on the other side.

- **Leg Relieve** Before BISAM can take a step when walking in stride the leg which is about to swing (l_{swing}) has to be relieved which is insured by this reflex. The force in l_{swing} is used as criterion r, the output u of the reflex is a shift of the body in the y direction and an appropriate bending of the leg opposite to l_{swing}.

4.3 Standing and Walking Behaviour

For the experiments on BISAM two more complex behaviours have been designed: one behaviour ensuring a stable standing on four legs despite perturbences from the environment; and a behaviour implementing a statically stable stride. The complete network of the used behaviours is shown in figure 8. All behaviours described in this chapter have been implemented in MCA[2], a realtime capable modular controller framework (Scholl et al., 2000), (Scholl et al., 2001).

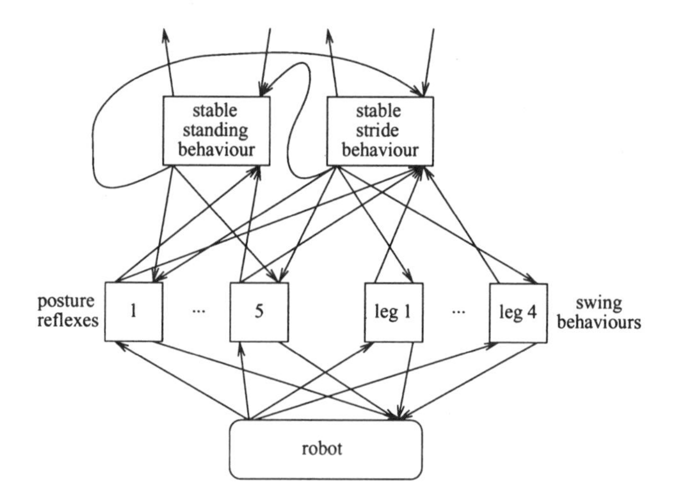

Figure 8: The network of behaviours for posture control.

Using the posture reflexes described above it is relatively easy to achieve a behaviour for stable standing. An apropriately weighted activation of all the reflexes except, of course, the leg relieving reflex will accomplish satisfying results.

Unfortunately just turning on all the reflexes won't suffice to assure a statically stable walk as some reflexes work against each other. One such reflex conflict occurs while trying to relieve the leg that is about to swing: the Leg Relieve reflex will move the machine body to one side and bend the leg opposite the swing leg, whereas the Even Force X/Y, Even Force Z and Body Inclination reflexes will try to move the body back to the most

[2]Visit http://mca2.sf.net/ for more information

stable posture but with $\frac{1}{4}$ of the load on the swing leg. To resolve these conflicts a reflex weighting mechanism depending on the current stance phase has to be included in the walking behaviour. Using the information passed from the posture reflexes, i.e. their target rating r, a suitable activation can be calculated, for example by a simple finite state automaton. The resulting activation ι of the posture reflexes during one swing of a leg is shown in figure 9. Note that during the intermittent phase (cycle 20 to 310) the reflexes for stabilizing the machine are activated. In the swing phase (cycle 310 to 380) the Leg Relieve reflex is the significant one, only Even Force X/Y is switched on a bit to ensure the stability of the robot in case of disturbances. Furthermore the behaviour has to trigger the swing behaviour of the next leg when the corresponding state is reached.

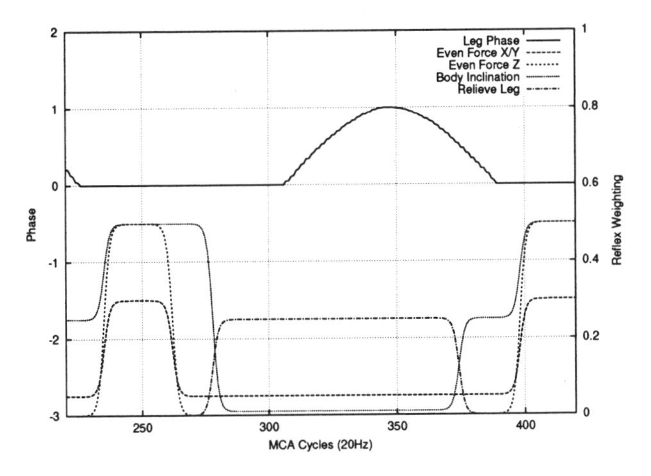

Figure 9: The activation of the different reflexes using the input i of the reflex units.

4.4 Experimental Results

Several experimental setups have been used to verify the functioning of the initial behaviour based control using the simple posture reflexes. Three of them are presented here. One most basic test for the reflexes is changing BISAM's load distribution (figure 10). A weight of about 10 pounds is placed on the front part of the robot's central body. At once the activity of the Even Force X/Y reflex increases and the posture of the machine is adapted. The same procedure is repeated when the weight is lifted again.

To further study the standing behaviour BISAM is placed on a platform of about 10cm in height with one leg reaching over the platform's rim (figure 11). When toppling down the platform (at about $t \sim 35$ in figure 12) the machine's stability (visualized by the stability margin) will decrease until the lower leg touches the ground ($t \sim 65$). Without any further posture control the fourth leg won't reach the ground and the stability will remain

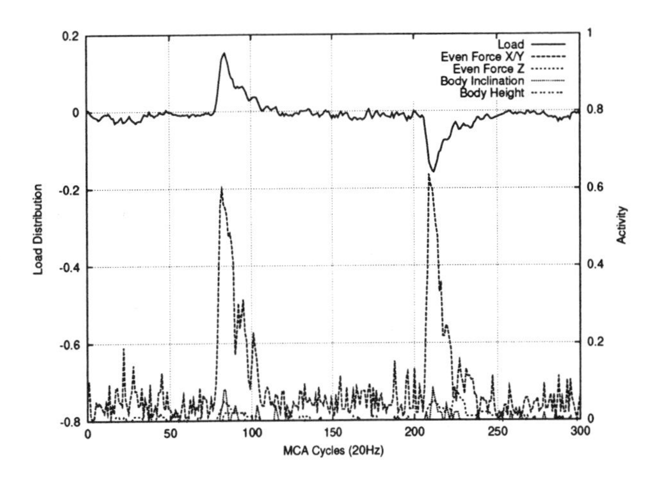

Figure 10: Changing the load distribution.

on a low level (solid line), whereas it will dramatically increase when the reflexes are activated (dashed line): after similarly toppling down it takes the reflexes only a short moment to get all four legs on the ground. Afterwards the reflexes will continue to optimize the machine's posture which explains their ongoing activity.

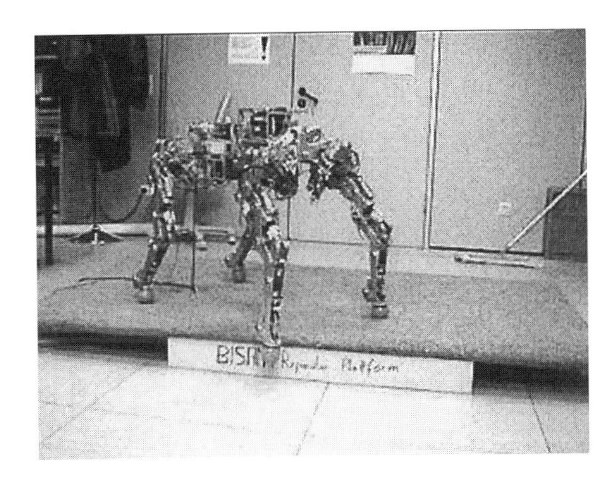

Figure 11: BISAM stepping down a platform.

To demonstrate the walking abilities BISAM has to walk on even ground as well as over obstacles. Figure 13 shows the activity of the posture reflexes during a swing cycle. The upper, solid line represents the swing and stance phases of the legs. At the end of the previous swing phase the reflexes responsible for reaching a stable stand on four leg are heavily active. When they are satisfied with the machine's posture their target rating r will drop low and the walking behaviour will activate the Leg Relieve reflex (the double-dotted line). With the swing leg relieved its swing behaviour is triggered and the machine will take another step.

If the swinging leg hits an obstacle (figure 14) the reflexes will hide its occurrence from the rest of the control

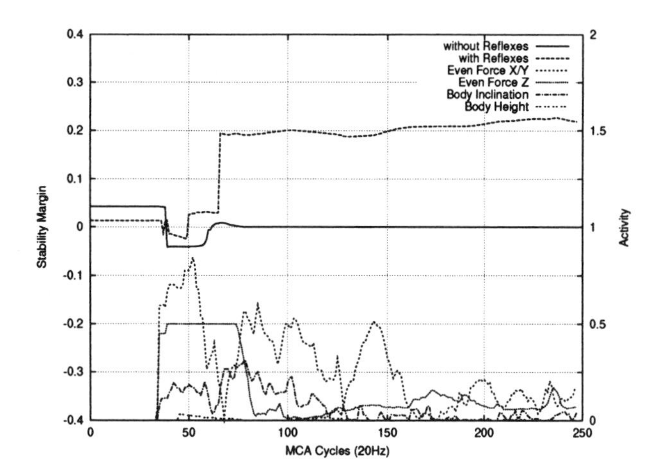

Figure 12: Comparing stability with and without activated fuzzy reflexes. The stability margin and the reflex activity are plotted against time.

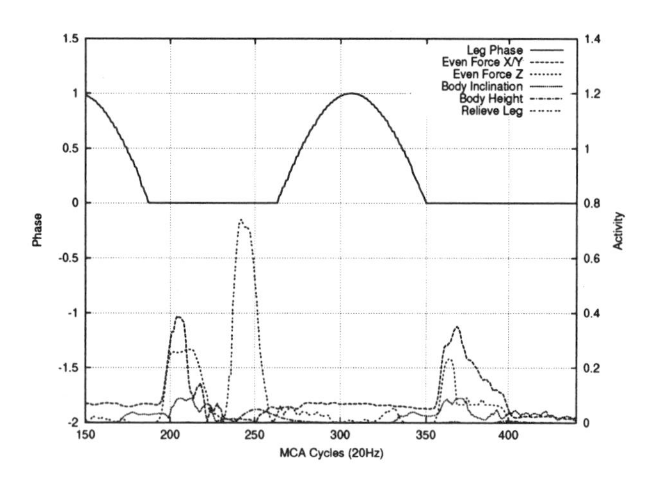

Figure 13: The reflex activity during a swing cycle on even ground.

system: mainly the Even Force Z and the Body Inclination reflex will show more activity to adapt the robot's posture to the obstacle (see figure 15, the dotted line). This way BISAM can walk over obstacles like steps or bricks of up to 10cm in height.

5 Conclusion and Outlook

As initial experiments showed, the proposed behaviour architecture is capable to generate a statically stable gait, to control the posture while standing and to allow a flexible transition between standing and walking. The same set of behaviours/reflexes which allows to adapt to unstructured ground while standing can also be applied to adapt to the terrain while walking by only using a different set of values for the activation inputs. Additionally we demonstrated that it is possible to smoothly

Figure 14: BISAM stepping on an obstacle.

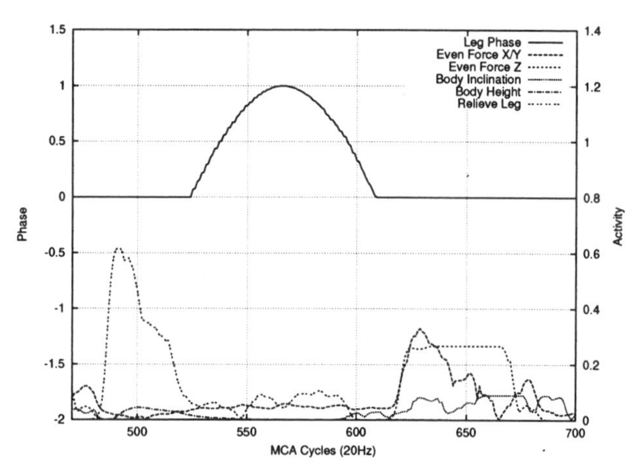

Figure 15: The reflex activity during a swing cycle when hitting an obstacle.

shift between these to sets.

Future work includes the extension of BISAM's control architecture to allow trotting and moving in extremely unstructured terrain, as well as the improvement of the transition between different higher behaviours (e.g. trotting, walking, standing). Furthermore we will transfer the behaviour based archtiecture on a mammal-like (Witte et al., 2001a), pneumatic-muscle activated leg and the six-legged walking machines Lauron III (B. Gassmann, 2001) and AirBug (Berns et al., 2001). To accomplish this a unified and robust design methodology for constructing the proposed behaviour networks has to be developed. Another field of interest is to try to pre-activate behaviours or reflexes based on additional sensor or control information, like extra tactile sensors for obstacle detection (Dürr and Krause, 2001) or a navigation layer. Additional it is planned to optimise the fuzzy-based behaviours, by transferring them to radial basis function networks and train them with reinforcement learning methods. Machine learning algorithms will also be applied on other promising parameters of the behaviour control.

References

Albiez, J., Ilg, W., Luksch, T., Berns, K., and Dillmann, R. (2001a). Learning reactive posture control on the four-legged walking machine bisam. In *International Conference on Intelligent Robots and Systems (IROS)*, Hawaii, USA.

Albiez, J., Luksch, T., Ilg, W., and Berns, K. (2001b). Reactive reflex based posture control for a four-legged walking machine. In *Proceedings of the 4th International Conference on Climbing and Walking Robots (CLAWAR)*, Karlsruhe, Germany. FZI.

Arkin, R., Kahled, A., Weitzenfeld, A., and Cervantes-Prez, F. (2000). Behavioral models of the praying mantis as a basis for robotic behavior. *Journal of Autonomous Systems.*

B. Gassmann, K.U. Scholl, K. B. (2001). Locomotion of lauron iii in rough terrain. In *International Conference on Advanced Intelligent Mechatronics*, Como, Italy.

Berns, K., Albiez, J., Kepplin, V., and Hillenbrand, C. (2001). Control of a six-legged robot using fluidic muscle. In *International Conference on Advanced Robotics*, Budapest, Hungary.

Berns, K., Ilg, W., Deck, M., Albiez, J., and Dillmann, R. (1999). Mechanical construction and computer architecture of the four-legged walking machine BISAM. *IEEE Transactions on Mechatronics*, 4(1):1–7.

Berns, K., Ilg, W., Eckert, M., and Dillmann, R. (1998). Mechanical construction and computer architecture of the four-legged walking machine bisam. In *CLAWAR '98 - First International Symposium*, pages 167–172.

Brooks, R. (1986). A robust layered control system for a mobile robot. *IEEE Journal of Robotics and Automation*, RA-2(1):14–23.

Buehler, M., Cocosco, A., Yamazaki, K., and Battaglia, R. (1999). Stable open loop walking in quadruped robots with stick legs. In *Proceedings of the IEEE International Conference on Robotics and Automation*, pages 2348–2354, Detroit.

Cruse, H. and Bartling, C. (1995). Movement of joint angles in the legs of a walking insect, carausius morosus. *J. Insect Physiol.*, (41):761–771.

Cruse, H., Dürr, V., and Schmitz, J. (2001). Control of a hexapod walking - a decentralized solution based on biological data. In *Proc. of the 4th International Conference on Climbing and Walking Robots (CLAWAR)*, Karlsruhe, Germany.

Dürr, V. and Krause, A. (2001). The stick insect antenna as a biological paragon for an actively moved tactile probe for obstacle detection. In *Proc. of the 4th International Conference on Climbing and Walking Robots (CLAWAR).*

Endo, Y. and Arkin, R. (2001). Implementing tolman's schematic sowbug: Behaviour-based robotics in the 1930's. In *Proceedings of the 2001 IEEE International Conference on Robotics and Autonomous Systems.*

Ferrell, C. (1995). Global behavior via cooperative local control. volume 2, pages 105 – 125.

Gienger, M., Löffler, K., and Pfeiffer, F. (2001). In *Proc. of the IEEE International Conference on Robotics and Automation (ICRA).*

Hosoda, K., Miyashita, T., and Asada, M. (2000). Emergence of quadruped walk by a combination of reflexes. In *Procceedings of the International Symposium on adaptive Motion of Animals and Machines*, Montreal.

Ilg, W. (2001). *Eine biologisch motivierte adaptive Bewegungssteuerung für eine vierbeinige Laufmaschine.* Infix-Verlag. Dissertationsschrift (in german).

Ilg, W., Albiez, J., and Dillmann, R. (2000). Adaptive posturecontrol for a four-legged walking machine using some principles of mammalian locomotion. In *Int. Symposium on Adaptive Motion of Annimals and Machines.*

Ilg, W., Albiez, J., and Jedele, H. (1998a). A biologically inspired adaptive control architecture based on neural networks for a four-legged walking machine. In *Proceedings of the 8th International Conference on Artificial Neural Networks*, pages 455–460, Skoevde.

Ilg, W., Berns, K., Jedele, H., Albiez, J., Dillmann, R., Fischer, M., Witte, H., Biltzinger, J., Lehmann, R., and Schilling, N. (1998b). Bisam: From small mammals to a four legged walking machine. In *Proceedings of the Fifth International Conference on Simulation of Adaptive Behaviour*, pages 400–407, Zürich.

Ilg, W. and Scholl, K.-U. (1998). Cqran: Continuous q-learning resource allocation network. towards reinforcement learning with multi-dimensional continuous action spaces. In *Proceedings of the 8th International Conference on Artificial Neural Networks, Skoevde, Sweden*, pages 455–460, Skoevde.

J. Ayers, J. Witting, C. O. N. M. D. M. (2000). Lobster robots. In Wu, T. and Kato, N., (Eds.), *Proceedings of the International Symposium on Aqua Biomechanisms.*

J. Ayers, J. Witting, C. W. P. Z. N. M. and Massa, D. (2000). Biomimetic robots for shallow water mine countermeasures. In *Proc. of the Autonomous Vehicles in Mine Countermeasures Symposium.*

Kimura, H., Fukuoka, Y., Hada, Y., and Takase, K. (2001). Three-dimensional adpative dynamic walking of a quadruped robot by using neural system model. In *Proc. of the 4th International Conference on Climbing and Walking Robots (CLAWAR)*, Karlsruhe. FZI.

Kimura, H. and Nakamura, H. (2000). Biologically inspired dynamic walking on irregular terrain - adaptation based on vision. In *International Symposium on Adaptive Motion of Animals and Machines*, Montreal.

Kimura, H., Shimoyama, I., and Miura, H. (1990). Dynamics in the dynamic walk of a quadruped robot. *Advanced Robotics*, 4(3):283–301.

Löffler, K., Gienger, M., and Pfeiffer, F. (2001). Simulation and control of a biped jogging robot. In *Proceedings of the 4th International Conference on Climbing and Walking Robots (CLAWAR).*

Likhachev, M. and Arkin, R. (2000). Robotic comfort zones. In *Proceedings of the SPIE: Sensor Fusion and Decentralized Control in Robotic Systems*, volume 4196, pages 27–41.

Likhachev, M. and Arkin, R. (2001). Spatio-temporal case-based reasoning for behavioral selection. In *Proceedings of the 2001 IEEE International Conference on Robotics and Automation (ICRA)*, pages 1627–1634.

Mataric, M. J. (1997). Behavior-based control: Examples from navigation, learning, and group behavior. *Journal of Experimental and Theoretical Artificial Intelligence*, Special issue on Software Architectures for Physical Agents, 9(2-3):323–336.

Mataric, M. J. (1998). Behavior-based robotics as a tool for synthesis of artificial behavior and analysis of natural behavior. *Trends in Cognitive Science*, 2(3):82–87.

Matsumoto, O., Ilg, W., Berns, K., and Dillmann, R. (2000). Dynamical stable control of the four-legged walking machine bisam in trot motion using force sensors. In *Intelligent Autonomous Systems 6.*

Pearson, K. (1995). Proprioceptive regulation of locomotion. *Current Opinions in Neurobiology*, 5(6):768–791.

Pirajanian, P. (1999). Behaviour coordination mechanisms - state-of-the-art. Technical Report IRIS-99-375, Institute for Robotics and Intelligent Systems, School of Engineering, University of Southern California.

Raibert, M. H. (1986). *Legged Robots That Balance.* MIT Press, Cambridge, MA.

Scholl, K.-U., Albiez, J., and Gassmann, B. (2001). Mca - an expandable modular controller architecture. In *In proceedings of the 4th Linux Real Time Workshop*, Milano.

Scholl, K.-U., Kepplin, V., Albiez, J., and Dillmann, R. (2000). Developing robot prototypes with an expandable modular controller architecture. In *ICIAS2000.*

Vukobratovic, M., Borovac, B., Surla, D., and Stokic, D. (1990). *Biped Locomotion.* Springer–Verlag, Heidelberg, Berlin, New York.

Witte, H., Hackert, R., Fischer, M. S., Ilg, W., Albiez, J., Dillmann, R., and Seyfarth, A. (2001a). Design criteria for the leg of a walking machine derived by biological inspiration from quadruped mammals. In *Proc. of the 4th International Conference on Climbing and Walking Robots (CLAWAR)*, Karlsruhe, Germany.

Witte, H., Hackert, R., Lilje, K., Schilling, N., Voges, D., Klauer, G., Ilg, W., Albiez, J., Seyfarth, A., Germann, D., Hiller, M., Dillmann, R., and Fischer, M. (2001b). Transfer of biological priciples into the construction of quadruped walking machines. In *Second International Workshop On Robot Motion And Control*, Bukowy Dworek, Poland.

Yoneda, K. and Hirose, S. (1992). Dynamic and Static Fusion Gait of a Quadruped Walking Vehicle on a Winding Path. In *Proceedings of the IEEE/RSJ International Conference on Intelligent Robots and Systems*, pages 143–148, Nizza.

Relating Behavior Selection Architectures to Environmental Complexity

Orlando Avila-García   Elena Hafner   Lola Cañamero
Adaptive Systems Research Group, Department of Computer Science, University of Hertfordshire
College Lane, Hatfield, Herts AL10 9AB, UK
O.Avila-Garcia@herts.ac.uk, draguita@hotmail.com, L.Canamero@herts.ac.uk

1 Introduction

Assessing the performance of behavior selection architectures is a complex task that depends on many factors. Comparative studies such as (Maes 1991, Tyrrell 1993, Bryson 2000) have tended to focus on the respective merits and drawbacks of hierarchical (structured) versus flat (parallel) architectures. In our opinion, however, the gap between those types of architectures is too big, and finer-grained and more systematic analyses of both architectures and environments, and of how these relate to each other, are needed to obtain a deeper understanding of the adequacy of different architectures for different types of tasks and contexts. This paper outlines an initial study comparing four motivated behavior-based architectures, all of them variants of (Cañamero 1997), performing in different worlds with varying degrees and types of complexity, and analyzes performance results relating architectural elements to environmental complexity. The criteria used to measure and compare performance are based on Ashby's viability theory (Ashby 1952).

2 Architectures for Behavior Selection

The architectures studied are neither strictly flat (parallel) nor hierarchical (structured), but consist of two internally parallel layers—motivational and behavioral—performing a two-step computation of intensity: motivational intensity must be computed prior to the calculation of behavioral intensity. They have been implemented in Khepera robots using the Webots 3.0.1 simulator (www.cyberbotics.com). All the architectures have the same elements, but vary in the way these are combined (their arbitration mechanisms):

Sensors: *external* (IR proximity and collision, a radio emitter/receiver used to transmit and detect the attack of another robot, and a RGB color camera used to detect direction and discriminate objects) and *internal* (measuring physiological variables).

Physiology: essential variables (e.g. energy, pain) that must remain within a range of viable values so that environmental changes do not put the robot's life in danger.

Motivations: urges to action (e.g. hunger, thirst) that implement a homeostatic process to maintain an essential physiological variable within a certain range. A feedback detector generates an error signal—a drive—when the value of this variable departs from the ideal value (setpoint), and this assigns an intensity (activation level) to the motivation, and triggers the execution of behaviors to adjust the variable in the adequate direction.

Behaviors: coarse-grained subsystems implementing different competencies. They can be activated and executed with different intensities[1] set by motivations (and in some cases external stimuli).

The four architectures differ along three parameters in the way these elements are interconnected, as summarized in Table 1: (a) Link between motivations and behaviors; (b) Locus of influence of external stimuli; and (c) Point of main selection decision—motivations (a winner motivation selects the behavior that best satisfies it) or behaviors (a behavior may allow to satisfy several motivations at the same time).

Arch.	Links	Stimuli computed	Decision point
A1	fixed weights	on behaviors	behaviors
A2	fixed weights	on motivations	behaviors
A3	physiology	on motivations	motivations
A4	physiology	on motivations	behaviors

Table 1: Characterization of the four architectures.

3 Experiments

We have created a typical behavior selection environment (Valimar) in which our robots (Nessas) must select among and perform different activities in order to survive. The world is populated by cylindrically shaped objects of different colors: food and water sources, nests, obstacles, dull blocks, and Enemies. We varied Valimar to explore the effects of three sources of environmental complexity: (a) amount of objects, (b) availability of resources, and (c) dynamism (Enemies that can at-

[1] Behavioral intensity influences the speed of the wheels and the modification (increment or decrement) of specific physiological variables (and hence the duration of the behavior).

tack and kill Nessas, and also hamper foraging activities). We created five Valimar settings: V1, V2 and V3 progressively decremented (a) and (b) (and dynamism was absent), while we added dynamism in V4 (one Enemy) and V5 (two Enemies). Five sets of experiments[2] were performed, each of them testing the four architectures in one of the five worlds. To measure and compare the performance of the four architectures, we have used three indicators: (1) Viability (V), defined as the complementary of the normalized sum of errors of the robot's physiological variables; (2) Life span (S_{life}) in simulation steps; and (3) Quality of life, $Q_{life} = V \times S_{life}$.

Results for A1, A2 and A4 were very similar in the first three (static) worlds, where A3 obtained the worst results in terms of viability, as it deals worse with extreme situations in which at least one variable is near its limit. In terms of life span, survival was very good in the static worlds. The introduction of Enemies in V4 and V5 leads to significant changes with respect to the three indicators. Viability and life span become considerably worse for all the architectures, due to the negative impact of Enemies' attacks on the physiology of Nessas. In these dynamic worlds, A3 outperforms the other architectures because its winner-takes-all policy makes it more reactive to external changes, dealing better with situations of self-protection, while the opposite is the case for A4. Table 2 ranks our architectures regarding some of the phenomena that (Maes 1991) proposes as requirements to achieve flexible behavior selection.

Phenomenon	A1	A2	A3	A4
Openness (reactivity)	3	2	1	4
Stability	2	2	4	1
Opportunism	3	1	3	2
Displacement behaviors	–	–	–	1

Table 2: Ranking of architectures (1 means best, 4 poorest).

Displacement behaviors were only observed in A4[3], as its "maximum profit" policy can lead to *mutual inhibition* (cancellation) of two motivations with high intensity. This policy is also responsible for the lower activation levels that behaviors receive in A4, compared to the other architectures.

Stability of a sequence of behaviors occurs when their intensities are (nearly) similar. A non-stable sequence results in sudden changes in the robot's velocity and modification of its variables. A3, being more reactive, was the least stable architecture, while A4, due to its "maximum profit" policy, showed the higest stability.

[2]Each set consisted of 40 runs, 10 for each architecture, with a total of 200 runs (about 40 hours).

[3]Displacement behaviors would also have been possible in A1 and A2 if we had not considered only positive weights between motivations and behaviors, but are not possible in A3, where only one motivation drives behavior selection.

Opportunism management varies considerably in the four architectures as a consequence of the way in which the influence of external stimuli is taken into account. Architectures A1 and A3 are less opportunistic than A2 and A4, since the influence of external stimuli is computed only once to calculate the intensity of the winner behavior. Although opportunism is in general a desirable feature that provides flexibility, too much opportunism can present disadvantages in environments with few resources located far from each other.

Situations of self-protection are more difficult to deal with when Nessas are executing a consummatory behavior near a resource, as they are more exposed to Enemies, which can attack them on the back and block them against resources. A4 is more often trapped in these situations than the other architectures due to its lower reactivity and to the fact that its lower intensity levels make Nessas spend more time next to resources. A3, being more reactive (like a simple emotional system), is the best in these situations.

4 Conclusion and Future Work

The results of these experiments show that small variations in the way in which the same architectural components are combined greatly influence the way in which behavior selection is performed, and therefore the adaptivity of the robot to different environmental conditions. To continue this study we envisage several directions for future work, namely: (1) complementing our viability and life quality indicators with a measure of the internal equilibrium achieved in terms of the standard deviation of the motivations' error; (2) adding different sources of dynamism such as extinction and mobility of resources; (3) adding basic emotions to perform behavior selection, in particular in highly dynamic worlds.

References

Ashby, W.R. (1952). *Design for a Brain: The Origin of Adaptive Behavior.* London: Chapman and Hall.

Bryson, J. (2000). Hierarchy and Sequence vs. Full Parallelism in Action Selection. In J.A. Meyer *et al.*, eds., *Proc. 6th SAB*, 147–156. MIT Press.

Cañamero, L.D. (1997). Modeling Motivations and Emotions as a Basis for Intelligent Behavior. In W.L. Johnson, ed., *Proc. First Intl. Conf. on Autonomous Agents*, 148–155. New York: ACM Press.

Maes, P. (1991). A Bottom-Up Mechanism for Behavior Selection in an Artificial Creature. In J.A. Meyer and S.W. Wilson, eds. *Proc. First SAB*, 238–246. Cambridge, MA: MIT Press.

Tyrrell, T. (1993). The Use of Hierarchies for Action Selection. *Adaptive Behavior*, 1(4): 387–419.

INTERNAL WORLD MODELS AND PROCESSES

Global localization and topological map-learning for robot navigation

David Filliat*
*DGA/Centre Technique d'Arcueil
16 bis, av. Prieur de la Côte d'Or
94114 Arcueil Cedex - France
david.filliat@etca.fr

Jean-Arcady Meyer**
**AnimatLab-LIP6
8, rue du capitaine Scott
75015 Paris - France
jean-arcady.meyer@lip6.fr

Abstract

This paper describes a navigation system implemented on a real mobile robot. Using simple sonar and visual sensors, it makes possible the autonomous construction of a dense topological map representing the environment. At any time during the mapping process, this system is able to globally localize the robot, i.e. to estimate the robot's position even if the robot is passively moved from one place to another within the mapped area. This is achieved using algorithms inspired by Hidden Markov Models adapted to the on-line building of the map. Advantages and drawbacks of the system are discussed, along with its potential implications for the understanding of biological navigation systems.

1 Introduction

The word navigation refers to all the strategies that may be used by a robot to purposely move in its environment. Such strategies range from simple visible goal heading behavior to complex map-based navigation that allows the planification of movements to arbitrary distant goals (Trullier et al., 1997). Using the latter strategies basically raises three sub-problems : *map-learning*, which concerns the construction of a map representing the environment, *localization*, which concerns the estimation of the robot's position inside this map and *planification*, which concerns the design of a plan to reach a given goal.

Every navigation strategy may call upon two sources of information. The first is the idiothetic source that provides information about the robot's movements using internal sensors such as accelerometers. This information can be directly expressed in a metrical space. The second one is the allothetic source that provides information about the robot's position inside its environment using external sensors such as sonar sensors or a camera. The characteristics of these two sources are complementary : while idiothetic information suffers from cumulative errors that make it unreliable for long-term position estimation, allothetic information suffers from the *perceptual aliasing* problem that prevents the robot from distinguishing between two places. Therefore, the efficiency of a navigation system usually relies on its capacity to efficiently combine these two types of information.

It is important to note that allothetic information can be used in two different ways. The first makes use of a metrical model of the sensors, which permits the allothetic data to be expressed in the metrical space of idiothetic information. This is, for example, the case for sonar data used to estimate the position of obstacles in a metrical map of the environment (Moravec and Elfes, 1985). The second way avoids any use of metrical models of the sensors and directly resorts to allothetic information to compare and recognize different positions. This is, for example, the case when the colors of the environment are used to recognize a position in a topological map (Ulrich and Nourbakhsh, 2000). This paper will limit itself to methods that use allothetic sensors without any associated metrical model. Indeed, this choice makes a much more general use of allothetic data possible, as it does not require sensors measuring metrical properties of the environment. This way, information like a color, an odor or a temperature can be used to map the environment. Moreover, such simple use of allothetic data seems more representative of the way an animal like a rat builds an internal model of its environment.

Without metrical models of the sensors, however, a navigation system will have to cope with some limitations. Most of these limitations stem from the fact that it is impossible to infer what should be perceived at a distant position without actually going there. For example, it is easy, with a metrical sensor model, to infer that a wall perceived two meters away will be perceived as being one meter away if the robot moves one meter in the direction of this wall. On the contrary, such an inference is impossible without using a metrical sensor model. Consequently, a map-learning system will only provide information about positions that have already been visited at least once. As will be shown in the remainder of this paper, this limitation must be dealt with by the map-learning and localization procedures.

The main issue with map-based strategies lies in the necessity of simultaneously tackling localization and map-learning problems. The difficulty arises from the *chicken and egg* status of these problems (Yamauchi et al., 1999). In other words, a map is necessary to estimate the position, while knowing the position is necessary to update the map. It is true that the localization problem when a map is given a priori has been given efficient solutions (Thrun et al., 1999). Notably, some models are able to tackle the *lost robot problem*, i.e. the estimation of the robot's position without any initial cues about its position. Unfortunately, the corresponding models that are able to *globally localize* a robot are difficult to extend to on-line map-learning.

In the context of animat research, strong emphasis is placed on autonomy. A map-based navigation system should therefore make it possible to accurately localize an animat in any, eventually initially unknown, environment without human intervention. These requirements are met by global localization models that build environmental maps on-line. The model described in this paper affords solutions to such requirements. Moreover, for the reasons stated above, this model does not make use of any metrical sensor model. It draws inspiration from the literature on bio-mimetic navigation systems, on the one hand, and from purely robotic navigation systems, on the other hand. Several improvements to the simulation model presented by Filliat and Meyer (2000) will be described here, together with new results that were obtained with a real robot implementation.

2 Global localization and map-learning

Localization models described in the literature basically pertain to three categories called respectively *direct-position inference*, *single-hypothesis tracking* and *multiple-hypothesis tracking* (Filliat and Meyer, 2002).

2.1 Direct-position inference

These models (e.g., Franz et al., 1998, Gaussier et al., 2000) call upon environments and sensory capacities that are not subject to perceptual aliasing. Allothetic information is supposed to directly provide an unambiguous estimate of the position, without the need to use any idiothetic information. These models therefore heavily rely on perceptual systems that are able to discriminate between a great number of positions. However, such an hypothesis about the absence of perceptual aliasing within a whole environment is hard to assume *a priori* in any initially unknown environment.

2.2 Single-hypothesis tracking

These models (e.g., Smith et al., 1988, Dedeoglu et al., 1999) take the perceptual aliasing issue into account and solve it by using idiothetic information to disambiguate positions. This information is used to estimate the current position relative to the previous one, and this estimate is used to limit the search space of the position that corresponds to current allothetic data. Assuming that the restrained search area no longer exhibits perceptual aliasing, the corresponding position is unique. This mechanism allows a single position hypothesis to be tracked, as the alternative positions that would correspond to the same allothetic data are simply discarded.

This method is *local* in the sense that the current position is searched for only in the vicinity of the previous position estimate and not over the whole map. As a consequence, an initial position estimate has to be provided to the system either by a separate direct position inference mechanism or by an operator. This requirement limits the robot's autonomy and moreover precludes future correct position estimation if the current estimate should accidentally prove false.

2.3 Multiple-hypothesis tracking

A solution to avoid the dependence on an initial position estimate in perceptually aliased environments is to track multiple hypotheses of the robot's position. According to this scheme, instead of discarding the positions corresponding to current allothetic data that do not match the previous position estimate, these positions are memorized as alternative hypotheses of the robot's position. All these hypotheses are subsequently tracked in parallel and their relative credibilities are monitored. At every moment, the most credible hypothesis is considered as the robot's current position.

This approach allows a *global localization* that is not tied to an initial position estimate. Moreover, the set of concurrent hypotheses may be empty and may be initialized with all the positions that correspond to the first allothetic information gathered in the environment. Therefore, this approach solves the *lost robot problem*, and it affords a high degree of autonomy to the localization process.

The corresponding implementation may call upon the explicit process of monitoring several possible positions in parallel (Piasecki, 1995), or it may call upon Partially Observable Markov Decision Processes (Simmons and Koenig, 1995, Fox et al., 1998). These latter solutions may be viewed as implicit multiple-hypothesis tracking, where each possible position in the map is considered as a position hypothesis. This solution already yielded highly successful robots operating in challenging environments (Thrun et al., 1999).

2.4 Map-learning

From a recent review of map-learning strategies in robots (Meyer and Filliat, 2002), it appears that combining map-learning with direct position inference is relatively straightforward as it simply entails adding to the map allothetic situations that have never been seen before.

A lot of models also combine map-learning with single-position tracking methods (Arleo and Gerstner, 2000, Dedeoglu et al., 1999) because this approach still works when the robot gets outside the area already mapped. Indeed, in such case, it is straightforward to insert a new position in the map, because it is defined relatively to a previously known position.

On the contrary, combining map-learning with multiple-hypothesis tracking algorithms is more difficult. The reason is that these algorithms rely heavily on the completeness of the map to estimate the relative credibilities of the different position hypotheses. This estimation entails comparing what the robot currently perceives with what it should perceive in each of the possible positions monitored. Therefore, when the map is incomplete - which is the case during map-learning - this estimation is difficult, as the robot may be either inside or outside the currently mapped area. If it is inside, the global localization procedure can estimate the robot's position; if it is not, this procedure cannot be used.

Various attempts have been made to overcome this difficulty while nevertheless combining global localization with map-learning. A first method is to use off-line mapping algorithms that build a map corresponding, with the highest possible probability, to a set of data gathered by the robot (Shatkay and Kaelbling, 1997). However, this method does not meet our requirement of autonomy because localization and map-learning are to be separated.

A second method that works on-line is to use powerful distance sensors, along with associated metrical models, in order to prevent the robot from traveling outside the mapped area (Thrun et al., 2000). Indeed, as argued in the introduction, metrical sensor models make it possible to build a map that extends beyond the current robot's position. Accordingly, frequently estimating the robot's position guarantees that it always remains within the mapped area.

A third method will be used here, which combines global localization and map-learning without resorting to any metrical sensor model. This method entails frequently checking whether the robot is in the mapped area or not. If such is the case, a global localization algorithm can be used directly. If not, a single hypothesis tracking method based on the previous positions is used temporarily, until the robot re-enters the mapped area. To decide between these two alternatives, Filliat and Meyer (2000) proposed to simply use the credibility of the most credible among the concurrent position hypotheses. Should this credibility fall below a given threshold, the robot would be considered to be outside the mapped area. However, additional experiments with such a procedure showed it to be brittle, because the corresponding threshold needed to be changed according to the particular environment mapped. Moreover, large uncertainties in the robot's position, which lead to low credibilities of the concurrent hypotheses, always led to believe that the robot was outside the mapped area, thus rendering the mapping process quite unstable.

This paper describes an updated model where the decision between the two cases calls upon an heuristic based on the variation of the sum of credibilities of the various hypotheses. This heuristic, that will be described later on, efficiently detects when the robot exits the mapped area, thus affording the model a substantial gain in robustness, notably because the corresponding parameters become independent of the environment.

3 The-model

This section outlines a simplified version of the model that assumes that panoramic sensors are used. Experimental results presented further were obtained with directional sensors and active perception strategies described in Filliat and Meyer (2000) and Filliat (2001).

3.1 Structure

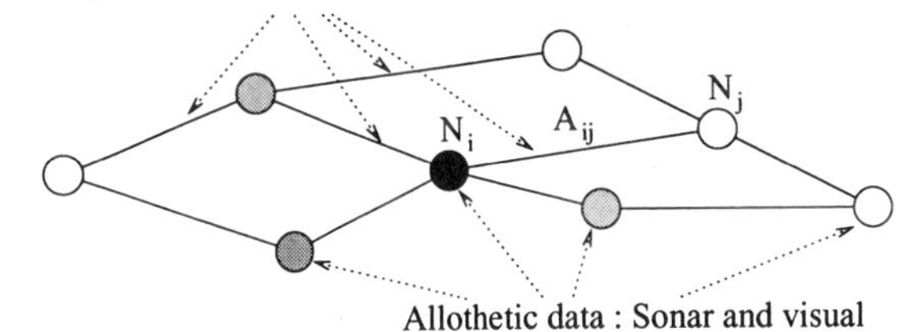

Figure 1: The topological map used in the model.

The map built by the system is a dense topological map, the nodes of which represent close positions in the environment (with a mean spacing of 25 cm). Each node stores the allothetic data that the robot can perceive at the corresponding place in the environment. A link between two nodes memorizes at which distance and in which direction the corresponding places are positioned relatively to each other, as measured by the robot's idiothetic sensors (Figure 1). All the directions used in the model are absolute directions, assuming a fixed reference direction given by a magnetic compass. The robot's position is represented by an activity distribution over the nodes : activity A_i of node i represents the probability that the robot is at the corresponding position. These probabilities are estimated using allothetic and idiothetic data gathered by the robot, as will be described in section 3.3.

The model iterates the following steps that are explained in the paragraphs below :

- Update the activity of each node in the map;
- Recognize a node as corresponding to the robot's current position or create a new one;
- Update visual and sonar data stored in the recognized node using the current allothetic data;
- Update the idiothetic data stored in the links;
- Choose the direction of the next move in order to explore the environment or to reach a goal.

3.2 *Model inputs*

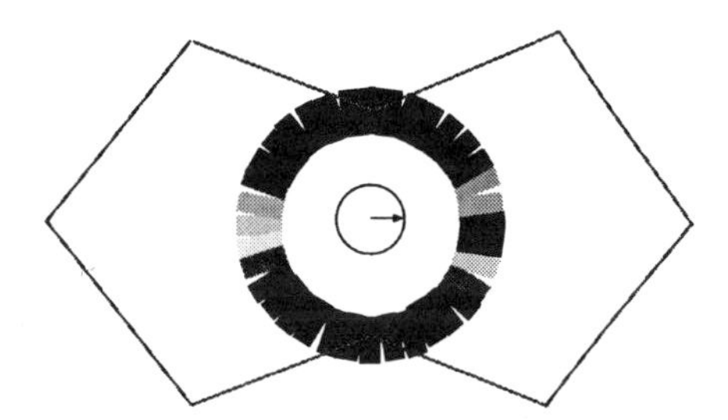

Figure 2: Schematics of allothetic data used in the model. The broken line joins the points detected by sonar sensors in eight absolute directions. The rectangles arranged on a circle indicate the mean grey-level perceived in the corresponding direction by the camera.

Two series of allothetic data are used in the model : sonar data and visual data (Figure 2). Sonar data are gathered through a 16-sonar belt and aggregated into eight virtual sensors that provide distances to obstacles in eight absolute directions. Visual data are gathered by an omnidirectional camera and down-sampled to the values of 36 virtual sensors that measure the mean grey-level of the environment in 36 absolute directions.

Both sonar and visual allothetic data are associated with a procedure P_O that compares two perceptions O_M and O_P. This procedure, which returns 1 if the two perceptions are identical, and decreases to 0 more quickly the more the perceptions are different, is used to estimate the probability that the robot is at a position characterized by data O_M, given the currently perceived allothetic data O_P. In the experiments described below, we used the following function[1] :

$$P_O(O_M/O_P) = \sqrt[l]{\prod_{k=1}^{l} F(O_M^k - O_P^k)}$$

[1]This procedure is adapted to the case of partial data when a directional camera is used. See Filliat (2001) for details.

where O_M^k and O_P^k are the values of allothetic data in the absolute direction k, l is the total number of directions for the considered sensor - i.e., eight for sonar data and 36 for visual data - and F is a Gaussian function given by $F(x) = e^{-x^2/K^2}$. The parameter K is chosen empirically for each sensor so as to give $P_O = 10^{-6}$ for maximally different sensor values. The model seems robust with respect to this parameter, since the same value was efficiently used for all simulated and real experiments.

Idiothetic data are used to estimate the probability that the robot has moved from one node in the map to another. Given a displacement of direction θ_{od} and length r_{od} measured by the robot's odometry, the probability of having moved from node A to node B is :

$$P_D(AB/od) = E_1 \times E_2$$

with :

$$E_1 = exp\left(\frac{-(\theta_{od} - \theta_{AB})^2}{L^2}\right)$$

$$E_2 = exp\left(\frac{-(r_{od} - r_{AB})^2}{M^2}\right)$$

where θ_{AB} and r_{AB} are the direction and length of the link between nodes A and B, L and M are empirically set to $L = 30$ *degrees* and $M = 20$ *cm* through statistics gathered on the moves interspersed with activity updates. Here also, the same values have been used in all simulated and real experiments.

3.3 *Activity updates*

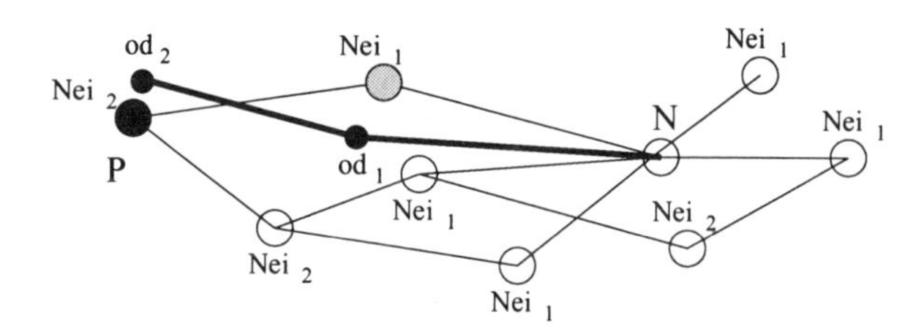

Figure 3: Illustration of the use of idiothetic data for activity updates. Nei_k is the set of all the nodes linked to node N by k connections, and od_k is the position of the robot at time $t-k$ as measured by the odometry relatively to node N. In this example, the activity of node N will be a function of the activity of node P, at time $t-2$ (see text for details).

The activity of each node is updated each time the robot has moved by a given distance (50 cm in the experiments). Such updates are directly inspired by the equation used in POMDP-based navigation models (Simmons and Koenig, 1995) and are adapted to the irregular structure of our model. Idiothetic data are first integrated using the equation :

$$A_i(t) = \max_{k \in [1..K]} \left(\max_{j \in Nei_k(i)} (A_j(t-k) \times P_D(ij/od_k)) \right)$$

where $Nei_k(i)$ is the set of all the nodes linked to node i by k connections, $A_j(t-k)$ is the activity of node j at time $t-k$, and od_k is the position of the robot at time $t-k$ as measured by the odometry relatively to node i.

The effect of this equation is to estimate the probability of the robot's being at node i, taking into account the node j that best fits the robot's path over K past time-steps (see Figure 3). The sum S_a of the activities of all the nodes is then calculated. It will be used to decide whether the robot is in the mapped area or not (see next section).

Then, allothetic data O_P are integrated using :

$$A_i(t+1) = A_i(t) \times P_O(O_i, O_P)$$

The effect of this equation is to increase the activities of nodes characterized by allothetic data that match the current perceptions, and to decrease the activities of the other nodes. Activities are then normalized such that their sum equals 1.

3.4 Position estimation

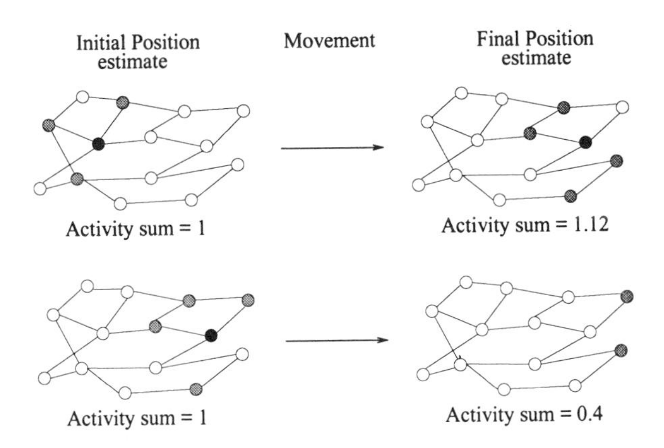

Figure 4: Illustration of the heuristic used to detect whether the current position is in the mapped area or not. When the robot is in the mapped area, the sum of the activities remains approximately constant (top half of the figure) while, if the robot exits the mapped area, the sum decreases (bottom half of the figure).

The model presented so far estimates the robot's most probable position, assuming that this position is part of the map. However, during map-learning, the robot can get out of the mapped area. To decide if the robot has exited the mapped area, an heuristic based on the variation of the sum of the activities before and after the integration of idiothetic cues is used. The idea underlying this heuristic is that, when the robot exits the mapped area, the sum S_a of the activities should suddenly decrease (see Figure 4). If the robot remains in the mapped area, on the contrary, this sum should either increase or remain stable.

Taking into account that the sum of activity is 1 before idiothetic cue integration, the algorithm used to estimate the position is then :

- If $S_a \geq 1$, the node with the highest activity is recognized as the current position.
- If $S_a < 1$, the robot's position is estimated using odometry information gathered since the last recognized node. If this position falls close to an existing node, this node is recognized; otherwise, a new node is added to the map[2]. Such a procedure amounts to temporarily using a single-position tracking method.

3.5 Map updates

Once the node corresponding to the current position has been determined, the allothetic data that characterize it are updated using the newly perceived data.

The direction and distance that correspond to the link between the previously recognized node and the current one are also updated using the newly measured displacement. To achieve map consistency, the values of all the links in the map are then updated using the relaxation algorithm of Duckett et al. (2000). In this context, a map is considered to be consistent if, when two different paths link two nodes, the relative positions of these nodes, calculated by summing the connection data along these two paths, are identical. Basically, the relaxation algorithm "shakes" the relative positions of all the nodes in the map so as to make these relative positions as close as possible to their measured values, thereby resulting in a globally coherent map.

3.6 Exploration strategy

Once the map has been updated, the exploration of the environment resumes. The exploration strategy used in the model aims at limiting localization errors and at ensuring exhaustive exploration. As global localization is efficient only when the robot is in the mapped area, the exploration strategy limits the distance that the robot may travel in an unmapped area. This is implemented thanks to a mechanism that retraces the recent route backwards if the model consecutively creates five nodes, i.e. if the heuristic mentioned above detects that the robot is outside the mapped area during five consecutive

[2]It should be noted that the heuristic thus used has a tendency to over-estimate the novelty of a position, which results in having any unmapped position always being correctly recognized as new. However, it also often causes a position previously mapped to be classified as new. This over-estimation is compensated for by verifying the existence of a node close to the position estimated before creating a new one.

time-steps. When this mechanism is not active, on the contrary, the direction of movement is chosen towards the less explored area, i.e. the direction free of obstacles where there are fewer nodes in the map, so as to ensure exhaustive exploration.

3.7 Path planning

If a goal is assigned to the robot, a movement is planned towards this goal. To achieve this, a *policy*, determining in which direction D_i to move from each node i of the map to reach the goal, is calculated using a simple spreading-activation algorithm starting from the goal. The direction of the next move is then chosen according to a voting method (Cassandra et al., 1996). A score is accordingly calculated for 36 sectors of 10 degrees surrounding the robot. This score is the sum of the activities of the nodes whose associated direction falls in this sector :

$$V(d) = \sum_{d-5<D_i<d+5} A_i$$

where $V(d)$ is the score of the sector of direction d, D_i is the direction of the goal associated with node i, and A_i is the activity of node i. The direction to be taken by the robot corresponds to the sector that achieves the highest score.

A detour mechanism may also be triggered when the planned trajectory to the goal turns out to be blocked by an unforeseen obstacle (Tolman, 1948). In such a case, the contradiction between planned movements that would lead the robot to cross the obstacle and the local obstacle-avoidance procedures that repel the robot from this obstacle generates an oscillatory behavior in front of the obstacle. These oscillations are detected by a continuous check of the robot's progression and a threshold is used to detect when too low a progression indicates it is probably impossible to reach the goal. The nodes that are close to the robot's position are then excluded from the planning process, which is entirely repeated. This results in a new policy that avoids the blocked position and leads the robot to the goal by a different route whenever possible (Filliat, 2001).

4 Experimental results

The model has been implemented on a Pioneer 2 mobile robot (see Figure 5). This robot is equipped with 16 sonar sensors and a directional camera. Although a magnetic compass could be used to estimate the absolute direction, this sensor turned out to be inefficient in our environment because of numerous magnetic disturbances. In the current system, the direction is therefore estimated using the robot's odometry, and its error is periodically compensated for by manually aligning the robot with a reference direction. This correction is made every 50 time-steps, i.e. approximately every 10 minutes. A set of low-level procedures allows local obstacle-avoidance during navigation.

Figure 5 shows a map obtained by the system in the corridors of our laboratory, this map being superimposed on an architectural sketch of the environment. It was created in 2000 time-steps in approximately six hours of operation, most of this time being consumed in stopping and starting the robot and in orienting the camera at each time-step. This time could be significantly reduced by the use of an omnidirectional camera that would allow the system to operate without stopping the robot at each time-step. Be that as it may, the map thus obtained correctly reproduces the structure of the laboratory and permits the robot's position to be estimated precisely. Figure 6 shows part of the robot's trajectory, as estimated either by the whole localization system or by a sub-part of this system that called upon the robot's odometry only. The trajectory estimated by the whole system is closer to the real trajectory, because it remains in the open area and does not cross any wall, thus demonstrating that the localization system is efficient.

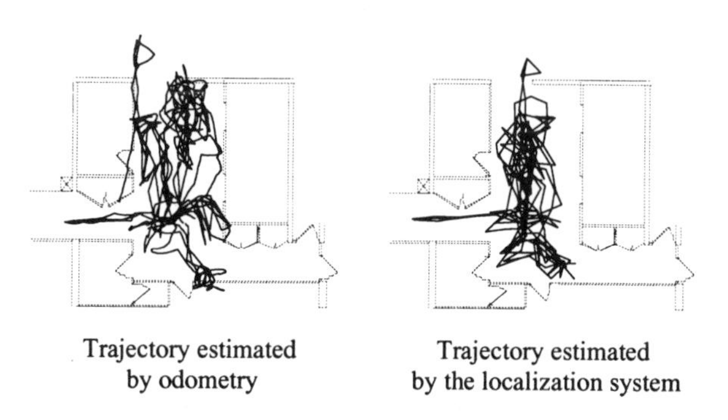

Figure 6: Comparison of two procedures to estimate the robot's trajectory. Left: results obtained with odometry alone. Right: results obtained with the full navigation system.

Moreover, the localization algorithm effectively achieves global localization most of the time. Indeed, it frequently computes the robot's position using node activities instead of using the position-tracking method that is temporarily triggered when the navigation system detects that the robot is outside the mapped area (Figure 7).

We carried out specific experiments to demonstrate this global localization capacity. In particular, we stopped the localization system when the robot was correctly localized at position A and subsequently manually moved it to position B in the environment of Figure 5. The standard localization and exploration process were then resumed without providing the system any cue about this displacement. Figure 8 shows the error in the estimation of the position during the subsequent lo-

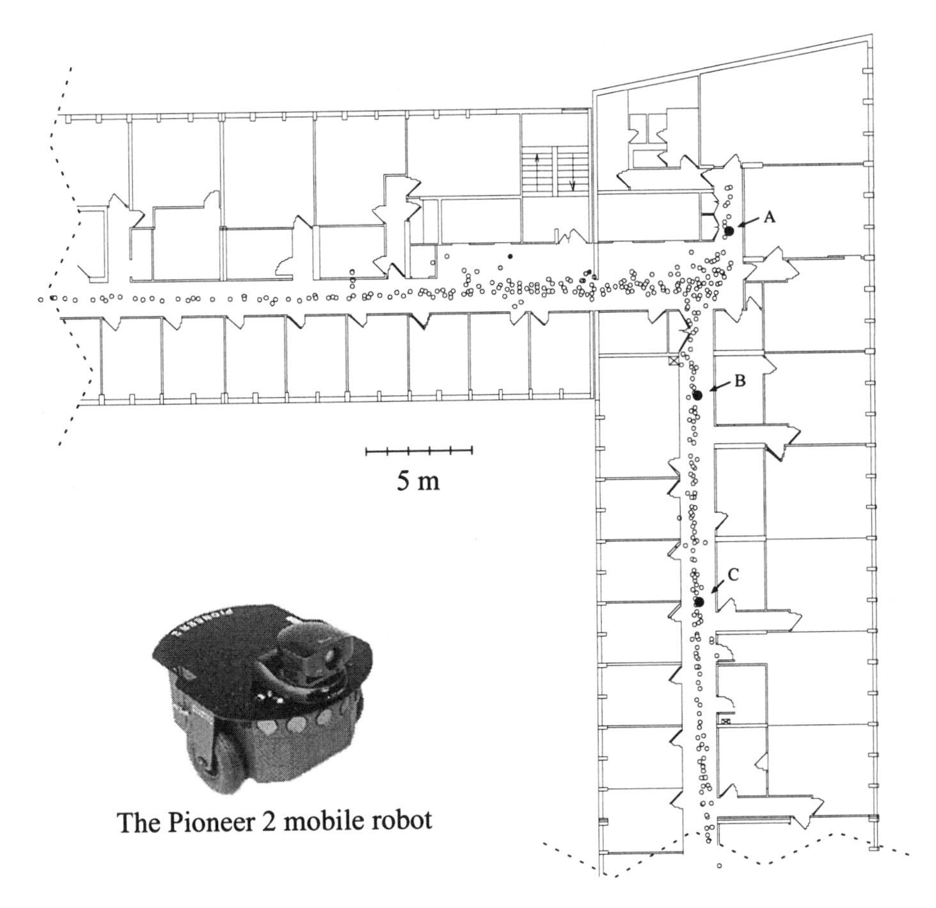

Figure 5: An example of a map created in the corridors of our laboratory. The map is superimposed on an architectural sketch of the environment.

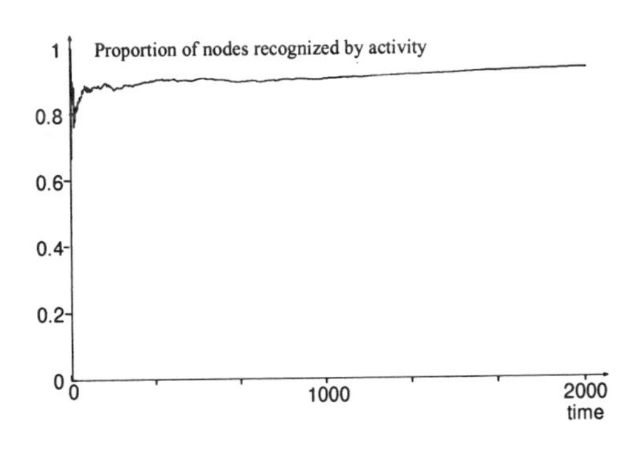

Figure 7: Proportion of the number of nodes that are recognized by the global localization system using node activities relatively to the total number of recognized nodes.

calizations. It thus turns out that the robot succeeds in getting correctly re-localized after 10 time-steps, when the localization error returns to its initial value, typically equivalent to the robot's diameter (50 cm). The large augmentation of the error between the third and seventh time-steps is caused by perceptual aliasing that causes the environment near position B to look very similar to the environment near position C. Consequently, while the robot is effectively positioned near position B, the system wrongly estimates that there is a high probability of its being near position C. Such an incorrect inference gets corrected after 10 time-steps when the robot is far enough from position B for the environment to be sufficiently different from what it looks like near position C.

It is important to note, however, that, contrary to what was demonstrated in simulation in a previous paper (Filliat and Meyer, 2000), such a re-localization capacity may temporarily prove to be inefficient. The main reason is that the real vision system is much noisier than the simulated one, which enhances perceptual aliasing difficulties. As a consequence, information provided by sonar sensors and by idiothetic cues about the structure of the environment is assigned much greater importance in actual case than in simulation. This causes the re-localization procedure to become inefficient on the real robot when, for instance, a wrongly estimated position belongs to the same corridor as the real one. In this case, re-localization is not effective until the robot has entered an open area or a different corridor. Unfortu-

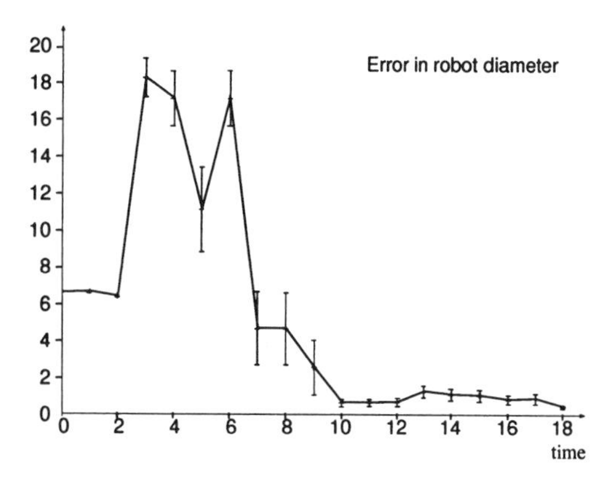

Figure 8: Evolution of the error in the estimation of the position after a passive displacement of the robot from point A to point B (Figure 5).

nately, the exploration strategies employed in the model emphasize strong local exploration in order to avoid localization errors. When such re-localization issues are encountered, local exploration prevents movements that would rapidly lead the robot out of a corridor and that would make prompt re-localization possible.

A solution to this problem would be to implement an active navigation strategy that would guide the robot toward areas where re-localization would be efficient. This suggestion is supported by the fact that, in the current system, manually assigning a goal to the robot when it is temporarily lost entails getting out of the corridor in question and permits a rapid re-localization.

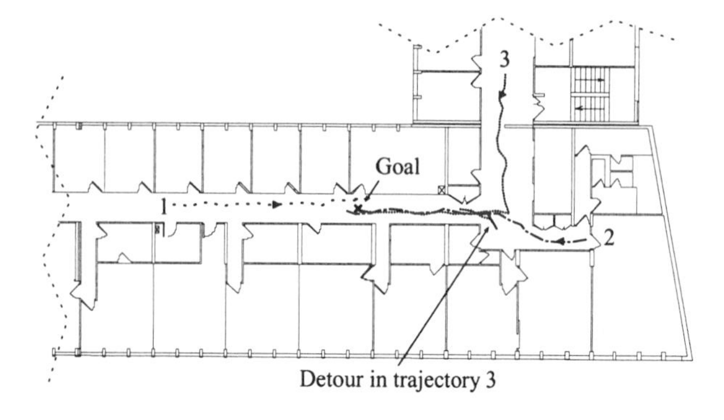

Figure 9: Three examples of goal-directed trajectories starting from three different positions. Trajectories 1 and 2 are direct, while trajectory 3 entails a re-planification leading to a slight detour.

Finally, the model makes it possible to efficiently reach any goal position in the environment. To demonstrate this, we performed ten trials to reach a fixed goal, starting from different positions. Among these trials, one failed due to the robot's getting trapped into a narrow dead-end. The nine other trials were successful, because the robot either directly reached the goal in five trials, or after the use of the above-mentioned re-planning procedure in the last four ones (Figure 9). The mean precision of the final positions in these nine trials was 50 cm, all trials ending less than 80 cm from the goal. These data are representative of the performance obtained with any other goal in the environment.

5 Discussion

The navigation system presented here therefore affords important autonomy capacities to mobile robots by combining global localization with map-learning. Its performances are achieved using relatively simple sensors and without resorting to metrical models for these sensors. The localization precision thus obtained (50 cm) is sufficient for most navigation tasks in common office environments[3]. In cases where it wasn't, the existing procedures could be supplied with additional short-range visual guidance algorithms, as demonstrated by current research efforts (Gourichon and Meyer, 2002).

The absence of any metrical model for the sensors is compensated in our model by the need for an exhaustive exploration of the environment. Indeed, the navigation system strongly relies on careful exploration to avoid localization instabilities during the map-learning process.

The capacities of the system have been demonstrated on a real robot in an environment mostly made up of hallways. Experiments in simulation indicate that navigating in open environments will be possible without any loss of precision using an omnidirectional camera. However, when a directional camera is used, as is the case in this paper, the system could present instabilities in the mapping process due to the higher rate of localization failures caused by the incompleteness of available data. In this case, the structure of the environment provided by the corridor is important, as shown by the mentioned limitations to the re-localization capacity. Further experiments in wider environments and using an omnidirectional camera will be conducted in the context of a new application within the AnimatLab, the Psikharpax project.

Complete autonomy of the system would be achieved if the robot were able to monitor its direction, along with its position. Indeed, the current method - which entails estimating the direction through odometry and periodically correcting the resulting error through an external reference - could be automated if the robot were able to learn how to associate the relative positions of some landmarks with its current orientation. Encouraging results have already been obtained with a preliminary implementation of such a capacity. This implementation entails first detecting colored landmarks from the initial position (using the method described in Gourichon and Meyer, 2002) and memorizing the directions

[3]For example, it allows a door to be reached correctly.

of these landmarks in the first map-node. The robot is then periodically guided by our navigation system toward this initial position where its direction estimate is reset using the perceived direction of these landmarks. Improvements to this scheme should entail memorizing such landmarks in several nodes of the map, so as to be able to reset the direction estimate in several positions and to avoid recurrent visits to the start node.

As mentioned in the previous section, the system could also be improved by the implementation of active navigation strategies to enhance the re-localization capacity. Such strategies could, for example, guide the robot, according to current position hypotheses, toward areas where the positions corresponding to the various hypotheses would be easy to differentiate.

With respect to other navigation models, this one shares several features with the ELAN model presented by Yamauchi and Beer (1996). However, the authors report that the latter model, which was functional in simulation, failed in real robot experiments. We believe that three main differences with respect to ELAN allow our model to work on a real robot and that they are therefore important for robustness :

- the regular correction of the direction by an external procedure that avoids large direction estimation errors and allows meaningful activity estimation,
- the use of vision instead of range sensors to reduce perceptual aliasing,
- the use of a dedicated heuristic to decide when to add a new node to the map.

This third point is particularly interesting, as the use of the heuristic mentioned by Yamauchi and Beer, i.e., a threshold on the most activated node, leads in our model to a severe loss of robustness. Comparisons of our model with other approaches can be found in Filliat and Meyer (2000).

Finally, this model is highly reminiscent of several biologically inspired navigation models described in the literature (Trullier et al., 1997). Indeed, nodes that have been used herein may be viewed as counterparts of *place-cells* found in the hippocampus of the rats. Our approach, however, relies on global localization, while most existing biologically-inspired models (e.g., Balakrishnan et al., 1999; Arleo and Gerstner, 2000) simply call upon single-hypothesis tracking and upon special procedures for the initial estimation of the position. Nevertheless, there are some indications that rats might in fact resort to global localization procedures also. For example, Zemel et al. (1997) describe a method to encode arbitrary probability distributions in the activities of a population of neurons. This technique potentially allows multiple-position hypotheses to be encoded in place-cell activities in a way very similar to what is done in our model. Another paper (Zhang et al., 1998), also demonstrates that deducing the position of a rat in a maze from place-cell recordings is much more precise when a probabilistic framework similar to that underlying this model is used, instead of resorting to a standard method like population vector coding.

In other words, such cues suggest that it might be useful to interpret the functioning of the hippocampus of rats during navigation within a probabilistic framework similar to the one used in this article.

6 Conclusion

The navigation system presented herein allows a high degree of autonomy by integrating global localization and map-learning processes with minimal human intervention. Moreover, this integration has been achieved using simple sensors, without resorting to any metrical sensor model, through the implementation of dedicated heuristics. Its capacities have been demonstrated on a real mobile robot operating in an unmodified office environment. Current research efforts to further enhance the autonomy of the system already provided encouraging results.

There is also good reason to think that the inner workings of the model could bear some resemblance to their biological counterparts found in the rat.

7 Acknowledgements

This work was supported by Robea, an interdisciplinary program of the French Centre National de la Recherche Scientifique.

References

Arleo, A. and Gerstner, W. (2000). Spatial cognition and neuro-mimetic navigation : A model of hippocampal place cell activity. *Biological Cybernetics, Special Issue on Navigation in Biological and Artificial Systems*, 83:287–299.

Balakrishnan, K., Bousquet, O., and Honavar, V. (1999). Spatial learning and localization in rodents : A computation model of the hippocampus and its implications for mobile robots. *Adaptive Behavior*, 7(2):173–216.

Cassandra, A. R., Kaelbling, L. P., and Kurien, J. A. (1996). Acting under uncertainty : Discrete bayesian models for mobile-robot navigation. In *Proceedings of IEEE/RSJ International Conference on Intelligent Robots and Systems.*

Dedeoglu, G., Mataric, M., and Sukhatme, G. S. (1999). Incremental, online topological map building with a mobile robot. In *Proceedings of Mobile Robots XIV - SPIE.*

Duckett, T., Marsland, S., and Shapiro, J. (2000). Learning globally consistent maps by relaxation. In *Proceedings of the International Conference on Robotics and Automation (ICRA'2000).*

Filliat, D. (2001). *Cartographie et estimation globale de la position pour un robot mobile autonome (in french).* PhD thesis, Universitè Pierre et Marie Curie.

Filliat, D. and Meyer, J. A. (2000). Active perception and map-learning for robot navigation. In *From Animals to Animats 6. Proceedings of the Sixth Conferences on Simulation of Adaptive Behavior.* The MIT Press.

Filliat, D. and Meyer, J.-A. (2002). Map-based navigation in mobile robots - I. A review of localization strategies. *Submitted for publication.*

Fox, D., Burgard, W., Thrun, S., and Cremers, A. B. (1998). Position estimation for mobile robots in dynamic environments. In *Proceedings of the Fifteenth National Conference on Articial Intelligence (AAAI-98).*

Franz, M., Scholkopf, B., Georg, P., Mallot, H., and Bulthoff, H. (1998). Learning view graphs for robot navigation. *Autonomous Robots*, 5:111–125.

Gaussier, P., Joulain, C., Banquet, J., Lepretre, S., and Revel, A. (2000). The visual homing problem : An example of robotics/biology cross-fertilisation. *Robotics and autonomous systems*, 30(1-2):155–180.

Gourichon, S. and Meyer, J.-A. (2002). Using coloured snapshots for short-range guidance in mobile robots. *Special Issue on Biologically Inspired Robots - International Journal of Robotics and Automation*, Submitted for publication.

Meyer, J.-A. and Filliat, D. (2002). Map-based navigation in mobile robots - II. A review of map-learning and path-planing strategies. *Submitted for publication.*

Moravec, H. and Elfes, A. (1985). High resolution maps from wide angular sensors. In *Proceedings of the IEEE International Conference On Robotics and Automation (ICRA-85).* IEEE Computer Society Press.

Piasecki, M. (1995). Global localization for mobile robots by multiple hypothesis tracking. *Robotics and Autonomous Systems*, 16:93–104.

Shatkay, H. and Kaelbling, L. P. (1997). Learning topological maps with weak local odometric information. In *Proceedings of the Fifteenth International Joint Conference on Artificial Intelligence.*

Simmons, R. and Koenig, S. (1995). Probabilistic navigation in partially observable environments. In Mellish, S., (Ed.), *Proccedings of IJCAI'95.* Morgan Kaufman Publishing.

Smith, R., Self, M., and Cheeseman, P. (1988). Estimating uncertain spatial relationships in robotics. In Lemmer, J. F. and Kanal, L. N., (Eds.), *Uncertainty in Artificial Intelligence*, pages 435–461. Elsevier.

Thrun, S., Bennewitz, M., Burgard, W., Cremers, A. B., Dellaert, F., Fox, D., Haehnel, D., Rosenberg, C., Roy, N., Schulte, J., and Schulz, D. (1999). MINERVA : A second generation mobile tour-guide robot. In *Proceedings of the IEEE International Conference on Robotics and Automation (ICRA-1999).*

Thrun, S., Burgard, W., and Fox, D. (2000). A real-time algorithm for mobile robot mapping with applications to multi-robot and 3D mapping. In *Proceedings of the IEEE International Conference on Robotics and Automation (ICRA-2000).*

Tolman, E. C. (1948). Cognitive maps in rats and men. *Psychological Review*, 55:189–208.

Trullier, O., Wiener, S., Berthoz, A., and Meyer, J. A. (1997). Biologically-based artificial navigation systems : Review and prospects. *Progress in Neurobiology*, 51:483–544.

Ulrich, I. and Nourbakhsh, I. (2000). Appearance-based place recognition for topological localization. In *Proceedings of the IEEE International Conference on Robotics and Automation (ICRA-2000).*

Yamauchi, B. and Beer, R. (1996). Spatial learning for navigation in dynamic environments. *IEEE Transactions on Systems, Man, and Cybernetics-Part B,Special Issue on Learning Autonomous Robots*, 26(3):496–505.

Yamauchi, B., Schultz, A., and Adams, W. (1999). Integrating exploration and localization for mobile robots. *Adaptive Behavior*, 7(2):217–230.

Zemel, R. S., Dayan, P., and Pouget, A. (1997). Probabilistic interpretation of population codes. In Mozer, M. C., Jordan, M. I., and Petsche, T., (Eds.), *Advances in Neural Information Processing Systems*, volume 9, page 676. The MIT Press.

Zhang, K., Ginzburg, I., McNaughton, B., and Sejnowski, T. (1998). Interpreting neuronal population activity by reconstruction : A unified framework with application to hippocampal place cells. *Journal of Neurophysiology*, 79:1017–1044.

Cortico-Hippocampal Maps and Navigation Strategies in Robots and Rodents

Jean Paul Banquet[+] Philippe Gaussier[++] Mathias Quoy[++]
Arnaud Revel[++] Yves Burnod[+]
+ IMSERM 483 Neuroscience et Modélisation, UPMC
9, quai St Bernard 75252 PARIS, France. banquet@ccr.jussieu.fr
++ CNRS 2235 ETIS-Neurocybernétique UCP/ENSEA,ENSEA,
6, Av. Ponceaux, Cergy-Pontoise, France. gaussier@ensea.fr

Abstract

A biologically inspired integrated model of different hippocampal subsystems makes a distinction between place cells (PC) within entorhinal cortex (diffuse) or dentate gyrus (segregated), and transition cells (TC) in CA3-CA1 that encode transitions between events. These two types of codes support two kinds of hippocampo-cortical cognitive maps: -A context-independent map in subiculum and EC encodes essentially the spatial layout of the environment thanks to a local dominance of ideothetic movement-related information over allothetic (visual) information; -A task-and-temporal-context dependent map based on the TCs in CA3-CA1 allows encoding, in higher order structures, maps as graphs resulting from combination of learned sequences of events. The dominantly spatial and the temporal-task-dependent maps are permanently stored in parietal cortex and prefrontal cortex respectively. On the basis of these two maps two distinct goal-oriented navigation strategies were designed in experimental robotic paradigms: -one based on a (population) vector code of the location-actions pairs to learn and implement to reach the goal; another based on linking TCs together as conditioning chains that will be implemented under the top-down guidance of drives and motivations.

1 Introduction

Submitting large-scale biological models of functionally integrated brain structures to robotic paradigms, navigation in particular, results in mutual benefits for robotics and biology, and imposes also constraints on the modeller. The constraints of the physical laws (temporal and spatial in particular) must be complied, and behavioural and dynamic factors have to be taken into account. Besides a vertical integration between different explanatory levels (behavioral, neurobiological,...), robotic experiments suppose also an horizontal mechanistic integration between the different networks of the architecture of the model. This horizontal integration is rarely achieved in nowadays network modelling characterized by a tendency to build dedicated architectures for specific tasks without caring of their functional integration in a system.

Figure 1: Photo of our six-wheeled Koala robot (size 32 cm x 32 cm) from KTeam SA. The CCD camera is mounted on a servo-motor to control the direction of gaze.

In this work, a mathematical model achieves a coherent synthesis of the neurobiological and behavioural data on hippocampus proper (HS) and the parahippocampal regions of the cortex, and is implemened as a control architecture for an autonomous robot (Fig. 1). In the classical framework of hippocampal function in relation with navigation, we will demonstrate that a robotic implementation of a network model brings the modelling effort a step further in various ways and with several benefits. The animat approach (using autonomous artificial animals) changes the way modelization is classically viewed. Embodied models are more than just another way to represent and/or explain biological and psychological results. The test of the model in a complex physical setup, ruled by the perception-action paradigm, generates new perspectives. Contradictions can be unexpectedly resolved, or conversely may emerge. Model-grounding in real world by compliance to physical laws more than a constraint can become a lever to implement simpler and more efficient solutions. Interdependance between the control system and the agent morphology is revealed. Under the pressure for survival, neurobiological systems have adapted simple fast solutions based on learning, anticipation and planning, at the detriment of purely reactive adaptations limited to learning phases. The adaptation and implementation of these functions can bring solutions to critical issues of robotic control, often in a simpler way than classical algorithms plagued by a combinatorial explosion in the dimensionality of the solution space.

2 An overview of the hippocampal function:goals and hypotheses

Even in rodents, some aspects of hippocampal function cannot be easily characterized as dominantly spatial. Some complex cells respond to trace conditioning or reflect stages of behavioural tasks, as animals occupy multiple locations (Wiener and Korshunov, 1995). Our fundamental hypothesis supposes that navigation and declarative-episodic memory are both grounded on a basic hippocampal processing providing elements for the computation of temporospatial sequences of complex, non-repetitive events. Indeed, procedural learning of repetitive sequences made of simple sensory stimuli and elementary motor responses involve essentially basal ganglia (caudate nucleus in particular) and their cortical projections in supplementary motor area (SMA) (Sakai et al., 1999). Nevertheless, when sequence elements are complex, non reproducible events, as in multimodal temporo-spatial contexts or place learning and recognition, hippocampal contribution becomes essential (Maguire et al., 1997, Ghaem et al., 1997).

This sequencing function itself relies on elementary associative and pattern-encoding-recognition processes in hippocampus and cortex. Then, our first goal is to delineate the complementary contributions of perirhinal (PR), parahippocampal (PH) and entorhinal (EC) cortices versus hippocampal systems such as dentate gyrus (DG), CA3-CA1, and subiculum (SUB), inside a coherent frame of spatial processing. Our second goal is to provide a unified interpretation of behavioral and spatial firing correlates of complex cells. The proposed model elucidates whether hippocampal cells encode place independently of behavioral context, or vice versa by showing that hippocampus (HS) can do both in different neural maps ("universal" versus contextual) elaborated by distinct hippocampal subsystems. Our final goal is to bridge the gap between spatial function and declarative-episodic memory function of HS. Place cell (PC) activity is obtained by generic computations (association and pattern recognition) applicable to other types of information that involve temporo-spatial sequences (episodic memory, language), thus unifying spatial and non-spatial processing.

This work expands previous work on homing (Gaussier and Zrehen, 1995, Gaussier et al., 2000), sequence learning (Banquet et al., 2001, Gaussier et al., 1998, Andry et al., 2001) and planning (Gaussier et al., 2002, Banquet et al., 2001, Quoy et al., 2001) involving robotic experiments and neurobiological modelisation focusing on hippocampo-cortical functions. First, in our model the cortical gateways to hippocampus proper(perirhinal (PR), parahippocampal (PH) and entorhinal (EC) cortices)are not just considered as simple relay stations, but participate to spatial information processing. Second, the types of associations performed in parahippocampal region and in HS proper are contrasted. Association-fusion performed by local recurrent cortical nets of pyramidal cells and/or dendritic matrices are supposed different from association-correlation, preserving the compositionality of the items (Cohen and Eichenbaum, 1993) performed in CA3 associative network. These different properties are related to the distinct architectures of hippocampal and cortical association and recurrent networks. Selective lesions of the cortical-hippocampal subsystems support this distinction between different types of associations (cortical and hippocampal in particular). They point to a hierarchy of associations from simple object-location associations up to place cells and self-location.

At the level of neural populations, these computations provide in our model at least two types of maps coexisting in cortico-hippocampal systems, that can be associated with distinct navigation strategies: -an EC neural population with position-dependent activity computes a context-independent "universal" map as learned "landscapes" of potentials particular to each location; -the CA3 hetero-association and recurrent networks compute transitions between place fields that form the building-blocks of sequences, graphs, and contextual maps. These two types of maps are finally stored in parietal (PT) and prefrontal (PF) cortex, respectively. Both of them are under the influence of the head direction system. This modulation achieves external coherence (put the maps in register with the external world), and also internal coherence (rendering the positional firing independent of the orientation of the agent by putting in register the different views taken from multiple directions). Because the model is implemented as a control system for the navigation of a mobile robot it comprises, beyond the hippocampal system, an intermediate level (basal ganglia) linking hippocampal output to motor response in relation with drives and reinforcement; a cortical prefrontal level stores sequences to form maps and graphs supporting planning in relation with goals and motivations (Banquet et al., 2001, Banquet et al., 2002, Gaussier et al., 2002).

Our model integrates in a coherent way in a single architecture two classes of models representing spatial processing: - population-vector models associating locations to specific movements; - associative models creating graphs as place-field chains. The implementation of place and transition learning as particular cases of association and pattern learning and recognition provides versatility to the system that can also be used for purely temporal sequence learning, and imitation (Gaussier et al., 1998, Banquet et al., 2001).

3 Anatomical and experimental motivations

The navigation paradigm after selective brain lesions reveal the specific contributions of the parahippocampal region and of HS to spatial processing.

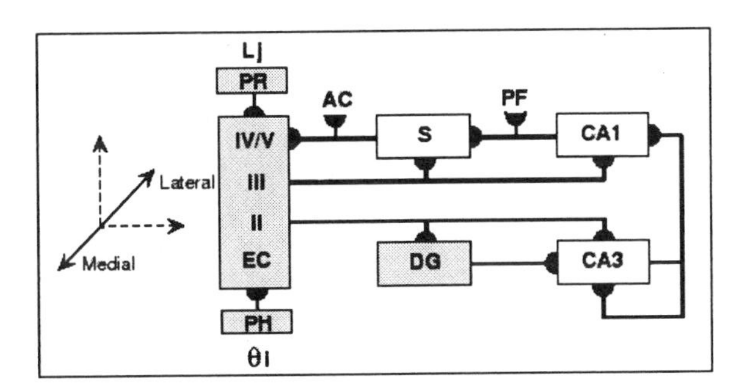

Figure 2: Schematic representation of the hippocampal circuits. Superficial entorhinal cortex (EC) receives information from associative cortical areas. Information from EC layer II is transmitted through the perforant pathway to DG and CA3 pyramidal cells terminal dendrites. CA3 proximal dendrites receive mossy fibers from the DG granules. CA3 pyramidal neurons connect to other CA3 neurons by recurrent collaterals and to CA1 by Schaffer collaterals. Terminal CA1 dendrites receive direct connections from EC layer III. CA1 connects to subiculum (SU) and also directly to deep EC layers and to prefrontal cortex. Subicular connections close both the intrahippocampal and cortico-hippocampal loops.

3.1 Perirhinal-parahippocampal cortices and simple associations

PR and PH, first level in the hierarchy of association networks of the cortico-hippocampal loop constitute convergence zones for neocortex unimodal and polymodal association areas (Fig 2). Anatomical projections still appear segregated at their level. PR (receiving dominantly from inferior temporal lobe area TE) and PH (receiving from posterior parietal cortex and V4) are dedicated respectively to novel item recognition and to spatial arrangement of items. Selective lesions of these structures induce mild deficits qualitatively different from those resulting from hippocampal damage, or even no deficit at all. In contrast with this absence of effect on self-location and navigation, PR-PH removal disrupts the animal's ability to detect when a specific object has changed position(Aggleton and MWBrown, 1999). Accordingly, DMS/DNMS tasks based on object-location associations are affected by these lesions. Thus, PR-PH seems more specifically dedicated to object location (landmark), rather than subject location (place field) in space. PR-PH areas remain cortically-oriented since (visual) stimulus-responsive cells are more frequently encountered there than in EC. They generate a first order association giving rise to a multimodal but still sensory representation. In the model pattern-encoding PR and direction-encoding PH, both converge to form a landmark by a product operation (Pi) on the PR/PH network.

3.2 Entorhinal cortex and place encoding

Subject position codes appear for the first time in EC which performs a second wave of association. Joint activation from PR, PH and other polysensory areas activate EC superficial layers which receive also unidirectional connections from deep EC layers (Fig 2). Superficial layers (II, III) send inputs to the different hippocampal structures. Deep layers send external outputs to cortical areas, and also internal outputs to superficial EC layers. At variance with PR and PH, EC plays an important role not only in object-location, but also in self-location and navigation. Extensive EC lesions: -reduce the fraction of hippocampal cells presenting location specific firing, and the stability of the place fields following maze rotation; -cause spatial deficits comparable to hippocampus-related deficits. Further, Place cell-like activity has been recorded in the superficial and deep layers of the medial EC. Accordingly, in the model, PR-PH implementing intermodal associations is singled out from EC implementing further associations but integrating also hippocampal feedback. This convergence is thought to be essential for a stable and robust spatial coding.

3.3 Hippocampus proper and transition-cells

Well delimited, typical place cells are initially recorded in DG, on the trisynaptic loop. First, probably due to input divergence, DG is the first hippocampal stage where the anatomical segregation still prevailing in EC is lost. Second, the temporal integration of inputs may be performed by excitatory loops between DG granule cells and interneuronal mossy cells. In the model, a temporal delay in DG helps to create event-transitions by rendering two successive events simultaneous in time. Third, EC

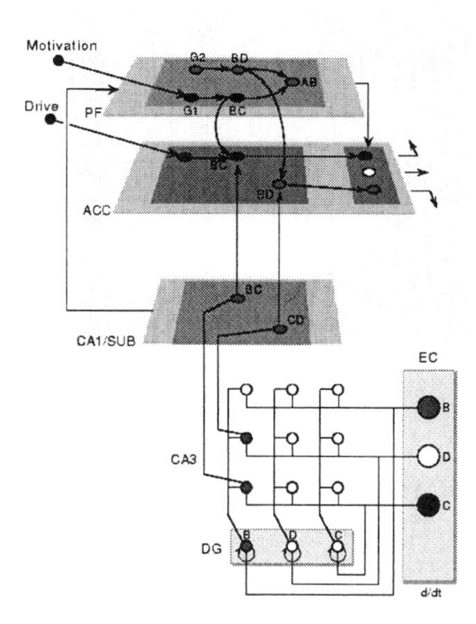

Figure 3: In this architecture for navigation control, DG is characterized by delay neurons (precise spectral timing is not necessary to learn simple sequences). CA3-CA1 learn transitions. Accumbens (ACC) and Prefrontal Cortex (PF) learn, store, and perform temporospatial sequences, as conditioning chains

positional firing cells become place-cells with clearly delimited borders and minimal overlap. The strong inhibitory network and a lack of any significant direct recurrent connectivity seem to be responsible for this "orthogonalization" that provides a uniform partition of space into place-fields. Finally, the strong connections, through the mossy fibers, between DG and CA3 imply that DG is capable (alone or in conjunction with the direct perforant path to CA3) to enforce its activation pattern on the CA3 network.

The third wave of association-pattern recognition taking place in CA3-CA1 computes and learns transitions between events whatever their nature (here places). These transitions are at the basis of sequence learning in higher structures (Banquet et al., 2001). Hippocampal contribution to sequence learning could be essential in several conditions, when complex, non repetitive events have to be integrated in an unified representation. First, event sequences contribute unique *episodes* to our personal history, as illustrated by anterograde amnesia, characterized by an incapacity to acquire such new event sequences. Second, HS could be involved at an initial phase of *procedural* learning, prior to reaching ceiling performance. During conditioning, septal lesions (producing an amnesic syndrome similar to HS lesions) greatly delay (but do not prevent), eyeblink conditioning. HS becomes also necessary if conditioning requires a unified representation of a complex CS set as in contextual conditioning,....Third, during *navigation* in a complex environment visual inputs cooperate with proprioceptive and inertial inputs to learn place field sequences, buiding blocks of graphs and maps. A variable overlap at the border between two neighbour place fields suggests these place field chains. As the animal becomes more familiar with its environment, the overlap increases by retrograde expansion of the upcoming field. We interpret that as earlier anticipation of the upcoming field as learning proceeds. Finally, complex cell firing reflects the *successive stages* of a behavioral task as animals occupy multiple, but physically and functionally similar locations(Wiener and Korshunov, 1995).

4 Neural network model, simulation, and robotic experiments

An integrated architecture featuring the different hippocampal and cortical subsystems served as a control system for robot navigation in open or maze environment.

4.1 EC Place Cells, spatial maps and vector-based navigation

4.1.1 Network input

For a given landmark l, the effect of lateral diffusion on activity of neuron j on the "where" layer was expressed as a non-normalized Gaussian activity profile $\Theta_j = \exp - \frac{((\theta_k^l - \frac{2\pi}{N}j) mod 2\pi)^2}{2\sigma^2}$ where θ_k^l represents the azimuth of the lth landmark and $\frac{2\pi}{N}j$ the preferred direction of neuron j. The influence on Θ_j of the activity related to l^{th} landmark decays exponentially as a function of the angular distance between neuron j preferred direction and the azimuth of the lth landmark. If this difference is nil $\Theta_j = 1$. The diffusion of activation implied that a neuron did not need to be precisely tuned for the direction of a given landmark in order to become active.

4.1.2 Sigma-Pi networks

The core of the system, the two-dimensional merging array PR-PH performed a fusion of "what" and "where" streams by a product between the two inputs (AND operator); a sum of these products was performed at the EC stage. Thus, PR-PH and EC cooperatively computed a typical SigmaPi operation. AND operations in biological networks can be performed by the staged merging of excitatory synapses on dendritic trees. These fusions form rigid monolithic representations lacking the property of compositionality. The dynamical equations of the PR-PH network are

$$\frac{dX_{kl}}{dt} = (B_X - X_{kl}).I_{kl} + \lambda_1.X_{kl} - X_{kl}.\sum_m In_m.W_{m,kl} \quad (1)$$

The *excitatory* component of equation (1) includes:
$=I_{kl}$ a global input to neuron kl detailed below
Activity X_{kl} of a kl neuron in the PR-PH matrix fluctuating between 0 and B_X such that $X_{kl}(t) = [X_{kl}(t-dt) + dX_{kl}]^+$
$[.]^+ = max(.,0)$ denotes a threshold nonlinearity that makes EC activities nonnegative
$=\lambda_1.X_{kl}$ correspond to a positive feedback (Working Memory) on the neural activity allowing the build up of a landmark constellation. The inhibitory term in equation (1) induces a reset of the representation of a learned landmark constellation: $=W_{m,kl}$ represents fixed inhibitory weights between the inhibitory interneuron m and a PR-PH pyramidal cell kl.
$=In_m$ represents the activity of mth inhibitory interneuron triggered by a sensori-motor reset signal at T, 2T, 3T,... nT, where T is a constant period for a visual panoramic exploration of the scenery.
The learning of a new set of landmarks is triggered by a sufficient drift of the input.
$=I_{kl}$ is the global input to neuron kl of the PR-PH matrix, computed as a product.

$$I_{kl} = \left(\sum_{i \in N_{l i}} L_i.W_{i,kl}\right) \cdot \left(\sum_{j \in N_{l j}} \Theta_j.W_{j,kl}\right) \quad (2)$$

$=W_{i,kl}$ ($W_{j,kl}$) are the weights of the connections between any ith landmark (jth azimuth) input to the kl EC neuron. $=L_i$ and Θ_j represent the "What" and "Where" network inputs, respectively. The product therefore instantiates a logical AND operation. Hebbian learning took place between the different systems, but between PH and the PR-PH neurons it was restricted to maximally active input neurons.

4.1.3 Self-organizing cooperative competitive pattern encoding

The emergence of pseudo-place cells in EC is accounted for in the model by a summation. An "OR" operator complements the "AND" operator of the PR-PH network to globally perform as a Sigma-Pi. The *activity* Y of an EC pyramidal neuron coding for places is given by:

$$Y_j = f_{\overline{D_j}}\left(\sum_{kl \in N_{kl}} W_{j,kl}.X_{kl}\right) \quad (3)$$

$f_{\overline{D}}(x)$ represents an exponential output function of EC neuron that performs a learning-dependent tuning to spatial parameters. The output function depends on a tuning factor $\overline{D}$, a vigilance parameter, and a scaling factor. Before learning EC neuron response is weak, with very little specificity. After learning, the response is stronger but only to specific inputs.

A local competition is implemented:

$$Y_j^* = \begin{cases} Y_j & \text{if } Y_j = \max_{i:|i-j|<d_1 max} Y_i \\ 0 & \text{otherwise} \end{cases}$$

$d_1 max$ is a parameter determining the distance on which neurons compete.

4.1.4 Navigation with EC place cells

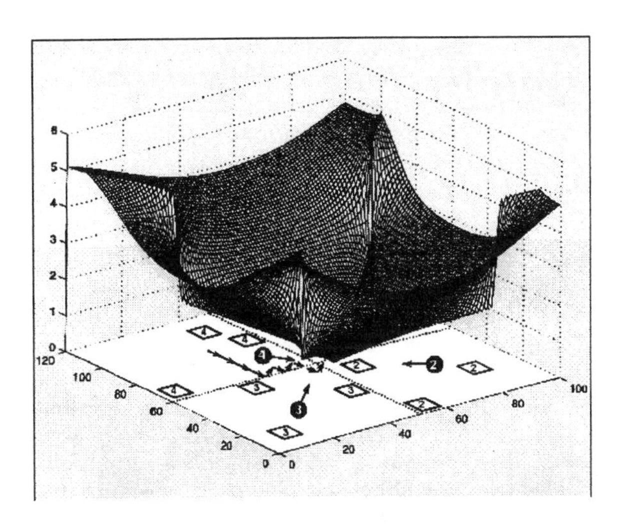

Figure 4: The combination of several overlapping PC activities builds a potential lanscape (not represented on the figure), proper to every location. The x,y plane illustrates the partitioning of space by 4 place cells, the learning of 4 location-movement associations around the goal, and the path used by the robot to reach the goal from an unlearned location, thanks to generalization. Any location in the environment is associated to the closest learned view(s). The trajectory to the goal in this figure comes from a real robot experiment. Therefore the movement (or the weighted combination of movements) associated to these views is performed. This strategy amounts to a gradient descent. The associated theoretical attractor bassin illustrated in the figure results from sensori-motor learning

Spatial maps based on EC diffuse place fields (Fig 5, top) allow navigation to a goal based on the activation of a neural population.That several strategies can be used for navigation results from experimental paradigms using Morris watermaze. Rats with HS lesion are still but only able to use a heading vector from an intraarea landmark to the platform. With improved performance from session to session, this strategy gives an edge to hippocampal rats when platform is moved between sessions. Comparable results confirm that hippocampal rats are not impaired in the Morris watermaze when the platform is visible or a cue points to the platform location. A taxon-like strategy can then be used. Yet, besides this strategy, normal rats also use a supposed cognitive map which provides for a further improvement between trials within the same session.

A population code has been implemented on the basis of the potential map derived from the different levels of coactivation of a neural population of overlapping PCs. Each PC is associated with a directly learned (or secondarily associated by generalization) movement direction that will lead to the goal. Learning a few places around the goal, and the actions to reach the goal from these places, provides for a generalized goal-reaching strategy equivalent to a gradient-descent(Gaussier et al., 2000). In our model, the direction to move can be given by a winner-take-all cell (Fig 4) or by a weighted summation of the directions corresponding to the most active cells. The difference with other population codes comes from the capacity of generalization such that just a few places around the goal need to be learned. Indeed, these strategies usually need to reach the goal many times from different points, in order to build up an attractor. When the goal is moved a new attractor surface must be computed. Nevertheless, the performance of these models improves dramatically when a diffuse (instead of a punctuate) spatial representation like PCs is formed during exploration. Using position-dependent cell activity of the type encountered in EC or subiculum, as in our model, avoids the access problem related to localized information. Further, multiple goals can be accommodated as multiple attractors if planification is not required. Motivation induces a bias in place recognition. From EC, direct and indirect (through DG) inputs reach the CA3-CA1 system through the perforant path.

4.2 Transition Cells, Temporo-Spatial Sequences and Graph-Maps

We assume that transition cells are computed in CA3-CA1 under the control of DG.

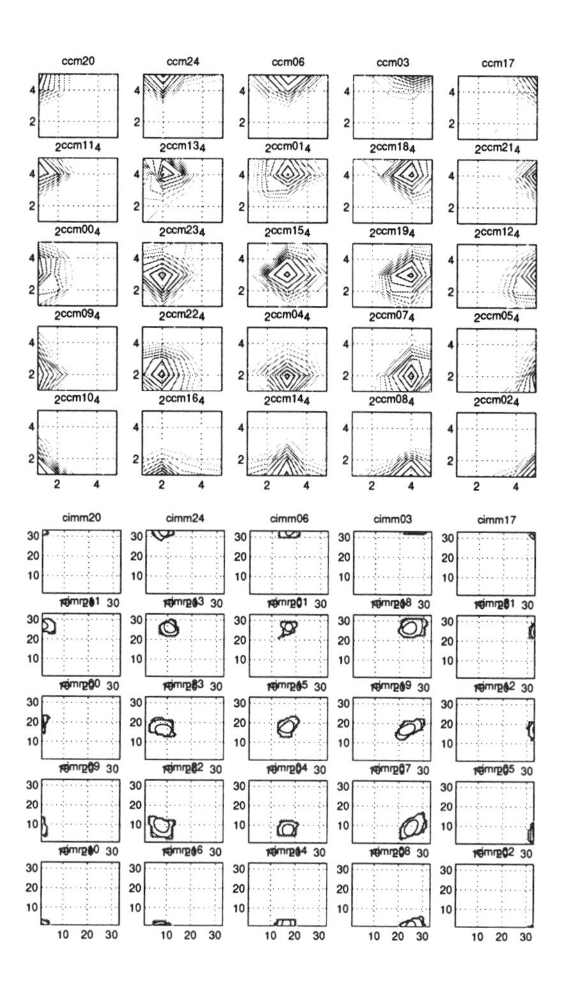

Figure 5: Top: 25 EC neurons, 25 places, 5 measures averaged per place, diffusion 35 degrees. Neurons are considered to be isolated (activation only comes from the direct input). Learning was supervised: each neuron is associated with a particular location in a 5×5 paving of the room in which the experiments have been carried out. Each rectangle is a map of the room. The curves show the activity of the neurons corresponding to the learned locations in the room. Down: same setup as above. A competition between DG neurons is introduced. Only the winner neuron remains activated while the others are set to 0.

4.2.1 DG system

DG was modeled as a self-organizing competitive network performing input orthogonalisation. A strong inhibitory interneuron network, a lack of any direct recurrent connectivity along with a strong input divergence induced input orthogonalization in DG. A strong competition (strict WTA) extended to the entire DG domain implemented this characteristic. Mossy cells as excitatory interneurons could provide a local recurrent activation of granule cells and therefore a WM function.

A simplified granule cell equation is:

$$U_{j,l}(t) = \frac{m_0}{m_j} \exp - \frac{((t - \tau_l) - m_j)^2}{2\sigma_j} \qquad (4)$$

where m_j denotes a particular time constant, with standard deviation σ_j, m_o being the faster referential time-constant; l the label of the recognized input pattern that triggers on its associated time battery, and τ_l the instant of activation of the lth battery. When DG granule cells are implemented as simple delay neurons (instead of spectral timing neurons), that is the case in navigation, m_j, τ_l, σ_j are the same for all cells, and σ_j is large. A strong competition and input convergence reduced the overlap between clearly delimited place fields (Fig 5, down)

4.2.2 CA3-CA1 system and Transition cells

CA3-CA1 learned event-transitions as building-blocks for temporo-spatial sequences. The computation of event-transitions solved the problem of the implementation of a cognitive graph-map in the sensori-motor system by an appropriate movement selection. An internal representation of an event-transition (not just a scene recognition) can only be linked with a single movement. The neural implementation (Fig 3) features a group of CA3 neurons linked (distal dendrites) to the derivative of the current place recognition in EC, and also linked (soma-proximal dendrites) to the memory of the previous input. The activity of transition-prediction neurons, $V_{i,j}$, results from a summation of the activity of the two inputs. Appropriate threshold and initial weights make the activity of a single input unsufficient to trigger the activity. The transition prediction is achieved by reinforcing (e.g. by Hebbian learning) the link between the delayed input and CA3 assembly coding the transition, such that the delayed activity becomes by itself sufficient to activate CA3 neurons. Correct predictions reinforce learning; otherwise weights are depressed, and another (learned or new) node becomes active.

$V_{i,j}$ the activity of CA3 TCs is:

$$V_{i,j} = \sum_j W_{i,j}^{j,l}.U_{j,l} + W_{i,l}^{i}.Y_l \quad (5)$$

$W_{i,j}^{j,l}$ is the strength of the link between the (j,l) DG neuron and the (i,j) CA3 neuron; $U_{j,l}$ is the activity of the j cell of the l DG battery; Y_l is the activity of the EC neuron connected to a (i,j) CA3 neuron, and $W_{i,l}^{i}$ the strength of the link between them. A CA3 neuron fires only when its potential reaches its maximum value(the sign of the derivative of $V_{i,j}$ goes from positive to negative). The firing condition is computed according to Eqs (6).

$$V_{i,j}^{a} = \mathcal{F}(V_{i,j}) \quad (6)$$

$$\mathcal{F}(x(t)) = \begin{cases} 1 & \text{if } \frac{dx(t)}{dt} < 0 \text{ and } \frac{dx(t-1)}{dt} > 0 \\ 0 & \text{otherwise} \end{cases}$$

The weight modification rule between a DG granule cell and a CA3 pyramidal cell is:

$$W_{i,j}^{j,l} = \begin{cases} \frac{U_{j,l}}{\sum_{j,l}(U_{j,l})^2} & \text{if } Y_j^* \neq 0 \\ \text{unmodified} & \text{otherwise} \end{cases}$$

4.2.3 Graph-based Navigation Strategy

The alternative strategy resulted from chaining "transition cells" computed by CA3/CA1 and forming chains in a structure (Accumbens and/or Prefrontal cortex) that stored combination of these chains to form graph-maps

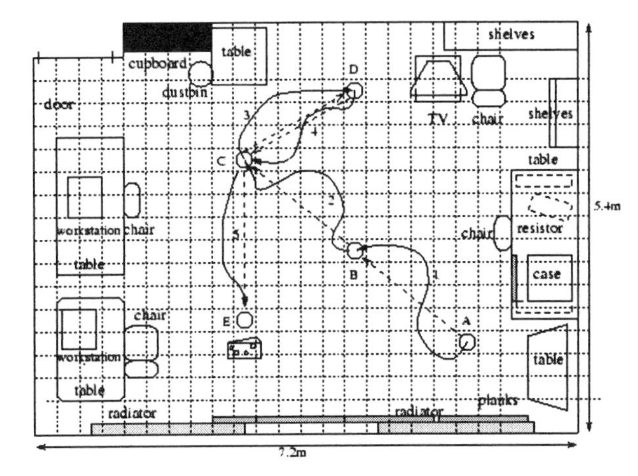

Figure 6: Exploration and latent learning of a graph-map by the robot, in an indoor environment. The curved lines delineate the real robot trajectory. The straight lines represent the ideal trajectories derived from path integration between two adjacent locations, and associated to the transitions between these locations. In spite of the absence of direct connections between nodes coding transition BC and CE (not experienced during exploration), the combination of the transition-prediction system and of top-down motivation allows to select this shortcut.

related to specific goal-locations.During a phase of random exploration (Fig 6) the robot learns in the EC-DG system places A, B, C, D, E.contributing to the EC map. When the level of activation corresponding to an explored or recognized place falls below some threshold, a new PC takes over (Banquet et al., 2000). Simultaneously, in CA3-CA1, new nodes are created for each new transition AB, BC, CD, DC, CE linking nodes as they are encountered by simple Hebbian latent learning independent of any explicit reinforcement. The ideal movement associated to a given transition is computed by path integration. This path integration is reset when the robot reaches a new place. Simple paths can thus be learned and combined to form graph-maps. These maps built by simple latent learning, can be used in a reactive mode if a modulation of the graph is introduced by a reward-based reinforcement (Accumbens). These graph-maps are also necessary for the implementation of a proactive planning mode (prefrontal cortex), based on incentive motivation (Fig 3). The combination of a transition-anticipation system with a motivation-related top-down activation allows not only to invent shortcuts in the graph-map (Fig 7), but also to manage in a consistent way several simultaneously active goals.

The two strategies of navigation presented here are based on two different types of documented PCs. This remark is congruent with the fact that both of the strategies belong to hippocampus-based locale navigation in the taxonomy designed by O'Keefe. The other two types,

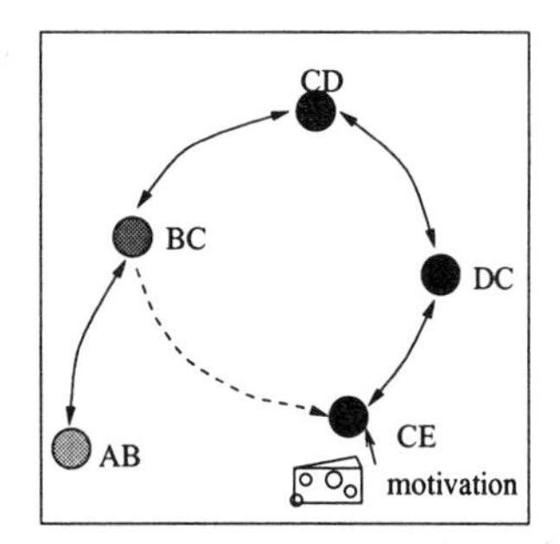

Figure 7: Cognitive map built after the exploration of the environment shown fig. 6. Note that there is no direct connection between the neuron coding the transition BC and the neuron coding the transition CE (because it was not experienced by the robot during the exploration). With the transition prediction system, the robot can nevertheless choose the transition CE instead of CD,after the transition BC because of its opportunistic behavior (recognition of C allows to predict CD and CE, and top-down motivation bias the choice in favor of the shortest path).

taxon and route strategies are less dependent of the integrity of the hippocampal system, and more related to an S-R strategy.

5 Discussion

The model made a distinction between different levels of association. First parahippocampal region, PR-PH and EC performed Sigma-Pi operations (Rumelhart and McClelland, 1986) mimicking cortical local association-fusions to give rise to landmarks (object-location association) that were combined as constellations forming local views characterizing entorhinal PCs. These diffuse PCs (probably in conjunction with path-integration) supported a context-independent "universal" map formed in SU-EC and supposedly stored in parietal cortex. Second, widespread flexible CA3 hetero-associations, preserving the compositionality of the items (Cohen and Eichenbaum, 1993)were used to link well-delimited DG placefields into event transitions and sequences forming graphs and context-dependent maps that could be stored or at least reactivated in prefrontal cortex. At each level, decoding these associations implied specific pattern recognition stages in EC, DG, CA1/SUB selforganizing competitive networks. The model suggest that local associations support context-independent maps, and global associations context-dependent ones.

On the basis of these two kinds of associations, two different maps context-independent and context-dependent, are hypothesized, respectively in EC-SU and CA3-CA1: -first, the scope and range of cortical fusion-associations is too limited to form comprehensive contexts without the hippocampus mediation; -second, the convergence in SU and/or EC (Redish and Touretzky, 1997) of genuinely spatial, movement-generated informations (path-integration) could gate the activation of a PC population encoding a unique location corresponding to several local views or even task contexts(Frank et al., 2000). A strong positional signal could dominate the weak directional information provided by panoramic views.

In the robotic experiment and in the simulation learning and performance were not distinct stages. A vigilance parameter active during both stages decided after a quantum motion whether the new panorama was sufficiently different to be learned. Thus, the mismatch-dependent decrement of PCs activity combined to the tuning of the output function induced the transition from one place code to the next. This feature of the model is suggested by acethyl-choline/GABA septal modulation of HS, that favors the encoding of novel events (rather than the expression of previously learned patterns) when a mismatch-dependent low level activity triggers a selective ACh depression of synaptic expression in hippocampal subsystems (Hasselmo et al., 1995).

The inputs to the model represented a "what" pattern temporal stream, and a "where" directional parietal stream. This last one combined object-related directional visual information with subject-related (movement-dependent) head-body axis of the robot referred to a stable external reference. The relation of the "where" azimuth information to an external reference represented in the model the contribution of the head direction system to put the maps in register with the distant environment and make them independent of the robot orientation. An extension of the model incorporated path integration (Babeau et al., 2000). The model assumed that there was no crucial need of a precise evaluation of distance to distant landmarks in order to compute the subject's precise location. Landmarks' bearing with respect to an external reference, combined with pattern recognition, was enough. Conversely, an evaluation of the distance as well as azimuth of the boundaries of the arena could be important for the computation of the shape and size of the place fields (Hartley et al., 2000).

On the basis of this information a planification based locale navigation was implemented (Banquet et al., 2001, Gaussier et al., 2002), as opposed to the *gradient-descent* navigation implemented by the PC population in EC. Furthermore, these basic operations of the model could be adapted to any type of information in different formal spaces (e.g., word list learning). Thus, the model contributes to bridge the gap between the two main theories of hippocampal function, cognitive mapping versus episodic memory.

The situations where the goal, or a cue directly related to the goal, are in view do not require the hippocampus, but need the contribution of parahippocampal region. In this view, hippocampal models of navigation can be divided in two general classes. A first class uses associatives capacities of the CA3 network to create chains of events(McNaughton, 1989), maps or graphs (Muller et al., 1991, Trullier and Meyer, 2000), or attractor networks(Samsonovich and McNaughton, 1997). Another class of models maps space as potential fields (Dayan and Sejnowski, 1994), vectors, or vector fields (O'Keefe, 1991, Burgess et al., 1994, McNaughton et al., 1996). Each of these approaches seems to refer to a particular aspect of hippocampal processing. Our tryptic model implements, through different stages, the two types of approaches.

The model makes several original contributions. First, it takes into account and gives a functional significance to the existence of at least two types of PCs, diffuse EC place cells which adapt to task and geometrical context rather than change their code, and DG-CA3-CA1 well-delimited PCs that depend on task context, and therefore change their code if context is changed. Both types of PCs seem to be under the control of head direction cells, since a rotation of the landmarks induces a commensurate rotation of the place fields. The EC PCs would encode the spatial layout of the environment independently of task and context constraints, on the basis of purely spatial dominant movement-related information used for path-integration. Conversely, CA3-CA1 transition cells would encode temporo-spatial sequences dependent on the task context, in particular. In this contextual encoding the temporal, or at least sequential aspect of learning during task performance would prevail on the purely spatial aspect. Both types of maps are complementary. Distinct navigation strategies in order to capture a goal have been associated to each of them. We make the distinction between reactive (reflex or conditional) and proactive (planning) modes of implementation of the maps. The context-dependent map can be used in both modes, but seems particularly appropriate for the planning mode.

Second, the encoding of transitions in CA3-CA1 (instead of simple locations as in other models) allows an unambiguous association of a single movement to any situation since path-integration computes the ideal trajectory between two locations, whatever the exploration path taken between them. This is not the case when simple locations are associated with many possible displacements. Besides a local desambiguation of trajectories at strategic locations, a straightforward link is establised between sensori-motor associations, and performance of action sequences as in planning.

Third, these high-level functional properties derive from the basic distinction between the local associations performed by the different cortices, with a limited (even if increasing with the hierarchies of associations) scope, and the global all-inclusive associations performed by the CA3 system, which by the sole criteria of processing complexity and integration, could well be located at the top of the processing hierarchy.

Fourth, the submission of the model to the test of a robotic paradigm of navigation has made possible the straightforward integration of originally two distinct models of hippocampal function, namely PC computation, and timing-sequence learning. The biological model has provided the idea of transition codes, and thanks to learning, avoids the combinatorial explosion of the codes associated to a sigma-pi operation, making it plausible, and useful for robotic application.

Finally, the functional integration of the system illustrates two dimensions of neural organisation, relevant for robots and humans as well. These dimensions account respectively for the nature of the functions performed (e.g. spatio-temporal processing and navigation, linguitic processing, episodic learning,...) and the level of performance (reflex, conditioning, planning). The first, horizontal dimension specifies according to the combined perceptuo-motor modalities, the nature of the parallel processes performed, independently of their performance level. Each particular combination of modalities specifies a function. The second vertical dimension specifies the performance level (reflex, conditioned behaviour, planned controlled behaviour) independently of the nature of the function performed. This second sequencial or rather iterative aspect unfolds in cortico-subcortical loops characterized by the dual process of convergence-contraction and divergence-expansion of information.

References

Aggleton, J. and MWBrown (1999). Episodic memory, amnesia, and the hippocampal-anterior thalamic axis. *Behav. Brain Sci.*, 22:425–489.

Andry, P., Gaussier, P., Moga, S., Banquet, J., and Nadel, J. (2001). Learning and communication in imitation: An autonomous robot perspective. *IEEE transactions on Systems, Man and Cybernetics, Part A*, 31(5):431–444.

Babeau, V., P.Gaussier, C.Joulain, A.Revel, and Banquet, J. (2000). Merging visual place recognition and path integration for cognitive map learning. In *SAB2000*, pages 101–110, Paris France.

Banquet, J., Gaussier, P., Quoy, M., and Revel, A. (2002). From reflex to planning: Multimodal, versatile, complex systems in biorobotics. *Behavioral and Brain Science*, in press.

Banquet, J., Gaussier, P., Revel, A., and Burnod, Y. (2000). A neural network model of entorhinal and hippocampal place cells in rodents and primates. In

International Joint Conference on Neural Networks, Como, Italy.

Banquet, J., Gaussier, P., Revel, A., Moga, S., and Burnod, Y. (2001). Sequence learning and timing in hippocampus, prefrontal cortex and accumbens. In *International Joint Conference on Neural Networks*, pages 1053–1058, Washington, DC.

Burgess, N., Recce, M., and O'Keefe, J. (1994). A model of hippocampal function. *Neural Networks*, 7(6/7):1065–1081.

Cohen, N. and Eichenbaum, H. (1993). *Memory, amnesia, and the hippocampal system.* MIT Press, Cambridge, MA.

Dayan, P. and Sejnowski, T. (1994). *td*(λ) converges with probability 1. *Machine Learning*, 14(3):295–301.

Frank, L., Brown, E., and Wilson, M. (2000). Trajectory encoding in the hippocampus and entorhinal cortex. *Neuron*, 27:169–78.

Gaussier, P., Joulain, C., Banquet, J., Leprêtre, S., and Revel, A. (2000). The visual homing problem: An example of robotics/biology cross fertilization. *Robotics and Autonomous Systems*, 30:155–180.

Gaussier, P., Moga, S., Banquet, J., and Quoy, M. (1998). From perception-action loops to imitation processes: A bottom-up approach of learning by imitation. *Applied Artificial Intelligence*, 12(7-8):701–727.

Gaussier, P., Revel, A., Banquet, J., and V.Babeau (2002). Fom view cells and place cells to cognitive maps: processing stages of the hippocampal system. *Biol. Cybern.*, 86:15–28.

Gaussier, P. and Zrehen, S. (1995). Perac: A neural architecture to control artificial animals. *Robotics and Autonomous System*, 16(2-4):291–320.

Ghaem, O., E.Mellet, Crivello, F., Tzourio, N., Mazoyer, B., Berthoz, A., and Denis, M. (1997). Mental navigation along memorized routes activates the hippocampus, precuneus, and insula. *Neuroreport*, 8(3):739–44.

Hartley, T., Burgess, N., Lever, C., Cacucci, F., and O'Keefe, J. (2000). Modeling place fields in terms of the cortical inputs to the hippocampus. *Hippocampus*, 10(4):369–79.

Hasselmo, M., Schnell, E., Berke, J., and Barkai, E. (1995). A model of the hippocampus combibing self-organization and associative memory function. In Tesauro, G., Touretzky, D., and Leen, T., (Eds.), *Advances in Neural Information Processing Systems*, volume 7, pages 77–84, Washington DC. The MIT Press.

Maguire, E., Frackowiak, R., and Frith, C. (1997). Recalling routes around london: activation of the right hippocampus in taxi drivers. *Journal of Neuroscience*, 17(18):7103–10.

McNaughton, B. (1989). Neuronal mechanisms for spatial computation and information storage. In L.Nadel, Cooper, L., Culicover, P., and Harnish, R., (Eds.), *Neural connections and Mental Computations*, pages 285–349, Cambridge, MA. MIT Press.

McNaughton, B., Barnes, C., Gerrard, J., Gothard, K., Knierim, M. J. J., Kudrimoti, H., Qin, Y., Skagges, W., Suster, M., and Weaver, K. (1996). Deciphering the hippocampal polyglot: the hippocampus as a path integration system. *Journal of Experimental Biology*, 199:173–185.

Muller, R., Kubie, J., and Saypoff, R. (1991). The hippocampus as a cognitive graph. *Hippocampus*, 1(3):243–246.

O'Keefe, J. (1991). The hippocampal cognitive map and navigational strategies. In Paillard, J., (Ed.), *Brain and Space*, pages 273–295, Oxford. Oxford University Press.

Quoy, M., Banquet, J., and Dauce, E. (2001). Sequence learning and control with chaos:from biology to robotics. *Behavioral and Brain Science*, 24(5).

Redish, A. and Touretzky, D. (1997). Cognitive maps beyond the hippocampus. *Hippocampus*, 7:15–35.

Rumelhart, D. and McClelland (1986). *Parallel Distributed Processing.* MIT Press, Cambridge.

Sakai, K., Hikosaka, O., Miyauchi, S., Sasaki, Y., Fujimaki, N., and Pütz, B. (1999). Presupplementary motor area activation during sequence learning reflects visuo-motor association. *J Neurosci:RCI*, 19(10).

Samsonovich, A. and McNaughton, B. (1997). Path integration and cognitive mapping in a continuous attractor neural network model. *Journal of Neuroscience*, 17(15):5900–20.

Trullier, O. and Meyer, J. (2000). Animat navigation using a cognitive graph. *Biol Cybern*, 83:271–85.

Wiener, S. and Korshunov, V. (1995). Place-independent behavioural correlates of hippocampal neuro nes in rats. *Neuroreport*, 7(1):183–8.

Learning to Autonomously Select Landmarks for Navigation and Communication

Jason Fleischer **Stephen Marsland***
Dept. of Computer Science
University of Manchester
Manchester M13 9PL, UK
{jfleischer,smarsland}@cs.man.ac.uk

Abstract

Selecting landmarks for use by a navigating mobile robot is important for map-building systems. However, it can also provide a way by which robots can communicate route information, so that one robot can tell another how to find a goal location. A route through an environment can be described by the landmarks encountered along the path, and a robot following the same path must identify the perceptions corresponding to the actual landmarks in the description in order to localise itself. This paper presents an algorithm to automatically select landmarks, choosing as landmarks places that do not fit into a model of typical perceptions acquired by the robot. Four methods of aligning the landmarks between different runs on the same route are also presented. The different alignment methods are evaluated according to both how well they produce matching landmarks and how suitable such alignment methods would be for use in a route communication system.

1 Introduction

Perceptual landmarks – navigational landmarks based on the sensory perceptions of the robot – form the basis of many successful mobile robot navigation systems, see for example (Yamauchi and Beer, 1996, Duckett and Nehmzow, 1998). These landmarks are used because they avoid the problem of drift error that is inherent in odometry measurements.

The selection of landmarks is important for other applications, too. The problem of robot communication has become increasingly important over the last few years, as multiple robot systems have become more and more common. One question that has begun to be addressed is how a group of robots can share navigational information, for example so that one robot can tell another how to find a goal location. A possible way to approach this question is to send information about the landmarks that are seen along the route and the actions that were taken at these landmarks. Such a system would require the following capabilities:

- A landmark selection method
- A mapping from selected landmarks to words to be communicated
- An inverse mapping from words to landmarks

It is the first of these requirements that is addressed in this paper. Wherever navigational landmarks are used it is important that they are detected consistently, so that the same landmarks are found on every run through an environment. Otherwise, the route that should be taken is unclear and, for communication, the physical grounding of the symbols that the robots were communicating with would be very different.

In previous work (Marsland et al., 2001) it was shown that a robot could adaptively learn a model of its environment by using its current sensory perceptions to predict the next set of perceptions. It was argued that the places where the prediction was not accurate – those places that were not adequately described in the acquired model – were suitable choices for landmarks. These places were found by using a Kalman filter to monitor the error curve of the output of the neural network.

In that work, which is described in more detail in section 3.1, the robot travelled a constant distance (20 cm) between sensor scans of the environment, and stopped in order to make the scans. This meant that interesting landmarks could be missed if they were placed between the sampling steps. Here we investigate a more complex version of that task, where the robot samples the environment continually as it explores, even while turning corners. This means that the only limit on the number of samples taken by the robot as it travels is the operating capabilities of the robot.

We compare the quality of landmark matches between successive runs in an environment by looking at how well

*Authors are given in alphabetical order, both being principal authors

the landmarks match in terms of preserving landmark sequence, distance, and category between the two runs. The results are promising.

2 Relevant Literature

2.1 Landmark Selection

The most common method of selecting landmarks has been to take sensory scans at regular intervals with every scan being used as a landmark that is put into the map. This avoids the problem of ensuring that landmarks should be detected consistently, but since in many environments – such as corridors – a large number of perceptions are almost identical, the map is filled up with information that does not aid navigation.

A number of researchers have considered selecting suitable landmarks for mobile robot navigation. One technique is to ask the user to define the landmarks before the robot explores. Humans typically select objects that they find easy to recognise, such as doors, or line segments extracted from camera images recorded as the robot travels (Kortenkamp and Weymouth, 1994). However, there is no guarantee that these objects are easy for the robot to recognise.

(Thrun, 1998) approached the problem through Bayesian learning, aiming to select an optimal set of landmarks for performing self-localisation in one specific environment. The landmarks were made up of a projection from the robot's raw sensory perceptions (camera images) onto vectors in a lower-dimensional space. This optimisation was performed by minimising directly the quantity of interest, namely the robot's error in self-localisation. Thrun showed that his method produced better performance than localisation using designer-determined landmarks including doors and ceiling lights.

A similar, but computationally cheaper technique, was developed by (Vlassis et al., 2000), who showed that their optimisation method produced better results than principal component analysis. Our approach differs in that we do not carry out any analysis of the utility of the landmarks selected, but instead use a self-acquired model of 'typical' sequences of perceptions, which is independent of any particular task or environment.

(Zimmer, 1996) considered the problem of selecting landmarks in a topological map through a process of 'life-long learning', where the robot's map was continuously adapted on-line during exploration. This approach used global statistical information, based on comparison of accumulated error statistics at each of the nodes, to decide where to add and delete nodes in the map. A related idea can be found in (Bourque and Dudek, 2000), who addressed the 'vacation snapshot' problem of deciding in which locations to take camera images in order to obtain a set of images that best represent an entire environment. This approach kept running statistics on what is a 'typical' perception, together with backtracking to previously visited locations that were subsequently found to be 'atypical'.

2.2 Landmark Communication

In the route communication task, landmarks can be considered as categories that the robot recognises and can communicate about. The problem of communicating landmarks between robots then becomes one of learning a mapping between the different categorisations that are privately held by each robot in the communicating community. Although it is not difficult for agents to communicate once they posses the same categorisations of the environment, the methods by which agents can acquire matching categorisations and learn to communicate them is still the topic of ongoing research. The most common method of learning to communicate categorisations is to assume that categorisation is a process that is private to each individual agent, and communication consists of learning a public code, or set of symbols, that maps meanings between the internal categories of each agent. Early work by (Yanco and Stein, 1993) showed how two robots could use reinforcement learning to create an encoding that enables the robots to communicate the simple categories 'turn left' and 'turn right'. (Billard and Dautenhahn, 2000) demonstrated that imitation is a plausible mechanism for robots to learn a common symbol-meaning system describing various perceptions in the environment. (Steels, 1996) proposed a formalism to both learn perceptually grounded meanings of such encodings and a means by which agents can modify their internal categorisations to increase their success rate in communication. An alternative approach is to produce a non-adaptive system that maps categories onto symbols. For instance, (Skubic et al., 2001) showed how a set of fuzzy logic spatial relationships can be combined with a sensor prototyping method called the histogram of forces to generate linguistic descriptions of a robot's environment that are understandable to human users.

The problem of learning symbols to describe a robot's route through an environment has not, to our knowledge, yet been addressed in the literature. The closest related work is that done on segmenting a robot's sensory flow into categories by (Tani and Nolfi, 1999) and (Linaker and Niklasson, 2000). Both of these works mention the possibility of communicating routes between robots as sequences of categories, but do not actually attempt the task. Our own work (Fleischer and Nehmzow, 2001) has demonstrated that it is possible for two robots to learn individual categorisations of locations in the environment and then an encoding from their categorisations onto a set of public symbols, enabling the robots to agree on symbols representing locations in the environment.

3 Approach

3.1 *Overview*

We suggest that the perceptions that are most suitable as landmarks are those that differ in some way from the main run of perceptions. This means that we are attempting a form of novelty detection, highlighting perceptions that are in some way unusual. The first part of any novelty detection process (Marsland, 2001) is to learn the model of normality. This is done by allowing the robot to travel through the environment using a wall-following behaviour and collecting data from its sonar sensors. These readings are used as inputs to train a single-layer neural network, described in section 3.2. Landmark detection is performed by finding peaks in the error curve of the single-layer network. This is a fairly difficult problem as the residual noise level is high. A one-dimensional Kalman filter is used, as is described in section 3.3.

3.2 *The Sensor Prediction Network*

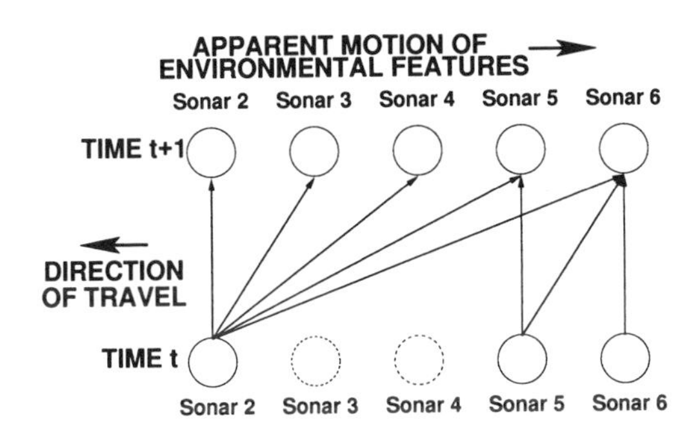

Figure 1: The single-layer network used to learn the mapping between the current sensory perception and the next. To aid clarity, connections for nodes drawn dotted are not shown.

A single-layer neural network with a sigmoidal activation function, shown in figure 1, is trained to acquire a model of the relationship between successive sensory perceptions. The network structure is not fully connected; the perceptions of sensors facing forwards are used as inputs to model rear-mounted sensors, but not vice-versa. This reflects the fact that the robot drives forward, and so the future perceptions of any one sensor depend only on the sensors in front of it. While this assumption about the sensors breaks down when the robot turns corners, this does not seem to be a problem. In addition to the network inputs shown in the figure, a bias input that is permanently set to -1 is used.

The network weights, W, are adapted during a training phase, where the robot explores an environment and records its sensory perceptions. The sensory perceptions at each time step, $p(t)$, are used as both network inputs (at the current time step) and training examples for the next time step. The weights are adapted using the standard Widrow-Hoff learning rule:

$$\Delta W(t) = \eta \left(p(t) - W(t)p(t-1)\right)^{\mathrm{T}} p(t-1)\,. \quad (1)$$

The single-layer network was trained using early stopping on a training set made of runs through a number of corridor environments, using a learning rate of $\eta = 0.2$. Further details are given in section 4.

Once the training is completed the landmark selector can be used on-line. The error curve of the neural network is monitored, where the error is the squared difference between the predicted output, $o_k(t)$, and the actual perceptions of the robot, $p_k(t)$, for each sensor k:

$$E = \sum_k \left(o_k(t) - p_k(t)\right)^2. \quad (2)$$

3.3 *Landmark Selection Using a Kalman Filter*

The criteria that is used for selecting a landmark is that the output error of the sensor prediction network is sufficiently high. A Kalman filter (Kalman, 1960) provides a principled method for determining when this is true. The Kalman filter is a method for recursively estimating the state of a discrete-time controlled process that is governed by a linear stochastic difference equation. In our case we have a scalar variable, the error of the network, E, which would be a constant value, hopefully near zero, if the sensor values remained unchanging. However, there is noise in E from both sensor repeatability (measurement noise) and robot orientation and position (process noise).

The Kalman filter attempts to optimally re-estimate the variable E at each step along the way so at to remove the effects of both measurement and process noise, leaving behind only the prediction error actually produced by variation of the environment. The filter equations compute a gain, K, that is used to recursively update estimates of the true error, $\widehat{E}$, and its variance, v. The Kalman filter equations for our simple, one-dimensional case are:

$$K(t) = \frac{v(t-1)}{v(t-1)+R}, \quad (3)$$

$$\widehat{E}(t) = \widehat{E}(t-1) + K(t)\left(E(t) - \widehat{E}(t-1)\right), \quad (4)$$

$$v(t) = (1-K)\left(v(t-1)+Q\right), \quad (5)$$

where Q is the process noise variance and R is the measurement noise variance. The value of Q is estimated as

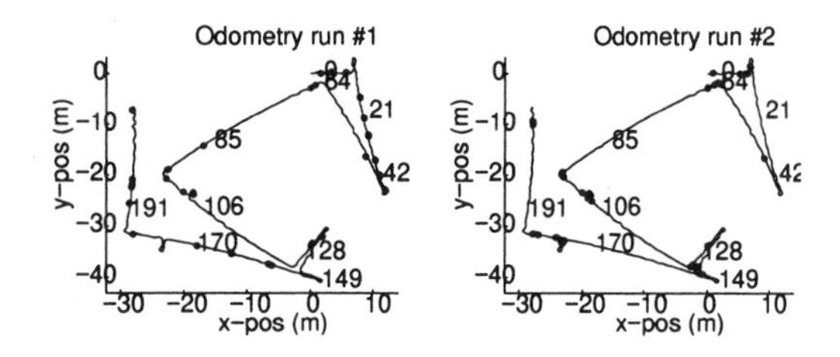

Figure 2: The difference between a constant valued measurement noise variance (run 2, right) and the time-varying function of variance calculated via the Delta method (run 1, left). The line shows the robot path as measured by the robot odometry, and the points are the places that were selected as landmarks. The numbers refer to their distance from the start of the run. The time-varying noise model shows only 74 landmarks while the constant model finds 181 using the same data. The mean distance between landmarks in the time-varying model was 2.86 m but only 1.16 m in the constant one.

the variance of $\widehat{E}$ during training, and R can either be assigned to be the same value, set to 0, or approximated by measuring the variance of the sonar sensors as the robot travels in a short, perfectly straight stretch of featureless hallway, and propagating this variance through the nonlinear transfer functions of the neural network and sum-of-squares error using the Delta Method (Rice, 1994) to generate a time-varying noise model. Figure 2 shows the effects that this choice makes on a typical set of sonar data.

The time-varying noise model produces fewer, more distinct landmarks than either $R = 0$ or $R = Q$ noise models. Most of the extra landmarks found by the constant-valued noise model come from many successive perceptions at the same location being counted as landmarks, whereas the time-varying model often prevents this problem from occurring. We used the time-varying noise model in all of the experiments reported here.

As well as trying to remove noise from the error estimate, the Kalman filter also conveniently maintains an estimate of the error variance. This variance estimate can be used to determine if $\widehat{E}(t)$ is greater then some number n of standard deviations away from the mean of the Kalman estimated error at the current time, $\bar{E}(t)$. Therefore, we can define a landmark as any perception where,

$$\widehat{E}(t) > \bar{E}(t) + n\sqrt{v(t)}\,. \quad (6)$$

The parameter n provides a method for adjusting the required level of conspicuousness and therefore relative frequency of landmarks. Typically, we find that values of n between four and five work well.

3.4 Categorising Landmarks

Two of the four methods of aligning landmarks between different runs that we used employed landmark categories. Landmark categories are also of more general interest because they can form the basis for communication about landmarks between different robots (Fleischer and Nehmzow, 2001). We used Kohonen's Self-Organising Map (Kohonen, 1982), an unsupervised neural network based on vector quantisation, to perform categorisation of the landmarks. The Self-Organising Map (SOM) is a neural network that maps a set of inputs onto activations of a set of output nodes arranged on a lattice. The network categorises the robot's sensor perceptions by feeding them as inputs into the network and taking the identity of the output node with the highest activation level (the 'best-matching node'), as the category of the input. The SOM is trained by presenting it with a data set and adapting the network weights such that, for each input in the data set, the region in the output map around the best-matching node is moved closer to the presented input.

We used the SOM Toolbox (Vesanto et al., 2000) to produce the SOMs used in this work. The toolbox will automatically select the size and shape of the map and the training parameters based on the number of inputs, the number of data points, and the principal components of the training data set. This feature was employed here. For the training data sets used in this work, two-dimensional, toroidal SOMs with between 24 and 66 output nodes were produced, with one dimension of the map being roughly twice as long as the other.

4 Experiments

4.1 Description

The experiments in this paper were performed on a Nomad Scout, a differential drive robot with 16 Polaroid sonar sensors capable of giving range information on objects between 15 cm and 6 m away from the robot. The sensor values were updated as quickly as the processing speed of the PC controller would allow, giving less temporal structure to the data than there would be if the sensors were only updated at fixed distance or time intervals. Each experiment run consisted of between 30 and 200 metres of travel in normal office building corridors that have not been modified for the robot's use in any way. These runs were made during daytime hours with normal use being made of the corridors; no attempt was made to remove the anomalies in the data created by people walking through the sensor range, as these would be the conditions in which the system would have to work as part of a navigation or communication system.

The robot used a hard-wired wall-following program to follow the left-hand side wall at a constant distance.

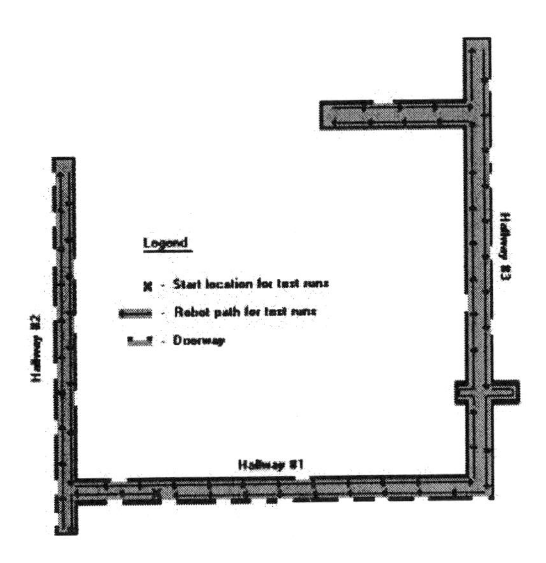

Figure 3: The hallway environment. The robot was trained in either any one of these hallways or in all three. During testing runs the robot travelled through all three hallways. Each run consists of a complete circuit of the hallway(s); the robot executed a left-hand side wall-following program until it arrived back at the location it had started from.

Noise in the system comes principally from inaccuracies in the wall-follower and error in the sonar sensors.

Three different, adjacent, hallway sections were used for training the landmark detection algorithm (see figure 3). The robot travelled through each hallway three times, thus producing a total of nine different data sets for training the algorithms. The training data were used to do three things: (1) train the sensor prediction model, (2) calculate values of Q and R for the Kalman filter and (3) train the SOM for landmark categorisation. The testing data was produced by propping open the doors between the three hallways and collecting a further three data runs using a 200 m long route through all three hallways. To compare the alignment between landmarks the three test sets were paired in all six possible permutations (since order of comparison matters) and the measures described in the next section were calculated.

4.2 Analysis

We investigated four different methods of producing alignments between landmarks in two runs. These different alignment methods reflect some possible ways that a robot might use to align its own landmark perceptions with a landmark-based route description. The alignment methods used match landmarks between runs using sequence, distance travelled, and landmark category, or a combination of them. In each alignment the first run is taken as the route description received by the robot, which is trying to match up the the second run (the landmarks it has perceived) against the original. The alignment methods can be described as:

Sequential Each landmark in the second run is aligned with the next landmark in the first run. If there is a mismatch between the number of landmarks there is no alignment for the excess landmarks.

Distance The $x - y$ position odometry is transformed into a single dimension of distance travelled from the beginning of the run. Landmarks in the second run are aligned to the landmarks in the first run that are closest to them in distance travelled.

Category and distance Each landmark in the second run is aligned to the landmark in the first run that is closest to the same distance travelled and is also a member of the same landmark category. If there is no category match then no alignment is made for that landmark.

Category and distance with limited range As previously, but matches are only allowed that are at the same distance travelled plus or minus an error term representing odometric drift. In these experiments the allowed misalignment, ν_k increases with distance travelled, d (in metres), $\nu_k = 0.25 + 0.05d$, for each landmark k.

Alignments between test runs can be simply evaluated by looking at actual landmark matches. Examples are shown in figures 4 to 7. In addition to these figures three metrics were devised to evaluate the alignments according to the criteria they were based on:

Category score The number of assignments where the aligned landmarks in both runs share the same category, divided by the number of alignments.

Distance score The mean of $\exp \frac{-\delta_k^2}{2\nu_k^2}$ for all aligned landmarks, where δ_k is the difference between the recorded odometry distance for landmark k and its aligned partner.

Sequential score The fraction of alignments where the aligned partner of landmark $k + 1$ is the same or further distance than the aligned partner of landmark k. Landmarks without assigned alignments are not counted as a break in the sequence.

Note that these metrics do not penalise alignments that produce fewer matches of better quality. This is based on the assumption that navigation with many landmarks of uncertain quality is more difficult than navigation with fewer, better quality landmarks, i.e., landmarks that are found consistently.

Alignment	Score	Left Lmks μ	Left Lmks σ	All Lmks μ	All Lmks σ	Left All μ	Left All σ
Sequence	Cat	0.24	0.16	0.15	0.10	0.20	0.14
	Dist	0.30	0.19	0.30	0.19	0.30	0.19
	Seq	1.00	0.00	1.00	0.00	1.00	0.00
Distance	Cat	0.27	0.18	0.19	0.11	0.23	0.16
	Dist	0.82	0.14	0.82	0.14	0.82	0.14
	Seq	1.00	0.00	1.00	0.00	1.00	0.00
Category	Cat	1.00	0.00	1.00	0.00	1.00	0.00
+	Dist	0.64	0.12	0.63	0.12	0.62	0.12
Distance	Seq	0.87	0.05	0.81	0.04	0.87	0.05
Category,	Cat	1.00	0.00	1.00	0.00	1.00	0.00
Distance,	Dist	0.85	0.11	0.86	0.10	0.85	0.10
Rng lim	Seq	0.94	0.05	0.90	0.05	0.95	0.04

Table 1: Alignment scores for a landmark selector ($n = 5$), training the SOM with three different types of input: 'Left Lmks' left-side sonars from landmark perceptions, 'All Lmks' all sonars from landmark perceptions, or 'Left All' left-side sonars from all perceptions. The scores represent mean and standard deviation values of the 54 possible permutations in the training/testing sets. Two-sample t-tests show that the 'Left All' and 'Left Lmks' parameters produce better scores than 'All Lmks', but are indistinguishable from each other.

5 Results

We investigated how suitable the four alignment methods described previously were for use in the route description task using the three metrics. The way in which the SOM was trained was also varied. We tested using all of the 16 sonar sensors as inputs, or just the five left-hand sensors that were used as inputs to the single-layer network. We compared using a SOM trained only on the sensor inputs that were selected as landmarks with one trained on every set of sensor readings. Finally, we compared training in hallways shorter then the test environment with training in hallways of the same size.

5.1 *Alignments and Scores*

It is useful to note some general trends and features of the alignments and scores that are presented in this section. Table 1 shows mean values for each score and alignment combination. As would be expected, the sequence and distance alignments have perfect sequential scores, but they produce very poor categorisation scores. This is presumably because they are matching landmarks that are close together, either in the sequence or by distance, but that are caused by different features. The sequential alignment also produces the worst distance score. This occurs because the number of landmarks can vary by 100%, producing misalignments of up to 80 m, as can be seen in figure 7. The category-based algorithms, naturally, produce perfect category scores. While the category and distance alignment produces good alignments in many cases, it can periodically match up landmarks at opposite ends of the environment. The range-limited version was devised to overcome this problem, and is very successful in doing so; it produces many fewer landmark alignments, an average of 50, as compared with 122 for the other alignments. It succeeds in producing the highest quality and most desirable matches when there are small clusters of landmarks in the same location. Another benefit is that the alignments it produces are relatively insensitive to the order of comparison, unlike the other alignment methods.

5.2 *Effects of SOM Inputs*

Table 1 and figures 4 to 6 show the alignments and scores for three choices of SOM training inputs. We compared the scores with a two-sample t-test ($\alpha = 0.05$) to determine what effects SOM training inputs had on alignment performance. Using only the left-side sonars produced a detectable improvement in the sequential score on both category-based alignment methods. It is possible that the slightly improved performance for left-side only sonars is because there was not space for someone to walk between the robot and the wall it was following (left), and so the sonar returns were not confused by people walking past the robot. We also compared training the SOM with only landmark perceptions versus all sonar scans. In this case, there were no statistically significant differences ($\alpha = 0.05$) in any of the scores.

5.3 *Effects of Training Environment*

The landmark selection algorithm sets parameters related to the noise models and the SOM learns to categorise perceptions based on the training environment. If there is enough difference between training and testing environments we might expect poor landmark selection and alignment. Of course it is unlikely that it would be possible to train in the entire environment that the robot will need to navigate in; often at best a subset, or a related environment might be all that was available. Therefore, the assumed standard in table 1 is to train in only one hallway and test in runs spanning all three hallways. However, comparing those values to training on three-hallway runs using the same training parameters we can see no statistically significant differences in any score for any alignment ($\alpha = 0.05$). Qualitatively, the difference can be seen by comparing the alignments shown in figure 6, trained in one hallway, with the alignments in figure 7, trained in all three. Both figures show alignments generated on the same test data. But there are fewer landmarks detected when training occurs in the same hallways as testing, and while there are fewer landmark alignments numerically, a higher proportion of those landmarks are assigned matches in the category-based alignment methods.

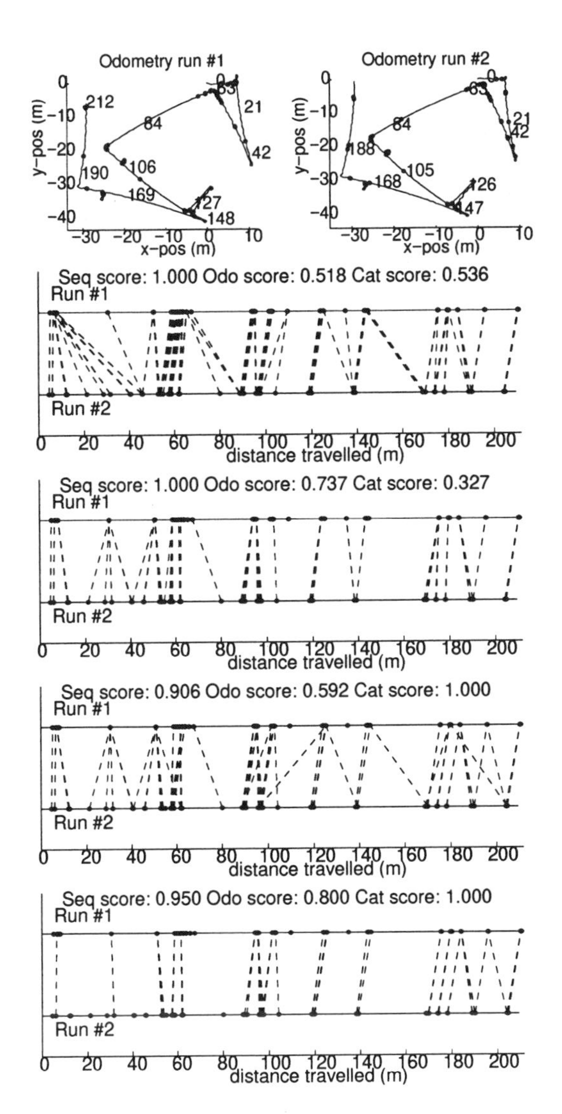

Figure 4: Example alignments for a landmark detection system trained on a run in hallway 1 ($n = 5$, using left-side sonars at landmark perceptions for SOM training). The system was then tested on two separate runs that spanned all three hallways. This set of SOM training parameters produces the best alignment scores, along with the ones in figure 6, which are statistically indistinguishable from this set (see table 1).
Robot path is represented by solid lines, landmarks by black dots, and alignments by dashed lines. *Top:* The robot's internal odometry showing $x - y$ position. The numbers on the path are the total distance travelled in the to that point. *Underneath:* The $x - y$ odometry is projected onto a distance-travelled axis, and aligned landmarks have dashed lines drawn between them. The alignments are (from 2nd from top to bottom) sequential, distance, distance & category, and distance & category with range limit.

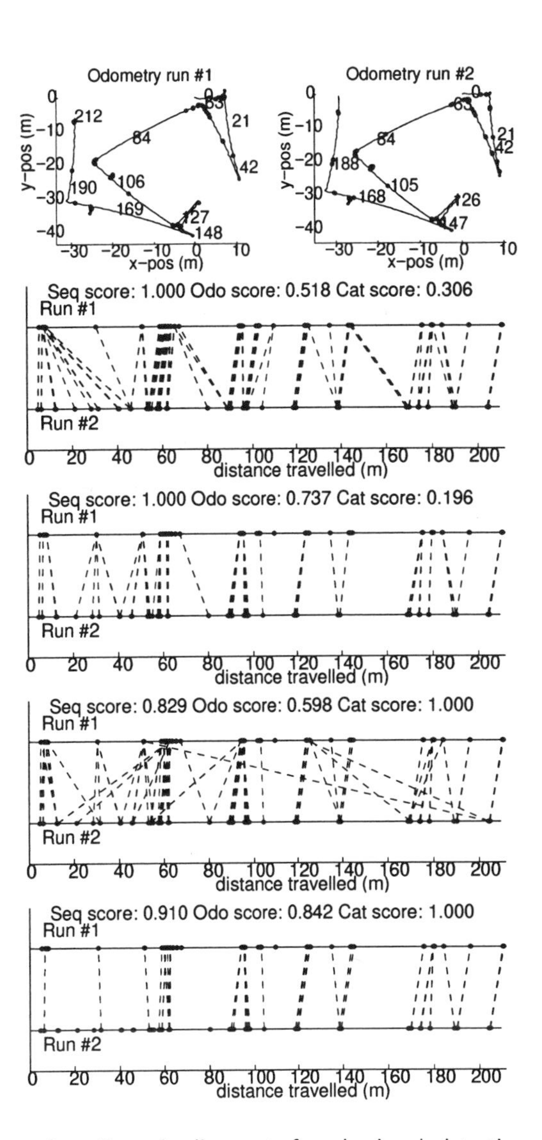

Figure 5: Example alignments for a landmark detection system trained on a run in hallway 1 ($n = 5$, using all 16 sonars at landmark perceptions for SOM training). The system was then tested on two separate runs that spanned all three hallways. Using all the sonars like this is less successful than using only the sonars on the left of the robot, as can be seen in table 1.
Robot path is represented by solid lines, landmarks by black dots, and alignments by dashed lines. *Top:* The robot's internal odometry showing $x - y$ position. The numbers on the path are the total distance travelled in the to that point. *Underneath:* The $x - y$ odometry is projected onto a distance travelled axis, and aligned landmarks have dashed lines drawn between them. The alignments are (from 2nd from top to bottom) sequential, distance, distance & category, and distance & category with range limit.

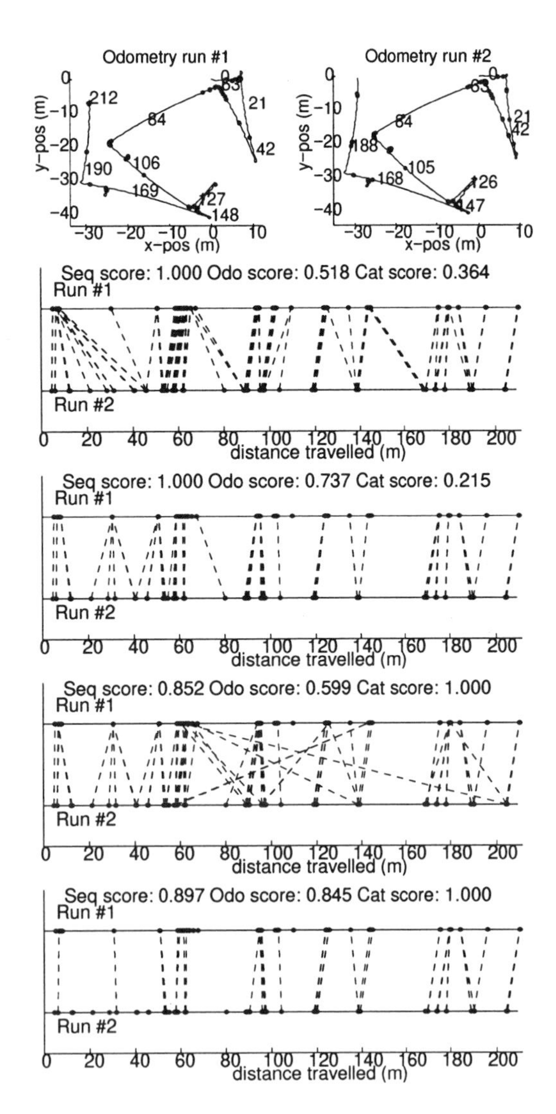

Figure 6: Example alignments for a landmark detection system trained on a run in hallway 1 ($n = 5$, using left-side sonars at every perception for SOM training). The system was then tested on two separate runs that spanned all three hallways. This set of SOM training parameters produces the best alignment scores, along with the ones in figure 4, which are statistically indistinguishable from this set (see table 1).
Robot path is represented by solid lines, landmarks by black dots, and alignments by dashed lines. *Top:* The robot's internal odometry showing $x - y$ position. The numbers on the path are the total distance travelled in the to that point. *Underneath:* The $x - y$ odometry is projected onto a distance travelled axis, and aligned landmarks have dashed lines drawn between them. The alignments are (from 2nd from top to bottom) sequential, distance, distance & category, and distance & category with range limit.

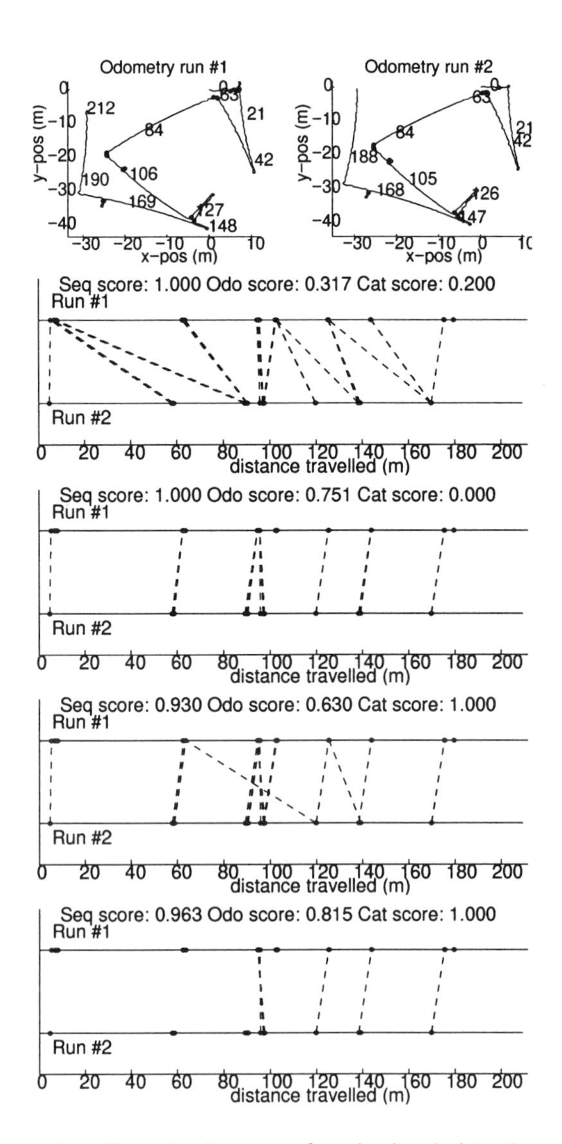

Figure 7: Example alignments for a landmark detection system trained on a run spanning all three hallways ($n = 5$, using left-side sonars at every perception for SOM training). The system was then tested on two separate runs that also spanned all three hallways. Note that there are considerably fewer landmarks in each alignment, as compared with the other figures. Robot path is represented by solid lines, landmarks by black dots, and alignments by dashed lines. *Top:* The robot's internal odometry showing $x - y$ position. The numbers on the path are the total distance travelled in the to that point. *Underneath:* The $x - y$ odometry is projected onto a distance travelled axis, and aligned landmarks have dashed lines drawn between them. The alignments are (from 2nd from top to bottom) sequential, distance, distance & category, and distance & category with range limit.

5.4 *Too Complex?*

Given how much computational machinery is at work in this algorithm, it is important to ask how much of it is actually necessary. There are three major components: a predictor network, a Kalman filter operating on the predictor network error to detect landmarks, and a SOM producing categorisations of the resulting landmarks. Our previous work (Marsland et al., 2001) has shown that simple threshold detection of the error curve is insufficient, and that a Kalman filter is necessary. This paper also demonstrated that the single-layer network was sufficient for the sensor prediction. The SOM is used because it is a well-known self-organising algorithm, but many other algorithms could be substituted to produce landmark categorisations.

But could the Kalman filter operate directly on the sonar inputs, removing the need for the prediction network? The problem with this approach is that the R values would then be the variance of the sonar values due to measurement error. The sonar that is directly perpendicular to the wall will have an R value of almost zero, since sonars have very little measurement error for hard surfaces that have small angles of incidence to the beam. By the Kalman update equations the gain will become unity and the variance estimate will become zero. The filter will see every perception as a landmark unless some clever voting scheme is introduced that removes the effects of the perpendicular sonar.

6 Conclusions and Future Work

We have presented a system that is capable of extracting landmarks from realistic, continually updated, robot sonar data. The landmarks are selected based on the principle of unexpected perceptions – places where the robot's predictive sensor model breaks down are unusual and therefore conspicuous and distinctive. Previous work has shown that the system works well in a tightly structured and discretely sampled environment (Marsland et al., 2001), but this is the first time that the method has been applied to more realistic, noisy, continually sampled data. In spite of the added difficulties, some slight modifications to the algorithm produced a robust and efficient landmark detector capable of reliably reproducing most of the same landmarks at each pass through an environment. While it is not perfect – it will typically produce uneven numbers of landmarks at certain locations or occasional landmarks that appear in one run but not another – from a qualitative standpoint, the landmark selection is very satisfactory.

We evaluated the use of the landmarks generated by this system for aligning landmarks between two different trips through the same environment. We produced alignments between landmarks in pairs of runs based on sequence, distance, and both categorisation and distance. These alignments were scored for their quality in three different measures and inspected visually for suitability. In general, alignment using categorisation and distance with a limited range produced the most consistent scores and most pleasing alignments visually. It is clear that combining different types of landmark alignments (e.g., category and distance) produces better results than using just one type of alignment.

Training the SOM using only left-side sonars produced statistically better scores than using all sonars — an effect that might be attributable to reducing the influence of humans walking through the hallway on the robot perceptions. Training in a subset of the testing hallways produced scores which were not detectably worse than training in the entire hallway area. This is useful since a robot may not be able to train in every environment in which it might need to operate, nor is it desirable to have to perform so much time-consuming training.

We believe that the results to date serve as an strong indication of the suitability of this method for the route communication task. We have demonstrated in this paper that the same robot can align landmarks with high accuracy between different runs of the same environment using information about distance travelled and the categories of the landmarks. In previous work (Fleischer and Nehmzow, 2001), we showed that two robots can learn to reliably and consistently link symbols with perceptual categories of the environment. The communication system can provide a link between the internal categories of two robots, thus enabling one robot to interpret a set of landmarks and categories describing a route followed by another robot. Alternatively, the landmark alignment system could be seen as a method by which two robots might be able to learn symbols representing perceptions that they have both encountered on runs in the same environment. The next step in our investigations will be to combine these two components into a system capable of aligning landmarks between two runs in the same environment performed by different robots. We will be investigating both the possibility of using the landmark detection and alignment algorithms to enable the robots to learn consistent symbol-landmark mappings, and the effects of symbol-category consistency on the performance of matching landmarks between two runs by different robots in the same environment.

7 Acknowledgements

Part of this work was supported by a Canadian & U.S. Citizens Scholarship and an ATLAS grant from the Department of Computer Science at the University of Manchester.

References

Billard, A. and Dautenhahn, K. (2000). Experiments in social robotics: Grounding and use of communication in autonomous agents. *Adaptive Behavior*, 7(3/4).

Bourque, E. and Dudek, G. (2000). On-line construction of iconic maps. In *Proceedings of the IEEE International Conference on Robotics and Automation (ICRA'00)*, pages 2310–2315.

Duckett, T. and Nehmzow, U. (1998). Mobile robot self-localisation and measurement of performance in middle scale environments. *Robotics and Autonomous Systems*, 24(1–2):57–69.

Fleischer, J. and Nehmzow, U. (2001). Towards robots that give each other navigational directions: Learning symbols for perceptual categories. In *Towards Intelligent Mobile Robotics: Proceedings of the 3rd British Conference on Autonomous Mobile Robotics*, Manchester.

Kalman, R. (1960). A new approach to linear filtering and prediction problems. *Journal of Basic Engineering*, 82:34 – 45.

Kohonen, T. (1982). Self-organised formation of topologically correct feature maps. *Biological Cybernetics*, 43:59–69.

Kortenkamp, D. and Weymouth, T. (1994). Topological mapping for mobile robots using a combination of sonar and vision sensing. In *Proceedings of the Twelth National Conference on Artificial Intelligence (AAAI'94)*, pages 979–984, Seattle, Washington.

Linaker, F. and Niklasson, L. (2000). Extraction and inversion of abstract sensory flow representations. In *From Animals to Animats: The 6th International Conference on Simulation of Adaptive Behaviour (SAB'00)*, pages 199 – 208. MIT Press.

Marsland, S. (2001). *On-line Novelty Detection Through Self-Organisation, With Application to Inspection Robotics.* PhD thesis, Department of Computer Science, University of Manchester.

Marsland, S., Nehmzow, U., and Duckett, T. (2001). Learning to select distinctive landmarks for mobile robot navigation. *Robotics and Autonomous Systems*, 37:241 – 260.

Rice, J. (1994). *Mathematical Statistics and Data Analysis.* Duxbury, 2nd edition.

Skubic, M., Matsakis, P., Forrester, B., and Chronis, G. (2001). Generating linguistic spatial descriptions from sonar readings using the histogram of forces. In *Proceedings of the 2001 IEEE International Conference on Robotics and Automation.*

Steels, L. (1996). Emergent adaptive lexicons. In *Animals to Animats 4*, pages 562 – 567. MIT Press.

Tani, J. and Nolfi, S. (1999). Learning to perceive the world as articulated: an approach for heirarchical learning in sensory-motor systems. *Neural Networks*, 12:1131–1141.

Thrun, S. (1998). Bayesian landmark learning for mobile robot localisation. *Machine Learning*, 33(1):41 – 76.

Vesanto, J., Himberg, J., Alhoniemi, E., and Parhankangas, J. (2000). SOM Toolbox for Matlab 5. Report A57, Helsinki University of Technology, Neural Networks Research Centre, Espoo, Finland.

Vlassis, N., Motomura, Y., and Krose, B. (2000). Supervised linear feature extraction for mobile robot localization. In *Proceedings of the IEEE International Conference on Robotics and Automation (ICRA'00)*, pages 2979–2984.

Yamauchi, B. and Beer, R. (1996). Spatial learning for navigation in dynamic environments. *IEEE Transactions on Systems, Man and Cybernetics Section B*, 26(3):496 – 505.

Yanco, H. and Stein, L. A. (1993). An adaptive communication protocol for cooperating mobile robots. In *Animals to Animats 2*, pages 478–485.

Zimmer, U. R. (1996). Robust world-modelling and navigation in a real world. *Neurocomputing, Special Issue*, 13(2–4):247 – 260.

Localization of Function in Neurocontrollers

Lior Segev* **Ranit Aharonov**** **Isaac Meilijson*** **Eytan Ruppin***
*Schools of Computer and Mathematical Sciences
Tel-Aviv University, Tel-Aviv, Israel
{liorseg, isaco, ruppin}@post.tau.ac.il
**Center for Neural Computation
The Hebrew University, Jerusalem, Israel
ranit@alice.nc.huji.ac.il

Abstract

This paper presents the Functional Contribution Algorithm (FCA) that addresses the fundamental challenge of localizing functions in artificial and natural neural networks. The FCA is based on an assignment of contribution values to the elements of the network, such that the ability to predict the network's performance in response to multi-lesions is maximized. The algorithm is thoroughly examined on evolved neurocontrollers, which are simple enough, but not too simple. We demonstrate that the FCA portrays a stable set of contributions and accurate multi-lesion predictions, which are significantly better than those obtained based on the classical single-lesion approach. Our results demonstrate the potential of the FCA to provide insights into the organization of both animat and animate nervous systems.

1 Introduction

Recent years have witnessed a growing interest in the study of neurally-driven evolved autonomous agents (EAAs). Much progress has been made in finding ways to evolve autonomous agents which successfully cope with diverse behavioral tasks [1, 2, 3, 4]. *However, relatively few studies have dealt with the important and challenging question of understanding how the neurocontrollers perform the tasks successfully.* This is a fundamental milestone in using EAAs as a neuroscience research tool. EAAs are *less biased* than conventional neural networks used in neuroscience modeling as their architecture is not pre-designed in many cases. They are the *emergent result* of a simplified and idealized process that models the evolution of intelligent, neurally-driven life forms. This fundamental property naturally raises the possibility of using these agents as a vehicle for studying basic questions concerning neural processing. *In this paper, we present a rigorous and quantitative method for localizing functional tasks in EAA neurocontrollers.*

Several recent studies have dealt with analyzing evolved systems. A series of studies have developed a rigorous, quantitative analysis of the dynamics of central pattern generator (CPG) networks evolved for locomotion [5]. The activity of the internal neurons in the network as a function of a robot's location and orientation was charted via a simple form of receptive field measurement in [1]. Others have systematically clamped neuronal activity and studied its effects on the robot's behavior [6]. Single-lesion analysis was used to discover "command" neurons in EAAs [7]. A more "procedural" kind of ablations to the network, where different processes are systematically cancelled out was used recently in [8]. Overall, these studies have provided only glimpses of the processing in these networks.

One of the difficult challenges in analyzing neuronal systems is to identify the roles of the network elements, and to assess their contributions to the different tasks. In neuroscience, this is traditionally done either by assessing the deficit in performance after lesioning a specific area, or by recording the activity in the area during behavior. These classical methods suffer from two fundamental flaws: first, they do not take into account the probable case that there are complex interactions among elements in the system. For example, if two neurons have a high degree of redundancy, lesioning of either one alone will not reveal its influence. Second, they are mostly qualitative measures, lacking the ability to precisely quantify the contribution of a unit to the performance of the organism and to predict the effect of new, multiple-site lesions. The relative simplicity and the availability of full information about the network's structure and dynamics make EAA models an ideal testbed for studying neural processing. *In this framework, we have presented a rigorous, operative definition for the neurons' contributions to the organism's performance in various tasks and a novel Functional Contribution Algorithm (FCA) to measure them* [9]. This definition permits an accurate prediction of the performance of EAAs after multi-lesion damage, and yields a precise quantification of the distribution of processing in the network, a fundamental open question of neuroscience [10, 11].

A description of an FCA of general agent performance was presented in [9]. In this paper, we extend the FCA to study distinct tasks and to compute their localization in the network. We also present a new adaptive algorithm for selecting efficient lesioning experiments. Moreover, we introduce the high-dimensional FCA, which is a conceptual extension of the basic FCA, allowing to find important compound elements in the system. The rest of the paper is organized as follows: section 2 describes the Functional Contribution approach and provides the necessary definitions. In section 3, we use the FCA to analyze evolved agents, and in section 4, we tackle two fundamental challenges that arise. We conclude with a discussion.

2 The Functional Contribution Approach

Consider an agent (either natural or artificial) with a controller network of N interconnected neurons (or more generally, units) that performs a set of K different functional tasks. Addressing the question of which elements contribute to which tasks, it is natural to think in terms of a contribution matrix, where C_{ik} is the contribution of element i to task k, as shown in Figure 1.

Element	Task 1	Task 2	$\cdots$	Task K	
1	C_{11}	C_{12}	$\cdots$	C_{1K}	
2	C_{21}	C_{22}	$\cdots$	C_{2K}	$\rightarrow S_2$
$\vdots$	$\vdots$	$\vdots$	$\ddots$	$\vdots$	
N	C_{N1}	C_{N2}	$\cdots$	C_{NK}	
	$\downarrow L_1$				

Figure 1: *The contribution matrix*

The data analyzed for computing the contribution matrix is gathered by inflicting a series of multiple lesions onto the agent's network. Under each lesion, the resulting performance of the agent in different tasks is measured. Given this data, the FCA finds the contribution values C_{ik} that provide the best performance prediction on average for all possible multi-site lesions. If a task is completely distributed in the network, the contributions of all neurons to that task should be identical (full *equipotentiality* [12]). Thus, we define the *localization* L_k of task k as a deviation from equipotentiality along column k (e.g., L_1 in Figure 1), and similarly, S_i, the specialization of neuron i is the deviation from equipotentiality along row i of the matrix (e.g., S_2 in Figure 1). Below we rigorously define these concepts and describe the algorithm to compute the contributions.

2.1 Definitions

We consider each of the functional tasks separately. Suppose that a multi-lesion experiment is performed where a set of neurons in the network is lesioned and the neurocontroller network then performs a certain task. The result of this experiment is described by the pair $\{\mathbf{m}, p_m\}$ where the lesion configuration vector $\mathbf{m}$ has $m_i = 0$ if neuron i was lesioned and 1 if it was left intact. p_m is the corresponding performance of the lesioned network on that task, divided by the baseline performance of the fully intact network. For a system of N units there are 2^N such possible multi-lesion experiments, i.e., 2^N possible lesioning configurations $\mathbf{m}$.

The underlying idea of our definition is that the contributions of the units are those values which allow the most accurate prediction of performance following lesions of any degree to the system. Given a configuration $\mathbf{m}$ of lesioned and intact neurons, the predicted performance of the network is the sum of the contribution values of the intact neurons ($\mathbf{m}\cdot\mathbf{c}$), as evaluated by a performance prediction function f. Formally, we seek to find the pair $\{\mathbf{c}, f\}$, which minimizes

$$E = \frac{1}{2^N} \sum_{\{\mathbf{m}\}} [f(\mathbf{m}\cdot\mathbf{c}) - p_m]^2, \tag{1}$$

where f is a smooth non-decreasing[1] function and $\mathbf{c}$ is a normalized column vector such that $\sum_{i=1}^{N} |c_i| = 1$. The resulting vector $\mathbf{c}$ is taken as the *contribution vector* for the task tested (i.e., the task's column in the contribution matrix), and the corresponding f is its adjoint *performance prediction function.* Thus, the optimal contribution vector $\mathbf{c}$ and its accompanying prediction function f minimize the Mean Squared Error (MSE) of predicted vs. actual performance, over all possible lesioning configurations.

2.2 Computing the Contributions

The contribution vector and the prediction function for a task are computed using a training set of lesioning configurations and their corresponding performance levels in the task at hand. The training error, E', is defined as in Equation 1, but averaging only over the configurations present in the training set. The FCA is a gradient descent search algorithm which iteratively updates f and $\mathbf{c}$ until a local minimum of the training error, E', is reached. The steps of the FCA are:

1. **Choose** an initial normalized contribution vector $\mathbf{c}$ for the task, and compute f as in step 4.
2. **Compute** $\mathbf{c}$. Using the current f compute new values of $\mathbf{c}$ by performing a gradient descent on the error E', searching for $\mathbf{c}$ values that minimize E' using the current f, via Eq. 2:

$$E' = \frac{1}{n} \sum_{\mathbf{m}\in M} [f(\mathbf{m}\cdot c) - p_m]^2, \tag{2}$$

where M is the training set, and n is its size.

[1] Complying with the basic assumption about the definition of a contribution, it is assumed that as more important elements are lesioned ($\mathbf{m}\cdot\mathbf{c}$ decreases), the performance (p_m) decreases, and hence the postulated monotonicity of f.

3. **Re-normalize c**, such that $\sum_{i=1}^{N} |c_i| = 1$.

4. **Compute f.** Given the current $\mathbf{c}$, perform isotonic regression on the pairs $\{\mathbf{m} \cdot \mathbf{c}, p_m\}$ in the training set. Use a smoothing spline on the result of the regression to obtain the new f.

   Repeat steps 2 through 4 until the error converges or a maximal number of steps has been reached.

In the results presented in this paper, lesioning of a neuron was performed by making its firing pattern random, rather than by completely silencing its output. More precisely, at every time step a lesioned neuron fires with probability equal to its overall mean firing rate, independent of its input field. This ensures that the lesioning does not effect the mean field of other neurons, only the information content received form the lesioned neuron (we return to this important point in the discussion). In addition, when lesioning motor neurons, we do not alter the activity transmitted to the motors themselves. This enables us to isolate the role of the motor units in the computation of the recurrent controller networks (i.e. their contribution to other neurons), without immobilizing the agent.

2.3 Localization and Specialization

Returning to the contribution matrix (Figure 1), we define the *localization* L_k of task k as a deviation from equipotentiality. Formally, L_k is the standard deviation of column k of the contribution matrix divided by the maximal possible standard deviation,

$$L_k = \frac{\mathrm{std}(C_{\cdot k})}{\sqrt{(N-1)/N^2}}. \tag{3}$$

Note that L_k is in the range $[0, 1]$ where $L_k = 0$ indicates full distribution and $L_k = 1$ indicates localization of the task to one neuron alone. Similarly, if neuron i is highly specialized for a certain task, $C_{i\cdot}$ will deviate strongly from a uniform distribution, and thus we define S_i, the *specialization* of neuron i, as

$$S_i = \begin{cases} 2 \cdot \mathrm{std}(|C_{i\cdot}|) & \text{if } K \text{ is even} \\ \frac{2 \cdot \mathrm{std}(|C_{i\cdot}|)}{\sqrt{(K^2-1)/K^2}} & \text{otherwise.} \end{cases} \tag{4}$$

Note that S_i uses the absolute value of the contributions to reflect the intuition that the specialization of the unit is determined more by the magnitude of its contribution than by its sign. Again, $S_i = 1$ indicates maximal specialization of neuron i. We note, however, that in principle other measures can be defined. The crucial point is that the contribution matrix enables one to define such quantitative measures to capture qualitative notions.

3 Analyzing Evolved Agents

Using evolutionary simulations, we have previously developed autonomous agents controlled by fully recurrent artificial neural networks [7]. High performance levels were attained by agents performing simple life-like tasks of foraging and navigation. Classical neuroscience methods were used to analyze the neurocontrollers. Here, the FCA is applied to the analysis of those neurocontrollers.

3.1 The Evolved Agents

The agents in the model live in a grid arena of size 30×30 cells surrounded by walls (Figure 2). "Poison" items are randomly scattered all over the arena, and "food" items are randomly scattered in a restricted 10×10 "food zone" in the corner of the arena. The agents' behavioral task is to eat as much of the food as they can while avoiding the poison. The complexity of the task stems from the partial sensory information the agents have about their environment. The agents are equipped with a set of sensors, motors (given and constant), and a fully-recurrent neurocontroller which is coded in the genome and evolved.

The initial population consists of 100 agents equipped with random neurocontrollers. Each agent is evaluated in its own environment, which is initialized with 250 poison items and 30 food items. The life cycle of an agent (an *epoch*) is 150 time steps, in each of which one motor action takes place. At the beginning of an epoch the agent is introduced to the environment at a random location and orientation. The agent's general performance (fitness) is the total amount of food eaten minus the amount of poison consumed, normalized by the total number of food items available. Selecting the best agents of each generation, and reproducing with variability, successful agents emerge after 5000-30000 generations.

Each agent is controlled by a fully recurrent binary neural network consisting of 10 to 45 neurons, excluding the input neurons (the number is fixed within a given simulation run). Of these, four are output motor neurons which command the agent's motors. In each step a sensory reading occurs, network activity is then updated synchronously, and a motor action is taken according to the resulting activity in the designated output neurons.

The agents are equipped with a basic sensor (the somatosensor) consisting of five probes. Four probes sense the grid cell the agent is located in and the three grid cells immediately ahead of it (Figure 2). These probes sense the difference between an empty cell, a cell containing a resource (with no distinction between food and poison), and the wall. The fifth probe can be thought of as a *smell probe*, which discriminates between food and poison if either is present in the cell occupied by the agent, but gives a random reading otherwise. In some simulations the agents were equipped with an extra sensor returning the location of the agent in the arena. The

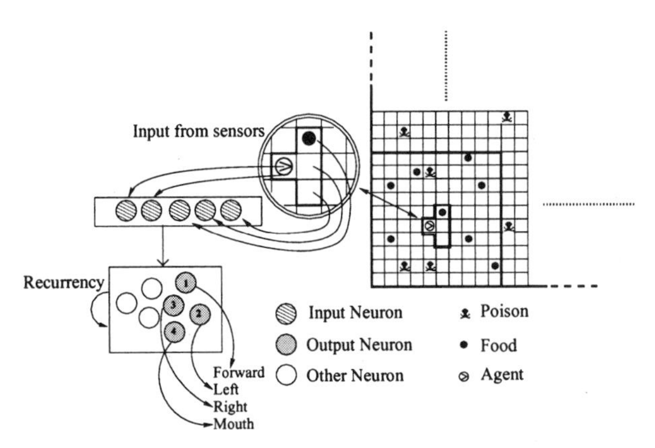

Figure 2: *An outline of the grid arena* (corner) and the agent's neurocontroller. For illustration purposes the borders of the food zone are marked, but they are invisible to the agent. The agent is marked by an arrow, whose direction indicates its orientation. The T-shape marking in front of the agent denote grid cells which it senses. Output neurons and interneurons are all fully connected to each other.

motor system allows the agent to go forward, turn left or right, or attempt to eat (a costly action, as it requires a time step with no movement).

We studied two types of agents: an SP-type agent possessing both a somatosensor and a position sensor, and an S-type agent possessing a somatosensor only. We found that for both types the most successful strategy relies upon a switch between two behavioural modes – *exploration* and *grazing*. The exploration mode consists of moving in straight lines, ignoring resources in the sensory field which are not under or in front of the agent, and turning at walls. The grazing mode consists of turning to resources to the right or left to examine them, turning at walls, and maintaining the agent's location in a relatively restricted region. In both modes eating occurs when stepping on a food item. Exploration mode is mostly observed when the agent is out of the food zone, allowing it to explore the environment and find the food zone. Inside the food zone, however, the agents tend to display grazing behaviour, which results in efficient consumption of food.

Examining the networks of successful agents equipped with a position sensor (SP-type) reveals an important common feature: certain interneurons have a position-sensitive response. One of these neurons typically fires outside the food zone and remains quiescent inside the food zone and in its close vicinity. Clamping this *"position-sensitive cell"* and measuring the behavioural mode of the agent, reveals that it acts as a *command neuron*. When this command neuron is active the agent assumes an exploration behaviour, *regardless* of where it is in the environment, whereas clamping it to a silent state results in grazing behaviour. In the following sections we apply the FCA to one such agent, whose network consists of 10 neurons (SP10).

Similar to the SP case, we have identified at least one neuron whose activity was position-dependent, in agents possessing only the somatosensor (S-type), and lacking a position sensor. In some agents, one such neuron alone serves as a command neuron. Here, we study such an agent, whose network consists of 22 neurons (S22). In other agents a combination of such neurons effectively assume the role of a command system, determining the behavioral mode of the agent. One such example, with a 10 neuron network (S10) is also studied in subsequent sections. These command neurons emerge even though no explicit information about location is available to S-type agents. Moreover, their activity is not locked to the absolute position in the arena, but rather to the location of the food zone, which may vary. We have demonstrated that this computational ability is based on a spontaneously emerging stochastic memory mechanism. Thus, the "knowledge" of location is not a real position-dependent activity but rather activity produced by a form of memory of the time elapsed since the last eating event[7]. As will become clear from the analysis below, this memory mechanism relies on feedback from the motor neurons, supplying information about eating events of the agent.

3.2 FCA Analysis of General Performance

We consider first the performance obtained in the general task, i.e., maximizing fitness by consuming food and avoiding poison. We study three agents, two S-type agents (S10 and S22) and one SP-type agent (SP10), as described above. We apply the FCA to these networks by performing lesioning experiments and testing the performance levels of the agents. The performance levels are normalized so that the performance of the intact agent is 1. For each agent, a small training set was chosen (see section 4.1), and the FCA was applied 50 times, each with a different random initial condition (section 2.2). The result of the FCA is the contribution vector $\mathbf{c}$ and the performance prediction function f. As explained above, given any lesioning configuration $\mathbf{m}$, the predicted performance of the agent on the task is $f(\mathbf{m} \cdot \mathbf{c})$. The prediction capability of the FCA is measured by computing the MSE on a test set consisting of all 2^{10} configurations in S10 and SP10, and of 50000 random configurations in S22[2]. The MSE on these test sets were all below 3.5×10^{-4}, demonstrating that the FCA succeeded in finding a pair $\{\mathbf{c}, f\}$, that brings the error in Eq. 1 to a very low value, even when trained on a very limited subset of the configuration space.

Each panel in Figure 3 depicts the mean and standard deviation of the contributions of the different neurons of the agent, over the 50 different initial conditions. Importantly, the FCA converges to similar minima of the error,

[2]Since obtaining the full 2^{22} set is impossible, we chose a large enough set, such that any set of that size results in the same error.

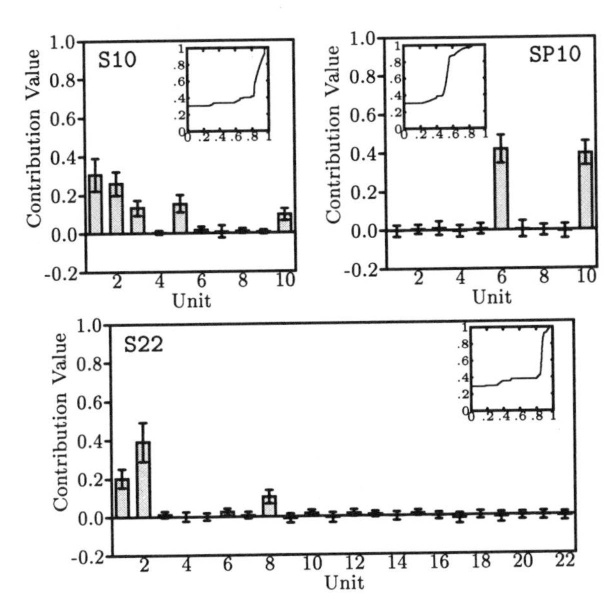

Figure 3: *Contribution values of the agents' neurons:* the mean and standard deviation of the contribution values computed by the FCA. A, B. Agent S10 (A) and agent SP10 (B). FCA trained on 30 lesioning examples, obtained by the adaptive algorithm (section 4.1). C. Agent S22. FCA trained on 99 random lesioning examples. The inset in each panel depicts the performance prediction function.

yielding consistent measures of the contributions. As expected, we find that the command neurons (section 3.1) receive positive contributions in all three agents. Interestingly, in the two S-type agents, some of the motor neurons (neurons 1-4) receive high contribution values, testifying to the importance of the recurrency from the output units[3]. This is consistent with the notion that the feedback from the motor units is required for signaling eating events, enabling the agents to switch to and remain in grazing mode. Note that in SP10, to which location information is available, the motor neurons are not significant. The inset in each panel plots the performance prediction function obtained by the FCA[4].

3.3 Localization and the Contribution Matrix

In the previous section we have discussed the results of applying the FCA to the overall performance measure of the agent. In this section, we present an in-depth analysis of one agent, S10, and compute its contribution matrix. We consider two tasks which correspond to two emergent behaviors of successful agents, exploration and grazing (section 3.1). Exploration performance, p^e, is measured by randomly placing the agent in the arena and measuring the number of steps t that elapsed before the agent reaches the food zone for the first time (maximal allowed t is 1000). The performance is computed by

$$p^e = \frac{T-t}{T} - \frac{p/F}{t/T}, \tag{5}$$

where p is the number of poison items consumed before reaching the food zone, $F = 30$ is the total number of food items in the arena, and $T = 150$ is the original number of steps in an epoch. The first term is concerned with the speed of reaching the food zone, and becomes negative for $t > T$. The second term penalizes poison consumption which should be avoided in any behavioral mode, and is normalized to be consistent with the general performance normalization. Grazing performance, p^g, is measured by placing the agent in the arena for T time steps. Denote by f the total number of food items eaten, and by p_a the number of poison items consumed after reaching the food zone for the first time, then,

$$p^g = \frac{(f - p_a)/F}{(T-t)/T}, \tag{6}$$

where t is measured as before. The enumerator is consistent with the general performance measure, and the denominator normalizes by the time inside the food zone.

The FCA is performed separately for each task yielding a contribution vector for each the task (Figure 4). As is evident, the distribution of the two tasks within the network differs quite considerably. The most evident difference is in the contributions of neurons 5 and 10, both participating in determining the behavioral mode of the agent. Whereas both neurons are important for grazing behavior, neuron 5 does not contribute to exploration behavior, and neuron 10 even hinders it.

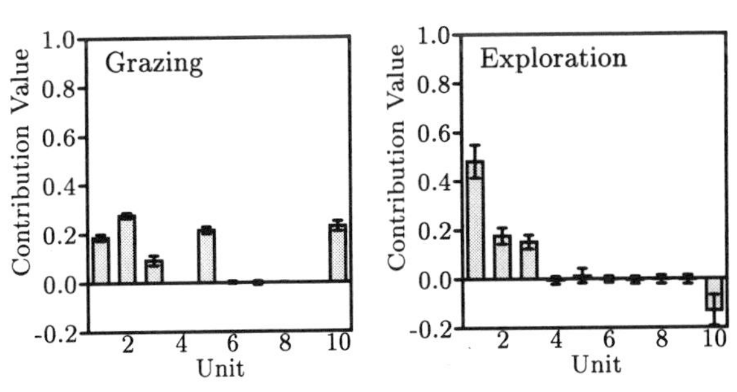

Figure 4: *Contribution values for grazing and exploration:* each is mean and standard deviation of 50 FCA runs. Training on the full 2^{10} set.

Table 1 depicts the contribution matrix of the agent, along with the measures of localization and specialization computed from it (section 2.3). Since the evolved neurocontrollers are arbitrarily large, some neurons do not participate in any task the agent performs. This should be taken into account when considering the localization, and thus, we compute the effective localization $\tilde{L}_k$ by considering only the significant neurons

[3]Recall that when lesioned, the motor action is kept intact; only the information transfer to the rest of the network is severed.

[4]All 50 functions are similar, one was arbitrarily chosen.

($c_i > 0.01$), as shown for the different tasks performed by agent S10 in Table 1. Returning to the other agents, the effective localization index for agent SP10 is, as expected, very low (0.0220), since the two significant neurons have very similar contributions. In contrast, the three significant neurons of agent S22 have quite different values of contribution, resulting in high effective localization (0.7589).

N	Grazing	Exploration	S_i
1	**0.1864**	**0.4819**	**0.2955**
2	**0.2745**	**0.1758**	**0.0987**
3	**0.0911**	**0.1508**	**0.0597**
4	-0.0000	-0.0065	0.0065
5	**0.2141**	**0.0118**	**0.2023**
6	0.0010	-0.0028	0.0038
7	-0.0017	-0.0058	0.0042
8	0.0010	-0.0031	0.0041
9	0.0001	-0.0047	0.0048
10	**0.2280**	**-0.1319**	**0.3599**
L_k	0.1083	0.1610	
$\tilde{L}_k$	0.1524	0.5107	

Table 1: Localization and specialization in agent S10. Contribution values are mean values of 50 runs. $\tilde{L}_k$ denote effective localization (see text).

3.4 *Besting the Single Lesion Approach*

An important aspect of the FCA is using information from multiple lesions, extending the limited information available in single lesions alone. To study the significance of using multiple lesions, we compute the null hypothesis contributions, i.e., relying only on the performance levels when single units were lesioned. The contribution of neuron i is taken to be the decrease in the performance due to lesioning it alone, normalized as before[5]. The black bars in Figure 5(A) depict the contributions computed from single lesions for the general performance of agent S10, where the light bars describe the contributions computed by the FCA. As is evident, the contributions assigned by the FCA differ from those obtained with knowledge of single lesion results. Are the FCA-computed contributions better? In order to compare contribution estimations, we take each contribution vector, and fit the optimal prediction function for the test set (the full 2^{10} configuration set). This ensures that any difference in prediction capabilities stems from a difference in the contributions themselves. Panels B and C of Figure 5 compare the predicted vs. actual performance, using single lesions (B) and the FCA (C). Clearly, single lesions do not reveal the true contributions of the neurons, testifying to the existence of nonlinear interactions between the units. Further, Figure 6 compares, for each agent and task, the MSE on the test set using single lesion information, and using the FCA derived contributions. The results demonstrate again that in some cases nonlinear interactions exist, yielding suboptimal contribution estimations when considering only single lesions. Note that for agent SP10, the effect is less significant, testifying to the simpler functional organization of the network.

We emphasize that the nonlinear interactions are still modeled well by the FCA, which is a linear model that is generalized by the nonlinear prediction function. The fact that single lesions do not suffice to reveal the underlying contributions, is not in itself a cause to discard a linear approach, as is clear from the above analysis. In section 4.2, we deal with a case where the basic 1-D FCA fails, even when multiple lesions are considered, and offer a suitable solution.

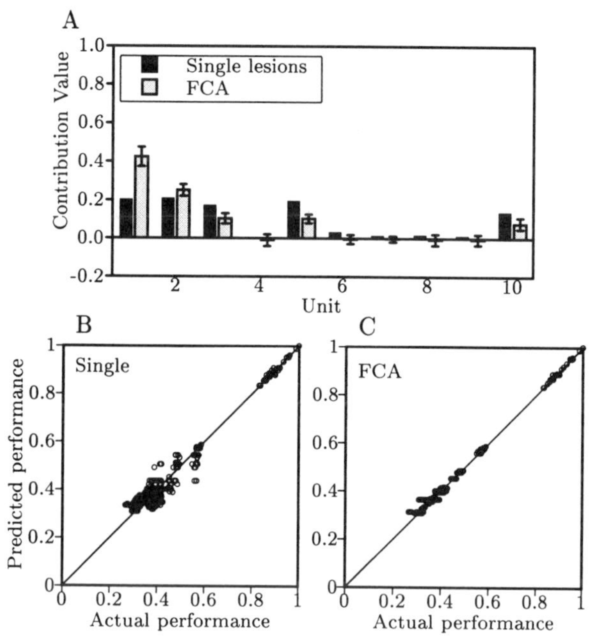

Figure 5: *Comparison between the single lesion approach and the FCA on agent S10:* A. contribution values obtained by the two methods. B, C. predicted vs. actual performance on the test set using single lesions information (B) and the FCA trained on the full set of multiple lesions (C).

4 Fast and High Dimensional FCA

Section 3 demonstrated that the FCA enables analysis of networks performing different functional tasks even when the interactions between the units are nonlinear. However, these were relatively small and simple, even if recurrent, networks. Thus, two important questions arise. First, can we do better than randomly inflicting damage to the network? *That is, can we judiciously select a small subset of lesions that will maximize the FCA accuracy for a given training size?* Section 4.1 describes

[5] $c_i = (1-p_i)/(\sum_{i=1}^{N} |1-p_i|)$, where p_i denotes the performance when neuron i is lesioned alone.

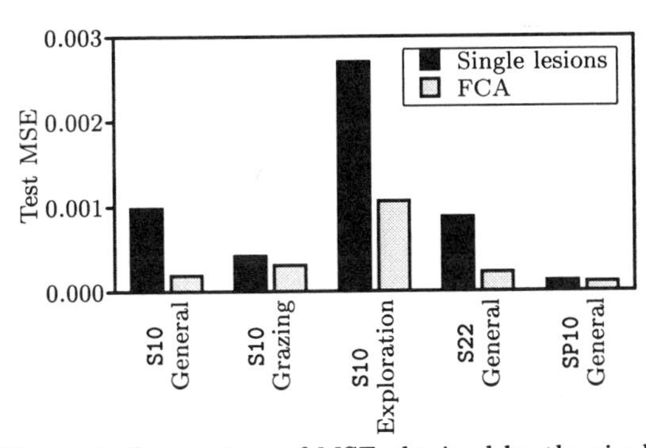

Figure 6: *Comparison of MSE obtained by the single lesion approach and the FCA.* Test sets are the full 2^{10} set for S10 and SP10, and the 50000 configuration set for S22. All predictions used an optimally fitted prediction function for the test set (see text).

such an algorithm for lesions selection. Second, *can one apply the FCA to networks with more complex interactions?* In section 4.2 we present an extension of the FCA which deals with an important type of complexity, the paradoxical lesioning effect.

4.1 Fast FCA

The most time consuming and demanding part of the FCA analysis is testing the agent's corresponding performance for a given lesioning configuration. If the analyzed system is biological, it is impractical to obtain a large set of lesioning configurations. Even when the system is an agent's neurocontroller, it might still be impossible to obtain the full 2^N set of configurations, and might be time consuming to obtain sufficient volume of lesions/performance data.

To reduce the number of configurations needed, it is essential that the most informative set of configurations is selected for training. For this purpose, we have developed the adaptive configuration selection algorithm, which iteratively selects the next lesioning configuration to be evaluated, based on the current configuration set:

1. Create a random initial core set of N configurations[6]. Compute $\mathbf{c}$ and f using the FCA.

2. From all possible configurations, find the configuration whose *estimated* performance is farthest from the *known* performance values of the configurations currently in the training set. In other words, given the current configuration set T, choose the next lesioning configuration $\mathbf{m}$ such that

$$\mathbf{m} = \arg\max_{\mathbf{m}' \notin T} \left\{ \min_{\mathbf{m}'' \in T} |p_{\mathbf{m}''} - f(\mathbf{m}' \cdot \mathbf{c})| \right\} \tag{7}$$

3. Add $\mathbf{m}$ to T, recompute $\mathbf{c}$ and f using the FCA, and return to step 2.

Steps 2 and 3 are repeated until either a predefined number of training examples is reached, or the change in the test prediction error falls below a threshold criterion.

As can be seen in Figure 7, for any size of training set, the adaptive algorithm results on average in a lower MSE on the test set, than randomly choosing the training set (note that for both methods the training set always includes the all-intact and the all-lesioned configurations). Moreover, the adaptive algorithm is much more consistent in finding good training sets. Random sets are prone to badly representing the full set, and thus more often result in large MSE on the training sets (see error bars in figure). The efficiency of the adaptive algorithm stems from the fact that by uniformly sampling the values of p, it concentrates on the areas of f in which a small perturbation in $\mathbf{m} \cdot \mathbf{c}$ results in a large variation in the prediction. Thus, it focuses on the areas in which an error is most costly.

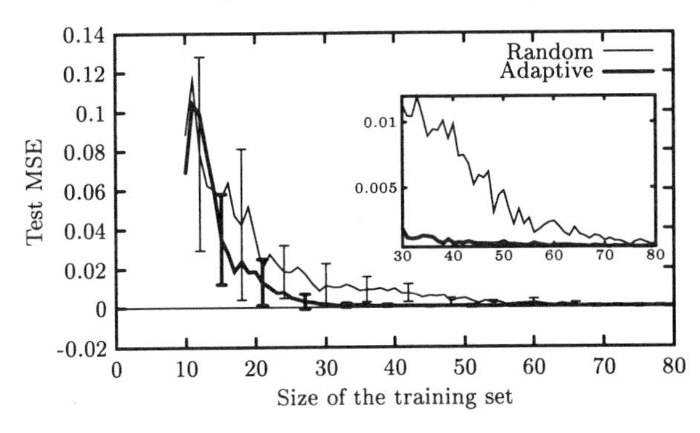

Figure 7: *Comparison of different configuration selection algorithms:* mean and standard deviation of the test MSE vs. number of training configurations on agent S10. Test MSE is measured on the full 2^{10} configuration set. The inset focuses on a subset of the same data, portraying the number of training configurations leading to significant, low test MSE values.

4.2 High Dimensional FCA

In a classical paper, Sprague [13] has shown that orienting deficits resulting from large cortical visual lesions can be reversed through additional removal of the contralateral Superior Colliculus, or section of the collicular commissure. This has been known as the Sprague effect, or the paradoxical lesioning effect, extensively studied (e.g., [14, 15]). Such effects pose a great conceptual challenge: if an area is important for a behavior (lesioning it hinders performance), how can its lesioning under a different condition improve the behavior? The main conclusion is that the contribution of such a unit *depends* on the state of the other units.

[6] Since $\mathbf{c}$ has N parameters, the training set must consist of at least N configurations. The set always includes the all-intact and the all-lesioned configurations.

Lesioining configuration	Performance
○ ○ ○ ○ ○ ○ ○ ○ ○ ○	0.76
⊗ ○ ○ ○ ○ ○ ○ ○ ○ ○	0.64
○ ⊗ ○ ○ ○ ○ ○ ○ ○ ○	0.68
⊗ ⊗ ○ ○ ○ ○ ○ ○ ○ ○	0.72

Figure 8: *A simple example of paradoxical lesions:* the system contains 10 elements, with varying contributions. The first two elements exhibit paradoxical effects. Lesioning both of them results in a higher performance (0.72) than lesioning each one alone (0.64, 0.68). These are 4 of 2^{10} configurations.

To study this case, we have devised a synthetic example, in which 2 of 10 elements exhibit the paradoxical lesioning effect (Figure 8). Not surprisingly, the 1-D FCA described so far fails to find a pair $\{\mathbf{c}, f\}$, which brings the prediction error to a reasonably low value. This is because it is based on the assumption that there exists a notion of a contribution of a single element irrespective of the state of the other elements. In this scenario, this is clearly an incorrect assumption. Here, precise multi-lesion prediction requires the consideration of contributions from additional, functionally important *conjunctions* of the basic units. We thus introduce the concept of a high dimensional FCA, where compound elements of order 2 and above are considered. Once the new compound elements are chosen the rest of the definitions and concepts remain the same. For example, in a *full* 2-D FCA, the performance is predicted by

$$p(\mathbf{m}) = f\left(\sum_{i=1}^{N}\sum_{j=i}^{N} m_i m_j c_{ij}\right), \tag{8}$$

where c_{ii} corresponds to the contribution of the simple element i. Clearly, such an approach introduces the problem of finding the minimal optimal set of conjunctions of elements.

The simplest approach to this problem, and most time-consuming, is to iteratively try each conjunction up to a certain order. Then, select the one that yields the lowest prediction error, and add it to the current set of compound elements. A faster approach, suitable for 2D-FCA (which includes only two element conjunctions), is to add all conjunctions of two elements, and select those which have contribution values above a threshold. However, this approach requires $O(N^2)$ conjunctions, and thus is intractable when N is large. We have devised a more efficient algorithm, called Compound Element Error Reduction (CERE) which uses *estimates* of the reduction in the prediction error that each conjunction yields. This algorithm is much faster than the previous two, and proved almost as successful (data not shown).

CERE estimates the prediction error when a compound element $\pi = (\pi_1, \pi_2, \ldots, \pi_k)$ is added to the current set of elements by

$$\Delta E_\pi = \left(\frac{1}{2^N}\frac{N_\pi}{n_\pi}\right) \sum_{\mathbf{m}\in T_\pi} \Big([f(\mathbf{m}\cdot\mathbf{c}) - p_m]^2 - [f(\mathbf{m}\cdot\mathbf{c} + c_\pi) - p_m]^2\Big), \tag{9}$$

where T_π is the set of lesion configurations from the train set that match π, and n is the size of the training set. A lesion configuration $\mathbf{m}$ is said to *match* a compound element π if $m_{\pi_i} = 1$ for all $i = 1, 2, \ldots, k$. The term $N_\pi/(2^N \cdot n_\pi)$ normalizes the error to the expected error over the complete set of 2^N configurations, where N_π is the number of possible configurations that match π, and n_π is the number of such configurations in the train set. c_π is the contribution value assigned to π, computed to maximize ΔE_π by gradient descent. The CERE algorithm is comprised of the following steps:

1. Initialize the set of compound elements to the empty set. Compute $\{\mathbf{c}, f\}$ using the FCA on single elements.
2. For each candidate compound element π, compute c_π to maximize ΔE_π via gradient descent (using Eq. 9).
3. Select the compound element π that maximizes the reduction in the prediction error ΔE_π and add it to the set of compound elements.
4. Recompute $\{\mathbf{c}, f\}$ using the FCA, using all N simple elements and all compound elements selected so far.

Steps 2-4 repeat until a stopping criteria is met.

We apply a high-dimensional FCA to the example illustrated above. The FCA automatically finds the correct compound element[7] as can be seen in Figure 9. This leads to an MSE of practically zero. Note that the first two elements receive a negative contribution, but that the element composed of their conjunction has a positive contribution, larger than the sum of those single contributions. Thus, when both elements are intact their overall contribution to performance ($c_1 + c_2 + c_{12}$) is positive (lesioning both of them indeed reduces performance from 0.76 to 0.72). However, having one element intact results in a lower overall contribution than when both are lesioned (both c_1 and c_2 alone are negative).

Importantly, The usage of multiple lesions in the basic 1-D FCA is important for revealing nonlinear interactions between the units (e.g. redundancy), as was clearly demonstrated in section 3.4. However, 1-D FCA is adequate only if the notion of a constant contribution of an element is a consistent one. If an element augments performance in one network configuration and hinders it in another, compound elements must be introduced. Thus, the issue of the dimension of the FCA is inherently different from the question of using information obtained from lesioning of high orders.

[7] In this example, all three conjunctions-selection algorithms always selected the same compound element.

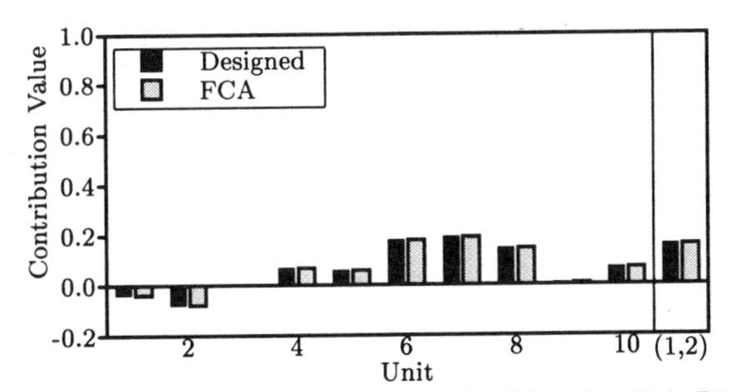

Figure 9: *Contribution values obtained by the High Dimensional FCA:* element $(1, 2)$ is a compound element composed of the conjunction of the first two elements. The black bars depict the original contributions assigned for constructing the example. All FCA runs resulted in the same compound element. Standard deviations are too small to be observed.

5 Discussion

This paper presents a novel functional contribution algorithm that addresses the fundamental challenge of localizing functions in neural networks. The FCA is based on a rigorous assignment of contribution values to the elements of the network analyzed, such that the ability to predict the network's performance in response to multi-lesions is maximized. The algorithm is thoroughly examined in an EAA environment, which is simple enough for studying it but not too simple.

From a methodological standpoint, in the recurrent yet simple EAA neurocontrollers studied here, the basic 1-D version of the FCA suffices to portray a stable set of contributions and to yield fairly accurate performance predictions under multi-lesions, which are significantly better than those obtained with a single-lesion analysis method. The FCA may be trained on a relatively small subset of all multi-lesion configurations. Its computational efficacy may be further enhanced by the Fast FCA, which performs a smart and incremental selection of future lesions based on the information extracted from the lesions it has already been trained on.

From a thematic perspective, the application of the FCA to the analysis of an agent performing multiple tasks reveals that many neurons receive vanishing overall contribution values (both with respect to the general performance and with respect to the specific tasks). Such neurons can obviously be discarded, since they play no part in network processing. They are the mere consequence of the fact that the agents are evolved with arbitrarily large neurocontroller networks. Removing these neurons reveals the true, effective size of the controller network, which in turn reflects the complexity of the task in hand. Thus, the FCA analysis may serve both to estimate the difficulty of the EAA environment, and to uncover the functional localization of the tasks in the network. After removing the insignificant neurons, one may now correctly compute the true, effective localization and specialization indices.

The analysis of more complex networks requires FCA with a dimension higher than 1. This is demonstrated on a basic example of paradoxical lesioning, where the addition of a single 2-dimensional unit suffices to make the analysis tractable. The concept of high-dimensional FCA is a very important one. It may well be that the analysis of future complex animat neurocontrollers and certainly their animate counterparts will require fairly high-dimensional FCA, with basic atomic units spanning a varying number of dimensions. On one hand, such a decomposition may reveal basic functional units in the system. On the other hand, it will probably call for a revisit of the way we think about localization of function. EAA studies provide, in our minds, an essential simple conceptual tool for thinking about and studying these issues and the questions they naturally raise.

These results should be viewed as an encouraging but first step in the quest for understanding neural information processing in animate and animats. Two important issues arise from this work, which deserve considerable attention in the future. First, indeed, the analysis of complex networks will most likely require high-dimensional descriptions. However, the number of parameters that should be determined increases exponentially with the dimension of the model. Hence, we should find smart ways to heuristically focus on just a "relevant" subset of these parameters, i.e., compound combinations of the atomic units composing the network (the current high-dimensional FCA already presents an initial step in this direction). Second, the FCA makes one think about the "correct" or "best" way to perform the lesions in the network. Our studies indicate that the conventional way of lesioning employed in neuroscience, that is, completely silencing the lesioned unit(s), is inadequate for general multi-lesion studies. A discussion of the issue is beyond the scope of this paper. A rather intuitive explanation is that since silencing a neuron interferes also with its function as a constant bias to other neurons, the damage inflicted is propagated through the system, creating dispersed effective lesions. In this study we have chosen one informative way for lesioning, but is it the best one? Importantly, the FCA provides important and useful information about function localization and the true effective network size, but it does not otherwise advance our understanding of how neural information processing takes place; how is information represented, encoded and manipulated on the neural, assembly and network levels.

The FCA has potential significance beyond the scope of EAAs. Multi-lesion analysis algorithms like the FCA will become an essential tool in neuroscience for the analysis of reversible inactivation experiments. Reversible inactivation methods alleviate many of the problematic aspects of the classical lesion technique (ablation), enabling the acquisition of reliable data from multiple lesions of different configurations [16]. It is most likely that a plethora of data will accumulate in the near future. The sensible integration of such data will require quantitative methods to complement the available qualitative ones. FCA like algorithms could also be used for the analysis of transcranial magnetic stimulation studies which aim to induce multiple transient lesions and study their cognitive effects [17]. Another possible use is the analysis of functional imaging data by assessing the contributions of each element to the other, i.e., extending previous studies which employed linear models to study the network's effective connectivity (e.g., [18]).

In summary, applying algorithms such as the FCA should prove useful in obtaining insights into the organization of both animat and animate nervous systems, providing a rigorous method for determining function localization and settling the long-lasting debate about local versus distributed computation. Yet, as it happens time and again in science, providing some answers raises even more challenging questions.

Acknowledgments

We acknowledge the valuable contributions and suggestions made by Hanoch Gutfreund, Nira Dyn, Oran Singer and Matan Ninio. This research has been supported in part by the FIRST grant of the Israeli Academy of Sciences, by the Adams Super Center for Brain Studies in Tel Aviv University, and by the Israel Science Foundation.

References

[1] D. Floreano and F. Mondada. Evolution of homing navigation in a real mobile robot. *IEEE Transactions on Systems, Man, and Cybernetics - Part B*, 26(3):396–407, 1996.

[2] C. Scheier, R. Pfeifer, and Y. Kunyioshi. Embedded neural networks: Exploiting constraints. *Neural Networks*, (7-8):1551–1569, 1998.

[3] J. Kodjabachian and J.A. Meyer. Evolution and development of neural controllers for locomotion, gradient-following and obstacle-avoidance in artificial insects. *IEEE Transactions on Neural Networks*, 9(5):796–812, 1998.

[4] F. Gomez and R. Miikkulainen. Incremental evolution of complex general behavior. *Adaptive Behavior*, 5(3/4):317–342, 1997.

[5] R.D. Beer, H.J Chiel, and J.C. Gallagher. Evolution and analysis of model CPGs for walking II. general principles and individual variability. *Journal of Computational Neuroscience*, (7):119–147, 1999.

[6] I. Harvey, P. Husbands, and D. Cliff. Seeing the light: Artificial evolution, real vision. In D. Cliff, P. Husbands, J.A. Meyer, and S. Wilson, editors, *From Animals to Animats 3.* MIT Press, 1994.

[7] R. Aharonov-Barki, T. Beker, and E. Ruppin. Emergence of memory-driven command neurons in evolved artificial agents. *Neural Computation*, (13):691–716, 2001.

[8] K.O. Stanley and R. Miikkulainen. Evolving neural networks through augmenting topologies. *Technical Report AI01-290*, Department of Computer Sciences, University of Texas at Austin 2001.

[9] R. Aharonov, I. Meilijson, and E. Ruppin. Understanding the agent's brain: A quantitative approach. In *The Sixth European Conference in Artificial Life. (ECAL 2001).* 2001.

[10] J. Wu, L. B. Cohen, and C. X. Falk. Neuronal activity during different behaviors in aplysia: A distributed organization? *Science*, 263:820–822, 1994.

[11] S. Thorpe. Localized versus distributed representations. In M. A. Arbib, editor, *Handbook of Brain Theory and Neural Networks.* MIT Press, 1995.

[12] K. S. Lashley. *Brain Mechanisms in Intelligence.* University of Chicago Press, Chicago, 1929.

[13] J.M. Sprague. Interaction of cortex and superior colliculus in mediation of visually guided behavior in the cat. *Science*, (153):1544–1547, 1966.

[14] C.C. Hilgetag and S.G. Lomber abd B.R. Payne. neural mechanisms of spatial attention in the cat. *Neurocomputing*, 38:1281–1287, 2000.

[15] S.G. Lomber and B.R. Payne. Task-specific reversal of visual hemineglect following bilateral reversible deactivation of posterior parietal cortex: A comparison with deactivation of the superior colliculus. *Visual Neuroscience*, 18(3):487–499, 2001.

[16] S. G. Lomber. The advantages and limitations of permanent or reversible deactivation techniques in the assesment of neural function. *J. of Neuroscience Methods*, 86:109–117, 1999.

[17] V. Walsh and A. Cowey. Transcranial magnetic stimulation and cognitive neuroscience. *Nature Reviews Neuroscience*, (1):73–79, 2000.

[18] K.J. Friston, C.D. Frith, and R.S.J. Frackowiak. Time-dependent changes in effective connectivity measured with PET. *Human Brain Mapping*, (1):69–79, 1993.

Articulation of Sensory-Motor Experiences by "Forwarding Forward Model": From Robot Experiments to Phenomenology

Jun Tani
The Brain Science Institute, RIKEN
2-1 Hirosawa, Wako-shi, Saitama, 351-0198 Japan

Abstract

This paper introduces the so-called "Forwarding Forward Model" network that explains how complex behavior can be learned and generated while its sensory-motor flow is hierarchically articulated. This model is characterized by a distributed representation of behavior primitives at each level, which contrasts with our prior models utilizing localist views. The model was examined through experiments using a 4-degrees of freedom arm robot with a vision system. The experimental results showed that behaviors can be generated both robustly and flexibley going through the bottom-up and the top-down interactions between levels. The characteristics of the distributed representation are discussed. Our discussion is further extended to the phenomenological issue of subjective time perception. A novel idea for explaining the sense of "nowness" is derived by applying our idea of articulating experiences to Husserl's notions of retention and protention.

1 Introduction

It is generally understood that higher-order cognition involves structural information processing which deals with the level of abstraction. For motor systems, it is generally assumed that the lower level system stores motor primitives and the higher level manipulates them for generating complex behaviors [1] [2]. Previously, we proposed a neural network model [3] which is characterized by modular and level structures. Through the learning processes, each primitive for the sensory-motor representation and their abstract representation is self-organized in a local module in the lower and the higher levels, respectively.

In contrast with this localist scheme, the current study introduces a novel scheme, namely "Forwarding Forward model" (FF-model), which emphasizes its distributed representation scheme. The FF-model is characterized by two levels of forward models which interact with each other. The lower level forward model learns to generate various sensory-motor sequences with self-associating values of the so-called parametric bias. Multiple sensory-motor spatio-temporal patterns are distributedly represented in the lower level forward model where each sensory-motor profile can be evoked by switching of the parametric bias values. When the parametric bias is changed, the dynamic pattern generated in the lower level forward model changes structurally. This is analogous to the way that dynamical structures of nonlinear systems bifurcate as their parameters change. (Note that the time constant of the parametric bias change is much slower than that of the sensory-motor flow.) The higher level forward model, on the other hand, learns to generate the sequential changes of the parametric bias by which the desired sensory-motor sequence can be produced in the lower level.

Our experiment will clarify how complex behaviors are both recognized and generated as "articulated" through learning in this distributed representation scheme. Here, "articulated" means that the sensory-motor flow is recognized/generated as segmented/combined with reusable pieces or chuncks. It will be also shown that the bottom-up and the top-down interactions between levels are essential to generate both robust and flexible behaviors.

In the discussion, we attempt to make correspondences between our technical discussions and phenomenology, as we have been inspired by the neurophenomenology research program proposed by Varela [4]. More specifically, we will apply our ideas of the sensory-motor articulations to the phenomenological discussions of subjective time. By applying our dynamical systems views to Husserl's ideas of retention and protention in the formulation of a sense of "nowness", a novel view of nowness is introduced.

2 Model

We explain our model briefly. Figure 1 shows our proposed neural network architecture. The main architecture on the left-hand side consists of two Jordan-type recurrent neural nets (RNNs) [5] which correspond to the lower and the higher level networks. These RNN networks are operated through utilizing the working

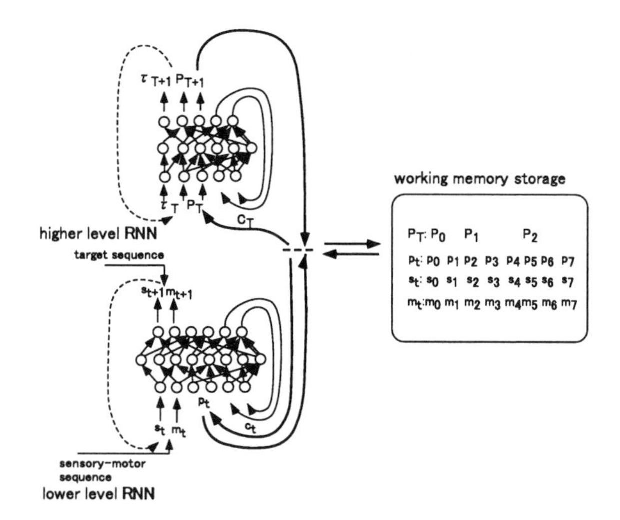

Figure 1: The FF-model utilizing two levels of RNNs.

memory storage shown on the right-hand side of the figure. In the main architecture, the lower level RNN receives two types of input. One type is the vector of the current sensory-motor values (s_t, m_t); the other is the vector of the current parametric bias p_t. This RNN outputs the prediction of the sensory-motor values at the next time step $(\hat{s_{t+1}}, \hat{m_{t+1}})$. On the other hand, the higher level RNN receives p_t as inputs, then outputs its prediction at the time step $t+1$. It is also noted that the connection of p_t between the lower and the higher levels is bi-directional depending on the operational processes. The working memory storage is used to store the sequences of the parametric bias and the sensory-motor inputs/outputs where computation of regression as well as motor planning take place as will be described later.

In the top-down process, the sequence of p_t is generated in the higher level RNN by means of its forward dynamics and its sequence is fed into the parametric bias units in the lower level RNN. Then, the lower level RNN generates the sensory-motor sequence as corresponding to the inputs of p_t. As will be described later, p_t tends to change stepwisely from time to time in the sequence. Such stepwise changes in p_t cause dynamic changes of the sensory-motor profile generated in the lower level. It is said that command-like signals of p_t sent from the higher level trigger to generate detailed sensory-motor flows in the lower level. The higher level is said to play a role of the 2nd order forward model, as its forward model learns to predict how characteristics of the lower level forward model changes in terms of the parametric bias. With closing the loop between the outputs of the sensory-motor state and its inputs, lookahead prediction is made for future sensory-motor sequences in the lower level. This mechanism is utilized for generating motor program.

The bottom-up processes are utilized in the processes of recognizing its own sensory-motor sequence experienced from certain steps before to the current step. Now, let us consider that the system experiences a sensory-motor sequence while its arm is moved with a specific patten through manual guidances. If the system already learned this pattern previously, the lower level RNN can re-generate this target sequence with adapting the parametric bias sequence to adequate one. The sequence of p_t is searched by means of the inverse problem of minimizing the errors between the target and output sequences with the smoothness constraints on the p_t sequence. Actually, p_t is obtained by back-propagating the error obtained during the regression and the delta error in the parametric bias node is utilized to update p_t. Here, the details of the update mechanism is described more specifically. The temporal profile of p_t in the sequence is computed via the back-propagation through time (BPTT) algorithm (Rumelhart, Hinton, & Williams, 1986), utilizing the sequence of the internal values of the parametric bias ρ_t, the target and the regenerated outputs of the sensory-motor sequences in the working memory storage. The total number of steps of these sequences in the working memory is L_p. For each iteration, the forward dynamics of the RNN are computed for L_p steps through establishing closed sensory-motor loops. Once the L_p steps of the sequence are regenerated, the errors between the regenerated outputs and the target ones are computed and then back-propagated through time in order to update the values of the parametric bias at each step in the sequence. The update equations for the ith unit of the parametric bias at time t in the sequence are:

$$\delta\rho_t{}^i = k_{bp} \cdot \sum_{t-l/2}^{t+l/2} \delta_t^{bp^i} + k_{nb}(\rho_{t+1}^i - 2\rho_t^i + \rho_{t-1}^i) \quad (1)$$

$$\Delta\rho_t^i = \epsilon \cdot \delta\rho_t{}^i + \eta \cdot \Delta\rho_{t-1} \quad (2)$$

$$p_t^i = sigmoid(\rho_t/\zeta) \quad (3)$$

In Eq. (1) the δ force for the update of the internal values of the parametric bias ρ_t^i is obtained from the summation of two terms. The first term represents the delta error δ_t^{bp} back-propagated from the output nodes to the parametric bias nodes which is integrated over the period from the $t - l/2$ to the $t + l/2$ steps. By integrating the delta error, the local fluctuations of the output errors will not affect the temporal profile of the parametric bias significantly. The parametric bias should vary only corresponding to structural changes in the sensory-motor sequences. The second term plays the role of a low pass filter through which frequent rapid changes of the parameter values are inhibited. ρ_t is updated by utilizing the delta force obtained from the steepest descent method, as shown in

Eq. (2). Then, the current parameters p_t are obtained by means of the sigmoidal outputs of the internal values ρ_t. A parameter ζ is employed such that the gradation of the parametric bias can be controlled. With setting ζ as relatively small values, the parametric bias tends to have more extreme values of either near to 0 or near to 1. Its value changes stepwisely only when the profile of the sensory-motor flow changes significantly.

When the robot actually behaves, the motor plan for the future L_f steps is generated in the top-down process while the past L_p steps sensory-motor sequence is regressed in the bottom-up process as described above. These computation of regression and planning are conducted as on-line iteratively for the window of $L_p + L_f$ steps which is shifted to forward at each time step while the robot actually behaves. Since this regression of the past can re-interpret and update p_t sequence, the on-line planning for future, which depends on the p_t sequence, can be generated as contextually depending on the past. The actual p_t from the past to the current step is determined utilizing both forces of the top-down prediction and the bottom-up regression with k_{top} as an arbitration coefficient between these two.

Finally, the learning is a process to search for optimal synaptic weights in both of the lower and higher level RNN and the parametric bias sequences by which the target teacher sensory-motor sequences (given through the manual guidances) can be regenerated in the outputs with minimum errors. Firstly, the lower level RNN is trained with a set of target teacher sequences. After determination of corresponding p_t sequences and the synaptic weights, the higher level RNN is trained to be able to regenerate these p_t sequences. For the reason of simplifying the learning scheme, p_t sequence is segmented with their stepwisely changing points and the resultant segmented sequence of P_T is learned associated with each segment step length τ_T in the higher level. The update of the synaptic weights are conducted by utilizing the back-propagation through time algorithm [6]. The update of the parametric bias sequence follows Eq. (2).

3 Experiments

We utilized an arm robot with 4 degrees of freedoms equipped with a vision system as shown in Figure 2. The arm sweeps horizontally on the surface of the task table on which a colored object is located. The positions of the object as well as the arm hand are perceived by a real time vision system. A handle is attached in the arm top by which the trainer can guide the arm for specific behaviors.

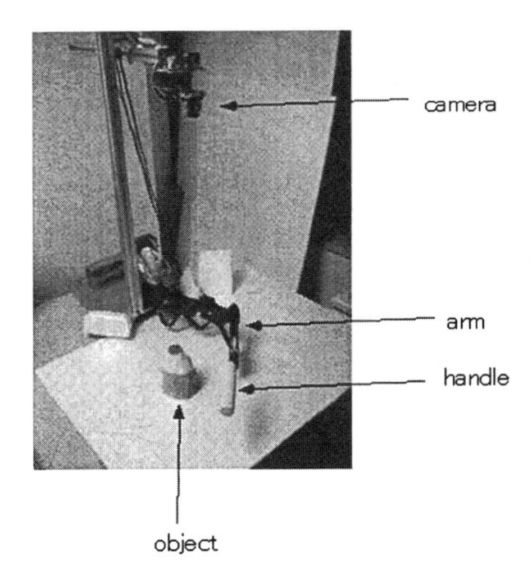

Figure 2: The arm robot.

We conduct three types of experiments. In the first experiment we examine how the robot learns to generate a set of behavior patterns by focusing on the ways of self-organizing primitive representations in the network. Nextly, we examine how the system can imitate to generate novel behavioral patterns by combining the pre-learned primitives. In the last experiment, we investigate how the motor plans can be dynamically modified in the course of their execution in response to situation changes in the environment. This experiment provides us with an opportunity to examine the roles of the bottom-up and the top-down interactions in on-line behavior generation.

These experiments are conducted using the proposed neural network model of the same specification. The network sizes as well as the parameters were determined by try and error base in order to find robust conditions for the robot experiments. The lower level RNN has 8 input nodes which are allocated for the 4 motor positions of the arm and the two dimensional cartesian positions of the hand and of the object obtained through the video camera image processing, for the current time step. The output nodes are allocated in the same way as the input nodes, but with the values for the next time step. All values are normalized to between 0.0 and 1.0. The lower level RNN has 20 hidden nodes and 8 context nodes. It also has 4 parametric bias nodes in the input layer. The higher level RNN has 4 input and output nodes which are allocated to the parametric bias of the current and the next time step respectively; it also has 10 hidden nodes and 6 context nodes.

3.1 Learn to articulate

Before the first learning experiment, the trainer prepares a set of primitive behaviors , as shown in in Figure 3(a), for which he himself practices to guide the

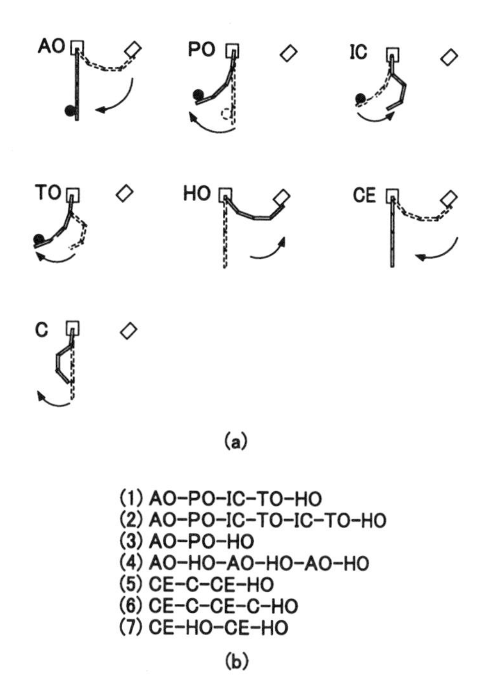

Figure 3: (a)A set of primitive behaviors and (b) the behavior sequences learned to imitate.

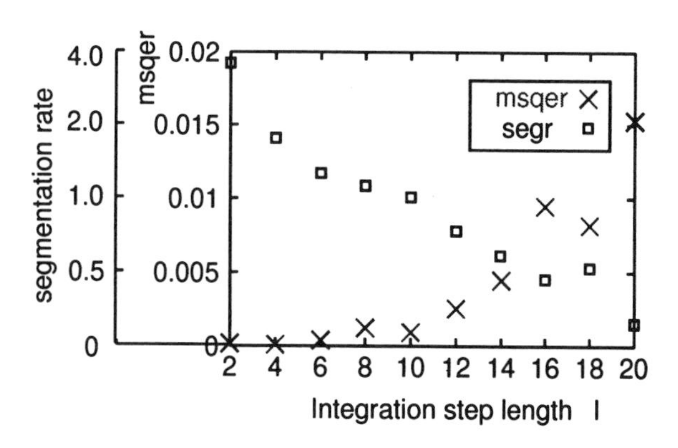

Figure 4: The mean square error (msqer) and the segmentation rate (segr, logscale) plotted as a function of l in the repeated learning trials.

robot arm until his manual guidances become stable enough. Those primitives are AO: approach to object in the center from the right-hand side, PO: push object from the center to the left-hand side, IC: perform inverse C shape, TO: touch object, HO: back to home position, CE: go to the center from the right-hand side, and C: perform C shape. Then, the robot is guided with seven sequences as shown in Figure 3(b) each of which is generated by combining the primitives prepared in sequences. Note that the training sequences are carefully designed such that they do not include any deterministic sub-sequences of the primitives. For example, after AO either PO or HO can follow. If PO were always to follow after AO, the AO-PO sequence could be regarded as an alternative primitive. The objective of this experiment is to investigate how the system seeks the segmentation structures hidden in the training sequences by self-organizing primitives in the proposed network architecture.

The learning experiments are repeatedly conducted for various integration step lengths l since this parameter is assumed to affect significantly the ways of segmenting the sensory-motor flow. We examined how the behavioral primitives are acquired as articulated in the training patterns by observing the relationship between the training error and the segmentation rate with parameter l variation. The segmentation rate is calculated as the average ratio of the actual number of the segments generated in the learning processes to the actual numbers of primitives combined in the training sequence patterns. The results are plotted in Figure 4 in which the mean square error and the segmentation rate (on a log scale) are plotted as a function of the integration step length. It is observed that the mean square error becomes higher and the segmentation rate becomes lower as the integration step length increases. This means that the learning error can be minimized if fragmentation of the segmentation is allowed to be generated, while the error with typically increase if such fragmentation is not permitted through control of the parameter l.

We inspect the results in more detail for the case in which l is set to 6, as a representative example. Figure 5 shows how the parametric bias is activated in the learning results, for three of representative training sequences. The plots in the top row in this figure show the activation of four parametric bias units as a function of the time step; the activation values from 0.0 to 1.0 are represented using the gray scale from white to black, respectively. The plots in the second and the third rows represent the temporal profile of motor and sensor values for each training sequence. The vertical dotted lines indicate the occurrence of segmentation when the behavior sequence switches from one primitive to another in generating the training sequence. The capital letters associated with each segment denote the abbreviation of the corresponding primitive behavior. In this figure, it is observed that the switching of bit patterns in the parametric bias takes place mostly in synchronization with the segmentation points known from the training sequences although it is observed that some segments are fragmented. Our examinations for all the trained sequences showed that the bit patterns in the parametric bias correspond uniquely to primitive behaviors in a one-to-one relationship in most cases.

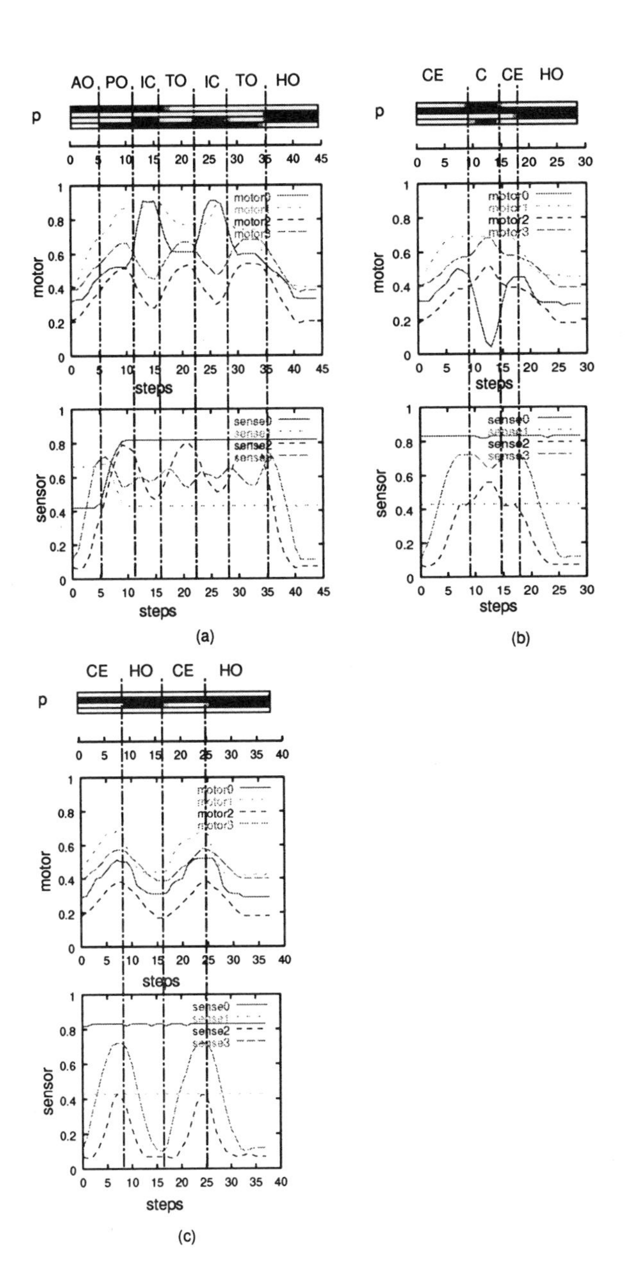

Figure 5: For the three representative training sequences (a)-(c), the temporal profiles of the parametric bias, the motor outputs are plotted in the second row and the sensor inputs are plotted in the third row. The vertical dotted lines denote occurrence of segmentation when the primitive behaviors switched. The capital letters associated with each segment denote the abbreviation of the corresponding primitive behavior.

3.2 Imitate novel combinations of the primitives

Next, we examine how the robot can imitate behavioral patterns which are prepared as novel combinations of the pre-learned primitive behaviors. Two behavioral patterns are prepared. Each behavioral pattern is taught in which only the connective weights in the higher level RNN are allowed to adapt, while those in the lower level RNN are unchanged, assuming that the internal representations for primitives are preserved in the lower level RNN. More specifically, the sequences of the parametric bias p_t are obtained by iterative computation using Eq. (1) for the lower level RNN in which the learning rate of the connective weights ϵ is set to 0.0. Subsequently, the higher level RNN is trained using the articulated sequences of the parametric bias P_T.

After the learning in the higher level RNN converges, the robot attempts to regenerate each behavioral pattern. The actual behavior of the robot is generated based on the scheme of the motor planning and regression processes described previously. Figure 6 shows the comparison of the temporal profiles between the taught patterns (the top row) and the regenerated patterns, as the robot actually behaves (the bottom row) associated with the sequence of the parametric bias (the middle row), for each sequence. It is observed that the sensory-motor profiles are successfully regenerated from the patterns taught without any significant discrepancies. These results suggest that the robot successfully learned to generate the guided behavior sequences as articulated.

3.3 On-line motor plan modification

In the next experiments, the characteristics of on-line motor plan adaptation are examined by focusing on the bottom-up and the top-down interactions. The robot is trained with two different behavioral patterns each of which is associated with a specific environmental situation. The training is conducted only for the higher level RNN while the synaptic weights in the lower level RNN are preserved as acquired in the previous experiment. The first behavior is that the arm repeats a sequence of approaching to touch the object and then returning home while the object is located in the center of the task space. The second behavior is that the arm repeats a sequence of centering, making a C-shape, and then returning home while the object is located to the left-hand side of the task space.

The test is then to examine how the behavioral patterns are switched between when the position of the object is suddenly moved from the center to the left-hand side of the task space in the middle of executing the first behavior. As the position of the object is

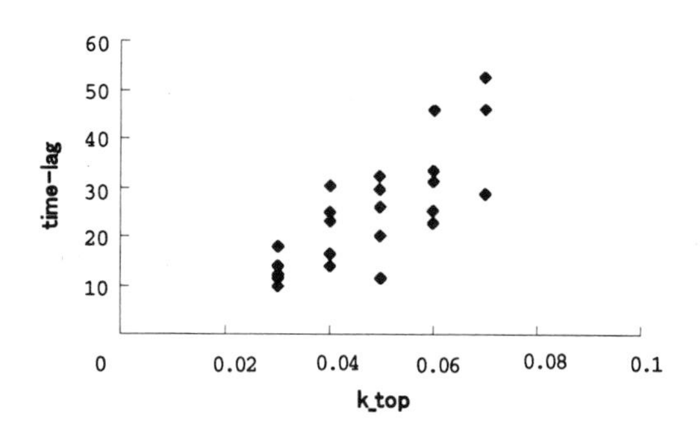

Figure 7: The time lag in the behavior switching plotted as a function of the top-down coefficient k_{top}.

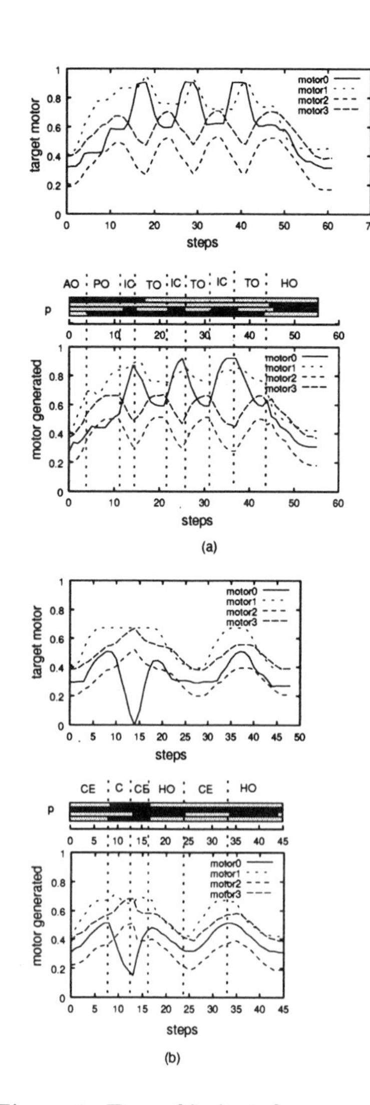

Figure 6: Two of imitated sequences.

moved, certain errors are generated in the prediction of the visual sensory inputs in the lower level RNN as a result of which the parametric bias tends to be modulated in the bottom-up way while the higher level RNN continues to proceed with the current behavior pattern, generating the same top-down signal for the parametric bias. Here, we expect to observe a transient dynamic during the switching as caused by interactions between the bottom-up and the top-down processes for determining the parametric bias.

As it is assumed that the balance between the top-down and the bottom-up processes affects the system's behavior to a large extent, the experiments on behavior switching are conducted repeatedly, changing the strength of the top-down effects by varying the coefficient k_{top} from 0.0 to 0.1 with 0.01 increment. In particular, we examine the smoothness of the behavior switching by observing the time lag from the moment the object is moved to the moment when the second behavior is activated. The switching trial was repeated for 5 times for each setting of k_{top} value. Other conditions such as the timing of the object move were set as the same for all the trials. Figure 7 shows the plot of the time lag versus the coefficient k_{top}.

It was observed that the first behavior pattern was not accomplished when k_{top} was set to less than 0.03: the parametric bias was not activated as learned since the top-down force was too weak. When the top-down behavior plan in terms of the parametric bias sequence is executed as on-line, the sequence was easily affected by even small predcition error in the lower level which was caused by noise in the robot operation. When k_{top} was set to between 0.03 and 0.07, behavior switching from the first behavior to the second behavior took place. There was a tendency that the time lag increased as k_{top} was increased. This indicates that the motor plans tend to be less sensitive to the sensation of the situation changes in the environment when the top-down effects become larger, as expected. An important observation, however, is that there is relatively large distribution of the time lag for each k_{top} value.

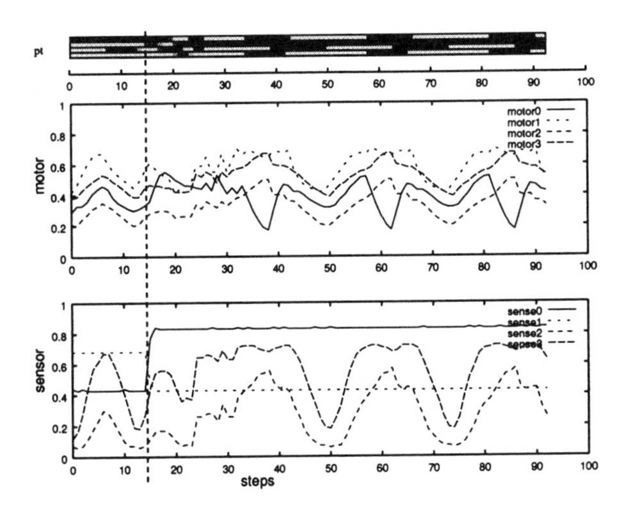

Figure 8: The temporal profiles of the parametric bias, the motor outputs and the sensory inputs in the behavioral switching trial.

This suggests that the transient behavior patterns are generated diversely during the switching. When k_{top} was set to larger than 0.07, no behavior switching was observed. The parametric bias was not affected anymore by the bottom-up sensations since the top-down influence on the parametric bias was too strong.

Figure 8 shows the temporal profile of the behavior generated in the case where k_{top} was set to 0.05. The profiles of the parametric bias, the motor outputs and the sensor inputs are plotted in the top, in the second and in the third rows, respectively. The vertical dotted line denotes the moment when the object is moved from the center to the left-hand side of the task space. It is observed that it takes 20 steps until the second behavior pattern is initiated after the object is moved to the left-hand side. It is also observed that the parametric bias, the motor outputs as well as the sensory inputs fluctuate during this transition period. The fluctuation is initiated because of the gap generated between the top-down prediction of the parametric bias and the reality as it appears in the bottom-up process. The fluctuations in the parametric bias result in generating complex motor patterns in the lower level by means of the top-down pathway which turns out to generate the sensory prediction error that is again fed-back to the parametric bias by means of the bottom-up pathway. In the 5 times of repeated trials in this parameter setting, the profiles of the traisient patterns were never repeated, as have been suggested by the large distribution of the delay time. The observed fluctuation seems to play the role of a catalyst in searching for the diverse transition path from the steady attractor associated with the first behavior pattern to that for the second behavior pattern. Once the transition is complete, the second behavior pattern proceeds steadily.

4 Discussion

4.1 Distributed or local representation

There are continuing discussions about whether primitives should be represented locally or distributedly in the networks. Tani and Nolfi [3] as well as Wolpert and Kawato [7] showed localist view models in which complex behaviors could be decomposed into a set of reusable behavioral patterns each of which is stored in a specific local neural network module. Our FF-model contrasts with this localist view in a sense that various behavior primitives are represented in a distributed manner in a single RNN at each level. A specific difference between two schemes is that number of the primitives in the local representation scheme is constrained by the number of local modules while that in the distributed one is by the number of possible bit combination in the parametric bias. It is, furthermore, assumed that an infinite number of different patterns could be generated if the parametric bias is allowed to take a graded value. In such situation, it would be furthermore interesting to ask what type of mapping is generated between space of the parametric bias and that of the behavior pattern. It is intuitively assumed that patterns could be generated by linearly interpolating learned patterns by changing the parametric bias. However, our recent studies [8] showed that the attractor patterns bifurcates nonlinearly with the linear change of the parametric bias. It was found that the nonlinear interferences among trained patterns results in generating a distorted mapping between the parametric bias and the patterns to be generated. Therefore, diverse behavior patterns could be generated which cannot be explained by the linear interpolations of learned patterns. Such example was shown in the behavior generation during the transient period in the experiment of the behavior switching. It can be said that the diversity is gained by taking advantage of the nonlinear interferences caused by the distributed representation in the network.

Which is better between the distributed and the local representation scheme? This is supposed to be a trade-off problems between the diversity and the stability in generating and learning patterns. In the distributed representation, if a novel pattern is learned in the network, it cannot be avoided that this pattern interferes with the memory patterns stored previously to some extents, since each pattern shares the same resources in the network. On the other hand, although number of patterns generated could be limited to that of local networks in the localist scheme, learning of novel patterns would not affect the memory of previously learned patterns since they do not share the resources in the networks. Our future research goal is to explore possible scheme which reside in between these two extremes.

4.2 The subjective time

In this sub section, we attempt to apply our ideas of sensory-motor articulation to the phenomenological problem of subjective time perception. First of all, we would like to explain why our research, which mostly focuses on technical issues, has been extended to the region of phenomenology. The reason is that current disciplines such as neuroscience, cognitive modeling, and phenomenology seem to be not yet powerful enough to obtain satisfactory answers for various questions of cognition. For example, our modeling of sensory-motor articulation might seem to be plausible in the view of nonlinear dynamics cognitive modeling but we cannot expect to obtain neurobiological evidence for the scheme so soon. In such a situation, Varela [4] proposed the so-called neurophenomenological hypothesis which states as follows: *Phenomenological accounts of the structure of experience and their counter parts in cognitive science relate to each through reciprocal constraints.* He considered that the following three ingredients play equally an important role: (1) the neuro-biological basis, (2) the formal descriptive tools from nonlinear dynamics, and (3) the nature of lived temporal experience studied under phenomenology. He expected that mutual interactions among those three, where effects of constraint and modification can circulate effectively, could induce substantial novelty in the exploration of lived cognition. Following this idea, we conducted reciprocal analysis between the neuro-mechanism of articulation and the experiences of subjective time.

Our question is that how we sense the temporality in our experiences. The major agreement concerning the time perception in phenomenology is that time does not flow like linear sequence as defined in physics, but having complex texture and structures [9]. Such complex structures are apparent in our awareness of the present. William James [10] pointed out the apparent paradox of human temporal experience as: on the one hand there is the unity of the present, an aggregate we can describe where we reside in basic consciousness, and on the other hand this moment of consciousness is inseparable from a flow, a stream. Intuition is that although "nowness" is perceived as a static concrete object, it is still a part of flow. Consider video, where each frame is a distinct object but also part of the larger sequence.

Husserl [11] introduced an idea of retention and protention for explaining this paradoxical nature of "nowness". He used an example of hearing a sound phrase

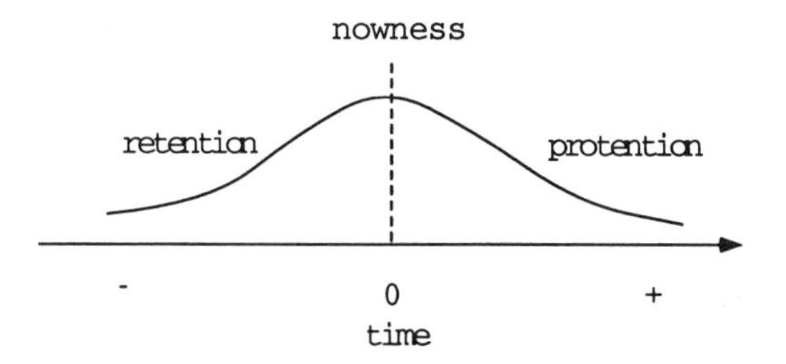

Figure 9: Husserl's idea of retention and protention. The nowness includes fringes of the immediate past and future while the physical present time is a point denoted by the dotted line.

as like "Do Mi So" for explaining the idea. When we hear the note of "Mi", we would still perceive lingering impression of "Do" and at the same time we would anticipate to hear the next note of "So". The former is called as retention and the latter as protention. Those terms are to designate the dynamics that follows impression in the present which intends the just-past and the immediate future. Those effects are a part of automatic processes and they cannot be controlled consciously. Husserl considered that the subjective experience of "nowness" are extended to include fringes both in the past and the future in terms of the retention and pretension. (Figure 9 shows a sketch for this idea.)

After coming to understand Hussearl's idea of "nowness", a question is arisen. We ask where the "nowness" is bounded? Husserl seems to argue that immediate past does not belong to a *re-presentational* conscious memory, but just to an impression. Then, how can the immediate past which is experienced just as impression slip into a distant past which can be evoked only through conscious memory retrieval? What kind of mechanism qualitatively changes an experience as from just impression to consciously retrieval episodic memory? Here, we consider that the idea of the articulation could be a key to answer the question. Our main idea is that the "nowness" can be bounded where the flow of experience is articulated. A sequential notes of "Do Mi So" constitutes a chunk within which a perfect coherence is organized in the coupling between the neural dynamics and the sound stimulus flow. Within the chunk, everything proceeds smoothly, automatically and unconsciously. However, when we hear a next phrase of "Re Fa La" after "Do Mi So", a temporal incoherence emerges in the transition between two phrases since this second phrase is not necessarily predictable from the first one. In our neural network model, the prediction error is generated at this moment which triggers the stepwise change in the parametric bias, then after a coherent state is achieved again in the system. It is supposed that at the very moment of this transition from one chunk to another, the system becomes conscious about passing of time in terms of event transition.

Our argument could be clarified further more when the issue is discussed as related to the notion of the momentary self [10]. William James [10] considered that we are conscious about ourselves not continuously in time but only discontinuously. This observation would correspond to Strawson's [12] view in which he suggests the image of a string of pearls, as an image of a self. He claims that each self should be considered as a distinct existence, an individual thing or object, yet discontinuous as a function of time. This unity of momentary self can be interpreted as the dynamical state of coherence in our system, as have been discussed also in [13]. And when the system state become incoherent where the flow is segmented, we first time becomes conscious about this unity which has already become a part of retrieval episodic memory. It can be said that the nowness is noticed only in a posterior manner after that part of flow is articulated into consciously manipulable object.

One of Husserl's goal was to explain the emergence of the objective time from the immanent time of the retention and protention level [11]. Husserl seems to consider that the sense of objective time would emerge as a natural consequence when each experience has been organized into one line of consistent sequence. The idea seems to be applicable also to a question how a consistent self can be sensed not only for momentary experiences but also those extended in time [14]. An idea of so-called narrative self explains that a coherent self emerges through making his or her story by interweaving momentary episodes experienced in the past [15]. This idea, in an abstract sense, would be explained by the memory consolidation [16] known in neuroscience. It is said that each momentary experience once stored in the short-term memory are later transfered into the long-term memory where it is consolidated into a web of organized episodic memory. Our system level explanation for this is that the chunks segmented in the lower level are learned to be a line of sequence in the higher level network.

However, if we say that the higher level deals with the formation of consciously retrieval episodic memory, another problem would arise. Let us think that we repeatedly hear "Re Fa La" after "Do Mi So" as a sequence. In such case, we can think of generating a new chunk in the further higher level which ties these two phrases into a familiar sequence. Then, the problem is that when we hear the phrase of "Do Mi So", we would have retention of "Re Fa La". Question is that if the nowness is bounded inside of "Do Mi So", or "Re Fa La" is also included in the nowness in this situation. For this problem, we assume that nowness would be sensed hierarchically as like the experience can be articulated in multiple levels in our scheme. Let us suppose a situation that we hear a phrase "Ci Re So" instead of "Re Fa La" after "Do Mi So". Then, we would feel a sense of incompatibility in this new phrase which was not anticipated after "Do Mi So" and we would say that "now" I hear a strange phrase. However, it will be different if we hear a phrase like "Re Re Fa" in which the second note is generated by a mistouch. The sense of incompatibility comes from the note level in this situation and we would say that "now" I hear a strange note. Our discussion is that our sense of nowness could be directed in different levels depending on of which level coherence is broken. Since we recognize our experience as hierarchically articulated, the subjective time should have the corresponding hierarchical structures.

In the end of this section, our arguments are summarized. We considered that the phenomenology of the subjective time perception is deeply related to the sensory-motor level behavioral structure. In order to make sense of the world, the sensory-motor experience should be articulated structurally utilizing a set of reusable schema or primitives. The experience is consciously recognized at the very moment of switching between different schema. In other words, the present experience becomes a consciously manipulable object after it is articulated into an event. Therefor, we can sense "nowness" for such event only as a posterior manner. It is also natural to assume that the sense of nowness is organized in a hierarchy since the experiences are recognized utilizing the sensory-motor level hierarchical structures.

5 Conclusions

In this paper, we have proposed the FF-model which is characterized with its distributed representation scheme. Our experiment demonstrated that complex behavior can be learned and generated as articulated by self-organizing behavior primitives by utilizing the distributed representation scheme in the lower level. It was also shown that the bottom-up and the top-down interactions are essential for the system to adapt environment both robustly and flexiblely. We discussed that the proposed mechanism of the sensory-motor articulation would explain the phenomenology of the subjective time. It was explained that the nowness is sensed hiererchically as we recognize our experiences as articulated in multiple levels.

References

[1] E. Bizzi, N. Acornero, W. Chapple, and N. Hogan, "Posture control and trajectory formation during arm movements", *J. Neurosci.*, vol. 4, pp. 2738–2744, 1984.

[2] A. Feldman, "Superposition of motor programs, I. Rhythmic forearm movements in man", *Neuroscience*, vol. 5, pp. 81–90, 1980.

[3] J. Tani and S. Nolfi, "Learning to perceive the world as articulated: an approach for hierarchical learning in sensory-motor systems", in *From animals to animats 5*, R. Pfeifer, B. Blumberg, J. Meyer, and S. Wilson, Eds. Cambridge, MA: MIT Press., 1998.

[4] F. Varela, "Neurophenomenology: a methodological remedy for the hard problem", *J. of Conscious Studies*, vol. 3, no. 4, pp. 330–350, 1996.

[5] M.I. Jordan, "Attractor dynamics and parallelism in a connectionist sequential machine", in *Proc. of Eighth Annual Conference of Cognitive Science Society*. 1986, pp. 531–546, Hillsdale, NJ: Erlbaum.

[6] D.E. Rumelhart, G.E. Hinton, and R.J. Williams, "Learning internal representations by error propagation", in *Parallel Distributed Processing*, D.E. Rumelhart and J.L. Mclelland, Eds. Cambridge, MA: MIT Press, 1986.

[7] D. Wolpert and M. Kawato, "Multiple paired forward and inverse models for motor control", *Neural Networks*, vol. 11, pp. 1317–1329, 1998.

[8] J. Tani, "Self-organization of behavioral primitives as multiple attractor dynamics: a robot experiment", in *Proc. IJCNN'2002*, 2002, pp. ??–??

[9] F. Varela, "Present-Time Consciousness", *J. of Conscious Studies*, vol. 6, no. 2-3, pp. 111–140, 1999.

[10] W. James, *The principles of psychology*, Dover Publ. (reprinted 1950), 1890.

[11] E. Husserl, *The phenomenology of internal time consciousness, trans. J.S. Churchill*, Indiana University Press, Bloomington, IN, 1964.

[12] G. Strawson, "The self", *J. of Consciousness Studies*, vol. 4, no. 5-6, pp. 405–428, 1997.

[13] J. Tani, "An interpretation of the "self" from the dynamical systems perspective: a constructivist approach", *Journal of Consciousness Studies*, vol. 5, no. 5-6, pp. 516–42, 1998.

[14] S. Gallagher, "Philosophical conceptions of the self: implications for cognitive science.", *Trends. CogSci.*, vol. 4, no. 1, pp. 14–21, 2000.

[15] D. Dennet, *Consciousness explained*, Little Brown Co., 1991.

[16] L.R. Squire, N.J. Cohen, and L. Nadel, "The medial temporal region and memory consolidation: A new hypothesis", in *Memory consolidation*, H. Weingartner and E. Parker, Eds., pp. 185–210. Erlbaum, Hillsdale, N.J., 1984.

Learning default mappings and exception handling

Fredrik Linåker*,** **Nicklas Bergfeldt***
*Department of Computer Science, University of Skövde
P.O. Box 408, SE-541 28 Skövde, Sweden

**Department of Computer Science, University of Sheffield
Regent Court, 211 Portobello Street, Sheffield S1 4DP, United Kingdom

{fredrik.linaker, nicklas.bergfeldt}@ida.his.se

Abstract

We show how a layered neural adaptation process comes to learn a simple default mapping from inputs to outputs. Then, as a succeeding step, it learns to use context for detection and handling of exceptions from this default mapping. Thereby tasks can be learnt in a step-wise manner, and the system exhibits graceful degradation if higher structures in the system are damaged.

1 Introduction

Learning correlations between multidimensional noisy inputs and outputs is a very difficult task, especially if the mapping requires that remote past inputs are taken into account. The layered architecture in (Bergfeldt and Linåker, to appear), was created to deal with this problem. It is based on two separate layers. The lower layer, layer 1, operates on the time-step level, directly mapping sensor inputs to motor outputs in a purely reflexive manner. The top layer, layer 2, consists of a recurrent neural network which can maintain a trace of the stimuli through its recurrent connections. The state of this layer is *encapsulated* through an input and output filter or *gate*, drawing on ideas from the very powerful Long Short-Term Memory (LSTM) neural networks (Hochreiter and Schmidhuber, 1997). The input gate only lets inputs flow on to layer 2 when drastic changes are detected in the inputs. At such points in time, a new pattern of activation on the output units of layer 2 is formed. This pattern is applied as *modulation* of the layer 1 mapping until the next input change occurs, then once again leading to an opened input gate. An overview of the architecture is presented in Figure 1.

A problem with this approach is that *every* internal change of layer 2 (caused by an opened input gate) will affect the layer 1 input-to-output mapping, regardless of the situation. However, all internal context updates should not necessarily require such a behavioural change, and thus an output gate was added to the system. However, an even more selective modulation could be of great

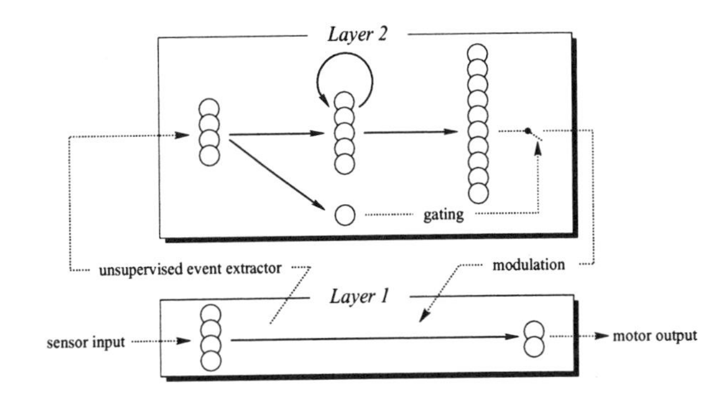

Figure 1: Overview of the layered architecture, consisting of two interconnected layers working on different time-scales.

use. Such a scheme is presented in the following, and some major advantages of this new approach are presented.

2 Experiments

Consider a simple T-maze environment in which the robot passes a light source on either the left or the right side, indicating the correct turning direction at an upcoming junction. The system should learn to apply different types of modulation at the junction, depending on the earlier light stimulus. In the previous architecture, the system thus opened the output gate whenever "junction inputs" were encountered. Working solutions were however difficult to find for the evolutionary algorithm. We here show how connecting the output gate to the *internal* nodes on layer 2 as well as the input nodes on that layer, the learning becomes much simpler. Consider the case where the right-side lights occur more frequently (70% of trials) than left-side lights (30%). In that case, the adaptation process can now be characterized as going through the following four stages (Figure 2):

a. Random movement - initially the weights on all layers are just randomly assigned values, producing

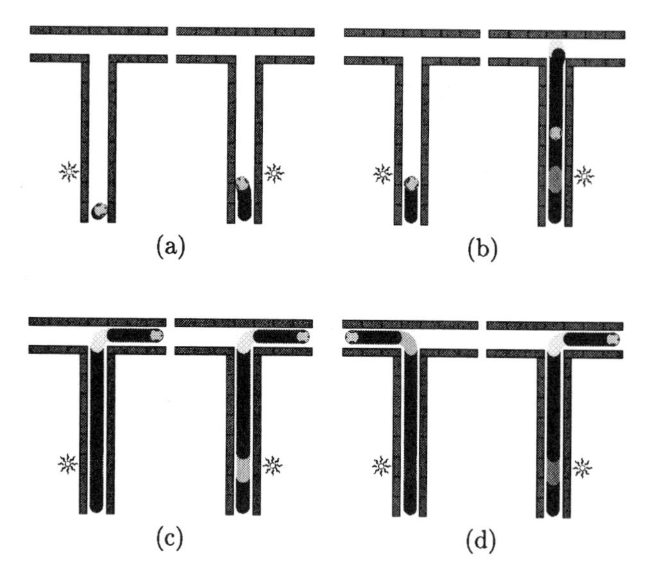

Figure 2: The four stages emerging in layered adaptation.

erratic behaviour with wall crashes leading to aborted runs and very low performance scores.

b. Obstacle avoidance - basic obstacle avoidance appears, allowing the robot to stay active for a longer time-period, but instead often are "timed out" as neither of the end-zones (in the arms of the T-maze) are reached. Most often, the obstacle avoidance behaviour is manifested in layer 1, and modulation is never applied during runs, as learning the correct modulation output in layer 2 involves many more weights than in layer 1 and there are many internal states in layer 2 which may cause the robot to crash into a wall if they are used for modulation.

c. Default mapping - some of the obstacle avoiding robots will reach the end-zones and get substantial fitness rewards if they happen to stray into the *correct* end-zone (depending on the placement of the light source they passed in the corridor). The simple reflexive layer 1 is adequate for establishing a *default mapping* which always makes the robot turn to the most frequently correct direction at the junction. The robot will thus always end up in the same end-zone, and can be said to have developed a *position habit* (always goes to the same position regardless of the context). These robots have 70% of the maximum fitness; quite good considering they still only use the simple feed-forward network in layer 1.

d. Exception handling - after a while, some robots will develop an *exception* for those cases when the light stimulus is on the less frequent side (in this example left). As long as the light stimulus is on the right-hand side, the robot will just use the default mapping as discussed earlier. However, when the robot encounters a light stimulus on the left side this is noticed but not acted upon immediately. (Right-hand light stimuli are also noticed but never explicitly acted upon, even at a later stage.) Then, at the junction, the robot modulates its own behaviour in order to change the default mapping of turning right into the desired left turning behaviour. With this simple exception handling the robot is able to complete the task to 100%, using the simple default mapping 70% of the time and modulating the behaviour during the other 30%. That is, involving further resources (higher layers) only becomes necessary when exceptions occur. When the regular, most frequent, stimuli associations are encountered, the default mapping handles the situation and the higher processing levels do not need to intervene.

There are some other advantages having this structure in the system (i.e. having a default mapping handling the most frequent decisions and then extending the system with exception handling to also deal with any irregularities). If damage is inflicted to higher processing levels (i.e. layer 2 here), then the system reverts to the best non-informed decision available, namely the default mapping, and thus still successfully handles the task most of the time (here 70% performance). If damage is inflicted to the low-level processing (layer 1), the whole system will however remain unable to act, even though the higher level processing systems will receive events and generate modulation as previously.

3 Conclusions

We have showed how a layered system can be constructed which is able to appropriately apply selective modulation. In this system, the lower reflexive layer automatically comes to implement a *default mapping* from inputs to outputs, corresponding to the best (most frequently rewarding) available action the system can take without taking any context into account. The higher layer stores massive amounts of context and uses this to detect *exceptions* where a behaviour different from the default mapping could be employed to receive an even higher reward. Then the lower layer is modulated to generate the alternative behaviour instead. In this manner, complicated long-term stimuli-response mappings can be learnt in a step-wise manner, first establishing a default mapping in the lower layer for the most commonly rewarding behaviour, then the higher layer can learn exceptions from this default mapping as a separate stage.

References

Bergfeldt, N. and Linåker, F. (to appear). Self-organized modulation of a neural robot controller. In *International Joint Conference on Neural Networks (IJCNN2002)*.

Hochreiter, S. and Schmidhuber, J. (1997). Long short-term memory. *Neural Computation*, 9(8):1735–1780.

SELF-ORGANIZATION AND LEARNING

Self organisation in a simple task of motor control

Christian R. Linder

Department of Biological Cybernetics, University of Bielefeld, PO Box 100131, 33501 Bielefeld, Germany.
Email: christian.linder@biologie.uni-bielefeld.de

Abstract

This paper provides a mechanism that is able to self adjust a network that is used as a feedback controller in a motor control system. The following disturbances are considered: (1) synaptic drift, (2) a change in the slope of the sensor characteristics and (3) a shift along the ordinate in the sensor characteristics. I show that these disturbances can be counterbalanced, if the following conditions are fullfilled: (1) The sign of each synapse, i. e. being an excitatory or an inhibitory one, but not its exact value, is genetically determined. (2) The input characteristics are described by straight lines. (3) A mechanism is required that changes the input synapses of a cell in a way that the longterm influence of positive (excitatory) and negative (inhibitory) synapses cancel each other. (4) Both input values have to be distributed in a way that they vary around the same mean value. No explicit knowledge, neither concerning the shape of the trajectory nor concerning the involved sensor characteristics is required. This solution is possible because in the architecture proposed here each neuron is endowed with two inputs of opposite sign.

1 Introduction

In physically realized or in simulated animats that are controlled by neuronal networks, the latter are usually considered to have fixed synaptic connections, because the learning processes necessary to train the networks are performed offline in most cases.

In biology, the situation is quite different. Neurons are dynamically changing entities, cell membranes are semifluid bimolecular layers. New ion channels are produced continously and old ones are disposed. Hormones and other substances alter these dynamics and take influence on the information processing and synaptic plasticity. Furthermore, sensors and effectors can be destroyed or deteriorated by mechanical influences, as for example sensory hairs may brake, transparent tissue may become dull and the body as a whole gets weaker and deficient with age. In addition, body size including sensor and effector characteristics may change drastically during

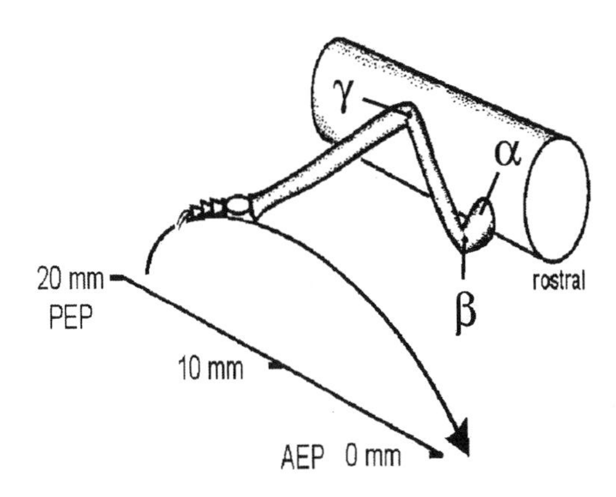

Figure 1: The Swing Movement

development. Through all these changes, the functionality of the neuronal system has to be maintained. This can not be accomplished by synaptic weights that are coded in the genome and then are built up once and kept constant during lifetime. There have to be some kind of self-organising principles that lead to homeostatic properties and stabilize the system against such disturbances.

Generally, the following three possible disturbing influences are possible: On the input side, the sensor characteristics can change in various ways. For linear characteristics, a change in slope, a shift along the ordinate and partial or complete dropouts of a sensory field could occur. Synaptic weights can change over time as the overall number of ion channels per synapse varies when new channels are produced and old ones are disposed. On the effector side the characteristic of the muscles, i. e. the translation from neuronal signals to force might change.

The long term goal of this investigation is to find self-organising principles that could be applied to biological as well as artificial neural networks. Using simulation as a research tool, my work concentrates on the WALKNET, an ANN that has been proposed to control six legged walking (Cruse et al., 1998). In this architecture the problem of six legged walking is divided into two tasks, one concerns the coordination between the legs, the other concerns the coordination of the movement of the three joints of each individual leg. The latter task can be divided into two subtasks: the generation of stance movement and the generation of a swing trajectory. In this article, I focus on one specific module of the WALKNET, the SWINGNET, a simple, one-layered, Perceptron-like neuronal network with linear activation func-

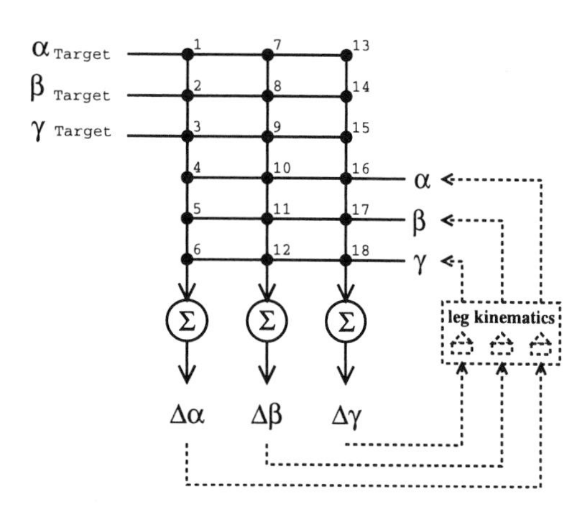

Figure 2: The SWINGNET (Cruse et al., 1998)

tions responsible for the the swing movement of the leg of a walking hexapod (Fig.1). This movement starts when the tip of the leg is lifted off the ground at the posterior extreme position (PEP). The tip is then moved forward until it touches the ground again at its anterior extreme position (AEP). To perform this movement, three joints, called α, β and γ have to be moved in a coordinated way.

The SWINGNET (Fig. 2) consists of three neurons in one layer. Their input is composed of six values. Three values code for the target angle of the alpha, beta and gamma joint. This is given by a module called TARGETNET, a three layered feed forward network which receives input from sense organs of the neighbouring leg. This input codes for the desired AEP. The TARGETNET will not be considered further here. The three other input values deliver the proprioceptive information, which corresponds to the actual angle of each joint of the leg to be controlled. The output codes for the angular change in the next timestep.

2 Results

2.1 Genetic Algorithms

In earlier investigations the weights of the SWINGNET were either hand coded or found using different types of supervised learning. As a first step and as a basis for later investigation, the solution space of the network will be considered using traditional GA.

In the genome, the weights are directly encoded as floating point values between -1 and 1. The mutation rate was set to 0.3, which means that every weight in every generation had a probability of 0.3 to undergo a point mutation. This point mutation either corresponded to the addition of a random number between -0.01 and 0.01 (probability 0.2) or set the weight to 0 (probability 0.1). Additionaly, I introduced a 10% probability for a genome mutation, which corresponds to the multiplication of every weight by a random number between 0.5 and 1.5. The crossover possibility was defined as the likelihood of skipping the reading process to the other

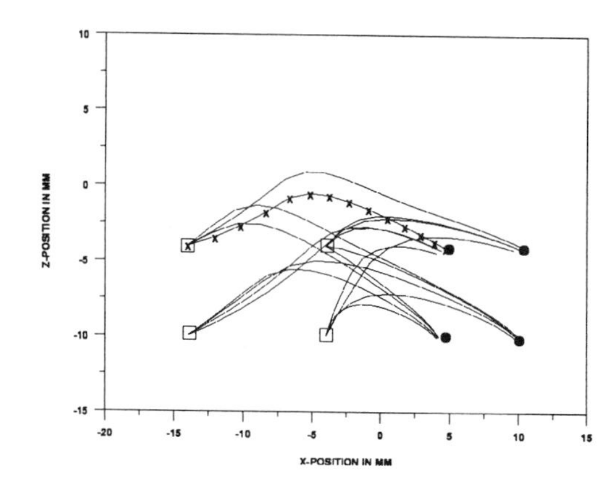

Figure 3: Movement of the tibia endpoint from four different start positions (squares) to four different target positions (filled circles) viewed from the side. Depicted are 16 trajectories produced by a weight distribution determined with GA. For one trajectory, timesteps are marked with crosses.

parents genome (after having read one weight) and was set to 0.5. After evaluation, the genomes were sorted according to their overall fitness, and parental genomes from the upper third were chosen to produce offspring that replaces the lower two-thirds.

The SWINGNET as shown in Fig. 2 is completed by a simulated external world that transforms the motor output of each unit into an angular position for the corresponding joint. This is simply done by a linear integration of the output signal (see dotted box in Fig. 2). In the simulation, the swing movement is finished if either a given time is reached or if the tip of the leg has crossed the line connecting starting position (PEP) and the desired end position (AEP). As fitness value I used the distance between final position and desired position, i.e. no information concerning the form of the trajectory has been used (in contrary to earlier investigations). In a second approach, the environment was made somewhat more complex in the way that obstacles of different height and position were placed between starting position and desired end position.

Using this simple fitness function, without application of the obstacles, the simulation produces very flat movements different to that shown in Fig. 1. Using external knowledge, for example concerning the upper extreme position of the swing trajectory, and applying this knowledge to the fitness function, would easily lead to better results. However, this would not reflect the situation of the biological system, because it is very impropable that the neuronal system has explicit knowledge concerning the form of the leg trajectory. Instead, I applied an approach closer to biology with simply changing the environment in a realistic way: I introduced different obstacles to impede the movement if the trajectory was not high enough and then applied the same fitness function as above. This procedure indeed led to trajectories that followed the biological example much better (see Fig. 3 for a selection of trajectories). If the height of the obstacles used is low-

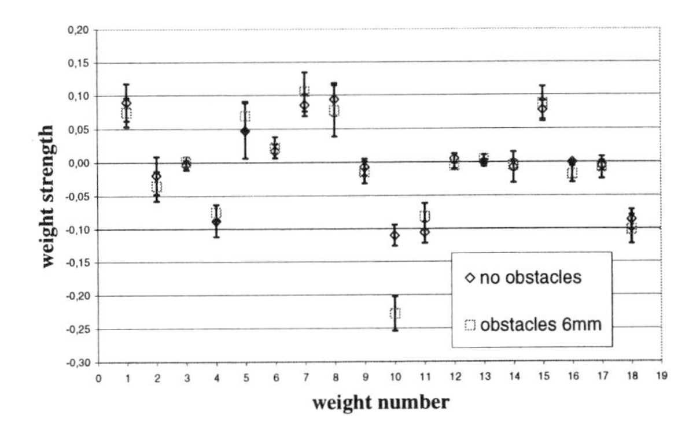

Figure 4: Weight distributions for the SWINGNET obtained in different runs of the genetic algorithm. Mean values and standard deviation of 7 runs without obstacles and 7 runs with obstacles are shown.

ered, the leg quickly adapts to this lower height. This is due to a trade-off between targetting accuracy and swing height. For higher swing trajectories there is also a higher mean targetting error. This means that the system can automatically adapt to the form of ruggedness of the environment, an effect that could be interpreted as a way to save energy.

To investigate the solution space of the task, the GA procedure was repeated 14 times. Mean values and standard deviation of the resulting weights were determined. The results plotted in Fig. 4 show that all weights have quite narrow distributions indicating that there is only one solution for the task. Many of the weights are always close to zero. Interestingly, the non-zero weights mainly refer to the three connections within one joint (Fig. 4, weights 1, 4, 8, 11, 15 and 18) plus two weights (Fig. 4, weight 7 and 10) that connects the controller of the beta joint with the input from the alpha joint and the alpha target. The value of weight no. 10 varies systematically with the height of the swing movement.

As a number of weights turned out to approach values close to zero, in a second run the GA procedure was repeated with these weights being fixed to exactly zero. The results were esentially the same (not shown).

These results led to the approach deployed in the following sections. The task of generating a swing movement is devided into a subtask responsible for the targetting behaviour and another subtask controlling the lifting of the leg. In the following section, Only the first of these subtasks is explained in detail. Possible mechanisms regarding the latter subtask are briefly described in section 2.3.

2.2 *Online self organisation for the targeting behaviour*

The SWINGNET has two types of synapses, synapses that directly connect input and output of a single joint, e. g. target angle α, sensory input for joint α and motor output to that joint. This refers to the synapses 1, 4, 8, 11, 15 and 18 in Fig. 2. Other synapses interconnect the three joint controllers. As has been shown in 2.1, with one exception all synapses of the latter type are not necessary for the SWINGNET. To start with a simple case, I will therefore first concentrate on a single joint system, i. e. a controller that contains only synapses of the first type.

Such a simple single joint controller is depicted in Fig. 5. The system can be considered as the neuronal implementation of a negative feedback controller. There are two inputs. One represents the target angle provided by a neuronal system, which in WALKNET is simulated by TARGETNET. The latter in turn depends on sensory input describing the position of the rostrally neighbouring leg. As mentioned above, the details of this net will not be considered here. The other input is received from sensors that monitor the actual angle of the joint to be controlled. The sensors transform an actual angle value, termed real world angle (RWA), into a neuronal signal, the sensory output SO. I assume here, that the characteristic of the sensor can be described by a line with constant slope. Therefore the transformation can be written as $SO_A = m_A * \mathrm{RWA}_A + t_A$, with slope m and intercept t as constant values. This signal is then given to the motor unit via a synapse w_2. Therefore the motor unit receives from the sensor an internal signal

$$IS_A = w_2 * m_A * \mathrm{RWA}_A + w_2 * t_A.$$

Correspondingly, the input given by the target net is transmitted via a synapse with weight w_1 and can therefore be described by $IS_T = w_1 * m_T * \mathrm{RWA}_T + w_1 * t_T$, as long as we stay with the simplification of approximating the characteristic of the TARGETNET by a line with constant slope, too.

I assume further, that the following three conditions are fulfilled for this system:

1. The sign of each synapse, i. e. being an excitatory or an inhibitory one, but not its exact value, is genetically determined.

2. A neuron can control its longterm mean activity in the following way: There exists an internal metabolic mechanism for long term integration of the cell's activity that is able to scale up or down the overall strength of excitatory and inhibitory synapses such that the long term average of the cell activity remains constant. In our case the desired ratio between long term excitatory input and long term inhibitory input is assumed to be 1:1, i.e. the excitatory input integrated over time being the inverse value of the inhibitory input integrated over time. In other words this means that the strength of the excitatory input and that of the inhibitory input cancel each other in the long term average. There are biological investigations showing that this can be acomplished by metabolic processes of a single neuron (Turrigiano et al., 1998).

3. On a long term scale, the input of each pathway is distributed axially symmetric to the midline over the whole

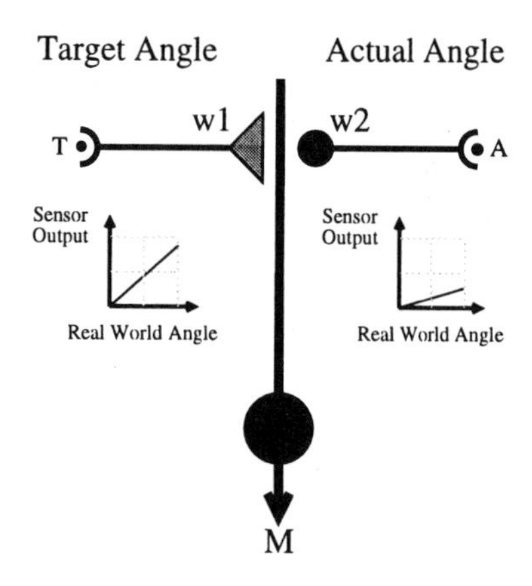

Figure 5: A simple model for a single joint controller: There are only two inputs. One coding for the target angle and one for the actual angle (proprioceptive input). The output (=activation of the motorneuron) is coding for angular change. The characteristics are assumed to be described by $SO = m * \text{RWA} + t$. For the inset figures shown here, t=0 in both cases.

sensor range. This is fulfilled, for example, if the sensory input follows a Gaussian distribution or if it is equally distributed. The latter assumption is often used to describe systems that are self organising, for example in the development of Kohonen networks. However, as will be explained in the discussion, this condition may often not be given in real life and special assumptions are necessary to fulfill this condition.

2.2.1 *The simplest possible model of a single joint controller*

Applying condition 1 to the system shown in Fig. 5, I assume that the synapse described by weight w_1 has a positive sign and that synapse w_2 has a negative sign. Therefore the output value is given by

$$M = IS_T + IS_A = w_1 * m_T * \text{RWA}_T + w_1 * t_T +$$

$$+ w_2 * m_A * \text{RWA}_A + w_2 * t_A \quad (1)$$

with w_1, m_A, m_T, t_A and t_T being zero or positive and w_2 being negative. The output is send to the effector and codes for an angular change. For this first example, we assume the encoding of negative values as possible. A solution for biological neurons with only positive firing rates is explained only for the extended architecture of the following section, but can be easily transferred to the system at hand. The task of the system is to bring the actual angle closer to the target angle. Therefore, the output M should be zero, if real world target angle RWA_T and real world actual angle RWA_A are equal (i. e. for $\text{RWA}_T = \text{RWA}_A$), since then the target angle is reached and any further movement would worsen the situation. In other words, the task of the system is to provide a positive output M whenever the real world target angle RWA_T is higher than the actual real world angle RWA_A, and a negative output for the reverse case, respectively.

The question adressed here now is: Can the system determine the values of the parameters w_1 and w_2 in a self organised way such that the task of the system can be fulfilled, even under influence of disturbances?

For an easier explanation, let us first start with the special assumption $t_T = t_A = 0$. This assumption will later be released to meet the general case. For $t_T = t_A = 0$, (1) simplifies to

$$M = w_1 * m_T * \text{RWA}_T + w_2 * m_A * \text{RWA}_A \quad (2)$$

If $w_1 * m_T \neq -w_2 * m_A$, the movement will not stop as required when $\text{RWA}_T = \text{RWA}_A$. This means that the task of $M = 0$ is only solved, if $w_1 * m_T = -w_2 * m_A$. Therefore, in the special case of $t_T = t_S = 0$ and arbitrary, but fixed values of m_T and m_A, our question reduces to find a mechanism that is able to achieve the condition of $w_1 * m_T = -w_2 * m_A$. This goal can be reached, if we apply the conditions 2 and 3 mentioned above: Application of condition 2 means that the neuron has a built-in mechanism that self regulates the long term average activity $\overline{M}$ by scaling up or down its input synapses w_1 and w_2. If, for example, $w_1 * m_T * \overline{\text{RWA}_T} > -w_2 * m_A * \overline{\text{RWA}_\text{A}}$, the neuron will be overexcited and a metabolic process will be triggered that scales down the excitatory synapse w_1 and/or scales up the inhibitory synapse w_2. Correspondingly, if $w_1 * m_T * \overline{\text{RWA}_T} < -w_2 * m_A * \overline{\text{RWA}_\text{A}}$, w_1 will be scaled up and w_2 will be scaled down. This process continues until the system reaches its equilibrium at $\overline{M} = 0$, which means that $\overline{w_1 * m_T * \text{RWA}_T + w_2 * m_A * \text{RWA}_\text{A}} = 0$.

Supposing that the rate of weight change is very slow and therefore a magnitude below the time interval used to calculate the long term average $\overline{\text{RWA}}$, and if we believe m_T and m_A to be constant in these intervals, this can be written as

$$w_1 * m_T * \overline{\text{RWA}_T} + w_2 * m_A * \overline{\text{RWA}_\text{A}} = 0 \quad (3)$$

If we assume, that during the procedure of adjustment, the whole range of physiologically relevant positions can be represented, RWA_T and RWA_A both cover the same range. Furthermore, following condition 3, both the RWA_T and RWA_A values vary symmetrically around their respective long term average, $\overline{\text{RWA}_T}$ and $\overline{\text{RWA}_\text{A}}$. With these two assumptions, it follows that $\overline{\text{RWA}_T} = \overline{\text{RWA}_\text{A}} = \overline{\text{RWA}}$ and therefore Equation 3 simplifies to

$$w_1 * m_t + w_2 * m_A = 0$$

This means that, on the long run, these mechanisms lead to an equilibrium where $w_1 * m_T = -w_2 * m_A$ as desired, given that the change of weight values is much slower than the change of the RWA values, or, in other words, if the learning rate is small enough.

In the case of $t_T = t_A = 0$ and with the sign of the synapses being determined genetically, the system can stabilize against arbitrary changes of the weights w_1 and w_2 and

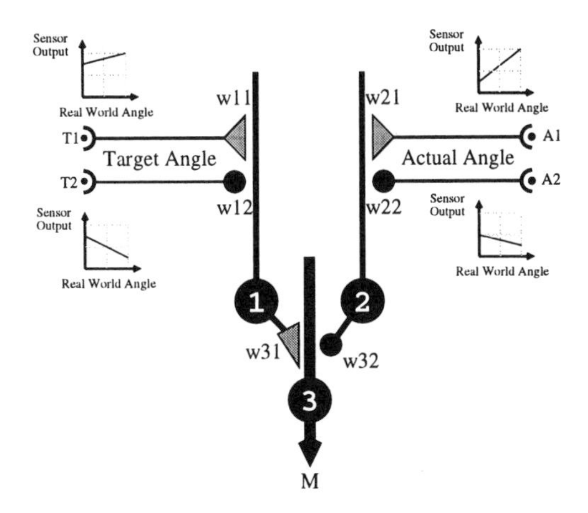

Figure 6: A model of a single joint controller using 3 neurons, two interneurons and a motorneuron. Each interneuron receives input via two antagonistically organised channels as indicated by the input characteristics..

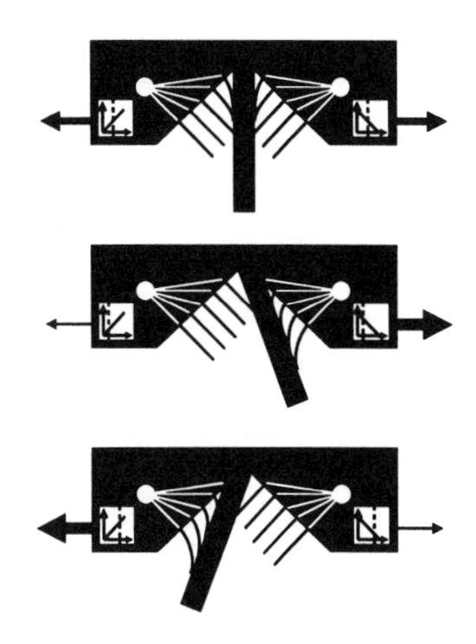

Figure 7: In the stick insect Carausius morosus, there is not a single sensor gauging a joint angle, but fields of sensory hairs on opposite sides of the joint. Therefore, in the improved architecture proposed here, there are inversely proportional inputs for each pathway, which resembles much better the situation in the stick insect.

at the same time counteract changes in the slope of the sensor characteristics m_T and m_A in order to maintain the behavior of a sensible feedback controller.

2.2.2 *Taking into account redundant sensory information*

Is it also possible to stabilize the system if the condition $t_T = t_A = 0$ is not anymore fulfilled? To achieve this goal, a new architecture will be introduced. The single neuron shown in Fig. 5 is now replaced by three neurons of the same type, i. e. neurons that receive one excitatory and one inhibitory input. These neurons, called Interneuron 1, Interneuron 2 and Motorneuron 3, are connected as shown in Fig. 6. What happens if we apply condition 2 to interneuron 1? The latter receives input only from the system providing the target angle, Interneuron 2 receives input only from the sense organs measuring the actual joint angle. The important new aspect is that each interneuron does not receive one signal, but instead receives two inputs providing redundant information. As it is often found in biology, these two input channels represent an antagonistically structured system. For example, the joint angle may be monitored by two sense organs. One increases excitation, when the joint angle increases. The other sensor works in the opposite way: its activation decreases when the joint angle increases. In Fig. 7 this is illustrated by a sketch showing two hairfields arranged at either side of the joint (This situation can be found in the coxal joint of insects (Wendler, 1964). Also chordotonal organs are known to have sensory cells that respond to elongation, and other cells that respond to relaxation of the organ (Burrows, 1992). It has been shown that interneurons in the motor system of insects are arranged antagonistically, too (Bueschges and Schmitz, 1991). Again, I use the simplifying assumption that the translation for each sensory system can be approximated by a line with constant slope $SO = m * \text{RWA} + t$. The two internal signals received by Interneuron 1 are thus $IS_{T1} = w_{11} * (m_{T1} * \text{RWA}_T + t_{T1})$ and $IS_{T2} = w_{12} * (m_{T2} * \text{RWA}_T + t_{T2})$. The antagonistic structure means that m_{T1} and m_{T2} have different sign. Let us assume without loss of generality that m_{T1} has a positive sign and m_{T2} has a negative one. Likewise for the weights we assume that w_{11} has a positive and w_{12} a negative sign. The overall excitation of Interneuron 1 follows as

$$X_{IN1} = IS_{T1} + IS_{T2} = w_{11} * (m_{T1} * \text{RWA}_T + t_{T1}) + \\ + w_{12} * (m_{T2} * \text{RWA}_T + t_{T2}) \quad (4)$$

or

$$w_{11} * t_{T1} + w_{12} * t_{T2} + (w_{11} * m_{T1} + w_{12} * m_{T2}) * \text{RWA}_T \quad (5)$$

with w_{11}, m_{T1}, t_A and t_T being positive and w_{12} and m_{T2} being negative. According to condition 2 on a long term scale X_{IN1} should average to zero by scaling up or down the weight w_{11} and w_{12} until

$$\overline{X_{IN1}} = 0 \quad (6)$$

As $\overline{(m * \text{RWA} + t)} = m * \overline{\text{RWA}} + t$, equation 4 and 6 can be written as

$$w_{11} * (m_{T1} * \overline{\text{RWA}_T} + t_{T1}) + \\ + w_{12} * (m_{T2} * \overline{\text{RWA}_T} + t_{T2}) = 0 \quad (7)$$

The expressions $(m_{T1} * \overline{\text{RWA}_T} + t_{T1})$ and $(m_{T2} * \overline{\text{RWA}_T} + t_{T2})$ are not negative since they code for a neuronal spike frequency (the activation of the sensor T1 and T2, respectively). Therefore equation 7 is always solvable for an appropriate

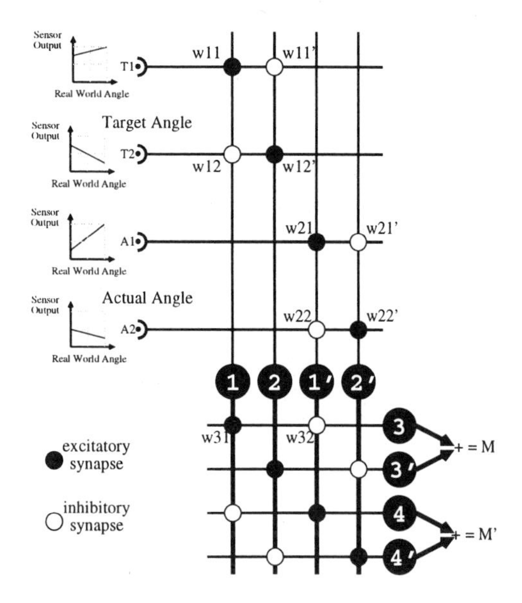

Figure 8: A model closer to biology, where only positive values are encoded. Output is given via two motorneurons converging on each of the antagonistic muscles. Motorneuron 3 is activated for forward movements in the second half of the input space, motorneuron 3' for forward movements in the first half. For movements originating in the first half and aiming to the second, both neurons are activated. The same holds for motorneuron 4 and 4' in backward movements.

pair of a positive w_{11} and a negative w_{12} as demanded. It can be transformed to

$$(w_{11} * t_{T1} + w_{12} * t_{T2}) = -(w_{11} * m_{T1} + w_{12} * m_{T2}) * \overline{\mathrm{RWA}_T}$$

If we substitute the first term of equation 5 by the latter term, the overall excitation of Interneuron 1 for a specific real world angle becomes

$$X_{IN1} = (w_{11} * m_{T1} + w_{12} * m_{T2}) * (\mathrm{RWA}_T - \overline{\mathrm{RWA}_T}) \quad (8)$$

What have we achieved by applying condition 2? Equation 8 can be seen as a linear equation of the form y=m*(x+k), which corresponds to a straight line with slope m and an x-axis intercept of k. This means, we have obtained a defined abscissa intercept at $k = \mathrm{RWA}_T$. The corresponding considerations can be applied to IN2. Accordingly, the overall excitation of Interneuron 2 can be written as

$$X_{IN2} = (w_{21} * m_{A1} + w_{22} * m_{A2}) * (\mathrm{RWA_A} - \overline{\mathrm{RWA_A}}) \quad (9)$$

Since according to condition 3, RWA_T and RWA_A cover the same range and therefore $\overline{\mathrm{RWA}_T} = \overline{\mathrm{RWA}_A} = \overline{\mathrm{RWA}}$, equation 8 and 9 can be written as

$$X_{IN1} = m_{IN1} * (\mathrm{RWA}_T - \overline{\mathrm{RWA}})$$

and

$$X_{IN2} = m_{IN2} * (\mathrm{RWA}_A - \overline{\mathrm{RWA}})$$

with $m_{IN1} = w_{11} * m_{T1} + w_{12} * m_{T2}$ and $m_{IN2} = w_{21} * m_{A1} + w_{22} * m_{A2}$. This means that both lines may vary in slope, but are shifted by the same value ($\overline{\mathrm{RWA}}$) along the abscissa. For their biological meaning, of course, X_{IN1}, X_{IN2} and M can only be positive. From now on, we therefore restrict the consideration to cases where $\mathrm{RWA}_T > \mathrm{RWA_A} > \overline{\mathrm{RWA}}$. This results in a restriction of the motor output to positive values, and also a restriction to movements in the second half of the input (and output) space. In the complete model (Fig. 8), the encoding of the negative excitation is handled by additional neurons: The motorneuron 3' for movements in the first half of the input space and an antagonistic pathway with interneuron 1' and 2' and motorneuron 4 and 4' for backward movements in the first and second half of the input space, respectively. All the synapses of this parallel, but antagonistic system are connected inversely (i. e. an excitatory synapse would be an inhibitory one and vice versa). Hence, although these channels use as well only positive excitation, they enable the system as a whole to code for movements in both directions. The logic of this antagonistic output pathway corresponds to the system explained here, and is therefore not considered further.

In the following step, the rectified output values of both interneurons X_{IN1} and X_{IN2} are given as input to the motorneuron, using a positive weight w_{31} for X_{IN1} representing the target angle and a negative weight w_{32} for X_{IN2} representing the actual joint angle. As both functions intercept the x-axis at the same point, also the sum of both, i. e. the activation of the motorneuron, has the same x-axis intercept, but an arbitrary slope. The output of the motorneuron can be described as:

$$M = w_{31} * X_{IN1} + w_{32} * X_{IN2} =$$
$$= w_{31} * m_{IN1} * (\mathrm{RWA}_T - \overline{\mathrm{RWA}}) +$$
$$+ w_{32} * m_{IN2} * (\mathrm{RWA}_A - \overline{\mathrm{RWA}})$$

If we now again apply condition 2 to the motorneuron, the weights w_{31} and w_{32} will change over time until the long term average $\overline{M} = 0$, which means that

$$w_{31} * m_{IN1} * \overline{(\mathrm{RWA}_T - \overline{\mathrm{RWA}})} =$$
$$= -w_{32} * m_{IN2} * \overline{(\mathrm{RWA}_A - \overline{\mathrm{RWA}})} \quad (10)$$

According to condition 3, RWA_T and $\mathrm{RWA_a}$ are distributed axially symmetric over the whole range, which means that $\overline{\mathrm{RWA}_T} = \overline{\mathrm{RWA}_A}$ and therefore also $\overline{\mathrm{RWA}_T - k} = \overline{\mathrm{RWA}_A - k}$, if both input channels cover the same range. Note that $\mathrm{RWA}_T - \overline{\mathrm{RWA}} \neq 0$ in this case, since we are still restricted to positive neuronal excitation ($\mathrm{RWA}_T > \overline{\mathrm{RWA}}$ and $\mathrm{RWA}_A > \overline{\mathrm{RWA}}$). Therefore, with application of condition 2 and 3, equation 10 simplifies to $w_{31} * m_{IN1} = -w_{32} * m_{IN2}$. The desired behaviour is established and the actual angle approximates the target angle asymptotically.

As seen in the previous section, self organisation in the motorneuron is possible, if the characteristics of the input channels vary in slope, but not in their respective abscissa intercept. This in turn can be accomplished by interneurons using exactly the same mechanisms: If in compliance with condition 2 and 3, their output represents the target (and actual) angle, shifted along the abscissa for a common value (the long term average $\overline{\text{RWA}}$). The final behaviour of the network is therefore independent of arbitrary shifts and slope changes in any input characteristics and can be attained with randomized weight values in the beginning, always assumed that the inputs are evenly distributed for the time of adjustment and that the characteristics are linear over the whole sensory range.

2.2.3 *Simulation results*

For better understanding, a simulation of the system shown in Fig. 6 was conducted for different changes in the input characteristics. Generally, the following cases are imaginable:

1. synaptic drift in one, many or all of the involved synapses
2. a change in the slope of a characteristic
3. a shift along the ordinate
4. a partial dropout of sense organs, i. e. one of the input channels does not cover the whole range
5. a complete dropout of an entire pathway
6. non linear characteristics

A SWINGNET with fixed synaptic weights is completely distorted by any of these changes: The actual angle will not approximate the target angle anymore. In the simulation, I show that with the mechanisms described in the preceeding section, distortions of type 1 through 3 can be counterbalanced. The results are depicted in Fig. 9.

The input values for both target and actual angle were scattered randomly (and therefore evenly distributed) between 0 and 180. The long term average of excitation was calculated over the 10 precedent movements. In case of a long term over- or underexcitement, both positive and negative weights were adjusted by adding or subtracting the value 0.001. This corresponds to the learning rate, which was purposely chosen to be very low to prevent oscillation or indefinite growth. In the following, the output signal in the case of $\text{RWA}_T = \text{RWA}_\text{A}$ will be called error signal. As mentioned before, the output signal should be zero in this case, because the target is reached already.

Dealing with distortions of type 1

For the beginning, the input channels for target and actual angle were chosen to have the same characteristics, and the antagonistic pathways were chosen to have inversely proportional characteristics. To test for distortions of type 1, all weights w_{ij} of the network shown in Fig. 6 are randomized. The randomized weight distribution results in an initially high error signal. On a longer time scale, however, the weights change due to the neuronal conditions and the

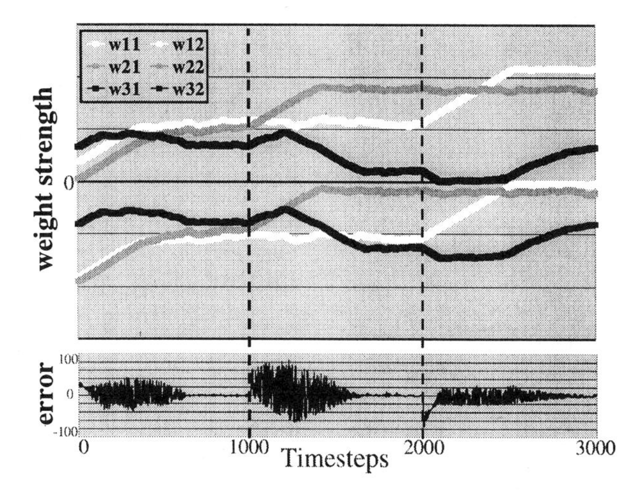

Figure 9: Simulation of the dynamics of the network depicted in Fig. 6, showing the recovery from different distortions. At the beginning, all weights are randomized. From timestep 1000, the gain of the sensor A1 is multiplied by 0.1. In addition to this, from timestep 2000 T1 is multiplied by 0.1, T2 is multiplied by 4 and shifted by $T2_{Max}$.

error signal decreases to zero. Since the antagonistic pathways were chosen to have inversely proportional characteristics, the weights take on pairwise inversely proportional values as well ($w_{11} = -w_{12}, w_{21} = -w_{22}, w_{31} = -w_{32}$ in Fig. 6).

Dealing with distortions of type 2 and 3

As stated above, the three neuron system in Fig. 6 should be able to counterbalance for additional distortions of the type 2 and 3, too.

To verify this, the parameters of the initial characteristics were changed at two different points in time (the dashed lines in Fig.9). First, the gain of the input channel A1 (see Fig. 6) was multiplied by a factor of 0.1 (see Fig. 9, t=1000). As a consequence, the error signal increases drastically, since the joint angle comes to a rest at 30 degrees for a real world target angle of 3 degrees, for example. In the long run, however, the weigths adjust and the error signal drops to zero. The weights w_{21} and w_{22} (and, as a consequence, w_{31} and w_{32}) are now no longer inversely proportional, since the corresponding input channels are not inversely proportional anymore. To simulate a more complex distortion, I additionally changed the slope for the sensor characteristics of T1 and T2, the latter also being shifted for by the maximum value of T2 (Fig. 9, t=2000). The result is similar to that of the situation described before: The initial increase of the error signal is followed by a gradual re-adjustment and is eventually completely neutralized.

2.3 *Lifting the leg off the ground*

The preceeding section led to a functional targetting movement robust against synaptic drift and independent of the slope and intercept of the input characteristics. What is still missing is a mechanism to lift the leg off the ground at the beginning and to bring it down at the desired AEP. From the evolutionary solution described in section 2.1 we know, that one interjoint connection is sufficient to establish the desired behaviour. The connection might be established in the following way:

The existence (not the weight!) of this interjoint connection could be coded in the genome. Then, a mechanism implementing a neuronal realisation of the evolutionary principle as explained in section 2.1 might be applied to this single interjoint connection. This possibility is explained in some detail in the discussion section.

As an alternative solution, an architecture introduced as the "antagonist feedback structure" (AFS) could be used. It is based on mutual inhibition between the two antagonistic output channels. During normal walking, as a result of the contraction of the depressor during stance and the sudden switch at the transition to swing, the levator is activated due to a rebound effect and thereby the leg is lifted off the ground automatically (Cruse, 2002).

3 Discussion

WALKNET is a biologically motivated controller for six legged walking. It represents a distributed neuronal net with the same architecture for each leg and different modules for the different tasks of a single leg. So far, the modules of the WALKNET were trained using algorithms for supervised learning. This is sufficient to demonstrate the ability of the system to produce biologically relevant behaviour. Nevertheless, for a biological system it is unclear, where the required teaching input should come from. In other words it is still unknown how a sensible weight distribution can be obtained without explicit knowledge of the task to be optimized.

Here I first show that genetic algorithms are principally sufficient to explain self organisation in a single joint controller of the SWINGNET. No external knowledge concerning the form of the trajectory is needed to find weight distributions that produce trajectories very similar to the biological model. The only information required is the distance to the anterior leg at the end of the swing movement. This can be obtained from sense organs at the legs and is known to be used by stick insects during walking (Cruse, 1979, Brunn and Dean, 1994). However, realistic swing movements could only be obtained if a realistic environment, i. e. one endowed with obstacles, is used. In the literature, genetic algorithms are either used as an optimisation procedure without referring to any biological meaning, or are used to simulate self organisation on a phylogenetic level. However, under specific assumptions this principle could also be applied to explain self organising processes on an ontogenetic level. This is possible when the neural system is endowed with two types of synapses: "Static" synapses with fixed weights and "dynamic" synapses. The latter are formed spontaneously in random fashion and may be transformed to static synapses by a reward signal applied to the neural network (e. g. in the form of a chemical substance). If within a given time period no reward signal is received, the dynamic synapses disappear again.

As an alternative to explain online self organisation, I propose the exploitation of local synaptic plasticity rules for neuronal homeostasis. Such mechanisms are known to be existent in biological systems (Abbott and Marder, 1995). Information from long term integration of excitatory and inhibitory input pathways is used to readjust the corresponding weights. Recently, the detailed dynamics of such a mechanism have been revealed (Turrigiano et al., 1998, Turrigiano, 1999). So far, the functional value of these mechanisms has mainly been the subject of speculation. Now, more and more possible fields of application are discussed. Stemmler and Koch proposed the usage of these mechanisms to maximize coding efficiency (Stemmler and Koch, 1999). Also it has been proposed that the modification of intrinsic properties of neurons during homeostasis may function as a form of memory mechanism (Marder et al., 1996). Furthermore, it has been stated, that "effective associative learning with Hebbian synapses alone is biologically implausible and that Hebbian synapses must be continuously remodeled by neuronally driven regulatory processes in the brain" (Citation from (Chechik et al., 2000), textual similar in (Turrigiano, 1999)).

In the system at hand, the qualitative information about the character of a synapse, i.e. being an excitatory or an inhibitory one, is assumed to be coded in the genome. Furthermore, the input of each pathway has to be distributed axially symmetric to the midline over the whole range of physiologically relevant input values, i. e. leg positions. This last assumption was not inspired by biology but necessary for the desired functionality. Hence, it is the most problematic assumption. For biological systems, this assumption does not hold in general. For example, the target input to the swing module during normal walking is almost always in the anterior quartile of the whole range. However, there could be a biological solution for the problem: If there exists a mechanism to record maximal and minimal excitations for particular neurons over a time (e.g. in periods of activity), a sort of "playback" could be used to generate the desired distribution. This of course is only possible when the connection to motor output is inhibited, for example during periods of no activity.

The mechanism proposed here is able to stabilize the system against synaptic drift (homeostasis) and at the same time modifies the system to cope with changes in the input characteristics (plasticity). In the simplest possible model of a single joint controller, a one-neuron system depicted in Fig. 5, the system can cope with changes in the slope of input characteristics. This is possible because in the system shown in Fig. 5 both inputs have opposite sign. In the interneurons of the extended system (Fig. 6) this was possible by exploiting antagonistic input channels, which can therefore also be applied with weights of opposite sign. The mechanisms proposed here cope with shifts in the sensor characteristics and provide signals coding for target and actual angles that coincide in their abscissa intercept. The remaining variation in slope can be counterbalanced by the motorneurons using the same mechanisms.

For adjustment of the targeting behaviour, only the relative strength of a single synapse (compared to its counterpart of opposite sign) is important. A multiplication of both weights by arbitrary values does not interfere with the desired behaviour, but solely affects the velocity of movement. Hence, the velocity of movement has to be controlled separately.

If only positive encoding is used (Fig. 8), a proportionately higher number of interneurons and motorneurons can cope with the abovementioned distortions. However, in this architecture each of the 4 motorneurons converges to its own long term equilibrium independently, resulting in different velocities for different situations (i.e. moving forward vs. moving backward). The harmonisation of these different velocities can be handled by additional mechanisms that compare the mean excitation of two neurons. Then, through chemical signals, the process of (for example) upscaling excitatory and downscaling inhibitory synapses is biased to one ore the other. The details of this mechanism are examined in an ongoing project.

The proposed architecture is transferring complexity from the genotype-phenotype mapping into neuronal characteristics. While the virtual genome was so far assumed to code for the strength of each single synapse, it can now be seen as a blueprint for metabolic processes, which is of course much closer to biology. It is also highly reusable since many neurons can be build according to the same principle and therefore decreasing complexity on a holistic perspective.

In a very similar study, Di Paolo has shown that with local plasticity rules for neuronal homeostasis a robot evolved for phototactic behaviour can cope with inversions of the visual field (Di Paolo, 2000). As a controller, he used a fully connected, eight neuron, dynamic neural network. Different rules for local plasticity were coded in the genome, for each synaptic connection separately. Through selection for long term stability of the phototactic behaviour, a complex network of different rules for local plasticity is evolved, showing robustness against inversions of the visual field and against changes in the gain values as an emergent behaviour. For the evolutionary process, however, left/right symmetry of the gain values for sensory and motor pathways was forced.

The architecture proposed here represents an even more simplistic model showing the properties of self organisation. In this way, the fundamental logic of self organisation through local plasticity rules for neuronal homeostasis can be understood on a step by step basis. In the system on hand, all synaptic connections and all neurons show the same, simple properties. Neither threshold values nor different plasticity rules have to be coded in the genome, no evolution with an enforced symmetry between the input channels is necessary. It is therefore an extreme case of self organisation without any a priori value system. Nevertheless, the system is not only able to cope with changes in gain, but also with shifts in the input characteristics, a distortion Di Paolos robot was not tested for. This is possible since, due to antagonistic input channels and equally distributed inputs, adaptivity for the suggested joint controller is reached for a long term activity of zero, which renders the definition of distinct reference or threshold values unnecessary.

Acknowledgements

Thanks to Holk Cruse for ideas and support. This work was funded by DFG grant no. Cr 58 / 9-3 and the GK "Verhaltensstrategien und Verhaltensoptimierung".

References

Abbott, L. F. and Marder, E. (1995). Activity-dependent regulation of neuronal conductances. *In Arbib, M. (Editor), Handbook of Brain Theory and Neural Networks, MIT Press*, pages 63–65.

Brunn, D. and Dean, J. (1994). Intersegmental and local interneurons in the metathorax of the stick insect carausius morosus that monitor middle leg position. *Journal of Neurophysiology*, 72:1208–1219.

Bueschges, A. and Schmitz, J. (1991). Nonspiking pathways antagonize the resistance reflex in the toraco-coxal joint of stick insects. *Journal of Neurobiology*, 22:224–237.

Burrows, M. (1992). Local circuits for the control of leg movement in an insect. *TINS*, 15:226–232.

Chechik, G., Meilijson, I., and Ruppin, E. (2000). Neuronal normalization provides effective learning through ineffective synaptic learning rules. *Neurocomputing*, 32-33:345–351.

Cruse, H. (1979). The control of the anterior extreme position of the hindleg of a walking insect carausius morosus. *Physiol. Entomol.*, 4:121–124.

Cruse, H. (2002). The functional sense of "central oscillations" in walking. *Biological Cybernetics*, in press.

Cruse, H., Kindermann, T., Schumm, M., Dean, J., and Schmitz, J. (1998). Walknet - a biologically inspired network to control six-legged walking. *Neural Networks*, 11:1435–1447.

Di Paolo, E. A. (2000). Homeostatic adaptation to inversion of the visual field and other sensorimotor disruptions. In Meyer, J.-A., Berthoz, A., Roitblat, D. F. H., and Wilson, S. W., editors, *From animals to animats 6*, pages 440–449, Cambridge, Massachusetts. Proceedings of the

Sixth International Conference on Simulation of Adaptive Behavior, MIT Press.

Marder, E., Abott, L. F., Turrigiano, G. G., Liu, Z., and Golowasch, J. (1996). Memory from the dynamics of intrinsic membrane currents. *Proceedings of the National Academy of Sciences of the United States of America*, 93:13481–13486.

Stemmler, M. and Koch, C. (1999). How voltage-dependent conductances can adapt to maximize the information encoded by neuronal firing rate. *Nature Neuroscience*, 2:521–527.

Turrigiano, G. G. (1999). Homeostatic plasticity in neuronal networks: the more things change, the more things stayy the same. *TINS*, 22(5):221–227.

Turrigiano, G. G., Leslie, K. R., Desai, N. S., Rutherford, L. C., and Nelson, S. B. (1998). Activity-dependent scaling of quantal amplitude in neocortical neurons. *Nature*, 391:892–896.

Wendler, G. (1964). Laufen und Stehen der Stabheuschrecke Carausius morosus: Sinnesborstenfelder in den Beingelenken als Glieder von Regelkreisen (Walking and standing of the stick insect Carausius morosus: sensory hair fields in the leg joints as elements of control loops). *Z. vergl. Physiologie*, 120:369–385.

Timed Delivery of Reward Signals in an Autonomous Robot

William H. Alexander
Department of Psychology
Indiana University
Bloomington, IN 47405
wialexan@indiana.edu

Olaf Sporns
Department of Psychology
Indiana University
Bloomington, IN 47405
osporns@indiana.edu

Abstract

In this paper, we implement a computational model of a neuromodulatory system in an autonomous robot. The model is based on a set of anatomical and physiological properties of the mammalian dopamine system, one of several diffuse ascending systems of the brain. The output of this system acts as a value signal, which modulates widely distributed synaptic changes in sensory and motor areas. During reward conditioning, the model learns to generate tonic and phasic responses, which are consistent with potential roles as reward predictions and prediction errors. Different sets of neural units generate precisely timed signals that exert positive effects (predictive) and negative effects (if a predicted reward is omitted) on neuroplasticity. We test the learning and behavior of the robot in different environmental contexts, and observe changes in the development of neural connections within the neuromodulatory system that depend on the robot's interaction with the environment. Simulation of a computational model responsive to rewarding stimuli leads to the emergence of conditioned appetitive behaviors. These studies represent a step towards investigating the behavior of autonomous robots controlled by biologically based neuromodulatory systems.

1. Introduction

Robot learning is an active and rapidly progressing area of research (Touretzky and Saksida, 1997; Sharkey, 1997; Nolfi and Floreano, 1999; Schaal, 2002; Sporns, 2002). In many of these approaches, learning and plasticity depend upon behavioral actions of an agent (animal, robot), which is embedded in an environment. One broad set of functional roles played by neuromodulators in the brain is in mediating the effects of behavior on plasticity. This pivotal role of neuromodulation provides a clear rationale for cross-level computational studies incorporating plasticity and behavior.

Throughout the nervous system, neuromodulators have a variety of functions ranging from the regulation of neuronal excitability and plasticity to effects on gene expression and structural modifications in neural circuits. In computational models, functional effects of neuromodulation have been implemented as an influence on neuronal response functions, learning rates, or other model parameters (Servan-Schreiber et al., 1990; Hasselmo, 1995; Pennartz, 1996; Fellous and Linster, 1998; Doya, 2000; Hasselmo et al., 2002). Several models have focused on the potential functional roles of diffusely projecting neuromodulatory systems in influencing the magnitude and direction of synaptic plasticity, ultimately resulting in behavioral change. In this paper we present one possible computational implementation of a neuromodulatory system and we investigate the behavior of an autonomous robot in which this system has been embedded.

In order for behavior to be adaptive, the behaving agent (organism or robot) must be sensitive to the consequences of its own actions. Salient environmental stimuli and events are important in influencing learning and plasticity. Saliency, defined as "predictive power" or "relevance" to the agent, is derived from the intrinsic structure of the model as well as from its action within an environment. Mechanisms mediating plastic changes in an autonomous agent must be unsupervised (or self-supervised), allowing the agent to learn as a result of its own actions and in the absence of an external teacher. In recent years, reinforcement learning, and in particular temporal difference learning, has emerged as a useful computational framework for this type of adaptive change (Sutton, 1988; Sutton and Barto, 1990; Montague et al., 1996; Sutton and Barto, 1998). In a related approach, "value" systems have been proposed as internal mediators of environmental saliency and have been studied in a number of computer simulations and robot models (Reeke et al., 1990; Edelman et al. 1992, Friston et al., 1994, Verschure et al., 1995, Scheier and Lambrinos, 1996, Rucci et al., 1997, Almassy et al., 1998; Pfeifer and Scheier, 1999, Sporns et al., 2000). Value systems serve to modulate neural activity or plasticity by delivering a diffuse, globally acting signal. Value systems are incorporated in the neural network architecture of the agent, and typically become active after the occurrence of specific sensory stimuli, often as a consequence of behavioral actions of the agent. The activation of these value systems is short-lasting (phasic) and constitutes a timing (gating) signal for synaptic modification. Certain aspects of the value signal may be regarded as an "innate" feature of the neural network, the result of evolutionary adaptation (such as the effects on behavior of food), while other aspects are "acquired" due to the experience or behavior of an agent. In computational models, innate value is determined by fixed (hard-wired) connections, while acquired value is the result of synaptic modifications within the value system itself (Friston et al., 1994; Sporns et al., 2000).

The anatomical and physiological properties of multiple diffuse ascending systems of the vertebrate brain implicate them in the mediation of neuromodulatory effects on synaptic plasticity. These systems respond to a variety of salient sensory stimuli, including food rewards (Ljungberg et al., 1992), as well as stimuli, which attract attention, signal novelty, or trigger aversive responses (Jacobs, 1986, Aston-Jones et al., 1991). Widespread areas of the brain receive anatomical projections from these systems, and their immediate physiological effects include modulation of the "signal-to-noise" ratio of cortical neuronal activity (Hasselmo et al., 1997) as well as modulation of synaptic efficacy (Bear and Singer, 1986; Hasselmo and Barkai, 1995).

Recent studies of the function of mammalian midbrain dopamine system implicate one of its components, the ventral tegmental area (VTA), in reward conditioning (Schultz, 1998). Dopaminergic neurons within the VTA project to parts of the striatum, the amygdala and to widespread cortical areas, including frontal and prefrontal cortex. Prior to conditioning, VTA dopamine neurons show a phasic response to food rewards. In the course of learning, the response pattern changes in a characteristic manner. When a reward is preceded by stimuli, which reliably predict the occurrence of the reward, dopamine neurons no longer respond to the occurrence of the (already predicted) reward. Rather, their phasic activation is "transferred" to the onset of the reward-predicting stimulus, while activation at the occurrence of the primary (now completely predicted) reward becomes attenuated or disappears entirely. Moreover, if a reward is fully predicted but does not occur, there is a transient depression in the baseline firing rate of dopamine neurons at the time of the expected occurrence of the reward. Several computational models of the midbrain dopamine system have been proposed (Montague et al, 1996; Schultz et al., 1997; Schultz, 1998; Suri and Schultz, 2001), forging a strong connection between dopaminergic responses and temporal difference learning (Sutton and Barto, 1990). An alternative interpretation empahasizes the potential role of a phasic dopamine signal following unexpected behaviorally significant stimuli in switching attentional control (Redgrave et al., 1999).

In this paper, we focus on the question of how relatively unconstrained behavior executed by an autonomous agent may be used to study the operation of neuromodulatory systems, in particular the precise timing of their signals. We first present a neural implementation of a neuromodulatory system mediating reward that shares many structural and functional characteristics with the mammalian midbrain dopamine system (see also Sporns and Alexander, 2002). We then demonstrate the operation of this neuromodulatory system in simple robot experiments, emphasizing the relationship between behavior, environment and neural change.

2. Methods

2.1 General

All experiments reported in this paper were carried out using neural simulations implemented in Matlab 6.0 (Mathworks, Natick, MA), run on Linux 6.2 or 7.1 workstations (ASL, Newark, CA) and interfaced with autonomous khepera robots (K-Team, S.A., Préverenges, Switzerland), equipped with a color CCD video turret and a gripper module. Basic Matlab scripts and serial line communication modules for Linux are available from the authors upon request. Some additional material including short movies of robot behavior can be found at php.indiana.edu/~osporns/lab.htm.

2.2 Robot Design

The robotic system described in this paper is named "Monad" (after Giordano Bruno and Gottfried Wilhelm Leibniz). Monad consisted of a mobile circular platform (Ø ≈ 5 cm) capable of translation and rotation limited to an approximate speed of 5 cm/s. An arm/gripper module allowed a one degree-of-freedom gripper to be raised and lowered, as well as closed and opened, thus permitting physical contact with objects. Rotational movements of the robot wheels as well as movements of the arm/gripper were triggered by the activation of neural units within the simulation. Robot sensors consisted of a color CCD camera (fl 3mm, 59.8 deg × 42.6 deg field of view) mounted on top of the mobile platform and angled forward, 8 infrared (IR) sensors mounted around the periphery of the platform, and resistivity sensors on the inner surfaces of the gripper. The color camera continuously transmitted RGB images (320 × 240 pixels), which were used as input to the visual part of the neural simulation. A low-resolution gray level image was derived from the RGB signal and used as input to the

Figure 1: (Left) Monad in robot environment. Environment consists of an enclosed platform (approx. 90 x 90 cm) containing objects (small cubes). Monad's tether, consisting of power and communication lines connected through a rotating contact, allows unrestricted movements. In the picture, multiple objects are distributed at random throughout the environment. (Right) Monad, consisting of mobile base, arm/gripper module and camera, shown approaching an object (small cube).

visual approach system. The 8 IR sensors acquired infrared reflectance readings at 20 ms intervals that were converted into motor signals using a fixed motor map. Synaptic weights were set so as to effectively steer Monad away from obstacles (see below). The resistivity sensors recorded the conductivity across the surface of objects, a signal that served as a measure of object "taste" (low conductivity = appetitive taste, high conductivity = aversive taste). Resistivity was recorded as a scalar variable with 8-bit resolution; for simplicity, all readings were converted to binary outputs, with high conductivity activating an aversive taste receptor (T_{av}) and low conductivity activating an appetitive taste receptor (T_{ap}).

2.3 Robot Environment

Monad was tethered via a flexible wire bundle and a rotating contact, which was mounted directly above an illuminated environmental enclosure with white floor and walls, containing various stimulus objects. These objects were black 1-inch cubes, with a single colored face at the top. Objects were visually indistinguishable except for their color, which was red or blue. The black surfaces of the objects were either electrically non-conductive or conductive, a physical property analogous to "taste". Objects were sufficiently light to be easily manipulated by Monad's gripper and were either presented manually by the experimenter or placed at random positions within the environment at the beginning of an experiment.

2.4 Robot Behavior

At each point in time, Monad was in one of several behavioral modes, forming a simple behavioral hierarchy. In the absence of overt visual targets, Monad was traversing the environment at fairly constant speed (≈ 3-5 cm/s) while sensing IR reflectance and avoiding obstacles or walls (*mode = navigate*). Upon detection of a high-contrast visual target, Monad approached the target under the guidance of the visual approach system (*mode = approach*). Activity of units in a visual map was translated by this system into motor (speed) commands relayed to Monad's two high-precision DC motors, via a fixed motor map. While approaching a target, no IR sensor readings were taken and all steering movements were under visual control. Approach terminated if a visual target loomed large (i.e. was physically close) in the center of Monad's visual field (fovea). Before learning, Monad attempted to establish physical contact with all targets located in close physical proximity (*mode = interact*). Interaction consisted of lowering the arm and closing the gripper. After obtaining sensor readings of the conductivity of an object, the object was released (*mode = withdraw*) by opening the gripper and returning the arm to a raised position. Then, Monad turned away from the released object to resume navigation or approach another target.

The behavioral sequence outlined above constituted Monad's "default" or innately specified behavioral pattern. All objects of high-visual contrast were approached and "tasted", regardless of their visual appearance. When objects were encountered, Monad emitted different unconditioned responses (UR) depending upon the nature of the unconditioned stimulus (US), i.e. "taste". If an object was found to be appetitive, a prolonged gripping response ensued. After learning (essentially amounting to an instrumental conditioning paradigm), the visual appearance (color) of objects was sufficient to trigger conditioned responses, a reward-related conditioned response (CR_R) consisting of immediate approach and gripping for appetitive stimuli. This conditioned response was triggered by activation of motor unit M_{ap}, as soon as its activation difference exceeded a behavioral threshold β (β=0.3). In all robot experiments, connections driving M_{ap} activation were subject to value-dependent learning (see below).

2.4 Neural Simulation

All visual images and sensor readings were relayed to a neural simulation, and motor commands were initiated from the simulation and relayed to Monad via simple serial line commands. Neural and behavioral states were continuously recorded and saved for off-line display and analysis. On average, a single simulation cycle required about 250 ms of CPU time.

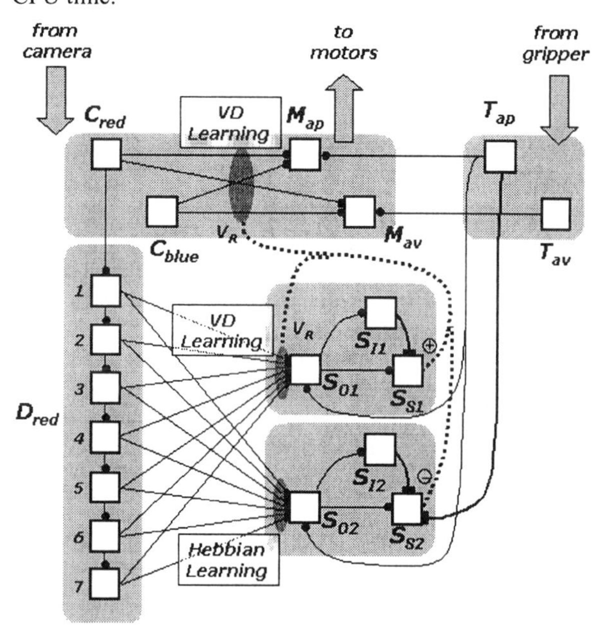

Figure 2: Schematic diagram of Monad's reward system (neural structures used for IR-guided obstacle avoidance, visually guided approach and color vision are not shown). Major functional components are shaded, neural units and networks are indicated as square boxes, excitatory connections are shown as thin lines connecting boxes, inhibitory connections are shown as thick lines, the output signals of the neuromodulatory system are indicated as thick hatched lines, and their modulatory targets are indicated as ellipsoid shaded areas. Some feedback inhibitory connections between D_{red} units that are used for generating temporally specific responses are not shown for clarity. VD learning = value-dependent learning. See text for a more detailed description of anatomy and physiology.

All neural units were implemented using a continuous firing rate model with a single saturating non-linearity, according to

$$s_i(t+1) = \phi[A(t) + \Omega s_i(t)]$$

where $s_i(t)$ is the activity of unit i at time t, $A(t)$ is the total synaptic input to unit i at time t, Ω is the unit's temporal persistence ($0<\Omega<1$), and $\phi[.]$ is the saturating nonlinearity (both hyperbolic tangent and sigmoidal functions are used). $A(t)$ was calculated as the linear sum of all inhibitory and excitatory inputs, i.e. $\Sigma c_{ij} s_j(t)$.

Schematic diagrams of the neural model and its constituent networks are shown in Figure 2 (exact parameter values are given in Sporns and Alexander, 2002). IR sensors are connected to motor units driving the two wheels of Monad via a fixed motor map. A raw image delivered by Monad's CCD camera provides input to the visual approach system, after being converted to a low-resolution format, thresholded and contrast-inverted. The resulting visual array is converted to steering motions of the two wheels using two symmetrical fixed motor maps. The central (foveal) part of the raw image is converted to red, green, blue and yellow (red+green/2) arrays, which are further processed using red-green and blue-yellow center-surround convolutions. The resulting neural arrays (R+G-, B+Y-) are used to drive color-selective units in C_{red} and C_{blue} that discount visual topography and report the presence or absence or red or blue color within the visual field. We note that the emphasis of the present model was on the computational aspects of the neuromodulatory system. Therefore, no attempt was made to implement complex categorization or sensorimotor mappings (for earlier work on visual categorization see Almassy et al., 1998 and Krichmar et al., 2000). Stimulus categories were limited to the object's color, specifically "red" and "blue" and category-dependent behavioral outputs were simple "appetitive" or "aversive" behaviors.

The main component of the robot's neuromodulatory system mediated effects of appetitive (reward) stimuli. The structure of the reward component is shown in Figure 2. Color selective units (C_{red}/C_{blue}) provided sensory inputs to a network transforming this input into a continuous temporal representation. As a result of excitatory and inhibitory interactions within this network (essentially forming a delay chain), stimulus-specific units D_{red}/D_{blue} within this network became active after a specific amount of time had elapsed since the onset of their preferred stimulus. These units had fairly broad temporal tuning with significant mutual overlap in terms of their "temporal receptive field" (see Figure 3 for example activity traces). A more realistic implementation of such a network would consist of intermixed populations of neurons that show complex spatio-temporal activation patterns. D_{red} units projected to two subcomponents of the reward system mediating different aspects of reward stimuli through two distinct sets of modifiable connections. Both of these subcomponents consisted of an "integrator" unit (S_{O1} and S_{O2}, respectively), which activated a feedforward inhibitory unit (S_{I1} and S_{I2}, respectively) and a phasic response unit (S_{S1} and S_{S2}, respectively). S_{S1} and S_{S2} emitted two phasic components of the reward-related neuromodulatory output signal. Initially, the connections linking temporal delay units and S_{O1}/S_{O2} were weak, but in the course of learning they became strengthened in specific patterns and capable of driving responses in S_{O1}/S_{O2}. S_{O1}/S_{O2} were also activated by the primary reward (appetitive taste), through an "innate" and excitatory connection. Due to their combined excitatory and inhibitory inputs, S_{S1} and S_{S2} both showed a phasic response profile, a short burst of activation followed by inhibition from S_{I1} and S_{I2}. The S_{S2} phasic response is inhibited ("cancelled") by the simultaneous occurrence of a primary reward, due to an "innate" and inhibitory connection from T_{ap} to S_{S2}. Essentially, activation in S_{S1} and S_{S2} constituted a temporal derivative of increases (but not decreases) in S_{O1}/S_{O2} activation. The firing level of S_{S1} was taken to be proportional to an increase in the level of neuromodulator released at projection targets over a stationary baseline. In turn, the firing level of S_{S2} was taken to be proportional to a decrease in the level of neuromodulator below the same stationary baseline. The overall level of neuromodulator released by the reward system, taken to be a "value signal", was calculated as $V_R = S_{S1}-\lambda S_{S2}$, with λ setting the overall difference in gain between the two components (λ=1 in this paper). V_R was a signed scalar variable, which was positive if the level of neuromodulator increased above baseline and negative if it decreased below baseline (i.e. we assume that baseline = 0).

Two types of learning were employed in the present model, value-dependent learning and Hebbian learning. Through value-dependent learning, the value signal V influenced synaptic modification in sensorimotor connections linking color-selective units in $C_{red/blue}$ to motor units in M_{ap}, as well as in connections linking the temporal delay units $D_{red/blue}$ to S_{O1} and S_{O2}. Thus, value exerted a dual influence, by directly modifying behavior through changes in sensorimotor linkages, and by modifying the response characteristics of parts of the neuromodulatory system itself. Connections that were subject to value-dependent learning were updated according to a ternary learning rule:

$$c_{ij}(t+1) = (1-\varepsilon)c_{ij}(t) + \eta s_j(t)F(s_i(t))V$$

where c_{ij} = connection weight from unit j to unit i, ε = incremental decay rate of connection weight per iteration, η = learning rate, F(.) = nonlinear function applied to postsynaptic activity, V = value signal. F(.) determined if a connection weight increased or decreased, depending upon the level of postsynaptic activity. F(.) was a continuous saturating function ($-1<F(.)<1$), here modeled as

$$F = \kappa(1-\tanh(\varphi_1 s_i(t)-\xi_1))+\tanh(\varphi_2 s_i(t)-\xi_2).$$

Parameter values κ, φ_1, φ_2, ξ_1 and ξ_2 were set by the experimenter solely to determine the shape of F(.) and were not thought to have direct physiological analogues. The shape of F modeled the observed dependence of the sign of synaptic modification on postsynaptic activity (cf. Bienenstock et al., 1982; Bear et al., 1987). For plastic connections that are not value-dependent (i.e. connections between D_{red}/D_{blue} and S_{O2}), the value component of the above equation is set to 1, and thus synaptic modification depends entirely on pre- and postsynaptic activity only (Hebbian learning).

3. Results

The functional characteristics of the neuromodulatory system were explored in two sets of experiments. In the first set, objects associated with reward (appetitive taste) were presented manually to the robot, in order to ensure reproducible timing of reward delivery within individual learning trials. In the second set of experiments, objects were placed at random throughout the environment and all robot behavior and learning proceeded in a fully self-guided and autonomous fashion. This tended to degrade the consistent timing of reward delivery.

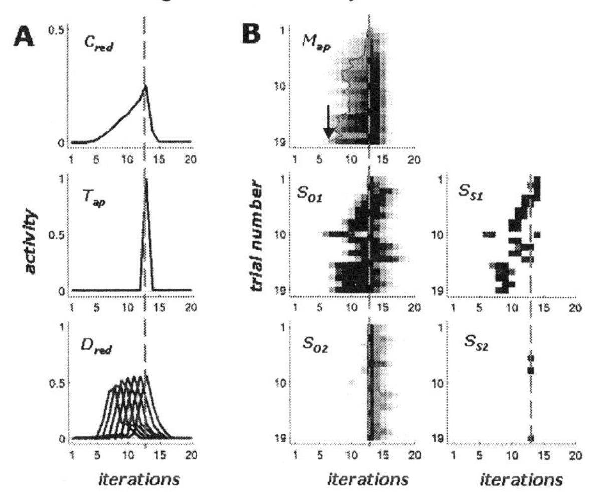

Figure 3: Neural activation patterns for various sensory, motor and neuromodulatory networks. All data shown is from one representative experiment, lasting 1000 iterations (19 trials or object encounters). (A) Averaged activity of non-plastic neural units. Top to bottom, plots show neural activity in C_{red}, T_{ap}, and D_{red} units, averaged over all 19 trials and temporally aligned to the onset of taste (hatched vertical line at iteration 13). (B) Raster plots of neural activity for plastic neural units, shown for each of the 19 trials (trial 1 is at the top of each plot; black indicates high activity, white indicates low or absent activity). Contour line in M_{ap} activity plot marks the time when activity exceeds the behavioral threshold β (β=0.3), resulting in the triggering of a reward-related conditioned response (CR_R, arrow). S_{O1}/S_{S1} panels show that the phasic reward-related value signal is initially triggered by the onset of taste and, after learning, is triggered by the onset of object vision (C_{red}). S_{S2} is usually inactive, indicating that the predicted reward arrives at the expected time after a red object is detected. Again, in all plots, the hatched vertical line at iteration 13 indicates the onset of appetitive taste (reward delivery).

3.1 Timed Reward Trials

Our first goal was to investigate the development of neural connectivity and activation patterns in Monad's neuromodulatory system in a task setting in which there was consistent relative timing of object vision and subsequent reward (appetitive taste), controlled by the experimenter. The experimenter placed objects in Monad's navigational path, at positions just outside of the visual field. This produced a fairly stereotypic behavioral sequence, beginning with initial visual acquisition, followed by guided visual approach, interaction with the object through physical contact with the gripper and the sensing of taste. After each trial the object was released and navigation through the environment resumed. Individual trials, from first visual contact to taste sensing, took about 1-2 seconds of real time, or 6-8 iterations. About 20-25 trials were conducted in a typical experiment lasting 1000 iterations (4-5 min). Figure 3 shows data from a representative experiment with a total of 19 trials (stimulus encounters). The figure shows typical activation patterns of plastic and non-plastic components of the neural network.

Throughout the experiment, all red objects were approached, resulting in activation of red-selective units (C_{red}) as well as a temporal representation of the stimulus in D_{red} units. On completion of approach, the object was gripped and the primary reward (appetitive taste) was delivered, resulting in activation of T_{ap}. Figure 3 shows neural activation patterns for motor and neuromodulatory components of the neural network at various stages of the learning process. Prior to learning (i.e. trial 1 in Figure 3), motor (M_{ap}) and S_{O1}/S_{O2} activation was triggered only by the delivery of the primary reward (T_{ap}), here at iteration 13 (Figure 3). Color-selective cells had no effect on motor or neuromodulatory responses. Activity in S_{S1} indicated a positive temporal difference in S_{O1}, while S_{S2} activation was cancelled due to delivery of the primary reward. During learning (trials between 1 and 19), selectively strengthened inputs from C_{red} began to trigger M_{ap} while value-dependent modifications of connections between D_{red} and S_{O1} resulted in activation of S_{O1} due to activity in temporal units. S_{S1} showed phasic activation during visual approach, while tonic activity in S_{O1} prevented a second response when the primary reward was delivered. Activation of S_{O2} continued to be triggered by the primary reward in addition to being partially driven by strengthened connections from D_{red} units whose "temporal receptive field" coincided with the delivery of the reward. S_{S2} remained suppressed for the majority of trials due to reward delivery at the "expected" time. In contrast, activation in S_{S2} is seen for trials in which the primary reward is delayed or withheld. After learning (trial 19 in Figure 3), M_{ap} units were strongly driven by C_{red}. S_{O1} and S_{S1} were activated immediately after a red objects was visually acquired, with S_{O1} remaining active for a prolonged time period, while S_{O2} was driven by both primary reward and timing signals from D_{red}.

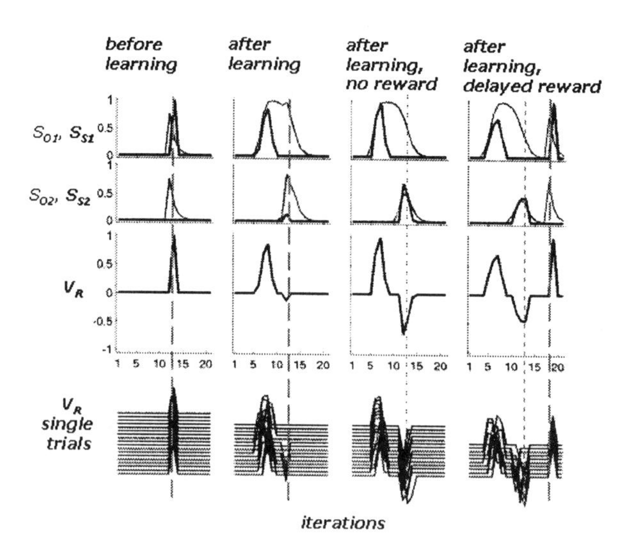

Figure 4: Activation patterns of reward-related neuromodulatory networks, averaged over multiple trials before and after learning, as well as over trials for which a predicted reward was omitted or delayed (left to right). Plots summarize data from 18, 15, 15, and 10 individual trials, respectively. Top to bottom, plots show S_{O1}/S_{S1}, S_{O2}/S_{S2}, V_R and a plot of V_R signals recorded in individual trials. Thick hatched vertical line indicates the actual onset of appetitive taste (reward delivery), thin hatched line indicates expected time of predicted reward (no reward delivery).

Figure 4 shows activation levels, averaged over several trials, of the reward component of the neuromodulatory system, as well as average value signals used in synaptic modification. Panels in Figure 4 show S_{O1}/S_{S1} and S_{O2}/S_{S2} activations, as well as V_R, both as an average over multiple trials and for individual trials. Before learning, S_{O1}/S_{S1} were activated by the primary reward (delivered at iteration 13), accounting for the single spike in the value signal. After learning, S_{O1}/S_{S1} were triggered by visual input from C_{red} and D_{red} units. S_{O2} responded to the primary reward in most trials and no S_{S2} activity resulted. Note that in two of the trials the reward was accidentally slightly delayed (as measured from the onset of red visual input) and some S_{S2} activation occurred. On average, the value signal consisted of a single spike temporally aligned to the onset of the reward-predicting stimulus (color red). After learning, if the reward was withheld (by presenting red objects to Monad and then quickly withdrawing them as the robot approached), S_{O1}/S_{S1} activation was unaffected, but S_{O2}/S_{S2} activation was entirely driven by the temporally specific expectation of the reward, mediated by selectively strengthened connections between D_{red} and S_{O2}. This reward expectation signal was not cancelled by an actual reward and thus contributed a negative spike to the value signal. If, after learning, the reward was delayed (by pulling the object away from Monad during visual approach), S_{S2}

activation occurred at the expected time of reward followed by a second S_{SI} activation to the primary reward as soon as it was delivered. This produced a tri-phasic value signal with an initial positive spike due to the appearance of the reward-predicting stimulus (color red), followed by a negative spike due to reward omission at the expected time, followed by a second positive spike due to the final, now unpredicted, delivery of the primary reward.

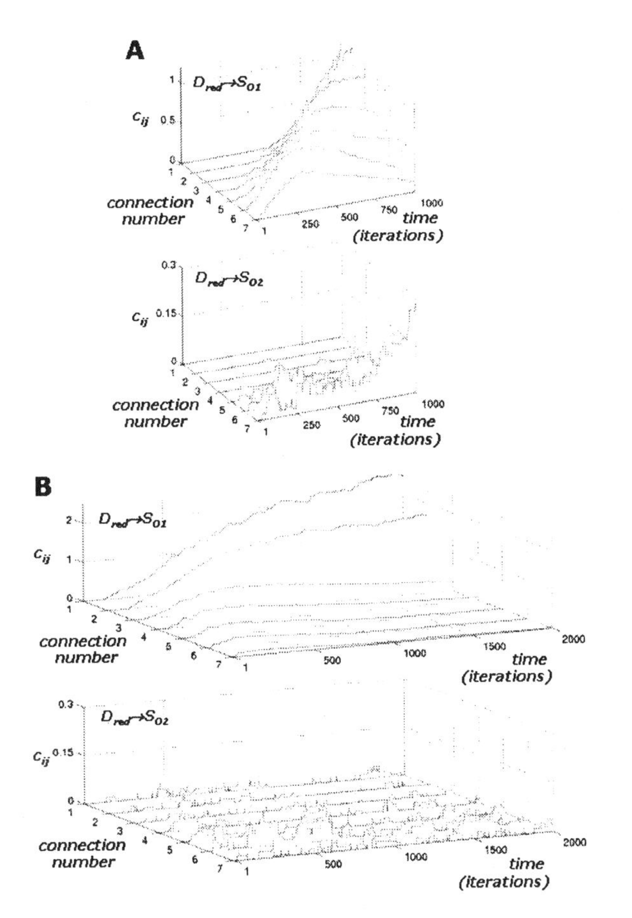

Figure 5: Time evolution of synaptic weights for connections mediating reward-related neuromodulatory responses. (A) Individual connection weights (c_{ij}, numbered 1 through 7, compare Figure 2) linking units in D_{red} and S_{O1} (top panel) and D_{red} and S_{O2} (bottom panel), plotted as a function of time (1000 iterations). Weights are averages over 4 individual experiments. (B) Same as in (A), but weights are obtained from 4 individual experiment involving fully autonomous behavior (2000 iterations each).

Figure 5A shows connection weights between temporal representation units D_{red} and S_{O1}/S_{O2}, obtained from representative learning experiments. Both sets of weights started at values near zero, but showed characteristic patterns at the end of learning. The connections terminating on unit S_{O1} first showed strengthening for longer temporal delays (connections numbered 5-7; compare Figure 2). This is due to the initial phasic activation of S_{SI} by primary reward stimuli and the fact that the value signal derived from this S_{SI} activation was at first temporally coincident with the primary reward. As a result of the broad "temporal receptive field" of D_{red} units, earlier temporal components gradually became capable of driving S_{O1} activation and caused S_{SI} to be active prior to the primary reward. The retrograde transfer of the temporal onset of the S_{O1} response was completed when the earliest onset of C_{red} and D_{red} triggered both S_{O1} and S_{SI} responses. The weight profile shows high synaptic weights for D_{red} units that became active immediately after the onset of C_{red} activity. The connections terminating on S_{O2} were modified using a Hebbian rule, without the modulatory action of the value signal. This ensured that the association between the primary reward and the appropriate temporal delay units remained "fixed in time". If the timing between the reward-predicting stimulus and the actual reward was consistent across trials, a weight pattern emerged that drove responses in S_{O2} at the expected time of reward. As shown, in Figure 5A, consistent timing of visual approach and appetitive taste resulted in the selective strengthening of connections 6 and 7.

3.2 Autonomous Behavior

A second goal was to investigate neural development of the model in an unconstrained context, and to compare the resulting activation patterns with those obtained in controlled experiments. The second set of experiments allowed Monad to behave autonomously without manual manipulation of targets by the experimenter. Prior to the experiment, targets were placed in the environment at random. Density of the objects was either low or high. In conditions of low object density, neural activation and synaptic patterns closely resemble those that emerged during manual training (see above). Behavioral sequences involving visual approach and contact with objects were fairly stereotypic, and precise timing between visual acquisition of a target and reward delivery was preserved. Densely crowded objects, however, upset the consistent timing of reward delivery. In many cases, objects were encountered "suddenly", leading to immediate interaction with the object and delivery of the reward. Other objects were lost during visual approach as other, more salient, targets interfered. Figure 5B shows connection patterns between stimulus representation and neuromodulatory components of the neural network. Connection weight data were obtained over four separate learning experiments (2000 iterations each). Connection weights between D_{red} and S_{O1} closely resemble those obtained after manual training, while connection patterns between D_{red} and S_{O2} were flat, reflecting the lack of consistent timing between onset of visual input and reward.

4. Discussion

In many higher organisms, neuromodulatory systems are essential in linking behavior and neuroplasticity. In the present computational model, we implemented a biologically based neuromodulatory system in a behaving robot. Experiments with the model embedded in an autonomous robot captured several characteristics of neuromodulatory systems involved in mediating the effects of rewarding sensory stimuli.

The "value signal" used to change synaptic connections in the present paper, as well as in previous work (Friston et al., 1994; Sporns et al., 2000), implements the functional effects of neuromodulators in neural plasticity. Value acts as a global signal affecting widespread projection areas, and value signals are often phasic and short-lasting. Incorporating a combination of temporal specificity and spatial uniformity, value operates as a modulating third factor, in addition to traditional Hebbian-type synaptic rules containing terms for pre- and postsynaptic activity only. Value signals change their own response properties through modulation of sensory afferents to the value system, and link sensory and motor units, resulting in adaptive behavioral change. The phasic nature of values signals acts as a temporally specific gate for plasticity, while the projections of the value system to a wide range of areas affects plasticity in a more or less uniform manner. These properties of value systems correspond closely with the anatomical and physiological characteristics of mammalian neuromodulatory systems, including the dopamine system. This correspondence provides a promising rationale for the incorporation of such systems in computational models implemented in autonomous robots. The action of the value system in the present model, through changes in sensorimotor connections and inputs to the neuromodulatory system itself, allows an autonomous agent to learn and adaptively change its behavior without an external "teacher."

Figures 3 and 4 summarize the temporal characteristics of Monad's neuromodulatory system. After learning, the neuromodulatory system contained specialized units, which emitted phasic and tonic responses, comparable to responses found in various neural structures and task contexts (reviewed in Suri and Schultz, 2001). The appearance of sensory stimuli, which predict reward, but themselves are unpredicted, is signaled by the activity of S_{S1}. Negative errors in reward prediction are signaled by S_{S2}, which carries precise temporal information about the occurrence of expected reward. These two (positive and negative) phasic response components are related to formulations of temporal difference learning. In these formulations the prediction error is represented as a single (positive and negative) first derivative of the reward prediction (e.g. Montague et al., 1996; Suri and Schultz, 2001). In contrast, in our model, the positive and negative components are distinct elements of the neural network, implemented by different cell populations. The positive prediction error (S_{S1}) is derived as a positive change in the prediction (S_{O1}), while the negative prediction error is derived separately through S_{O2}/S_{S2}. Activation of S_{S2} depends on the development of coincidence detectors associating the occurrence of a primary reward with the representation of a salient stimulus. In combination, S_{O1}/S_{S1} and S_{O2}/S_{S2} yield a signal (V_R) that is similar to formulations of temporal difference (TD) learning. The generation and maintenance of temporal specificity by S_{O2}/S_{S2} required the use of Hebbian learning rather than value-dependent learning. This suggests that substrates for positive and negative "prediction errors" may in fact be segregated within the midbrain. In fact, there is neurophysiological evidence that midbrain dopaminergic nuclei contain functionally heterogeneous populations of cells (Kiyatkin and Rebec, 1998).

Several other temporal characteristics of the present model are of interest in comparison to formulations of TD learning. In TD learning, transference of the reward prediction to the onset of a reward-predicting stimulus is propagated through intervening time steps between the delivery of the reward and the stimulus onset. In our model, this is achieved through value-dependent synaptic modification of connections linking S_{O1} to D_{red} units with temporal receptive fields that are intermediate between stimulus onset and the occurrence of the primary reward. Once the value signal has been transferred to the earliest onset of the reward-predicting stimulus, D_{red} units that become active at later times tend to again decrease their synaptic inputs to S_{O1}, including units that are coincident with the primary reward (see Figure 5). Interestingly, in TD learning formulations a negative prediction error (as would occur in the case of violated predictions, e.g. due to omitted rewards) undergoes the same kind of transference to the onset of a stimulus, where it ultimately leads to extinction. A similar approach in our model would necessitate both excitatory and inhibitory connections from temporal units to the value system. This may suggest that additional complementary (possibly inhibitory) subsystems in the midbrain are responsible for extinction of learned behaviors.

In addition to phasic signals in S_{S1} and S_{S2}, tonic activity is emerging in the "integrator" unit S_{O1}. After learning, this tonic activity remains elevated in the period between the initial occurrence of a predictive stimulus and the predicted primary reward. This anticipatory activity is dependent on the participation of the phasic response in synaptic changes between the temporal stimulus representation and the integrator unit. Neurons with tonic activity patterns have been observed in parts of the striatum and various cortical areas (Suri and Schultz, 2001) and are thought to represent anticipatory signals.

The present model needs further testing and extension in order to capture more of the rich neurophysiology and behavioral effects found in reward conditioning. Interesting differences emerge in our model when comparing neural development during timed trials (in which behavior is fairly stereotypic and constrained) and during autonomous behavior (Figure 5). The system is no longer capable of precise temporal predictions, which leads to a degradation (or failure to develop) of the S_{O2}/S_{S2} ("negative prediction error") component of the model. The S_{O1}/S_{S1} component, on the other hand, functions more as a saliency-based gating signal, responding preferentially to novel, unexpected and behaviorally relevant stimuli. Thus, the function of the dopamine system in autonomous behavior seems to be consistent with interpretations based on temporal prediction as well as novelty and saliency.

References

Almassy, N., Edelman, G.M., and Sporns, O. (1998) Behavioral constraints in the development of neuronal properties: A cortical model embedded in a real world device. *Cerebral Cortex*, 8, 346-361.

Aston-Jones, G., Chiang, C., and Alexinsky, T. (1991) Discharge of noradrenergic locus coeruleus neurons in behaving rats and monkeys suggests a role in vigilance. *Progress in Brain Research*, 88, 501-520.

Bear, M.F., Cooper, L.N., and Ebner, F.F. (1987) A physiological basis for a theory of synaptic modification. *Science*, 237, 42-48.

Bear, M.F., and Singer, W. (1986) Modulation of visual cortical plasticity by acetylcholine and noradrenaline. *Nature*, 320, 172-176.

Bienenstock, E.L., Cooper, L.N., and Munro, P. (1982) Theory for the development of neuron selectivity: Orientation selectivity and binocular interaction in visual cortex. *Journal of Neuroscience*, 2, 23-48.

Doya, K., (2000) Metalearning, neuromodulation, and emotion. In: *Affective Minds*, Hatano, G., Okada, N., and Tanabe, H. (eds.), pp. 101-104, Elsevier.

Edelman, G.M., Reeke, G.N., Gall, W.E., Tononi, G., Williams, D. and Sporns, O. (1992) Synthetic neural modeling applied to a real-world artifact. *Proceedings of the National Academy of Sciences USA*, 89, 7267-7271.

Friston, K.J., Tononi, G., Reeke, Jr., G.N., Sporns, O., and Edelman G.M. (1994) Value-dependent selection in the brain: Simulaition in a synthetic neural model. *Neuroscience*, 59, 229-243.

Fellous, J.-M., and Linster, C. (1998) Computational models of neuromodulation. *Neural Computation*, 10, 771-805.

Hasselmo, M.E., Wyble, B.P., and Fransen, E. (2002) Neuromodulation in mammalian nervous systems. In: Arbib, M. (ed.). *Handbook of Brain Theory and Neural Networks, 2nd Edition*. Cambridge, MA: MIT Press.

Hasselmo, M.E., Linster, C., Ma, D., and Cekic, M. (1997) Noradrenergic suppression of synaptic transmission may influence cortical "signal-to-noise" ratio. *Journal of Neurophysiology*, 77, 3326-3339.

Hasselmo, M.E. and Barkai, E. (1995) Cholinergic modulation of activity-dependent synaptic plasticity in rat piriform cortex. *Journal of Neuroscience*, 15, 6592-6604.

Hasselmo, M.E. (1995) Neuromodulation and cortical function: Modeling the physiological basis of behavior. *Behavioral Brain Resarch*, 67, 1-27.

Jacobs, B.L. (1986) Single unit activity of locus coeruleus neurons in behaving animals. *Progress in Neurobiology*, 27, 183-194.

Kiyatkin, E.A., and Rebec, G.V. (1998) Heterogeneity of ventral tegmental area neurons: single-unit recording and iontophoresis in awake, unrestrained rats. *Neuroscience* 85, 1285-1309.

Krichmar, J.L.; Snook, J.A.; Edelman, G.M.; Sporns, O. (2000) Experience-dependent perceptual categorization in a behaving real-world device. In: Animals to Animats 6: *Proceedings of the Sixth International Conference on the Simulation of Adaptive Behavior*, Meyer, J.A.; Berthoz, A.; Floreano, D.; Roitblat, H.; Wilson, S.W., (Editors), MIT Press: Cambridge, MA. p. 41-50.

Ljungberg, T., Apicella, P., and Schultz, W. (1992) Responses of monkey dopamine neurons during learning of behavioral reactions. Journal of *Neurophysiology*. 67, 145-163.

Montague, P.R., Dayan, P., and Sejnowski, T.J. (1996) A framework for mesencephalic dopamine systems based on predictive hebbian learning. Journal of *Neuroscience*, 16, 1936-1947.

Nolfi, S. and Floreano, D. (1999) *Learning and Evolution. Autonomous Robots*, 7, 89-113.

Pennartz, C.M.A. (1996) The ascending neuromodulatory systems in learning by reinforcement: comparing computational conjectures with experimental findings. *Brain Research Reviews*, 21, 219-245.

Pfeifer, R., and Scheier, C. (1999) *Understanding Intelligence*. MIT Press, Cambridge, MA.

Redgrave, P., Prescott, T.J., and Gurney, K. (1999) Is the short-latency dopamine response too short to signal reward error? *Trends in Neurosciences* 22, 146-151.

Reeke, G.N., Jr., O. Sporns, and G.M. Edelman (1990) Synthetic neural modeling: The "Darwin" series of recognition automata. *Proc. IEEE* **78**:1498-1530.

Rucci, M., Tononi, G., and Edelman, G.M. (1997) Registration of neural maps through value-dependent learning: Modeling the alignment of auditory and visual maps in the barn owl's optic tectum. *Journal of Neuroscience*, 17, 334-352.

Schaal, S. (2002). Robot Learning. In: Arbib, M. (ed.). *Handbook of Brain Theory and Neural Networks, 2nd Edition*. Cambridge, MA: MIT Press.

Scheier, C. and Lambrinos, D. (1996) Categorization in a real-world agent using haptic exploration and active perception. In *From animals to animats: Proceedings of the Fourth International Conference on Simulation of Adaptive Behavior*, pp. 65-75, Eds. Maes, P., Mataric, M., Meyer, J.-A., Pollack, J., Wilson, S.W., MIT press, Cambridge, MA.

Schultz, W., Dayan, P. and Montague, P.R. (1997) A neural substrate of prediction and reward. *Science*, 275, 1593-1599.

Schultz, W. (1998) Predictive reward signal of dopamine neurons. *Journal of Neurophysiology*, 80, 1-27.

Servan-Schreiber, D., Printz, H., and Cohen, J.D. (1990) A network model of catecholamine effects: gain, signal-to-noise ratio, and behavior. *Science*, 249, 892-895.

Sharkey, N.E. (1997) The new wave in robot learning. *Robotics and Autonomous Systems*, 22, pp.

Sporns, O., Almassy, N., and Edelman, G.M. (2000) Plasticity in value systems and its role in adaptive behavior. *Adaptive Behavior*, 8, 129-148.

Sporns, O. (2002) Embodied cognition. In: *Handbook of Brain Theory and Neural Networks, 2nd Edition.*, Arbib, M. (ed.), Cambridge, MA: MIT Press.

Sporns, O.,and Alexander, William H. (2002) Neuromodulation and plasticity in an autonomous robot. *Neural Networks* (in press).

Suri, R.E., and Schultz, W. (2001) Temporal difference model reproduces anticipatory neural activity. *Neural Computation*, 13, 841-862.

Sutton, R.S. (1988) Learning to predict by the methods of temporal difference. *Machine Learning* 3, 9-44.

Sutton, R.S., and Barto, A.G. (1990) Time derivative models of Pavlovian reinforcement. In *Learning and Computational Neuroscience: Foundations of Adaptive Networks*. Gabriel, M., and Moore, J. (eds.), pp. 539-602, Cambridge, MA: MIT Press.

Sutton, R.S., and Barto, A.G. (1998) Reinforcement Learning. Cambridge, MA: MIT Press.

Touretzky, D.S. and Saksida, L.M. (1997) Operand conditioning in skinnerbots. *Adaptive Behavior*, 5, 219-247.

Verschure, P.F.M.J., Wray, J., Sporns, O., Tononi, G., and Edelman, G.M. (1995) Multilevel analysis of a behaving real world artifact: An illustration of synthetic neural modeling. Robotics and Autonomous Systems, 16, 247-265.

Using Markovian decision problems to analyze animal performance in random and variable ratio schedules of reinforcement

Jérémie Jozefowiez* **Jean-Claude Darcheville*** **Philippe Preux****

*Unité de Recherche sur l'Evolution
des Comportements et des Apprentissages
Domaine du Pont de Bois, B.P. 149
59653 Villeneuve D'Ascq, France
{jozefowiez, darcheville}@univ-lille3.fr

**Laboratoire d'Informatique du Littoral
50 rue Ferdinand Buisson
62228 Calais, France
preux@univ-littoral.fr

Abstract

Markovian decision problems are a kind of optimization problems in which an agent must learn how to optimize the amount of reward it can collect during its interaction with its environment. We use them to analyze the task faced by an animal in random and variable schedules of reinforcement. Predictions of the model derived from this analysis are compared to three sets of data obtained in men, rats and pigeons and are contrasted with the ones of its main challenger in psychology, Herrnstein's equation. This reveals the existence of two response strategies in ratio schedules, one which corresponds to our model, the other which is closer to Herrnstein's equation.

1 Introduction

1.1 Operant conditioning

Pavlovian and operant conditioning are the two main procedures for the study of animal learning in experimental psychology (Skinner, 1938, Pavlov, 1927). Pavlovian conditioning allows the study how an animal changes its behavior when it is passively exposed to a succession of stimuli while operant conditioning deals with the way animals change their behavior as a function of their consequences.

Developing computational models accounting for the data collected in these two procedures could lead to advances in the design of artificial adaptive systems. Indeed, the first works on reinforcement learning (Sutton and Barto, 1981) took their inspirations from Pavlovian conditioning. This is a paradox because reinforcement learning algorithms are solving Markovian decision problems (Sutton and Barto, 1998), a class of optimization problems which bears striking similarities with operant conditioning. But, no attempt to analyze operant conditioning within the framework of Markovian decision problems (MDP) have been made up to this day[1].

This is what we try to do in this paper. The next section provides a brief of overview of MDP. Then, we use them to analyze animal performance in so-called random and variable ratio schedules of reinforcement.

1.2 Markovian decision problems

A MDP is composed of an environment, characterized by a set of states $s \in S$ and of an agent characterized by a set of controls $u \in U$. The agent is further characterized by a policy π which attributes to each $s \in S$ a control $\pi(s) \in U$. The agent and the environment interact together according to the following sequence: (a) at time t, the environment is in state $s(t) = s$, (b) the agent emits control $u(t) = \pi[s(t)] = u$, (c) the environment goes into state $s(t+1) = s'$, (d) the agent collects an immediate amount of primary value $r(t) = r$.

The dynamics of the environment is Markovian: $s(t+1)$ only depend upon $s(t)$ and $u(t)$. So, we can define the state transition probability $p(s'|s,u)$ which is the probability that $s(t+1) = s'$ when $s(t) = s$ and $u(t) = u$. The Markov property also holds for $r(t)$ whose mean value is determined by the return function $f[s(t), s(t+1), u(t)]$. This allows the computation of $V^\pi(s)$, the situation value of state s for policy π. It is the total (discounted) amount of primary value that an agent can expect to collect if, while the environment is in state s, it begins to follow policy π. We have

$$V^\pi(s) = \sum_u p(u|s) \sum_{s'} p(s'|s,u)[f(s,s',u) + \gamma V^\pi(s')] \quad (1)$$

[1] A few models of operant conditioning combining artificial neural networks with reinforcement learning have recently been proposed (see, for instance, Donahoe et al., 1993). But, none of them explicitly exploit the MDP framework.

This is Bellman's equation for policy π. γ is a free positive parameter always smaller than 1 called the discount factor which controls the impact of delayed rewards on the value of $V^{\pi}(s)$. $p(u|s)$ is of course dependent on π.

The goal of the agent is to find an optimal policy π^* which maximizes the total (discounted) amount of primary value that the agent can collect in any state of the environment e.g. it is to find a policy π^* whose value function is V^*, the optimal value function with

$$\begin{aligned} V^*(s) &= \max_{\pi} V^{\pi}(s) \\ &= \max_{u} \sum_{s'} p(s'|s,u)[f(s,s',u) + \gamma V^*(s')] \end{aligned} \quad (2)$$

This is Bellman's optimality equation (see Bertsekas, 1995 or Sutton and Barto, 1998 for further details).

1.3 Ratio schedules of reinforcement

One of the major research field in operant conditioning is the one of reinforcement schedules e.g. the way the consequence (which is most of the time food) is scheduled as a function the animal's responding (Fester and Skinner, 1957).

In ratio schedules, for instance, a certain number of responses (the ratio of the schedule) must have been emitted for the reward to be delivered. In variable-ratio schedules (VR), it varies around a mean determined by the ratio of the schedule. The idealization of a VR is a random ratio (RR) schedule. In such a schedule, the probability for a response to be rewarded is constant and equal to the inverse of the ratio.

From the point of view of MDP, this is a very interesting property. In a MDP model of operant conditioning, $s(t+1)$ should discriminate between the fact that the animal has or has not been reinforced and, eventually, between the fact that a response has or has not been emitted. In RR schedules, these two events are dependent only upon u and the ratio of the schedule. That means that, no matter the nature of the set of states S, Bellman's optimality equation can always be written. Moreover, experimental studies (Crossman et al., 1987, Mazur, 1983) have failed to find significant differences between RR and VR schedules and, in theoretical analysis, RR schedules are often used instead of VR schedules (see, for instance, Killeen, 1994). For these reasons, we will start our analysis of operant conditioning within the framework of MDP by RR schedules. We provide below the detailed analysis of what is maybe the simplest possible MDP model of RR performance.

2 A Markovian decision problem model of ratio schedule performance

2.1 Description of the model

The agent is controlling the probability of emission of the response and keeps track of the local rate of reward and of the local rate of responding through exponential moving averages. Every ϕ seconds, it checks the state $s(t)$ of the environment and selects a probability of responding u. This results in a cost $g(u)$ and in the collect of a reward $h(r)$, where r is the actual amount of reward collected. These two quantities are then used to update two exponential moving averages. One keeps track of the local rate of reward while the other one deals with the local rate of responding. These updated averages determine the next state $s(t+1)$ and the actual amount of primary value collected. g is of course a strictly decreasing function of u while h is a strictly increasing function of r.

Let's put this formally. At time t, the state of the environment is $s(t)$. It is composed of two components: first, $v(t)$, the local rate of reward and second, $c(t)$, the local rate of responding (which is a measure of the cost of responding). The agent emits control u with a cost $g(u)$ and this allows it to collect r amount of reward which is passed through a reward function h. Hence, we have $s(t+1) = [v(t+1), c(t+1)]$ with

$$\begin{aligned} v(t+1) &= ah(r) + (1-a)v(t) \\ c(t+1) &= bg(u) + (1-b)c(t) \end{aligned} \quad (3)$$

where a and b are two strictly positive parameters smaller than or equal to 1 which govern the rate of growth and decay of the two exponential moving averages. Moreover, the immediate amount of primary value collected is simply

$$R(t) = v(t+1) + c(t+1) \quad (4)$$

So, as it is often assumed by operant conditioning theorists (see, for instance, Baum, 1981), it is the local rates of reward and of responding that matter to the agent, not the actual amount of reward collected or the immediate cost of responding. Since u and r have already been filtered, it seems unnecessary and inelegant to filter again $v(t+1)$ and $c(t+1)$ to get $R(t)$. Finally, note that r can only take two values in the current model: 1 if the agent is rewarded, 0 otherwise. By convention, we have $h(0) = 0$ and $h(1) = 1$ so that we do not really have to matter about the shape of h.

So, at time t, the environment is in state $s_0 = [v, c]$ and the agent emits control u. If it is rewarded, the environment goes into state $s_1 = [a + (1-a)v, bg(u) + (1-b)c]$ and the agent collects $a + (1-a)v + bg(u) + (1-b)c$ amount of primary values. If, on the other hand, it is not rewarded, the environment goes into state

$s_2 = [(1-a)v, bg(u) + (1-b)c]$ and the agent collects $(1-a)v+bg(u)+(1-b)c$ amount of primary values. To be rewarded, the agent must emit a response and a reward must have been scheduled. Hence, since u is the probability responding at time t and since, in a RR schedule, the probability for a response to be rewarded is equal to the inverse of the ratio, we have $p(s_1|s_0, u) = \frac{u}{N}$ where N is the ratio of the schedule. So, $p(s_2|s_0, u) = 1 - \frac{u}{N}$. Hence, Bellman's optimality equation for this MDP is

$$\begin{aligned} V^*(v,c) &= \max_u \frac{u}{N}\{a + (1-a)v + bg(u) + (1-b)c \\ &\quad +\gamma V^*[a + (1-a)v, bg(u) + (1-b)c]\} \\ &\quad +\frac{N-u}{N}\{(1-a)v + bg(u) + (1-b)c \\ &\quad +\gamma V^*[(1-a)v, bg(u) + (1-b)c]\} \\ &= \max_u Q^*(v,c,u) \end{aligned} \quad (5)$$

To complete the analysis of this model, we need to find the optimal policy and to study the impact of g on it. We already know that g is a strictly decreasing function of u. We will also assume two additional constraints e.g. that $g(0) = 0$ and $g(1) = -1$. Moreover, we will require that g is differentiable over $[0;1]$. Even with these constraints, it can assume several shapes which all have a particular meaning in term of motivation:

1. $g(u)$ could be negatively accelerated, that is to say $g''(u) < 0$. An example would be $g(u) = -(u^d)$ where d is a positive real number greater than 1. Such a function obeys the law of diminishing marginal value supposed to be a fundamental property of motivational systems. That means that if $u_0 < u_1 < u_2$ and $u_1 - u_0 = u_2 - u_1$, then $g(u_1) - g(u_0) < g(u_2) - g(u_1)$. So, the cost of increasing u becomes more and more important as u increases: the more the agent has to work, the harder it is for it to work more.

2. $g(u)$ is positively accelerated, that is to say $g''(u) > 0$. An example would be $g(u) = -(x_0+1)\frac{x}{x_0+x}$ where x_0 is a positive free parameter. Such a function has the reverse property of a negatively accelerated one e.g if $u_0 < u_1 < u_2$ and $u_1 - u_0 = u_2 - u_1$, then $g(u_1) - g(u_0) > g(u_2) - g(u_1)$ and so, the cost of increasing u becomes less and less important as u increases: the more the agent has to work, the less hard it is for it to work more.

3. $g(u)$ could also be linear. No matter the value of u, increasing it has always the same cost e.g. if $u_0 < u_1 < u_2$ and $u_1 - u_0 = u_2 - u_1$, then $g(u_1) - g(u_0) = g(u_2) - g(u_1)$.

2.2 Optimal policy

As it is shown in the appendix, $g(u)$ determines the nature of the optimal policy

- If $\frac{\partial V^*}{\partial u} = 0$ has no root over $[0;1]$ then the optimal policy is either 0 ($\frac{\partial V^*}{\partial u} < 0$) or 1 ($\frac{\partial V^*}{\partial u} > 0$). This is especially the case when $g(u)$ is linear. So, the function relating ratio to response rate (number of responses per unit of time) is a step function, going suddenly for a critical ratio value from the maximum response rate to no responding. This all-or-none response pattern is totally at odds with the data about VR and RR which shows that response rate continuously varies with the ratio.

- If $g(u)$ is positively accelerated, then although $\frac{\partial V^*}{\partial u} = 0$ has a solution over $[0;1]$, it is a local minimum. So, the optimal policy is 1 if $Q^*(v,c,1) > Q^*(v,c,0)$ and 0 otherwise. This is the same kind of all-or-none policy as for a linear $g(u)$ and so, this class of models is ruled out because it is at odds with the data.

- If $g(u)$ is negatively accelerated, then $\frac{\partial V^*}{\partial u} = 0$ has a solution and it is a local maximum. Its value is the optimal probability of responding. Hence,

$$u^* = -g'^{-1}\left\{\frac{a[1-\gamma(1-b)]}{Nb[1-\gamma(1-a)]}\right\} \quad (6)$$

For instance, if $g(u) = -(u^d)$,

$$u^* = \sqrt[d-1]{\frac{a[1-\gamma(1-b)]}{dNb[1-\gamma(1-a)]}} \quad (7)$$

Equation (6) is of the form $f(\frac{q}{N})$ where q is a real number whose value is determined by the free parameters a, b and γ. Then, it can be shown that if $f(\frac{q_1}{N_1}) = f(\frac{q_2}{N_1})$ then $f(\frac{q_1}{N_2}) = f(\frac{q_2}{N_2})$ for any N_2 as long as $N_2 = nN_1$ where n is some rational number. So, although the model can approximate any data for a given ratio by setting its free parameters to the appropriate value, it always predicts the same variation of response rate as a function of the ratio. This makes it a very valuable model since it will be easy to see if it fits or not the actual data about VR. On the other hand, in case of success, its predictions must be taken seriously because they are dependent upon a very limited number of free parameters, all related to the cost function.

2.3 From optimal probability to response rate

We use a version of the model with $g(u) = -(u^d)$ to test its empirical validity. Since most studies only report mean response rates, we need to know how to pass from the optimal probability predicted by our model to response rate. First, remember that, for a given ratio, the optimal probability u^* is constant. Second, we have assumed that the animal emits a response with probability u^* every ϕ units of time. ϕ could be interpreted as the minimum possible interresponse time. A session

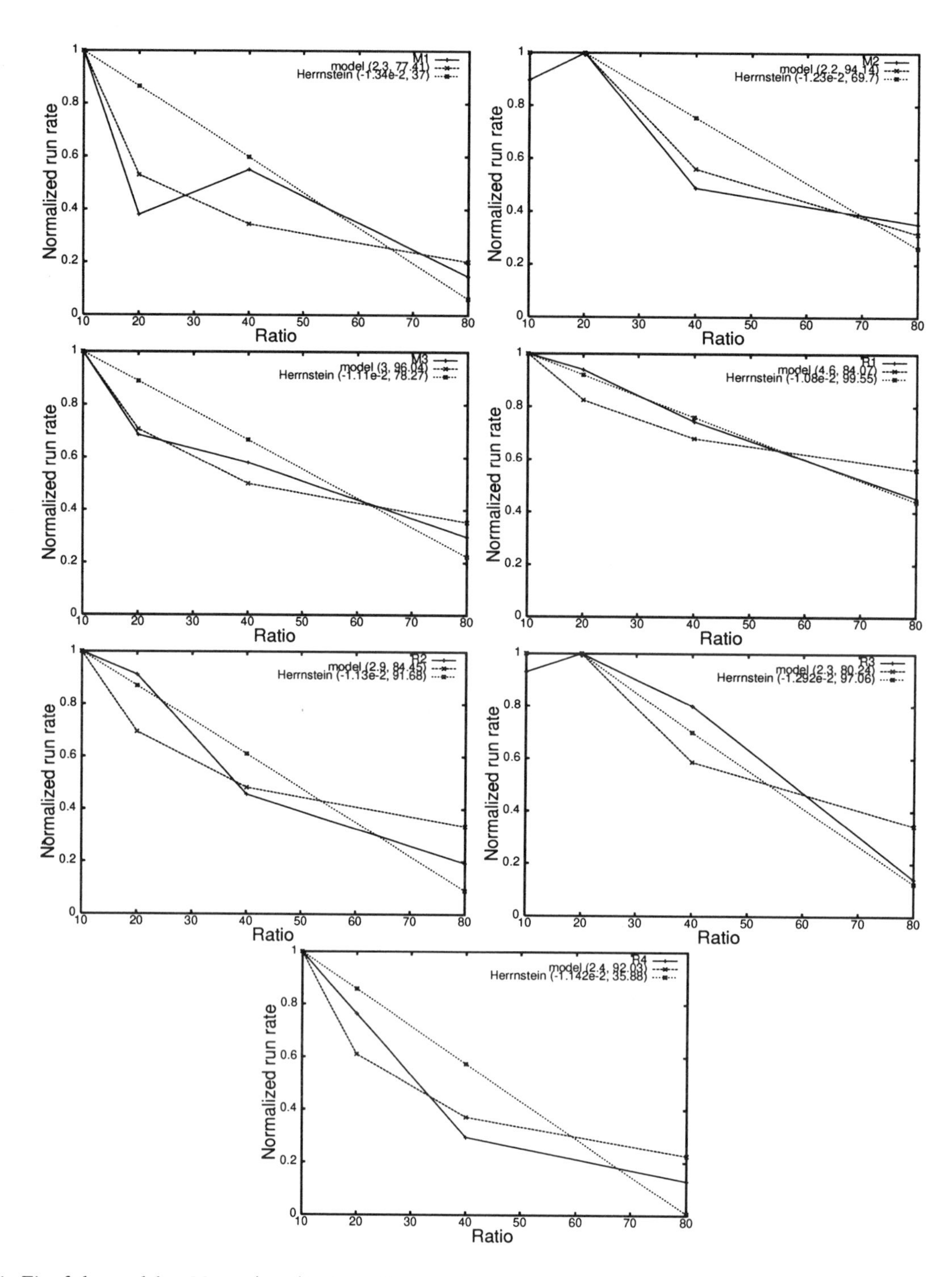

Figure 1: Fit of the model to Mazur (1984)'s data. Numbers in parentheses are first the free parameter of the equation, second the percentage of variance explained by that equation. Subjects are rats.

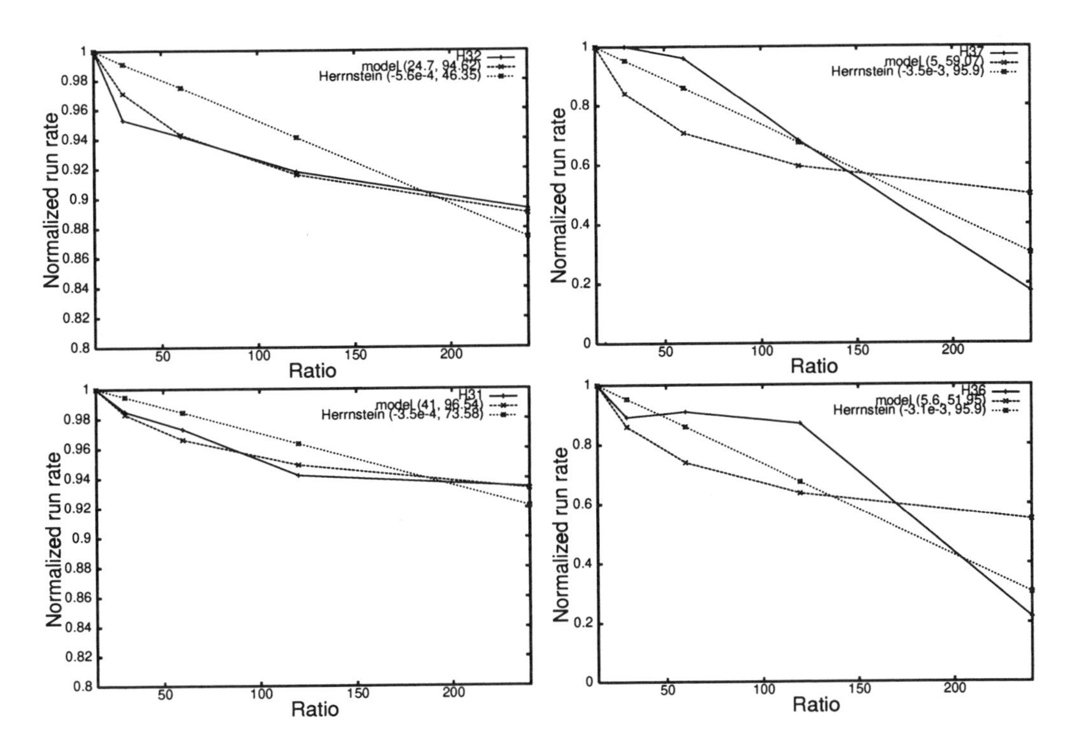

Figure 2: Fit of the model to McDowell and Wixted (1986)'s data. Numbers in parentheses are first the free parameter of the equation, second the percentage of variance explained by that equation. Subjects are humans.

ends after $x\phi$ units of time have elapsed where x is a given positive real number.

So, during the whole session, the agent would have emitted xu^* responses and so, its response rate is simply $\frac{u^*}{\phi}$. So, the function relating response rate to the ratio has exactly the same shape as the one relating the optimal probability to the ratio. Since it is the shape of this function which is really predicted by the model, we can directly use the optimal probability to compare the predictions of the model to real response rates.

2.4 *Qualitative comparison with the data*

Experimental research on VR and RR schedules both with rats (Mazur, 1983) and pigeons (Baum, 1993, Bizo and Killeen, 1997, Brandauer, 1958, Green et al., 1982) have found that response rate is a bitonic function of the ratio: first, response rate increases with the ratio until it reaches a critical ratio value. Then, response rate decreases with the ratio. This seems to discard any version of our model.

But, Baum (1981) has argued that the bitonic ratio/response function is an artifact due to the "postreinforcement pause" (PRP)[2] whose impact on the computation of response rate is more important for low ratios than for high ratios because the interreinforcement time is shorter. So, the decrease in response rate for low ratio values would actually be totally artificial and a more appropriate dependent variable would be the run rate (number of responses divided by the overall duration of the session minus the time filled by the PRP).

Moreover, according to Baum (1993), the PRP would be more a kind of reflexive activity dependent upon the reward rate and only the run rate would demonstrate a real modulation of responding as a function of its consequences. Hence, in a study where he submitted pigeons to two different types of reinforcement schedules (so-called variable interval schedules versus RR schedules) with equalized reward rate, Baum (1993) observed that, for a given reward rate, the run rate was different from one schedule to the other while the PRP had the same duration (see Baum, 1993 for more details as well as for further discussions on this topic).

Indeed, when run rate is used instead of response rate as the dependent variable (Baum, 1993, Mazur, 1983, McDowell and Wixted, 1986), it appears that it is a decreasing function of the ratio, as predicted by our model (see also McDowell and Wixted, 1988 for a reanalysis of data showing a bitonic ratio/response rate and which tries to evaluate the impact of the PRP).

Finally, except sometimes for very high ratio values, run rate is highly regular (Fester and Skinner, 1957) which is coherent with the constant optimal probability of responding predicted by our model.

[2] After a rewarded response, pigeons and rats usually wait before responding again. This time is filled by species-specific activities related to food consumption such as grooming in rats.

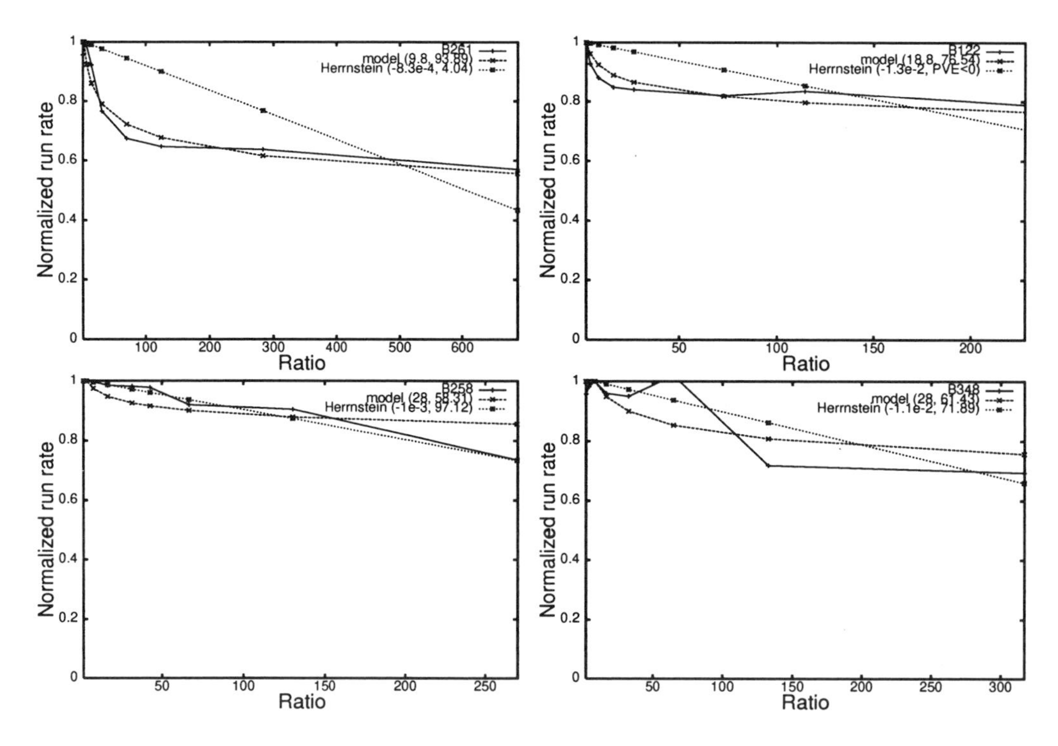

Figure 3: Fit of the model to Baum (1993)'s data. Numbers in parentheses are first the free parameter of the equation, second the percentage of variance explained by that equation. Subjects are pigeons.

2.5 Quantitative comparison with the data

We compared the quantitative fit of the model to three sets of data: Mazur (1983)'s study on rats, McDowell and Wixted (1986)'s study on humans and Baum (1993)'s study on pigeons. Rats and pigeons were trained to press a lever or to peck a key for food while human subjects pressed a button to earn points that could lately be exchanged for money. Since it is the correspondence between the shape of the predicted ratio/run rate function and the empirical one that we will test, we have normalized all the data described below so that the highest possible response rate is 1.

To fit the data below, d, the only free parameter of the model, which controls the curvature of the cost function $g(u)$, will be adjusted so as to minimize the average least mean square error between the prediction of the model and the data. This error is given by the formula

$$e(d) = \frac{1}{2n} \sum_{i=1}^{n} (B_i - u_i^d)^2 \tag{8}$$

where n is the number of ratio values, B_i is the empirical normalized run rate observed for the ith ratio value and u_i^d is the optimal normalized run rate predicted by the model for the ith ratio value when the free parameter of the cost function is fixed to d. To quantify the fit of the model, we will compute the proportion of variance explained (PVE) using the following equation

$$PVE = 1 - \frac{\sum_{i=1}^{n} (B_i - u_i^d)^2}{\sum_{i=1}^{n} (B_i - \overline{B})^2} \tag{9}$$

where $\overline{B}$ is the mean empirical normalized run rate. If the PVE is negative, that means a horizontal line crossing the y-line at $\overline{B}$ is a better fit to the data than the model.

To assess the quality of these fits, we need to compare our model with other models of performance under VR schedules. Actually, we only know three of these models in the literature. One of them, Killeen (1994)'s "mechanical" model of operant conditioning, predicts the response rate and not the run rate so, it is irrelevant here. The two other models are predicting the run rate. The first one is Herrnstein's equation (Herrnstein, 1970) which, in the context of ratio schedules (Killeen, 1994, Pear, 1975), is simply a linear function with a negative slope. Hence,

$$u_n^d = k - dn \tag{10}$$

where n is the value of the ratio and k the asymptotic rate of responding (see also Baum, 1993 for a discussion about the relevance of Herrnstein's equation to ratio responding). Another VR model is McDowell and Wixted (1988)'s application of the linear system theory to VR schedules which can actually be approximated by Herrnstein's equation. So, we will just compare our model to equation (10). Just like our model, Herrnstein's equation will be adjusted to the model so as to minimize equation (8).

Figure 1, 2 and 3 shows the results of these fits. Two conclusions can be draw from them.

1. The predictions of the model are on the whole better than the ones of Herrnstein's equation. So, the mean PVE for the model is 78.74 per cent while it is only 65.36 per cent for Herrnstein's equation. Moreover, Herrnstein's equation failures are worse than the ones of the model. The PVE for the model never falls below 50 per cent while it is several times the case for Herrnstein's equation (note particularly B261 and B122 in Baum's study).

2. But, in each one of the studies, we always find two groups of subjects: one which is better fitted by our model than by Herrnstein's equation and another one which is better fitted by Herrnstein's equation than by our model. When our model is better, there is also a good qualitative fit with the data since the empirical ratio/run rate function is the kind of negatively decelerated function predicted by the model (and this fit is sometimes quite impressive: see, for instance, H31 and H32 in the McDowell and Wixted (1986)'s study and B261 in Baum (1993)'s study) while this is not always true for Herrnstein's equation.

3 Conclusion

We provided an analysis of RR and VR schedule performance within the framework of MDP. The results correspond quite well to some of the empirical data collected on three different species although the subjects are clearly following two response strategies, a fact that has not been noted before in the literature. Only one of these strategies correspond to our model, the other being a kind of decreasing S-shaped function closer to the straight line described by Herrnstein's equation.

In further studies, we will try to account for these two strategies. An obvious research direction would be to use a cost function $g(u)$ whose curvature change as a function of u. For instance, we have observed that if $g(u)$ is a sigmoid ($g(u) = -(x_0^2+1)\frac{x^2}{x_0^2+x^2}$), the model predicts the existence of two response strategies, one corresponding to the negatively decelerated function corresponding to $g(u) = -u^d$, the other to the all-or-none response function corresponding to $g(u) = -(x_0+1)\frac{x}{x_0+x}$. The trick would be to find a cost function for which the second response strategy would be the decreasing S-shaped function observed in the data.

Acknowledgements

This work was supported by a grant from the Conseil régional du Nord-Pas-de-Calais accorded to the first author (contract n 98 49 0262). We would like to thank William Baum for having given access to his raw data about VR performance in pigeons.

References

Baum, W. M. (1981). Optimization and the matching law as the basis for instrumental behavior. *Journal of the experimental analysis of behavior*, 36:387–403.

Baum, W. M. (1993). Performance under ratio and interval schedules of reinforcement: data and theory. *Journal of the experimental analysis of behavior*, 59:245–269.

Bertsekas, D. P. (1995). *Dynamic programming and optimal control (2 volumes)*. Athena Scientific, Belmont, MA.

Bizo, L. and Killeen, P. R. (1997). Models of ratio schedule performance. *Journal of experimental psychology: animal behavior processes*, 23:351–367.

Brandauer, C. M. (1958). *The effect of uniform probabilities of reinforcement on the response rate of the pigeon.* Doctoral dissertation, University Microfilms N0 59-1478.

Crossman, E. K., Bonem, E. J., and Phelps, B. J. (1987). A comparison of response patterns on fixed, variable and random ratio schedules. *Journal of the experimental analysis of behavior*, 48:395–406.

Donahoe, J. W., Burgos, J. E. and Palmer, D. C. (1993). A selectionist approach to reinforcement *Journal of the experimental analysis of behavior*, 60:17–40.

Fester, C. B. and Skinner, B. F. (1957). *Schedules of reinforcement.* Appleton Century Croft, New York.

Green, L., Kagel, J., and Battalio, R. C. (1982). Ratio schedules of reinforcement and their relation to economic theories of labor supply. In Commons, M. L., Herrnstein, R. J., and Rachlin, H., (Eds.), *quantitative analysis of behavior, vol 2: matching and maximizing accounts*, pages 395–429. Ballinger, Cambridge, MA.

Herrnstein, R. J. (1970). On the law of effect. *Journal of the experimental analysis of behavior*, 13:243–266.

Killeen, P. R. (1994). Mathematical principles of reinforcement. *Behavioral and brain sciences*, 17:105–172.

Mazur, J. E. (1983). Steady state performance on fixed, mixed and random ratio schedules. *Journal of the experimental analysis of behavior*, 39:293–307.

McDowell, J. J. and Wixted, J. T. (1986). Variable ratio schedules as variable interval schedules with linear feedback loops. *Journal of the experimental analysis of behavior*, 46:315–329.

McDowell, J. J. and Wixted, J. T. (1988). The linear system theory's account of behavior maintened by variable ratio schedules. *Journal of the experimental analysis of behavior*, 49:143–169.

Pavlov, I. P. (1927). *Conditioned reflexes.* Oxford University Press, New York.

Pear, J. J. (1975). Implications of the matching law for ratio responding. *Journal of the experimental analysis of behavior*, 23:139–140.

Skinner, B. F. (1938). *The behavior of organisms.* Appleton Century Croft, New York.

Sutton, R. S. and Barto, A. G. (1981). Toward a modern theory of adaptive network: expectation and prediction. *Psychological review*, 88:135–170.

Sutton, R. S. and Barto, A. G. (1998). *Reinforcement learning: an introduction.* MIT Press, Cambridge, MA.

Appendix

In this appendix, we show how to compute the optimal policy for the model.

According to Bellman's principle of optimality (Bertsekas, 1995)

$$V^*_{n+1}(s) = \max_u \sum_{s'} p(s'|s,u)[r + \gamma V^*_n(s')]$$

where V^*_n is the optimal value function for the MDP with an horizon of n. That means that the agent and the environment interact only for n time steps. This equation can be used to find V^* since

$$V^*(s) = \lim_{n\to\infty} V^*_n(s)$$

We will use Bellman's principle of optimality to derive an expression of V^*_n and of its associated optimal policy π^*_n. To get V^* and π^* from them, we just need to see what these expressions become when $n \to \infty$.

We proceed by induction. Consider the MDP with an horizon of n. Suppose that for $j \in [0;n]$, the optimal policy π^*_n consists of a single optimal control u^*_n. Moreover, suppose that Suppose that

$$\begin{aligned} V^*_n(v,c) &= v\sum_{i=0}^{n}\gamma^i(1-a)^{i+1} + c\sum_{i=0}^{n}\gamma^i(1-b)^{i+1} \\ &+\frac{a}{N}\sum_{i=0}^{n}\gamma^{n-i}\sum_{j=0}^{i}\gamma^i(1-a)^i u^*_j \\ &+b\sum_{i=0}^{n}\gamma^{n-i}\sum_{j=0}^{i}\gamma^i(1-b)^i g(u^*_j) \end{aligned} \quad (11)$$

Let's shown that this expression is also true for $n+1$. We have

$$\begin{aligned} V^*_{n+1}(v,c) &= \max_u \frac{u}{N}\{a + (1-a)v + bg(u) + (1-b)c \\ &+\gamma V^*_n[a+(1-a)v, bg(u)+(1-b)c]\} \\ &+\frac{N-u}{N}\{(1-a)v + bg(u) + (1-b)c \\ &+\gamma V^*_n[(1-a)v, bg(u)+(1-b)c]\} \\ &= \max_u \frac{au}{N} + bg(u) + (1-a)v + (1-b)c \\ &+\frac{\gamma u}{N}\Delta V^*_n(v,c) \\ &+\gamma V^*_n[(1-a)v, bg(u)+(1-b)c] \end{aligned}$$

where

$$\begin{aligned} \Delta V^*_n(s,c) &= V^*_n[a+(1-a)v, bg(u)+(1-b)c] \\ &-V^*_n[(1-a)v, bg(u)+(1-b)c] \\ &= a\sum_{i=0}^{n}\gamma^i(1-a)^{i+1} \end{aligned}$$

and, hence

$$\begin{aligned} V^*_{n+1}(v,c) &= \max_u \frac{au}{N} + bg(u) + (1-a)v + (1-b)c \\ &+\frac{\gamma au}{N}\sum_{i=1}^{n}\gamma^i(1-a)^{i+1} \\ &+\gamma(1-a)v\sum_{i=0}^{n}\gamma^i(1-a)^{i+1} \\ &+\gamma bg(u)\sum_{i=0}^{n}\gamma^i(1-a)^{i+1} \\ &+\gamma(1-b)c\sum_{i=0}^{n}\gamma^i(1-b)^{i+1} \\ &+\frac{\gamma a}{N}\sum_{i=0}^{n}\gamma^{n-i}\sum_{j=0}^{i}\gamma^i(1-a)^i u^*_j \\ &+\gamma b\sum_{i=0}^{n}\gamma^{n-i}\sum_{j=0}^{i}\gamma^i(1-b)^i g(u^*_j) \\ &= \max_u[\frac{au}{N} + \frac{au}{N}\sum_{i=0}^{n}\gamma^{i+1}(1-a)^{i+1} + bg(u) \\ &+bg(u)\sum_{i=0}^{n}\gamma^{i+1}(1-b)^{i+1}] \\ &+(1-a)v + v\sum_{i=0}^{n}\gamma^{i+1}(1-a)^{i+2} \\ &+(1-b)c + c\sum_{i=0}^{n}\gamma^{i+1}(1-b)^{i+2} \\ &+\frac{a}{N}\sum_{i=0}^{n}\gamma^{n+1-i}\sum_{j=0}^{i}\gamma^i(1-a)^i u^*_j \end{aligned}$$

$$+b\sum_{i=0}^{n}\gamma^{n+1-i}\sum_{j=0}^{i}\gamma^i(1-b)^i g(u_j^*)$$

$$= \max_u\{\frac{au}{N}\left[1+\sum_{i=0}^{n}\gamma^{i+1}(1-a)^{i+1}\right]$$

$$+bg(u)\left[1+\sum_{i=0}^{n}\gamma^{i+1}(1-b)^{i+1}\right]\}$$

$$+v\left[(1-a)+\sum_{i=0}^{n}\gamma^{i+1}(1-a)^{i+2}\right]$$

$$+c\left[(1-b)+\sum_{i=0}^{n}\gamma^{i+1}(1-b)^{i+2}\right]$$

$$+\frac{a}{N}\sum_{i=0}^{n}\gamma^{n+1-i}\sum_{j=0}^{i}\gamma^i(1-a)^i u_j^*$$

$$+b\sum_{i=0}^{n}\gamma^{n+1-i}\sum_{j=0}^{i}\gamma^i(1-b)^i g(u_j^*)$$

$$= \max_u[\frac{au}{N}\sum_{i=0}^{n+1}\gamma^i(1-a)^i$$

$$+bg(u)\sum_{i=0}^{n+1}\gamma^i(1-b)^i]$$

$$+v\sum_{i=0}^{n+1}\gamma^i(1-a)^{i+1}+c\sum_{i=0}^{n+1}\gamma^i(1-b)^{i+1}$$

$$+\frac{a}{N}\sum_{i=0}^{n}\gamma^{n+1-i}\sum_{j=0}^{i}\gamma^i(1-a)^i u_j^*$$

$$+b\sum_{i=0}^{n}\gamma^{n+1-i}\sum_{j=0}^{i}\gamma^i(1-b)^i g(u_j^*)$$

So, we have

$$\frac{\partial V_{n+1}^*}{\partial u} = \frac{a}{N}\sum_{i=0}^{n+1}\gamma^i(1-a)^i + bg'(u)\sum_{i=0}^{n+1}\gamma^i(1-b)^i \quad (12)$$

Neither v nor c appear in this equation. So, π_{n+1}^* consists of a single optimal control u_{n+1}^*. To find it, we need to solve $\frac{\partial V_{n+1}^*}{\partial u} = 0$ and then to check if the roots of this equation are local minima or maxima. Only local maxima are optimal policies. Two cases are possible

1. $\frac{\partial V_{n+1}^*}{\partial u} = 0$ has no root over $[0;1]$. That means that either V_{n+1}^* is an increasing function of u over $[0;1]$, either it is a decreasing function of of u over $[0;1]$. The optimal policy is 1 in the first case, 0 in the other one. Given the above expression of $\frac{\partial V_{n+1}^*}{\partial u}$, it is clear that this is almost always the case when $g(u)$ is linear.

2. $\frac{\partial V_{n+1}^*}{\partial u} = 0$ has at least one root over $[0;1]$. These roots are the solution of the equation

$$u_{n+1}^* = -g'^{-1}\left[\frac{a\sum_{i=0}^{n+1}\gamma^i(1-a)^i}{bN\sum_{i=0}^{n+1}\gamma^i(1-b)^i}\right]$$

The ones for which $\frac{\partial^2 V_{n+1}^*}{\partial^2 u} > 0$ are local minima. If the function has only local minima, then it is a kind of inverted U shape function. In such cases, the optimal policy is 1 if $Q_{n+1}^*(v,c,1) > Q_{n+1}^*(v,c,0)$, 0 otherwise (except in the case where $Q_{n+1}^*(v,c,1) = Q_{n+1}^*(v,c,0)$. In such cases, any point on $[0;1]$ is an optimal policy). The roots for which $\frac{\partial^2 V_{n+1}^*}{\partial^2 u} < 0$ are local maxima and are optimal policies. Since we have

$$\frac{\partial^2 V_{n+1}^*}{\partial^2 u} = bg''(u)\sum_{i=1}^{n+1}\gamma^i(1-b)^i$$

the sign of the second derivative of V_{n+1}^* is entirely determined by $g''(u)$. So, when g(u) is positively accelerated ($g''(u) > 0$), $\frac{\partial^2 V_{n+1}^*}{\partial^2 u} > 0$ while $\frac{\partial^2 V_{n+1}^*}{\partial^2 u} < 0$ when $g(u)$ is negatively accelerated ($g''(u) < 0$).

So

$$\max_u\left[\frac{au}{N}\sum_{i=0}^{n+1}\gamma^i(1-a)^i + bg(u)\sum_{i=0}^{n+1}\gamma^i(1-b)^i\right]$$

$$= \frac{au_{n+1}}{N}\sum_{i=0}^{n+1}\gamma^i(1-a)^i + bg(u_{n+1})\sum_{i=0}^{n+1}\gamma^i(1-b)^i$$

$$= \frac{a}{N}\sum_{i=0}^{n+1}\gamma^i(1-a)^i u_{n+1} + b\sum_{i=0}^{n+1}\gamma^i(1-b)^i g(u_{n+1})$$

Since

$$\frac{a}{N}\sum_{i=0}^{n+1}\gamma^i(1-a)^i u_{n+1} + \frac{a}{N}\sum_{i=0}^{n}\gamma^{n+1-i}\sum_{j=0}^{i}\gamma^i(1-a)^i u_j^*$$

$$= \frac{a}{N}\sum_{i=0}^{n+1}\gamma^{n-i}\sum_{j=0}^{i}\gamma^i(1-a)^i u_j^*$$

and

$$b\sum_{i=0}^{n+1}\gamma^i(1-b)^i g(u_{n+1}) + b\sum_{i=0}^{n}\gamma^{n+1-i}\sum_{j=0}^{i}\gamma^i(1-b)^i g(u_j^*)$$

$$= b\sum_{i=0}^{n+1}\gamma^{n-i}\sum_{j=0}^{i}\gamma^i(1-b)^i g(u_j^*)$$

we have show that

$$V_{n+1}^*(v,c) = v\sum_{i=0}^{n+1}\gamma^i(1-a)^{i+1} + c\sum_{i=0}^{n+1}\gamma^i(1-b)^{i+1}$$

$$+\frac{a}{N}\sum_{i=0}^{n+1}\gamma^{n-i}\sum_{j=0}^{i}\gamma^i(1-a)^i u_j^*$$

$$+b\sum_{i=0}^{n+1}\gamma^{n-i}\sum_{j=0}^{i}\gamma^i(1-b)^i g(u_j^*)$$

So, we have shown that, if equation (11) holds for the MDP with an horizon of n, it also holds for the MDP with an horizon of $n+1$. Moreover, as a corollary, the optimal policy is determined by the roots of equation (12).

To complete the demonstration, we need to show that equation (11) also holds for $n = 0$. Bellman's equation is

$$V_0^*(v, c) = \max_u \frac{au}{N} + bg(u) + (1-a)v + (1-b)c$$

The optimal policy is given by the following equation

$$\frac{\partial V_0^*}{\partial u} = \frac{a}{N} + bg'(u) = 0$$

This is just equation (12) for $n = 0$. So, u_0 is 0 or 1 if that equation has no root or if all these roots are local minima or are the roots of this equation that correspond to local maxima. So,

$$V_0^*(v, c) = v(1-a) + c(1-b) + \frac{a}{N} + bg(u_0)$$

which is equation (11) for $n = 0$. Hence, the property is true for $n = 0$ and so, it is true for any $n \in \aleph$.

Testing the Ecological Rationality of Base Rate Neglect

Peter M. Todd*
*Center for Adaptive Behavior and Cognition
Max Planck Institute for Human Development
Lentzeallee 94, 14195 Berlin, Germany
ptodd@mpib-berlin.mpg.de

Adam S. Goodie**
**Department of Psychology
University of Georgia
Athens, GA 30602-3013 USA
goodie@egon.psy.uga.edu

Abstract

What simple learning rules can allow agents to cope with changing environments? We tested whether a rule that neglects base rates of events in the world — something that is usually considered irrational — could be as successful as Bayesian inference that combines base rates and cue accuracies — the usual standard of rationality — in making cue-based predictions about events in time-varying environments. We focused on environments in which base rates change more frequently than cue accuracies, a condition that, we argue, is common in the real world. Five strategies (Bayesian combination, cue accuracy alone, adjusted cue accuracy, base rates alone, and a Least Mean Square learning rule) were compared across "lifetimes" of 10,000 predictions, in which base rates and cue accuracy independently changed every 10, 50, 100, 500, 1000, or 5000 events. The results confirmed that simple strategies that are typically deemed irrational (base rate neglect and its opposite, conservatism) can rival the typical standard of rationality, Bayesian combination of information, by producing ecologically rational decisions in appropriately varying environments.

1 Introduction

The best predictor of the future is the past. But what aspects of the past should be used to predict the events of the future? The information that could possibly be used includes the prevalence of particular past events — that is, their base rates — and cues that more or less reliably preceded past events. Given this information, making predictions based on combining both sources according to Bayesian updating is usually considered to be the best (most rational, even optimal) approach. However, in a wide array of prediction tasks, people have been found to underweight (or "neglect") base rates and overweight situation-specific cues relative to Bayesian prescriptions. As an example of a situation in which people would be likely to neglect base rates (Nisbett & Ross, 1980), consider choosing between two car brands to buy, the first of which has a better long-term repair record. But then you hear from a friend with that brand that their car just died on the highway — now which brand would you be likely to pick? Here, a single new piece of information should not overwhelm the evidence of cumulative repair records (base rates) — but in similar sorts of experimental tasks it often does. There are many manipulations that mitigate this effect (Koehler, 1996), but it is difficult to eliminate it entirely in many laboratory experiments. Indeed, in summing up a wide range of research, Bar-Hillel (1980) stated that "The genuineness, the robustness, and the generality of the base-rate fallacy [i.e., base rate neglect] are matters of established fact."

Why do people sometimes neglect base rates in this way? More generally, why do people (as well as other species) systematically make decisions like these that are typically described as "irrational"? A large literature explores the conditions under which "reasoning errors" can be elicited (e.g., Kahneman, Slovic, & Tversky, 1982), while an opposing tradition demonstrates how such errors can be eliminated or substantially reduced (e.g., Gigerenzer, 1991). One view that strives to explain both sets of findings posits that human decision mechanisms are designed to work with specific structures of information in the environment. When these mechanisms are used in inappropriate environments, irrational decisions can result. But when they are applied to decision problems in appropriately-structured environments, humans can make decisions with *ecological rationality* — that is, appropriate choices that result from a match between a decision mechanism and the environment in which it is applied (Gigerenzer & Todd, 1999).

In this paper, we argue that learning and decision mechanisms that produce a pattern of behavior usually held to be irrational, base rate neglect, can actually be ecologically rational by producing appropriate decisions in environments with appropriate structure. We demonstrate this match between decision strategies and environments through Monte Carlo simulations. Furthermore, we argue that the environment structure most suited for base-rate-neglecting mechanisms is actually common in the real world. Thus, the fact that people can be shown to inappropriately neglect base rates in laboratory experiments may be an indication that the experimental setups violate people's reasonable expectations about environment structure, rather than demonstrations of deep-seated irrationality.

As an example of a situation in which base rate neglect may be reasonable, Gigerenzer (1991) introduced a variation on the dying-car problem that casts some doubt on the universal applicability of the traditional Bayesian approach. Consider deciding whether to let your child play in the local river or in the nearby forest. Crocodiles have not been seen in the river for the past several years — establishing a safe base rate for choosing the river over somewhat risky tree climbing — but yesterday one attacked somebody at the river's edge. Where should you let your child play today? In this case, it would seem the wiser strategy to emphasize the recent single case and discount (or, better, discard) the overall base rate.

The crocodile problem illustrates the importance of considering decisions in the context of the dynamic environment in which they are likely to occur. For the car-buyer, the base rates of brand reliability may be stable over the course of a few years, and a single breakdown might be merely an erroneous cue. For the parent, however, it is clear that the base rates of crocodile infestation can vary from year to year, while a recent report of a crocodile attack remains a reliable cue to the present situation. The cue of yesterday's crocodile attack may serve not only its traditional role in the likelihood ratio of Bayes' formula, but may also indicate that the prior probabilities — the base rates — have changed.

In light of such considerations, it has been proposed (Gigerenzer, 1991; Goodie & Fantino, 1999) that in environments where base rates change more frequently than cue accuracies[1] (including the probably most common cases where base rates change to some extent and cue accuracies change little if at all), neglecting base rates when making predictions may not impair performance to any great extent. Thus, observations of "irrational" base rate neglect may indicate that an otherwise-ecologically rational mechanism is being used in an environment for which it is not intended. Here we test this idea, by comparing in simulation the performance of five prediction strategies derived from the literature, including partial and complete base rate neglect, Bayesian integration, base rate use alone and connectionist learning, in a range of environments with varying cue accuracies and event base rates.

2 Past work on ecologically rational learning

The question of which particular learning strategies are ecologically rational in which environments is closely related to the question of when (i.e., in what environmental circumstances) different forms of learning are expected to evolve. While there is wide agreement that rapidly changing environments are favorable to the evolution of learning in general (e.g., Belew, 1990; Miller & Todd, 1990; Stephens, 1991), there has been relatively little exploration of specific environment structures that favor the evolution of particular types of learning. Todd & Miller (1991) evolved simple recurrent neural networks that exhibited the ability to sensitize and habituate in "clumpy" environments where important stimuli were encountered in (more or less noisy) patches over time. In effect, the networks experienced changing base rates, but there were no explicit cues (other than the presence or absence of the eliciting stimuli) to use along with base rates, greatly limiting the range of learning strategies that could be employed. Here we extend this early work by explicitly comparing a set of clearly defined (rather than messily evolved) learning strategies deployed in a wider range of environment structures to determine when the two fit, and thus produce ecological rationality.

McKenzie (1994) took a similar direct approach by simulating intuitive strategies and more traditional rational models making decisions in a broad range of environments. He found high correlations in performance between the intuitive strategies and the rational models in many instances, indicating the ecological rationality of the former. (Gigerenzer & Hoffrage, 1995, extended the catalog of intuitive near- and non-Bayesian strategies that people use when faced with Bayesian decision tasks.) Our work in this paper expands on McKenzie's efforts by exploring the important case of strategies tuned to changing, rather than fixed, environments.

3 Simulating cue and outcome learning

To test the efficacy of different prediction strategies in different settings, we created a simulation that can be viewed as comparing the behavior of organisms following various prediction strategies in particular environments. The "lifetime" of each simulated organism consisted of 10,000 occasions on which either of two events, E_0 or E_1, could occur, preceded in each case by one of two cues (or, equivalently, two states of one cue, e.g. absent or present), C_0 or C_1. The organism used its particular strategy to predict the event after "perceiving" the preceding cue, and the success of each strategy was measured as the average number of events correctly predicted over 50 lifetimes.

3.1 The environments

To create environments in which event base rates and cue accuracies can vary over time, we specified two parameters, the base rate change interval ($BRCI$) and the cue accuracy change interval ($CACI$). Each parameter was assigned intervals of 10, 50, 100, 500, 1000, or 5000 events, yielding 36 environments. For each organism's lifetime, an initial base rate and cue accuracy were selected randomly from a uniform distribution between 0.0 and 1.0. Then, after $BRCI$ events, the base rate was changed to a new value chosen at random from the same interval. Similarly, after $CACI$ events, a new cue accuracy was selected uniformly between 0.0 and 1.0. So for example an environment characterized by a $BRCI$ of 50 and a $CACI$ of 1000 has base rates that change rapidly while cue accuracies remain relatively stable — the kind of situation in

[1] We use the term cue accuracy to mean Bayesian likelihoods, in other words $p(cue|outcome)$.

which base rate neglect might be adaptive.

The simulated environment as we have set it up here includes two assumptions that are important to note. First, we have used the widest possible range of variation for both cue accuracies and base rates, selecting values uniformly between 0.0 and 1.0. This may not reflect the real distributions of values that hold in different domains in the natural world. Cue accuracies may often be distributed clustering around 0.5 on the one hand — very inaccurate, nearly useless cues — and around 0.0 and 1.0 on the other hand — very accurate and hence useful cues (pointing in opposite directions). But we do not yet have good data on this point, so our simulations take a neutral stand by using a uniform distribution. Similar arguments apply to the distribution of base rates of important events in the world.

Second, we use an equal range of rates of change for both cue accuracies and base rates. Again we expect that this is not an accurate reflection of natural states of affairs — cue accuracies will typically change much less often than base rates, as discussed in section 5. (But when they do, it is obviously very important for agents to respond appropriately, usually by switching to more accurate, or simply using more, cues.) Again, though, in the absence of more data regarding this aspect of environment change, we wanted our exploration to cover a wide range of possibilities.

The current base rate and cue accuracy values controlled the generation of the events and cues an organism would experience on a given trial. First, event E_0 or E_1 was selected randomly with a probability distribution defined by the current base rate. For example, if the base rate were .7, E_0 would be selected with 30% probability and E_1 with 70% probability. Second, based on the selected event, a cue would be generated according to the cue accuracy. (In these simulations, we make the simplifying assumption that $p(C_0|E_0) = p(C_1|E_1)$.) For example, if E_1 was chosen and the cue accuracy was .4, then C_1 would be generated with 40% probability, and otherwise C_0 would be generated. In this way, 50 lifetime sequences of 10,000 cues and events were generated for each of the 36 $BRCI/CACI$-specified environments. Each strategy described in the next section was tested on the same set of 50 sequences per environment.

3.2 The strategies

We compared the performance of five strategies in two classes: four that kept track of cue-event pairs in a decaying memory table, and one that learned connection strengths between cues and events. Each strategy computed the probability p of event E_1 given the observed cue (C_0 or C_1) and its knowledge of the past, and then predicted E_1 if $p > 0.5$ and E_0 if $p \leq 0.5$.

3.2.1 Contingency table-based strategies

Four of our strategies are based on memory for the frequencies of cue-event pairings as represented in the 2×2 table

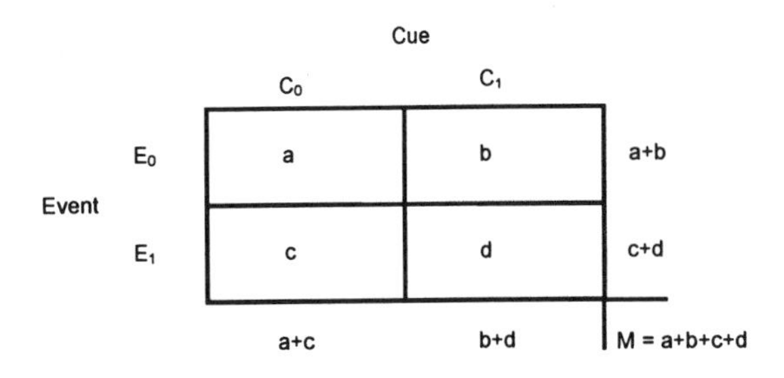

Figure 1: The structure of the memory for strategies 1 to 4. A decaying memory trace was kept for the four cells a, b, c, and d, and different strategies used some or all of these values.

shown in Figure 1. The most straightforward application of this approach would imply a full and perfect memory of M events, with complete amnesia for all observations $M + 1$ events ago and earlier. To use a more realistic model of memory that incorporates the assumption that old information is less relevant than recent information, we employed a table in which stored events decay exponentially as further events occur. For instance, entry a in the table is updated as follows: $a_{t+1} = f \cdot a_t + 1$ if the cue-event pair C_0, E_0 was just seen, and $a_{t+1} = f \cdot a_t$ otherwise, where f is the decay rate parameter ($0 < f < 1$). In this way, a, b, c, and d maintain a weighted count of the past event-cue pairs seen, with the most recent observations getting the most weight. To paraphrase Douglas MacArthur[2], old events never depart from memory; they just fade away. Each of the four table-based strategies we investigated then computed its prediction on the basis of the values a, b, c, and d, as described below and summarized in Table 1. We varied the memory decay rate f, but to allow memory length to be compared more directly to rates of environment change, we give our results in terms of "half-life," which is the number of trials it takes a given event-cue pair entered into memory initially with a strength of 1.0 to decay to a strength of 0.5 (half-life $\approx -1/\log_2 f$). The four contingency table-based strategies were defined as follows:

1. Bayesian combination. This strategy consists of Bayesian integration of both base rates and cue accuracy, defined as the conditional probabilities of E_1 given the observed cue. This strategy employs all four pieces of available information, and is usually considered the benchmark for rationality in this domain (Castellan, 1977).

2. Cue accuracy alone (base rate neglect). This strategy considers only the probability that the cue matches the upcoming event. Because it neglects base rate information, this strategy should be outperformed by the Bayesian combination strategy 1 in environments where base rates are stable and useful, but its relative performance should improve when

[2] Address to a Joint Meeting of Congress, April 19, 1951.

	prediction after C_0	prediction after C_1
1. Bayesian combination	$R\left(1-\frac{a}{a+c}\right)$	$R\left(1-\frac{d}{b+d}\right)$
2. Cue accuracy	$R\left(1-\frac{a+d}{M}\right)$	$R\left(\frac{a+d}{M}\right)$
3. Adjusted cue accuracy	$R\left(1-\frac{a+d}{M}+.05\right)$ if $\frac{a+b}{M}<.5$	$R\left(\frac{a+d}{M}+.05\right)$ if $\frac{a+b}{M}>.5$
	$R\left(1-\frac{a+d}{M}-.05\right)$ if $\frac{a+b}{M}>.5$	$R\left(\frac{a+d}{M}-.05\right)$ if $\frac{a+b}{M}<.5$
4. Base rates	$R\left(\frac{c+d}{M}\right)$	$R\left(\frac{c+d}{M}\right)$

Table 1: Formulas for calculating the predictions of the contingency-table-based strategies (1-4) from the information stored in the 2×2 table shown in Figure 1. $R(x)$ denotes the function of rounding a number to the nearest integer. Because all values represented here are bounded by 0 and 1, all rounded values are 0 or 1, corresponding to predicting E_0 or E_1, respectively. Some values to be rounded are subtracted from 1 because, for example, the cue accuracy strategy should predict E_0 when $a + d$ is close to 1, not when it is close to 0.

base rates change frequently.

3. Adjusted cue accuracy. Absolute base rate neglect as embodied in strategy 2 has seldom been reported in the literature (Koehler, 1996; but see Gigerenzer & Hoffrage, 1995, for instances). Rather, humans and other animals seem to incorporate base rate information partially in their decisions, as is modeled by this strategy. The cue-based probability used in strategy 2 was adjusted by a small amount (5 percentage points) in the direction prescribed by the current base rate. This strategy is consistent with the substantial but incomplete base rate neglect observed by Goodie and Fantino (1995; 1999).

4. Base rates alone. The complement to using cue accuracy exclusively is using base rates exclusively. Where the former neglects base rates entirely, this strategy neglects cue-based information entirely, employing only the probability with which each event occurs. This strategy is similar to an extreme form of conservatism in decision making (Edwards, 1982). While conservatism has been described as erroneous reasoning, this strategy should perform well when cue accuracy changes frequently.

3.2.2 *The connectionist learning strategy*

The fifth strategy arises from a connectionist model of categorization that has been applied to base rate phenomena. This strategy dispenses with directly counted memories of events and instead learns continuously shifting association strengths between experienced cues and events. A set of input nodes receives activation according to the current cue, and an output node makes an event prediction based on the summed input activation weighted by the learned connection weights between input and output nodes. Learning takes place after each cue-event pair is seen, adjusting the connection weights to minimize the difference between predictions and actual events. The fifth strategy is defined as follows:

5. Least Mean Squares (LMS) learning rule. Gluck and Bower (1988) trained people to diagnose fictitious diseases from symptoms in a situation analogous to the cue-to-event prediction we consider here. They found base rate neglect that could be modeled with a simple neural network trained according to the LMS learning rule. Our strategy used a similarly trained network with three input nodes: one activated by C_0, one by C_1, and a third that was always activated, simulating background cues. We followed Gluck and Bower's formulas for updating activation weights and converting them to response probabilities.

3.3 *Predictions and questions*

These five strategies were compared by simulating their predictions over 50 lifetimes of 10,000 cue-event pairs in each of the 36 environments[3]. The most immediate question addressed by this comparison is whether base rate neglect (strat-

[3] We performed a similar comparison across these three environment structures for two additional types of strategies: a set corresponding to strategies 1-4 that used an exact count of cue-event pairs in a sliding memory window rather than a decaying memory; and a set corresponding to strategies 1-5 that predicted outcome events using probability matching rather than exclusively choosing the more likely event. In both cases, the patterns of results closely resembled those reported here.

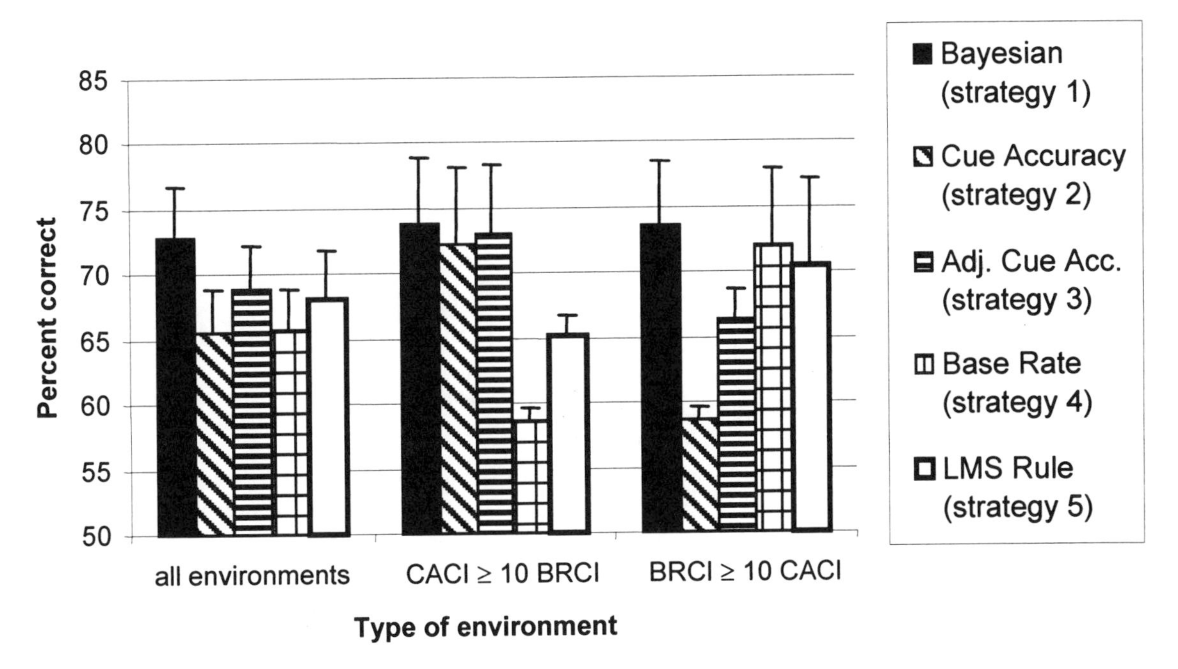

Figure 2: The performance of the five prediction strategies averaged across all 36 environments (left panel), across those environments with stable cue accuracies and rapidly varying base rates (middle panel), and across environments with stable base rates and rapidly varying cue accuracies (right panel). Error bars show standard deviations across the environments averaged together in each case.

egy 2) can successfully compete with Bayesian integration (strategy 1) when base rates change more frequently than cue accuracy. Of course, if the Bayesian strategy has accurate knowledge of the base rates and cue accuracies in the current environment at every moment, it will have an advantage in using this knowledge. But such knowledge does not come from a guidebook — it must be acquired through experience. If base rates change so frequently that they are often estimated incorrectly, then a strategy that uses these estimates may have little advantage over a strategy that ignores base rate information. We expected this to be the case when the Bayesian combination strategy attempted to incorporate rapidly changing base rates. Particularly when cue accuracy changed seldom, we expected base rate neglect strategies (2 and 3) to perform about as well as Bayesian integration (strategy 1).

We correspondingly expected that "cue-accuracy neglect" (strategy 4) would rival Bayesian integration (strategy 1) in environments with rapidly-shifting cue accuracies and stable base rates. Finally, we were not sure how a connectionist strategy would fare in this comparison; in principle, it could learn and use both base rate and cue accuracy information, like Bayesian integration. But because such models have been used to predict (correctly) base rate neglect in comparable situations, we thought strategy 5 might behave similarly to strategies 2 and 3.

4 Comparing rule performance in different environments

The prediction performance of all five strategies, using a memory half-life of 100 cue-event pairs for strategies 1-4, is shown in Figure 2. Averaged across all 36 environments, the Bayesian integration strategy proved the most successful, making correct predictions for about 73% of the cue-event pairs (see Figure 2, left panel). The adjusted cue accuracy strategy (3), which must keep track of three of the cells in the 2×2 table, combined this information in a slightly less effective manner, reaching about 69% correct predictions. This was followed closely by the LMS strategy with 68%. The two simplest strategies, using only two of the cells in the 2×2 table (cue accuracies or base rates alone), each scored about 66% correct overall.

However, when environments with different structures are examined separately, the ordering of successful strategies changes systematically and dramatically. First, consider those environments where base rates are much less stable than cue accuracies, defined as those where the interval between cue accuracy changes is at least ten times as long as the interval between base rate changes ($CACI \geq 10 \cdot BRCI$). Ten of the 36 environments have this structure. Averaged across these environments (see Figure 2, middle panel), Bayesian integration still makes correct predictions most often, even exploiting this environment structure to improve performance slightly to 74%. However, base rate neglect now nearly matches Bayesian performance, with 72% correct. Including the base rate adjustment (strategy 3) improves performance by another percentage point (73% correct). The LMS rule's performance is reduced by the rapidly changing base rates to 65%. Finally, strategy 4, in relying solely on base rates, not surprisingly does worst of all, but is still able to extract enough information to make correct predictions 59% of the time.

In environments where cue accuracies are much less stable than base rates ($BRCI \geq 10 \cdot CACI$; Figure 2, right panel), we find a mirror image of these results. Base rates alone (strategy 4) now nearly matches Bayesian performance (72% vs. 74%), while base rate neglect performs worst (59%). Partial base rate neglect (strategy 3) achieves an intermediate accuracy (66%). The LMS rule also benefits from stable base rates (improving to 70%).

Because our primary question was when base rate neglect (strategy 2) would approach the accuracy of the Bayesian combination strategy, the difference in performance between these two strategies is portrayed across all environments in Figure 3. When cue accuracies changed slowly and base rates changed rapidly (front left corner), base rate neglect was about as accurate as Bayesian integration. As cue accuracies also changed more and more rapidly (moving along the front edge toward the right), the two strategies continued to perform similarly, in part because all strategies fared less well in environments with rapidly changing cue accuracies and base rates (front right corner). Only when base rates became much more stable ($BRCI \geq 500$) did a large difference between the strategies appear, particularly when cue accuracies were less reliable than base rates (rear right corner).

In comparing the LMS learning rule with the contingency-table strategies, three unpredicted but robust effects emerged. First, the LMS rule did a better job of capitalizing on stable base rates than on stable cue accuracies (see Figure 2, difference between right and middle panels, respectively). Second, this strategy performed better than any other strategy in the most variable environments ($BRCI, CACI = 50$; see Figure 4). Third, however, it lags behind other strategies in environments of frequent base rate change where base rate neglect does well (Figure 2, middle panel), perhaps because it does relatively poorly at taking advantage of stable cue accuracies. This is at first surprising in view of Gluck and Bower's (1988; see also Shanks, 1990) demonstration that their model using the LMS rule produces base rate neglect. This situation may be clarified by noting the distinction between base rate neglect as an empirical phenomenon, as the term is most commonly used, and base rate neglect as a decision process, as was employed here — the LMS rule can produce base rate neglect phenomena in some circumstances, but our results make it clear that it does so in a way that differs from the cue-accuracy strategy (2).

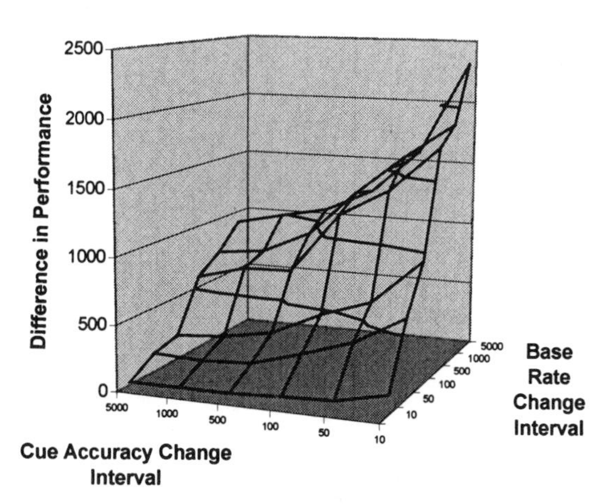

Figure 3: Comparison of the performance of Bayesian integration and base rate neglect across all 36 environments specified by $BRCI$ and $CACI$. The z-axis shows the difference in number of correct predictions (out of 10,000 predictions made) between the two strategies (Bayesian integration minus base rate neglect).

4.1 Effects of memory length

Do these effects hold with other assumptions about the length of memory for cue-event pairs? We tested the robustness of our results to changes in memory length by comparing the performance of strategies 1-4 with half-lives of 20, 50, 100, 200, 500, 1000, 2000, and 5000 events. We used four environments combining slow and fast changes: $BRCI/CACI = 1000/1000, 1000/50, 50/1000$, and $50/50$. The results appear in Figure 4.

Memory length has a strong effect on overall performance: Longer memory traces almost always impair performance. This is because longer memories blur current and outdated base rate and cue accuracy information. Furthermore, the main patterns seen before remain robust across different memory lengths: When base rates change much more rapidly than cue accuracy ($BRCI = 50, CACI = 1000$), base rate neglect (strategy 2) performs nearly as well as Bayesian integration (strategy 1) for all but the shortest memories. Likewise, when cue accuracy changes more rapidly than base rates ($BRCI = 1000, CACI = 50$), base rate use (strategy 4) performs nearly as well as Bayesian integration for all but the shortest memories. These relationships are buttressed by the facts that the error ranges indicate greater variability within each strategy than between them, and poor strategies, having little variability, are unlikely to ever beat the better strategies by chance. Finally, because short memory lengths do not seem likely (in light of, e.g., Goodie & Fantino's 1999 learning studies), the ability of simple strategies to match Bayesian performance in appropriate environments appears robust under reasonable and not very restrictive assumptions concerning memory length.

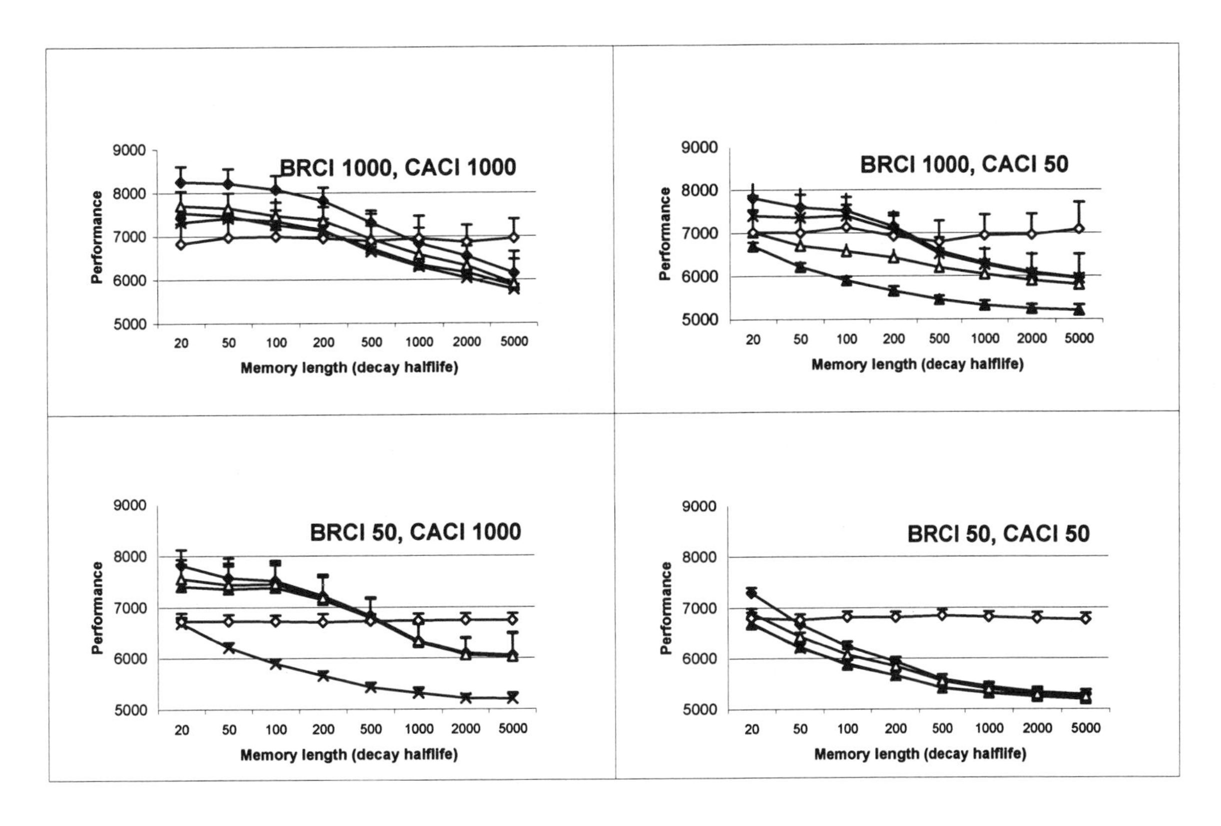

Figure 4: Performance (number of correct predictions out of 10,000) of strategies 1 to 4 with different memory decay parameters, tested in four environments characterized by different rates of base rate and cue accuracy change. The performance of the LMS rule (which is not affected by variations in the decay parameter) is shown in each case for comparison. Error bars show standard error across lifetimes averaged for each data point; error bars in the negative direction are omitted in every case for clarity.

5 Forms of environmental variation

Overall, these results show that simple predictive strategies matched to environments with a specific structure can rival "optimal" models in prediction performance. We compared the performance of a variety of decision strategies across a range of environments, and found that although some of these strategies violate traditional norms of rationality, for instance by ignoring or underweighting base rates, they produce good decisions in environments to which they are matched, and thereby yield ecological rationality (Gigerenzer & Todd, 1999).

But *do* base rates change more often than cue accuracies in general? We contend that they do, because base rates of natural events are usually caused by complex and interacting factors, while good cues often bear a relatively simple, and therefore stable causal relation to the events they predict. Consider the problem of predicting picnic weather. The base rate of rain on a particular day in a particular place is complexly determined by a number of time-varying factors that are often difficult to assess, such as approaching fronts, humidity, microclimate, and time of year, making accurate assessment of the prior probability difficult and time-consuming. However, these multifarious causal variables often combine to create not only rain, but also rain clouds. Dark clouds are a good cue to impending rain, and because they are mostly caused by the same forces, the likelihood of storm clouds when rain is imminent is expected to stay constant, even as the base rates of both rain and rain clouds may vary widely.

Similarly, the incidence of measles may vary from time to time, or from one locale to another, but red spots are always a reliable cue. Birth rates are different in different countries and in different years, but pregnancy tests remain equally reliable. Another way of thinking of this is that when things in the world are clumped in space or time, organisms will experience changing base rates of those things as they move through space or time. Just how clumped various things are compared to how stable (over space and time) their cues are will determine when base rates and cue accuracies are informative.

In contrast, we can ask what environments exist where cue accuracies vary more rapidly than base rates. In such environments, we could expect people to rely relatively more on base rate information, and thus not show (strong) base rate neglect. Situations in which cues are socially constructed, and so may vary rapidly, are one candidate for rapidly varying cue accuracies, especially when there are conflicts of interest involved between signalling and receiving individuals. Similarly, when entering a new arbitrarily constructed situation (as social environments can be), cue accuracies may be expected to change more, or more frequently, than base rates.

Goodie and Fantino (1995, 1996) found evidence that people may expect such changes in cue accuracy, but not base rates, when encountering novel situations. They created conditions of direct experience in which, unlike in past studies, participants were not required to process or comprehend any statistical information, and did not have to report any quantitative judgments. They asked participants to predict events consisting of the occurrence of blue or green rectangles on the basis of cues that were either other blue or green rectangles (Goodie & Fantino, 1995) or the words "blue" or "green" (Goodie & Fantino, 1996, experiment 2). In both cases, participants showed standard base rate neglect, behaving as though they expected the cues to be stable, accurate predictors while the base rates could be variable and hence not useful. This is understandable: We commonly match colors and match names to colors (thus experiencing reliable cue accuracies), and we also see some colors more often than others in different situations (thus experiencing variable base rates). However, when the cues were instead unrelated to the events, such as vertical or horizontal lines preceding the blue or green rectangles, base rate neglect disappeared entirely (Goodie & Fantino, 1996, Experiment 1), suggesting that participants were now using a strategy consistent with the belief that both event base rates and cues could be important in this novel prediction setting.[4]

6 Implications and conclusions

While the events and cue-event associations of one's past are an important guide to the future, knowledge of the past can prove untrustworthy if it springs from processes that are no longer in place. And in the real world, change is constant. The changing seasons, the shifting populations of interacting predator and prey (or parasite and host) species, the new surroundings encountered by migrating creatures, all contribute to reducing the sameness of past and future (Potts, 1996).

With the present simulations we have demonstrated that a supposedly normative, Bayesian combination strategy is often little more successful than less "rational" strategies at predicting the events in particular changing environments. Thus, the fact that people can be made to act "irrationally" in underusing base rates (or conversely being too conservative in changing them) appears to say more about the environment structure that people expect (whether via learning or evolution) to encounter than about the purported shortcomings of our cognitive abilities. There are several extensions to this work that must be explored before definitive statements can be made. The simulations should be enhanced, for instance eliminating the constraint of equal accuracies for both cues, using more extreme base rates and cue accuracies in the environments, and testing the efficacy of a wider range of strategies. We will also assess how well different strategies account for the decision behavior of individual people presented with different changing environments in experimental settings. Nonetheless our initial results here indicate that observed human behavior in many learning situations can be ecologically rational, rather than irrational.

In addition to the near-Bayesian-integration performance of strategies that neglect base rates under presumably common shifting environmental conditions, two more of our findings bear on questions of standards of rationality. First is the finding that longer memories lead to poorer performance. It is perhaps surprising when more information (here stored in memory) results in reduced performance, but such "less is more" effects may not be so uncommon (e.g., Goldstein & Gigerenzer, 1999). This instance arises from a compromised ability to respond rapidly to changes in the environment.

The other notable outcome is the performance of the LMS rule relative to the Bayesian combination strategy: sometimes better, sometimes worse. Both LMS learning (Widrow & Hoff, 1960) and Bayesian decision making (Tversky & Kahneman, 1982) are widely viewed as optimal — but only under certain conditions. Indeed outside those conditions, as in the environments explored here, the two strategies do different things. How then can we choose an appropriate norm of rationality to use as a benchmark when judging the performance of difference decision mechanisms (and of human decision making; see Hertwig & Todd, 2000)? One could argue that whichever approach performs better is the truly rational model. But the fact that Bayesian integration does better in some environments while the LMS rule does better in others shows that norms, like decision mechanisms, must be chosen relative to a particular environment. Thus, the structure of the environment proves to be the final arbiter of ecological rationality.

Acknowledgements

We thank Valerie M. Chase, Gerd Gigerenzer, Ralph Hertwig, and two anonymous reviewers for detailed comments on earlier versions of the text, Thomas Dudey for help with the figures and formatting, and Martin Dieringer for valuable assistance in programming.

References

Bar-Hillel, M. (1980). The base-rate fallacy in probability judgments. *Acta Psychologica, 44,* 211-233.

Belew, R.K. (1990). Evolution, learning, and culture: Computational metaphors for adaptive algorithms. *Complex Systems, 4,* 11-49.

[4]Furthermore, when participants in the familiar matching situation were allowed to acquire a great amount of experience about the particular experimental environment — 1600 predictions — they began by neglecting event base rates as indicated above. Over time, though, their performance changed in a way that indicated the participants had learned that within this context a new and stable set of base rates held and hence a strategy incorporating both base rates and cues could again be used.

Castellan, N. J. (1977). Decision making with multiple probabilistic cues. In N. J. Castellan, D. P. Pisoni, & G. R. Potts (Eds.), *Cognitive theory*, Vol. 2 (pp. 117-147). Hillsdale, NJ: Erlbaum.

Edwards, W. (1982). Conservatism in human information processing. In D. Kahneman, P. Slovic, & A. Tversky (Eds.), *Judgment under uncertainty: Heuristics and biases* (pp. 359-369). New York: Cambridge University Press.

Gigerenzer, G. (1991). How to make cognitive illusions disappear: Beyond "heuristics and biases". *European Review of Social Psychology, 2,* 83-115.

Gigerenzer, G., & Hoffrage, U. (1995). How to improve Bayesian reasoning without instructions: Frequency formats. *Psychological Review, 102,* 684-704.

Gigerenzer, G., & Todd, P. M. (1999). Fast and frugal heuristics: The adaptive toolbox. In G. Gigerenzer, P. M. Todd, and the ABC Research Group, *Simple heuristics that make us smart* (pp. 3-34). New York: Oxford University Press.

Gluck, M. A., & Bower, G. H. (1988). From conditioning to category learning: An adaptive network model. *Journal of Experimental Psychology: General, 117,* 227-247.

Goldstein, D. G., & Gigerenzer, G. (1999). The recognition heuristic: How ignorance makes us smart. In G. Gigerenzer, P. M. Todd, and the ABC Research Group, *Simple heuristics that make us smart* (pp. 37-58). New York: Oxford University Press.

Goodie, A. S., & Fantino, E. (1995). An experientially derived base rate error in humans. *Psychological Science, 6,* 101-106.

Goodie, A. S., & Fantino, E. (1996). Learning to commit or avoid the base rate error. *Nature, 380,* 247-249.

Goodie, A. S., & Fantino, E. (1999). What does and does not alleviate base rate neglect under direct experience. *Journal of Behavioral Decision Making, 12,* 307-335.

Hertwig, R., & Todd, P. M. (2000). Biases to the left, fallacies to the right: Stuck in the middle with null hypothesis significance testing. Commentary on Krueger on social-bias. *Psycholoquy 11*(28). http://www.cogsci.soton.ac.uk/psyc-bin/newpsy?11.028

Kahneman, D., Slovic, P., & Tversky, A. (1982). *Judgment under uncertainty: Heuristics and biases.* New York: Cambridge University Press.

Koehler, J. J. (1996). The base rate fallacy reconsidered: Descriptive, normative and methodological challenges. *Behavioral and Brain Sciences, 19,* 1-53.

Lovett, M. C., & Schunn, C. D. (1999). Task representations, strategy variability and base rate neglect. *Journal of Experimental Psychology: General, 128,* 107-130.

McKenzie, C. R. M. (1994). The accuracy of intuitive judgment strategies: Covariation assessment and Bayesian inference. *Cognitive Psychology, 26,* 209-239.

Miller, G.F., & Todd, P.M. (1990). Exploring adaptive agency I: Theory and methods for simulating the evolution of learning. In D.S. Touretzky, J.L. Elman, T.J. Sejnowski, & G.E. Hinton (Eds.), *Proceedings of the 1990 Connectionist Models Summer School* (pp. 65-80). San Mateo, CA: Morgan Kaufmann.

Nisbett, R. E., & Ross, L. (1980). *Human inference: Strategies and shortcomings of social judgment.* Englewood Cliffs, NJ: Prentice-Hall.

Potts, R. (1996). *Humanity's descent: The consequences of ecological instability.* New York: William Morrow.

Shanks, D. R. (1990). Connectionism and the learning of probabilistic concepts. *Quarterly Journal of Experimental Psychology, 42,* 209-237.

Stephens, D.W. (1991). Change, regularity, and value in the evolution of animal learning. *Behavioral Ecology, 2,* 77-89.

Todd, P.M., & Miller, G.F. (1991). Exploring adaptive agency III: Simulating the evolution of habituation and sensitization. In H.-P. Schwefel & R. Maenner (Eds.), *Proceedings of the First International Conference on Parallel Problem Solving from Nature* (pp. 307-313). Berlin: Springer-Verlag.

Tversky, A., & Kahneman, D. (1982). Evidential impact of base rates. In D. Kahneman, P. Slovic, & A. Tversky (Eds.), *Judgment under uncertainty: Heuristics and biases* (pp. 153-160). New York: Cambridge University Press

Widrow, B., & Hoff, M. E. (1960). Adaptive switching circuits. In *IRE WESCON Convention Record* (pp. 96-104). Reprinted in J. A. Anderson & E. Rosenfeld (Eds.), *Neurocomputing: Foundations of research* (pp. 126-134). Cambridge, MA: MIT Press.

Asynchronous Learning by Emotions and Cognition

Sandra Clara Gadanho* **and Luis Custódio**
Institute of Systems and Robotics
Torre Norte, IST, Av. Rovisco Pais 1, 1049-001 Lisbon, Portugal
{sandra,lmmc}@isr.ist.utl.pt

Abstract

The existence of emotion and cognition as two interacting systems, both with important roles in decision-making, has been advocated by neurophysiological research (LeDoux, 1998; Damasio, 1994). Following this idea, this paper proposes the ALEC agent architecture which has both emotion and cognition learning capabilities to adapt to real-world environments. These two learning mechanisms embody very different properties which can be related with those of natural emotion and cognitive systems.

Experimental results show that both systems contribute positively for the learning performance of the agent.

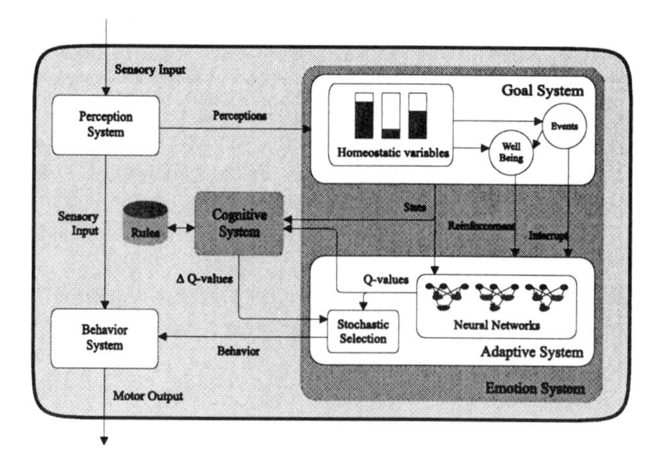

Figure 1: The ALEC architecture.

1 Introduction

Gadanho and Hallam (2001) and Gadanho and Custódio (2002) proposed an emotion-based architecture which uses emotions to guide the agent's adaptation to the environment. The agent has some innate emotions that define its goals and then learns emotion associations of environment state and action pairs which determine its decisions. The agent uses a Q-learning algorithm to learn its policy while it interacts with its world. The policy is stored in neural networks which allows limiting memory usage substantially and accelerates the learning process, but can also introduce inaccuracies and does not guarantee learning convergence.

The ALEC (Asynchronous Learning by Emotion and Cognition) architecture proposed here aims at a better learning performance by augmenting the previous emotion-based architecture with a cognitive system which complements its current emotion-based adaptation capabilities with explicit rule knowledge extracted from the agent-environment interaction.

2 The ALEC Architecture

The ALEC architecture is an extension of the emotion-based architecture presented in (Gadanho and Hallam, 2001; Gadanho and Custódio, 2002). Inspired by literature on emotions, Gadanho and Hallam (2001) have shown that reinforcement and deciding when to switch behavior[1] can be successfully addressed together by an emotion model. The justification for the use of emotions is that, in nature, emotions are usually associated with either pleasant or unpleasant feelings that can act as reinforcement (Tomkins, 1984) and frequently pointed to as a source of interruption of behavior (Sloman and Croucher, 1981). Later the emotion model was formalized into a goal system with the purpose of establishing a clear distinction between motivations (or goals) and emotions (Gadanho and Custódio, 2002). In this system, emotions take the form of simple evaluations or predictions of the internal state of the agent. This goal system is based on a set of homeostatic variables which it attempts to maintain within certain bounds. The emotion-based architecture is composed by two major systems: the goal system and the adaptive system. The goal system evaluates the performance of the adaptive system in terms of the state of its homeostatic variables and asynchronously determines when a behavior should be interrupted. On such interruptions, the adaptive sys-

*Post-doctorate sponsored by the Portuguese Foundation for Science and Technology.

[1] Behavior-switching may be motivated by several factors: the behavior has reached or failed to reach its goal, the behavior has become inappropriate due to changes in circumstances, the behavior needs to be rewarded or punished. The correct timing of behavior-switching can be vital (Gadanho and Hallam, 2001).

tem learns which behavior to select using reinforcement-learning techniques which rely on neural-networks to store the utility values.

The ALEC architecture adds a cognitive system to the emotion-based architecture described previously. The function of the cognitive system is to provide an alternative decision-making process to the emotion system. The cognitive system collects knowledge independently and can step in to correct the emotion system's decisions because it relies on a more exact memory representation based on a collection of important individual events which is not prone to inaccuracies due to over-generalization. The cognitive system is based on the rule-based level of the CLARION model (Sun and Peterson, 1998). One of the main reasons for selecting CLARION's rule system is that it does not derive rules from a pre-constructed set of rules given externally, but extracts them from the agent-environment interaction experience.

The cognitive system maintains a collection of rules which allow it to make decisions based on past positive experiences. Rule learning is limited to those few cases for which there is a particularly successful behavior selection and leaves the other cases to the emotion system which makes use of its generalization abilities to cover all the state space. If the rule is often successful the agent tries to generalize it by making it cover a nearby environmental state; otherwise if the rule's success rate is very poor it attempts to make it more specific (same as in Sun and Peterson, 1998). In ALEC a behaviour is considered successful if it leads to a positive transition of the agent's internal state, or more specifically, of its homeostatic variables.

If the cognitive system has a rule that applies to the current environmental state, then it makes the selection of the behaviors suggested by the rule more probable.

3 Experiments

The experiments tested ALEC within an autonomous robot which learns to perform a multi-goal and multi-step survival task when faced with real world situations such as continuous time and space, noisy sensors and unreliable actuators.

Results show that ALEC not only learns faster than the original emotion-based architecture (Gadanho and Custódio, 2002) but also achieves a better final performance level.

The cognitive and the emotion systems together perform better that either one on its own. On the one hand, the cognitive system of ALEC improves learning performance by helping the emotion system to make the correct decisions. On the other hand, the cognitive system cannot perform well without the help of the emotion system because it only has information on part of the state space.

4 Conclusion

The ALEC approach implies that while emotion associations may be more powerful in its range capabilities, they lack explanation power and may introduce errors of over-generalization. Cognitive knowledge, on the other hand, is restricted to learning about simple short-term relations of causality. Cognitive information is more accurate, but at a price — since it's not possible to store and consult all the single events the agent experiences, it selects only a few instances which seem most important.

The way the emotion system influences the cognitive system is akin to Damásio's somatic-marker hypothesis (Damasio, 1994). In his hypothesis, Damásio suggested that humans associate high-level cognitive decisions with special feelings which have good or bad connotations dependent on whether choices have been emotionally associated with positive or negative long-term outcomes. If these feelings are strong enough, a choice may be immediately followed or discarded. Interestingly, these markers do not have explanation power and the reason for the selection may not be clear. In fact, although the decision may be reached easily and immediately, the person may feel the need to subsequently use high-level reasoning capabilities to find a reason for the choice. Meanwhile, a fast emotion-based decision could be reached which depending of the urgency of the situation may be vital.

ALEC shows similar properties, when it uses emotion associations to guide the agent. Furthermore, the cognitive system can correct the emotion system when this reaches incorrect conclusions. Knowing the exceptions from previous experiences, it may choose to ignore the emotion reactions, which although powerful can be more unreliable.

References

Damasio, A. R. (1994). *Descartes' error — Emotion, reason and human brain.* Picador, London.

Gadanho, S. C. and Custódio, L. (2002). Learning behavior-selection in a multi-goal robot task. Technical Report RT-701-02, Instituto de Sistemas e Robótica, IST, Lisboa, Portugal.

Gadanho, S. C. and Hallam, J. (2001). Robot learning driven by emotions. *Adaptive Behavior*, 9(1).

LeDoux, J. E. (1998). *The Emotional Brain.* Phoenix, London.

Sloman, A. and Croucher, M. (1981). Why robots will have emotions. In *IJCAI'81*, pages 2369–71.

Sun, R. and Peterson, T. (1998). Autonomous learning of sequential tasks: experiments and analysis. *IEEE Transactions on Neural Networks*, 9(6):1217–1234.

Tomkins, S. S. (1984). Affect theory. In Scherer, K. R. and Ekman, P., (Eds.), *Approaches to Emotion.* Lawrence Erlbaum, London.

Behaviour Control Using a Functional and Emotional Model

Ignasi Cos **Gillian Hayes**
IPAB, Division of Informatics
University of Edinburgh
Scotland, UK
ignasi@dai.ed.ac.uk, gmh@dai.ed.ac.uk

Abstract

Animals are capable of pursuing long term goals while coping with the ever-changing events in their scenario. In the quest of animats towards autonomy, we propose to use a functional model biased by emotions to drive their behaviours. In addition to that, we suggest that emotions can be used as source of reinforcement to build functional description of objects, on the grounds of which to decide what behaviour to take next.

1 Introduction

An autonomous agent should be able to pursue a variety of goals while coping with daily reality. To that aim, Gibson (1966) and von Uexküll (1921) assert that the structure of the body continuously evolves to adapt to changes in the environment. According to Gibson, to perceive implies the activation of the perceptual centres in our nervous system, that are sensitive to certain sensations, e.g. to a particular object, action or event. In his theory of perception he introduced the term *affordance* to designate the functionality objects offer to the agent. The affordances of an object derive from the constant features of the object under the view of that particular agent —shape, weight, size, etc. (Gibson, 1966)[1].

To close the perception-action loop, we suggest that affordances could function as bridge between the perception of an object and a goal-directed action. If the corresponding affordance is active, the goal related to its action or behaviour may be possible. This is supported by the research of Rizzolatti et al. (2000), according to which some groups of neurons in the pre-motor cortex of macaque monkeys are active when perceiving some objects and the actions these afford to them.

So far we have discussed how to relate goal-directed actions with the possibilities that objects in the environment offer to the agents. Nevertheless, triggering that particular functional neurological centre to execute a particular action also involves willingness to do it.

Humans are capable of coping with overall goals, but in fact, we do it in a very fuzzy way. Our relation to the environment is based on our perception, that gives rise to physiological changes in our body. This is one of the elements that influence the arousal of emotions (Izard, 1993). Emotions are a very controversial subject, but there seems to be some consensus in asserting that they play a role in evaluating situations and triggering one or another behaviour (Frijda, 1995).

An autonomous animat with several goals should encompass both *possibility* and *state of mind* to trigger a decision —functional and emotional stance.

2 Emotion and function together

We propose to build an animat model following the aforementioned ideas. Its **perception** will consist of two modules, the first capable of extracting affordances from objects in the scene and the second capable of providing an emotional stance (measure of its state of mind) on the basis of the agent's experiences and circumstances.

According to Gibson's theory, *affordances* are bound to features of objects and to the morphology of the agent. However, because affordances are *a priori* unknown, the animat will have to learn to associate correctly the perception of the object to its functionalities from the whole available set. This will point out the *possible actions* to the agent.

The *mood* of the agent will be instead provided by the *emotional stance* through a model derived from the work of Gadanho (1998), who developed a model for behaviour control using basic emotions, inspired in the research of Damasio (1994). The model includes a representation for feelings (hunger, pain, restlessness, temperature and eating), which monitor the body[2] and for fundamental emotions (happiness, sadness, fear and anger), which can

[1] Depending on the morphology of the agent, an object will have different affordances. E.g. a cup does not afford the same to a human (grasp) as to a dog (bite?).

[2] The definition of feeling is controversial. Though we have respected the name given by Damasio, this denomination has been widely discussed. Other authors refer to it as "drives".

be considered as indicators of the *mood* of the agent. The emotional model will be further grounded by including relevant neurologic processes, such as the double response circuit described by LeDoux (1997).

The emotional stance plays two different roles. Firstly to estimate the probability of an affordance being attached to an object. Once affordances have been first estimated, a decision to take a behaviour will follow. According to any variation in the emotional stance the estimation will be further refined. In order to make this reinforcement possible, we have **hypothesised** that *the agent's use of an affordance should not produce a negative emotional impact.*[3] Secondly, emotions play a role in biasing the decision for one or another behaviour. The decision making integrates both the *possibilities* of the scenario (affordances) and the *mood* provided by the emotional stance into a reinforcement learning scheme that by apprehending from past events, we expect will improve the autonomy of the agent.

3 The Animat Model

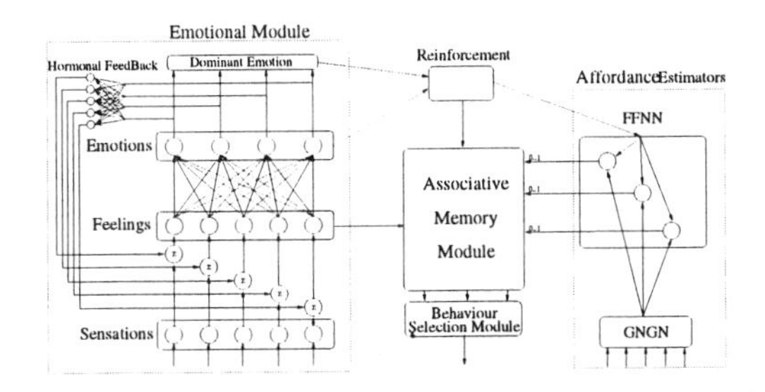

Figure 1: Behaviour Control Architecture

A model for an animat according to the aforementioned ideas is introduced in figure 1. Its three main modules are, from left to right: the emotional model, which represents feelings and emotions[4]; the associative memory, that integrates emotional and functional information for the decision making, and the affordance extractor. The latter consists of a set of estimators each of them associated to a particular affordance, that receive as input a fuzzy description of the sensed object, and an emotional reward.

The animat includes a reinforcement learning module to learn to choose behaviours (associative memory), whose state vector is defined by its feelings and emotions plus the affordances of the sensed objects. To close the circle, the associative memory calculates the expected outcome of performing each behaviour on the grounds of the state, and according to that decides which of them to execute. After some recursion, if there is some significant change in the emotional stance (feelings or emotions), we can use this to improve the decision process, i.e. what to decide according to the circumstances, and to refine the estimators of affordances of the objects in the scenario.

[3]If we try to sit on a large cactus it will surely be painful. This may indicate that it does not offer that functionality.

[4]E.g. the anger node gets input from all the feelings nodes (hunger, pain, restlessness), which contribute in some measure to the anger.

4 Approach

In a first stage we will perform experiments with the perceptual capabilities of Khepera robots for building the affordance estimator independently of the emotional stance. The first approach will be to point to very simple goals (survive) in a *toy* scenario composed of several easily distinguishable objects and an appositely built oracle to test the validity of the estimator.

On a second stage, experiments will be conducted to make a comparative study of the decisional process. Functional and emotional information can be used separately for the decision making. The hypothesis mentioned in section 2 can be then tested.

Finally, both sources of information can be merged together into a single state for testing the autonomy of the complete agent. Results can be compared with the ones obtained in earlier stages.

References

Damasio, A. (1994). *Descartes' Error: Emotion, Reason and the Human Brain.* Picador, New-York.

Frijda, N. H. (1995). Emotions in robots. In Roitblat, H. and Meyer, J., (Eds.), *Comparative Approaches to Cognitive Science*, pages 510–516. Cambridge, MA: The MIT Press.

Gadanho, S. C. (1998). *Reinforcement Learning in Autonomous Robots: An Empirical Investigation of the Role of Emotions.* PhD thesis, University of Edinburgh.

Gibson, J. J. (1966). *The Senses Considered as Perceptual Systems.* Houghton Mifflin Company, Boston.

Izard, C. (1993). Four systems for emotion activation: Cognitive and noncognitive processes. *Psychological Review*, 100(1):68–90.

LeDoux, J. E. (1997). Emotion, memory and the brain. *Scientific American*, 7(1). Special Issue: Mysteries of the Mind.

Rizzolatti, G., Fogassi, L., and Gallese, V. (2000). Cortical mechanisms subserving object grasping and action recognition: A new view on the cortical motor functions. In Gazzaniga, M., (Ed.), *The New Cognitive Neurosciences*, pages 539–552. MIT Press.

von Uexküll, J. (1921). *Umwelt und Innenwelt der Tiere.* Julius Springer, Berlin.

The Road Sign Problem Revisited: Handling Delayed Response Tasks with Neural Robot Controllers

Mikael Thieme & Tom Ziemke
Department of Computer Science
University of Skövde
Box 408, 54128 Skövde, Sweden
{mikaelt|tom}@ida.his.se

Abstract

The 'road sign problem' is a class of delayed response tasks in which an agent's correct turning direction at a T-junction is dependent on a stimulus it has encountered earlier. Neural robot controllers of four different architectures have been evaluated in experiments with six different variations of the problem. The highest reliability was achieved by Extended Sequential Cascaded Networks, a higher-order recurrent neural network architecture.

1. Introduction & Background

The 'road sign problem' is a class of delayed response tasks that recently has received much attention in the adaptive behavior and artificial neural network (ANN) research community (e.g. Ulbricht, 1996; Jakobi, 1997; Rylatt & Czarnecki, 1998, 2000; Linåker & Jacobsson, 2001, in press; Bergfeldt & Linåker, in press). In the simplest version, illustrated in Figure 1, an agent's correct turning direction at a T-junction depends on where (on which side) it earlier has encountered a certain stimulus (e.g. a light).

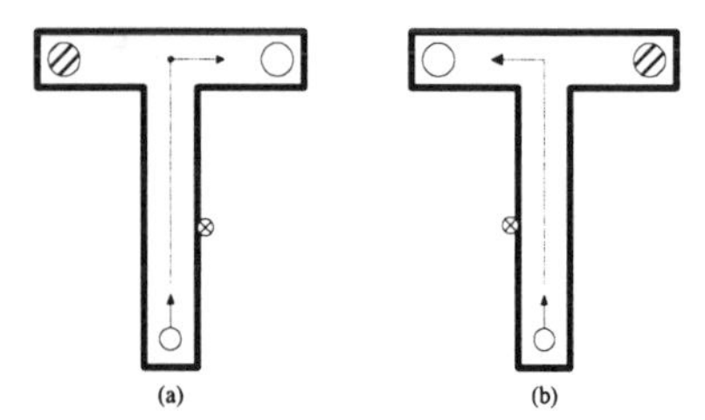

Figure 1: The two situations in the simple T-maze environment. The agent should turn to the side on which it has encountered the light. Adapted from Ulbricht (1996).

Rylatt & Czarnecki (2000) showed that standard recurrent neural networks (of the multilayer perceptron type) trained by error backpropagation, have problems dealing with long delays. In their experiments they found that networks could typically not deal with delays of more than 12-13 time steps, i.e. they would usually already have forgotten about the stimulus by the time they reached the T-junction. Linåker & Jacobsson (2001, in press) showed that the problem can be solved by abstracting from the sensorimotor values and individual time steps, using unsupervised competitive learning, to simple *concepts* like 'corridor' and 'corner' and the transitions at that conceptual level, which then could be learned more easily. In a similar vein Bergfeldt & Linåker (in press) constructed a two-level ANN architecture in which the sensorimotor mapping could dynamically be modulated by a higher-level mechanism triggered by events/transitions at the conceptual level.

2. Experiments

In contrast to the above previous work our intuitions have been that (1) relatively simple recurrent neural networks can in fact very well deal with the task when trained with evolutionary algorithms, and (2) no conceptual abstraction in the above sense is required. These questions have been evaluated systematically in experiments with a simulated Khepera robot, illustrated in Figure 2, facing the six different variations of the road sign problem illustrated in Figure 3. ANNs of the four different architectures illustrated in Figure 4 were used as robot controllers. They were trained, for each of the six environments separately, by evolving the connection weights using a fairly standard evolutionary algorithm. For details on these experiments and their cognitive-scientific background see Thieme (2002).

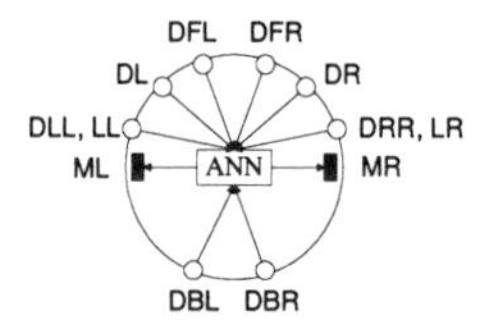

Figure 2: Simulated Khepera robot, receiving input from eight distance sensors (D) and two light sensors (L), and using two motors (M) to control left and right wheel.

The experimental results show that our intuitions were correct. When the best individuals of the final generations were tested on all possible combinations of light and goal locations, all architectures achieved 100% reliability in the original task (a) (cf. Figure 3). Even the feed-forward nets solved the task perfectly, making the robots move reactively towards the sensed light stimulus and then simply follow the wall to the goal (cf. also Bergfeldt & Linåker, in press).

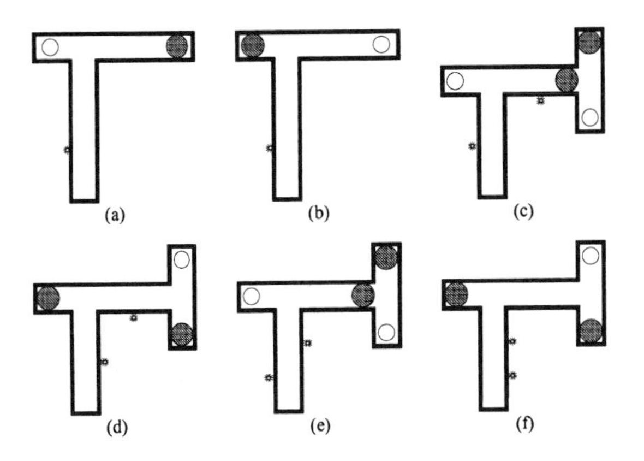

Figure 3: Example situations in six variations of the road sign problem (exact start position, orientation and light locations vary randomly; goal locations vary accordingly). Empty circles indicate goal locations (maximum fitness), gray circles areas in which the agent 'dies' immediately. In (a) the agent should turn to the side where the stimulus appeared, whereas in (b) it should turn to the other side. Both (c) and (d) are *repeated T-maze problems*: in (c) the agent should turn to the side where the stimulus appeared, possibly twice, whereas in (d) the lights have different meanings, i.e. in the second case the agent should turn towards the other side. Both (e) and (f) are *multiple T-maze problems*, i.e. *both* stimuli come before the first T-junction. In (e) both turns should go towards the side where the respective stimulus appeared, whereas in (f), as in (d), the meaning of the second light is reversed.

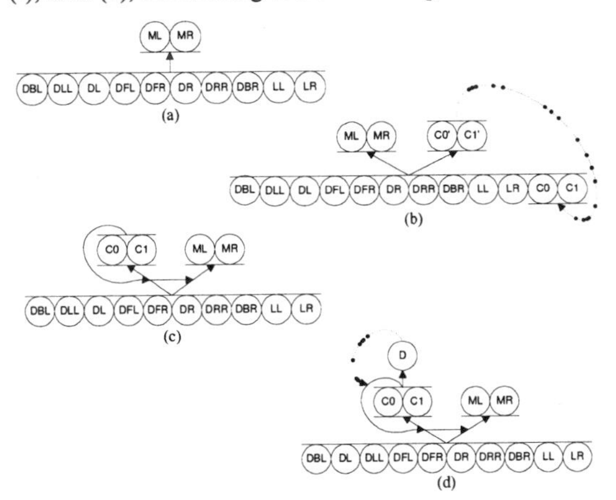

Figure 4: ANN robot controllers mapping sensor inputs to motor outputs (cf. Figure 2): (a) feed-forward net, (b) very simple recurrent net, (c) Sequential Cascaded Network (SCN; Pollack, 1991), and (d) Extended SCN (ESCN; Ziemke, 1999, 2000). C0/1 denote context or state units. In both SCN and ESCN the sensorimotor mapping (input-output weights) can be changed dynamically (in the ESCN depending on the activation of decision unit D).

Averaging over all six variations of the task, the Extended Sequential Cascaded Networks (ESCNs) performed best by far, achieving an impressive 99.4% reliability, while the other architectures lay around 90% (feed-forward: 87.1%, simple recurrent: 92.5%, SCN: 91.2%). Analysis of the ESCN-controlled robots (cf. Thieme, 2002) shows that they solve the task by selectively adapting their sensorimotor mapping through dynamic changes of the input-output weights. For example, when encountering the light stimulus on a particular side the ESCN would re-set its weights such that later, independent of the actual length of the delay, the activation of the frontal sensors at a junction triggers the correct turning response. This allows context-dependent adaptation of behavioral dispositions (cf. Ziemke, 1999, 2000), functionally similar to Bergfeldt & Linåker's architecture, but in a simpler and more elegant fashion.

References

Bergfeldt, N. & Linåker, F. (in press). Self-Organized Modulation of a Neural Robot Controller. In *Proc. of the Intl. Joint Conf. on Neural Networks* (IJCNN02).

Jakobi, N. (1997) Evolutionary robotics and the radical envelope of noise hypothesis. *Adaptive Behavior*, 6(2), 325-368.

Linåker, F. & Jacobsson, H. (2001). Mobile Robot Learning of Delayed Response Tasks through Event Extraction. In *Seventeenth Intl. Joint Conf. on AI*, pp. 777-782.

Linåker, F. & Jacobsson, H. (in press). Learning Delayed Response Tasks through Unsupervised Event Extraction. *International Journal of Computational Intelligence and Applications*, to appear.

Pollack, J. B. (1991). The induction of dynamical recognizers. *Machine Learning*, 7, 227-252.

Rylatt, R. M. & Czarnecki, C. A. (1998). Beyond physical grounding and naive time: Investigations into short-term memory. In *From animals to animats 5*, pp. 22-31. Cambridge, MA: MIT Press.

Rylatt, R. M. & Czarnecki, C. A. (2000). Embedding connectionist autonomous agents in time: The 'road sign problem'. *Neural Processing Letters*, 12, 145-158.

Thieme, M. (2002). *Intelligence without hesitation*. Masters Dissertation HS-IDA-MD-02-001, Department of Computer Science, University of Skövde, Sweden.

Ulbricht, C. (1996) Handling time-warped sequences with neural networks. In *From Animals to Animats 4*, pp. 180-189. Cambridge, MA: MIT Press.

Ziemke, T. (1999). Remembering how to behave: Recurrent neural networks for adaptive robot behavior. In Medsker & Jain, editors, *Recurrent Neural Networks: Design and Applications*, pp. 355-389. New York: CRC Press.

Ziemke, T. (2000). On 'Parts' and 'Wholes' of Adaptive Behavior: Functional Modularity and Diachronic Structure in Recurrent Neural Robot Controllers. In *From animals to animats 6*, pp. 115-124. Cambridge, MA: MIT Press.

Development of the First Sensory-Motor Stages: A Contribution to Imitation.

Pierre Andry* **Philippe Gaussier *** **Jacqueline Nadel****
* Neurocybernetic team, ETIS Lab, UPRES A 8051 UCP-ENSEA, France
** Equipe Développement et Psychopathologie UMR CNRS 7593, France
andry@ensea.fr, gaussier@ensea.fr

Abstract

We present the first stages of the developmental course of a robot using vision and a 5 degree of freedom robotic arm. During an exploratory behavior, the robot learns autonomously the visuo-motor control of its mechanical arm. We show how a neural network architecture, combining elementary vision, a self-organized algorithm, and dynamical Neural Fields is able to learn and use proper associations between vision and arm movements, even if the problem is ill posed. We show as a robotic result that the architecture can be used as a basis for simple gestural imitations.

1. Introduction

Far from complex recognition systems, we propose a neural network architecture able to learn and store association between 2-D elementary vision and the 3-D working space of a robot's arm. We show how the use of a new type of associative coding neurons coupled with dynamical equations solves the reaching of visual parts of the working space, independently of the number of degrees of freedom of the arm (and their potential redundancy). We will also emphasize the importance of the on-line learning process, where the robot performs sensory-motor associations thru a random exploration of its own motor dynamics and physics. As a robotic validation, we will present how dynamic and perceptive properties of such a generic architecture allow to exhibit real time imitations gestures.

2. A Neural network architecture for Visuo-Motor development

With complex arms (such as the one we use), the same position of the extremity corresponds to multiples vector positions of the joints. To allow the association of multiples proprioceptive vectors with a single visual perception, we use a new kind of sensory-motor map, composed of *clusters* of neurons (fig 1). Each cluster associates a single connection from one neuron of the visual map with

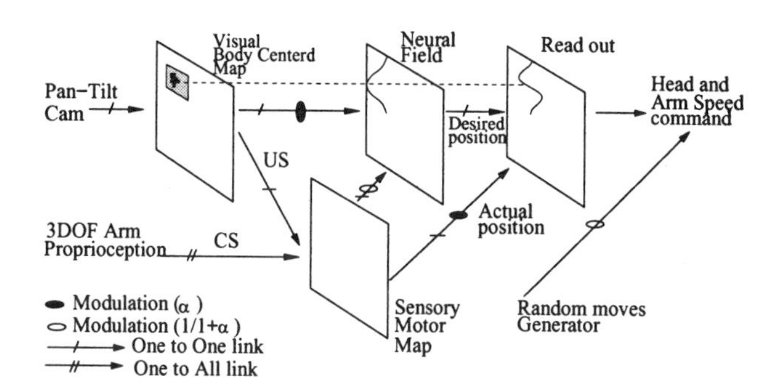

Figure 1: The simplified architecture.

multiples connections from the arm's proprioception. Visual information is considered as a 2-D Unconditional Stimuli (US) that controls the learning of particular pattern of a 3-D proprioceptive input(Conditional Stimuli -CS). Thus, this sensory-motor map has the same global topology as the visual map. A cluster i is composed of (fig 2): *(1)*one input neuron X_i linked the visual map. This neuron responds to the US and triggers learning. *(2)*A small topological *submap* of Y_i^k neurons ($k \in [1, n]$) which learn the association between proprioceptive vectors and a given visual position. *(3)* One output Z_i neuron, merging information from X_i and $submap_i$. The winner neuron Z will represent the "visual' response associated to the proprioceptive input presented. Thus, many proprioceptive configurations are able to activate the same "visual feeling", while close visual responses can be induced by very different proprioceptions (thanks to the independence between each cluster). The learning phase is performed during an exploratory behavior. Both head and arm are controlled by the sensory-motor map, providing random excitations at the begining of the learning phase. During the learning process, visuo-motor associations are slowly learned, inducing progressively a coherent control of the head and arm.

A 2D dynamic for 3D arm movements

To reach a perceived target, the error between the desired position (the visual position of the target) and the current position of the device has to be minimized. The error has to be converted in an appropriate movement

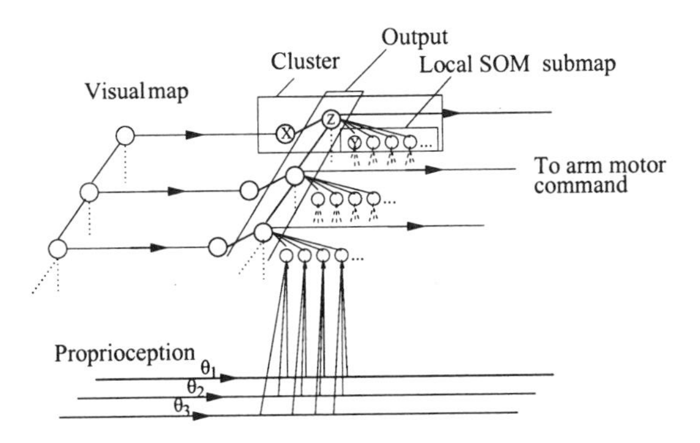

Figure 2: Organization of the sensory-motor map. The activity of one neuron of the vision map will trigger the learning of the corresponding cluster (one to one links).

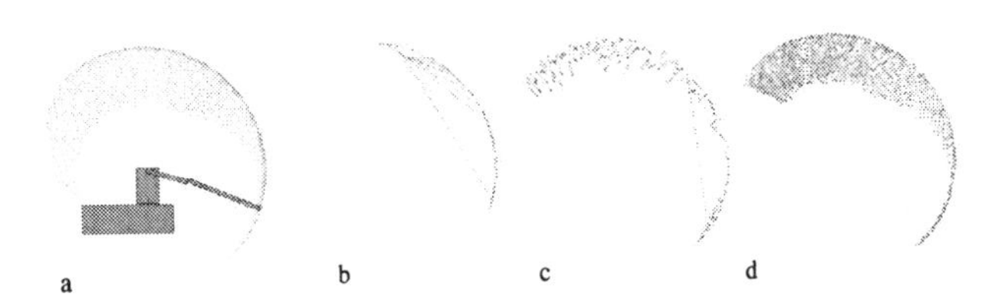

Figure 3: A sensory-motor vector of 146 clusters learning vertical movements (redundancy between θ_2 and θ_3 joints). Each cluster is composed of a submap of 6 Y neurons. A point represent the values of the Y neuron's weights learning the θ_2 and θ_3 values. a): theoretical positions. b,c,d): Progressive dissociation of Y neurons.

vector to move each device's joint toward the target. This process is done by computing a dynamical attractor centered on the target stimuli. The computation is made on maps of neurons having the same topology that the visual map, using neural field equations (eq 1, (Amari, 1977)):

$$\tau \cdot \frac{f(x,t)}{dt} = -f(x,t) + I(x,t) + h + \int_{z \in V_x} w(z) \cdot g(f(x-z,t))\, dz \quad (1)$$

The spatial derivate of the NF activity is interpreted as two 1-D desired speed vectors (horizontal and vertical), to reach the attractor target (*read-out* mechanism (Gaussier et al., 1998)). Thus each joint will then contribute simultaneously to the global move of the arm toward the visual objective, the shape of the NF activity on the target ensuring convergent moves. The robot behavior appears as a minimization between visual and proprioceptive representations (homeostatic principle).

A Contribution to imitation behaviors

An elementary imitative behavior can be triggered by exploiting the *ambiguity* of the perception. Using only movement detection, the system can't differentiate it's extremity from another moving target, such as a moving hand. The generated error will induce movements of the robotic arm, reproducing the moving path of the human hand: an imitative behavior emerges. The experimenter was naturally moving its arm in front of the robot. The camera rapidly tracked the hand (the most moving part of the scene) and the arm, reproduced in real time the hand's perceived trajectory. The use of neural fields ensures a reliable filtering of movements and a stable, continuous tracking of the target by the head and the arm of the robot.

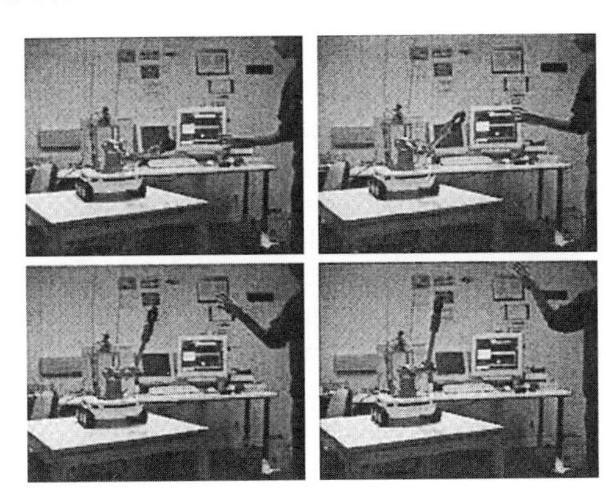

Figure 4: Real time imitation of simple vertical gesture.

3. Conclusion

Far from building a control architecture dedicated to imitation tasks, we showed how a generic system with learning capabilities and dynamical properties, can easily exhibit low-level imitations. These imitations of arm movements are performed without any internal model of the human arm, and can easily be transposed to imitation of robotic arm movement, whatever the morphology of the arm is. We have also shown how a very simple vision system, only built on movement recognition, gets for free end point tracking, without the use of any recognition system of the hand. The "low-level" characteristic of the imitative behavior is here related to the basic sensory-motor repertory whose accuracy is dependant of the duration of the learning proces. But independently of the accuracy of the sensory-motor associations, we have shown how the use of dynamical equations ensures stable and coherent behavior.

References

Amari, S. (1977). Dynamic of pattern formation in lateral-inhibition type by neural fields. *Biological Cybernetics*, 27:77–87.

Gaussier, P., Moga, S., Banquet, J., and Quoy, M. (1998). From perception-action loops to imitation processes: A bottom-up approach of learning by imitation. *Applied Artificial Intelligence*, 7-8(12):701–727.

Introducing chronicity: a quantitative measure of self/non-self in immune response

Yoshiki Kashimori, Yoshihiro Ochi, Mei Hong Zheng, and Takeshi Kambara

Division of Bioinformatics,Department of Applied Physics and Chemistry,
The University of Electro-Communications, Chofu,Tokyo,182-8585 Japan
e-mail:kashi@pc.uec.ac.jp

Abstract

This paper presents a new theory of how lymphocyte-antigen interaction is governed. We present `chronicity', a quantitative record of previous lymphocyte-antigen interactions, which is used to regulate lymphocyte behavior. A simplified model of the immune response, incorporating chronicity, is described. When the chronicity of a lymphocyte increases with the interaction and gets beyond the lower threshold, the lymphocyte can proliferate. Simulations using this model show that non-self antigens cause lymphocyte proliferation which destroys the antigen and results in the creation of a `memory' of this antigen in the form of a pool of lymphocytes of correct specificity. Equally importantly, `self' antigens are not destroyed. The lymphocytes attacking specifically the antigens increase quite rapidly their chronicity through the continuous interaction with the antigens that are rare of change in concentration. When the chronicity gets beyond the upper threshold, the lymphocytes get in the tolerance state ensuring non destruction of self antigens. The lymphocytes secrete actively negative cytokines which reduce the activation of other lymphocytes. This self/non-self discrimination ability and overall system behavior is similar to that observed in real immune systems.

1. Introduction

Simplified descriptions of the interaction between lymphocyte and antigen often use `lock-and-key' imagery, with the lymphocyte as a self-propelled lock testing antigens for fit, and destroying those which match. Burnet (1959) proposed in his `clonal-selection' theory that the enormous range of lymphocyte specificities (the lock shapes) needed is created before birth, that lymphocytes are pre-committed to their target antigens. Lymphocytes which match self antigens were believed to be eliminated in the same period. This `neonatal tolerance' model was proposed by Burnet (Burnet, and Fenner, 1949; Lederberg, 1959) and supported experimentally by Medawar and colleagues (Billingham, Brent, Medawar, 1953; Billingham,1956).

However, recent experiments (Forsthuber, Yip, and Lehmann,1996;Pennisi, 1996; Ridge, Fuchs, and Matzinger, 1996; Sarzotti, Robbins, and Hoffman, 1996; Hertz, Kouskoff, Nakamura, and Nemazee, 1998; Rajewsky, 1998) have revealed that the discrimination between self and non-self is not made in early life, extending our understanding of the mechanisms involved in the discrimination process. New ideas for the discrimination process have therefore been proposed. Zinkernagel and Doherty (1974, 1979) proposed the `second signal' model in which not only stimulation from a MHC-peptide complex but also a second signal from an antigen-presenting cell are needed for a T lymphocyte to become active. Marzinger (1994) proposed the `danger signal' model in which the immune system attacks antigens only when an antigen-presenting cell has antigen on its surface and it receives a danger signal from damaged host tissue. Although these models differ in their details, they both assume that the discrimination between self and non-self is made based on a qualitative measure, the binary presence or absence of some signals.

The existence of a quantitative measure and the importance of considering the dynamic properties of the immune system in self/non-self discrimination was suggested by Grossman (1984). However, it was not clear how this quantitative measure could systematically control the discrimination process.

In this paper an alternative new theory of the discrimination process is proposed based on Grossman's idea that discrimination between self and non-self in the immune response could be quantitative. This quantitative value, which we have named `chronicity', is a lymphocyte state variable which represents the frequency of interaction of this lymphocyte with matching antigens. We believe that our model provides a clear answer to the interesting idea suggested by Grossman.

First, a minimal model of the immune system is described. This model contains antigens of specific type and generic lymphocytes each with a particular antigen specificity. The behavior of these lymphocytes when they encounter a matching antigen is determined partly by the levels of non-specific positive and negative cytokines nearby, but mostly by their internal chronicity value.

The results from running a grid-world simulation of this model using a)non-self and b) self antigens are described. Major differences in these results show that self/non-self discrimination can be systematically explained based on the changes in chronicity over time.

2. Model

2.1 Overview

Our cellular automata model of a simplified immune system comprises five kinds of immune elements: generalized lymphocytes, two kinds of antigens (foreign and self), and two kinds of cytokines (positive and negative). Lymphocytes keep track of their interaction with antigens through a quantitative measure introduced in this paper, which we have termed `chronicity'. Individual lymphocyte behavior is governed by its internal level of chronicity as well as by the presence of antigen of matching specificity. If chronicity is low, then the lymphocyte is in `rest' mode where it might proliferate but does little else. If chronicity is medium, the lymphocyte is in `active' mode, where it is likely to proliferate, likely to eliminate antigen, and will secrete positive cytokines. If chronicity is high enough, lymphocytes secrete negative cytokines. The thresholds determining these levels vary dynamically, depending upon the levels of the different types of cytokine. All interactions are probabilistic, so the overall system behavior is not deterministic.

2.2 Model components:

a) Lymphocytes

Our generalized lymphocytes have basic characteristics that are common to various different types of lymphocyte such as helper T cells, killer T cells, suppressor T cells, B cells, and antigen-presenting cells. They have a particular antigen specificity, they age and die, and under the right circumstances they proliferate. In our model, they also have a `chronicity level' quantifying recent antigen interaction and two variable thresholds determining whether that chronicity level counts as being low, medium, or high.

The range of possible values for each of the five variables is given in Table 1. Lymphocytes live until the expiration of the life span T_{LS}^{LP}. The specificity of both lymphocytes and antigens is given as a number. Although this should more properly be thought of as a type or even a shape, using a numerical value allows similarities and differences in specificity to be more easily represented. Lymphocytes and antigens with `similar' specificity are more likely to interact than those with differing specificity. We considered only sixty specificities in the model for simplicity, although there are tremendously many kinds of antigens in the real world and the immune system can provide lymphocytes which interact specifically with any single kind of antigen. Lymphocyte chronicity represents the recent frequency of that lymphocyte's interaction with antigens, so changes dynamically. Lower and upper thresholds of chronicity are demarcation values by which the range of chronicity is divided into three regions as shown in Fig.1. The values of both lower and upper thresholds change depending on the current state of immune system.

The behavioral mode of any lymphocyte changes depending on its current chronicity level and current chronicity thresholds as shown in Fig. 1. This mode affects lymphocyte proliferation, secretion, and interaction with antigens. The detailed behavior of lymphocytes in each mode is described in subsection 2.4.

The biological plausibility of the existence of some quantitative measure of lymphocyte/antigen interaction and of varying its threshold values is considered below. We describe briefly experimental results indicating and suggesting that the state of lymphocytes may be changed as their chronicity changes and that there exists one lower threshold and one upper threshold.

We may regard our artificial lymphocytes as approximating to T lymphocytes, because self/non-self discrimination, rapid proliferation and differentiation in response to foreign antigens, and an effective response to negative stimuli orchestrating the silencing of an immune response are all characteristics of T lymphocytes (Welsh, Selin, Razvi, 1995).

(i) The lower threshold: Viola and Lanzavecchia (1996) and Rothenberg (1996) have shown that T cells commit themselves to activation only when a certain threshold number of T cell receptors have been engaged by the peptide-MHC complex and down-regulated. If the number of T cell receptors engaged is proportional to or a simple function of chronicity, then the activation process in the T cells can be described based on variations in their chronicity.

(ii) The upper threshold: The process generating self-tolerance in the immune system consists of two stages, the central process due to the death of developing lymphocytes responding to self antigens in the generative lymphoid organs and the peripheral process necessary for the mature lymphocytes which escaped the selective death and entered into peripheral tissues. We consider here the mechanism terminating immune response only in the peripheral process. The mechanism proposed so far fall into two broad categories: passive control mechanisms due to the decline or absence of activating stimuli and active mechanisms induced by the presences of lymphocyte activation(Van Parijs and Abbas, 1998). We develop a model of immune system based on the activation induced tolerance model in which the activation of lymphocytes itself triggers feedback mechanisms that limit their proliferation and differentiation.

The termination of immune responses due to activation induced cell death(AICD) is most important for tolerance to widely disseminated and persistent antigens, such as self antigens(Van Parijs and Abbas, 1998). The termination process induced by Fas-mediated AICD is that when T cells are repeatedly stimulated by high concentrations of antigens, the death receptors(Fas) and ligands(FasL) of Fas are expressed on the surface of T cells, and Fas-FasL interactions may trigger off apoptotic death of the T cells.

Some experiments (Rocha, von Boehmer, 1991; Kyburt et al, 1993; Malalaki et al, 1993) have shown that resting T cells will respond to antigen in a productive proliferative manner, but many eventually respond to further stimulation by undergoing apoptosis, ie, programmed cell death. This indicates that developing sensitivity to AICD is correlated with the extent of cellular proliferation. In the process of generating tolerance, it has been also reported that the expression of cell cycle associated proteins is

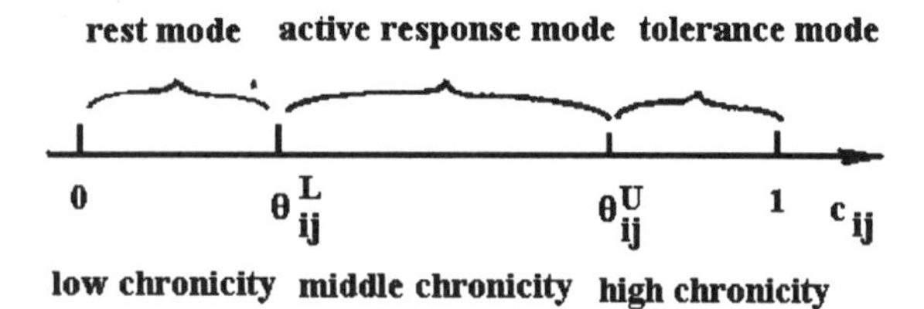

Fig.1 Three lymphocyte behavioral modes determined by the lymphocyte's chronicity level, and the position of its lower (θ_{ij}^{L}) and upper (θ_{ij}^{U}) chronicity thresholds.

elements	variables	notation	range
lymphocyte	age	T_{ij}^{LP}	$0 \sim T_{LS}^{LP}$
	specificity	s_{ij}	$1 \sim 60$
	chronicity	c_{ij}	$0 \sim 1$
	lower threshold of chronicity	θ_{ij}^{L}	$0 \sim 1$
	upper threshold of chronicity	θ_{ij}^{U}	$0 \sim 1$
antigen	age	T_{ij}^{Ag}	$0 \sim T_{LS}^{Ag}$
	specificity	s_{ij}^{Ag}	$1 \sim 60$
positive cytokine	concentration	ρ_{ij}^{P}	
negative cytokine	concentration	ρ_{ij}^{N}	

Table 1 State variables and their permitted ranges

rigorously controlled and down-regulated at different stages in the cell cycle in a timely fashion (Welsh et al, 1995). This work suggests that the existence of a certain upper limit of cell cycle number to trigger AICD is necessary for the induction of apoptosis of the T cells, because the lymphocytes are activated continuously by self antigens.

b) Antigens

In our model antigens have age and specificity in a similar way to lymphocytes. In the experiments reported here antigens of only two specificities, 6 and 20, were used. These numbers represent foreign and self antigen types, respectively.

Self antigens of many kinds are present in the body, most attached to body tissue. Their main characteristics are high concentration and persistence. This is not to say that self antigens are never destroyed, but that any destruction is normally insignificant in healthy tissue because of the huge reservoir of similar antigens in neighboring tissue cells. In order to make our model `self' antigens similar, we made them immobile, placed them in large, concentrated, blocks, and made their population unchanged, because of continuous supply compensating the destruction of them by lymphocytes.

Non-self antigens are invaders of some kind. They may replicate fast but do not have a large reservoir from which replacements can be drawn. Our model `foreign' antigens were therefore scattered through the model environment in clusters. They were given high replication rates but could be destroyed by lymphocytes of similar specificity.

c) Cytokines

Cytokines are growth factors which exist in solution and can diffuse throughout tissues. Although the natural immune system has different kinds of cytokines, such as INF, TNF, and interleukine, we combined these into two distinct types: positive cytokines which effectively activate lymphocytes, and negative cytokines which reduce lymphocyte activity. Lymphocytes interact specifically with antigens but non-specifically with cytokines.

2.3 The Model in Operation

The cellular automata model comprises many unit squares set in a grid. Each square can contain only one lymphocyte, one antigen, and some concentration of positive and negative cytokines. At each time step, we randomly choose one square to update. If there is a lymphocyte or an antigen in the square, their state variables are changed as described below. Foreign antigens may be destroyed. After possible proliferation, remaining elements may move to one of the eight neighboring squares with equal probability (1/8), but only if the site is vacant. When the destination site already contains an element of the same type, the element stays at the original site. Self antigens cannot move. It is quite possible for the model to include mobile self antigens, but this was not done here. The concentrations of positive and negative cytokines change by diffusion from one of the four directly neighboring sites.

The state variables of elements in all of the squares except for the square chosen are not changed. In the cellular automaton model used here, one grid square cannot contain more than one lymphocyte and one antigen. Therefore we can not move more than two elements simultaneously. However, repetition of this process allows other elements to change and move.

After 3000 time steps the first immune period finishes. The next immune period is started by supplying a preset number of new lymphocytes with random specificity to random squares on the grid. These new lymphocytes represent the supply from lymphoid organs which are not modeled here.

2.4 Temporal variation of state variables

The state variables of each element in the square (i, j) at time t are changed as follows:

a. Lymphocyte state variables
The age of lymphocyte in the square (i,j) is updated by one step: $T_{ij}^{LP}(t+1)=T_{ij}^{LP}(t)+1$.
The specificity S_{ij} is not changed.

$$c_{ij}(t+1)=c_{ij}(t)-\gamma_c c_{ij}^2+\Delta c_{ij}. \tag{1}$$

Lymphocyte chronicity level decays at a constant rate gc and is increased by interaction of lymphocyte with antigens. When the lymphocyte interacts with an antigen, its chronicity increases by $\Delta c_{ij}=\mu(1-c_{ij}(t))$, where μ is a constant rate parameter, $\mu=0.041$. The increment therefore becomes smaller as chronicity increases.

$$\theta_{ij}^L(t+1)=\theta_{ij}^L(t)+\alpha_L\frac{\theta_0^L-\theta_{ij}^L(t)}{\theta_0^L}-\beta_L\rho_{ij}^P(t)\frac{\theta_{ij}^L}{\theta_0^L}, \tag{2}$$

$$\theta_{ij}^U(t+1)=\theta_{ij}^U(t)+\alpha_U\frac{\theta_0^U-\theta_{ij}^U(t)}{\theta_0^U}-\beta_U\rho_{ij}^N(t)\frac{\theta_{ij}^U(t)}{\theta_0^U}. \tag{3}$$

The second terms in equations (2) and (3) represent the tendency of the lower and upper thresholds to return to their initial values, θ_0^L and θ_0^U, using constant rates α_L and α_U respectively. The third term in equation (2) represents the effect of positive cytokines, which decrease the lower threshold and so increase the likelihood of nearby lymphocytes being in the `active response' mode of Fig. 1. The third term in equation (3) represents the effect of negative cytokines, which destabilize the active response mode by decreasing the upper threshold of Fig. 1. Parameter values used in the experiments in this paper are: $\gamma_c=0.0038$, $\alpha_L=0.01$, $\beta_L=0.05$, $\alpha_U=0.01$, $\beta_U=0.02$, $\theta_0^L=0.3$, and $\theta_0^U=0.9$.

b. Antigen state variables

The age of any antigen in the square (i, j) is updated by one step: $T_{ij}^{Ag}(t+1)=T_{ij}^{Ag}(t)+1$.
Its specificity S_{ij}^{Ag} is not changed.

c. Concentrations of positive and negative cytokines in the square (i, j)

The concentrations of positive and negative cytokines, ρ_{ij}^P and ρ_{ij}^N, are determined by the following equations, respectively.

$$\begin{aligned}\rho_{ij}^P(t+1)&=\rho_{ij}^P(t)-\gamma_P\rho_{ij}^P(t)\\&+D\{\rho_{i+1,j}^P(t)+\rho_{i-1,j}^P(t)+\rho_{i,j+1}^P(t)+\rho_{i,j-1}^P(t)\\&-4\rho_{i,j}^P(t)\}-\nu\rho_{i,j}^N(t)+\Delta\rho_{i,j}^P,\end{aligned} \tag{4}$$

$$\begin{aligned}\rho_{ij}^N(t+1)&=\rho_{ij}^N(t)-\gamma_N\rho_{ij}^N(t)\\&+D\{\rho_{i+1,j}^N(t)+\rho_{i-1,j}^N(t)+\rho_{i,j+1}^N(t)+\rho_{i,j-1}^N(t)\\&-4\rho_{i,j}^N(t)\}-\nu\rho_{i,j}^P(t)+\Delta\rho_{i,j}^N.\end{aligned} \tag{5}$$

The second terms in equations (4) and (5) represent the natural decay of cytokines, the third terms represent the increment due to diffusion, the fourth terms represent the opposing interaction between positive and negative cytokines, and the final terms represent the increment due to lymphocyte secretion which is explained in d. Parameter values used here are: $D=0.2$, $\gamma_P=0.1$, $\gamma_N=0.2$, $\nu=1.0$.

d. Interactions between lymphocytes and antigens

When there exists a lymphocyte in the square (i, j), it may interact with antigens in the same and in the eight neighboring squares. This interaction occurs with probability P_{ij}^{Int} which is given by

$$P_{ij}^{Int}=\frac{1}{9}\sum_{k=i-1}^{i+1}\sum_{l=j-1}^{j+1}n_{k,l}^{Ag}\exp[-(\frac{S_{ij}^{LP}-S_{kl}^{Ag}}{\sigma})^2], \tag{6}$$

where n_{kl}^{Ag} is the number of antigens at (k, l), which is 0 or 1. The probability of interaction is proportional to the similarity in specificity between lymphocyte and antigen. The parameter σ determines the meaningful range of the difference and takes the value 3.0. In practice, interaction seldom occurs when the difference in specificity is larger than σ.

When a lymphocyte and an antigen interact,

lymphocyte chronicity is increased by Δc_{ij}. Some of the other state variables are also changed by the interaction, but how they are changed differs depending on the chronicity level of the relevant lymphocyte as follows.

(1) In `rest mode', when $0 \le c_{ij} \le \theta_{ij}^L$
The lymphocyte does not secrete any cytokine, that is, $\Delta\rho_{ij}^P = 0$ and $\Delta\rho_{ij}^N = 0$. Antigens are not affected by the interaction. The lymphocyte may proliferate according to the probability $P_{ij}^{Pro} = c_{ij}(t)/\theta_{ij}^L(t)$.

(2) In `active response mode', when $\theta_{ij}^L \le c_{ij} \le \theta_{ij}^U$
The lymphocyte secretes only positive cytokines. $\Delta\rho_{ij}^P$ is given by

$$\Delta\rho_{ij}^P = \xi_P \frac{c_{ij}(t) - \theta_{ij}^L(t)}{\theta_{ij}^U(t) - \theta_{ij}^L(t)} - \eta_P \frac{\theta_{ij}^L(t)}{\theta_0^L}, \quad (7)$$

where ξ^P and η^P are the constant rate parameters. Here, $\xi^P = 0.3$ and $\eta^P = 0.1$. The first term of equation (7) means that positive cytokine secretion is proportional to chronicity level. The second term may induce the correlative changes of $\theta_{ij}^L(t)$ and $\rho_{ij}^P(t)$ in cooperation with the third term of equation (2). When $\theta_{ij}^L(t)$ decreases(increases), $\Delta\rho_{ij}^P(t)$ increases(decreases) and as a result $\rho_{ij}^P(t)$ increases(decreases), and $\theta_{ij}^L(t+1)$ decreases(increases) according to equation (2). Therefore, the second term of equation (7) accelerates the change of $\theta_{ij}^L(t)$.

Since the lymphocyte does not secrete any negative cytokine, $\Delta\rho_{ij}^N = 0$. The lymphocyte may proliferate according to the probability
$P_{ij}^{Pro} = c_{ij}(t)/\theta_{ij}^L(t)$. If an antigen exists in the square (i, j), the antigen may be eliminated according to the probability,
$P_{ij}^{Eli} = [c_{ij}(t) - \theta_{ij}^L(t)][\theta_{ij}^U(t) - \theta_{ij}^L(t)]^{-1}$.

(3) In `tolerance mode', where $\theta_{ij}^U(t) \le c_{ij} \le 1$
The lymphocyte secretes only negative cytokines. $\Delta\rho_{ij}^P = 0$ and the increment in negative cytokine $\Delta\rho_{ij}^N$ is given by

$$\Delta\rho_{ij}^N = \xi_N \frac{c_{ij}(t) - \theta_{ij}^U(t)}{1 - \theta_{ij}^U(t)} - \eta_N \frac{\theta_{ij}^U(t)}{\theta_0^U}, \quad (8)$$

where ξ_N and η_N are the constant rate parameters, $\xi_N = 0.6$ and $\eta_N = 0.1$. The role of first term and second terms of equation (8) in the regulation of negative cytokines correspond to those of equation (7) for positive cytokines. The lymphocyte may proliferate according to the probability
$P_{ij}^{Pro} = [1 - c_{ij}(t)][1 - \theta_{ij}^U(t)]^{-1}$.

Antigens are not affected by interaction with lymphocytes in tolerance mode.

When the lymphocyte at the square (i,j) does not interact with any antigen present, and also when there is no lymphocyte at (i,j), nothing happens. $\Delta c_{ij} = 0$, $\Delta\rho_{ij}^P = 0$, $\Delta\rho_{ij}^N = 0$, the lymphocyte does not proliferate, and there is no effect on the antigen.

2.5 Variation in populations of lymphocyte and antigen

The population of lymphocytes is increased through two different processes, supply of new lymphocytes from the external lymphoid organs and proliferation of existing lymphocytes.

The supply from external organs is made continuously in our model with a constant rate λ^P (= 1/immune period). Each new lymphocyte supplied in this way is put on a vacant square (k, l) chosen randomly, where ``vacant" means that only that there is no lymphocyte present on that square.

The state variables of each newly supplied lymphocyte are given as follows $T_{kl}^{LP}=0$, $c_{kl}=0$, $\theta_{kl}^L = \theta_0^L$, $\theta_{kl}^U = \theta_0^U$, and the value of S_{kl}^{LP} is chosen randomly in the range of $(1 \sim 60)$.

Lymphocyte proliferation occurs as follows. When the lymphocyte at (i, j) is to proliferate, we decide based on the proliferation probability P_{ij}^{Pro} whether a lymphocyte is replicated at each vacant site (k, l) among the eight squares neighboring (i, j). The values of state variables of newly born lymphocytes are $T_{kl}^{LP}=0$, $c_{kl} = c_{ij}$, $\theta_{kl}^L = \theta_{ij}^L$, $\theta_{kl}^U = \theta_{kl}^U$. The specificity S_{kl}^{LP} of the new-born lymphocyte may not be identical to its parent, but is chosen with equal probability among S_{kl}^{LP}, $(S_{kl}^{LP}-1)$, and $(S_{kl}^{LP}+1)$. We adopted this rather high probability of specificity deviation in the model for the following reason. The specificity of lymphocytes is determined by several kinds of receptor proteins in the membrane of lymphocytes. Therefore, mutation of the DNA encoding the relevant proteins can cause the specificity of the daughter lymphocyte to deviate from the specificity of the mother cell. Although the real probability of these mutations is very small, we adopted the high value 2/3 in order to reduce simulation time by reducing the time taken for lymphocyte specificity to match antigen specificity.

The new born lymphocyte at (k, l) may be given a long life span (30000) with probability of 0.02, when $0.8 \le cij < \theta_{ij}^U$. Otherwise, it has a normal life span, TLSLP (30). This long life span reverts to normal if lymphocyte chronicity increases above the upper threshold.

The lymphocyte population is reduced through automatic death of any cells whose life span is exceeded.

The population of foreign antigens is increased through proliferation. The population of self antigens remains constant through continuous resupply. Any new antigen has age zero and specificity dependent on the type of antigen considered. Proliferation of foreign antigens occurs according to replication rate r_{Ag}, given by

$$r_{Ag} = r_{min} + \frac{r_{max} - r_{min}}{1 + \exp(-(\rho_{Ag} - \rho_{Ag}^{th}) / \varepsilon)}, \quad (9)$$

where r_{min} and r_{max} is the minimum and maximum value of replication rate, respectively, and ρ_{Ag} is the density of antigens. In the experiments presented here, $r_{min} = 0.0$, $r_{max} = 0.1$, $\rho_{Ag}^{th} = 0.15$, and $\varepsilon = 0.05$. Antigen population is decreased through natural death due to old age $T_{LS}^{Ag} = 100$, or from artificial death due to elimination of antigen by lymphocytes in the active state.

2.6 Experiment initialisation

We initialized our system with a cellular automata model comprising a 100x100 square field. A periodic boundary condition was applied to the lattice field. 100 lymphocytes with random specificity in the range 0 to 60 were put on random squares. 1024 self antigens, where used, were put thickly on the central square region of the field as illustrated in Fig. 2c. Where foreign antigens were used, 1446 antigens were put as several clusters on the field, as illustrated in Fig. 2a. The model was then run with the parameters set as in their descriptions above.

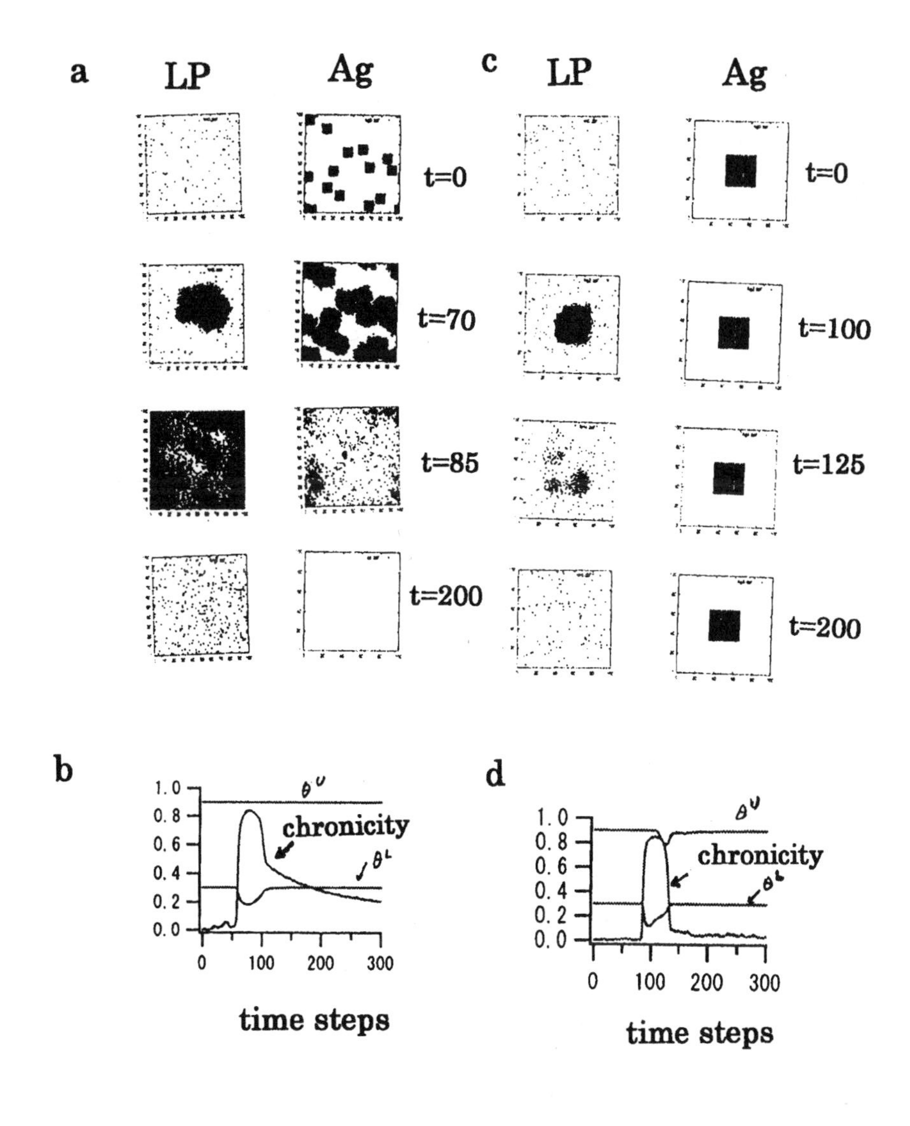

Figure 2 a, c, Snapshots showing spatial distribution of (a)lymphocytes (LP) and foreign antigens (Ag), and (c) lymphocytes (LP)and self antigens (Ag) at various times. b, d, Variations in lymphocyte chronicity and lower (θ^L) and upper (θ^U) thresholds when interacting with foreign (b) and self (d) antigens. The values in (b) and (d) are averaged over all lymphocytes. One unit of horizontal axis in (b) and (d) corresponds to 10 simulated immune periods (see Mode).

3. Results

Figure 2a presents snapshots of the spatial distribution of lymphocytes and foreign antigens at various times. The system is initiallised with a random distribution of lymphocytes whose specificity is given randomly and whose chroncity is zero. Antigens with specificity 6 and maximum replication rate 0.1 are put as several clusters on the field. As the time proceeds from 0 to 70, the population of antigen increases due to proliferation, becoming quite large. Lymphocytes whose specificity is close to antigen's specificity increase in chronicity during this period due to encounters with antigens. Many enter active mode so are more likely to proliferate themselves, producing daughter cells with similar specificity to themselves and therefore to the antigen. These active lymphocytes also secrete positive cytokines, which diffuse outwards and lower the activation threshold in neighbouring lymphocytes. This cyclic process causes the population of lymphocytes whose specificity is broadly tuned to the given antigens to increase rapidly, as can be seen from comparing the snapshots taken at t=70 and t=85. Thus, many antigens are eliminated by the increasing population of correct-specificity lymphocytes.

However, this destruction of antigen breaks the lymphocyte activation loop. New lymphocytes do not encounter sufficient antigens, so cannot interact with them and raise their chronicity above threshold level. Thus, most lymphocytes stay in the rest state where their chronicity is less than the lower threshold. These resting lymphocytes rarely proliferate and do not secrete positive cytokine. The reduction in positive cytokine concentration results in lower thresholds increasing back towards starting levels, reducing the chance of these lymphocytes becoming active still further. After their life span is over, they die without replicating, Therefore, the population of lymphocytes decreases as seen in Fig. 2a graph t=200. After the death of many lymphocytes, a small number of mature lymphocytes with a long life span remain. These do not require antigen stimulation. This state corresponds to a memory state, where the immune system contains a record of the specificity of antigen `invaders' and can respond faster to subsequent infection.

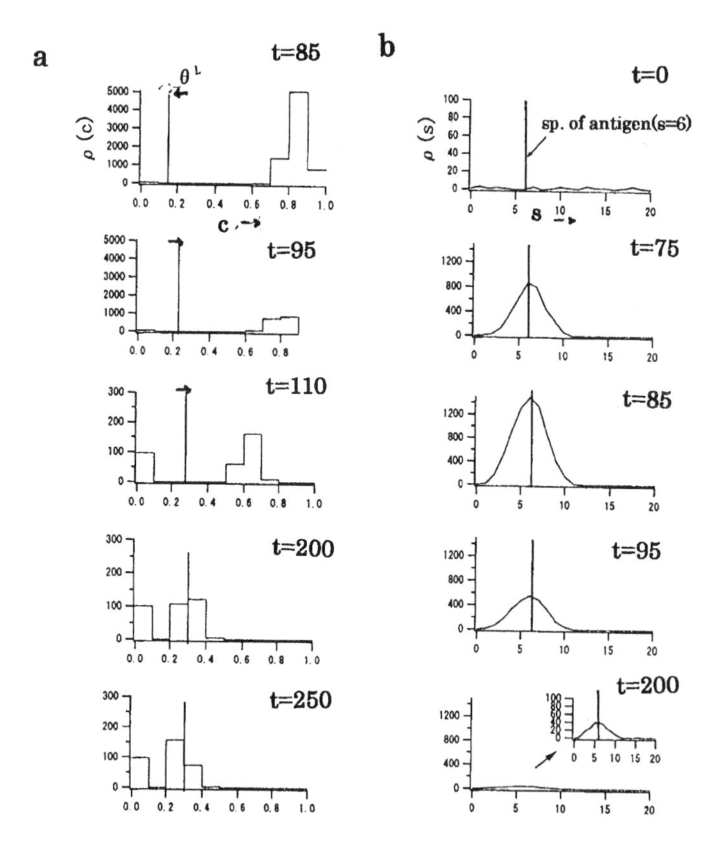

Figure 3 Immune response to foreign antigens with specificity 6. (a) variation in chronicity value, bin width 0.1. The movement of the lower threshold is also shown. The overall state of the immune system is active at times t=85 and t=95, half active and half rest mode at t=110, and rest mode after that. Note the change in vertical axis values during the lymphocyte population peak at t=85 and t=110. (b) specificity distribution (antigen specificity was 6 for this experiment). Lymphocytes with specificity from 20 to 60 were omitted because their numbers were so close to zero throughout.

The temporal variation in average chronicity and the changes in lower and upper thresholds with time, averaged over all lymphocytes, are shown in Fig. 2b. Note the large peak in average chronicity aroundt=90 when the lymphocyte population is enormous and most of these are in active mode. Note also the dip in the lower threshold which is responsible for the steepness of the chronicity increase through pushing more lymphocytes into active mode where they are more likely to interact with antigen.

Figure 2c shows similar snapshots for the immune system interaction with self antigens. Initialisation of lymphocytes is the same as in Fig. 2a. Self antigens with a specificity of 20 are put thickly onto the central square region of the field. Importantly, they are never eliminated. This mimics real self antigens such as those on healthy tissues which are permanent and whose cells get replaced as needed. Such antigens have a permanent dense and large population. As time proceeds from 0 to 100, lymphocytes with a matching specificity which contact these antigens interact with them, increase their chronicity level, and become active. The active lymphocytes proliferate and secrete positive cytokines as in the foreign antigen case above.

Figure 2d shows the temporal variation in chronicity and of the lower and upper thresholds averaged over all the lymphocytes. Up until around t=100 the pattern is much the same as for foreign antigens, with a steep rise in chronicity fuelled by a dip in the lower threshold. However, just after the chronicity peak a dip in the upper threshold can also be seen. This happens because the self antigens are not eliminated so their concentration remains high. This results in each lymphocyte of a self-matching specificity having so many interactions with antigens that its chronicity rises through active mode and into tolerance mode. As the time proceeds from 100 to 125, the number of lymphocytes in the tolerance state ($c_{ij} > \theta_{ij}^{U}$) increases so the concentration of negative cytokines

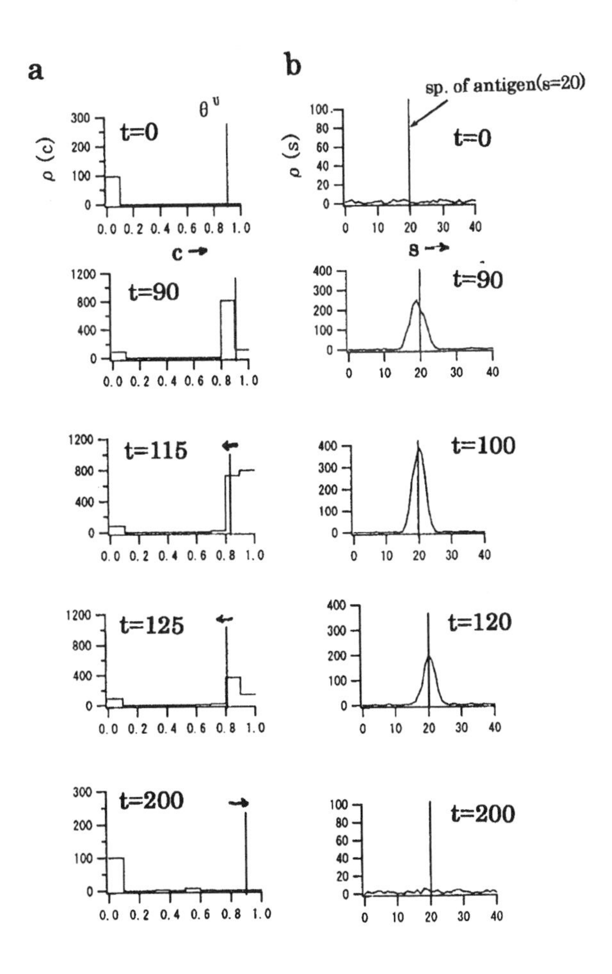

Figure 4 Immune response to self antigens with specificity 20. The graphs, (a) and (b) have the same meanings as those shown in Fig. 3. The overall immune states are: rest mode (t=0), active mode (t=90),tolerance mode (t=115), half tolerance mode half rest mode (t=125), and rest mode (t=200). The vertical line in (a) shows the upper threshold, with arrows to show the direction of upper threshold movement. Lymphocytes with specificity above 40 were omitted in (b), because their numbers were close to zero.

follow. This causes the upper threshold to be decreased, increasing the proportion of lymphocytes in tolerance mode in a positive-feedback manner.

As the time proceeds from 125 to 200, the population of lymphocytes decreases drastically because the proliferation rate of lymphocytes in the tolerance state is small. Active lymphocytes which attack self antigens disappear almost completely, since any that do appear are quickly taken into tolerance mode, where they remain until they die.

Because the activation process of lymphocytes is common to both the foreign and self antigens, the discrimination between self and foreign antigens arises from the difference in the termination process. This is illustrated in Fig. 3a, which shows snapshots of the distribution of chronicity values across the lymphocyte population during the termination of the response to foreign antigens It can be seen from Fig. 3a that the most probable chronicity value in the immune system changes from around 0.85 (near the top of the active region ($c_{ij} > \theta_{ij}^L$) to around 0.25 (just into the rest region ($c_{ij} < \theta_{ij}^L$) as the system settles into the memory state. The average lower threshold increases noticeably during this period. This shows clearly that after lymphocytes suppress antigen proliferation, they stay in the rest state and die, that is, the immune response is terminated by passive cell death.

There are two main determinants of the immune response to foreign antigens and the self-organization of memory state: the dynamical change in chronicity and the self-organized maturation of specificity. The first was shown in Fig. 3a, the second in Fig. 3b. It can clearly be seen from Fig. 3b that the diverse specificity of t=0 is tuned to the specificity of the foreign antigens as time proceeds. This indicates that a subset of mature lymphocytes is dynamically selected when a diverse population of cross-reactive lymphocyte clones, characterized by different interaction strengths, is driven by antigen specificity and subjected to regulation by chronicity values. The Darwinian selection of specificity has been experimentally shown for T cells (Zheng, Xue, and Kelsoe, 1994; Armed and Gray, 1996) and B cells (Hertz et al., 1994; Rajewsky, 1998). In memory state (t=200), lymphocyte specificity is well tuned to antigen specificity. This maturation of their affinity to antigens is the origin of the lymphocyte's function as memory cells. A second infection with antigens of the same specificity caused a rapid lymphocyte response in our system (results not shown).

The termination process of self antigens is quite different from that of foreign antigens. This is illustrated in Fig. 4, which shows the same data as Fig. 3 but for self antigens, ie Fig. 4a is a histogram showing the distribution of chronicity values in the lymphocyte population, and Fig. 4b gives lymphocyte specificity. It can be seen in Fig. 4a that the median chronicity value in the system changes rapidly from the active region ($\theta^L < c < \theta^U$) to the tolerance state ($\theta^U < c$). Fig. 4b shows lymphocyte specificity rapidly matching antigen specificity during the activation phase. Active lymphocytes exist for only a short period. Because self antigens are permanent and have a very large population, active lymphocytes specifically interacting with self antigens can not stay in the active mode but are rapidly pushed into tolerance mode due to their constant interaction with self antigens. The majority of lymphocytes stay in tolerance mode for a while, normally dying before they proliferate. As a result the total population of lymphocytes plummets. A small residual number of lymphocytes are in rest mode ($\theta^L > c$). These are mostly daughters of the original lymphocyte seeding, so have almost random specificity. This means that very few of the remaining lymphocytes attack the self antigens.

We also investigated the immune response in the case where self and foreign antigens coexist in the field. At t=0, self antigens were put thickly on the central square region and foreign antigens in clusters in the surrounding region. The population of lymphocytes interacting specifically with self antigens and that interacting with foreign antigens increased mainly on the central and surrounding regions, respectively, because the self and foreign antigens live separately from each other in the field. The actions of both types of lymphocytes are coupled through positive and negative cytokines secreted by the lymphocytes. The chronicity of both types of lymphocytes changes in opposite directions as before. Therefore, the termination processes for foreign antigens and for self antigens proceed rather independently in the two regions despite cytokine diffusion and even lymphocyte movement between the two.

Simulations using different parameter values give results with similar tendencies to those mentioned above. The behavior of our model is primarily dependent on parameters determining the dynamics of chronicity, and secondarily on parameters determining the dynamics of cytokines. When these parameter values are slightly changed, the rate of adaptation of the immune system to antigens becomes relatively faster or slower, but the overall behavior remains qualitatively unchanged.

4. Concluding remarks

Our model of the immune system was successful in distinguishing between self and non-self antigens. Furthermore, the difference in model behavior was biologically plausible in the two cases: foreign antigens were destroyed leaving a scattering of `memory cells' of the right specificity ready to respond quickly to any second infection; and self antigens were largely ignored after an initial lymphocytic investigation despite the continuing presense of antigen. We believe the distinction between self and non-self antigens in the model to be realistic. Therefore, we have shown that chronicity is sufficient to explain self/non-self discrimination in lymphocytes.

The present model proposes that lymphocytes interacting specifically with self-antigens increase their chronicity quite rapidly through continuous interaction with self-antigens that rarely change in concentration. Because the upper threshold θ^U is decreased considerably at the same time, the number of lymphocytes whose chronicity becomes larger than θ^U is increased almost discontinuously. Those lymphocytes attain tolerance mode ensuring non-destruction of self antigens. Furthermore, those lymphocytes actively secrete negative cytokines which reduce the θ^U of new lymphocytes. This result is consistent with the Activation Induced Cell Death process and with the cytokine-mediated regulation process proposed for self antigens by Van Parijs and Abbas (1998). The present model can be also applied to the other hypothesis of memory-maintenance (Zinkernagel, 1996) that immune system memory is maintained by lymphocytes continuously stimulated by the small numbers of residual antigens.

Our immune model is consistent with the linear differential pathway in memory cells proposed for foreign antigens by Ahmed and Gray (1996).

This present study presents a new viewpoint for the understanding of the immune system, that is, that immunoregulation such as self/non-self discrimination is regulated quantitatively and dynamically. The results obtained are complementary to those from experiments based on the structural approach.

References

Ahmed, R. and Gray, D. (1996). Immunological memory and protective immunity: Understanding their relation. *Science.*

Billingham, R. E. (1956). Actively acquired tolerance of foreign cells. *Proc. Soc. London Ser. B.*

Billingham, R. E., Brent, L. and Medawar, P. B. (1953). Quantitative studies of tissue tranplantation immunity. *Nature.*

Burnet, F. M. (1959). *The Clonal Selection Theory of Acquired Immunity*. University Press, Cambridge.

Burnet, F. M. and Fenner, F. (1949). *The Production of Antibodies*. Macmillan, Melbourn/London.

Forsthuber, T., Yip, H. C., and Lehmann, P. V. (1996). Induction of T_H1 and T_H2 immunity in neonatal mice. *Science.*

Grossman, Z. (1984). Recognition of self and regulation of specificity at the level of cell populations. *Immunol. Rev.*

Hertz, M., Kouskoff, V., Nakamura T. and Nemazee, D. (1998). V(D)J recombinase induction in splenic B lymphocytes is inhibited by antigen-receptor signalling. *Nature*

Kyburz, D., Aichele, P., Speiser, D. E., Hengartner, H., Zinkernagel, R. M. and Pircher, H. (1993). T cell immunity after a viral infection versus T cell tolerance induced by soluble viral peptides. *Eur. J. Immunol.*

Lederberg, J. (1959). Genes and antibodies. Science.

Mamalaki, C., Tanaka, Y., Corbella, P., Chandler, P., Simpson, E. and Kioussis, D. (1993). T cell deletion follows chronic antigen specific T cell activation in vivo. *Int. Immunol.*

Matzinger, P. (1994). Tolerance, danger, and the extended family, *Ann. Rev. Immunol.*

Pennisi, E. (1996). Teetering on the brink of danger. *Science.*

Rajewsky, K. (1998). Burnet's unhappy hybrid. *Nature.*

Ridge, J. P., Fuchs, E. J., and Matzinger, P. (1996). Neonatal tolerance revisited: turning on newborn T cells with dendritic cells. *Science.*

Rocha, B. and von Boehmer, H. (1991) Peripheral selection of the T cell repertoire. Science.

Rothenberg, E. V. (1996). How T cells count. *Science.*

Sarzotti, M., Robbins, D. S., and Hoffman, P. M. (1996). Induction of protective CTL responses in newborn mice by a murine retrovirus. *Science.*

Van Parijs, L. and Abbas, A. K. (1998). Homeostasis and self-tolerance in the immune system: Turning lymphocytes off. *Science.*

Viola, A. and Lanzavavecchia, A. (1996). T cell activation determined by T cell receptor number and tunable thresholds. *Science.*

Welsh, R. M., Selin, L. K. and Razvi, E. S. (1995). Role of apoptosis in the regulation of virus-induced T cell responses, immune suppression, and memory. *J. Cellular Biol.*

Zheng, B., Xue, W. and Kelso, G. (1994). Locus-specific somatic hypermutation in germinal center T cells. *Nature.*

Zinkernagel, R. M. (1996). Immunology taught by viruses. *Science.*

Zinkernagel, R. M. and Doherty, P. C. (1974). Immunological surveillance against altered self-components by sensitized T lymphocytes in lymphocytic choriomeningitis. *Nature.*

Zinkernagel, R. M. and Doherty, P. C. (1979) MHC-restricted cytotoxic T cells: studies on the biological note of poplymorphic major transplantation antugens determining T cell restriction-specificity, function, and responsiveness. *Adv. Immunol.*

The Digital Hormone Model for Self-Organization

Wei-Min Shen and Cheng-Ming Chuong

University of Southern California
4676 Admiralty Way, Marina del Rey, CA 90292, USA
shen@isi.edu, chuong@pathfinder.usc.edu

Abstract[1]

This paper presents the Digital Hormone Model (DHM) as a new computational model for self-organization. The model is a generalization of an earlier distributed control system for self-reconfigurable robots [1-3] and an integration of reaction-diffusion model with stochastic cellular automata. In this model, cells secrete hormones, and hormones diffuse into space and influence the behaviors of other cells to self-organize into global patterns. In contrast to the reaction-diffusion models, cells movements are computed not by the Turing's differential equations [4] but by stochastic rules that are based on the concentration of hormones in the neighboring space. Different from simulated annealing, the stochastic rules are not the Metropolis rule [5], which will not produce the desired results here. Experimental results have shown that the simulated cells in the DHM produce results that match to the observations made in the biological experiments where homogenous skin cells self-organize into feather buds regulated by hormones [6]. With the unique properties of being distributed, scalable, robust, and adaptive, the DHM opens opportunities for new hypotheses, theories, and experiments for further investigating the nature of self-organization.

The Digital Hormone Model

The Digital Hormone Model consists of a space of grids and a set of moving cells. Each living cell occupies one grid at a time and a cell can secrete chemical hormones that diffuse into its neighboring grids to influence other cells' behaviors. Hormones have types and diffusion functions. Similar to the reaction-diffusion models, two types of hormones are most common: an *activator* hormone that will encourage certain cell actions, while an *inhibitor* hormone will prohibit certain cell actions. We assume that hormones may react to each other and may diffuse to the neighboring grids according to certain functions. Similar to the extensions used in [7], we allow anisotropic and spatially non-uniform diffusion. To extend the traditional reaction-diffusion models further, we view cells as autonomous and "intelligent" entities and allow them to react to hormones and perform a set of possible actions such as migration, differentiation, or adhesion.

At any given time, a cell selects and executes one or more actions according to a set of internal behavior rules. These rules are probabilistic in nature. At the moment, we assume that the rules are given, unchangeable, and consistent. At any time, the rules will not cause a cell to select actions that are in conflict with each other.

Given the grids, cells, hormones, actions, and rules, the Digital Hormone Model works as follows:

1. All cells select actions by their behavior rules;
2. All cells execute their selected actions;
3. All grids update the concentration of hormones;
4. Go to Step 1.

To illustrate the above definitions, let us consider a simple DHM, where cells migrate on a space of N^2 grids and have only two actions: secretion and migration. The secretion action always produces two hormones: the activator A and the inhibitor I. The diffusion rates of A and I are characterized by the Guassian distributions, and the concentration of hormones in any given grid can be computed by summing up "A"s and "I"s in that grid. A cell is always migrating. The moving direction of a particular migration is governed by the distribution of hormones in its neighboring grids, and the probability to migrate to a particular neighboring grid is proportional to the concentration of A in that grid, and is inverse proportional to the concentration of I in that grid. The questions are now: Will the DHM enable cells to self-organize into patterns? Will the individual hormone diffusion profiles affect the self-organization process and the size and shape of the final patterns? Will the patterns change with the cell population and density? To answer these questions, we have run two sets of experiments described below.

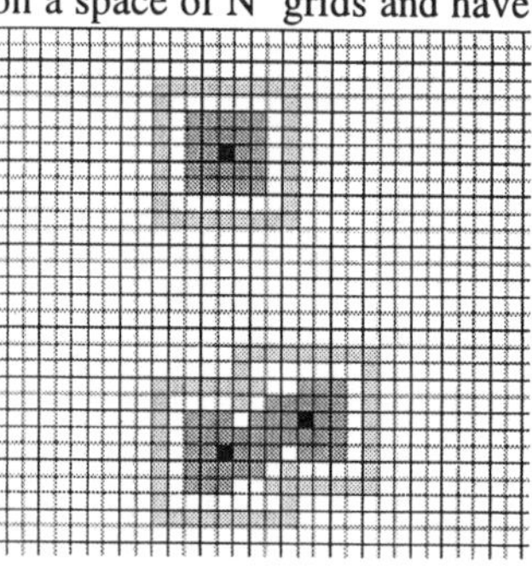

In the first set of experiments, we use the same hormone diffusion profile. For any single isolated cell, let the cell's n^{th} ring of neighbors be the neighboring grids at the distance of n grids away from the cell. Using this definition, we define the activator hormone level at the cell's surrounding grids as follows: 0.16 for the 0th ring

[1] This research is in part supported by the AFOSR contract F49620-01-1-0020 and the DARPA contract DAAN02-98-C-4032.

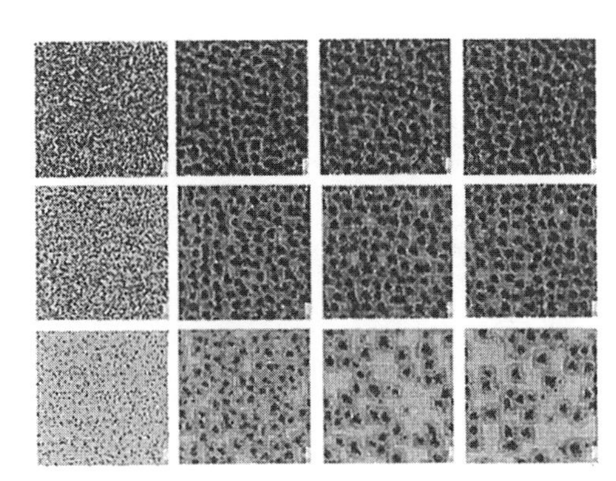

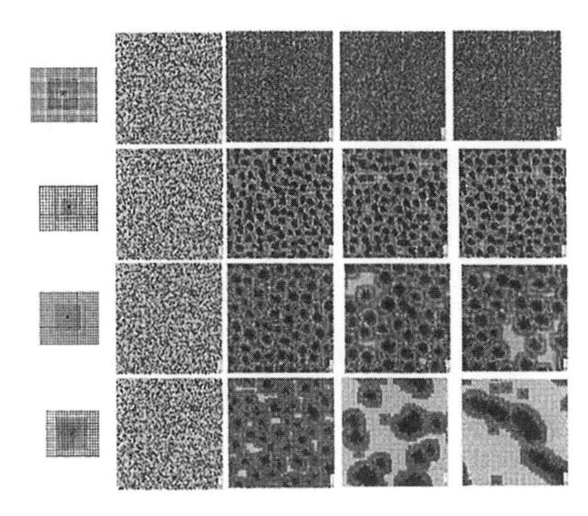

(i.e., the occupying grid), 0.08 for the 1st ring of neighbors, 0.04 for the 2nd ring of neighbors, 0.02 for the 3rd ring of neighbors, and 0 for the 4th and further rings of neighbors. For the inhibitor hormone, the levels for the 0th through 4th rings of neighbors are: -0.05, -0.04, -0.03, -0.02, and –0.01. Thus the combined levels of hormones at these rings are: 0.11, 0.04, 0.01, 0, and –0.01. As an approximation, we assume that the hormones from a cell will have no effect on the hormone level in grids that are beyond the 4th ring of neighbors.

Given this hormone diffusion profile, we have run a set of simulations on a space of 100x100 grids. To initiate the grids, we have randomly distributed cells on the grids with different cell population and density, ranging from 10%, 25%, through 50%. Each simulation run up to 1,000 action steps, and we recorded the configuration snapshots at the steps of 0, 50, 500, and 1,000. As we can see, in all the simulations the cells indeed form groups with approximately the same size. These results demonstrated that the digital hormone model indeed enables cells to form patterns. Furthermore, the results match the observations made in the biological experiments described in [6]. The size of the final clusters does not change with the cell population and density, but the number of clusters does. Indeed, we have seen that when the cell population is low, the number of final clusters is also low, but the size of clusters is roughly the same. When the population of cells increases, the size of final clusters remains the same, but the number of clusters increases.

In the second set of experiments, we have varied the hormone diffusion profiles and observed their effect on the results of pattern formation. As we expected, when a balanced profile of activator and inhibitor is given (see the second row), the cells will form final patterns as before. As the ratio of activator over inhibitor increases (see the third row), the size of final clusters also increases. However, when the ratio is so high that there are only activators and no inhibitors (see the fourth row), then the cells will form larger and larger clusters, and eventually become a single connected cluster. On the other hand, when the ratio is so low that there is only inhibitor and no activator, then the cells will never form any patterns (see the first row), regardless how long the simulation runs. From these experiments, we can conclude that both the activator and inhibitor hormones play important roles in self-organization and pattern formation, and neither alone can accomplish the job. Interestingly, we have also tried the Metropolis rule for migration in both experiments, but no patterns are formed no matter what temperature is set and how many steps it runs.

In summary, these two sets of experiments have demonstrated that the proposed digital hormone model is indeed an effective tool for simulating and analyzing self-organization phenomena. Furthermore, the results show that hormones play critical roles in self-organization. Finally, these results show that it is indeed possible for a set of locally interacted autonomous elements to form globally interesting patterns based on only local information and actions. This provides a departure point for new hypotheses, theories, and experiments for self-organization.

References

1. Shen, W.M., B. Salemi, and P. Will. *Hormones for Self-Reconfigurable Robots.* in *Proceedings of the 6th International Conference on Intelligent Autonomous Systems*. 2000. Venice, Italy.
2. Shen, W.M., Y. Lu, P. Will. *Hormone-based Control for Self-Reconfigurable Robots.* in *Proceedings of International Conference on Autonomous Agents*. 2000. Barcelona, Spain.
3. Salemi, B., W.-M. Shen, P. Will. *Hormone-Controlled Metamorphic Robots.* in *International Conference on Robotics and Automation.* 2001. Seoul, Korea.
4. Turing, A.M., *The chemical basis of morphogenesis.* Philos. Trans. R. Soc. London B, 1952. 237: p. 37-72.
5. Kirkpatrick, S., G.B. Sorkin, *Simulated Annealing*, in *The Handbook of Brain and Neural Networks*, M. Arbib, Editor. 1995, The MIT Press: Cambridge, MA.
6. Jiang, T.-X., Jung, H. S., Widelitz, R. B., and Chuong, C.-M., *Self organization of periodic patterns by dissociated feather mesenchymal cells and the regulation of size, number and spacing of primordia.* Development, 1999. 126: p. 4997-5009.
7. Witkin, A., M. Kass, *Reaction-diffusion textures.* Computer Graphics (also Proc. Siggraph 91), 1991. 25(3).

EVOLUTION

Active Vision and Feature Selection in Evolutionary Behavioral Systems

Davide Marocco
"Centro Interdip. della Comunicazione"
University of Calabria (UNICAL)
Arcavacata di Rende (CS), Italy
d.marocco@unical.it

Dario Floreano
Institute of Systems Engineering
Swiss Federal Institute of Technology (EPFL)
Lausanne, Switzerland
Dario.Floreano@epfl.ch

Abstract

We describe an evolutionary approach to active vision systems for dynamic feature selection. After summarizing recent work on evolution of a simulated active retina for complex shape discrimination, we describe in detail experiments that extend this approach to an all-terrain mobile robot equipped with a mobile camera. We show that evolved robots are capable of selecting simple visual features and actively maintaining them on the same retinal position, which largely simplifies the "recognition" task, in order to generate efficient navigation trajectories with an extremely simple neural control system. Analysis of evolved solutions indicates that robots develop a simple and yet very efficient version of edge detection and visual looming to detect obstacles and move away from them. Two evolved sensory-motor strategies are described, one where the mobile camera is actively used throughout the entire navigation and one where it is used only at the beginning to point towards relevant environmental features. The relationship between these two strategies are discussed in the context of the underlying visuo-motor mechanisms and of the evolutionary conditions.

1. Active Vision and Feature Selection

Visual processing for navigation in complex environments can be significantly simplified by selecting and paying attention only to a reduced set of environmental features. In computational vision, there are two major approaches to feature selection. One consists of equipping the behavioral system with a library of predefined feature extraction mechanisms, such as edge and motion detectors, depth from stereo-disparity, shape from shading, and landmark detection, to mention a few (Mallot, 2000, e.g.). The other consists of using some form of unsupervised learning in order to reduce the large amount of visual information to a subset of statistically invariant and optimal features[1] (Hinton and Sejnowski, 1999). The latter approach seems more powerful when the environment where a robot operates is largely unknown and may significantly change over time, but currently available algorithms produce stable results only for well-behaved probability distribution functions of the input data or require off-line learning on a large set of available visual data.

None of these two approaches takes into account the fact that the number and type of visual features that an organism is sensitive to depend also on the sensory-motor and behavioral characteristics of the organism in its environment (Gibson, 1979). From the perspective of a designer, an interesting complication of behavioral systems is that behavior is both determined by visual information and at the same time affects what type of visual information is gathered. In fact, this apparent "egg-and-chicken" issue can be exploited to turn hard sensory discrimination problems into simpler ones. For example, it has been shown that the fruit fly *Drosophila*, which is equipped with a very simple neural circuit, is capable of performing complex shape recognition by moving in order to shift the perceived image to a certain location of the visual field (Dill et al., 1993). The process of selecting by motor actions sensory patterns which are easy to discriminate is usually referred to as *active perception* (Bajcsy, 1988). In computer vision, the importance of active gaze control to facilitate object recognition in complex scenes has been pointed out by (Ballard, 1991) and (Rimey and Brown, 1994). The latter paper, in particular, explores the use of Bayes nets and decision theory to optimally position a vision sensor in an image, but it takes advantage of prior knowledge of environmental relations and geometrical structure. More recently, in a series of experimental results, Nolfi (1998) and Scheier et al. (1998) have shown that autonomous robots that au-

[1] So that the mean square root error between the original image and the *reconstructed* image is minimized.

tonomously evolve their behavior while freely interacting with the environment exploit active perception to turn hard classification problems into simpler ones. Using a problem classification theory developed by Clark and Thornton (1997), the authors showed that such evolved robots can indeed turn sensory classification problems of "type 2" (difficult ones) into problems of "type 1" (simple ones).

Within the context of vision processing, artificial evolution of behavioral systems is a powerful method to co-evolve feature-selection mechanisms and behavior of autonomous robots because it does not separate visual perception from behavior as in conventional system engineering methods (Cliff and Noble, 1997, Harvey et al., 1994). However, in all evolutionary experiments conducted so far the vision system is aligned with the body of the robot and cannot independently explore the environment while the robot moves around. A mechanically independent vision system complicates the sensory-motor coordination of the robot, but may provide improved navigation abilities by actively searching for simple features that can be useful for maintaining a smooth trajectory or maintaining in sight some cues in the visual field during navigation.

After summarizing the results of our recent work on evolutionary active vision for shape discrimination, we describe preliminary experiments that extend this approach to an all-terrain mobile robot equipped with an active vision system. We show that evolved robots are capable of selecting simple visual features and actively maintaining them on the same retinal position, which largely simplifies the "recognition" task of the system and generation of efficient navigation trajectories with an extremely simple neural control system.

2. Evolutionary Active Vision for Shape Discrimination

Recently, we have explored the idea of evolving an active retina capable of autonomously scanning the visual field in order to discriminate shapes with very limited computational resources (Kato and Floreano, 2001). Since the robotics experiments described in this paper represent an extension of that line of research, in this section we briefly summarize the main results.

Our system was inspired upon the evidence that humans and other animals asked to recognize shapes take their time to explore images with several rapid saccadic movements (figure 1) (Krupinski and Nishikawa, 1997, e.g.) instead of providing an immediate answer based on a single snapshot of the entire image (as most computer vision techniques do). These saccadic movements tend to sequentially foveate over salient areas of the image and we speculated that they might be useful (also) for simplifying the recognition problem by checking for the presence of simple features.

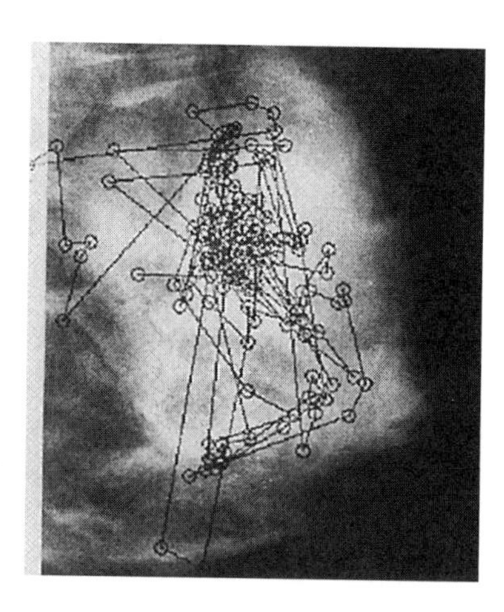

Figure 1: Patterns of eye movements of doctors scanning an X-ray image for the presence of breast cancer. Dots represent fixation points (Krupinski and Nishikawa, 1997). These and other images of human eye scans are available at `http://www.radiology.arizona.edu/ eye-mo/mainpage.htm`

We devised a simulated neuron-based vision system where a subset of output neurons can move the vision acquisition device around an image, zoom in and out, and dynamically select the filtering strategy used for preprocessing (figure2). The system consisted of a small simulated retina composed of a 3 by 3 matrix of visual cells whose receptive fields received input from a limited area of the image. Since the images were artificially created by the computer and had a limited size, an additional input neuron detected when the retina hit a border of the image (this neuron would not be necessary in a mobile robot immersed in an environment). The activations of the retinal cells were fed into a recurrent neural network without hidden units. The output units controlled the zooming factor (image area covered by the retina), the movements of the retina over the image (expressed as direction and distance from the current position), and the filtering strategy (average values of the pixels spanned by a vision cell or value of the the top leftmost pixel in the receoptive field of a vision cell). In addition, two output units coded the shape discrimination response of the vision system, in this case being the presence of a triangle or of a square.

The synaptic weights of the network were genetically encoded and evolved to discriminate between squares and triangles. Each individual was presented with 20 noisy images where a triangle or a square appeared at random positions and had a random size. The system was free to explore each image for a maximum of 50 cycles (input computation, network activation, retinal displacement, zooming and filtering, and recording of shape discrimination). The fitness of each individual was proportional to the number of correct responses for each im-

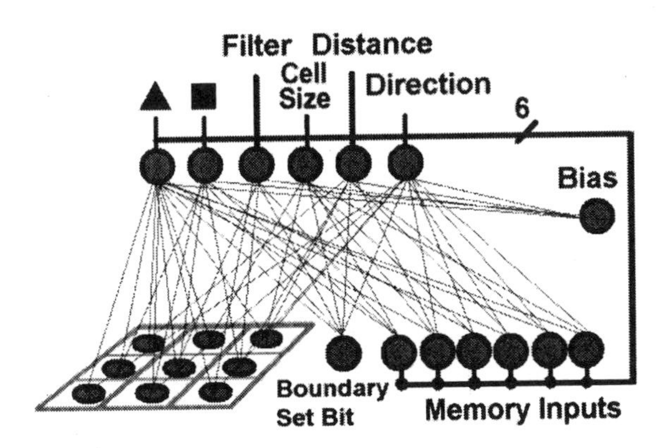

Figure 2: Neural architecture of the active vision system. Six output units receive signals from the retina cells and from a unit signalling whether the retina is against a border. The output units have recurrent connections, here represented as memory units that hold the activation of the output units at the previous time step (Elman, 1990).

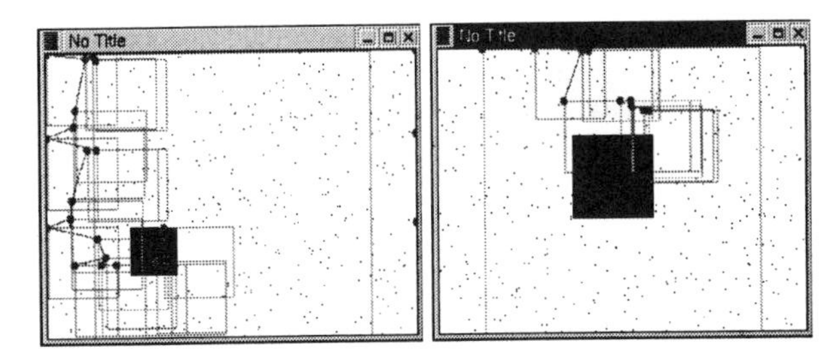

Figure 3: Examples of trajectories of an evolved individual. The retina always zooms and moves with respect to its top leftmost corner, here marked by a dot. The dots drawn after every retina movement are connected by a line. For graphical clarity, the values of the cells are not shown, only the retinal perimeter. *Left*: The retina starts with its initial size at the center of the image signalling "triangle". It then shrinks to the top left corner and moves down towards the square where it slides along its left edge and starts signalling "square". Finally, it explores the other three sides of the square maintaining the correct response. *Right*: The same individual begins signalling triangle and then moves towards the square where it visits the top and right edge changing the response into "square".

age. In other words, this fitness encouraged the retinal system to provide the correct answer as soon as possible while exploring the image.

Best evolved individuals reported 80% correct response (100% correct response was impossible because for each image it takes some time for the retina to find where the shape is) and managed to discriminate correctly all shapes. Evolved systems started with a fixed response (square or triangle, depending on the evolutionary run) and then moved towards the shape. Once over the shape, the retina slided back and forth along one of its vertical edges. If the edge was straight, it set its response to square, otherwise to triangle. Figure 3 shows the trajectories of the retina in the case of two squares and figure 4 shows the trajectories in the case of two triangles. A variation on this basic strategy consisted of scanning the corners of the shapes, instead of the edges, to detect whether it forms an acute or rectangular angle. Once the shape had been correctly recognized, sometimes the vision system moved away from the shape towards a border of the image maintaining the correct response (this behavior was made possible by the recurrent connections). These selected features, edge inclination and angle, turned out to be invariant to size and location of the shape (the angle feature is also invariant to rotation).

In addition, we showed that a feed-forward neural network (whose input was the entire image and whose output encoded the shape class) trained with the back-propagation algorithm on the same set of images could not solve the discrimination task, not even when equipped with a variable number of hidden units. Although we do not intend to claim that back-propagation networks with hidden units in general cannot solve this task, the experiments showed that evolved active vision and feature selection turned non-linearly separable problems (type 2) into linearly separable ones (type 1).

3. Robotic Experimental Setup

The aim of these experiments was to extend the approach described above to a mobile robot equipped with a CCD pan/tilt camera (figure 5). The robot was positioned in a square arena and asked to navigate as far as possible without hitting the walls. The robot was controlled by a recurrent neural network whose input was a filtered black-and-white image coming from its mobile camera and whose output determined the movement of the robot, of the camera, and the type of filtering technique applied to the camera image. The goals of this experiment were to *a)* investigate the type of visual features exploited by a simple evolutionary neurocontroller without hidden units; *b)* study the emerging type of sensory-motor coordination used to control navigation and movement of the vision system; *c)* explore the emerging interactions between active vision and feature selection.

We used a mobile robot Koala by K-Team SA equipped with a Sony EVI-D31 mobile video camera and on-board PC-104 computer board. The Koala robot (figure 6) has three soft rubber wheels on each side, but only two motors each connected to the wheel in the middle which is slightly lower in order to generate more traction. The remaining two wheels on each side provide only physical support. This mechanical solution allows nav-

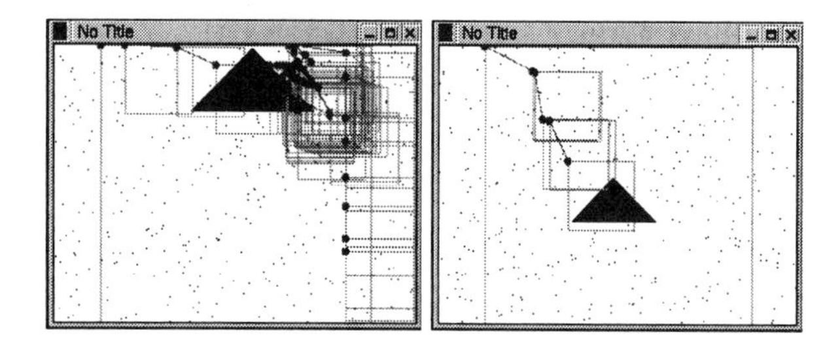

Figure 4: Examples of trajectories of an evolved individual, as in figure 3 above. *Left*: The recognition of a triangle is made by exploring its right corner and then drifting away while maintaining the correct response. *Right*: The recognition is performed by looking at the left edge of the triangle.

Figure 5: The Koala robot with the mobile camera and on-board computer board. The robot base is 30 cm w, 32 cm l, and 20 cm h; its weight is 4 kg and can carry a payload of 3 kg (camera and on-board computer are approximately 2 kg).

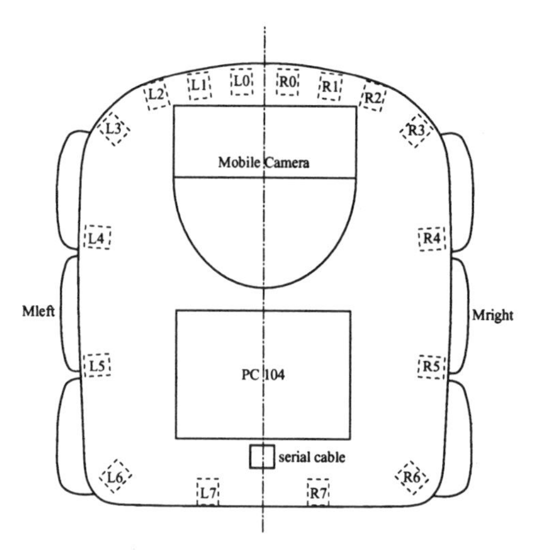

Figure 6: Sensory-motor layout of the Koala robot showing the position of the camera, of infrared sensors on left (L) and right (R) side, of the connection with the serial aerial line, and of the driving wheels (M_{left}, M_{right}).

igation on rough terrain while maintaining the simplicity of a two-wheeled robot like the Khepera. The video camera is equipped with two motors that allow both horizontal movement (*pan*) in the range $[-100°, 100°]$ and vertical movement (*tilt*) in the range $[-25°, 25°]$.[2] The camera returns to the on-board computer rectangular video frames which are then cropped to a square matrix of 240 by 240 pixels and color information is discarded. The on-board computer performs image pre-processing, activation of the neural network, control of the motors of the robot and of the camera, the evolutionary algorithm, as well as fitness computation and data storage for off-line analysis. Details of the communication protocol and software are given in the appendix. The Koala robot also has 16 infrared sensors distributed around the body that can detect obstacles at a distance of approximately 20 cm. These proximity infrared sensors were used to reposition the robot to a random location between trials, but their activations were not given to the evolutionary neural controller and not even used for fitness computation. In order to provide continuous electrical power without using time-consuming homing algorithms or heavy-duty batteries, the robot was connected to a power supply through an aerial serial cable attached to the rear side of the robot and suspended rotating contacts.

[2]The Sony EVI-D31 camera also allows motorized control of zoom and focus, but we did not use this option in current experiments.

The robot was evolved in a square arena measuring 200 cm by each side surrounded by 30 cm high white walls (figure 7). The arena was positioned in an office with a dark carpet floor and walls cover by posters, electric heaters, and windows. The office lights were kept on day and night. Near the arena was a desk with one or more researchers and visitors wandering around during the day while the robot evolved. All this visual information was potentially available to the mobile camera, but the robot was constrained to move only within its arena.

4. Neural Architecture and Evolution

Since a major goal of the experiment was to investigate the advantage provided by evolutionary active vision and feature selection with respect to computational complexity and resources, we employed a simple perceptron with recurrent connections to map visual input into motor commands. The neural network consists of 27 input units and 5 output units with discrete-time recurrent connections (figure 8). The input layer is an artificial retina of 5 by 5 visual neurons that receive input from a gray level image of 240 by 240 pixels. Visual neurons have non-overlapping receptive fields that receive

Figure 7: The evolutionary environment and the robot. The robot has visual access to the whole environment, but it can move only within the white arena. Lights were turned on day and night and researchers and visitors were free to come to the office during the evolutionary process. The pole on the back of the robot prevents the aerial cable from being trapped in the mobile camera. The other cable visible in this picture is used only after the evolutionary process to download data through ethernet from the PC-104 to a desktop computer for analysis.

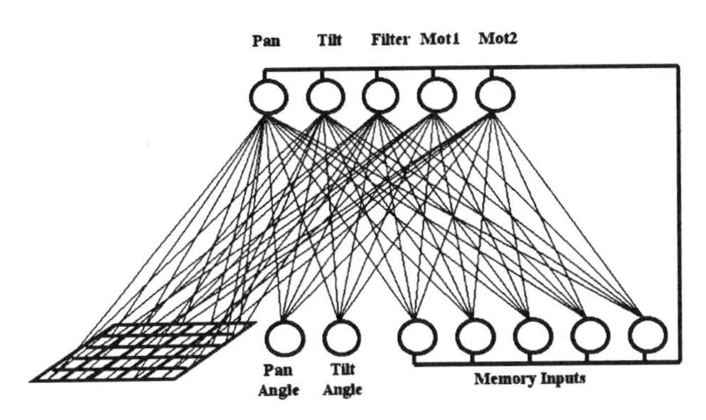

Figure 8: Architecture of neural network used in this experiment

information from a 48 by 48 pixels (240/5) area of the image. The state of visual neurons is determined by the activation of the *filter* output unit of the network. For activation values below 0.5, the state of each visual neuron corresponds to average grey level (scaled in the range $[0, 1]$) of the corresponding image patch (averaging filter). For activation values equal to or above 0.5, the state of each visual neuron corresponds to the grey level (scaled in the range $[0, 1]$) of the top leftmost pixel of the corresponding image patch (sampling filter). In addition, two proprioceptive input neurons encode the measured horizontal (pan) and vertical (tilt) angles of the camera. These values are in the interval $[-100, 100]$ and $[-25, 25]$ degrees for pan and tilt, respectively. Each value is scaled in the interval $[0, 1]$ so that activation 0.5 corresponds to 0 degrees (camera pointing forward parallel to the floor).

The activations of the output units are passed through the logistic function. Two output units determine the speeds of the wheels of the robot. In these experiments the speeds were set in the range $[-8, 8]$ cm/s. Activation values above 0.5 stand for forward rotational speed whereas activation values below 0.5 stand for backward rotational speed. Two output units encode the speed of the motor of the camera on the horizontal (pan) and vertical (tilt) planes in the same way described above. In this case, the maximum speed reachable in the horizontal plane is 80 degrees/sec and in the vertical plane is 50 degrees/sec. If the camera has reached a maximum allowed position ($-100, 100$ and $-25, 25$ degrees for pan and tilt, respectively), output speeds in the same direction have no effect. The remaining output unit encodes the filtering strategy, as described above. Discrete-time recurrent connections are implemented using 5 memory units that maintain a copy of the activations of output units at the previous sensory-motor cycle (Elman, 1990)

The architecture of the neural network is fixed and its connection strengths and neuron thresholds are evolved using a simple genetic algorithm described in (Goldberg, 1989). Connection strengths and thresholds can take values in the range $[-4.0, 4.0]$ and are each encoded on 5 bits. The architecture described above has 160 weights and 5 thresholds. A population of 40 individuals is evolved using truncated rank-based selection with a selection rate of 0.2 (the best 10 individuals make 4 copies each) and elitism (a randomly chosen individual of the population is replaced by the best individual of the previous generation). One-point crossover probability is 0.1 and bit-toggling mutation probability is 0.01 per bit.

Each individual of the population is tested on the same robot, one at a time, for 2 trials each consisting of at maximum 200 sensory-motor cycles. Each sensory-motor cycle lasts 300 ms (during which the wheels move at the latest computed speed) during which the following sequence of operations is performed: infrared sensor reading, image acquisition and pre-processing according to the value of the output filter (most of the time and computational resources are spent on this step), activation of the network, motion of the wheels and of the camera, real wheel speed reading and fitness computation.

At the beginning of each trial the robot is relocated in the environment at a random position and orientation by means of a motor procedure during which the robot moves forward and turns in a random direction for 20 seconds. During a trial, if the robot gets too close to a wall (at a distance of less than 10 cm), the trial is stopped and the robot is repositioned using the random motion procedure described above. This strategy is useful to prevent shocks that can damage the mechanics of the camera, the electronic contacts between the PC-104 and other components, and the head of the hard disk on the PC-104. In addition, it accelerates significantly the evolutionary process during early generations when most individuals tend to crash into walls and stay there.

The fitness function was conceived to select individuals capable of moving as fast forward as possible during the time allocated in each trial. Since the fitness function is computed and accumulated after every sensory-motor cycle (300 ms), robots whose trials are truncated earlier report lower fitness values.

The fitness criterion $\mathcal{F}(S_right, S_left, t)$ is a function of the measured speeds of the right wheel S_right and left wheel S_left, and of time t:

$$\mathcal{F}(S_right, S_left, t) = \frac{1}{E*T} \sum_{e=0}^{E} \sum_{t=0}^{T'} ((S^t_{right} + S^t_{left}) - |S^t_{right} - S^t_{left}|)$$

where S_{right}, S_{left} are in the range$[-8, 8]$ cm/s and $\mathcal{F}(S_right, S_left, t) = 0$ if S^t_{right} or S^t_{left} are less than 0 (backward motion of the robot); E is the number of trials (2 in these experiments), T is the maximum number of sensory-motor cycles per trial (200 in these experiments), and T' is the observed number of sensory-motor cycles (for example 34 for a robot whose trial is truncated after 34 steps to prevent collision with a wall).

5. Results

The entire evolutionary process has been carried out on the real robot, each generation taking about 1.5 hrs. Both average and best fitness values gradually increased and reached a stable level after only 8 generations (figure 9). We ended the evolution at 15th generation because the observed behaviors were very similar across all individuals of the population. After the initial 8 generations, we noticed an alternation of two behavioral strategies across generations, one where the robot camera and body movements and one where the camera is maintained at a fixed position. The former behavioral strategy disappears after 12 generations although its fitness performance is equal to the latter strategy. Since the former strategy is quite interesting, we will describe both of them and discuss why it disappears in the discussion section below. We will start describing the strategy that uses the motion of the camera because the other strategy is a reduced and less general one.

Figure 10 shows the trajectory, camera displacement, and visual input of the best individual of generation 12. The robot starts in the position marked by the star. The *horizontal direction* (pan) of the camera is shown by long arrows plotted at each sensory-motor cycle. The greyscale matrices show the activations of the visual neurons and, for sake of clarity, are plotted only before, during, and after avoidance of a wall and/or rotation of the camera to the opposite direction. This plot should be compared with figure 11 showing the horizontal (pan) and vertical (tilt) position of the camera as well as the output of the pan neuron and of the right motor neuron. The overall strategy consists of pointing and maintaining the camera downwards (thin continuous line in figure 11) so that the visual system can detect the edge between the dark floor and the white walls. In addition, the neural controller selects and maintains a *sampling* visual filtering (data not shown) so to enhance this brightness contrast (averaging filtering would blur out such contrast). This edge is clearly visible in the matrix plots of visual activations. The robot always follows a clockwise trajectory. The camera is moved to the left when the robot is approaching a wall on its left and then moved to right when the wall is sufficiently far away. While the camera is pointing to the right, it slowly scans back and forth and if a wall is detected at a certain distances, it moves to the left.

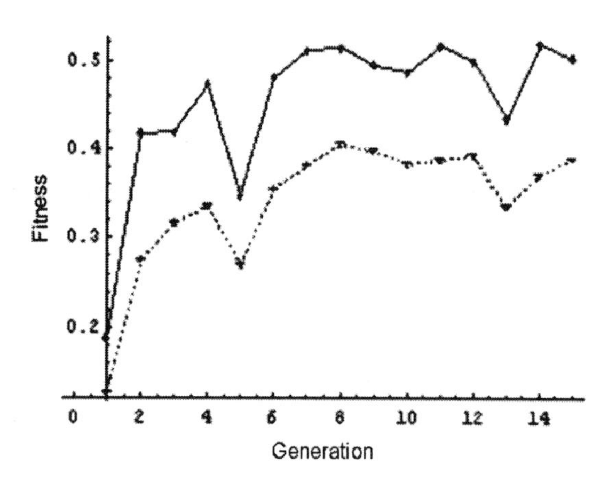

Figure 9: Fitness data scaled in the range [0, 1]. Fitness value is calculated from the speed of the each wheels of the robot according to function (1). The full line represent the performance of the best individual of generation. The dashed line represent the average performance of the entire population. Notice: the value 1.0 is not reachable, considering the characteristics of the fitness function, because the robot has to avoid the obstacles.

The values of output units encoding pan motor and right wheel motor are strongly correlated, as shown in figure 11, and are determined by the amount of white on the top two lines of the visual matrix. The closer the robot gets to a wall, the larger is the white area on the top portion of the visual field. This information is used to point the camera to the left facing the wall and slow down the rotation of the right wheel in order to turn right. As soon as the wall is sufficiently far away, the

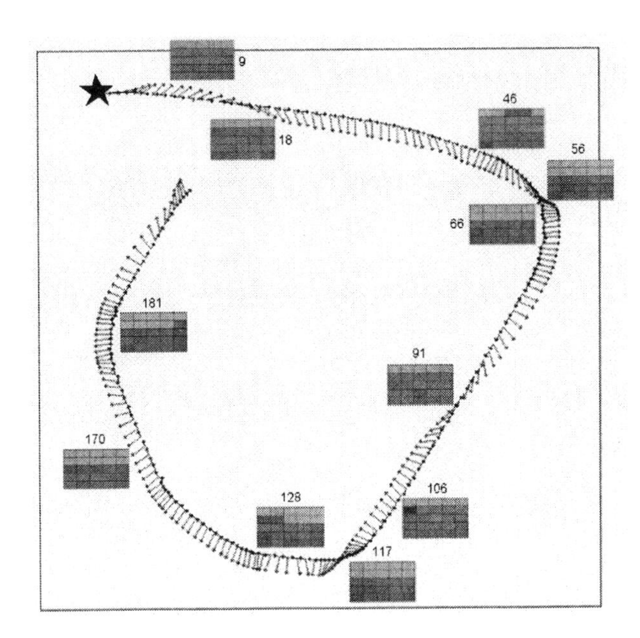

Figure 10: Robot trajectory (short arrows) and horizontal camera displacement (long arrows) of the best individual of generation 12 tested for one trial. Greyscale matrices represent the activations of the visual neurons (black = 0, white = 1) plotted before, during and after wall avoidance and camera movement. The numbers indicate the sensory-motor cycle corresponding to the visual plot and are should be compared with the graphs of figure 11.

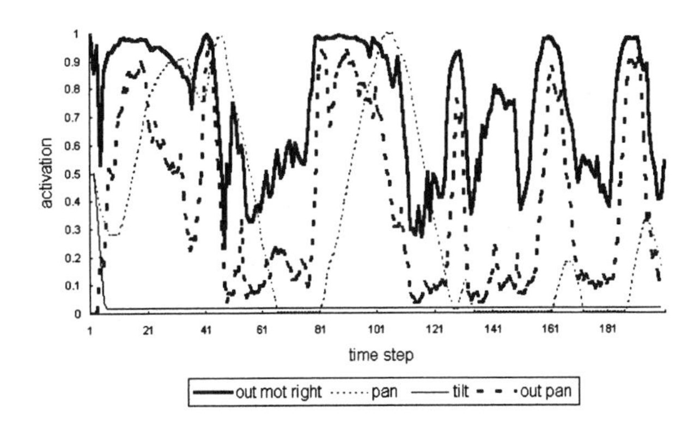

Figure 11: Pan and tilt angles scaled in the range [0, 1] (0.5 means 0 degrees), and output activations of the right wheel motor and of pan motor across one trial. Tilt angle is always negative ($-25°$), meaning that the camera points downward. The values of the output units encoding the right wheel motor and the pan motor are correlated with the retinal input provided by the camera (see figure 10).

robot shifts the camera to the front and then to right with slow scanning movements.

The alternative behavioral strategy (which becomes dominant after generation 12) is similar to that described above, but the camera is not actively used throughout the whole trial. At the beginning of the trial the robot points the camera downwards and to its left, and it keeps it there for the duration of the whole trial. The movement of the body is then sufficient to maintain the edge between the floor and the walls in sight and slow down the right wheel when it gets closer to a wall (signalled by the visual expansion of the white area on the top portion of the retinal image).

6. Discussion

The mobile camera allowed evolved robots to select two powerful, and yet computationally simple, visual features: *edge detection* and *visual looming*.

During the first couple of generations, robots were sensitive to a large variety of environmental features. For example, some of them watched the window while others were more "interested" in tall features (curious people watching the robot). However, selective attention to those features disappeared from the population because they were not sufficiently constant (window brightness changes throughout the day, people come and go). Instead, the edge between the dark floor and white arena walls remained constant across generations and was therefore quickly selected becoming a dominant feature for all individuals of the population.

Wall avoidance was performed by correlating the expansion of the white area on the retinal projection with the rotational speed of the right wheel, which moved the robot away from the walls. This simple visual feature is known as "visual looming", has been shown to be an innate feature of the avoidance behavior in rhesus monkeys (Caviness et al., 1962), and has been recently used as a computational strategy to measure distances in mobile robots (Sahin and Gaudiano, 1998).

Both edge detection and visual looming are instances of linearly-separable mappings between input and output space and therefore do not require multiple nonlinear transformations. These results add to the growing literature showing that behavioral systems can exploit sensory-motor self-organization to actively select simple sensory stimuli in order to remain operational (Nolfi and Floreano, 2000).

Finally, it is interesting to notice that whereas evolved active retinas for shape discrimination selected *sampling* visual filtering only in 61% of the cases (Kato and Floreano, 2001), all evolved robots always used it in these experiments. The reason is that in the former experiments the contrast between black shapes and white background was much higher than in the images captured by the robot camera. In this case, the sampling strategy provides a sharper contrast between the brown floor and the walls.

Evolved robots displayed two behavioral strategies

across generations, one where the camera is actively used throughout the whole navigation and one where the camera is maintained at a fixed position. Although both strategies exploit the mobile camera to select the area of the visual field, use the same visual features, and display similar fitness values, the latter strategy does not require sensory-motor coordination between the movement of the robot and that of the camera. There may be two reasons why it becomes the dominant strategy after 12 generations. The first reason is that the environment does not change and therefore evolved individuals can select a simpler navigation strategy that fits exactly the environment where they have been evolved. The second reason is that the set of possible connection strengths that support suitable coordination between the movements of the robot and of the camera may be much smaller than that of connection strengths that control only the movement of the robot. Therefore, small random mutations are more likely to shift selected individuals towards the simpler strategy.

The first strategy, where the camera is moved throughout the whole trial, is initially selected and maintained as long as the trajectory of the robot is not tuned to the geometry of the environment. Since during early generations some individuals turn more sharply and may end up encountering a wall on their right side, it pays off to visually check both sides of the body. This strategy appeared to be quite efficient for more complex environments, as we observed in a preliminary test of the robot with 8 cm high wood bricks scattered on the floor, although we didn't make a comparison respect of the strategy without camera movement, but as we discussed above the evolutionary environment did not impose sufficient selection pressure to maintain it. Interestingly, this behavioral strategy, whereby the robot points the camera towards the direction where it is going (and not simply in front), is similar to that used by humans when steering a car at a turn. In those cases, we do not watch straight ahead, but always in the direction the car will take us (Lee and Lishman, 1977).

7. Conclusion

We have described an evolutionary active vision system capable of dynamically selecting relevant visual features for generation of suitable behaviors. Using reverse engineering of evolved mechanisms, the strategies used by the robot described here could be translated into a very simple and efficient algorithm for robot indoor navigation.

The results indicate that sensory-motor coordination in a computationally simple system can turn difficult recognition problems into simpler ones. Notice that this holds also for the evolved strategy where the robot moves the camera only at the beginning of the trial. Indeed, if the camera is not mobile, it is up to the experimenter to design the environment or set the visual field so that the robot can perceive useful information.

We believe that the approach described in this paper may be quite powerful for more complex and rough terrains where the robot must actively search for and maintain selected features in sight as it moves. Preliminary tests on rough terrain with evolved individuals displaying camera/body coordination showed that robots successfully maintained the floor/wall edge in sight while moving up and down wood bricks. Rough terrains may also put stronger selection pressure on the development and maintainance of camera/body coordination.

Acknowledgements

The authors thank Claudio Mattiussi, Jesper Blynel, and Shloke Hajela for help with technical setup of the experiments. Thanks also to two anonymous referees for constructive questions and some bibliographic suggestion. This work was partially supported by the Swiss National Science Foundation, grant nr. 620-58049.

Appendix

The robot and the camera was controlled by an on-board PC-104 computer with 64 Mb of RAM and a CPU Pentium at 166 MHz. The operating system was Linux Mandrake 7.0. Low-level routines, such as PID motor control and sensor reading was performed by a Motorola 68332 microprocessor interfaced to the PC-104 by means of a MMA cable that allowed power supply and data exchange between the two processors. The command protocol used for motor control and sensor reading was provided by K-Team SA and is compatible with that used on the Khepera robot.

The camera was interfaced to the PC-104 board by means of an RS-232C serial communication cable using $VISCA^{TM}$ protocol ($VISCA^{TM}$ is an acronym of Video System Control Architecture). It is a network protocol designed to interface a wide variety of video equipment to computers. Under $VISCA^{TM}$, up to 7 EVI-D31 cameras can be connected to one controller using RS-232C communication. Protocol management was based on an informal technical report written by Thomas B. Moeslund (`http://www.vision.auc.dk/~tbm/Sony/`) because the documentation provided by the distributor of the camera was very poor. A framegrabber module installed on the PC-104 managed image acquisition and a modified version of Videodog software (`http://planeta.terra.com.br/informatica/gleicon/video4linux/videodog.html`) was used to format the data.

The entire algorithm and data storage was performed by the PC-104 board. At the end of an evolutionary run, the PC-104 was connected via ethernet cable to a desktop computer running Linux RedHat 6.0 in order to download the data for analysis.

Software code and video clips are available at `http://dmtwww.epfl.ch/isr/east/`.

References

Bajcsy, R. (1988). Active Perception. *Proceedings of the IEEE*, 76:996–1005.

Ballard, D. (1991). Animate vision. *Artificial Intelligence*, 48:57–86.

Caviness, J. A., Schiff, W., and Gibson, J. J. (1962). Persistent fear responses in rhesus monkeys to the optical stimulation of "looming". *Science*, 136:982–983.

Clark, A. and Thornton, C. (1997). Trading spaces: Computation, representation, and the limits of uniformed learning. *Behavioral and Brain Sciences*, 20:57–90.

Cliff, D. T. and Noble, J. (1997). Knowledge-based vision and simple vision machines. *Philosophical Transactions of the Royal Society of London: Series B*, 352:1165–1175.

Dill, M., Wolf, R., and Heisenberg, M. (1993). Visual pattern recognition in drosophila involves retinotopic matching. *Nature*, 355:751–753.

Elman, J. L. (1990). Finding Structure in Time. *Cognitive Science*, 14:179–211.

Gibson, J. J. (1979). *The Ecological Approach to Visual Perception.* Houghton Mifflin, Boston.

Goldberg, D. E. (1989). *Genetic algorithms in search, optimization and machine learning.* Addison-Wesley, Redwood City, CA.

Harvey, I., Husbands, P., and Cliff, D. (1994). Seeing the light: Artificial evolution, real vision. In Cliff, D., Husbands, P., Meyer, J., and Wilson, S. W., (Eds.), *From Animals to Animats III: Proceedings of the Third International Conference on Simulation of Adaptive Behavior*, pages 392–401. MIT Press-Bradford Books, Cambridge, MA.

Hinton, G. E. and Sejnowski, T. J., (Eds.) (1999). *Unsupervised Learning.* MIT Press, Cambridge, MA.

Kato, T. and Floreano, D. (2001). An evolutionary active-vision system. In *Proceedings of the Congress on Evolutionary Computation (CEC01)*, Piscataway. IEEE Press.

Krupinski, E. A. and Nishikawa, R. M. (1997). Comparison of eye position versus computer identified microcalcification clusters on mammograms. *Medical Physics*, 24:17–23.

Lee, N. D. and Lishman, R. (1977). Visual control of locomotion. *Scandinavian Journal of Psychology*, 18:224–230.

Mallot, H. A. (2000). *Computational Vision.* MIT Press, Cambridge, MA.

Nolfi, S. (1998). Evolutionary robotics: Exploiting the full power of self-organization. *Connection Science*, 10:167–183.

Nolfi, S. and Floreano, D. (2000). *Evolutionary Robotics: Biology, Intelligence, and Technology of Self-Organizing Machines.* MIT Press, Cambridge, MA.

Rimey, R. D. and Brown, C. M. (1994). Control of selective perception using bayes nets and decision theory. *International Journal of Computer Vision*, 12:2/3:173–207.

Sahin, E. and Gaudiano, P. (1998). Visual Looming as a range sensor for mobile robots. In Pfeifer, R., Blumberg, B., Meyer, J., and Wilson, S., (Eds.), *From Animals to Animats V: Proceedings of the Fifth International Conference on Simulation of Adaptive Behavior.* MIT Press-Bradford Books, Cambridge, MA.

Scheier, C., Pfeifer, R., and Kunyoshi, Y. (1998). Embedded neural networks: Exploiting constraints. *Neural Networks*, 11:1551–1596.

Genetic Programming for Robot Vision

Martin C. Martin
Artificial Intelligence Laboratory
Massachusetts Institute of Technology
mm@cmu.edu

Abstract

Genetic Programming was used to create the vision subsystem of a reactive obstacle avoidance system for an autonomous mobile robot. The representation of algorithms was specifically chosen to capture the spirit of existing, hand written vision algorithms. Traditional computer vision operators such as Sobel gradient magnitude, median filters and the Moravec interest operator were combined arbitrarily. Images from an office hallway were used as training data. The evolved programs took a black and white camera image as input and estimated the location of the lowest non-ground pixel in a given column. The computed estimates were then given to a hand-written obstacle avoidance algorithm and used to control the robot in real time. Evolved programs successfully navigated in unstructured hallways, performing on par with hand-crafted systems.

1. Introduction

Computer vision in unstructured environments, such as a typical office environment, is notoriously difficult. Different algorithms have their strengths and weaknesses, and no one algorithm is universally better or worse than the alternatives. In this work, Genetic Programming was used with a representation close to existing, hand-written algorithms to create an algorithm that worked empirically for a particular environment.

As in any field of research, one can find threads in the literature by following the evolution of a single idea. The idea that most inspired this work started with Ian Horswill's Ph.D. thesis on Polly the Robot [5]. Polly gave simple tours of the seventh floor of the MIT AI lab, which had a textureless carpeted floor. Obstacles, or at least their boundaries, could therefore be detected as areas of visual texture. The system had problems with other carpet patterns, or even sharp shadows. Liana Lorigo extended this work by assuming the bottom of each image represented floor, and searched for areas with different colors or texture than the floor [12]. However, an object near the robot could confuse it. Iwan Ulrich and Illah Nourbakhsh extended the work by taking the floor to be part of a previous image that had since been traversed [30].

In this work, Genetic Programming can be seen as automating this process. A traditional supervised learning framework was used. Images were collected from an office hallway, and the lowest non-ground pixel in six columns of each image was determined by hand. Programs were then evolved that, given an image and the location of a column, estimated the lowest non-ground pixel. An obstacle avoidance algorithm was then constructed by hand that used these estimates to guide the robot.

In [15] I presented an early implementation of this framework along with initial results. That work required that a hand written, poorly performing "seed" individual be inserted in the initial population. This paper describes several refinements that, among other things, obviate the need for the seed individual.

2. Previous Work

2.1 Evolutionary Robotics

Evolutionary Robotics is an emerging field that uses simulated evolution to produce control programs for robots. Recent work can be found in [7] and [23]. Most work uses bitstring Genetic Algorithms to evolve neural nets for obstacle avoidance and wall following using sonar, proximity or light sensors, e.g. [8, 20, 19]. Significantly, gradient based learning techniques consider recurrent neural networks much harder to train than feed forward networks, since gradient information typically isn't available. Evolutionary Computation doesn't use gradient information, and therefore even exploratory, toy problems can use recurrence.

Genetic Programming has been used occasionally. Lee et al. [11] use GP to evolve behavior primitives and arbitrators for a behavior based mobile robot. Nordin et al. [24] use a Genetic Programming variant that directly manipulates SPARC machine language. They use symbolic regression to predict the "goodness" of a state 300 ms in the future, based on the current sensor readings and action. For obstacle avoidance, the goodness is simply the sum of the proximity sensors, plus a term to reward moving quickly in a straight line. To choose a direction, the robot runs the best individual for all possible actions with the current sensor readings. The action with the highest estimated goodness is chosen. They use a population size of 10,000 and find that, in runs where perfect behaviour developed, it developed by generation 50.

Most work evolves in simulation, with the best individuals then run on a robot in the real world. Reynolds [27] has pointed out that without adding noise to a simulation, EC will find brittle solutions that wouldn't work on a real robot. Jakobi et al. [8] discovered that if there is significantly more noise in the simulation than on the real robot, new random strategies become feasible that also don't work in practice.

As well, to my knowledge, no one has tried to simulate CCD camera images, either using standard computer graphics techniques or morphing previously captured images. The Sussex gantry robot [2] uses a CCD camera, but the images are reduced to the average brightness over three circles. These are significantly easier to simulate than a full CCD image, especially when the only objects are pure white on a black background. Smith [28] simulated a 16 pixel one-dimensional camera with auto iris on a robot soccer field. The 16 pixels were actually derived from 64 pixels; Smith doesn't say how. This is an important step, but again much easier than simulating a CCD image at, say, 160 x 120 pixels or above.

A few research groups perform all fitness evaluations on the real robot. Floreano and Mondada [4] evolve recurrent neural networks for obstacle avoidance and navigation from infrared proximity sensors. It takes them 39 minutes per generation of 80 individuals, and after about 50 generations the best individuals are near optimal, move extremely smoothly, and never bump into walls or corners. Naito et al. [22] evolve the configuration of eight logic elements, downloading each to the robot and testing it in the real world. Finally, the Sussex gantry robot [2] mentioned earlier has used evaluation on the real robot. They used a population size of 30, and found good solutions after 10 generations.

The closest work to that reported here was done by Baluja [1], who evolves a neural controller that interprets a 15 x 16 pixel image from a camera mounted on a car. The network outputs are interpreted as a steering direction, the goal being to keep it on the road. Training data comes from recording human drivers.

In summary, Evolutionary Robotics has used low bandwidth sensors, such as sonar or proximity sensors, presumably to cut down the amount of information to process. There are typically less than two dozen such sensors on a robot, and each returns at most a few readings a second. However, much traditional work in computer perception and robotics uses video or scanning laser range finders, which typically have tens to hundreds of thousands of pixels, and are processed at rates up to 10 Hz or more. Evolutionary Robotics has much to gain by scaling to these data rich inputs.

In addition, most Evolutionary Robotics has designed algorithms for simplified environments that are relatively easy to simulate. While evaluating evolved programs on real robots is considered essential in the field, those environments are typically still tailored for the robot. The current work attempts to evolve algorithms to interpret video of an unmodified office environment in real time, to help a robot wander while avoiding obstacles.

2.2 Visual Obstacle Avoidance

Somewhat surprisingly, there have only been a handful of complete systems that attempt obstacle avoidance using only vision in environments that weren't created for the robot. Larry Matthies' group has built a number of complete systems, all using stereo vision [17]. They first rectify the images, then compute image pyramids, followed by computing "sum of squared differences", filtering out bad matches using the left-right-line-of-sight consistency check, then low pass filter and blob filter the disparity map.

Their algorithm has been tested on both a prototype Mars rover and a HMMWV. The Mars rover accomplished a 100m autonomous run in sandy terrain interspersed with bushes and mounds of dirt [16]. The HMMWV has also accomplished runs of over 2 km without need for intervention, in natural off-road areas at Fort Hood in Texas [18]. The low pass and blob filtering mean the system can only detect large obstacles; a sapling in winter, for example, might go unseen.

Ratler [10] used a stereo vision algorithm to do autonomous navigation in planetary analog terrain. After rectification, the normalized correlation is used to find the stereo match. The match is rejected if the correlation or the standard deviation is too low, or if the second best correlation is close to the best. Travelling at 50 cm/sec over 6.7km the system had 16 failures, for a mean distance between failures of 417m. No information on failure modes is available.

David Coombs' group at NIST has succeeded with runs of up to 20 minutes without collision in an office environment [3]. Their system uses optical flow from both a narrow and a wide angle camera to calculate time-to-impact, and provide feedback that rotationally stabilizes the cameras. Reasons for failure include the delay between perception and action, textureless surfaces, and hitting objects while turning (even while turning in place).

Liana Lorigo's algorithm [12, 13] assumes the bottom of the image represents clear ground, and searches up the image for the first window that has a different histogram than the bottom. This is done independently for each column. If the ground is mostly flat, then the further up the image an object is, the further away it is. The robot heads to the side (left or right) where the objects are higher up.

Failure modes include objects outside the camera's field of view, especially when turning. Other failure modes are carpets with broad patterns, boundaries between patterns, sharp shadows, and specularities on shiny floors.

Ian Horswill's algorithm [5, 6] is similar to the above. It assumes that the floor is textureless, and labels any area whose texture is below threshold as floor. Then, moving from the bottom of the image up, it finds the first "non-floor" area in each column, turning left or right depending on which side has the most floor.

The system's major failure mode is braking for shafts of sunlight. In addition, it cannot break for objects it has seen previously but doesn't see now. Textureless objects with the same brightness as the floor also cause problems, as does poor illumination.

Ulrich and Nourbakhsh [30] took the floor to be the part of a previous image that had since been traversed.

Illah Nourbakhsh has used depth from focus for robot obstacle avoidance [25]. Three cameras, focused at different distances (near, middle and far), image the same scene. Whichever image is sharpest is the most in focus, so the objects are roughly at that distance. Actually, the images are divided into 40 windows (8 across and 5 down), which are treated independently, giving an 8 by 5 depth map.

In hundreds of hours of tests, the robot has avoided stair cases as well as students, often running down its batteries before a collision. However, failure modes include areas of low texture and tables at the robot's head height.

3. Framework

Previous evolutionary robotics work has incorporated video as a sensor, but to do so has forced the image through a huge bottleneck, to either three or sixteen pixels. With such a low resolution, only problems in carefully constructed worlds were possible.

For the robot to be truly embodied in a real world environment, and to avoid all manner of obstacles using only vision, requires more complex algorithms that respond to more details in the image. Simulating real images to the required fidelity would be a research project in itself, if even possible at the speeds needed to support simulated evolution. However, evolving on the real robot is also not possible, since this limits the number of evaluations, and hence the complexity of what can be evolved. The research groups that have attempted this use population sizes of less than 100 individuals, whereas GP typically uses sizes of 2000 to 10,000.

Record Video	Robot, Real Time
Learn Offline	Genetic Programming
Build Navigation	By Hand
Validate Online	Robot, Real Time

Figure 1: The four phases of constructing the robot control program. First, a set of representative images were collected. Then an offline genetic algorithm created a vision subsystem. Next, a navigation subsystem was written by a human programmer. Finally, the combination of these two was used to control the robot.

In addition, without specifying any *a priori* structure for the evolved programs, the problem becomes many orders of magnitude harder than previous work. The input, essentially the amount of light in various directions, is simply too distantly related to the output, i.e. the direction to travel.

For both these reasons, the control program was divided into two components, the *vision subsystem* and the *navigation subsystem.* The interface between the two was completely specified: the vision algorithm takes in an image and a direction, and returns an estimate to the nearest obstacle in that direction. The navigation algorithm starts with a number of such estimates from the current image, and computes a direction to travel.

While specified, this interface is not arbitrary. Sonar and laser range finders, by far the most popular sensors, return distances, and the number of such estimates per image, six for training and twelve while controlling the robot, are a minimal representation of the environment compared to the dense depth map typically used in stereo based navigation. This minimal representation was inspired by arguments that representation should be used sparingly. The number six was chosen to be roughly equal to the number of sonar sensors in the camera's roughly 90° field of view. During online validation six coulmns was found to be too few, but twelve was sufficient. No other numbers were tried.

The vision subsystem was constructed first, independent of the navigation subsystem, see Figure 1. Then the navigation algorithm was constructed to be robust to the sorts of errors made by the vision algorithm. Lessons learned about which errors are easy to compensate for and which are more difficult could guide the design of the next version of the vision system.

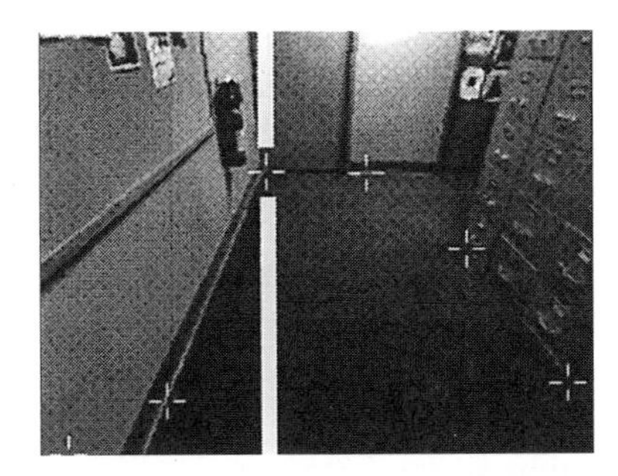

Figure 2: What the vision subsystem computed. For a given column of the image, the evolved vision subsystems computed the boundary between the ground (lower line) and a non-ground object (upper line). If there wass more than one such boundary, the lowest wass desired. This was done independently for six columns in the image.

For the person constructing the system, there is another asymmetry between the two subsystems: the vision subsystem is much harder to create. The output of the vision subsystem is the estimated distance to the first obstacle in various directions, and in this way similar to sonar. The navigation subsystem can therefore adapt designs created for sonar. When it comes to navigation, sonar is a much more common sensor than vision, so navigation from this type of data is well understood. In contrast, while there has been much work on computer vision, there has been little applying it to obstacle avoidance on a mobile robot.

For this reason, the core of this work was the development of the vision subsystem, which was constructed using Genetic Programming [9]. Potential programs were given an image and column location as input, and produced a single real-valued output, which was interpreted as the pixel location of the lowest non-ground pixel within that column. (See Figure 2.) Assuming objects touch the ground and that the ground is roughly flat, the height of the lowest non-ground pixel is a monotonic function of the distance from the robot. No state was maintained from one image to the next or between columns within an image, which means all programs were reactive.

A traditional supervised learning framework was used. To collect a training set, the robot navigated using sonar and passively collected a video stream. In each of six columns of each image, the correct answer, i.e. the location of the lowest non-ground pixel, was determined by the author and recorded.

An evolved program got a column "correct" if it was within 10 pixels of the hand labelled correct answer. Its fitness was simply the number of correct columns on the entire training set. This differed from previous work [15] where the fitness was the absolute value of the difference between what the individual returned and the hand labelled "correct" answer. The change discouraged the evolution from focusing differences of a few pixels, and instead focused it on getting more columns in the ballpark, which was much more important for obstacle avoidance..

Table 1: Functions and Terminals

root	`iterate-up`, `iterate-down`
rectangle sizes	`r22`, `r23`, `r32`, `r33`, `r24`, `r42`, `r44`, `r55`, `r26`, `r62`, `r36`, `r63`, `r66`, `r77`, `r28`, `r82`, `r38`, `r83`, `r88`
arithmetic	`*`, `+`, `%`, `-`, `sqr` and random constants
parameters	`x-obstacle`, the horizontal pixel location in which to find the obstacle; `area`, the area of the window in pixels; `image-max-x` (319): `image-max-y` (239); `first-rect`, one if this is the first rectangle of the iteration, zero otherwise; x and y, the center of the rectangle in pixels.
flow control	`prog2`; `prog3`; `break`, which halts the execution of the branch, returning immediately without any more iterations; `if-le`, the "if less than or equal to" operator.
memory	`set-a` ... `set-e`, `read-a` ... `read-e`
image statistics	average and average-of-squared over the window: raw, truncated median, median corner, Sobel magnitude, and four directional Moravec interest operators.

The most general and straight forward way of incorporating image information into GP is to simply allow it to access individual pixels and give it some looping constructs. However, with the limited computing power of today's computers, it would take a long time to find even a simple algorithm that examines the correct location in the image.

To give the representation a little more structure, successful visual obstacle avoidance algorithms were examined and a common building block was found. In all these existing system the bulk of run time was spent performing a computation over a rectangular window which iterated over a column or row of the image.

Thus, a new type of node was created, an *iterate* node, which moved a rectangular window over the image. In a simplification of previous work [15], it only moved vertically and took three arguments: the size of the rectangle, the column in which to iterate, and finally a piece of code to execute at each location. This last piece of code had access to the results of various image operators over the window.

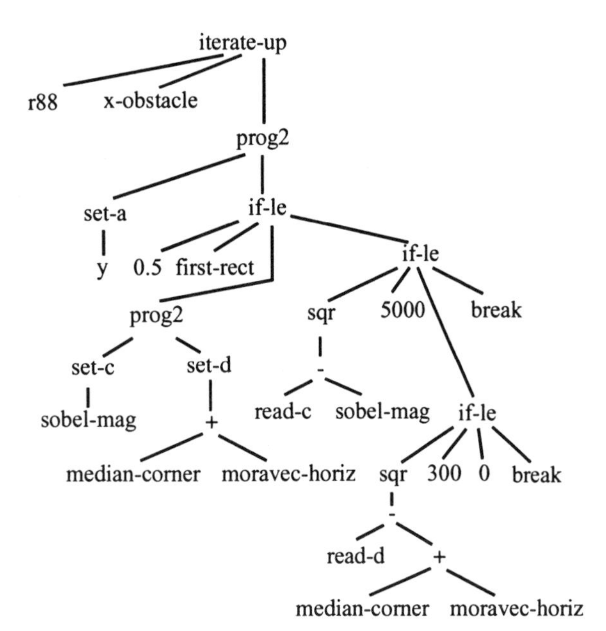

Figure 3: An example window iteration branch, implementing a Lorigo style algorithm [13].

In addition to the *iterate* node and its associated image operator nodes, standard mathematical functions were provided, as well as flow control and five floating point memory registers, with associated read and write nodes, similar to Teller and Veloso's work with PADO [29]. The complete list of functions and terminals in the window iteration branch is shown in Table 1. An example iterated window branch, implementing a Lorigo style algorithm, is shown in Figure 3.

There was only a single iterate node per tree, always appearing at the root. The creation and crossover operators enforced this constraint. Each individual had two such trees, to allow individuals to make two passes, e.g. to first compute an average intensity of pixels known to be floor, then on the second pass to compare each window to that average. The iterate node did not have a return value. Instead, one trees was designated as the *result producing branch*, and the final value of the first memory location was used at the return value for the entire individual. This tree could also use the final values of the memory locations of the other trees.

This paper extends work that was reported in [15]. There, a hand written, poorly performing "seed" individual was added to the initial population. This seed performed a little better than the best randomly generated individuals in that initial population, and evolution modified many aspects of it, improving its performance greatly.

While many aspects of the seed were modified, others were not. In particular, the successful evolved algorithms all iterated vertically, never horizontally, and from the bottom of the image to the top or vice versa, never starting part way up. Therefore, horizontal iteration was eliminated and the iterate node simplified to take only three arguments: the window size, the horizontal location in which to iterate, and the code to execute at every step. The result producing branch, which had simply returned the result of a single iteration branch, was also eliminated.

4. Experiments

4.1 Experimental Setup

All experiments were performed on the Uranus mobile robot, in the Mobile Robot Lab at Carnegie Mellon University. The robot has a three degree of freedom base with dead reckoned positioning. While forward/backward motion and turning in place are fairly accurate (~ 1% error), sideways motion isn't (about 10-20% error, significant rotation). For sensing it uses a b/w analog video camera and a ring of 24 sonar sensors. Processing was done by an off board 700 MHz Pentium III computer running BeOS.

Data was collected from two hallways in different buildings. The most problematic features were glossy, textureless grey walls which often confound local depth estimation techniques such as stereo, optical flow and depth from focus.

To collect images that were representative of what the cameras would see, the robot collected data while avoiding obstacles using sonar. While obstacle avoidance using sonar is considered easier than using vision, it still took many attempts to get a working system. The method that proved most successful determined speed from proximity to the nearest object, and determined direction of travel by fitting lines to points on the left and right sides of the robot. More details can be obtained from [14]. The data is summarized below.

The camera was calibrated using the system described in [21]. The camera was then mounted on the robot, pitched 51 degrees down from horizontal, and images of size 320 by 240 pixels were collected. These images became the training sets. Ground truth was then assigned by hand using a simple GUI. Typical images and ground truth are shown in Figure 4.

NSH Hallway

Total Number of Frames:	328
Frames In Training Set:	65 (every fifth)
Number of Fitness Cases:	$65 \times 6 = 390$
Elapsed Time:	75 sec
Frame Rate Of Training Set:	$65 \div 75 = 0.87$ fps

While the robot had to travel mostly straight down a hallway, it started out a little askew, so it approached one side. At one point, a person walks past the robot and is clearly visible for many frames. At the end of the hallway

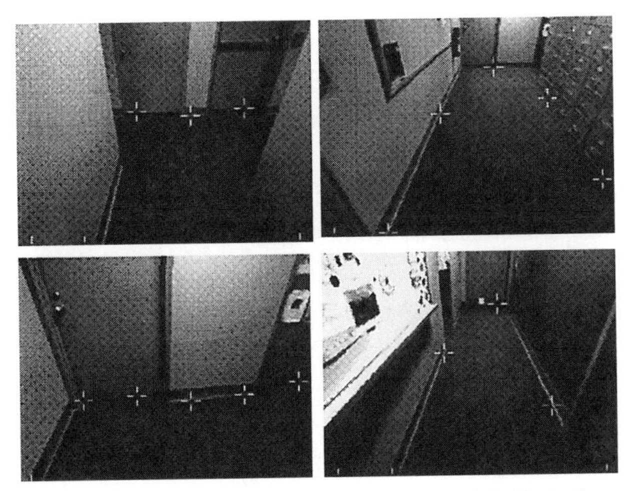

Figure 4: Training data, with ground truth indicated.

it turns right. The carpet is grey with a large black stripe.at one point. The shadow of the robot is visible at the bottom of most frames.

FRC Hallway

Total Number of Frames:	356
Frames In Training Set:	71 (every fifth)
Number of Fitness Cases:	71 × 6 = 426
Elapsed Time:	82 sec
Frame Rate Of Training Set:	71 ÷ 82 = 0.87 fps

Starts with robot in lab doorway. Moves straight until its in the hallway, then turns right, travels down hallway, at end turns right, then travels straight to dead end. All doors were closed. The fluorescent light bulbs at the start are burned out, so the intensity of the carpet varies widely. The shadow of the robot is visible at the bottom of most frames.

Combined

Total Number of Frames:	328 + 356 = 684
Frames In Training Set:	68 (every tenth)
Number of Fitness Cases:	68 × 6 = 408
Elapsed Time:	75 + 82 = 157 sec
Frame Rate Of Training Set:	68 ÷ 157 = 0.43 fps

This data set was simply the combination of the above two data sets, using every tenth frame instead of every fifth in order to keep the training set size approximately equal.

During offline learning, the images were first rectified to conform to an ideal perspective projection, and cropped to a horizontal field of view of 83 degrees using the above mentioned calibration information. The results of the operators were precomputed at every pixel, then the genetic programming run was started.

Genetic programming was performed on a dual 700 MHz Pentium III, and evaluation times varied widely, but averaged approximately 725 msec. The time for a simulated evolution run also varied widely, averaging approximately 20 hours. Ten runs were completed in each of the three experiments. A Koza style tableau is shown in Table 2.

Table 2: Koza Style Tableau

Objective	Given an image and a horizontal position within it, return the first non-ground pixel in that column.
Architecture of individuals	Two *window iteration* branches.
Function and Terminal Sets	See `Table 1`
Fitness cases	Six columns in each of 65-71 images. 390-426 total.
Raw fitness	The percentage of fitness cases "correct." A fitness case was "correct" if the pixel location estimated by the evolved program was within 10 pixels of the author's determination of the "correct" answer.
Standardized fitness	100% minus raw fitness.
Parameters	51 generations, population size of 10,000, tournament selection with tournament size of 7, ramped-half-and-half with min size 6 and max size 9.

Each evolved program took as input the image and the horizontal position of a column. It returned a single number, the vertical position, in pixels, of the first (i.e. lowest) non-ground pixel in that column of that image. It was run on six different columns per image. As mentioned earlier, a program's performance on a given column of a given image was deemed "correct" if its answer was within 10 pixels of the human-provided answer, and the total fitness was simply the number of correct columns in the training set. In contrast to earlier work, no seed individual was needed.

After the offline learning, a hand written navigation algorithm used the estimates to decide speed and direction to travel. This algorithm was very similar to the one used during data collection, except that its inputs came from vision, not sonar. The three different vision subsystems, one from each experiment, all used the same navigation subsystem. The best evolved algorithm was run on twelve columns in the image, twice as many as used in training.

The navigation algorithm classified estimates as either near (requiring an immediate halt), medium (slow to 2/3 speed to avoid collision) or far (avoid them before they

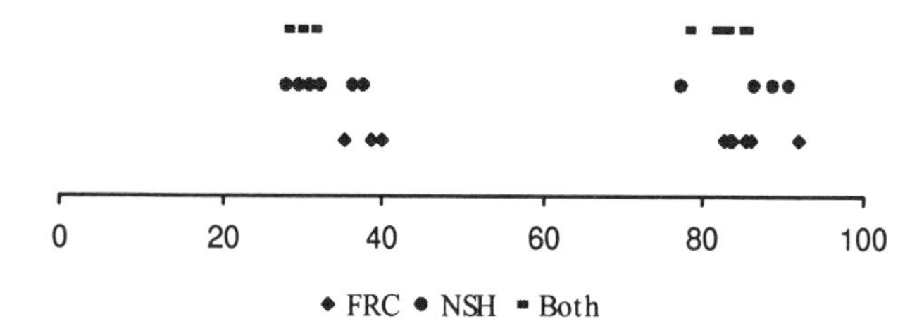

Figure 5: A scatter plot showing the fitness, i.e. the perentage of fitness cases correctly estimated, of the best-of-un individuals in all three experiments.

become a problem.) This case based approach is inspired by the Property Mapping approach of Nourbakhsh [26]. If any of the middle four estimates are in the lower fifth of the image, or either of the two readings outside are at the bottom, then the object is considered near and the robot immediately halts. Otherwise, it looks for objects within four feet to the left and the right of where the robot is and where it would be if it continued straight. To convert pixel height in the image to real world distances, it assumes that the floor is flat, and that the non-ground object touches the floor. If an object is sighted, on either the left or the right, a line is drawn through the readings on each side, and the robot turns to run parallel to the lines.

If there are no objects near or in medium-sides, the algorithm looks for objects straight ahead within the "far" boundary. Objects there cause it to respond by turning left or right, towards the largest gap. If all areas are clear other than far-sides, a line is fit to the side with the closest readings, and if the line is converging with the robots center line, the robot turns to move parallel. Finally, if there are no objects anywhere within the robot's field of view, it simply moves straight.

This algorithm was created by hand using traditional iterative design, and is still far from optimal. It is a natural application for simulated evolution, which is likely to do significantly better.

4.2 Training Results

This section is necessarily brief. For more details, the reader should consult [14]. A scatter plot of the final individuals in all conditions is show in Figure 5. All three experiments produced individuals which achieved a fitness of greater than 85%. They did this despite burned out lights and other effects that caused the carpet's average intensity to vary from zero to at least 140s out of 255; despite large gradients caused by imaging artifacts; despite moiré patterns of image noise; and despite the shadow of the robot.

In the Field Robotics Center, the best constant approximation, ignoring both the image and the desired column, was to return `image-max-y - 10`, i.e. 228, to get 101 answers correct for a fitness of 23.7%. In all but two of the runs, the best individual in the initial population did worse than that, returning `image-max-y - 3` and achieving 21.6%. The other two runs achieved 29.3% and 36.4% in the initial population. The other two conditions were similar, occasionally but rarely performing better than the best constant approximation.

Figure 6: Performance on the training data of the best individual from the FRC hallway experiment.

As can be seen in Figure 5, the results of all three experiments were bimodal. Interestingly, if an run was going to end up in the higher scoring mode, the fitness of the best-of-generation individual was already above 50% by generation 11.

The best individual from all FRC runs achieved 91.78%. Some examples from the training set are shown in Figure 6. While hard to see, the rectangles in the figure are the same size as used by the individual. In the two upper images, all six fitness cases were determined correctly. In the lower left, one of the two on the filing cabinets was higher than desired, although a difference this small would not affect obstacle avoidance significantly. In the lower right, the rightmost column should be at the bottom of the image instead of near the top.

The other two conditions did almost as well, with the best individual from all NSH runs achieving 90.51%, and the best from the combined data set 84.8%. In all three experiments the errors were transient with the exception of the stripe of red carpet in Newell Simon Hall, which was uniformly considered an obstacle, at least when near the camera.

The best individual from the FRC experiment was simplified using a combination of programming tools and by hand. The resulting program is quite readable, see

Result Producing Branch:

Iterate-Down, 3x3 window, centered on desired column:

```
if y ≤ 9 then
  a := y;

if second-branch-b > 9 then
  a := y;

if median > 35.4444 then
{
  if gradient > 413.96 then
    a := y;

  if gradient > 239 and (gradient / 239)^4 > diag-grad then
    a := y;
}
```

Second Branch:

b (initial value) := 3.40282e+38

Iterate-Up, 3x3 window, centered on desired column:

$$b = \frac{b(1+h)-(1+h+h^2+h^4)\text{desired-x}}{h^5}$$

Where h is horizontal gradient.

Figure 7: A simplified version of the best individal from the FRC experiment.

Figure 7. The two branches represent two very different styles of algorithm. The result producing branch was essentially a decision tree, although one of the decisions is a non-linear boundary in the space spanned by two image operators. The other branch, which detects obstacles at the bottom of the image, is a recurrent mathematical function involving both a single image operator and the location of the column.

This division into two cases—gradient based boundaries and objects that touch the floor—was discovered automatically. Nothing in the representation suggested the two different cases, nor that the two branches should each tackle a separate case. Similarly, ignoring the gradient when the image intensity was low, which ignored artificial gradients caused by a quirk in the imaging process, was not suggested by the representation either. These are examples of the genetic algorithm simultaneously exploiting regularities in both the problem domain and the representation.

The other best-of-experiment individuals were similarly simplified to discover how they worked. Evolution had exploited a number of techniques, including a sequence of if-then conditions similar to a decision tree but involving non-linear combinations of up to five different image terminals. In all cases, the bottom of the image was handled using different code than the rest of the image. This reflects a natural dichotomy in the images: at the bottom of the image a program must detect the presence or absence of an object, but in the middle of an image it must detect the *transition* between ground and non-ground. During a transition there is likely to be a significant gradient, whereas at the bottom of the image there isn't. The mechanisms for this varied; in the FRC experiment it used two different branches for the two conditions, whereas in the NSH and combined experiments a multitude of *if* statements were used to run different code at different locations. Interestingly, the raw image was never used, although the median filtered image was. All directional gradients of the image intensity were used except the vertical.

4.3 Validation on the robot

All three individuals generalized surprisingly well when run on new video from the same camera in the same hallway. They were relatively insensitive to the pitch and the horizontal location of the column in the image. With the camera set to automatic gain they also provided acceptable results over a wide range of iris settings. They detected objects that weren't present during training, such as chairs or people, with only a little less fidelity than they detected walls. They were also fast, running at about 10 Hz on a 700 MHz Pentium III.

The three algorithms were more than good enough for navigation. They produced occasional glitches, most often declaring that a pixel at the bottom of the image was non-ground when it was, in fact, ground. To stop these from causing too many panic halts, the hand written navigation algorithm filtered readings by taking the minimum (highest pixel location) of consecutive estimates.

The robot was then run in the same environment(s) it was trained in. Videos of this *online validation* can be found at www.metahuman.org/martin/Dissertation. These routes retraced the path of the training set and then continued much further. They included views of the same hallway from the opposite direction, as well as similar areas never seen during training. Objects were present that were not present during training, such as chairs, trash cans and people.

In general, the navigation system worked rather well. The evolved algorithms worked well despite months of wear & tear on the carpet. Most errors were transient, lasting 1 or 2 frames, even when the camera was stationary. It worked on a range of iris settings and camera tilts. It did not overfit to column location, and twice as many columns were used during navigation as were used for data collection.

When navigating under sonar, fourteen sonar sensors were used for a total field of view of approximately 215 degrees, seeing well to the sides. The area of awareness with vision was much smaller than with sonar, and entirely in front of the robot's base. In the reactive framework described here this makes it almost impossible to successfully navigate doorways, especially since the robot is only a few inches narrower than them. However, it performed very well at corridor following and avoiding obstacles such as people and chairs.

There were a few persistent errors. It was sensitive to small strips of paper or shiny pieces of metal. The red carpet, which was handled poorly even on the training data, was classified as an obstacle when nearby, causing the navigation to consistently treat it as a wall. Other errors would fool the navigation system only occasionally, although these were responsible for most collisions.

In 5 validation runs, the mean distance and time to collision for the FRC individual was 133 meters, 20:01 minutes, and the longest was 260 meters, 39:32 minutes. These are comparable to the published performance of the hand-written systems described in the Previous Work section.

Not surprisingly, the vision subsystem from the combined data set performed a little worse in each environment than the subsystem developed from only that data set. It did not have a new class of error, but instead was more likely to make one of the errors made by the other vision subsystems. In the FRC hallway, its average distance and time to collision were 72 meters and 10 minutes respectively.

5. Bibliography

[1] S. Baluja, Evolution of an Artificial Neural Network Based Autonomous Land Vehicle Controller. IEEE Transactions on Systems, Man and Cybernetics Part B: Cybernetics. 26, 3, 450-463. (1996)

[2] D. Cliff, P. Husbands and I. Harvey. Evolving Visually Guided Robots. In Proceedings of SAB92, the Second International Conference on the Simulation of Adaptive Behaviour. MIT Press, 1993.

[3] D. Coombs, M. Herman, T. Hong and M. Nashman Real-time Obstacle Avoidance using Central Flow Divergence and Peripheral Flow. Fifth International Conference on Computer Vision June 1995, pp. 276-83.

[4] D. Floreano and F. Mondada. Automatic Creation of an Autonomous Agent: Genetic Evolution of a Neural-Network Driven Robot. Proceedings of the Third International Conference on Simulation of Adaptive Behavior: From Animals to Animats 3. 1994.

[5] I. Horswill, Specialization of Perceptual Processes. Ph.D. Thesis, Massachusetts Institute of Technology, May 1993.

[6] I. Horswill, Polly: A Vision-Based Artificial Agent." Proceedings of the Eleventh National Conference on Artificial Intelligence (AAAI-93), July 11-15, 1993.

[7] P. Husbands and J.-A. Meyer (Eds.), Evolutionary Robotics, Proceedings, First Euopean Workshop, EvoRobot98, Paris, France, April 1998.

[8] N. Jakobi, P. Husbands and I. Harvey. Noise and the reality gap: The use of simulation in evolutionary robotics. In Advances in Artificial Life: Proceedings of the Third European Conference on Artificial Life, 1995.

[9] J. Koza, Genetic Programming, MIT Press, Cambridge, MA. 1992.

[10] E. Krotkov, M. Hebert & R. Simmons, Stereo perception and dead reckoning for a prototype lunar rover. Autonomous Robots 2(4) Dec 1995, pp. 313-331

[11] W.P. Lee, J. Hallam and H.H. Lund Applying Genetic Programming to Evolve Behavior Primitives and Arbitrators for Mobile Robots. In Proceedings of IEEE 4th International Conference on Evolutionary Computation, IEEE Press, 1997.

[12] L.M. Lorigo Visually-guided obstacle avoidance in unstructured environments. MIT AI Laboratory Masters Thesis. February 1996.

[13] L.M. Lorigo, R.A. Brooks, and W.E.L. Grimson Visually-guided obstacle avoidance in unstructured environments. IEEE Conference on Intelligent Robots and Systems September 1997.

[14] M.C. Martin, The Simulated Evolution of Robot Perception, Ph.D. Dissertation and Carnegie Mellon University Technical Report CMU-RI-TR-01-32. 2001, 159 pp.

[15] M. C. Martin Visual Obstacle Avoidance using Genetic Programming: First Results, Proceedings of the Genetic and Evolutionary Computation Conference, July 7-11, 2001, pp. 1107-1113

[16] L.H. Matthies, Stereo vision for planetary rovers: stochastic modeling to near real-time implementation. International Journal of Computer Vision, 8(1): 71-91, July 1992.

[17] L.H. Matthies, A. Kelly, & T. Litwin Obstacle Detection for Unmanned Ground Vehicles: A Progress Report. 1995.

[18] L.H. Matthies, Personal communication.

[19] L.A. Meeden, An Incremental Approach to Developing Intelligent Neural Network Controllers for Robots. IEEE Transactions of Systems, Man and Cybernetics Part B: Cybernetics. 26, 3, 474-485. (1996)

[20] O. Miglino, H. H. Lund and S. Nolfi, Evolving Mobile Robots in Simulated and Real Environments. Artificial Life , 2, 417-434 (1995).

[21] H.P. Moravec, DARPA MARS program research progress, http://www.frc.ri.cmu.edu/~hpm/project.archive/robot.papers/2000/ARPA.MARS.reports.00/Report.0001.html, Januray 2000.

[22] T. Naito, R. Odagiri, Y. Matsunaga, M. Tanifuji and K. Murase, Genetic Evolution of a Logic Circuit Which Controls an Autonomous Mobile Robot. Evolvable Systems: From Biology to Hardware. 1997.

[23] S. Nolfi and D. Floreano, Evolutionary Robotics. MIT Press / Bradford Books. 2000.

[24] P. Nordin, W. Banzhaf and M. Brameier, Evolution of a World Model for a Miniature Robot using Genetic Programming. Robotics and Autonomous Systems, 25, pp. 105-116. 1998.

[25] I. Nourbakhsh, A Sighted Robot: Can we ever build a robot that really doesn't hit (or fall into) obstacles? The Robotics Practitioner, Spring 1996, pp. 11-14.

[26] I. Nourbakhsh, Property Mapping: A simple technique for mobile robot programming. In Proceedings of AAAI 2000.

[27] C. W. Reynolds, Evolution of Obstacle Avoidance Behavior: Using Nosie to Promote Robust Solutions. In Advances in Genetic Programming, MIT Press, pp. 221-242. 1994.

[28] T. M. C. Smith, Blurred Vision: Simulation-Reality Transfer of a Visually Guided Robot. In Evolutionary Robotics, Proceedings of the First Euopean Workshop, EvoRobot98, Paris, France, April 1998.

[29] A. Teller and M. Veloso, PADO: A new learning architecture for object recognition. In Symbolic Visual Learning, Oxford Press, pp. 77-112. 1997.

[30] I. Ulrich and I. Nourbakhsh, Appearance-Based Obstacle Detection with Monocular Color Vision. In Proceedings of AAAI 2000.

Active Perception: A Sensorimotor Account of Object Categorization

Stefano Nolfi[1]
[1]Institute of Cognitive Science and Technologies CNR,
Viale Marx, 15, 00137, Rome, Italy
nolfi@ip.rm.cnr.it

Davide Marocco[2-1]
[2] University of Calabria, "Centro Interdipartimentale della Comunicazione"
87036 Arcavacata di Rende, Cosenza, Italy
davidem@ip.rm.cnr.it

Abstract

We describe the results of a set of experiments in which we evolved the control system of artificial agents that are asked to categorize objects with different shapes on the basis of tactile information. Agents are provided with a 3-segments arm with 6 degrees of freedom and extremely coarse touch sensors. As we will see, despite such a limited sensory systems, evolved individuals are perfectly able to solve the problem. The analysis of the obtained results shows that evolved individuals always develop a well defined behavioral strategy that allows them to easily and robustly discriminate different objects despite the limitation of their sensory apparatus. Moreover, we discuss the general advantage of the evolutionary method for the synthesis of effective artificial agents.

1. Introduction

The behavior of embodied and situated organisms is an emergent result of the dynamical interaction between the nervous system, the body, and the external environment Ashby, 1952; Beer, 1995). This simple consideration has several important consequences that are far from being fully understood. One important aspect, for instance, is the fact that motor actions partially determine the sensory pattern that organisms receive from the environment. By coordinating sensory and motor processes organisms can select favorable sensory patterns and thus enhance their ability to achieve their adaptive goals.

Examples of processes falling within this category have been identified in natural organisms. Dill *et al.* (1993) demonstrated that since the fruit fly drosophila cannot always recognize a pattern appearing at different locations in the retina, the insect solves this problem of shift invariance by moving so to bring the pattern to the same retinal location where it has been presented during the storage process. Franceschini (1997) demonstrated that flies use motion to visually identify the depth of perceived obstacles. Moreover, there is evidence that environmental feedback obtained through motor actions plays a crucial role in normal development (Chiel and Beer, 1997; Thelen and Smith, 1994).

With few notable exceptions (eg, Braitenberg, 1984; Franceschini, 1997; Scheier and Pfeifer, 1995), the possibility of exploiting sensorimotor coordination in the design of artificial systems has largely been left unexplored. This can be explained by considering that, as we said above, behavior is the emergent result of the interactions between the individual and the environment. Given that in dynamical systems there is a complex and indirect relation between the rules that determine the interactions and the emergent result of those interactions, it is very difficult to identify how the interactions between the organism and the external environments contribute to the resulting behavior. As a consequence, designing systems that exploit sensorimotor coordination is rather difficult (for an attempt to identify new design principles that might help to achieve this goal, see Pfeifer and Scheier [1999]).

From this point of view evolutionary experiments (Nolfi and Floreano, 2000) where artificial organisms autonomously develop their skills in close interaction with the environment represent an ideal framework for studying sensorimotor coordination (Nolfi and Floreano, 2002). Indeed, in most of the experiments conducted with artificial evolution one can observe the emergence of behavior exploiting active perception. The analysis of evolved robots and the identification of how they exploit the interaction with the environment is often very difficult and requires significant effort, but is generally much simpler than the analysis of natural organisms because the former are much more simple and can be manipulated much more freely than the latter. In addition, such analysis may allow the identification of new explanatory hypotheses that may produce new models of natural behavior that, later on, might be tested experimentally on the real organisms.

In the next section we describe the results of a set of experiments in which we evolved the control system of artificial agents that are asked to categorize objects with different shapes on the basis of tactile information. Agents are provided with a 3-segment arm with 6 degrees of freedom and extremely coarse touch sensors. As we will see, despite such a limited sensory systems, evolved individuals are perfectly able to solve the problem. The analysis of the obtained results shows that evolved individuals always develop a well defined behavioral strategy that allows them to easily and robustly discriminate different objects despite the limitation of their sensory apparatus. Finally, in the conclusions we discuss the general advantage of the evolutionary method for the synthesis of effective artificial agents.

2. Experimental Results

We evolved the control system of agents that are asked to categorize objects with different shapes on the basis of tactile information. Agents are provided with a 3-segments arm with 6 degrees of freedom (DOF) and extremely coarse touch sensors (see Figure 1).

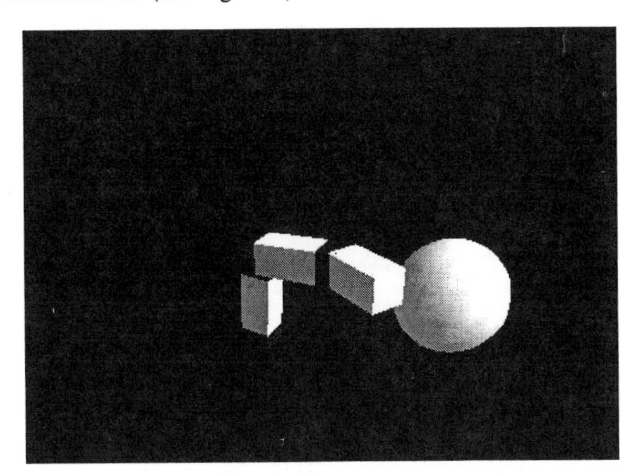

Figure 1. The arm and a spherical object.

To reduce the time necessary to test individual behaviors and model the real physical dynamics as accurately as possible we used the rigid body dynamics simulation SDK of Vortex^{TM} (see http://www.cm-labs.com/products/vortex/). This software allowed to build our robotic arm by means of several segments connected by joints and to run simulations faster than real time.

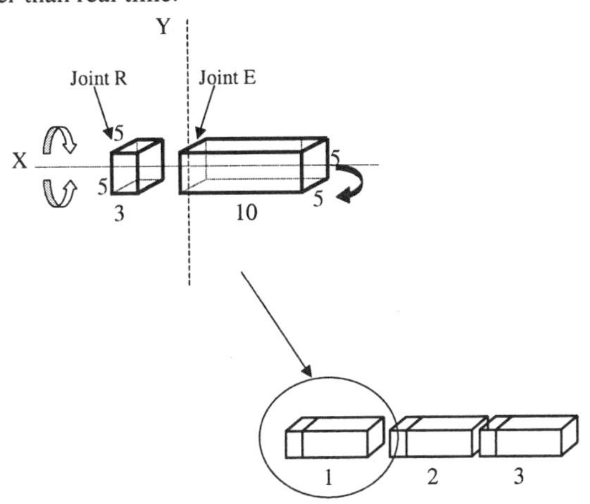

Figure 2. A schematic representation of the arm.

Given the specific characteristics of this tool, the implementation of the arm consists of a basic structure comprising two segments and two joints replicated three times (see Figure 2). The basic structure consists of a shorter segment of size {x=5, y=3, z=5} and a longer segment of size {x=5, y=10, z=5}. This two segments are connected by means of a joint (i.e. the *Joint E* in the Figure) that allows only one DOF on axis *Y* , while the shorter segment is connected at the floor, or at the longer segment, by means of a joint (i.e. the *Joint R*) that provides one DOF on axis *X*. In practice, the *Joint E* allows to elevate and lower the connected segments and the *Joint R* allows to rotate them in both direction. Notice that *Joint E* is free to moves only in a range between 0 and $\pi/2$, just like an human arm that can bend the elbow solely in a direction. The range of *Joint R* is $[-\pi/2, +\pi/2]$ Gravity is {0, -1,0, 0}. Each actuator is provided with a corresponding motor that can apply a maximum force of 50. Therefore, to reach every position in the environment the control system has to appropriately control several joints and to deal with the constraints due to gravity. Friction was set to 2.0.

The sensory system consists of a simple contact sensor placed on each longer segment that detects when this segment collides with an other object and proprioceptive sensors that provide the current position of each joint.

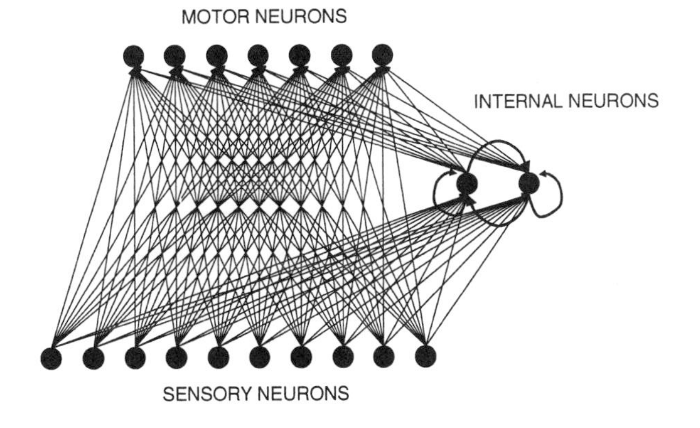

Figure 3. The architecture of the neural controller.

Each individual of the population was tested for 36 phases, each phase consisting of 150 timesteps. At the beginning of each phase the arm is fully extended and a spherical or a cubic object is placed in a random selected position in front of the arm (the position of the object is randomly selected between the following intervals: 20.0 >= X <= 30.0; 7.5 >= Y <= 17.5; -10.0 >= Z <= 10.0). The object is a sphere (15 units in diameter) during even phases and a cube (15 units in side) during odd phases so that each individual has to discriminate the same number of spherical and cubic objects during its "lifetime".

The controller of each individual consists of a neural networks with 10 sensory neurons directly connected to 7 motor neurons and 2 internal neurons receiving connections from the sensory neurons and from themselves and projecting connections to the motor neurons (see Figure 3). The first 9 sensory neurons encode the angular position (normalized between 0.0 and 1.0) of the 6 DOF of the joints and the state of the three contact sensors located in the three corresponding segments of the arm. The last sensory neuron is a copy of the last motor neuron that encode the current classification produced by the individual (see below). The first 6 motor neurons control the actuators of the 6 corresponding joints. The output of the neurons is normalized between $[0, +\pi/2]$ and $[-\pi/2, +\pi/2]$ in the case of elevation or rotational joints respectively and is used to encode the desired position of the corresponding joint. The motor is activated so as to apply a force (up to 50 units) proportional to the difference between the current and the desired position of the joint. The seventh motor neuron encodes the classification of the object produced by the individual (value below or above 0.5 are interpreted as classifications corresponding to a cubic or spherical object respectively).

The activation state of sensory and internal neurons was updated accordingly to the following equations (motor neurons were updated according to the logistic function):

$$A_j = t_j + \sum w_{ij} O_i$$
$$O_j = \tau_j O_j^{(t-1)} + (1 - \tau_j)(1 + e^{-A_j})^{-1} \quad \textbf{(1)}$$
$$0 \le \tau_j \le 1$$

With *Aj* being the activity of the *j*th neuron (or the state of the corresponding sensor in the case of sensory neurons), *tj* the bias of the *j*th neuron, *Wij* the weight from the *i*th to the *j*th neuron, *Oi* the output of the *i*th neuron. *Oj* is the output of the *j*th neuron, τj the time constant of the *j*th neuron. It should be noted that similar, although slightly worse performance, was obtained by using the standard logistic function for all neurons (result not shown).

The genotype of evolving individuals consists of 139 parameters that include 108 weights, 19 biases, and 12 time constants. Each parameter is encoded with 8 bits. Weights and biases are normalized between –10.0 and 10.0, time constants are normalized between 0.0 and 1.0. The fitness of individuals is computed by summing the number of phases in which the individuals correctly classify the current object. The classification is correct if at the end of the phase (i.e. after 150 cycles) the activation of the last motor units is below 0.5 and the object is a cube or is above 0.5 and the object is a sphere. By running 10 replications of the experiment and by evolving individuals for 50 generations we observed that in many of the replications evolved individuals display a good ability to classify the two objects and, in some cases, they produce close to optimal performance. Figure 4 shows the percentage of correct classifications measured through 100 trials for the best individual of each generation. As can be seen, in the case of the best replication (thin line), evolved individuals reach up to 98% of correct classifications.

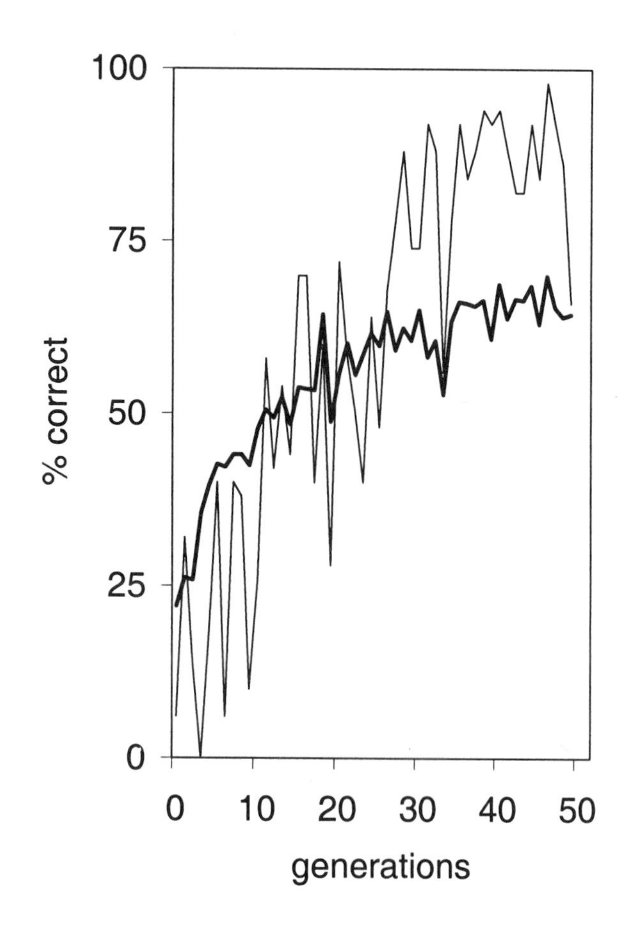

Figure 4. Percentage of correct classifications against generation for the best individual of each generation. The tick line represents the average performance of 10 replications. The thin line represents the performance of the best replication. Each individual has been tested for 100 phases.

By analyzing the obtained behaviors one can clearly see that in all experiments evolved individuals select a well defined behavior that assures that perceived sensory states corresponding to different objects can be easily discriminated and allows robust and effective categorizations. Figure 5 shows how a typical evolved individual behaves with a spherical and a cubic object (left and right side of the Figure respectively). As can be seen, first the arm bends on the left side and move to the right in order to start to feel the object with the touch sensor of the third segment. Then the arm moves so as to follow the curvilinear surface of the sphere or to keep touching one of the angles of the cubic object.

The fact that such behavior significantly simplifies the discrimination of the two objects can be explained by considering that the arm ends in very different conditions in the case of a sphere or of a cubic object. In particular, after a certain amount of time in which the arm is negotiating the object, it ends almost fully extended in the case of a spherical object and almost fully bent in the case of a cubic object. This implies that, given such a behavior, the state of the proprioceptive sensors after a certain amount of time can be used as a direct and straightforward indication of the category of the object. The fact that such behavior allows evolved individuals to produce robust and effective classifications can be explained by considering that the final classification is not the result of a single decision but is the end result of an interaction between the agent and the object that last several timesteps during which the agent keeps following the surface of the object so to ascertain whether it is curvilinear or not. Indeed, evolved individuals that display shorter negotiation periods with spherical objects also produce worse classification performance (result not shown).

The analysis of the activation state of the neurons during the behavior displayed in Figure 5 (see graphs H1 and H2 in Figure 6 and 7 that show the activation of the internal neurons when the arm has to discriminate the spherical or the cubic object respectively) shows that internal units are activated during the first phase (when the arm is looking for the object) and not activated during the second phase (when the arm starts to negotiate the object) for both spherical and cubic objects. Also notice how the activation state of unit C, that encodes the classification produced by the agent, starts low and then increases when the arm negotiates the sphere (Figure 6) while starts and remains low when the arm negotiates the cube (Figure 7). The fact that, at the end of the phase, the internal units tend to have the same activation states in the two cases shows that the classification is not accomplished on the basis of internal information extracted during the interaction between the arm and the object but rather on the basis of the final position of arm itself that, as claimed above, directly provides a clear indication of the category of the object with which the agent has previously interacted.

3. Discussion

Passive approaches to perception (e.g. Shapiro, 1987) assume that perception consists of the construction of a detailed representation of the external world. From this point of view the main challenge is that of transforming egocentric, incomplete, and noisy sensory information into an allocentric, complete, and precise representation of the external environment. To achieve this goal a large number of hard problems (in the case of vision, for example, infer 3D surfaces from 2D images or handle occlusions) have to be solved. Perception thus typically involve an intensive static analysis of passively sampled data. Within this view, motor behavior (i.e. the interaction with the external world) is not viewed as an opportunity but rather as a problem to be controlled --- the result of the perceptual process should be as independent as possible from the behavior displayed by the agent during the collection of sensory data.

Figure 5. Behavior of a typical evolved individual during an phase (150 cycles) in which the object consists of a sphere (left pictures) and of a cube (right pictures). For reason of space, the pictures show the position of the arm each 15 cycles.

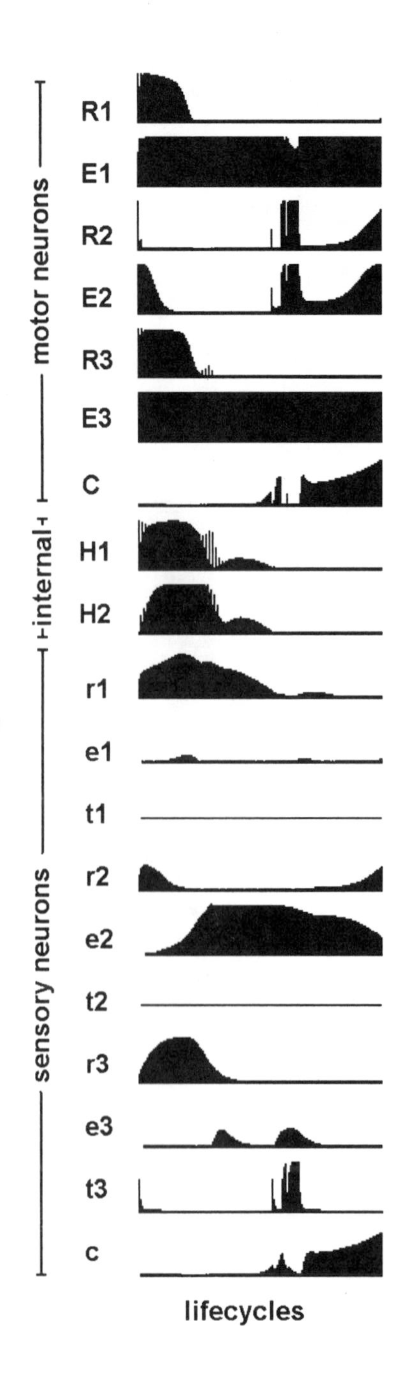

Figure 6. Activation state of the neurons during the behavior shown in the left side of Figure 4 through out 150 cycles. The height with respect to the baseline represents the activation state of the unit. *R1-R3* and *E1-E3* represent the activation state of the motor units that control the rotation and the elevation respectively of the three corresponding joints. *C* is the classification unit (value below and above 0.5 corresponds to spherical and cubic objects respectively). *H1* and *H2* represent the activation state of the two internal neurons. *r1-r3* and *e1-e3* represent the activation state of the sensory neurons that encode the current rotation and elevation of the three corresponding segments. *t1-t3* represent the activation state of the touch sensors located on the three corresponding segments. *c* is the copy of the *C* classification unit.

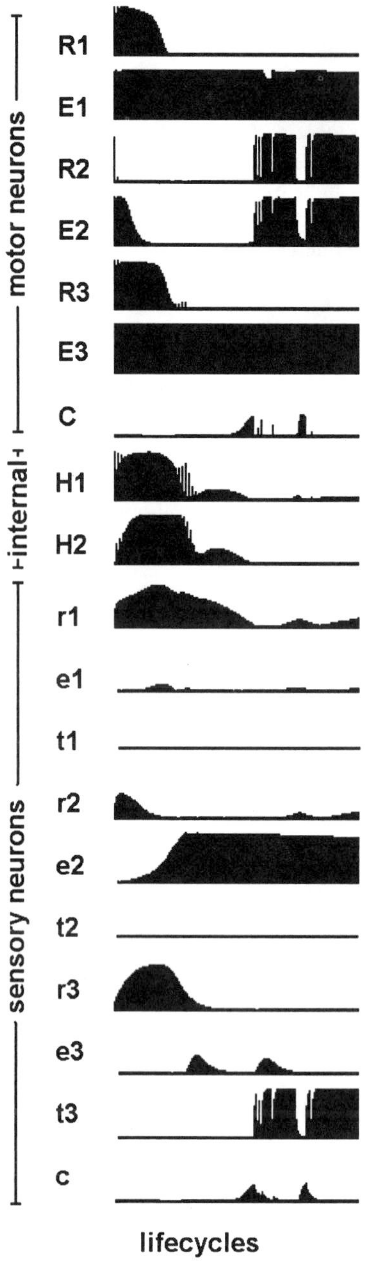

Figure 7. Activation state of the neurons during the behavior shown in the left side of Figure 4 through out 150 cycles. See legend of Figure 6.

Active approaches to perception (Bajcsy, 1988; Ballard, 1991), on the contrary, assume that the outside world serves as its own, external, representation and perception consists in mastering the regularities arising from sensorimotor interactions. From this point of view perception is a way of acting or, in other words, an exploratory activity of the environment. As pointed out by O'Reagan (2001, pp. 3) perception consists in identifying "the structure of the rules governing the sensory changes produced by various motor actions". Within this view perception and action cannot be separated and the produced behavior plays a crucial role in the outcome of the perception process.

From an engineering point of view active perception has several advantages. In many cases, solutions that exploit active perception, in fact, are much simpler than solutions that rely on passive approaches to perception (Bajcsy, 1988; Ballard, 1991; O'Reagan, 2001). In addition, active approaches, by not relying on a detailed internal representation of the external environment, are less affected by the problem of how to update such an internal representation when the environmental conditions change. On the other hand, active approaches require the designer to identify the appropriate behavior that in turn allows the agent to identify sensorimotor regularities that provide useful information. This task --- namely the identification of the appropriate way of interacting with the environment --- may be extremely difficult from the point of view of the designer given that, as we claimed in the introduction, behavior is the emergent result of the interaction between the agent and the environment. Therefore, from the point of view of the human designer that has to manually program the agent, the advantages of active perception might be counterbalanced by the difficulties of programming effective behaviors.

As we showed in this paper, evolutionary techniques in which individual agents are selected on the basis of the overall behavior emerging from the interaction between their control system and the environment (Nolfi and Floreano, 2000) represent an effective way to develop systems that are able to exploit active perception and, at the same time, to release the designer from the burden of identifying and programming the appropriate exploratory behaviors. The fact that similar results have been obtained by evolving wheeled robots, provided with different sensory systems ranging from infrared sensors to visual cameras, asked to categorize different type of objects (Scheier, C. *et al.*, 1998; Nolfi and Marocco, 2000; Nolfi 2002) demonstrates that the evolutionary method has a general validity and can be successfully applied to tackle different problems.

References

Ashby, W.R. (1952) *Design for a Brain*. London: Chapman and Hall

Bajcsy, R. (1988) Active perception *Proceedings of the IEEE* (76) 8, 996-1005

Ballard D.H. (1991) Animate vision. *Artificial Intelligence* 48:57-86.

Beer, R. (1995) A dynamical systems perspective on agent-environment interaction *Artificial Intelligence* 72, 173-215

Braitenberg, V. (1984) *Vehicles*. Cambridge, MA: MIT Press

Chiel, H.J. and Beer, R.D. (1997) The brain has a body: Adaptive behavior emerges from interactions of nervous system, body and environment *Trends in Neurosciences* 20, 553-557

Dill, M. *et al.* (1993) Visual pattern recognition in Drosophila involves retinotopic matching *Nature* 365, 751-753

Franceschini, N. (1997) Combined optical, neuroanatomical, electrophysiological and behavioral studies on signal processing in the fly compound eye. In C. Taddei-Ferretti (Ed.), *Biocybernetics of Vision: Integrative Mechanisms and Cognitive Processes*. London: World Scientific

Nolfi, S. (2002) Power and Limits of Reactive Agents. *Neurocomputing* 42:119-145

Nolfi, S. and Floreano, D. (2000) *Evolutionary Robotics: The Biology, Intelligence, and Technology of Self-Organizing Machines*. Cambridge, MA: MIT Press

Nolfi S. and Floreano D. (2002) Synthesis of autonomous robots through artificial evolution, *Trends in Cognitive Sciences* 1:31-37.

Nolfi, S. and Marocco D. (2000) Evolving visually-guided robots able to discriminate between different landmarks. In: J-A Meyer, A. Berthoz, D. Floreano, H.L. Roitblat, and S.W. Wilson (eds.) *From Animals to Animats 6. Proceedings of the VI International Conference on Simulation of Adaptive Behavior*. Cambridge, MA: MIT Press. pp. 413-419

O'Regan, J.K. and Noë A. (2001) A sensorimotor account of vision and visual consciousness. *Behavior and Brain Sciences*, 5: ?-?.

Pfeifer, R. and Scheier, C. (1999) *Understanding Intelligence*. Cambridge, MA: MIT Press

Scheier, C. and Pfeifer, R. (1995) Classification as sensorimotor coordination: A case study on autonomous agents. In F. Moran, A. Moreno, J.J. Merelo and P. Chacon (Eds.), *Advances in Artificial Life: Proceedings of the Third European Conference on Artificial Life*. Berlin: Springer Verlag

Scheier, C. *et al.* (1998) Embedded neural networks: exploiting constraints. *Neural Networks* 11, 1551-1596

Shapiro S. (1987) *Encyclopedia of Artificial Intelligence*. New York: Wiley Press.

Thelen, E. and Smith, L.B. (1994) *A Dynamics Systems Approach to the Development of Cognition and Action*. Cambridge, MA: MIT Press

Levels of Dynamics and Adaptive Behavior in Evolutionary Neural Controllers

Jesper Blynel **Dario Floreano**
Institute of Systems Engineering
Faculty of Engineering Science and Technology
Swiss Federal Institute of Technology (EPFL), CH-1015 Lausanne
Jesper.Blynel@epfl.ch, Dario.Floreano@epfl.ch

Abstract

Two classes of dynamical recurrent neural networks, Continuous Time Recurrent Neural Networks (CTRNNs) (Yamauchi and Beer, 1994) and Plastic Neural Networks (PNNs) (Floreano and Urzelai, 2000) are compared on two behavioral tasks aimed at exploring their capabilities to display reinforcement-learning like behaviors and adaptation to unpredictable environmental changes. The networks report similar performances on both tasks, but PNNs display significantly better performance when sensory-motor re-adaptation is required after the evolutionary process. These results are discussed in the context of behavioral, biological, and computational definitions of learning.

1 Introduction

In the great majority of experiments in Evolutionary Robotics (Nolfi and Floreano, 2000) evolved control systems consist of artificial neural networks most of which include recurrent connections. Recurrent connections potentially give a neural network rich temporal dynamics as well as the possibility to capture and exploit time-dependent events.

In recurrent neural networks, the activation states of the neurons can detect and maintain time-dependent activation patterns from sensors and/or other neurons that occur only briefly over time (Harvey et al., 1994). This information can modulate behavior (Beer and Gallagher, 1992), produce different actions for similar (or lacking) sensory information (Floreano and Mondada, 1996), and be used to detect and represent behavioral sequences (Tani, 1996).

Yamauchi and Beer (1994) showed that a particular class of Continuous-Time Recurrent Neural Networks (CTRNNs) (Hopfield, 1984) can be evolved to display reinforcement-learning-like behavior without modifications of the connection strengths. The authors evolved the connection strengths and neural time constants of networks that were asked to produce different output sequences in the presence of certain input patterns and reinforcement signals. The reinforcement signal consisted of toggling the value of an input neuron between zero and one. The authors analyzed the evolved CTRNNs using dynamical systems theory and showed that the reinforcement signals effectively modulated the state-space trajectories of the neuron activations in order to produce the desired output sequences.

Following a different approach, the dynamical and behavioral properties of Plastic Neural Networks (PNNs) was investigated in (Floreano and Urzelai, 2000, Urzelai and Floreano, 2001). The neuron sign, learning rate, and type of Hebbian learning (one of four possible learning rules) that was used to change on-line the strengths of all incoming connections to a neuron was evolved in networks with discrete-time recurrent connections.[1] In these experiments, the connection strengths were always initialized to small random values and modified after each neuron update using genetically-specified Hebb rules and learning rates. Analysis of experimental results showed that the evolved PNNs were capable of solving complex behavioral tasks that require precise sequences of actions by rapidly switching synaptic configuration whenever a new sequence of actions is required. Evolved PNNs also displayed remarkable on-line adaptability to new environmental conditions without requiring incremental evolution. However, no evidence was found that PNNs could actually learn and retain behavioral abilities over time or display reinforcement-learning-like properties.

These two investigation directions – CTRNNs and PNNs – provide apparently contrasting and non-intuitive results. On the one hand, neural networks with evolvable continuous-time dynamics can display learning-like behavior without synaptic plasticity. On the other hand,

[1] In networks with discrete-time recurrent connections the potential dynamics are limited compared to CTRNNs, but the system allows a simpler software implementation by maintaining a copy of the neural activities at the previous time step. These kinds of networks are sometimes referred to as having "memory units" and have been studied, among others, by (Elman, 1990).

networks with evolvable plastic connections can display rich behavioral dynamics without learning-like properties. This apparent contrast can be resolved if one considers that each model represents a specific implementation of a more general class of neural models with time-dependent states. In the case of CTRNNs it is the *neurons* which have a time-dependent state, whereas in the case of PNNs the *connection strengths* are time-dependent. Since the output of the network is a function of the product between neuron activations and connection strengths in both cases, one may arbitrarily decide where to apply the time-dependent property and, to a first approximation, the two models would be equivalent. Therefore, different abilities, such as reactive and learning-like behavior, could be explained purely in terms of different time-scales of dynamics.[2]

However, so far these two models have never been compared on a set of tasks which require time-dependent neural states. In this paper we begin to explore these issues by experimentally comparing evolutionary CTRNNs and PNNs on two sets of robotics experiments, one aimed at evolving reinforcement learning-like behaviors and the other aimed at testing the systems adaptability to various environmental changes.

2 Network Models

2.1 Continuous-Time Recurrent Neural Networks (CTRNNs)

In the continuous-time recurrent neural networks used in this paper the state of each neuron is governed by the following equation:

$$\frac{d\gamma_i}{dt} = \frac{1}{\tau_i}\Big(-\gamma_i + \sum_{j=1}^{N} w_{ij}A_j + \sum_{k=1}^{S} w_{ik}I_k \Big) \quad (1)$$

where N is the number of neurons, i $(= 1, 2, ..., N)$ is the index, γ_i describes the neuron state (cell potential), τ_i is the time constant, w_{ij} is the strength of the synapse from the presynaptic neuron j to the postsynaptic neuron i, $A_j = \sigma(\gamma_j - \theta_j)$ is the activation of the presynaptic neuron where $\sigma(x) = 1/(1 + e^{-x})$ is the standard logistic function and θ_j is a bias term. Finally, S is the number of sensory receptors, w_{ik} is the strength of the synapse from the presynaptic sensory receptor k to the postsynaptic neuron i and I_k is the activation of the sensory receptor $(I_k \in [0, 1])$. As in the work by (Yamauchi and Beer, 1994) the *Forward Euler* numerical integration method is used. The iterative update rule for the state of each neuron becomes:

$$\gamma_i(n{+}1) = \gamma_i(n){+}\frac{\Delta t}{\tau_i}\Big(-\gamma_i(n){+}\sum_{j=1}^{N} w_{ij}A_j(n){+}\sum_{k=1}^{S} w_{ik}I_k\Big) \quad (2)$$

where n is the iteration step number and Δt is the step size. This integration method is numerically stable when Δt is less than twice the smallest time-constant in the network(Hines and Carnevale, 1998). Initially the state of each neuron is $\gamma_i(0) = 0\ \forall i$, the step size set to $\Delta t = 1$. The range of the other parameters were the following:

$$\tau \in [1, 70],\ \ \theta \in [-1, 1]\ \ and\ \ w \in [-5, 5]$$

Notice from equation 2 that each neuron has an internal state and that the time constant τ controls its dynamics. Large time constants result in slowly changing neuron states while small time constants approximate reactive neurons.

2.2 Plastic Neural Networks (PNNs)

In the discrete-time, fully-recurrent, plastic neural networks used, each neuron activation A_i is updated using the following equation at every activation cycle:

$$A_i(n + 1) = \sigma\left(\sum_{j=1}^{N} w_{ij}A_j(n)\right) + I_i, \quad (3)$$

where the activation of the sensory receptor $I_i = 0$ for hidden and motor neurons. As a consequence, the range of A_i is $[0, 2]$ for input neurons and $[0, 1]$ for the hidden and output neurons. Each synaptic weight w_{ij} is randomly initialized in the range $[0, 0.1]$ and is updated after every sensory-motor cycle by the following rule:

$$w_{ij}(n + 1) = w_{ij}(n) + \eta\Delta w_{ij}(n) \quad (4)$$

where $0.0 < \eta < 1.0$ is the learning rate and Δw_{ij} is one of four Hebb rules specified in the genotype:[3]

1. *Plain Hebb rule*: strengthens the synapse proportionally to the correlated activity of the two neurons.

$$\Delta w = (1 - w)\, xy$$

2. *Postsynaptic rule*: behaves as the plain Hebb rule, but in addition it weakens the synapse when the postsynaptic node is active but the presynaptic is not.

$$\Delta w = w\,(-1 + x)\, y + (1 - w)\, xy$$

[2] This last point was raised during personal discussions with Inman Harvey and Ezequiel Di Paolo.

[3] Before applying a learning rule, the activation of input neurons are divided by 2 to scale them into the range $[0, 1]$.

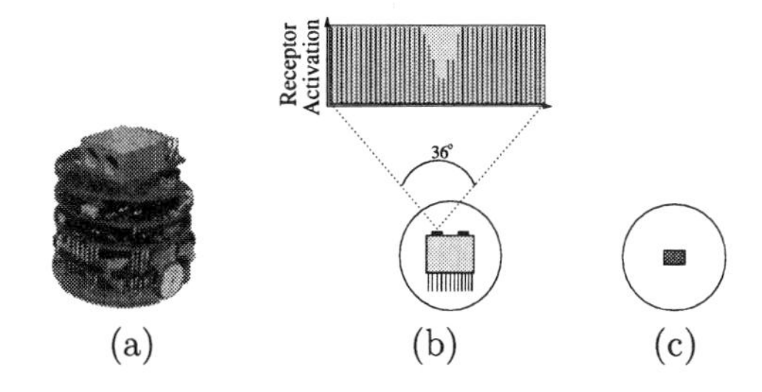

Figure 1: The Khepera robot used in the experiments. The robot has (a) 8 infrared sensors (small black rectangles) ditributed around the body that can measure object proximity and light intensity; (b) a linear vision module with 64 equally-spaced photoreceptors covering a visual field of 36°; and (c) a floor-color sensor under the body between the two wheels.

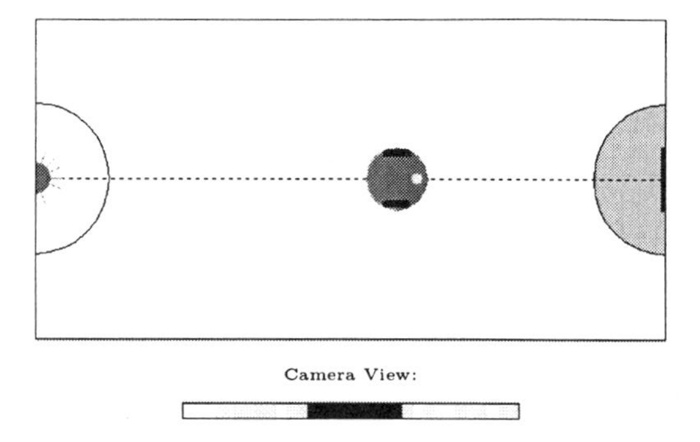

Figure 2: The environment used in the reinforcement learning task. To the left there is a light bulb and to the right there is black stripe on the wall. A gray reward-area can be randomly placed in either end (here to the right). The robot is constrained to move along the dashed line always facing the black stripe on the wall. Below the environment the view from the linear camera is shown.

3. *Presynaptic rule*: weakening occurs when the presynaptic unit is active but the postsynaptic is not.

$$\Delta w = wx\,(-1+y) + (1-w)\,xy$$

4. *Covariance rule*: strengthens the synapse whenever the difference between the activations of the two neurons is less than half their maximum activity, otherwise the synapse is weakened.

$$\Delta w = \begin{cases} (1-w)\mathcal{F}(x,y) & \text{if } \mathcal{F}(x,y) > 0 \\ (w)\mathcal{F}(x,y) & \text{otherwise} \end{cases}$$

where $\mathcal{F}(x,y) = \tanh(4(1-|x-y|)-2)$ is a measure of the difference between the presynaptic and postsynaptic activity. $\mathcal{F}(x,y) > 0$ if the difference is bigger or equal to 0.5 (half the maximum node activation) and $\mathcal{F}(x,y) < 0$ if the difference is smaller than 0.5.

The the self-limiting component $(1-w)$ is maintaining synaptic strengths within the range $[0,1]$. As a consequence, synapses do not change sign.

3 The "Reinforcement Learning" Task

In this first experiment CTRNNs and PNNs are compared on a task that requires acquisition and storage of knowledge on the basis of a reinforcement signal provided to the input units of the network. Given the simple experimental settings, the experiment is carried out in a realistic simulation by sampling infrared sensor activations (Miglino et al., 1995), computing geometric projections for the linear camera inputs, and adding 5% uniform noise to every value.

A simulated Khepera robot (figure 1) is positioned in a white rectangular arena with two areas of potential reward (figure 2), a light bulb to the left and a black vertical stripe on the right. The robot has 6 trials to find out where the reward area is located, go there and stay over it. At the beginning, the position of the reward area (grey-filled sector in figure 2) is randomly chosen, either below the light bulb or below the stripe, and remains the same for 3 consecutive trials. After 3 trials, the reward area is switched to the other end of the environment. At the beginning of each trial, the robot is randomly positioned within the center third of the dashed line shown in figure 2, always facing the black stripe. In order to make the task simpler, the robot can only move along this line back and forth at variable speed, but cannot rotate.

The reinforcement signal comes from a floor-color sensor (figure 1, c) which is *on* when the robot is inside the gray reward-zone and *off* otherwise. Notice that this information is a sensory input just like others, in contrast to conventional reinforcement learning systems where the reward signal plays a special role in the architecture and in the learning algorithm (Sutton and Barto, 1998).

3.1 Network Architectures and Genetic Encoding

In order to maintain consistency with previous results reported in the literature, the architectures and genetic encodings of the CTRNNs and PNNs used are slightly different.

The architecture difference is that sensory receptors in CTRNNs receive information only from the sensors of the robot (figure 3), whereas in PNNs sensory neurons receive information from all other neurons in the network, including other sensory neurons and self-connections (figure 4).

In addition, the PNNs have a bias neuron, with activation fixed to 1, whose plastic connections act as a variable threshold on post-synaptic neurons. Instead, the bias values (or thresholds) of the CTRNNs are fixed

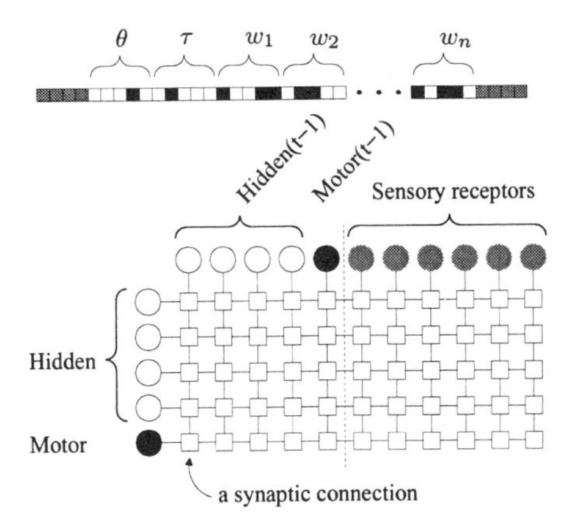

Figure 3: CTRNNs: Genetic encoding of the parameters for one neuron (top) and architecture of the CTRNN (bottom) used in the reinforcement learning task. *Genetic Encoding*: Each neuron parameter is encoded using 5 bits. θ is the bias, τ is the time constant, and $w_1 \ldots w_n$ are the strengths of the incoming synapses to this neuron. *Neural Architecture*: The network consists of 5 neurons (4 hidden + 1 motor output). Every neuron has synaptic connections from all neurons and all sensory receptors. In total, the network has 55 synaptic connections.

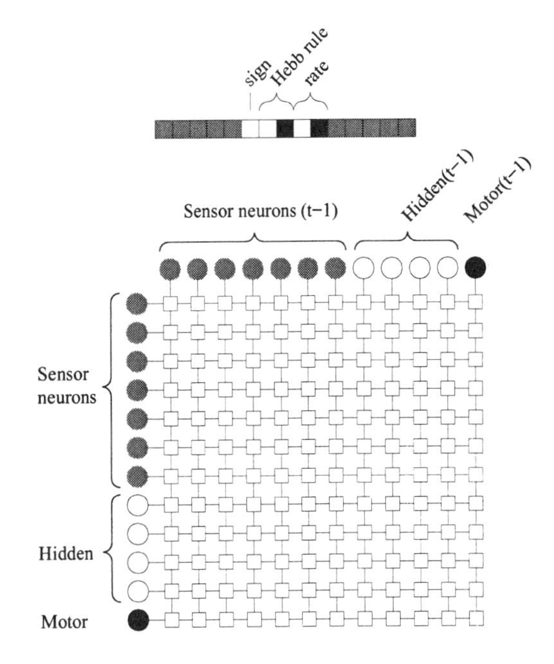

Figure 4: PNNs: Genetic encoding of the parameters for one neuron (top) and architecture of the PNN (bottom) used in the reinforcement learning task. *Genetic Encoding*: Each neuron is encoded using 5 bits. The first bit determines the sign of all outgoing synapses and the remaining four bits determine the Hebb rule (one out of four types) and learning rate (one out of four values) for all incoming synapses to that neuron. *Neural Architecture*: The network consists of 7 sensory input neurons, 4 hidden neurons and one motor output neuron. The network is fully recurrent giving a total of 144 synaptic connections.

and individually encoded in the genetic string (see description of genetic encoding below).

The networks have 6 sensory inputs, one from the each of the following receptors (figure 5):

- *2 Light receptors*: The robots infrared sensors in passive mode are used to measure the ambient light. Only the two sensors on the back of the robot are used.

- *3 Visual receptors*: The linear vision module has 64 equally spaced photoreceptors spanning a visual field of 36° (figure 1, b). The visual field is divided into 3 sectors and the average pixel-value (256 gray levels) in each sector is passed to the corresponding visual receptor.

- *1 Floor receptor*: An infrared sensor in the center of the robot pointing downwards (figure 1, c) measures the colour of the floor (in 256 gray levels). If the robot is inside the reward area the corresponding receptor is *on* and *off* otherwise.

The values from each receptor are scaled into the range $[0, 1]$. The activation of the motor neuron determines the speed of *both* wheels of the simulated robot. Activations above 0.5 correspond to forward motion, activations below 0.5 correspond to backward motion (the closer to 1 or 0, the faster the motion).

The parameters of both types of networks are encoded in genotype bitstrings. In the case of CTRNNs (figure 3, top), each neuron has 13 encoded parameters: a time constant (τ), a threshold (θ), and 11 synaptic strengths (w_{ij}). Each parameter is encoded using 5 bits giving a total genotype length of 325 bits.

In the case of PNNs (figure 4, top), each neuron has 3 encoded parameters: the sign bit determines the sign of all the outgoing synapses and the remaining four bits determines the properties of all incoming synapses to this neuron, (2 bits for one of four Hebb rules and 2 bits for one of four learning rates, namely $0.0, 0.3, 0.6, 1.0$). Consequently, all incoming synapses to a given node have the *same* properties. The total genotype length in this case is 60 bits.

3.2 Experiments

The experiments are carried out in simulation using a rank-based selection. A population of 100 neural controllers is evolved for 100 generations. At every generation the best 20 individuals make 5 copies each. One copy of the best individual remains unchanged (elitism). Single-point crossover with a 0.04 probability and bit-

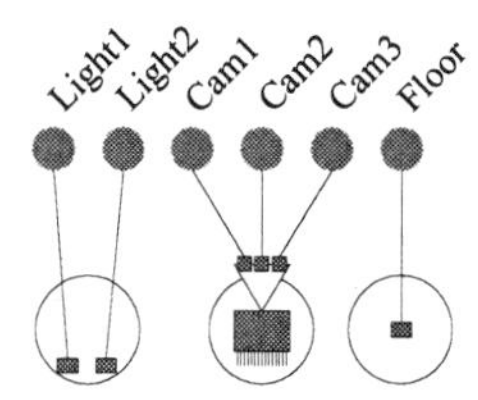

Figure 5: Configuration of the 6 sensory neurons used in the reinforcement learning task.

switch mutation with a 0.02 probability per bit are used. Every neural controller is tested for 3 times (epochs) of 6 trials each. Each trial lasts 150 sensory-motor cycles (one sensory-motor cycle corresponds to 100 ms on the real robot). At the beginning of each *epoch*, the neural controller is re-initialized. In the case of CTRNNs, the states of the neurons are set to 0 and in the case of PNNs the synaptic strengths are initialized to small random values in the interval $[0, 0.1]$ (recall that the sign of the synapse is given by the neuron from where the signal starts).

The fitness of the individual is proportional to the amount of time spent on the reward areas subtracted a penalty for spending time in only one of the two areas:

$$fitness = \frac{\sum_{i=1}^{epochs} f_1(i) + f_2(i) - |f_1(i) - f_2(i)|}{epochs \times trials\_per\_epoch \times steps\_per\_trial}$$

where f_1 is the number of steps spent in reward area in trials with the reward area to the left and f_2 is the number of steps spent in reward area in the trials with the reward area to the right. Notice that if f_1 or f_2 is zero in an epoch the total fitness is zero in that epoch. Without taking the absolute difference between f_1 and f_2, evolved controllers find the sub-optimal solution of only accumulating fitness points in one end of the environment.

For each type of neural controller (CTRNN and PNN) the experiment is repeated 10 times with different initializations of the pseudo-random number generator. The results of the experiments using CTRNNs are plotted in figure 6. Only 40 generations are required for the fitness values to reach a stable level.

The typical behavior of an evolved controller is visualized in figure 7. The x-position of the robot over time is plotted for an entire epoch of 6 trials [4].

Before the first step of each trial the robot is randomly placed within the center third of the x-axis. This evolved robot begins to move to the left where the reward area is positioned for the first 3 trials for this individual. After 3 trials the reward area is moved to the right side. The robot still moves to the left by default, but when it

[4]Each trial is shortened from 150 to 100 steps in figure 7 because the position of the robot always stabilizes within this timeframe for this individual.

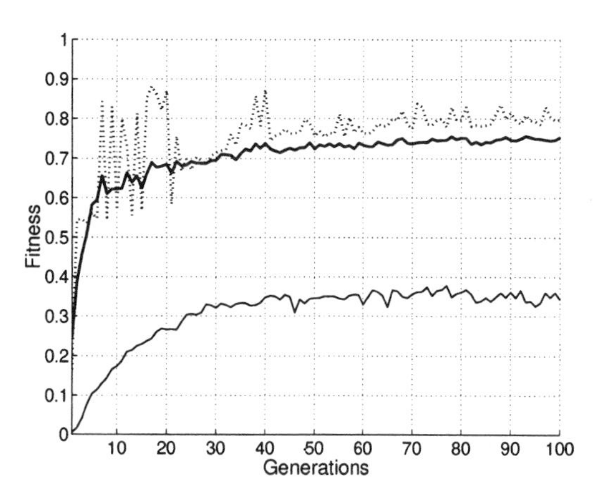

Figure 6: *Reinforcement Learning Task, CTRNNs.* Thick line = best individual; thin line = population average; dotted line = best individual of best replication.

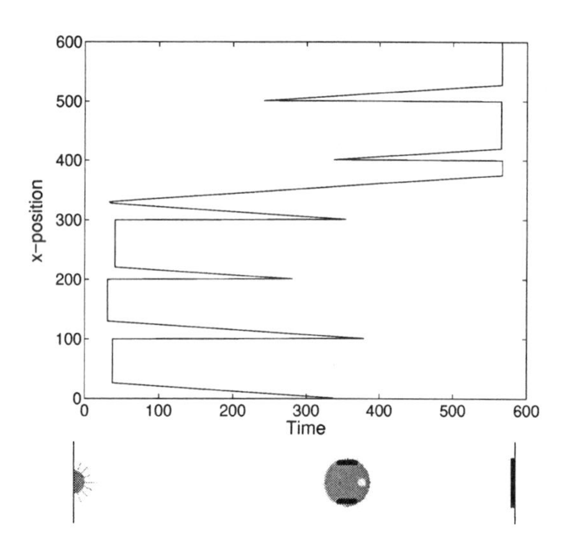

Figure 7: Typical behavior of an evolved robot in the reinforcement learning task. *Top*: Robot x-position is plotted against time over 6 trials. *Bottom*: Environment layout. See text for description.

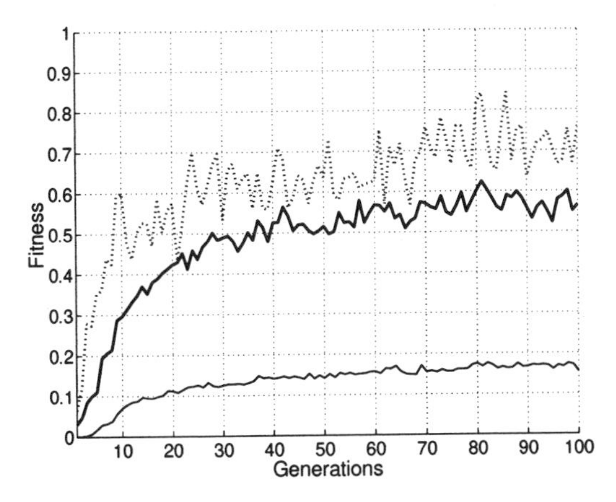

Figure 8: *Reinforcement Learning Task, PNNs.* Thick line = best individual; thin line = population average; dotted line = best individual of best replication.

discovers that the reward signal remains off, it reverses direction and moves towards the right side. For the remaining trials, it "remembers" that the reward area is on the right side and always moves in that direction.

The fitness data for the experiments with PNNs are plotted in figure 8 and the behavior of the best evolved individuals are similar to the one shown in figure 7. The fitness values in these experiments are lower for two reasons. The first reason is that PNNs must develop synaptic weights at the beginning of each epoch (which takes time and has an indirect fitness cost) whereas CTRNNs are functional from the very beginning of the epoch. The second reason is that only one replication of the experiment was successful at evolving individuals capable of solving the task reliably. In the other replications the succes of the best individuals depended on the initial position of the robot and the random initializtion of the synaptic weights resulting in a drop in performance under some *unlucky* initial conditions.

4 Adaptation to Changes

In order to further explore the adaptation capabilities of the two types of networks, an experimental setup investigated in (Floreano and Urzelai, 2000, Urzelai and Floreano, 2001) is now used. The environment (figure 9) is identical to the one used in the reinforcement learning task, but the position of the gray fitness area is now fixed under the light bulb and a black lightswitch area has been added under the black stripe. Initially the light is off, but it is switched on if the robot passes over the black area. The task of the robot is to spend as much time as possible under the light bulb when the light is on. Therefore, the robot must first discover how to switch on the light. Each individual of the population is tested for 3 trials of 500 sensory-motor cycles (100 ms) each. The experiments are carried out in simulation and some evolved individuals are then tested in the real environment (evolutionary experiments on the real robot with PNNs are reported in (Urzelai and Floreano, 2001) and the results are similar to those obtained in simulation).

Figure 9: *The Lightswitch Task*: The Khepera robot can gain fitness by staying on the gray area when the light is on. Initially the light is off, but the robot can switch it on by passing first over the black area. The neural controller has no information from the floor-sensor which is relayed to an external computer in order to automatically switch the light on and compute fitness values.

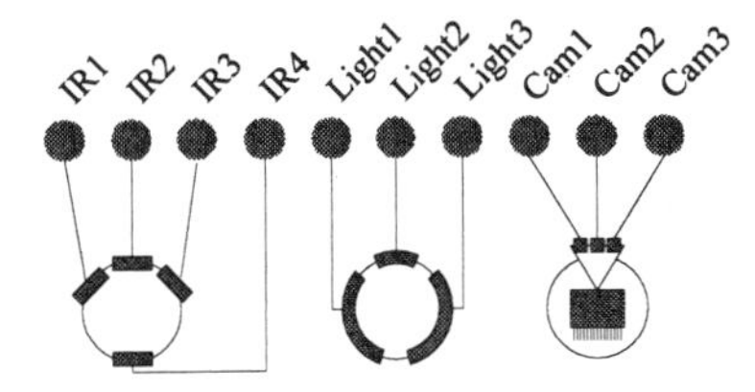

Figure 10: Configuration of the 10 sensory neurons used in the lightswitch task.

The fitness function is given by the number of sensory motor cycles spent on the gray area below the light bulb *when the light is on* divided by the total number of cycles available (500). In order to maximize this fitness function, the robot should find the lightswitch area, go there in order to switch the light on, and then move towards the light as soon as possible, and remain on the gray area.[5] Since this sequence of actions takes time (several sensory motor cycles), the fitness of a robot will never be 1.0. A robot that cannot manage to complete the entire sequence will get zero fitness.

The architectures of the two types of neural networks are identical to those described earlier, but the configuration of the sensory receptors and motor output neurons is adapted to this new task. The number of sensory receptors is extended to 10: 4 infrared proximity receptors to detect distance form walls (within a range of 5 cm),

[5] Notice that the fitness function does not explicitly reward this sequence of actions, but only the final outcome of the overall behavior.

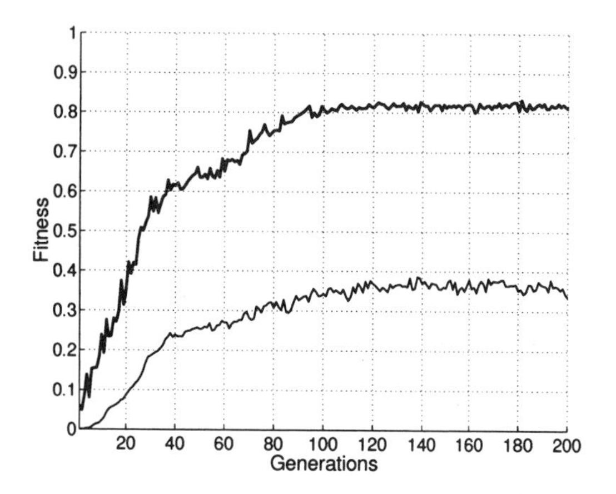

Figure 11: *Ligthswitch Task, CTRNNs.* Thick line = best individual; thin line = population average.

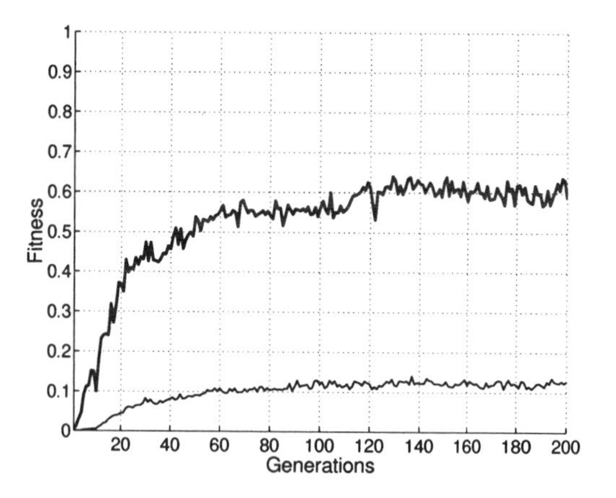

Figure 12: *Ligthswitch Task, PNNs.* Thick line = best individual; thin line = population average.

3 light receptors, and 3 visual receptors (figure 10).

4.1 Experiments

Two sets of experiments are run, one for each type of network. The results are shown in figures 11 and 12. In both cases, evolved controllers are capable of performing the entire sequence of actions. The reason why the PNNs report lower fitness values is that synaptic adaptation takes time and therefore has an implicit fitness cost. In addition, evolved PNNs tend to select lower motor speeds for the robot. However, all experimental runs end up with successful strategies.

A typical trace of a successful robot is shown in figure 13. The robot turns on the spot, moves directly towards the black area to turn on the light and finally moves towards the light bulb and remains over the gray fitness area. Small differences in the behaviors generated by the two types of networks are noticed. Robots controlled by CTRNNs generally turn and move directly towards the black area at first. Instead, robots controlled by PNNs often start out by first moving forward hitting the first wall and afterwards doing a backward turn while orienting towards the black stripe and approaching the black area. Recall the PNNs start out with randomly initialized synapses and must develop all abilities, including obstacle avoidance from scratch.

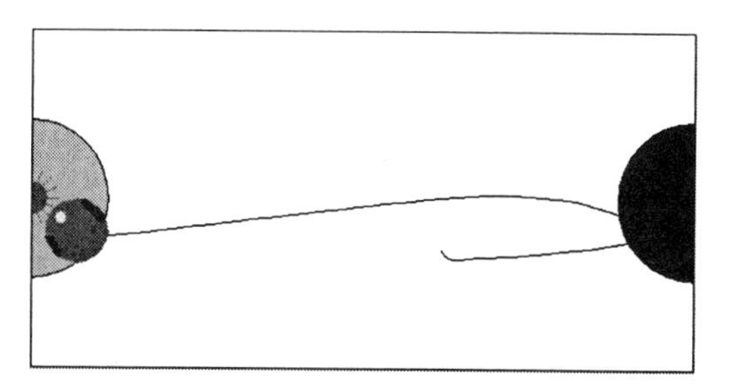

Figure 13: Typical trace of an evolved robot in the lightswitch task.

4.2 Environmental Changes

In contrast to the reinforcement learning task, this task requires more interactions between the robot and the environment and includes a larger number of sensory-motor correlations that must be taken into account by evolutionary neural controllers. Therefore, it is better suited for applying a number of modifications *after evolution* to test on-line adaptive abilities of evolved CTRNNs and PNNs.

In order to measure the performances of evolved controllers in environments with new characteristics, the best individual of the last generation for each of the 10 replications is tested in environments with white (as used during evolution), gray, and black walls. The responses of infrared proximity sensors are lower for darker walls, which requires the robot to avoid walls at lower sensory activation levels than with brighter walls.

Figure 14 shows the average fitness values of the best individuals in environments with white, gray and black walls for CTRNNs (left) and PNNs (right). In the CTRNN case, the performance drops significantly in the gray environment case and dramatically in the black environment. In the gray environment, evolved individuals are only able to complete the correct sequence of actions in one third of the trials and in the black environment this number drops to one out of 20 trials. In the PNN case a performance drop is observed in gray and black environments, but this is not so drastic when compared to the performance in white environments. Most importantly, PNN individuals can still solve the task but require longer time to do so because they interact with the walls for longer time. Evolved PNNs are capable of adapting to new sensory situations even in the black environment while CTRNNs cannot cope with it.

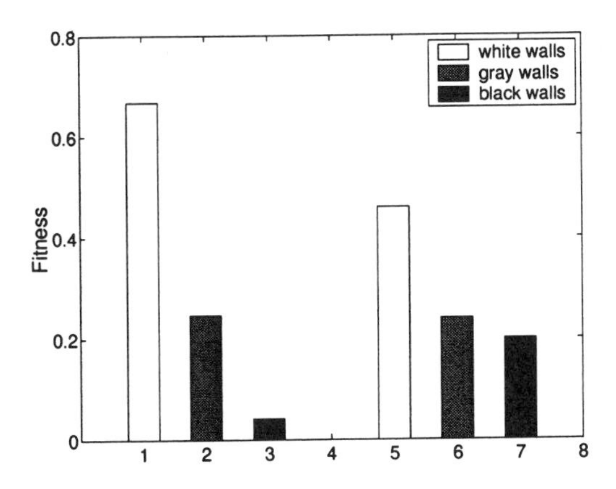

Figure 14: Performance of the evolved CTRNNs (left) and PNNs (right) tested in environments with white, gray and black walls. Each fitness value is the average over 100 tests (the best individual of each of the 10 replications is tested 10 times each).

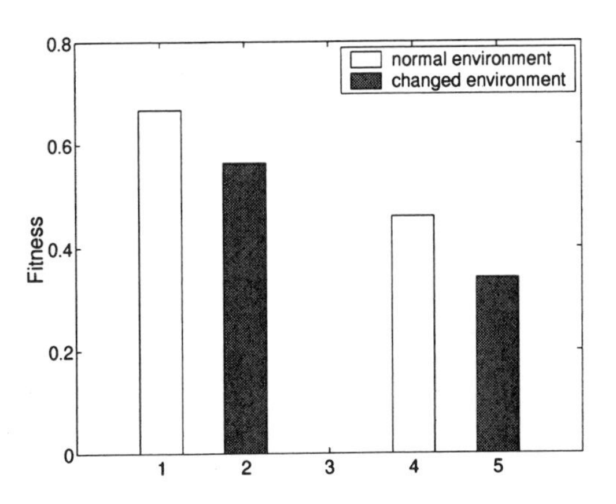

Figure 15: Performance of evolved CTRNNs and PNNs tested in the environment used during evolution (white bar) and in an environment where the lightswitch area and the fitness area are positioned randomly along the walls (gray bar). Each fitness value is the average over 100 tests (the best individual of each of the 10 replications is tested 10 times each).

In a second set of tests, the spatial relationships of the objects in the environment are modified. For each trial, the lightswitch area and the light bulb area is randomly located along the walls (discarding combinations with area overlap). The best individual of the last generation for each of the 10 replications is tested in 10 randomly generated environments. The results are reported in figure 15. To the left are the fitness values for individuals controlled by CTRNNs and to the right the fitness values for the individuals controlled by PNNs. In both cases there is a small drop in performance in the new environments, but both CTRNNs and PNNs succeed in solving the task. In other words, this modification does not show any difference in the adaptation capabilities of the two types of networks.

As a final test, the neural controllers (which were evolved in simulation) are tested in the real environment on a real Khepera robot (see figure 9). Despite the use of a realistic simulation, there is still a significant difference in sensory and motor properties of simulated and physical robots. The best individual of the last generation in each of the 10 replications is tested 3 times on the physical robot. Figure 16 shows that CTRNNs fail on the real robot, while PNNs are less affected by the transfer and still function in the real environment. The small performance drop in the PNN case is due to the fact that the robot sometimes performs looping trajectories around the fitness area without coming to rest. The main reason for the failure of the CTRNNs is that the robots not are able to reliably approach the black stripe on the wall. These networks are not capable of adapting to changed sensory-motor conditions and the robot often gets attracted to shadows or keeps spinning and bumping into walls.

5 Discussion

The experimental results presented in this paper question the definition of learning. From a behavioral perspective, learning is usually associated with acquisition of new knowledge, skills, and memory retention (Gallistel, 1990). From a biological and computational perspective, learning is associated to synaptic change (Churchland and Sejnowski, 1994). However, the experiments described in this paper show that both assumptions are not necessary and unique. On the one hand, dynamical neural networks without synaptic plasticity display reinforcement learning behavior (confirming earlier experiments described in (Yamauchi and Beer, 1994)). On the other hand, networks with plastic synapses do not necessarily retain previously acquired knowledge (data regarding this latter point are described in (Urzelai and Floreano, 2001)).

The experiments show that each type of network model – CTRNN and PNN – is capable of displaying learning-like abilities and can solve complex tasks which require sequential behavior. However, the PNNs display a clear advantage when confronted with environmental conditions not seen during the evolutionary process. In the two tests on unpredictable changes where PNNs scored significantly better than CTRNNs – walls of different brightness and transfer from simulated to physical robots – both sensory and motor adaptation is required. On the contrary sensory-motor adaptation is not necessary in the tests with spatial reconfiguration of the environment where the same stripe-directed navigation and light following behaviors as developed during evolution is efficient in several environments. In this latter test, both network models indeed display the same

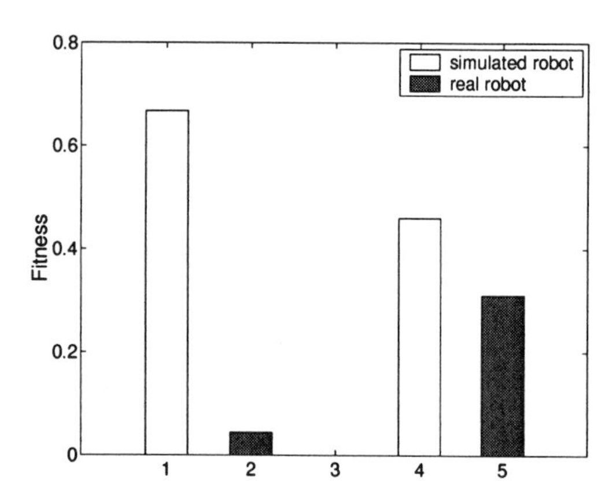

Figure 16: Performance of evolved CTRNNs and PNNs tested in the simulated robot used during evolution (white bar) and in the real Khepera robot (gray bar). Each fitness value is an average over 30 tests (the best individual of each of the 10 replications is tested 3 times each).

performance.

Correlation-based synaptic plasticity therefore seems useful to maintain coherent sensory-motor mappings in a variety of different environments. Considering the importance of sensory-motor coordination in even the most primitive organisms equipped with neural cells, one may (wildly) speculate that correlation-based (Hebbian) synaptic plasticity was discovered by evolution as a simple homeostatic mechanism (Ashby, 1960) to cope with partially changing and unpredictable environments. However, one may argue that the PNN architecture used in these experiments provides an advantage with respect to this because the sensory neurons are interconnected and therefore may better capture and maintain sensory-motor correlation than in the CTRNN case where correlations must be captured and maintained by hidden units.

In a previous work (Urzelai and Floreano, 2001), it was shown that PNNs rapidly change most synaptic strengths when the sensory information changes significantly. This fast re-wiring of the entire network is quite efficient in the simple sensory-motor tasks described in this paper, but may not be suitable for more complex situations where there is a survival advantage in acquiring and retaining complex skills or spatial configurations. Indeed, not all evolutionary runs with PNNs were capable of solving completely the reinforcement-learning task described here. A promising direction consists of adding mechanisms for enabling and disabling synaptic plasticity of the Hebb-rules described here. This idea has been recently explored to evolve reinforcement learning-like foraging (Niv et al., 2001), adapt to distorted optical information (DiPaolo, 2000), and adapting walking patterns to terrains of variable inclination (Fujii et al., 2001). However, the tasks explored in those works are not significantly more complex than those described here and it can be argued whether the enabling/disabling mechanism is necessary. Ironically, a serious problem seems to be the definition of sufficiently complex learning problems that require more than "minimally cognitive behaviors" and yet remain sufficiently simple for analysis.

As a final note, the size of the populations of CTRNNs in the experiments described here is one order of magnitude smaller than the one used in the experiments by (Yamauchi and Beer, 1994). Although the two types of networks are slightly different in the sensory-motor interface, the major difference is, that in the work presented in this paper, binary, instead of real-valued, genetic encoding is used. Consequently, the search space is much smaller, but this does not seem to limit the rich dynamics of the networks.

6 Conclusion

Two classes of dynamical recurrent neural networks, CTRNNs and PNNs, have been compared on two behavioral tasks aimed at exploring their capabilities to display reinforcement-learning like behaviors and adaptation to unpredictable environmental changes. Although both networks displayed similar performances (slightly offset in PNNs by the time required for synaptic development), PNNs displayed significantly better performance when sensory-motor re-adaptation is required after the evolutionary process.

These results have been discussed within the perspective of behavioral, biological, and computational definitions of learning. It has been argued that correlation-based synaptic plasticity may play a major role in developing and maintaining sensory-motor coordination in partially changing and unpredictable environments.

Although the combination of the two models may have some interest, we believe that more complex behavioral and cognitive abilities may be achieved by employing mechanisms for enabling and disabling synaptic plasticity in environmental conditions that require the development and retention of sensory-motor skills and spatial memories.

Acknowledgements

The authors thank Inman Harvey (Sussex University), Ezequiel Di Paolo (Sussex University), and Claudio Mattiussi (EPFL) for fruitful discussions and suggestions. This work was supported by the Swiss National Science Foundation, grant nr. 620-58049.

References

Ashby, W. R. (1960). *Design for a Brain: The Origin of Adaptive Behavior.* Chapman and Hall, London.

Beer, R. D. and Gallagher, J. C. (1992). Evolving dynamical neural networks for adaptive behavior. *Adaptive Behavior*, 1(1):91–122.

Churchland, P. M. and Sejnowski, T. J. (1994). *The Computational Brain.* MIT Press, Cambridge, MA.

DiPaolo, E. A. (2000). Homeostatic adaptation to inversion of the visual field and other sensorimotor disruptions. In Meyer, J., Berthoz, A., Floreano, D., Roitblat, H., and Wilson, S., (Eds.), *From Animals to Animats VI: Proceedings of the Fifth International Conference on Simulation of Adaptive Behavior*, pages 440–449. MIT Press-Bradford Books, Cambridge, MA.

Elman, J. L. (1990). Finding Structure in Time. *Cognitive Science*, 14(2):179–211.

Floreano, D. and Mondada, F. (1996). Evolution of homing navigation in a real mobile robot. *IEEE Transactions on Systems, Man, and Cybernetics-Part B*, 26:396–407.

Floreano, D. and Urzelai, J. (2000). Evolutionary robots with online self-organization and behavioral fitness. *Neural Networks*, 13(4–5):431–443.

Fujii, A., Ishiguro, A., Aoki, T., and Eggenberger, P. (2001). Evolving bipedal locomotion with a dynamically-rearranging neural network. In Kelemen, J. and Sosík, P., (Eds.), *Advances in Artificial Life (ECAL 2001)*, pages 509–518, Berlin. Springer Verlag.

Gallistel, C. R., (Ed.) (1990). *The Organization of Learning.* MIT Press, Cambridge, MA.

Harvey, I., Husbands, P., and Cliff, D. (1994). Seeing the light: Artificial evolution, real vision. In Cliff, D., Husbands, P., Meyer, J., and Wilson, S. W., (Eds.), *From Animals to Animats III: Proceedings of the Third International Conference on Simulation of Adaptive Behavior*, pages 392–401. MIT Press-Bradford Books, Cambridge, MA.

Hines, M. and Carnevale, N. T. (1998). Computer modeling methods for neurons. In Arbib, M. A., (Ed.), *The Handbook of Brain Theory and Neural Networks*, pages 226 - 230. MIT Press - Bradford Books, Cambridge, MA.

Hopfield, J. J. (1984). Neurons with graded response properties have collective computational properties like those of two-state neurons. *Proceedings of the National Academy of Sciences USA*, 81:3088–3092.

Miglino, O., Lund, H. H., and Nolfi, S. (1995). Evolving Mobile Robots in Simulated and Real Environments. *Artificial Life*, 2(4):417–434.

Niv, Y., Joel, D., Meilijson, I., and Ruppin, E. (2001). Evolution of reinforcement learning in uncertain environments: Emergence of risk-aversion and matching. In Kelemen, J. and Sosík, P., (Eds.), *Advances in Artificial Life (ECAL 2001)*, pages 252–261. Springer Verlag, Berlin.

Nolfi, S. and Floreano, D. (2000). *Evolutionary Robotics: Biology, Intelligence, and Technology of Self-Organizing Machines.* MIT Press, Cambridge, MA.

Sutton, R. and Barto, A. (1998). *Introduction to Reinforcement Learning.* MIT Press, Cambridge, MA.

Tani, J. (1996). Model-Based Learning for Mobile Robot Navigation from the Dynamical Systems Perspective. *IEEE Transactions on Systems, Man, and Cybernetics-Part B*, 26:421–436.

Urzelai, J. and Floreano, D. (2001). Evolution of adaptive synapses: Robots with fast adaptive behavior in new environments. *Evolutionary Computation*, 9:495–524.

Yamauchi, B. and Beer, R. D. (1994). Sequential behavior and learning in evolved dynamical neural networks. *Adaptive Behavior*, 2(3):219–246.

Evolving integrated controllers for autonomous learning robots using dynamic neural networks

Elio Tuci **Inman Harvey** **Matt Quinn**
Centre for Computational Neurosciences and Robotics,
School of Cognitive and Computing Sciences
University of Sussex, Brighton BN1 9QH, United Kingdom
phone: (+44) (0)1273 872945 - fax (+44) (0)1273 671 320
eliot, inmanh, matthewq@cogs.susx.ac.uk

Abstract

In 1994, Yamauchi and Beer (1994) attempted to evolve a dynamic neural network as a control system for a simulated agent capable of performing learning behaviour. They tried to evolve an *integrated* network, i.e. not modularized; this attempt failed. They ended up having to use independent evolution of separate controller modules, arbitrarily partitioned by the researcher. Moreover, they "provided" the agents with hard-wired reinforcement signals.

The model we describe in this paper demonstrates that it is possible to evolve an *integrated* dynamic neural network that successfully controls the behaviour of a khepera robot engaged in a simple learning task. We show that dynamic neural networks, based on leaky-integrator neuron, shaped by evolution, appear to be able to integrate reactive and learned behaviour with an *integrated* control system which also benefits from its own evolved reinforcement signal.

1 Introduction

In 1994, Yamauchi and Beer (1994) pointed out how research on learning robots typically drew a sharp distinction between the mechanisms responsible for an agent's behaviour and those responsible for learning (Kaelbling, 1993). They claimed that this distinction is difficult to defend biologically, because many of the same biological processes are involved in both reactive and learned behaviour. They based their claim looking at current theory in the study of learning in biology, which seems to suggest that animals' learning capabilities are exquisitely tuned by evolution, to the particular niche that they occupy (Garcia and Koelling, 1966; Wilcoxon et al., 1971; Davey, 1989).

Therefore, they set out an experiment in which they tried to evolve an autonomous learning robot by selectively integrating the necessary plasticity directly into the mechanisms responsible for the robot's behaviour. Their methodology was meant to guarantee that any structural or functional decomposition that may be found in the analysis of the behaviour or in the evolved control system should result entirely from the evolutionary contingencies (as it is in animals). Previous research on learning robots relied on explicitly hand-designed decomposition or modularization both of the behaviour and of the robot's control system (Kaelbling, 1993).

The approach undertaken by Yamauchi and Beer (1994), is part of a more general way of assuming an evolutionary perspective in designing control architecture for autonomous agents. This research area is generally referred to as Evolutionary Robotics (Harvey et al., 1997; Nolfi and Floreano, 2000). The potential benefits of an evolutionary approach to the design of simulated or real agents' control systems and possibly agents' morphologies, either for engineering purposes or as a new methodology for biology, are widely debated (see Webb, 2000; Nolfi, 1998). However, generally speaking, the appeal of an evolutionary approach to robotics is two-fold. Firstly, and most basically, it offers the possibility of automating a complex design task (Meyer et al., 1998; Nolfi and Floreano, 2000; Harvey et al., 1997). Secondly, since artificial evolution needs neither to understand, nor to decompose a problem in order to find a solution, it offers the possibly of exploring regions of the design space that conventional design approaches are often constrained to ignore (Harvey et al., 1992).

A growing amount of research in Evolutionary Robotics has been focusing on the evolution of controllers with the ability to modify the behaviour of a robot, in order to adapt to variation in its operating conditions. By variation we mean changes in the relationship between a robot's sensors and actuators, and its environment (e.g. Floreano and Urzelai, 2000, 2001b,a; Floreano and Mondada, 1996; Nolfi and Floreano, 1999; Eggenberger et al., 1999; DiPaolo, 2000). Yamauchi and Beer's approach clearly represents one of the first attempts in which continuous time recurrent neural networks have been exploited to integrate reactive, sequential and learned behaviour in a simulated robot.

Apart from Yamauchi and Beer (1994), current research in this area has been investigating the use of neural network controllers with some form of synaptic plasticity, that is, networks which incorporate mechanisms that change connection weights. Yamauchi and Beer (1994) did something distinctively different and unique, due to the characteristics of the control system and the singularity of the learning task employed. They employed continuous time recurrent neural networks (CTRNNs: described in their paper and also briefly in section 3.4 below), with genetically defined and *fixed* (as contrasted with plastic) connection weights, and leaky-integrator neuron (i.e. neurons whose activation decays with time). There is a further difference between the learning task described in their paper and the robot learning experiments of (e.g. Nolfi and Parisi, 1997; Floreano and Urzelai, 2000). The former specifically required the robot to learn the current relationship between a goal and a landmark, a relationship that changes from time to time; the latter require the robot to adapt to rather general variation concerning the relationship between a robot's sensors and actuators, and its environment — such as variation in the color of the arena walls (as in Nolfi and Parisi, 1997) or variation in lighting levels (as in Floreano and Urzelai, 2000).

However, Yamauchi and Beer (1994)'s attempts to evolve an *integrated* (i.e. not modularized) dynamic neural network, as a control system for a simulated agent capable of performing learning behaviour, failed. They ended up having to use independent evolution of separate controller modules, some of which are dedicated to reactive and sequencing behaviour and others dedicated to learning. Moreover, they hard-wired into the model a reinforcement signal that works as a feedback signal for the learning module. For reasons to be explained in section 2 we believe that is important to carry through the Yamauchi and Beer experiments as originally intended, even though they were forced to make compromises that we find unsatisfactory.

The model we are describing in this paper is intended simply as a proof of concept. We go beyond the Yamauchi and Beer (1994) approach, demonstrating that it is possible to evolve an *integrated* dynamic neural network, with fixed connection weights and leaky-integrator neurons that successfully control the behaviour of a khepera robot engaged in a learning task, similar to one proposed by (Yamauchi and Beer, 1994). We will bring evidence that the leaky-integrator neuron, if shaped by evolution, appears to be able to integrate reactive and learned behaviour within a single control system that is not explicitly modularized. The *integrated* control system we are describing also benefits from its own evolved reinforcement signals to generate the appropriate behavioural responses; this is in contrast to their explicit introduction of hard-wired reinforcement signals.

1.1 Structure of the paper

In what follows, we will firstly present a very brief description of the Yamauchi and Beer (1994) experiment (see section 2). We will then point out that, although a promising one, the Yamauchi and Beer (1994)'s approach is not completely satisfactory because the autonomy of the simulated agent is severely compromised by an external reinforcement signal and externally imposed modularization of the controller 2.1.

In section 3 we describe the methodology used, the task, the evaluation function used for evolution, the integrated network, the simulated robot and the Genetic Algorithm.

In section 4 we show experimental results. In section 5 we make comparisons with the original Yamauchi and Beer (1994) experiment. We discuss the differences that might have singularly contribute to make it possible for us to evolve an *integrated* dynamic neural network where we assume neither an *a priori* separated controller module nor an *a priori* external reinforcement signal as in Yamauchi and Beer (1994). Conclusions are drawn in section 6.

2 Yamauchi and Beer's model of a learning robot

Yamauchi and Beer set up an experiment where a simulated robot had to repeatedly find a goal using a landmark for guidance. The relationship between the position of the goal and the position of the landmark changed occasionally within the sequence of searches, so successful behaviour required the robot to "learn" when these changes occurred.

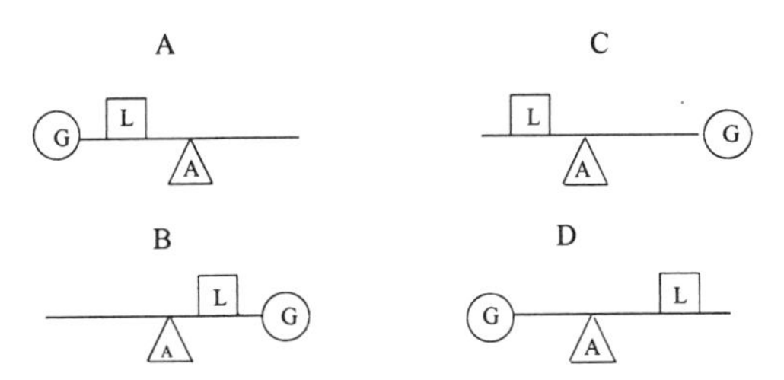

Figure 1: This picture is adopted from Yamauchi and Beer (1994). Environment for the one-dimensional navigation task. The triangle represents the agent, the circle represents the goal and the rectangle represents the landmark. The landmark-near environments (A and B) are on the left, the landmark-far environments (C and D) are on the right.

In the Yamauchi and Beer's model, an agent could move in either direction along a one-dimensional continuum environment (see figure 1). The environment contains a goal and a landmark. At the start of each trial, the agent was positioned in the center, and the goal was positioned randomly at either the left or right end. There were two possible way in which the landmark can be located within the environment. In the landmark-near environments 1A and 1B, the landmark was located on the same side of the agent as the goal. In landmark-far environments 1C and 1D, the landmark was located on the opposite side of the agent from the goal. The agent's task was to learn, over a series of successive trials, whether the current environment was landmark-far or landmark-near, and then reach the goal. Each trial lasted until either the agent reached the goal, the agent reached the non-goal end of the continuum, or the time limit was exceeded.

Attempts by Yamauchi and Beer to evolve a single fully CTRNN capable of solving the task were unsuccessful, so they used a modular, incremental approach instead, one in which the network architectures was split into three separately evolved subnetworks: one for goal-seeking behaviour in a landmark-far environment, one for goal-seeking behaviour in a landmark-near environment and one for environment classification. Each subnetwork consisted of a five-neuron fully interconnected CTRNN. All of these subnetworks had a sensor input for landmark detection, activated when the agents were directly in contact with the landmark, and two motor outputs (i.e. a left and a right motor neuron). In addition, the classifier network had a sensor input for reinforcement, which was provided every time the network reached the goal, and a classification output used to determine the environment type.

The two goal-seeking networks were specifically evolved to properly navigate in one or the other type of environment.

The classifier network were specifically evolved to control the behaviour of the agent during the first trial and to decide whether the agent was living in a landmark-far or in a landmark-near environment. The classification of the environment was landmark-far if the output of the classification neuron at the end of the first trial was less than 0.5, and landmark-near otherwise.

The evolved strategy was quite simple. The classifier network starts with the classifier output high, indicating a choice for landmark-near environment, and moving the robot towards the left side. Sensing a landmark causes the reversion of the direction of movement, while a reinforcement signal, hard-wired in to the model, makes the classifier output go low, indicating a choice for the landmark-far environment. If no reinforcement is provided, the classifier output remains high regardless of whether or not the agents had encountered a landmark.

These evolved responses make the classifier network able to correctly identify all four possible combination of environment type and goal location.

2.1 Thoughts and comments

We believe that the Yamauchi and Beer's model of evolved learning behaviour in artificial agents is valuable because it clearly represents one of the first attempts in which artificial evolution has been exploited to design the internal dynamics of a CTRNN to integrate the perceptions of the agents and to determine the agent's actions based upon these perceptions. In their model a fitness function rewards the behavioural responses that fulfil the conditions for learning.

However, they made use of an 'external' (i.e. not evolved) reinforcement signal, directly available to the robot, which signals to the robot when it performs correctly. This reward signal is clearly an externally imposed supervision, which severely compromises the autonomy of the system. Moreover, their attempts to evolve an *integrated* CTRNN, as a control system for a simulated agent capable of performing learning behaviour, failed. They ended up having to use independent evolution of separate controller modules, some of which are dedicated to reactive and sequencing behaviour and others dedicated to learning. This modularization of the controller reintroduces a structural and functional decomposition which is also externally imposed to facilitate the time-integration of different behavioural responses.

We argue that the Yamauchi and Beer (1994)'s model, relying as it does on independent evolution of separate controller modules (that are arbitrarily designed by the researcher and clearly dedicated to separate behavioural responses), and on an external reinforcement signal, failed to evolve an autonomous learning robot by selectively integrating the necessary plasticity directly into the mechanisms responsible for the robot's behaviour.

3 Evolving CTRNNs as controllers for autonomous learning robots

Drawing inspiration from the Yamauchi and Beer (1994)'s experiment, we set out to evolve an *integrated* dynamic neural network, with fixed synaptic weights and 'leaky integrator' neurons, as a control system for a robot engaged in a task where learning behaviour is required.

3.1 Description of the task

The task requires navigation within a rectangular arena in order to find a goal represented by a black stripe on the white arena's floor. The learning aspect requires the robots "to learn", over a series of successive trials, the relationship between the position of the goal with respect to a landmark that the robot must use for guidance to find the goal.

At the start of each trial, a robot is positioned in the left or right side of an empty arena (see the black

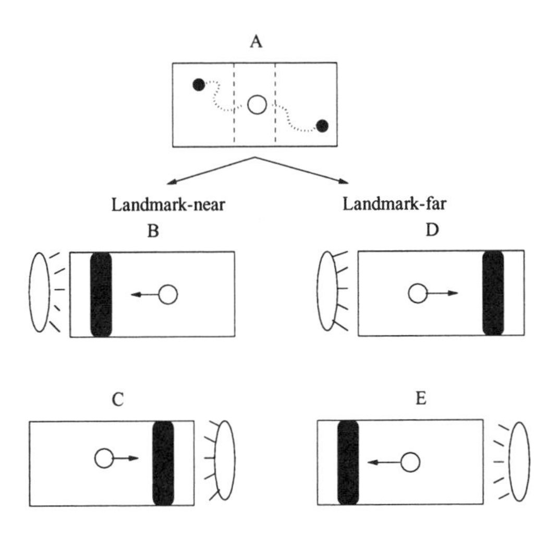

Figure 2: Depiction of the task. The small circle represents the robot. The white oval represents the landmark (i.e. a light source) and the black stripe in the arena is the goal. Picture A at the top represents the arena at the beginning of each single trial without landmark and goal. The black filled circles represent two possible starting points. The curved dotted lines represent two possible routes to the central part of the arena which is delimited by dashed lines. Pictures B and C represent the two possible arrangement of the landmark and goal within the landmark-near environment. The pictures D and E represent the two possible arrangement of the landmark and goal within the landmark-far environment. The arrows, in pictures B,C,D,E represent the directions towards which a successful robot should move.

filled circles in figure 2A). As soon as the robot reaches the central part of the arena (delimited in figure 2A by dashed lines), a goal (represented by a black stripe) is randomly positioned on the white arena's floor. This can be either in front of, or behind the robot, and is not perceivable by its proximity sensors (see figure 2B, 2C, 2D, 2E). The robot's behaviour will be evaluated according to a fitness function that rewards or penalises certain behaviours; this evaluation will be used to determine how many offspring it has within the evolutionary algorithm, but is not in any sense a reinforcement directly available to the robot. The robot is rewarded for leaving its current central position towards the goal until it finds it (see arrows in figure 2B, 2C, 2D, 2E); but it is penalized if it sets off away from the goal. The robot must "decide" which way to go; however, without any clue as to where the goal is, it can do no better than making a random decision.

The robot can use a landmark for guidance. The landmark is a light which lights up as soon as the goal is positioned, and it can be perceived by the robot from anywhere within the arena. The relationship between landmark and goal can vary; in some trials the landmark is close to the goal ('landmark-near environment' figure 2B and 2C), in other trials it is on the opposite side ('landmark-far environment' figure 2D and 2E). Both possibilities can be exploited by the robot to find the right way to the goal. If a robot approaches the landmark within the landmark-near environment, or moves away from the landmark within the landmark-far environment, it will consequently find the goal significantly more often than would be expected in a random walk. However, the current relationship between the landmark and the goal remains unknown to the robot until it has gathered enough experience to be able to disambiguate the environment between landmark-near or landmark-far.

Therefore, the task requires a robot to learn, over a series of successive trials, which of these two landmark-goal relationships currently holds true. The robot can disambiguate the environment between landmark-near or landmark-far by predicating its behaviour on its previous experience of the relationship between the location of the light and location of the goal. However, the simplicity of our scenario means that the light is only significant for the robot in this particular context. That is, the only adaptive reason for the robot to "pay attention" to the light is because the light is a cue to be exploited as part of the learning process. The robot has no reason to "pay attention" to the light in any other context. For a non-learning robot, the light presents no useful information because, on any given trial, there is no predictable relationship between the light and the goal. This presents a slight problem given that we wish to evolve the control mechanisms necessary for the robot to learn from the relative positioning of the light. It seems reasonable to suggest that unless the robot has *already* evolved to "pay some attention" to the light, it will be very difficult for it to evolve to learn from the relative relationship of the light position and the goal position. This problem can be addressed by given the light some significance other than as a learning cue. One way of doing this is to bias the proportion of each type of environment that the robots experience. For example, if robots are presented with landmark-near environment more often than the landmark-far environment, then the goal will be close to the light more often than it is far from it. In this way, the light can serve as a cue which can be exploited by even a purely reactive robot, since a robot which moves toward the light will be correct more often than it is wrong. Fully successful behaviour will nevertheless still require that the robot can learn to distinguish each type of environment and act appropriately. Rather than bias the proportion of times each environment is presented, we kept the proportion equal, but biased the relative weighting of the scores achieved in each environment, (as detailed in section 3.2, parameter $a = 1$ or 3.) in order to achieve the same effect.

The robot undergoes two test sessions in the

landmark-near and two more in the landmark-far environment. Each time the environment changes, the robot's control system is reset, so that there is no internal state or "memory" remaining from the previous environment. Within each of these four test sessions, a trial corresponds to the lapse of time that is given to the robot for reaching the middle of the arena and then finding the goal. The robot has 18 seconds to reach the middle of the arena. Then the landmark and the goal are positioned, and the robot has another 18 seconds to find the goal. Each trial can be terminated earlier either because the first time limit to reach the middle of the arena is exceeded; or because the agent reaches and remains on the goal for 10.0 seconds; or because the robot crashes into the arena wall. At the beginning of the first trial, and for every trial following an unsuccessful one the robot is positioned on either the left or the right part of an rectangular arena close to the short side. For every trial following a successful one, the robot normally keeps the position and orientation with which it ended the previous trial, but there is a small probability that it is replaced randomly. The simulation is deliberately noisy, with noise added to sensors and motors (see section 3.3); this is also extended to the environment dimensions. Every time the robot is replaced, the width and length of the arena, and the width of the central area that triggers the appearance of the landmark and the goal, are redefined randomly within certain limits. The robot undergoes 15 trials for each test session. During this time the relationship between landmark and the goal is kept fixed; the relationship remains unknown to the robot unless and until it can 'discover' it through experience. The position of the goal within the arena (i.e. left or right), is randomly determined every single trial and the landmark is subsequently appropriately positioned (depending on whether it is currently landmark-near or landmark-far). Over a full set of trials, all the 4 landmark/goal combinations (figure 2B, 2C, 2D, 2E) occur with equal likelihood.

3.2 Evaluation function

The robot is rewarded by the evaluation function F_{st} (per test session s, per trial t) for reaching the central area within the arena and then moving towards the goal until it find it (see arrows in figure 2B, 2C, 2D, 2E); but it is penalized either if it fails to reach the central area or if, once on the central area, it set off away from the goal.

$$F_{st} = (abcd)((3.0\frac{(d_f - d_n)}{d_f}) + \frac{p}{p_{max}})$$

with:

$s = [1, ..., 4]$, and $t = [2, ..., 15]$. The robot doesn't get any score during the first trial of each test session, because it is assumed that the robot doesn't know what kind of environment it is situated (i.e. landmark-near or landmark-far environment).

d_f represents the furthest distance that the robot reaches from the goal after the light is on. At the time when the light goes on, d_f is fixed as the distance between the centre of the robot body and the nearest point of the goal. After this, d_f is updated every time step if the new d_f is bigger than the previous one.

d_n represents the nearest distance that the robot reaches from the goal after the light is on. At the time when the light goes on, d_n is fixed as equal to d_f, and it is subsequently updated every time step both when the robot gets closer to the goal and when the robot goes away from the goal. d_n is also updated every time d_f is updated. In this case d_n is set up equal to the new d_f.

p represents the number of steps that the robot makes into the goal during its longest period of permanence. $p_{max} = 50$ is the maximum number of steps that the robot is allowed to make into the goal before the trial is ended;

a is a bias term for the type of environment; it is set to 3 in landmark-near environment, and to 1 in landmark-far environment.

b, c, d are the penalties. b is set to 0 if the robot fails to reach the central part of the arena, either because time limit is exceeded, or because it crashes into the arena wall; c is set to $\frac{1}{3}$ if the robot leaves the central area towards the opposite side of the arena respect to the goal (i.e unsuccessful behaviour); d is set to $\frac{1}{5}$ if the robot crashes into the arena wall; If the robot doesn't fall in any of these penalties, b, c, d are set to 1.

The total fitness of each robot is given by averaging the robot performance assessed in each single trail of each test session.

3.3 Simulated robot

A simple 2-dimensional model of a Khepera's robot-environment interactions within an arena was responsible for generating the base set aspects of the simulation (see Jakobi, 1997, for a definition of base set). The implementation of our simulator, both for the function that updates the position of the robot within the environment and for the function that calculates the values returned by the infra-red sensors and ambient light sensors, closely matches the way in which Jakobi designed his minimal simulation for a Khepera robot within an infinite corridor (see Jakobi, 1997, for a detailed description of the simulator). During the simulation, robot sensor values are extrapolated from a look-up table. Noise is applied to each sensor reading. Our robot sensor ability includes all its infra red sensors (Ir_0 *to* Ir_7 in figure 3), three ambient light sensors, positioned 45 degrees left and right and 180 degrees with respect to its face direction (A_1, A_4, A_6 in figure 3). The robot has a left and right motor, which can be independently driven forward

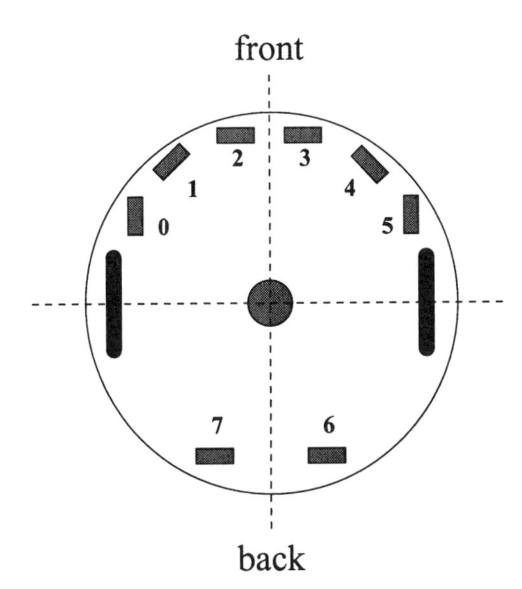

Figure 3: Plan of a Khepera mini-robot showing sensors and wheels. The robot is equipped with 6 infra red sensors (Ir_0 *to* Ir_5) and 3 ambient light sensors (A_1, A_4, A_6). It also has a floor sensor indicated by the central gray circle (F).

or in reverse, allowing it to turn fully in any direction.

The light is modelled as a bar that illuminates the whole arena with the same luminosity. Each ambient light sensor faces out from a point on the exterior of the robot and detects any light within plus or minus ±30 degrees from the normal to the boundary. The values returned by the ambient light sensors when impinged by the light, are set up to 1 if they exceed a fixed threshold otherwise they return 0. Our robot has an extra floor sensor, positioned on the belly of the robot, that can be functionally simulated as an ambient light sensor, that returns 0 when the robot is positioned over a white floor and 1 for a black floor. The simulation was updated the equivalent of 5 times a second.

3.4 The Network

Fully connected, 13 neuron CTRNNs are used. All neurons are governed by the following state equation:

$$\frac{dy_i}{dt} = \frac{1}{\tau_i}(-y_i + \sum_{j=1}^{k} \omega_{ji} z_j + gI_i)$$

$$with\ z_j = \frac{1}{1+exp[(y_j+\beta_j)]},\ i = 1, ., 13$$

where, using terms derived from an analogy with real neurons, y_i represents the cell potential, τ_i the decay constant, β_j the bias term, z_j the firing rate, ω_{ji} is the strength of synaptic connections from neuron j^{th} to neuron i^{th}, k corresponding to the number of input connections to neuron i both from other neurons and from itself, I_i the intensity of the sensory perturbation on sensory neuron i. 11 neurons receive input (I_i) from the robot sensors. These input neurons receive a real value (in the range [0.0 : 1.0]), which is a simple linear scaling of the reading taken from its associated sensor (i.e. neuron N_1 from infra red sensor Ir_0, N_2 from Ir_1, N_3 from Ir_2, N_4 from Ir_3, N_5 from Ir_4, N_6 from Ir_5, N_7 from $\frac{Ir_6+Ir_7}{2}$, N_8 from ambient light A_1, N_9 from A_4, N_{10} from A_6, N_{11} from floor sensor F). The 12th and the 13th neuron don't receive any input from the robot's sensors. Their cell potential y_i, mapped into [0.0 : 1.0] by a sigmoid function, and then linearly scaled into [−10.0 : 10.0], set the robot motors output. The strength of synaptic connections ω_{ji}, the decay constant τ_i, the bias term β_j, and the gain factor g are genetically encoded parameters. States are initialized to 0 and circuits are integrated using the forward Euler method with an integration step-size of 0.2. States are set to 0 every time the network is reset.

3.5 The Genetic Algorithm

A simple generational genetic algorithm (GA) was employed (Goldberg, 1989). Populations contained 100 genotypes. Each genotype is a vector of real numbers (length 196, given by 13 neurons, 169 connections, 13 decay constants, 13 bias terms, 1 gain factor). Initially, a random population of vectors is generated by initializing each component of every genotype to random values uniformly distributed over the range [0, 1]. Subsequent generations were produced by a combination of selection with elitism, recombination and mutation. In each new generation, the two highest scoring individuals ("the elite") from the previous generation were retained unchanged. The remainder of the new population was generated by fitness-proportional selection from the 70 best individuals of the old population. 100% of new genotypes, except "the elite", were produced by recombination. Mutation entails that a random Gaussian offset is applied to each real-valued vector component encoded in the genotype, with a probability of 0.1. The mean of the Gaussian is 0, and its s.d is 0.1. During evolution, vector component values were constrained within the range [0, 1].

Search parameters were linearly mapped into CTRNN parameters with the following ranges:

- biases $\beta_j \in [-2, 2]$;
- connection weights $\omega_{ji} \in [-4, 4]$;
- gain factor $g \in [1, 7]$.

Decay constants were firstly linearly coded in the range $\tau_i \in [-0.7, 1.3]$ and then exponentially mapped into $\tau_i \in [10^{-0.7}, 10^{1.3}]$

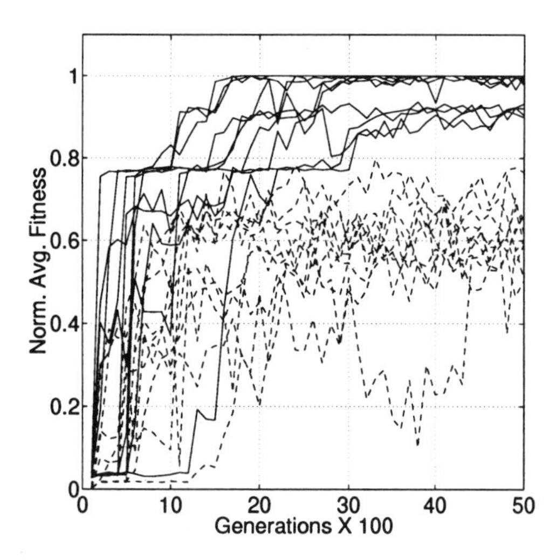

Figure 4: The graph shows the normalized fitness of the best robot (continuous line), and the normalized average fitness of the population (dotted line), for each generation for each run.

i	j					
	1	2	3	4	5	6
1	3.1	4	-1.7	3.6	0.2	-3.7
2	2.8	3.0	-0.6	-2.1	-2.0	0.4
3	-2.1	-3.3	-0.7	-1.8	-4	-4
4	-0.7	2.4	2.7	-4	0.7	2.8
5	-3.0	3.1	1.0	1.4	-1.2	-3.7
6	-1.5	-4	-0.0	-1.6	0.8	4
7	2.0	-1.6	2.3	3.4	-3.4	-2.3
8	4	1.6	-2.7	2.2	-2.2	-0.6
9	2.8	-0.5	-1.0	0.9	-3.9	-4
10	-3.7	-4	-2.8	4	4	-4
11	-4	2.9	-2.8	0.1	2.9	1.9
12	-2.9	-2.6	0.1	-4	-4	1.0
13	-0.4	0.8	2.5	-0.1	4	1.5

Table 1: The table above gives connection weights ω_{ij} from neuron N_i, (with $i = 1, ..., 13$) to neuron N_j (with $j = 1, ..., 6$) of the best network at generation 5000 of run n.1.

	i					
	1	2	3	4	5	6
τ_i	15.3	18.3	0.2	0.2	18.9	0.2
β_i	-0.9	-1.4	-1.4	-0.5	0.1	0.0

Table 2: Decay constants τ_i and the bias terms β_i, for neurons N_i, (with $i = 1, ..., 6$) of the best network at generation 5000 of run n.1.

i	j						
	7	8	9	10	11	12	13
1	-0.9	0.3	2.0	-3.3	1.4	3.4	2.7
2	2.1	4	4	-4	-1.8	-1.6	-3.4
3	-0.6	-2.3	3.2	-0.1	-1.9	-4	-3.6
4	-3.1	-1.4	0.3	1.3	-2.4	2.0	-4
5	-3.8	1.6	-1.3	-0.7	-1.8	1.5	-2.8
6	1.8	-4	3.8	3.4	-0.6	-2.5	-3.5
7	3.1	-0.7	1.4	1.2	4	3.0	-0.9
8	-0.7	-3.9	-1.1	-3.2	0.2	2.4	-4
9	-4	-4	-0.5	-0.0	-1.8	1.0	-0.5
10	-0.5	4	2.9	0.0	1.4	3.1	-1.9
11	1.9	-3.1	-3.5	2.6	2.9	-2.4	-4
12	-2.9	-4	2.4	-1.5	-3.9	2.6	4
13	-3.6	3.3	-3.2	1.4	3.0	3.0	0.4

Table 3: The table above gives connection weights ω_{ij} from neuron N_i, (with $i = 1, ..., 13$) to neuron N_j (with $j = 7, ..., 13$) of the best network at generation 5000 of run n.1.

	i						
	7	8	9	10	11	12	13
τ_i	3.4	0.6	0.2	0.3	0.2	1.3	0.2
β_i	-0.9	-2	-0.0	-1.4	0.7	0.8	-1.9

Table 4: Decay constants τ_i and the bias terms β_i, for neurons N_i, (with $i = 7, ..., 13$) of the best network at generation 5000 of run n.1. The gain factor g is equal to 6.88.

4 Results

Ten evolutionary simulations, each using a different random seed, were run for 5000 generations (see figure 4). We examined the best individual of the final generation from each of these runs in order to establish whether it had evolved an appropriate learning strategy. Recall that to perform its task successfully the robot makes use of the light source in order to navigate towards its goal. Thus, over the course of a test session, the robot must learn—and come to exploit—the relationship between the light source and the goal that exists in the particular environment in which it is placed. In order to test their ability to do this, each of these final generation controllers was subjected to 500 test sessions in each of the two types of environment (i.e. landmark-near and landmark-far). As before each session comprised 15 consecutive trials, and controllers were reset before starting a new session. During these evaluations we recorded the number of times the controller successfully navigated to the target (i.e. found the target directly, without first going to the wrong end of the arena).

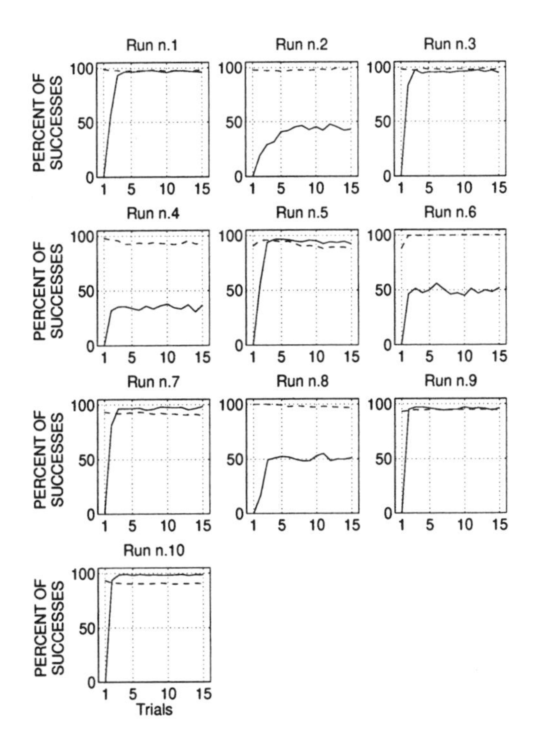

Figure 5: Each single graph refers to the performance of the best control system of the final generation of a single run, evaluated over 500 test sessions in both types of environment. Dashed lines indicate the percent of successful behaviour per trial in landmark-near environment. Continuous lines indicate the percent of successful behaviour per trial in landmark-far environment.

The results for the best controller of each of the ten runs can be seen in figure 5, which shows the percent of successful behaviour per trial during a session under each environmental condition. It is clear that in six of the runs (i.e. Run n.1, 3, 5, 7, 9 and 10) robots quickly come to successfully use the light to navigate toward the goal, irrespective of whether they are in an environment which requires moving towards or away from the light source. All these controllers employ a default strategy of moving toward the light, and hence are always successful in the landmark-near environment (see figure 5 dashed lines). Consequently they are initially very unsuccessful in the landmark-far environment (see figure 5 continuous lines). Nevertheless, as can be seen from figure 5, when these controllers are placed in the landmark-far environment they are capable of adapting their behaviour over the course of very few trials and subsequently come to behave very differently with respect to the light. Each of these controllers, when placed in the landmark-far environment, initially navigates toward the light, fails to encounter the target and subsequently reaches the wrong end of the arena. At this point it turns around and proceeds back toward the correct end of the arena where it ultimately encounters the target stripe. The fact that in subsequent trials the robot moves away from the light rather than towards it, demonstrates that the robot has learnt from some aspect, or aspects, of its earlier experience of the current environment. More prosaically, these controllers learn from their mistakes.

It should be noted that the controllers from the other five runs (i.e. 2, 4, 6, and 8), whilst performing the task with varying degrees of success, all modify their behaviour over the course of the initial trails; in each case this results in improved performance. For example, the controller from run 6 initially has a 80% success rate in landmark-near environment and a 0% success rate in landmark-far environment. In the landmark-near environment its success rate climb from around 80% to 100%. In the landmark-far environment its success rate will climb from 0% to around 50% within the course a few trials. Similarly, with the controller from run 2, 4 and 8, success in a landmark-far environment increases rapidly from 0% up to around 50% whilst success in the landmark-near remains consistently around 100%.

As reference, we provide connection weights (see table 1 and table 3), bias terms, decay constants and gain factor (see table 2 and table 4), of one of the best evolved networks, i.e. the best network at generation 5000 of run n.1. It should be notice that few weights are near to zero (i.e. 3 out of 169). This suggests that there is no immediately obvious evidence that the network can be meaningfully carved up into modularized structures of smaller number of neurons, functionally dedicated to particular sub-tasks. Unless and until such evidence is forthcoming, the default assumption is that functions are distributed across the network.

5 Discussion

Clearly there are strong similarities between Yamauchi and Beer (1994) and our experiment. We employed the same network style to control the behaviour of the robot, and we tested our robot in similar environmental conditions to those described in (Yamauchi and Beer, 1994). However there are also several differences which might have singularly contributed to our successful results.

First of all, contrary to (Yamauchi and Beer, 1994), we biased the relative weighting of the score achieved in each environment in order to give the landmark some significance other than as a learning cue. The effects of this feature of our fitness function can be clearly singled out both in figure 4 and in figure 5. From an evolutionary perspective, almost all the best robots in each run get to a stage in which they employ photo-taxis (see the plateau in continuous lines around fitness 0.8 in figure 4). This means that all the best robots make use of the light to navigate within the arena. Although the evolution of the mechanisms for photo-taxis contributes only three quarters of the total score, they might have significantly helped and guided the evolution towards the subsequent appearance of the learning controllers. The evolved learning robots consistently look for the light, and only when the conditions which reinforce the photo-taxis run out (i.e in the 'landmark-far environment') the robots change their behaviour to anti-photo-taxis (see figure 5).

Secondly, we didn't terminate the trial every time a robot finished up in the wrong end of the arena, far away from the goal, as it was for Yamauchi and Beer (1994)' agents. The robots, although slightly penalized for this, are allowed to recover from their mistakes. This produces a richer environment in which having more "time" and "freedom" to explore might help to evolve a proper reinforcement signal which makes the robot employ the right behaviour for each environment (i.e. landmark-near or landmark-far). However, the complexity and richness of our set up makes it harder to identify the condition of reinforcement as exploited by each learning robot. This is clearly an aspect of the evolved learning behaviour which needs to be clarified by further investigation.

Thirdly, our fitness function has not been designed in a discrete fashion either to strongly reward successful behaviour or to strongly penalize unsuccessful behaviour. In (Yamauchi and Beer, 1994) the agents get 1 for reaching the goal and 0 if they don't reach the goal. Instead of this 'black-and-white' approach, we arranged for 'shades of grey' by gradually increasing the reward starting from completely unsuccessful behaviours (e.g robots that don't manage to make the landmark and the goal appear on the arena) to very successful ones (e.g robots that reach the central part of the arena and then head successfully to the goal).

These are the three main differences between Yamauchi and Beer (1994) and our experiment which might have contributed to our successful results. However to assess the contribution that each of them brought to the achievement of our results, further tests and analysis need to be carried out.

6 Conclusion

In 1994 Yamauchi and Beer wanted to determine whether dynamic neural networks could provide an effective control mechanism for integrating reactive, sequential and learned behaviour in autonomous systems, without having to assume a sharp division between the components that are reactive and those that are in charge of sequencing or learning (Yamauchi and Beer, 1994). We gave a brief description of their experiment, which due to the characteristics of the control system and the singularity of the learning task employed, still is a distinctive and unique example in the area of autonomous learning robots (see section 2).

Their approach clearly represents one of the first attempts in which CTRNNs, with fixed connection weights and leaky-integrator neurons, have been exploited to integrate reactive, sequential and learned behaviour in a simulated robot. However, their attempts to evolve *integrated* CTRNNs, as controllers for simulated agents capable to perform learning behaviour, failed. They ended up having to use independent evolution of separate controller modules, some of which were dedicated to reactive and sequencing behaviour and others dedicated to learning. Moreover, they hard-wired into the model a reinforcement signal intended to be a feedback signal for the learning module.

The model we described in this paper demonstrated that it is possible to evolve *integrated* CTRNNs that successfully controls the behaviour of a simulated khepera mini-robot engaged in a learning task. We showed that dynamic neural network, if shaped by evolution, can integrate reactive and learned behaviour with a single control system that does not rely on independent evolution of separate controller modules, arbitrarily partitioned by the researcher, and does not rely on a hard-wired reinforcement signal. The evolved *integrated* control systems benefits from its own evolved reinforcement signals to generate the appropriated behavioural responses.

We hope that the evolutionary approach undertaken in our model might represent both a further methodological tool to test specific hypothesis about the selection pressures involved in the evolution of specific learning skills possessed by particular species, and a possible solution for engineering plastic control systems for learning robots.

Acknowledgments

The authors like to thank all the members of the Centre for Computational Neuroscience and Robotics (http://www.cogs.sussex.ac.uk/ccnr/) for constructive discussion and the Sussex High Performance Computing Initiative (http://www.hpc.sussex.ac.uk/) for computing support. ET likes to thank Elisa Niccolai, Andrea Vaccaro.

References

Davey, G. (1989). *Ecological learning theory.* Routledge, London, England UK.

DiPaolo, E. (2000). Homeostatic adaptation to inversion of the visual field and other sensorimotor disruptions. In Meyer, J.-A., Berthoz, A., Floreano, D., Roitblat, H., and Wilson, S., (Eds.), *From Animals to Animats VI: Proceedings of the 6th Interntional Conference on Simulation of Adaptive Behavior*, pages 440–449. Cambridge, MA: MIT Press.

Eggenberger, P., Ishiguro, A., Tokura, S., Kondo, T., and Uchikawa, Y. (1999). Toward seamless transfer from simulated to real worlds: A dynamically-rearranging neural network approach. In Watt, J. and Demiris, J., (Eds.), *Advances in Robot Learning, 8th European Workshop on Learning Robots, EWLR-8, Lausanne, Switzerland, September 18, 1999, Proceedings*, volume 1812 of *Lecture Notes in Computer Science*. Springer.

Floreano, D. and Mondada, F. (1996). Evolution of plastic neurocontrollers for situated agents. In Maes, P., Mataric, M., Meyer, J.-A., Pollack, J., and Wilson., S., (Eds.), *From Animals to Animats IV: Proceedings of the 4th International Conference on Simulation of Adaptive Behavior*. MA: MIT Press.

Floreano, D. and Urzelai, J. (2000). Evolutionary Robots: The Next Generation. In *The 7th International Symposium on Evolutionary Robotics (ER2000): From Intelligent Robots to Artificial Life*.

Floreano, D. and Urzelai, J. (2001a). Evolution of plastic control networks. *Autonomous Robots*. in press.

Floreano, D. and Urzelai, J. (2001b). Neural Morphogenesis, Synaptic Plasticity, and Evolution. *Theory in Biosciences*. in press.

Garcia, J. and Koelling, R. A. (1966). Relation of cue to consequence in avoidance learning. *Psychonimic Science*, 4:123–124.

Goldberg, D. E. (1989). *Genetic Algorithms in Search, Optimization and Machine Learning*. Addison-Wesley.

Harvey, I., Husband, P., Thompson, A., and Jakobi, N. (1997). Evolutionary Robotics: the Sussex approach. *Robotics and Autonomous Systems*, 20:205–224.

Harvey, I., Husbands, P., and Cliff, D. (1992). Issues in evolutionary robotics. In Meyer, J.-A., Roitblat, H., and Wilson, S., (Eds.), *From Animals to Animats II: Proceedings of the 2nd International Conference on Simulation of Adaptive Behavior*, pages 364–373, Cambridge MA. MIT Press/Bradford Books.

Jakobi, N. (1997). Evolutionary robotics and the radical envelope of noise hypothesis. *Adaptive Behavior*, 6:325–368.

Kaelbling, L. P. (1993). *Learning in Embedded Systems*. MIT Press.

Meyer, J.-A., Husbands, P., and Harvey, I. (1998). Evolutionary Robotics: A survey of Applications and Problems. In Husbands, P. and Meyer, J.-A., (Eds.), *Evolutionary Robotics: Proceedings of the 1st European Workshop, EvoRobot98*. Springer.

Nolfi, S. (1998). Evolutionary robotics: Exploring the full power of self-organization. *Connection Science*, 10:167–184.

Nolfi, S. and Floreano, D. (1999). Learning and evolution. *Autunomous Robots*, 7(1):89–113.

Nolfi, S. and Floreano, D. (2000). *Evolutionary Robotics: The Biology, Intelligence, and Technology of Self-Organizing Machines*. MA: MIT Press/Bradford Books.

Nolfi, S. and Parisi, D. (1997). Learning to adapt to changing environments in evolving neural networks. *Adaptive Behavior*, 5(1):75–98.

Webb, B. (2000). What does robotics offer animal behaviour? *Animal Behaviour*, 60:545–558.

Wilcoxon, H. C., Dragoin, W. B., and Kral, P. A. (1971). Illness-induced aversions in rat and quail: relative salience of visual and gustatory cues. *Science*, 171:826–828.

Yamauchi, B. M. and Beer, R. D. (1994). Integrating Reactive, Sequential, and Learning Behavior Using Dynamical Neural Network. In Cliff, D., Husbands, P., Meyer, J.-A., and Wilson, S. W., (Eds.), *From Animals to Animats III: Proceedings of the 3rd International Conference on Simulation of Adaptive Behavior*.

Using a net to catch a mate: Evolving CTRNNs for the Dowry Problem

Elio Tuci* **Inman Harvey*** **Peter M. Todd****
*Centre for Computational Neurosciences and Robotics,
School of Cognitive and Computing Sciences
University of Sussex, Brighton BN1 9QH, United Kingdom
eliot, inmanh@cogs.susx.ac.uk

**Center for Adaptive Behavior and Cognition
Max Planck Institute for Human Development, Lentzeallee 94, 14195 Berlin, Germany
ptodd@mpib-berlin.mpg.de

Abstract

Choosing one option from a sequence of possibilities seen one at a time is a common problem facing agents whenever resources, such as mates or habitats, are distributed in time or space. Optimal algorithms have been developed for solving a form of this sequential search task known as the Dowry Problem (finding the highest dowry in a sequence of 100 values); here we explore whether continuous time recurrent neural networks (CTRNNs) can be evolved to perform adaptively in Dowry Problem scenarios, as an example of minimally cognitive behavior [Beer, 1996]. We show that even 4-neuron CTRNNs can successfully solve this sequential search problem, and we offer some initial analysis of how they can achieve this feat.

1 Introduction

Life does not always present us with choices in the most convenient form. Rather than give us a supermarket of options to compare and choose between at our leisure — "pick a card, any card" — the world often conspires to confront us with a single possibility at a time — "take it or leave it". Another better option might come along later, but then again, it might not. This type of sequential choice problem can be encountered in searching for a job, or a place to live, or even a mate, when possibilities are seen one at a time and cannot be kept around for comparison to later options — the job, or house, or date you turn down today won't be waiting for you tomorrow. More generally, agents may have to deal with such sequential choice whenever resources they need are distributed in time or space: Should this prey item be pursued, or will a better one be found over the hill? Should this mate be courted, or will my genes stand a better chance with another individual later?

How can such a decision be made? This problem has been studied widely in statistics, economics, psychology, and other fields, and both optimal and non-optimal but reasonable approaches have been proposed, as we will discuss. But these proposals have all been made at the level of concrete symbolic information-processing algorithms: check this item, update that aspiration level. We wanted instead to find out whether lower-level subsymbolic dynamic systems of the sort proposed for biologically plausible agent architectures could also solve this problem, and explore what kinds of solutions evolution could find within this framework. Inspired by previous research on minimally cognitive behavior [Beer, 1996; Slocum et al., 2000], we set out to evolve continuous time recurrent neural networks (CTRNNs) that can deal with this important cognitive task of finding good options from sequences of possibilities.

1.1 The Dowry Problem

To study behavioral approaches to the problem of sequential choice, it is useful to take a well-specified form of the problem as a starting point. Probably the best-studied example of sequential search is captured in the Dowry Problem, in which the task is to select the woman with the highest dowry (money brought to a marriage) out of a sequence of 100, or its alternate form, the Secretary Problem, in which the goal is to select the best secretary (e.g. on typing speed) out of 100 applicants seen one at a time. In more detail (and with sex roles reversed and species changed for variety), the Dowry Problem goes like this:

A young female hangingfly wishes to find a mate, but she has many suitors to choose from. Male hangingflies attempt to woo her by offering her a nuptial gift (the equivalent of female dowries): an insect or other tasty morsel for her to dine on while the male goes about his inseminating business unnoticed. (From this point on we begin to take liberties with the true situation in nature for the purposes of formulating the problem properly — see Preston-Mafham and Preston-Mafham [1993] for the more peculiar real story.) She must choose a single male to mate with on the basis of the size of the nuptial gift he brings her. Her life is short, so she only has time to evaluate 100 of the eager males visiting her on her chosen leaf before she runs out of lifespan in which to mate successfully. Now her task in this problem is to select the one male, out of the 100 that she can possibly see, who has the biggest nuptial gift to offer her [1]. She can only see one male and his gift at a time, and then she must decide immediately if she thinks he is the one with the biggest nuptial gift out of all 100 males and accept his advances, or else rebuff him and go on to the next candidate. She cannot return to any male she has seen before — once she rejects them, they are gone forever. Moreover, the poor female has no idea of the range of gift sizes she might encounter, before she starts seeing the males — all of the males might be bringing fleas this season, or one might have bagged a whole rhino. What strategy can she possibly use to have the highest chance of picking the male with the biggest gift for her?

The Dowry Problem can be considered a formal model of many real-world decision problems, in which agents encounter options in temporal sequence, appearing in random order, drawn from a population with parameters that are partially or completely unknown ahead of time. What strategy can the agents possibly use to maximize the expected payoff or to minimize the expected cost associated with their choice? This general question has motivated research in statistics and probability theory [Ferguson, 1989], economics [Seale and Rapoport, 1997], and biology and psychology [Todd and Miller, 1999; Dudey and Todd, In Press]. Statisticians and economists tend to develop optimality theorems relevant to job search and consumer search. In biology and psychology, more psychologically plausible rules or heuristics for sequential choice have been proposed and investigated, decision algorithms that are not guaranteed to yield the highest likelihood of a best choice but which generally do a good job. Many of these studies focus on the problem of sequential mate search and mate choice, because of the evolutionary importance of this domain.

1.2 Structure of the paper

The Dowry Problem is a useful setting in which to explore the minimal cognitive mechanisms necessary to tackle the problem of sequential choice. To find such minimal cognitive mechanisms, we adopt the framework of evolutionary simulation, modelling the process of natural selection [Holland, 1975; DiPaolo et al., 2000] applied to dynamic neural networks as abstract instantiations of an agent's decision making structure. We take an evolutionary approach because it allows the exploration of possible cognitive architectures relatively unencumbered by *a priori* assumptions [Cliff et al., 1993; Cliff and Miller, 1996; Seth, 1998; Nolfi and Floreano, 2000]. As already pointed out by Slocum et al. [2000], these simple idealized models can serve as "frictionless planes" in which basic theoretical principles of the dynamics of agent-environment systems can be worked out.

In the next section, we present the analytic solution for the Dowry Problem and discuss the minimal cognitive requirements for this solution. In section 3 we describe our simulation for evolving Continuous Time Recurrent Neural Networks (CTRNNs) [Beer and Gallagher, 1992] with a genetic algorithm [Goldberg, 1989] to implement sequential choice strategies for the Dowry Problem. In section 4 the performance of the best evolved neural networks is analyzed and compared with the performance of some standard sequential choice algorithms. We demonstrate that the evolved CTRNNs can successfully deal with the slightly modified and more biologically plausible versions of the Dowry Problem in which perfection is not strictly required. Within these contexts, the neural network performance can actually surpass that of the standard strategies. Further analysis in section 5 gives some hints as to how the networks perform their search. Finally, we consider the implications of this work in section 6.

2 Solving the Dowry Problem

Roughly speaking, the Dowry Problem refers to a class of problems in which an agent has to maximize (or minimize) the expected payoff (or cost) given by choosing a single item from among a population of sequentially encountered items. The Dowry Problem is easy to state and has a striking solution. In its simplest form it has the following features[2]:

1. The agent (e.g., the female hangingfly) can make only one choice.

2. The number n of items in the population (e.g., number of males) is known.

3. The items are presented sequentially in random order, with each order being equally likely.

4. The items can be ranked from the best to the worst, without ties, according to a specific criterion (e.g.,

[1] Note that realistically she could do well to pick any male with a big-enough gift, and that evolution's goal would probably be to endow her with a way of picking large gifts on average. This is the payoff function we will explore in more detail later in the paper, but the original Dowry Problem is stated in terms of an all-or-nothing payoff, rewarding only selection of the single highest value.

[2] This formal description has been taken from [Ferguson, 1989] in which the problem is presented via its alter ago, the Secretary Problem.

size of nuptial gift). The decision to accept or reject an item must be based only on the relative ranks of those items presented so far.

5. Once an item is rejected, it cannot later be recalled (returned to).

6. The agent is very particular and will be satisfied with nothing but the very best item. (That is, the hangingfly's payoff is 1 — e.g., numerous offspring — if she chooses the best of the n males, and 0 — e.g., single death — otherwise).

7. If the agent does not make any choice before the end of the sequence, he or she must take the final option.

Agents facing the Dowry Problem seek perfection, with a payoff of one only for picking the very best item (highest dowry or biggest nuptial gift) and zero for picking anything else. They also ignore search costs such as time, ignore the problem of mutual choice (i.e., the possibility that the male hangingfly that the female selects will not agree to mate with her in return), and assume that they know the exact number of items that will be presented.

The solution to the Dowry Problem — that is, the strategy that gives the highest chance of selecting the single best option — requires sampling a certain initial subset of the population of items $r-1$ with $r \in [1, n-1]$, remembering the best of them, and then picking the next item that is even better [Ferguson, 1989]. It can be shown that for large n, it is optimal to wait until about 37% (i.e., $1/e$) of the population have been seen and then to select the next option encountered that is better than any seen previously. The probability of success in this case is also about 37%. For small values of n, the optimal r is the one that maximizes the probability $\phi_n(r)$, where

$$\phi_n(r) = (\tfrac{r-1}{n}) \sum_{j=r}^{n} \tfrac{1}{j-1}$$

One simple way to accomplish this optimal procedure is to set an aspiration level equal to the quality of the best option seen so far, updating this aspiration level as necessary as each new option is seen; then after 37% of the population has been encountered, fix this aspiration level and use it to stop the search with the next option seen that exceeds that aspiration. This approach is akin to Herbert Simon's notion of *satisficing* [Simon, 1990], and is what Seale and Rapoport refer to as a *cut-off rule* [Seale and Rapoport, 1997]. Thus, given that the searching agent knows the optimal 37% sample size — which assumes that the searcher first knows the total number of options n that could be encountered — following this *37% rule* is simple. The searcher must only keep in memory the optimal (37%) sample size and the quality of the best option seen so far. With each new option seen, a simple pair of comparisons (whether the option's quality exceeds the aspiration level, and whether the sequence-position number of that option exceeds the 37% value) is sufficient to decide whether to stop or continue search.

In a repeated version of the Dowry Problem where an agent sees multiple sequences and must choose the best value in each one independently, the agent must also be able to respond differently to the same item seen in different sequences, as a consequence of the fact that the item's relative ranking will vary across the sequences. This is an instance of the problem generally known as *sensory aliasing* [Nolfi and Marocco, 2001]. Sensory aliasing refers to the situation wherein two or more agent/environment states correspond to the same sensor pattern but require different responses. An agent that repeatedly plays the Dowry Problem game might encounter an item which is the best choice in one population and also the worst in a different population. An optimal strategy in this case should aim to select the item in the former case, and reject the same item in the latter. The analytic solution given above explicitly includes the memory and "behavioral plasticity" that is required for an agent to perform optimally when faced with multiple cases of the Dowry Problem. Our abstract model of the agent's decision making structure (i.e., CTRNNs) does not explicitly include these requirements. The challenge is for evolution to shape such a dynamic system so that its performance will compare favorably with that of the optimal strategy.

3 Evolving CTRNNs for the Dowry Problem

3.1 *Overview of the evolutionary scenario*

We evolve a population of CTRNNs to maximize the expected payoff given by choosing one item among a sequence of 20 items, where payoff is proportional to the relative rank (on some criterion) of the item chosen. (Here we use the term "rank" in the opposite of its usual sense — higher rank-numbers are better, and give higher payoffs.) The items are presented (see section 3.4) sequentially in random order, each order being equally likely. The network has to "decide" immediately after an item has been presented whether or not to select that item. A binary thresholded output neuron signals the decision of the network (see section 3.3). If, after the item is presented, the output neuron holds an activation value of 1, this item is chosen. If the output neuron holds 0, a new item is presented. Once rejected, an item cannot be recalled. The presentation of items is stopped either when the network selects an item or after the last of the 20 items is introduced; if the network does not select any item before the end of the sequence, the last item is taken as the network's choice. Then, the network is scored according to the relative ranking of the chosen item (see section 3.5). After that, the network is reset (see section 3.3), and a new sequence can be presented.

Our evolutionary scenario differs from the Dowry Problem mainly in the rank-proportional payoff function used (as opposed to the original all-or-nothing payoff). While picking the single best item yields the highest payoff, choosing the second best gives a somewhat reduced (but non-zero) payoff, picking the third gives slightly less than the second, and so on. This "friendlier" payoff function was used both because it is more realistic (see section 4.1), and because it proved to allow for much easier evolution of the networks. Aside from this, our scenario follows the Dowry Problem, with items simply being integer numbers randomly drawn from a pre-defined range (see section 3.2). Each network is evaluated on 60 different populations of items; we refer to each presentation of a sequence from a population of items as a trial. The 60 trials differ in the range from which their 20 numbers are drawn. Between any two trials there may be some sensory aliasing, in that the same items can occur in each trial but with a different ranking (a specific example is shown in the next section).

3.2 *How item values are drawn*

The networks are run through 60 independent trials (T_c, $c = 1, ..., 60$), each of which is made up of 20 integers. The integers are randomly drawn without replacement from a uniform distribution on the interval $[l_c, h_c]$, with the lowest number l_c and the highest number h_c of the range defined by the following rule:

$$c = 1, ..., 60 \left\{ \begin{array}{c} l_c = c; \\ h_c = l_c + 29 \end{array} \right.$$

While the networks see the trials in the simulation in order from the first to the sixtieth, this order is irrelevant, because the networks are reset (see section 3.3) at the beginning of every trial. (This of course cannot be done so easily for human subjects in Dowry Problem experiments, where great care must be taken to randomize the order of presentation of multiple trials, or to limit the experiment to a very small number of trials.)

Among all the trials, the lowest number that the networks can possibly experience is $l_1 = 1$ and the highest is $h_{60} = 89$. Values between 20 and 70 can be both the lowest within a trial and the highest within a different trial, while values between 2 and 88 can have a different ranking in different trials. The greatest amount of sensory aliasing between two trials occurs when they are consecutive (e.g., trials T_c and T_{c+1}); if two trials are more then 30 steps away from each other (e.g., T_c and T_{c+30}), they are certain to be made up of completely different items, so that no sensory aliasing exists between them.

Table 1 shows an example of two possible sequences of item-values that could occur on the first trial (T_1) and the 28th trial (T_{28}). The value 28 is an item in both populations, but in T_1 it is the best item, whereas in T_{28} it is the worst, yielding a very different rank-based score if it is chosen in the two trials. This is an example of sensory aliasing, because the network should select this item in T_1 and reject it in T_{28}.

Rank	**values**		**Rank**	**values**	
	T_1	T_{28}		T_1	T_{28}
1	4	28	11	16	44
2	5	33	12	17	46
3	6	35	13	18	47
4	7	36	14	21	48
5	9	37	15	22	49
6	10	39	16	23	50
7	11	40	17	24	51
8	12	41	18	25	52
9	13	42	19	27	53
10	14	43	20	28	57

Table 1: Two example trials of 20 values each, showing sensory aliasing owing to the value 28 being the highest-ranked item in the first trial and the lowest-ranked item in the second.

3.3 *The structure of the networks*

We model the agent decision-making mechanisms as fully connected, 4 neuron CTRNNs. All neurons are governed by the following state equation:

$$\tau_i \dot{y}_i = -y_i + \sum_{j=1}^{k} \omega_{ji} z_j + g I_i$$

$$\text{with } z_j = \frac{1}{1+exp[-(y_j+\beta_j)]},\ i = 1, ..., 4$$

where, using terms derived from analogy with real neurons, y_i represents the cell potential, τ_i is the decay constant, ω_{ji} is the strength of synaptic connection from neuron j to neuron i, k corresponds to the number of input connections to neuron i both from other neurons and from itself, z_j is the firing rate, I_i is the intensity of the sensory perturbation on sensory neuron i, g is the sensory gain factor, and β_j is a bias term. Only one neuron actually receives sensory perturbation in the networks we started with. There is also one output neuron that registers the network response. This neuron has a thresholded binary activation function: it will output 0 if its y-value ("cell-potential") is less than 0.5, otherwise it will output 1. The strength of synaptic connections ω_{ji}, the decay constant τ_i, the bias term β_j, and the gain factor g are genetically encoded parameters. Activation levels are integrated using the forward Euler method with an integration step-size of 0.2, which means that there are 5 updates of the network per second. To reset the network, all unit states are set to 0.

3.4 How items are presented to the network

Among the several possible ways in which an input can be presented to a network, we chose to use time-based input. While this is to some extent an arbitrary choice for our initial explorations described here, time (e.g. duration of sensory input) may correspond to magnitude in many natural situations. For instance, a bee can judge the size of a potential new hive site by the time it takes to fly from one side to the other, and a robot judging lengths of walls during navigations can use the time taken to pass along them. With such time-based input for the CTRNNs, the value of each item seen specifies the lapse of time in seconds (multiplied by 5 to give the number of network updates) during which the network external input I_i is set to 1 (see the network state equation in the previous section).

Thus, each trial proceeds as follows. At the beginning of each trial the network is reset. Then, before each item is introduced, the network is cleared for 2 seconds by setting the network external input I_i to 0 (where 2 seconds corresponds to 10 updates of the network's state). After that, the network is presented with the current item by setting the network external input to 1 for s seconds, with s equal to the value of the item being presented. After presentation, the network's output is checked to see if the item is accepted or not. If the item is rejected, the network is again cleared for 2 seconds, and then the next item is introduced.

3.5 How network fitness is scored

Each network is evaluated based on the items it has chosen in the 60 trials. Choices are made either by the network expressing its preference for some item, or by the network not expressing any preference and therefore being assigned the last presented item of the current trial as its choice. For each single trial, the payoff corresponds to the normalized relative ranking of the selected item. For example, assume that the network selects the item that holds value 13 among the population of items that made up trial T_1 in Table 1. The network's payoff for this trial corresponds to $P_1 = 9/20$ because 9 is the relative ranking of the selected number, 13, in that trial.

The total payoff $\bar{P}$ for each network is calculated by averaging the payoff obtained across the 60 trials:

$$\bar{P} = \frac{1}{60}\sum_{c=1}^{60} P_c, \quad c = 1, ..., 60.$$

3.6 The genetic algorithm

A real-valued steady-state genetic algorithm [Goldberg, 1989] was used to evolve the CTRNN parameters. A population of 50 individuals was maintained, with each individual's 4-neuron network encoded as a vector of 25 real numbers (16 connections, 4 decay constants, 4 bias terms, 1 gain factor). Initially, a random population of vectors was generated by setting each component of every individual to a random value uniformly distributed over the range [0, 1]. Individuals were selected for reproduction using a linear rank-based method with implicit elitism. Each vector component in a newly-created individual was mutated with a probability of 0.2 using a "creep" mutation operator which added a random number chosen uniformly from the range [-0.2, +0.2]. During evolution, vector component values could not move outside the range [0, 1]. Recombination when creating offspring was applied with a probability of 0.3.

i	1	2	3	4
τ_i	1.0	330.161170	1.071498	42.389673
β_i	-1.032616	0.187748	0.482778	-1.282664

Table 2: The table above gives the decay constants τ_i and the bias terms β_i, for each neuron of the best evolved network. The gain factor g is equal to 1.

Genetically encoded values were linearly mapped into CTRNN parameters with the following ranges: biases $\beta_j \in [-2, 2]$; connection weights $\omega_{ji} \in [-5, 5]$; and gain factor $g \in [1, 7]$. Decay constants were first linearly coded in the range $\tau_i \in [0, 2.8]$ and then exponentially mapped into $\tau_i \in [10^0, 10^{2.8}]$. These parameter ranges were chosen on the basis of having proven useful in other CTRNN experiments. The large range of the exponentially mapped decay constants is meant to allow for evolution to select both neurons that tend to change their state (cell potential) radically every time step (i.e., neurons with small decay constants), and neurons that tend to change their state only minimally every time step (i.e., those with large decay constants).

4 Results

Ten evolutionary simulations were run for 5000 generations each with 50 networks, and each network was assessed on all 60 sequences. We tested the best network of the final generation from each of these runs to establish how well these networks perform in the Dowry Problem and in another series of slightly different test problems. These problems differ from the standard Dowry Problem in the way in which payoffs are determined for chosen items, as explained in the next section. The very best network from all 10 simulation runs is shown in figure 1 with its connection weights ω_{ij}, while table 2) shows its decay constants τ_i, bias terms β_i, and gain factor g. This network's performance is compared with the average performance of the best networks from the 10 runs and with the performance of a set of more standard cutoff rules in section 4.2.

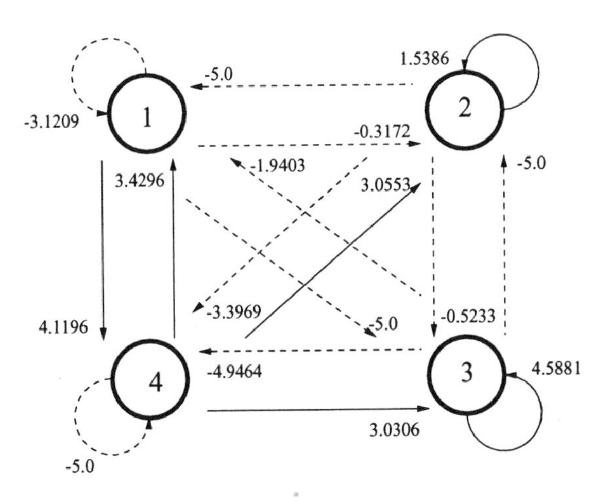

Figure 1: Morphology and connection weights of the best evolved network. The continuous lines indicate excitatory connections (positive weights), whereas dashed lines indicate inhibitory connections (negative weights).

4.1 Evaluation method

To evaluate the performance of the evolved networks, we developed an extended set of test problems because the Dowry Problem itself is probably not the most realistic or common form of sequential choice task that real agents could encounter. In the Dowry Problem, the single best option is the only good one — the only option that receives a non-zero payoff. But this very strict payoff function would be found in few natural situations.

In many species, many (if not most) animals find some mate, some food source, and some place to live, and thus receive some payoff in these search domains, even if they are not getting the highest possible payoff. Thus in these cases, a payoff proportional to the quality of the alternative chosen (e.g., proportional to the size of the nuptial gift selected, and as used in our fitness function for evolving the CTRNNs) is more appropriate than the all-or-nothing payoff of the standard Dowry Problem. In other species, the available alternatives (e.g., mates or habitats) may be limited so that not all individuals succeed in making a viable choice — for instance, only a quarter of the possible habitats in which one could settle may provide enough resources to raise offspring. For such cases, the search payoff function might fully reward only choices made in the top 25% of all available alternatives and give zero payoff to all other choices. Based on such considerations, we used the following set of test problems and associated payoff criteria [Todd and Miller, 1999]:

1. Dowry Problem — For each trial, the payoff is 1 for selecting the single best item, and 0 otherwise.
2. Top 10% Problem — For each trial, the payoff is 1 for selecting an item in the top 10% of the population of items for that trial, and 0 otherwise.
3. Top 20%, 30%, 40%, 50% Problems — As above, with payoff 1 for selecting an item in the top n% of the population, and 0 otherwise.
4. Bottom 20% Problem — For each trial, the payoff is 1 for selecting an item that is *not* in the bottom 20% of the population of items, and 0 otherwise.
5. Bottom 10% Problem — For each trial, the payoff is 1 for selecting an item that is *not* in the bottom 10% of the population of items, and 0 otherwise.
6. Maximize Mean Ranking Problem — For each trial, the payoff is proportional to the ranking of the value picked, calculated as explained in section 3.5.

Note that the last payoff function is the same as the fitness function that was used during network evolution; all of the other performance measures are testing the ability of the CTRNNs to generalize to new (though related) tasks. (Each network and strategy was evaluated on all performance measures simultaneously — that is, each choice made was evaluated for whether it was the single best, and in the top 10%, and bottom 20%, etc.)

To provide a benchmark for evaluating the performance of our best evolved networks, we also tested a range of cutoff rule strategies as introduced in section 2. These cutoff rules (as described in section 2) require checking a certain number $r-1$ of items from the population with $r \in [1, n-1]$ (where n is the number of items in the whole population), remembering the best of those, and then choosing the next item seen that is better than the best seen so far [Seale and Rapoport, 1997]. Varying the parameter r across its whole range, we get $n-1$ possible different cutoff rule strategies, or 19 different cutoff rules for our population of 20 items. For example, the cutoff rule defined by setting $r = 1$ samples 0 items (so that the best seen so far is also 0, or undefined), and hence always chooses the first presented item. This yields random performance in the Dowry Problem, which is a useful benchmark. The cutoff rule defined by setting $r = 11$ samples 10 items, remembers the best of them, and then picks the next seen (starting with item 11) that is even better. If no subsequent item is seen that is better than the best encountered during the initial sample (that is, no further item exceeds the aspiration level set), then the network "picks" the last item presented as its choice.

It is important to point out that these cutoff rules are not necessarily the optimal solution for each of our particular test problems, but they are a simple strategy that agents can use (and that people seem to use–see Seale and Rapoport [1997]). Thus, by finding the best cutoff rule for a particular test (i.e., by finding which value of r leads to the highest performance), we can establish a useful, and psychologically plausible, benchmark

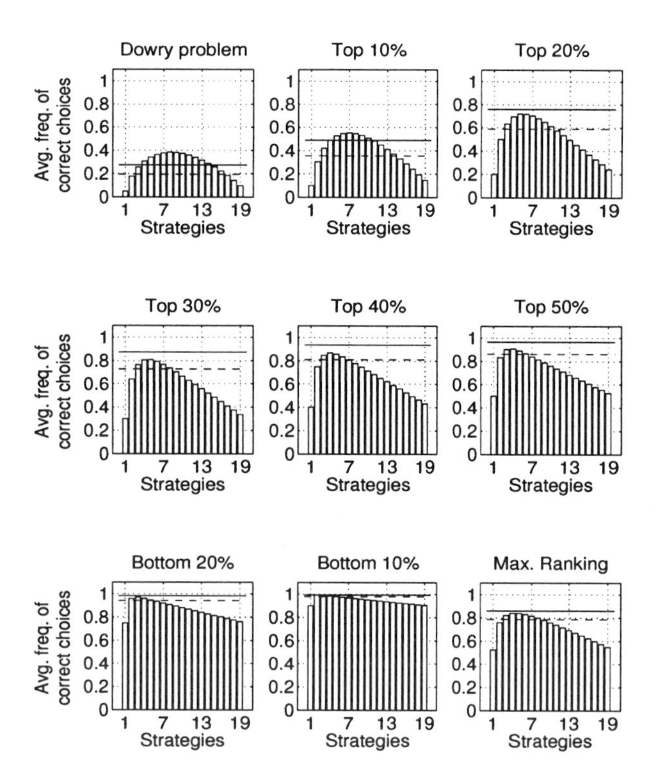

Figure 2: Performance on the different test problems for all cutoff rule strategies (white bars) compared with the single best evolved network (continuous lines) and the average performance of the best evolved networks from the 10 runs (dashed lines). For all graphs, the x-axis indicates the r value for the cutoff rules, while the y-axis shows the average payoff, calculated according to the particular test problem definition.

for comparison with the evolved networks. The optimal solutions are actually difficult to find; for the task of minimizing the mean rank selected (here with rank 1 being best), it is known that the optimal solution will find a rank of 4 or better on average [Chow et al., 1964], yielding another rough benchmark of 85% performance on the Maximize Mean Ranking Problem.

4.2 Comparing performance

Figure 2 shows the performance on the different test problems of the 19 cutoff rule strategies compared with that of the very best evolved network and the average performance of all the best evolved networks from the 10 simulation runs. In the Dowry Problem, the performance of the best evolved network does not match that of the best cutoff rule (for $r = 8$), achieving 27% success rate instead of the optimal 37%.

However, when perfection is not the goal (as it is for the Dowry Problem), the performance of our best evolved network gets more impressive. In the Top 10% Problem the performance of the best evolved network comes very close to the performance of the best cutoff rule strategy. Even more surprisingly, for all the other test problems the best network overcomes the performance of the best cutoff rule. How can this be?

5 Analyzing the best evolved network

It should be recognized that CTRNNs, with multiple recurrent connections and no a priori division into modules, are inherently difficult to analyze. There may be no direct translation of the network behavior into a strategy described in terms of simple rules. However, we can look into how the networks process their input information both behaviorally (e.g., by testing their choices in different experimental settings) and "neurologically" (e.g., by monitoring changes in internal activation levels over time) and see if this suggests any rules that can summarize their behavior.

To try to understand how the best evolved network can perform so well on search criteria related to the Dowry Problem, we should first consider how the problem we created differs from the true Dowry Problem. First, we always present values from a uniform distribution; second, we always present values in one sequence (trial) from within a 30-number span; third, the evolving networks only ever see numbers from 1-89. Given these restrictions, the evolutionary process can build a more specific solution to the search problem it faces. For instance, one approach would be to use the first few (even just two or three) values seen in a trial to compute an estimated mean of the range for this trial, and then add 12 (a bit less than half of the full range of 30) to this estimate to create the aspiration level for use in further search. We have not yet determined the strategy used by the best evolved network to this level of detail, but we have begun to take steps in that direction, as we now describe.

The first aspect of the network's behavior that we looked at is the percent of choices that are made at each relative rank within each of the 60 trials (see Figure 3), both for the best evolved network and for the best cutoff rule for the original Dowry Problem ($r = 8$). Choices here are both those that the network or strategy actively makes, and those that it is passively assigned when it must take the final presented option by default. Figure 3–Network shows that the best evolved network chooses the best item (rank 20) with the highest frequency throughout all the trials, regardless of the actual value of the item itself. The network also performs slightly better within the range of values that can be found in trials 10 to 40. Figure 3–Strategy shows that the cutoff rule distributes its choices across the relative ranks in a similar, but more consistent, manner, and with more of an emphasis on the single best value.

We focus in on explicitly expressed preferences (i.e., not including default final-value choices) in Figure 4,

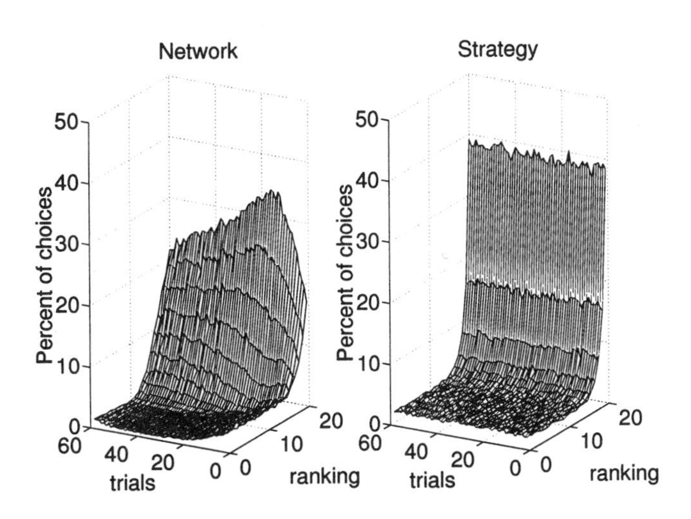

Figure 3: Percent of choices made per rank and per trial by the best evolved network (on the left) and by the cutoff rule with $r = 8$ (on the right). Ranking 1 refers to the item with the lowest (worst) value, while ranking 20 refers to the item with the highest (best) value.

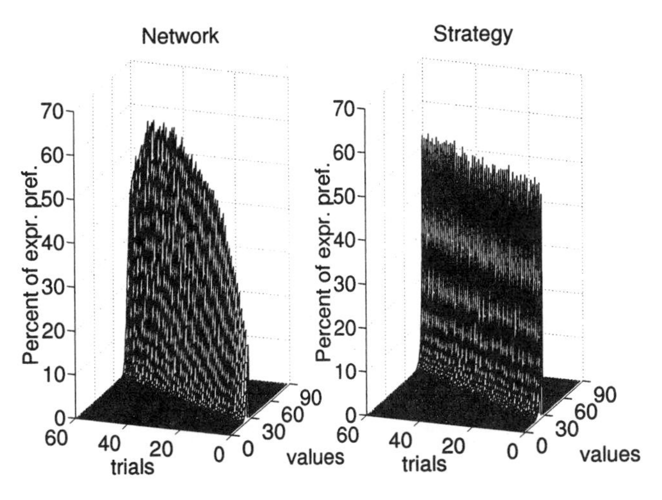

Figure 4: Percent of expressed preferences (actively-made choices) per item value and per trial by the best evolved network (on the left) and by the cutoff rule with $r = 8$ (on the right).

where we show the percentage of trials on which particular values (rather than ranks) are selected by the best network and cutoff rule. This graph makes it clear that the network tends to choose particular item values only when they are among the highest within the current population, thus overcoming the sensory aliasing problem. For example we can see from figure 4–Network that values around 30 were selected more frequently when they appeared in the first 10 trials than when they appeared in trials 20 and above.

Finally, figure 5 helps us to speculate on the possible approach that the network might have evolved. To understand the behavior of the network we refer to four different features of its decision making recorded during the evaluation test. Figure 5a shows the percent of expressed preference per trial. Figure 5b shows the average position at which a preference is expressed per trial. Figure 5c shows the average number of seconds of presentation of the ultimately-selected item in a given trial (that is, the amount of time that the input neuron receives input=1) until the activation level of the network's output neuron goes to 1. This is essentially the aspiration level or threshold that the network has set, based on the items it has seen so far in this trial, at the time that it makes its active choice. Graph d shows the average total number of seconds of input=1 from the beginning of each whole trial to the time when the activation level of the network's output neuron goes to 1 (making its choice) in each single trial. This is essentially how much "history" (in the sense of total summed input values) is required within a trial before the network makes an active choice.

Bearing in mind that the average value of the trial-population of items increases from trial 1 to trial 60, we can then describe and summarize the results shown in figure 5 in a few concepts: First, the network is not always expressing its preference with the same frequency (see Figure 5a). Looking at figure 5a, we can see that, from trial 1 to trial 15, the conditions triggering the network choice occur more often with later trials. After trial 15 these conditions seem always to be satisfied by the distribution of item values, until about trial 50, when the network starts getting more silent and expresses fewer explicit preferences. For the moment, we leave out any explanation of this phenomenon, but it will be treated more extensively in the next section.

Secondly, the response of the network is triggered by somewhat different conditions in each trial (see Figure 5b,c,d). These conditions are defined by the item values seen before a choice is made. Recall that the network experiences the item values in terms of 10 updates before each item with the external input set to 0, followed by a number of updates (5 times the item value) with the external input set to 1. Figure 5c then shows that the aspiration level of the network (in terms of amount of time or number of updates over which the selected item is presented before it is selected) goes up as the trials, and hence population values seen, go up. This is to be expected, in order for the network to perform well; but the increase in the aspiration level is not strictly proportional to the increase in the values in the population, with the first trial (values between 1 and 29) leading to an average aspiration level of about 25, while the last trial (with values between 60 and 89) leads to an aspiration level of about 80 (rather than 85). This can also explain why choices are made later in each trial for the early trials, and sooner in each trial for the later trials (Figure 5b). Finally, Figure 5d shows that, if we

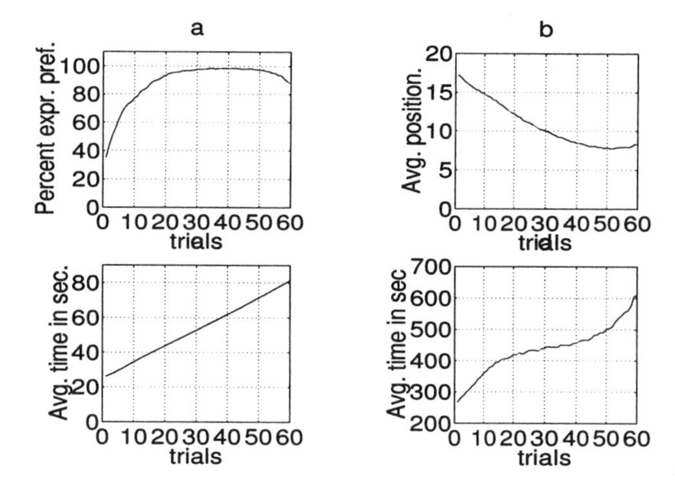

Figure 5: a. Percent of trials in which the best network actively expressed a preference, versus trial number. b. Average position at which an actively expressed choice is made per trial. c. Average number of seconds of input=1 from the beginning of the presentation of the selected item to the time when the activation level of the network's output neuron goes to 1. d. Average number of seconds of input=1 from the beginning of each whole trial to the time when the network output neuron's activation level goes to 1.

consider the sum of values seen before an active choice is made, this sum goes up with later trials; however, this is counteracted by the fact that fewer items are needed to create these higher sums in later trials (so again as shown in Figure 5b, with later trials fewer individual items must be seen before a choice is made).

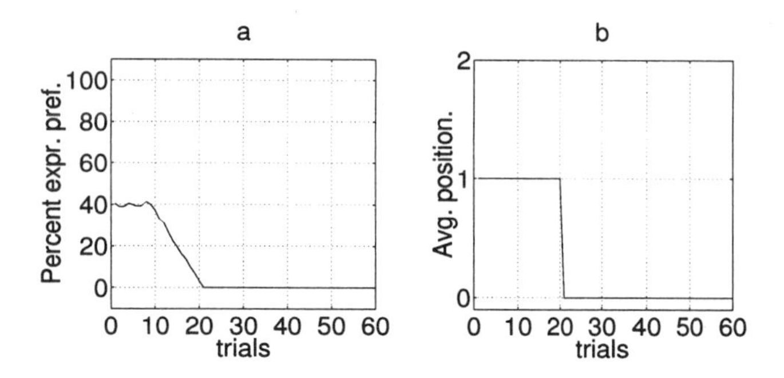

Figure 6: Robustness of the best evolved network. a. Percent of expressed preference per trial. b. Average position at which a preference is expressed in each trial.

5.1 *Robustness of the best evolved network*

We have carried out further tests to explore the behavioral robustness of the best evolved network. We ran the network 5000 times in an "unknown" Dowry Problem scenario, where the item values were taken from a distribution that had not been seen by the network nor by any of its ancestors. Values were drawn from a uniform distribution on the interval $T_c \in [l_c, h_c]$, with the lowest number l_c and the highest number h_c of the range defined as follows:

$$c = 1, ..., 60 \left\{ \begin{array}{l} l_c = 89 + c; \\ h_c = (l_c) + 29; \end{array} \right.$$

The message from the graphs in figure 6 is quite clear. The network starts expressing its preference with a rather low frequency (figure 6a), always selecting the first presented item (figure 6b). The frequency of expressed preference continues to decrease until trial 20 when it reaches 0 (figure 6a). At this point, the network stops making active choices (and then gets the last value as a default choice, yielding random performance on the task).

This message is clear, but it is also counterintuitive. We expected a different outcome, because we thought that any value that can trigger the network's response in trial 1 in this problem will appear more frequently on the following trials as the values get higher, so the rate of active choice should go up. Also, we thought that higher values should produce more active choice. Why was this not the case?

Part of the explanation may lie in the way that we interpret the output of the network. Bearing in mind that the network's output neuron — with which the network expresses its active choices — is checked only at the *end* of any item presentation, the behavior of the network may contradict our intuition because the output neuron's activation value need not rise to 1 and stay there during an item's presentation. That is, after the output neuron activation has been set to 1 by the conditions that trigger an active network choice, it may return to 0 if the network continues to be updated longer with the external input set to 1. Then, when the output level is checked at the end of the item's presentation, it may be back to 0, indicating that this item should not be chosen, even though it was at 1 earlier (which could have indicated active choice). Of course, we can get around this by having the unit act as a toggle, so that once output is 1, the choice is made immediately; this interpretation of the output unit must be tried during evolution to see what effect it has on the ability of these networks to generalize to other value ranges.

This same phenomenon may be responsible for the decrease in frequency of expressed preference shown by the best evolved network during the last 10 trials of the evaluation test (see figure 5a). In these trials, the item values are certainly the largest ever experienced by the network (and its ancestors). For particular orders of presentation, these items might cause the network to output 0 (do not choose) at the end of a particular item's presentation, even though the network might have output 1 (choose this item) during the presentation.

6 Conclusions

We have shown here that simple (or at least small) dynamic neural networks, shaped by evolution, can successfully solve the important adaptive problem of sequential choice, as embodied in variants of the Dowry Problem. This work demonstrates that CTRNNs are capable of exhibiting another form of minimal cognitive behavior [Beer, 1996; Slocum et al., 2000] — a behavior that agents in the real world face whenever they must find resources distributed in time or space. The surprising finding here is that the evolved networks can even outperform simple cutoff rules that usually do very well on such search problems; this performance advantage is pronounced when more biologically realistic forms of payoff (such as rank-proportional payoff) are employed. In addition, our results showed that the evolved CTRNNs can overcome a form of the sensory aliasing problem: the networks employ a "behavioral strategy" that is plastic enough to respond differently to the same value seen in different contexts determined by the current item population.

The reasons for this surprising performance must still be uncovered; so far, we have just begun to analyze how the networks achieve this feat. While our analyses to date have helped us to begin to see how the best evolved network is operating, several further tests are needed. In particular, we want to see how the network's aspiration level changes as a result of particular sets of input values it sees. For instance, does seeing the values {12, 18} result in the same aspiration level as {18, 12} or {12, 15, 18}? (It does not in human experiments.) How do the activation levels of the individual neurons change over time with presentation of a given value? Does the network seem to keep track of minimum or maximum values seen, or the average of all values, or the range of all values, or some combination of these? (We can try to determine this by a combination of manipulating the network's input and looking at what the neurons are doing.) By gathering these additional observations, we can home in on the kinds of strategies that this best network and other evolved networks are using. As a further step, we can then compare the mechanisms used by the networks with those used by people in similar experimental settings, as a way to try to understand more about how real biological agents deal with sequential search and choice problems.

Future work will explore the applicability of sequential choice CTRNNs for search problems influenced by other factors such as changing environments (e.g., where the population of values goes up or down in overall quality over time as the search is progressing) or multiple cues (e.g., where the agent learns about different features of each item rather than about each item's criterion value directly). These settings will allow our approach to be generalized to more biologically realistic settings including mate choice and habitat choice, enabling us to further extend the range of cognitive behaviors that CTRNNs can model beyond the merely minimal.

Acknowledgements

The authors wish to thank all the members of the Centre for Computational Neuroscience and Robotics (http://www.cogs.sussex.ac.uk/ccnr/), the ABC Research Group (http://www.mpib-berlin.mpg.de/abc) for constructive discussion and the Sussex High Performance Computing Initiative (http://www.hpc.sussex.ac.uk/) for computing support. We also thank two anonymous reviewers for their useful comments.

References

Beer, R. D. (1996). Toward the evolution of dynamical neural networks for minimally cognitive behavior. In Maes, P., Mataric, M., Meyer, J., Pollack, J., and Wilson, S., (Eds.), *From Animals to Animats IV: Proc. of the 4th Intl. Conf. on Simulation of Adaptive Behavior*, pages 421–429. Cambridge, MA: MIT Press.

Beer, R. D. and Gallagher, J. C. (1992). Evolving dynamic neural networks for adaptive behavior. *Adaptive Behavior*, 1(1):91–122.

Chow, Y. S., Moriguti, S., Robbins, H., and Samuels, S. M. (1964). Optimal selections based on relative ranks. *Israel Journal of Mathematics*, 2:81–90.

Cliff, D., Harvey, I., and Husbands, P. (1993). Explorations in evolutionary robotics. *Adaptive Behavior*, 2(1):73–110.

Cliff, D. and Miller, G. (1996). Co-evolution of pursuit and evasion II: Simulation methods and results. In Maes, P., Mataric, M., Meyer, J.-A., Pollack, J., and Wilson., S., (Eds.), *From Animals to Animats IV: Proc. of the 4th Intl. Conf. on Simulation of Adaptive Behavior*, pages 506–515. Cambridge, MA: MIT Press.

DiPaolo, E. A., Noble, J., and Bullock, S. (2000). Simulation Models as Opaque Thought Experiments. In *Artificial Life VII: The 7th Intl. Conf. on the Simulation and Synthesis of Living Systems.*

Dudey, T. and Todd, P. M. (In Press). Making good decisions with minimal information: Simultaneous and sequential choice. *Journal of Bioeconomics.*

Ferguson, T. S. (1989). Who Solved The Secretary Problem? *Statistical Science*, 4(3):282–296.

Goldberg, D. E. (1989). *Genetic Algorithms in Search, Optimization and Machine Learning.* Reading, MA:Addison-Wesley.

Holland, J. H. (1975). *Adaptation in Natural and Artificial Systems.* Ann Arbor, MI: University Press.

Nolfi, S. and Floreano, D. (2000). *Evolutionary Robotics: The Biology, Intelligence, and Technology of Self-Organizing Machines.* Cambridge, MA: MIT Press/Bradford Books.

Nolfi, S. and Marocco, D. (2001). Evolving robots able to integrate sensory-motor information over time. *Theory in Biosciences*, 120(3-4):287–310.

Preston-Mafham, R. and Preston-Mafham, K. (1993). *The encyclopedia of land invertebrate behavior.* Cambridge, MA: MIT Press.

Seale, D. A. and Rapoport, A. (1997). Sequential decision making with relative ranks: An experimental investigation of the "secretary problem". *Organizational Behavior and Human Decision Processes*, 69(3):221–236.

Seth, A. (1998). Evolving Action Selection and Selective Attention Without Actions, Attention, or Selection. In Meyer, J.-A. and Wilson, S. W., (Eds.), *From Animals to Animats V: Proc. of the* 5th *Intl. Conf. on Simulation of Adaptive Behaviour*, pages 139–147. Cambridge, MA:MIT Press.

Simon, H. A. (1990). Invariants of human behavior. *Annual Review of Psychology*, 41:1–19.

Slocum, A. C., Downey, D. C., and Beer, R. D. (2000). Futher Experiments in the Evolution of Minimal Cognitive Behavior: From Perceiving Affordances to Selective Attention. In Meyer, J.-A., Berthoz, A., Floreano, D., Roitblat, H. L., and Wilson, S. W., (Eds.), *From Animals to Animats VI: Proc. of the* 6th *Intl. Conf. on the Simulation of Adaptive Behavior*, pages 430–439. Cambridge, MA: MIT Press.

Todd, P. M. and Miller, G. F. (1999). From Pride and Prejudice to Persuasion: Realistic Heuristics for Mate Search. In Gigerenzer, G., Todd, P. M., and the ABC Research Group, *Simple Heuristics That Make Us Smart*, chapter 13. Oxford University Press, New York.

Fast homeostatic neural oscillators induce radical robustness in robot performance

Ezequiel A. Di Paolo
School of Cognitive and Computing Sciences, University of Sussex
Brighton, BN1 9QH, UK, ezequiel@cogs.susx.ac.uk

Abstract

A network of relaxation oscillators is evolved to produce phototaxis in a simulated robot. Oscillations are faster than the timescale of performance, and are designed to maintain the same average activation value independently of sensory or synaptic input. Neural activation cannot correlate with any action-relevant sensory information, but must be continuously modulated by sensorimotor coupling. Radical sensor robustness is shown by inverting the position of the sensors and also by removing either of them in turn – operations that do not alter the success of the strategy. Slowing down the timescale of oscillations results in less robustness.

This paper explores some of the issues that arise from evolving oscillatory neural controllers which operate timescales faster than that of performance, and whose elements compensate for long terms input patterns by keeping their average activation as close as possible to a middle range, thus making it difficult for action relevant information to be stored in individual elements.

Continuous-time recurrent neural networks (CTRNNs) (Beer, 1990) are extended to transform each neuron into a centre-crossing relaxation oscillator that maintains a constant average activation of 0.5. The equations describing traditional CTRNNs are:

$$\tau_i \dot{y_i} = -y_i + \sum_j w_{ji} z_j + I_i; \quad z_j = \frac{1}{1+\exp[-(y_j+b_j)]},$$

where y_i represents the membrane potential, τ_i the decay constant (range [0.4,4]), b_i the bias, z_i the firing rate, w_{ij} the strength of synaptic connection from node i to node j (range [-8,8]), and I_i the degree of sensory perturbation on the sensor node. These equations are extended by turning the neuron bias into a responsive variable that "keeps track" of the opposite value of the membrane potential with a τ_i^b greater than τ_i:

$$\tau_i^b \dot{b_i} = -(b_i + y_i); \quad \tau_i^b = \tau_i G_i;$$

where G_i ranges from 1 to T and is genetically set. The consequence of adding this equation is that the bias term is no longer constant but adapts so as to maintain a long term average firing rate of 0.5, regardless of all the other parameters and input pattern. The parameter T indicating the range of allowed values for τ_i^b was set for different series of runs at 2, 5, and 10.

Simulated robots are evolved to perform phototaxis on a series of light sources. Robots have circular bodies of radius $R_0 = 4$ with two motors and two light sensors. The angle between sensors is of 120 degrees. Motors can drive the robot backwards and forwards in a 2-D unlimited arena. Tests were run with fully connected networks of 4, 6, 8, 10 and 20 nodes with similar results. More details in (Di Paolo, 2002).

Robots are run for 4 independent evaluations, each consisting on the sequential presentation of 6 distant light sources. Only one source is presented at a time for a relatively long period T_S chosen randomly for each source from the interval [300,500]. The initial distance between robot and new source is randomly chosen from [60,80], the angle from $[0,2\pi)$ and the source intensity from [500,1500].

A population of 30 robots is evolved using a generational GA with truncation selection. All parameters are encoded in a real-valued vector. Only vector mutation is used with a standard deviation of vector displacement of 0.25. Fitness is calculated according to:

$$F = \frac{1}{T_S}\int f dt; \quad f = 1 - \frac{d}{D_i}$$

if the current distance to the source d is less than the initial distance D_i, otherwise $f = 0$.

In all cases the GA was run for 2000 generations. Control runs were performed using standard CTRNNs with genetically set but fixed bias terms from the range [-3,3]. The following results correspond to 6-node neural controllers. Each data point is taken by averaging the fitness over 10 trials of 5 independently evolved robots over a series of 50 light sources. The final distance to the source was in all cases between 3 and 10 units showing that all controllers perform phototaxis equally well.

A series of radical sensor perturbations has been performed. Figure 1 shows the proportional decay in performance for sensor inversion, consisting in swapping

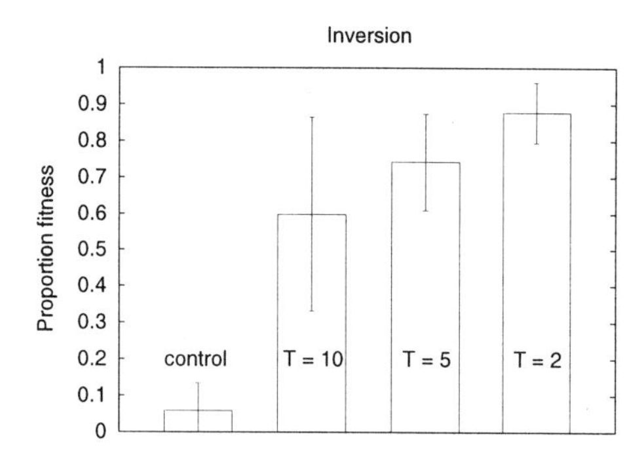

Figure 1: Average relative robustness (measured as proportion of unperturbed performance) for sensor inversion.

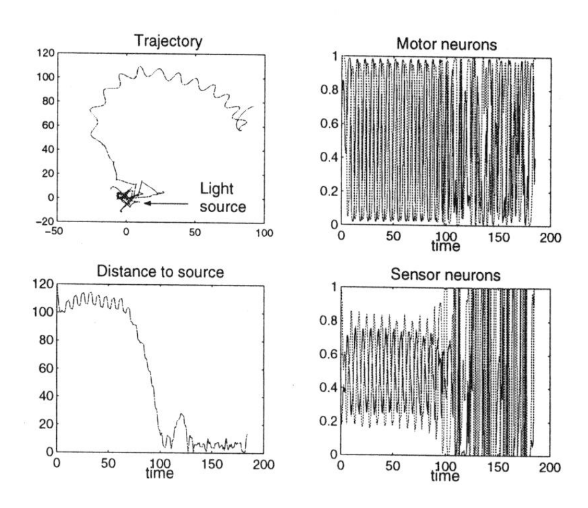

Figure 2: Robot behaviour, neural activation and distance to light source during phototaxis.

sensor positions left and right. Oscillating neural controllers show high robustness which increases as T is made smaller, i.e., for higher frequencies. Observed behaviour is unchanged by inversion for $T = 2$. Similar results have been obtained for sensor removal. Robustness is also shown in angular displacement tests where the position of both sensors is shifted by a same angle (Di Paolo, 2002). Control networks show a heavy reliance on a single sensor. Displacement of this sensor results in very different angular ranges for high fitness, and removal of the same results in negligible performance. In contrast, the oscillator-controlled robot shows the same angular distribution for all perturbations, indicating that the strategy used will work as long as there is at least one sensor (any sensor) facing in the forward direction of the robot.

Analysis of the evolved strategy has been performed for a 4-neuron network. All neurons behave as oscillators entrained to a common frequency of 0.11 cycles per unit of time. Neural response to step-changes in input is confined to less than one cycle of oscillation. After that the oscillation pattern is recovered with a possible phase shift. In all the cases tested the strategy used by the robot is a variation of the strategy shown in figure 2. Motor neurons oscillate maintaining a roughly constant phase difference. The trajectory far from the light can be described as alternating segments where the robot moves forwards, rotates on the spot about three quarters of a full turn, moves backwards, rotates on the spot, moves forwards, and so on. Because of the imposed limitations on the controller, the robot must use an active scanning strategy as sensor neurons will lose any instantaneous information they might acquire from the input currents.

The controller uses input currents to modulate oscillations and make the network switch between different regimes. This is clear from observing the behaviour of the sensor neurons whose oscillations are unstable in the presence of the intermittent pattern of input currents caused by rotating near the light source. Analysis of the transition between the near and far regimes throws some light into the origins of robust behaviour. Even though low input currents do not alter the oscillation mode of the neurons, they do modulate the phase of oscillation (Di Paolo, 2002). The effect of this phase modulation correlates with whether, during rotations, the robot faces the general direction of the light source or the opposite direction. In the first case, the angle of rotation is made smaller as a consequence of the increase in input current. In the second case, the angle is not affected, resulting in a deviation of the trajectory towards the source of light. For such a strategy the precise location of the sensors is not a very sensitive parameter, as long as they are placed towards the front of the body.

The results show that when single neurons are prevented from storing long-term information in their activation values, good solutions evolve that make use of the relative coordination between neural oscillations to solve the desired task. These solutions are extremely robust to sensory perturbation. Two essential components seem to be needed for this: the timescale of oscillation must be faster than the timescale of performance and the long-term average activation of each neuron must conserve an undifferentiated average value independently of the history of inputs.

The author wishes to acknowledge the support of the Nuffield Foundation, (grant no. NAL/00274/G).

References

Beer, R. D. (1990). *Intelligence as Adaptive Behavior: An Experiment in Computational Neuroscience.* San Diego: Academic Press.

Di Paolo, E. A. (2002). Evolving robust robots using homeostatic oscillators. Cognitive Science Research Paper 548, COGS, University of Sussex.

A Method for Isolating Morphological Effects on Evolved Behaviour

Josh C. Bongard **Rolf Pfeifer**
Artificial Intelligence Laboratory
Department of Information Technology
University of Zürich
Winterthurerstrasse 190, Zürich, Switzerland
{bongard|pfeifer}@ifi.unizh.ch

Abstract

As the field of embodied cognitive science begins to mature, it is imperative to develop methods for identifying and quantifying the constraints and opportunities an agent's body places on its possible behaviours. In this paper we present results from a set of experiments conducted on 10 different legged agents, in which we evolve neural controllers for locomotion. The genetic algorithm and neural network architecture were kept constant across the agent set, but the agents had different sizes, masses and body plans. It was found that increased mass has a negative effect on the evolution of locomotion, but that this does not hold for all of the agents tested. Also, the number of legs has an effect on evolved behaviours, with hexapedal agents being the easiest for which to evolve locomotion, and wormlike agents being the most difficult. Moreover, it was found that repeating the experiments with a larger neural network increased the evolutionary potential of some of the agents, but not for all of them. The results suggest that by employing this methodology we can test hypotheses about the behavioural effect of specific morphological features, which has to date eluded precise quantitative analysis.

1. Introduction

It has been over a decade since the idea of embodied AI was first introduced (for a review, see (Brooks, 1991)). Since that time, the belief that choices regarding an agent's or robot's body greatly affect its possible behaviours has come to be widely accepted, but relatively little quantitative data has been collected to support this view. One of the reasons for this is that embodied AI relies heavily on the synthetic methodology. That is, all aspects of agent design are interdependent, so building and then analyzing the behaviour of complete agents is the best way to generate autonomous, intelligent agents (Pfeifer and Scheier, 1999). However, it is then difficult to attribute the effect, if any, a part of an agent has on its resulting behaviour. This is complicated by the fact that designing, constructing and analyzing an autonomous, embodied agent takes a long time, even if computer simulation is employed.

Yet, with the recent advent and maturation of physical simulation, it has become possible to rapidly build and test the behaviours of embodied, situated agents. Such simulations are often coupled with artificial evolution. Some experiments focus on the evolution of controllers for a fixed agent design (Ijspeert and Arbib, 2000, Reil and Massey, 2001) or slight modification of a generic body plan (Bongard and Paul, 2001), or on the combined evolution of both the morphology and controller of the agent (Sims, 1996, Ventrella, 1996, Kikuchi and Hara, 1998, Lipson and Pollack, 2000, Komosinski, 2000, Bongard and Pfeifer, 2001, Taylor and Massey, 2001).

In the latter case, it has been demonstrated that agents with widely differing morphologies can accomplish the same task. However, little work has focussed on which properties of an agent's morphology make it suitable for a given task. An exception is the work by Lund *et al* (Lund et al., 1997), in which it was shown that for wheeled robots, a correlation between body size, wheel base and sensor range exists.

This paper investigates the behavioural effect of morphology by comparing a set of legged agents with the same number of sensors, actuated joints and neural network architectures, but differing body plans. Most papers published in the artificial life and adaptive behaviour literature study a single agent or robot, and attempt to draw conclusions from the resulting behaviour. However, Terzopoulos *et al* described learned controllers for three different fish morphologies in (Terzopoulos et al., 1996); Cruse *et al* described the commonalities and difference between locomotion strategies in real animals based on their biomechanical properties and environment (Cruse, 1991); and Cecconi *et al* (Cecconi and Parisi, 1991) evolved controllers for two grasping robots with different morphologies, but the second agent had a more complex neural network architecture.

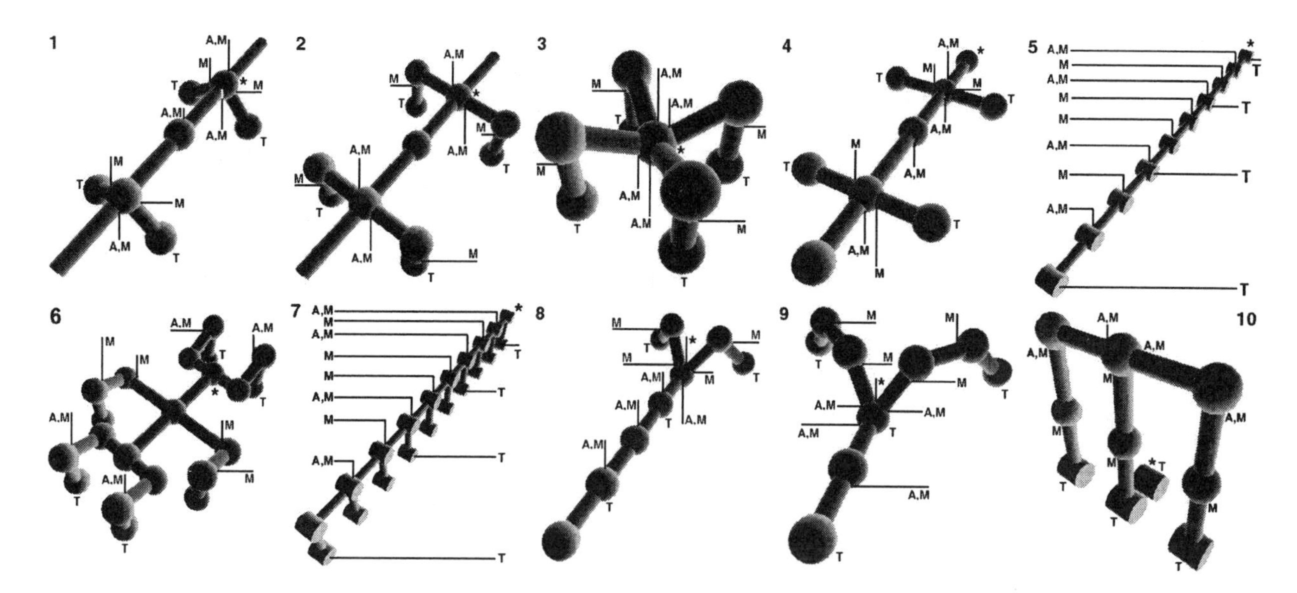

Figure 1: **The agents used for comparison.** Each agent contains four touch sensors (T), four angle sensors (A), and eight motors (M) actuating eight one degree-of-freedom joints. Fitness is based on the forward displacement of one of the body parts (indicated by *) contained in the agent over a fixed period of time.

The change in behaviour caused by different morphologies has been made clear by experiments in which evolved controllers are transferred from one type of robot to another (Floreano and Mondada, 1998), and from simulated agents to real robots (Miglino et al., 1995, Jakobi, 1997, Tokura et al., 2001). However, specific claims as to which aspects of the morphology cause the observed behavioural changes are not provided. Also, work has been done on heterogeneous robot groups (Parker, 1994), in which the actual morphologies of the robots differ, but there has been little or no mention of how particular aspects of the different robot morphologies affected the overall group task performance. Balch (Balch, 2000) has formulated a measure for determining the heterogeneity between robot groups, but this measure does not rely on, or clarify correlations between individual robot morphology differences, and differences in behaviour competencies.

In what follows, we introduce a methodology that can be used to isolate the effect of particular morphological properties—such as total mass, mass distribution (Paul and Bongard, 2001), size or stability—have on the evolution of agent behaviour. In the next section, this methodology is described in detail. In section 3, results are presented using this methodology. In section 4 the implications of this work for generalizing adaptive behaviour results to entire classes of agents are discussed. The final section provides some concluding comments and directions of future research.

2. Methods

In order to compare morphological effect on behaviour, 10 legged agents were constructed and tested in a physics-based, three-dimensional simulation toolkit developed by MathEngine PLC[1]. The morphologies of the ten agents are shown in Fig. 1. Each of the connecting cylinders has a radius of 10cm and a length of 50cm. Each of the spheres has a radius of 20cm. Each of the small cylinders contained in the two segmented agents and the triped (agents 5, 7 and 10 in Fig. 1) has a radius of 20 cm and a length of 40cm. All body parts have a mass of 1kg.

Each agent contains a total of four touch sensors, four angle sensors, and eight actuated, one degree-of-freedom joints, irrespective of its number of legs or body plan.

The touch sensors return a maximum positive signal if the body part in which they are contained is in the contact with the ground plane, and return a maximum negative signal otherwise. The angle sensors return a signal commensurate with the joint's current angle. For example, the sensors emit a maximum negative signal when the joint to which they are attached is at maximum flex, a zero value when the joint angle is equal to the original setting (shown in Fig. 1), and a maximum positive signal when the joint is at maximum extension.

The joints can rotate between $-\frac{\pi}{4}$ and $\frac{\pi}{4}$ radians of their original setting. Each of these joints is actuated by a torsional motor, which receives desired angle settings from the neural controller, and exerts torque proportional to the difference between the current joint angle and the desired angle using

$$\tau_{t+1} = max(I(\omega_t - k(\theta - \theta_d)), \tau_{max}),$$

where θ is the actual joint angle, θ_d is the desired joint angle, τ_{max} is the maximum torque ceiling, $\omega = \dot{\theta}$, and

[1]Final beta release of MathEngine SDK; `www.cm-labs.com`

I is the inertia matrix.

All motors in all the agents have the same maximum torque ceiling, as well as the same damping properties, which were tuned by hand to disallow extreme actions such as jumping or hopping in all 10 agents. However, combined motor action was sufficient for walking, and in some cases dynamic gaits in which the agent's centre of mass passed outside of the support polygon created by its contacts with the ground plane emerged.

The four motors actuating agent 1's four legs rotate through the transverse plane, while its four spinal motors rotate through the frontal plane. In agent 2, the four knee joints rotate through the transverse plane, and the four shoulder joints rotate through the frontal plane. The two joints on each leg of agent 3 rotate through the plane defined by that leg. The four shoulder joints in agent 4 rotate through the transverse plane, and the four spinal joints rotate through the frontal plane. The eight spinal joints in agents 5 and 7 rotate through the sagittal plane. All eight joints on the hexapedal agent (agent 6) rotate through the sagittal plane. The joints on the arms of agents 8 and 9 rotate through the plane defined by those arms; the spinal joints rotate through the sagittal plane.

The knee and hip joints on each of the three legs of agent 10 rotate through the sagittal plane, and the two pelvic joints rotate through the transverse plane.

All of the agents are controlled by a partially recurrent neural network, the architecture of which is shown in Fig. 2. The input and output layers correspond to the sensor and motor array, respectively. There is an additional bias neuron at the input and hidden layers that outputs a constant signal of 1. The input layer is fully connected to the hidden layer, and the hidden layer is fully connected to the output layer. In addition, the hidden layer is fully, recurrently connected.

At each time step of the simulation of an agent's behaviour, the eight sensor signals are scaled to floating-point values in $[-1.0, 1.0]$, and supplied to the input layer. The values are propagated to the hidden and output neurons. The hidden and output neurons scale their incoming values using the activation function

$$O = \frac{2}{1+e^{-a}} - 1,$$

where a is the summed input to the neuron.

A fixed length, generational genetic algorithm is used to evolve locomotion for the 10 agents. Genomes encode the 68 synaptic weights for the neural network as floating-point values, which can range between -1.00 and 1.00. For the experiments reported in the next section, each evolutionary run was conducted using a population size of 300, and was run for 200 generations. At the end of each generation, strong elitism was employed: the 150 fittest genomes were copied into the next generation. Tournament selection, with a tournament size of 3, is employed to select genomes from among this group to participate in mutation and crossover. 38 pairwise one-

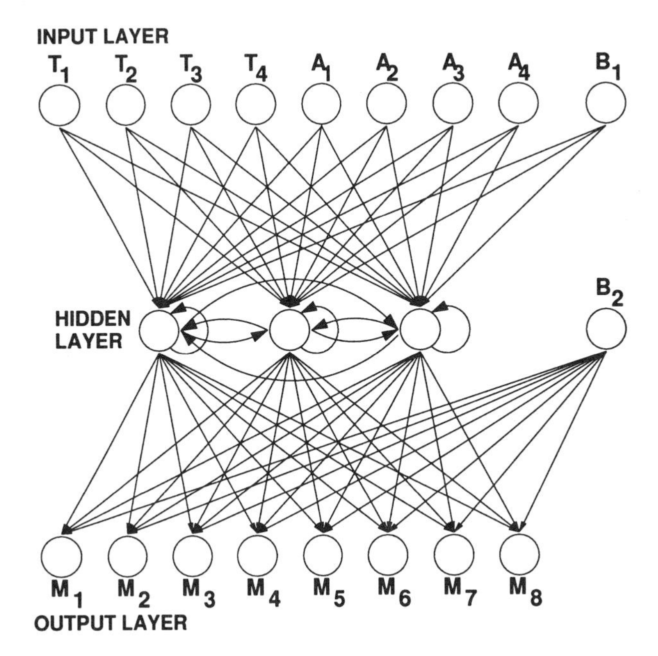

Figure 2: **The neural network architecture.** The four touch sensor signals are scaled and passed to input neurons T1–T4, and the angle sensors are scaled and passed to input neurons A1–A4. The output neuron values (M1–M8) are translated from desired angles into torque by the eight motors of the agent.

point crossings produce 76 new genomes. The remaining 74 new genomes are mutated copies of genomes selected from the previous generation: an average of three point mutations are introduced into each of these new genomes, using random replacement.

In the second set of experiments, the hidden layer was expanded to include five, instead of three hidden nodes. This increases the synapse count from 68 to 118, and thus the genome length from 68 to 118. However, except for the increase in genome length, no other genetic algorithm parameters were altered during this second set of experiments.

3. Results

For each of the 10 agents, 30 evolutionary runs were performed, in which fitness was set to the forward displacement of the selected body part (see Fig. 1) in the agent after 500 time steps of the physical simulation. During each time step of the evaluation, sensor readings are taken, the neural network is updated, the motor commands are translated into the torques. Also the body parts' positions, velocities and orientations are updated based on these torques as well as on external forces such as gravity, inertia, friction and collision or contact with the ground plane.

The highest fitness obtained in each generation was recorded, as well as the corresponding genome. For each agent, these fitness values from the 30 runs were averaged

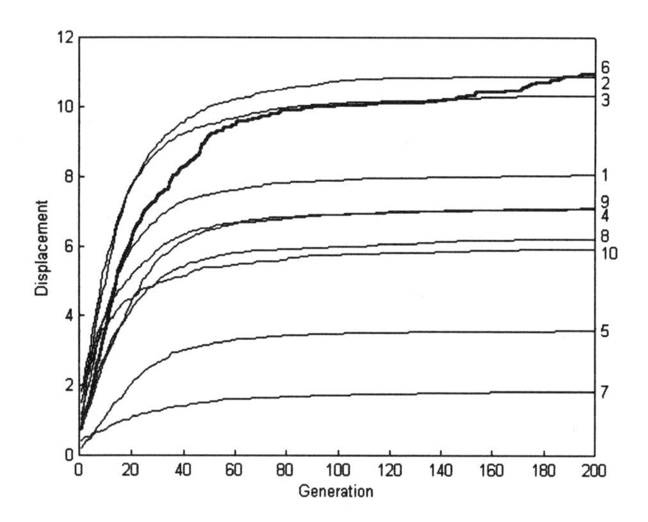

Figure 3: **Average evolutionary performance of the 10 agents.** The curves are averages of the best fitness curves taken over the 30 evolutionary runs for each agent. The numbers to the right indicate to which agent that curve belongs (ie., agents 6 and 2 performed the best). Displacement is in meters.

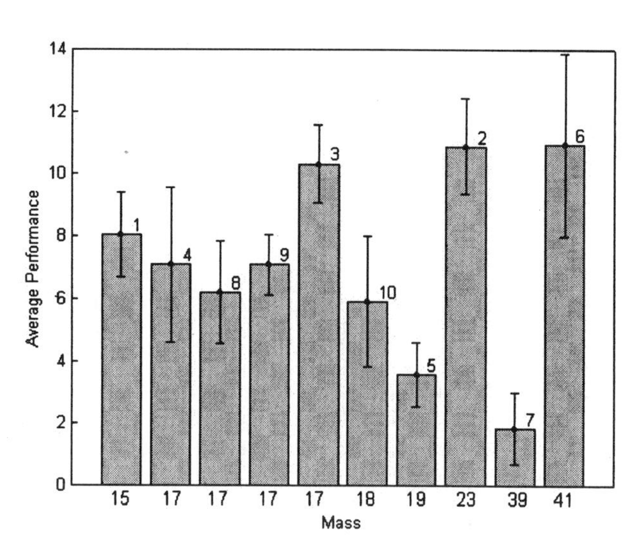

Figure 5: **Mass versus evolutionary performance.** The horizontal axis indicates the total mass of the agent, in kilograms. The vertical axis indicates the average displacement of the targetted body part for each agent type, in meters. The numbers above the bars indicate the agent index as given in Fig. 1. The error bars are two standard deviation units in length.

together, and are shown in Fig. 3.

Within each set of 30 evolutionary runs, the run which produced the fittest agent was found, and the time steps for which the agent's body parts were in contact with the ground plane were recorded. The footprint graphs for these agents from each agent type are are shown in Fig. 4.

In order to account for the performance differences indicated in Fig. 3, various morphological aspects of the agents were compared against their average evolutionary performance. Average evolutionary performance was computed by collecting the best fitness values achieved at the end of each of the 30 evolutionary runs, and averaging them. Fig. 5 plots the agent's total mass against average evolutionary performance. Fig. 6 plots the number of points of contact of the agent with the ground plane against average evolutionary performance.

Finally, a second set of experiments was conducted in which the hidden layer was expanded from three neurons to five neurons. Thirty evolutionary runs were again performed for each agent type, and the fittest genome was retained, and its fitness recorded, after each generation. The best fitness achieved at the end of each run was recorded and averaged within the set of 30 runs, for each agent. Fig. 7 plots the performance increase (or decrease) for each agent type realized by the increase in neural network size.

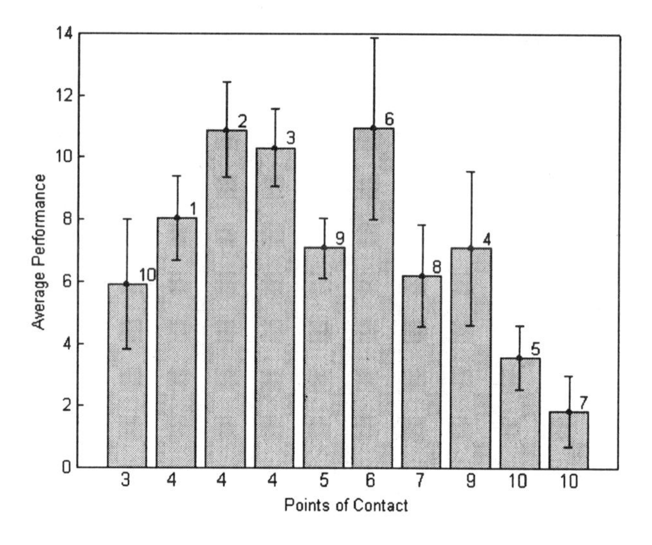

Figure 6: **Points of contact versus evolutionary performance.** The horizontal axis indicates how many body parts of the agent can contact the ground plane. The vertical axis indicates the average displacement of the targetted body part for each agent type, in meters. The numbers above the bars indicate the agent index as given in Fig. 1. The error bars are two standard deviation units in length.

4. Discussion

Fig. 4 shows that most of the agents achieve a relatively rhythmic gait during evolution, with the exception of the segmented agent 5. For example, the tripedal agent

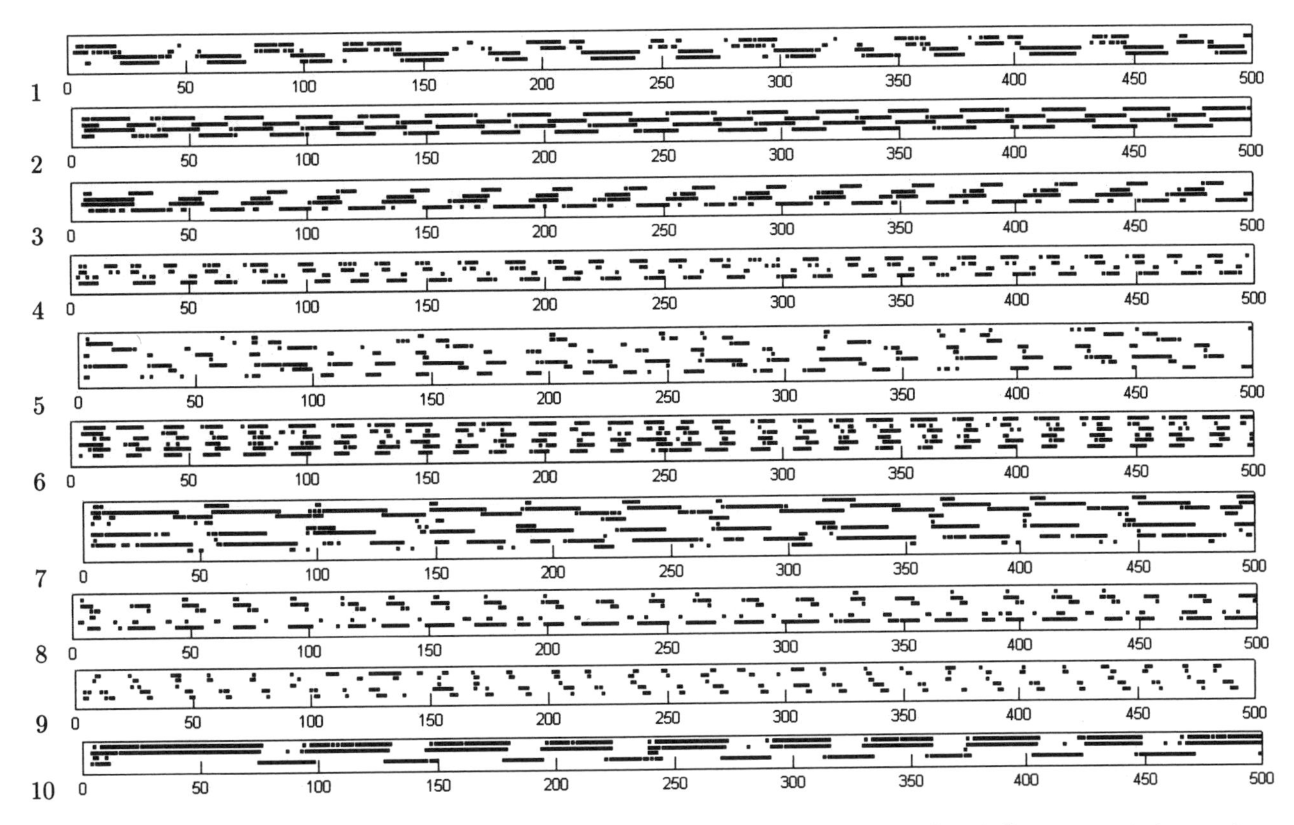

Figure 4: **Footprint graphs produced by the most fit agent of each type.** Numbers indicate agent index as given in Fig. 1. The horizontal axis indicates time; the rows arranged along the vertical axis correspond to one of the body parts comprising the agent that comes in contact with the ground plane for at least one time step during evaluation. Black bars indicate time periods for which the body part is in contact; the white gaps indicate periods in which it is not in contact with the ground plane.

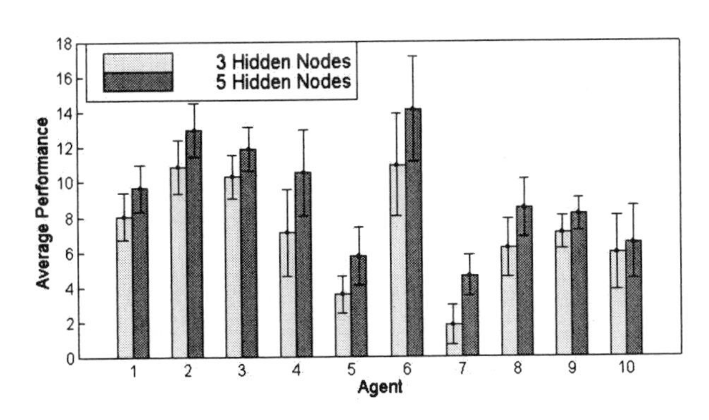

Figure 7: **Change in performance based on addition of hidden neurons.** The light coloured bars indicate the average evolutionary performance for that agent using three hidden neurons. The dark coloured bars indicate average performance for that agent using five hidden neurons. Numbers along the horizontal axis denote the agent's index number, as denoted in Fig. 1. The error bars are two standard deviation units in length.

(agent 10) keeps its left and right feet on the ground plane while its central leg swings into the air, and with the aid of the momentum of the return stroke falls into a regular gait where the left and right legs move in almost perfect synchrony (lowest panel, Fig. 4).

As can be seen from Fig. 3, it is much easier to evolve locomotion for two of the quadrupedal agents (agents 2 and 3) and the hexapedal agent (agent 6) using the genetic algorithm and neural network architecture reported here, than is the evolution of locomotion for the segmented agents (agents 5 and 7). Because the evolutionary method and neural controllers were kept constant for all agents, a morphological explanation must be found to account for this performance discrepancy.

One hypothesis is that the greater the number of legs an agent has, the more difficult it is to evolve a neural controller to coordinate them, or they generate more friction with the ground, and thus make locomotion more difficult.

However, Fig. 6 seems to refute this hypothesis, as there seems to be an inverse U-shape relationship between performance and leg number: performance increases from the tripedal agent (agent 10) up to the hexapedal agent, and then decreases again as the number of points of contact increases. This may be a general trend, and needs to be tested by including more agents in the group, such as bipedal and octapedal agents.

An alternative hypothesis is that the segmented agents have larger masses, and because the number of motors and the torque ceiling is kept constant across agents, it may simply be more difficult for the heavier agents to locomote. This hypothesis seems to be supported by the data reported in Fig. 5, because there is a partial negative correlation between mass and evolutionary performance. However, the three best agents (agents 2, 3 and 6) run against this apparent correlation, suggesting that mass is a necessary, but not a sufficient morphological explanation for the observed performance differences, and that six legs may be an optimal configuration for this experimental setup. We can envisage several other morphological explanations that could be tested using this method: static stability, dynamic stability, and orientation of joints are just a few possibilities.

Aside from isolating and testing hypotheses about specific morphological characteristics, by experimenting with sets of agents instead of just a single agent, we can measure the general effect of controller and evolutionary method choices, not just how they affect a particular agent. For example, Fig. 7 shows that by increasing the size of the neural network by adding hidden neurons is a great advantage for the segmented agents with many similar parts, helps somewhat with the other agents, but has no significant advantage for the tripedal agent. The advantage for the segmented agents is made more clear in Fig. 8, where the best evolved gait for agent 5 using three hidden neurons is contrasted against the best evolved gait using five hidden neurons. The first gait allowed the agent to travel 2.45 meters in 500 time steps; the second gait allowed the agent to travel 7.58 meters. It can be seen that in the first case the best gait has not yet achieved rhythmicity, whereas the second gait is much more rhythmic. Note also that despite hand-tuning the maximum motor torques, the second gait has achieved jumping: there are periods of time for which no part of the agent is in contact with the ground plane.

The large performance increase observed for the segmented agents lends support to the hypothesis that segmented animals require multiple, modular neural components in order to achieve travelling waves of muscular contraction in order to move (Ijspeert and Arbib, 2000). Our result also agrees with that of Gruau (Gruau and Quatramaran, 1997), who reported that a neural network with 16 hidden nodes was required, in his experimental setup, to evolve locomotion for an eight-legged robot.

By comparing new evolutionary techniques and controller architectures on different agents, it is possible to determine whether any observed gain in performance is general, or is useful only for particular agents. For example in this paper we have demonstrated that increasing network size is useful for segmented agents, but not for the tripedal agent. We hypothesize that if the triped, which is inherently unstable, were equipped with tilt sensors it may be possible to exploit the extra neural connections for balanced locomotion.

5. Conclusions

In this paper we have introduced a comparative methodology that serves two purposes. First, it can be used to measure how much a particular morphological characteristic will facilitate or hamper the evolution of behaviours for simulated agents. Moreover, because the methodology encompasses a group of agents, it could allow for predictions as to how easy or difficult it will be to evolve the behaviours for new agents, if the new agent shares one of the morphological characteristics with an agent from the original group. For example, because we have found that quadrupeds and hexapods are particularly good candidates for which to evolve locomotion, given our choice of neural network and evolutionary scheme, we predict that it would be relatively easy to evolve locomotion for new agents with quadupedal or hexapedal body plans.

Second, by modifying the evolutionary scheme or controller, re-evolving the agents in the set for the same behaviour, and then measuring performance changes, we can begin to understand how particular controller architectures or evolutionary schemes are appropriate—or inappropriate—for particular agents.

As physical simulation becomes more sophisticated and computational power continues to increase, it has become feasible to test hypotheses about adaptive behaviour on a whole class of agents, not just a single instantiation. Moreover, by gaining more specific insights into morphological effects on behaviour, it may become easier to transfer evolved agents from physical simulation to real world robots.

References

Balch, T. (2000). Hierarchic social entropy: An information theoretic measure of robot group diversity. *Autonomous Robots*, 8(3).

Bongard, J. C. and Paul, C. (2001). Making evolution an offer it can't refuse: Morphology and the extradimensional bypass. In Kelemen, J. and Sosik, P., (Eds.), *Sixth European Conference on Artificial Life*, pages 401–412.

Bongard, J. C. and Pfeifer, R. (2001). Repeated structure and dissociation of genotypic and phenotypic complexity in artificial ontogeny. In Spector, L. and Goodman, E. D., (Eds.), *Proceedings of The Genetic and Evolutionary Computation Conference, GECCO-2001*, pages 829–836.

Brooks, R. A. (1991). New approaches to robotics. *Science*, 253:1227–1232.

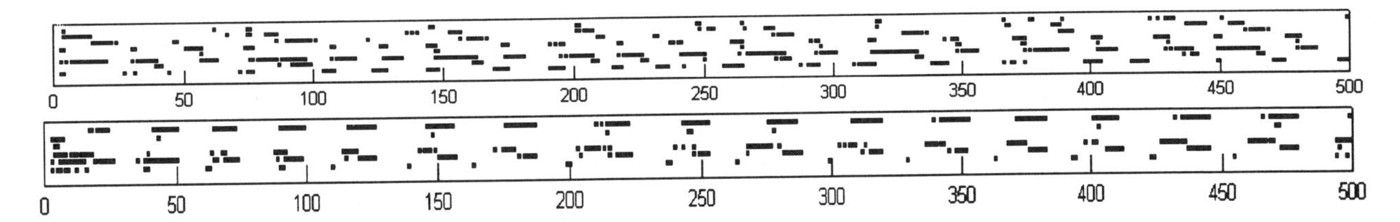

Figure 8: **The best evolved gaits using two different neural networks.** The upper panel shows the best evolved gait for agent 5 using a hidden layer with three neurons. The lower panel shows the best evolved gait for the same agent using a hidden layer with five neurons.

Cecconi, F. and Parisi, D. (1991). Evolving organisms that can reach for objects. In Meyer, J. A. and Wilson, S. W., (Eds.), *Proceedings, From Animals to Animats*, pages 391–399.

Cruse, H. (1991). Coordination of leg movement in walking animals. In Meyer, J. A. and Wilson, S. W., (Eds.), *Proceedings, From Animals to Animats*, pages 105–119.

Floreano, D. and Mondada, F. (1998). Evolutionary neurocontrollers for autonomous mobile robots. *Neural Networks*, 11:1461–1478.

Gruau, F. and Quatramaran, K. (1997). Cellular encoding for interactive evolutionary robotics. In Husbands, P. and Harvey, I., (Eds.), *Fourth European Conference on Artificial Life*, pages 368–377.

Ijspeert, A. J. and Arbib, M. (2000). Visual tracking in simulated salamander locomotion. In Meyer, J. A. and Berthoz, A., (Eds.), *Proceedings, From Animals to Animats 6*, pages 88–97.

Jakobi, N. (1997). Evolutionary robotics and the radical envelope of noise hypothesis. *Adaptive Behavior*, 6(1):131–174.

Kikuchi, K. and Hara, F. (1998). Evolutionary design of morphology and intelligence in robotic system using genetic programming. In Pfeifer, R. and Blumberg, B., (Eds.), *Proceedings, From Animals to Animats 5*, pages 540–545.

Komosinski, M. (2000). The world of Framsticks: Simulation, evolution, interaction. In *Proceedings, Second Intl. Conf. on Virtual Worlds (VW2000)*, pages 214–224.

Lipson, H. and Pollack, J. (2000). Automatic design and manufacture of artificial lifeforms. *Nature*, 406:974–978.

Lund, H. H., Hallam, J., and Lee, W. P. (1997). Evolving robot morphology. In *Proceedings of the Fourth International Conference on Evolutionary Computation.*

Miglino, O., Lund, H., and Nolfi, S. (1995). Evolving mobile robots in simulated and real environments. *Artificial Life*, 2:417–434.

Parker, L. E. (1994). Alliance: An architecture for fault tolerant, cooperative control of heterogeneous mobile robots. In *Proceedings of the IEEE/RSJ Intl. Conf. on Intelligent Robots and Systems (IROS)*, pages 776–783.

Paul, C. and Bongard, J. C. (2001). The road less travelled: Morphology in the optimization of biped robot locomotion. In *Proceedings of The IEEE/RSJ International Conference on Intelligent Robots and Systems (IROS2001).*

Pfeifer, R. and Scheier, C. (1999). *Understanding Intelligence.* MIT Press.

Reil, T. and Massey, C. (2001). Biologically inspired control of physically simulated bipeds. *Theory in Biosciences*, 120:1–13.

Sims, K. (1996). Evolving 3d morphology and behavior by competition. In Brooks, R. A. and Maes, P., (Eds.), *Proceedings, Artificial Life IV*, pages 28–39.

Taylor, T. and Massey, C. (2001). Recent developments in the evolution of morphologies and controllers for physically simulated creatures. *Artificial Life*, 7(1):77–88.

Terzopoulos, D., Tu, X., and Grzeszczuk, R. (1996). Artificial fishes with autonomous locomotion, perception, behaviour, and learning in a simulated physical world. In Brooks, R. A. and Maes, P., (Eds.), *Proceedings, Artificial Life IV*, pages 17–27.

Tokura, S., Ishiguro, A., Kawai, H., and Eggenberger, P. (2001). The effect of neuromodulations on the adaptability of evolved neurocontrollers. In Kelemen, J. and Sosik, P., (Eds.), *Sixth European Conference on Artificial Life*, pages 292–295.

Ventrella, J. (1996). Explorations in the emergence of morphology and locomotion behavior in animated characters. In Brooks, R. A. and Maes, P., (Eds.), *Proceedings, Artificial Life IV*, pages 436–441.

An Evolutionary Approach to Quantify Internal States Needed for the Woods Problem

DaeEun Kim* **John C.T. Hallam**
IPAB, Division of Informatics
University of Edinburgh
5 Forrest Hill
Edinburgh, EH1 2QL
Scotland, United Kingdom
{daeeun, john}@dai.ed.ac.uk

Abstract

The Woods Problem is a difficult problem for purely reactive systems to handle. The difficulties are related to the perceptual aliasing problem, and the use of internal memory has been suggested to solve the problem. In this paper a novel approach in evolutionary computation is introduced to quantify the amount of memory required for a given task. The approach has been applied to Woods Problems such as wood101, woods102, Sutton's gridworld and woods14.

Finite state machine controllers are used, as these permit easy measurement of the amount of memory in the controller. A concurrent evolutionary search for the minimal but optimal control structure in memory-based systems, using an evolutionary Pareto-optimal search mechanism, determines the best behavior fitness for each level of controller memory. This memory analysis demonstrates the effect of internal memory in evolved controllers for Woods Problems and is also used to investigate the relationship between the number of sensors available to an agent and the amount of memory necessary for effective behavior.

1 Introduction

Woods problems are goal-search problems for an agent to try to find a goal position, starting at random initial positions. The agent has eight sensors and it needs to find the shortest path to the goal. A variety of such problems have been defined in the literature (see below) and they are often cited as difficult problems for a memoryless strategy to solve.

McCallum defined a hidden state as any world state information not determined by the current immediate perception of a mobile agent (McCallum, 1996). The relevance of hidden states in robotics research can be observed in research on reactive systems (Brooks, 1986, Brooks, 1987, Kaelbling, 1986). Purely reactive agents choose their current motor action using only their current perception, and a significant amount of robotics research has concentrated on them (Maes and Brooks, 1990, Lee et al., 1997, Sutton, 1991, Chapman and Kaelbling, 1991, Lee, 1998, Mahadevan and Connell, 1991). However, in some non-Markovian[1] environments, purely reactive control cannot succeed in solving hidden state problems (Whitehead, 1992, Singh et al., 1994, McCallum, 1993). Hidden states often appear when sensors have a limited range of view of the surrounding environment or when there are a limited number of sensors. Such hidden state problems are often called perceptual aliasing problems (Whitehead and Ballad, 1991).

When an agent has only partial information about the surrounding environment through its sensory inputs, and the same perceived situation requires different actions in different contexts, the agent suffers from a perceptual aliasing problem. A solution to this problem is to find actions leading to a situation where the agent has an unambiguous sensory pattern (Nolfi and Floreano, 2000). However, this strategy is not a fundamental solution to perceptual aliasing problem and it is effective only when the agent can find at least one unambiguous sensory pattern. Some "reactive" robots employ internal state to deal with perceptual aliasing and their control systems are not actually purely reactive (Brooks, 1991). Even with primitive behaviors, memory internal to the controller is useful in some robotic experiments to overcome limitations of purely reactive systems (Kim and Hallam, 2001).

McCallum (McCallum, 1996) argued that internal memory should be added to solve perceptual aliasing

*Now with Max Planck Institute for Psychological Research, Amalienstr. 33, Munich, 80799, Germany

[1]The environmental features are not immediately observable, but they have dependencies with past states

problems in non-Markovian environments. He developed a reinforcement learning algorithm incorporating memory to prevent hidden states. When an agent suffers from hidden state problems, the perceptions cannot directly define the next motor action. Its decision will depend on the current perception and its internal memory about past perceptions and actions. Memory thus plays a role to disambiguate aliased perceptions.

Wilson used a zeroth-level classifier system (ZCS) for his animat experiments (Wilson, 1994). The original formulation of ZCS has no memory mechanisms, because the input-output mappings from ZCS are purely reactive, but Wilson suggested how internal temporary memory registers could be added. Adding an internal memory register consisting of a few binary bits can increase the number of possible actions in the system. Following Wilson's proposal, one-bit and two-bit memory registers were added to ZCS in Woods environments by Cliff and Ross (Cliff and Ross, 1995). They insisted ZCS manipulate and exploit internal states appropriately and efficiently in non-Markovian environments. Bakker (Bakker and de Jong, 2000) proposed a means of counting the number of states required to perform a particular task in an environment by extracting state counts from finite state machine controllers to measure the complexity of agents and environments. They applied their methodology to the Woods7 environment (Wilson, 1994) to estimate internal states. Colombetti and Dorigo developed ALECSYS, a classifier system to learn proper sequences of subtasks by maintaining internal state and transition signals which prompt an agent to switch from one subtask to another (Colombetti and Dorigo, 1994). Similar to the ZCS experiments (Wilson, 1994), Lanzi has shown that internal memory is effective with adaptive agents and reinforcement learning, when perceptions are aliased (Lanzi, 2000). Also there has been research using a finite-size window of current and past observations and actions (McCallum, 1996, Lin and Mitchell, 1992).

Woods Problems have often been mentioned in association with the perceptual aliasing problem. The environments include many ambiguous situations that require different actions. There have been many variations of the Woods environment: Sutton's gridworld (Sutton, 1990), McCallum's maze (McCallum, 1996), maze10 (Lanzi, 1998), woods102 and woods14 (Cliff and Ross, 1995). They range from simple to complex environments. They have been tackled with classifier systems or reinforcement algorithms (Lanzi, 1998, Lanzi, 2000). The reinforcement systems are mainly based on purely reactive systems (Kaelbling, 1986). Memory-encoding methods with reinforcement learning have been successful on complex Woods Problems (Cliff and Ross, 1995, Lanzi, 2000).

The incorporation of state information permits an evolved control structure to behave better, using past information, than a pure reaction to the current sensor inputs. Finite state machines have been used previously in evolutionary computation to represent state information (Fogel et al., 1966, Stanley et al., 1995, Miller, 1996). In this paper, control structures are based on finite state machines, with internal memory represented as a set of internal states. With the importance of internal states, a novel approach is introduced to quantify memory amount needed to solve Woods Problems. We discuss the potential of evolving memory-based control structures and how they work to improve the performance. Pareto optimization will be tools to recognize the relevance and importance of memory for given tasks.

2 Methods

To determine how many memory elements are required to solve a particular agent-environment interaction problem, evolutionary computation will try to optimize agent performance for each quantity of controller memory. By doing so, the trade-off between performance and quantity of memory can be explored. If sensors are discretized, a controller with internal memory can easily be expressed as a finite state machine (FSM). If a task can be completed with a purely reactive system, the controller can be represented as a 1-state FSM which is equivalent to memoryless strategy. Conversely, the amount on memory needed for a non-reactive task can be determined by counting the number of states in the FSM representing the minimal effective controller.

We have developed an evolutionary algorithm with Pareto optimization for variable sized finite state machines. Two objectives, behavior performance and memory size, are used in the Pareto optimization to try to maximize behavior performance and minimize the quantity of memory (number of controller states). The shape of the Pareto surface after a run indicates a desirable number of memory elements for a given performance level. As a result, one can often determine a threshold amount of memory needed to achieve a task. We assume that the quantity of memory in the optimal control structure for a given task represents the complexity of the problem faced by the agent.

The finite state machine we consider is a type of Mealy machine model, defined as $M = (Q, \Sigma, \Delta, \delta, \lambda, q_0)$ where q_0 is an initial state, Q is a finite set of states, Σ is a (sensor) input space $\{0,1\}^*$, Δ is the multi-valued output, δ is a state transition function from $Q \times \Sigma$ to Q, and λ is an output mapping from $Q \times \Sigma$ to Δ, i.e. $\lambda(q,a) \in \Delta$. $\delta(q,a)$ is defined as the next state for each state q and input value a, and the output action of machine M for the input sequence $a_1, a_2, a_3, ..., a_n$ is $\lambda(q_0, a_1), \lambda(q_1, a_2), \lambda(q_2, a_3), ..., \lambda(q_{n-1}, a_n)$, where $q_0, q_1, q_2, ..., q_n$ is the sequence of states such that $\delta(q_k, a_{k+1}) = q_{k+1}$ for $k = 0, .., n-1$.

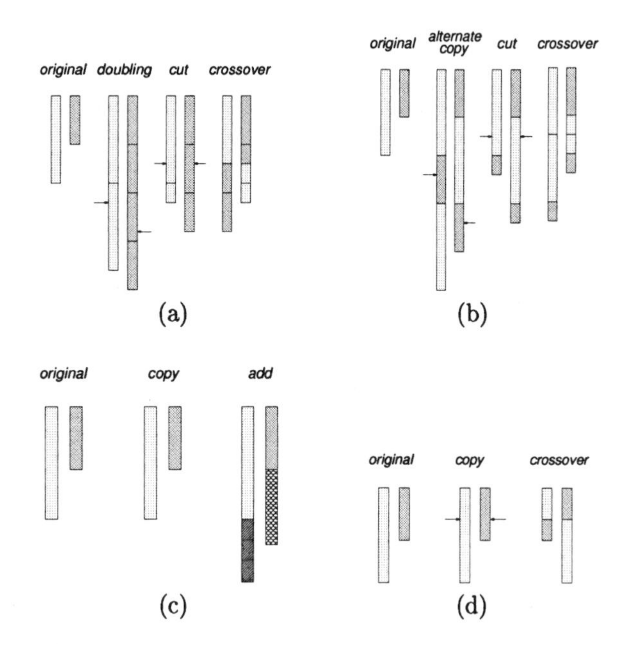

Figure 1: Genetic operators for variable state machines

The machine is encoded for the evolutionary algorithm as a sequence of pairs (state number, state output) on each sensor value in canonical order of state number. In our multi-objective optimization experiments, the genetic pool should allow variable length chromosomes for variable state machines; the size of finite state machines depends on the number of states and thus different members of the pool may have different length genetic representations.

In the experiments, tournament selection of size four is used for Pareto optimization. A population is initialized with random length chromosomes. The two best chromosomes are selected using a dominating rank method[2] over the two objectives, memory size and behavior performance. They reproduce themselves and the two worst chromosomes are replaced by new offspring produced using one point crossover followed by mutation. In the application of variable state machines, offspring with variable numbers of states should be generated to keep diversity in the genetic pool. Thus, a new size modifying genetic operator is introduced to maintain variable length coding. New offspring are thus produced with a size modifying operator, crossover and mutation by turns.

Three different methods for a size-modifying operator are shown in Figure 1. When offspring are produced, the number of memory states is randomly pre-selected for each new offspring. The chromosome size for each offspring will depend on this chosen amount of memory. The following size-modifying operators are then used to produce new offspring such that they have characteristics of their parents and have the pre-chosen chromosome length.

The first operator concatenates several copies of one of the parents, cutting the result at the pre-chosen length of the offspring's chromosome. This is repeated for the second offspring using the other parent – see Figure 1(a).

The second operator concatenates alternately a copy of one parent and a copy of the other parent, again cutting at the pre-chosen chromosome length. A second offspring is generated similarly starting with the other parent – see Figure 1(b).

The third operator takes a copy of one of the parents and if its size is larger than the pre-chosen chromosome length, it is cut; if not, random strings are added to build it up to the pre-chosen length. A second offspring is generated in a similar way, but instead it starts with a copy of the other parent – see Figure 1(c).

After applying the size-modifying operator, crossover is applied to the two offspring. The crossover point is selected inside both chromosome strings after aligning the prefixes of two strings. During this crossover process, bit flipping mutation can be used to change one integer value in the strings.

The size-modifying operator is applied to 75% of new offspring, in experiments with variable state machines. When it is not used, the offspring keep the size of their parents as shown in Figure 1(d).

3 Experiments with Woods Problems

Wilson suggested that temporary memory can be added to a zeroth-level classifier system (ZCS) to solve problems in non-Markovian environments (Wilson, 1994). It is implemented as a few bits in a memory register to be set or reset by actions. A Woods environment is a grid world where agents have discrete sensors and a limited number of motor actions. An agent explores the environment to find its food but there are many trees as obstacles to block moving forward. The agent can recognize the status of each of its eight neighboring cells through eight sensors and so processes only local information about its surrounding environment.

The performance of a strategy or solution to a Woods Problem is measured by the average number of time steps needed to reach a goal from all possible initial cells. An optimal solution has the minimum average steps to the goal. We suggest that an evolutionary approach as outlined above, with state transitions and memory, should be able to handle complex problems requiring long chains of actions and will find optimal solutions using minimal memory for Woods Problems. The complexity of the problems can then be defined in terms of necessary memory size. The proposed method for memory analysis, evolutionary multi-objective optimization (EMO), was applied to Woods Problems. The penalty

[2]The dominating rank method defines the rank of a given vector in a Pareto distribution as the number of elements dominating the vector. The highest rank is zero, for an element which has no dominators.

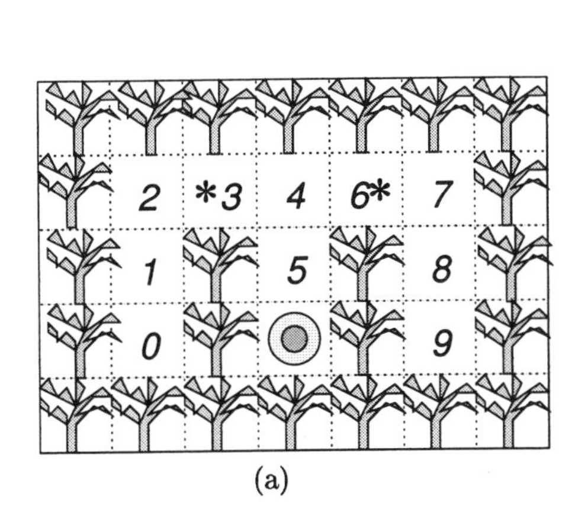

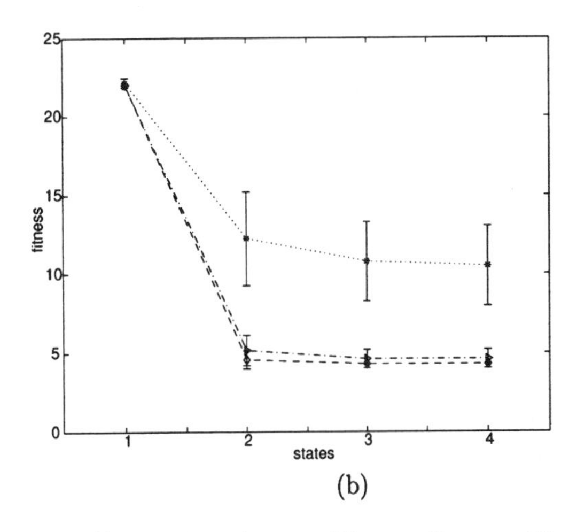

(a) (b)

Figure 2: Woods101 environment (a) grid world (b) memory analysis (* : 10 generations, ▹ : 30 generations, ◇ : 50 generations)

cell no.	trajectory							path length
0	D(1)	U(0)	U(0)	R(0)	R(0)	D(0)	D(1)	7
1	U(0)	R(0)	R(0)	D(0)	D(1)			5
2	R(0)	R(0)	D(0)	D(1)				4
3	R(0)	D(0)	D(1)					3
4	D(0)	D(1)						2
5	D(1)							1
6	R(0)	L(1)	L(0)	D(0)	D(1)			5
7	L(1)	L(0)	D(0)	D(1)				4
8	U(0)	L(1)	L(0)	D(0)	D(1)			5
9	D(1)	U(0)	U(0)	L(1)	L(0)	D(0)	D(1)	7

Table 1: A solution of woods101 problem (each symbol in the trajectory represents motor action and state number in FSM, U : up, D : down, L : left, R : right)

fitness function is defined as follows:

$$F_g = \frac{1}{|S|} \sum_{p \in S} \min(40, d(p, G)) \qquad (1)$$

where S is a set of initial cells, G is the goal position and $d(u, v)$ is the path length from position u to position v.

If the goal is not reachable from a cell, then the path length becomes infinity and a high penalty value (40) is taken; 40 time steps are assigned for each initial cell and if agents can reach the goal from an initial cell, the number of time steps spent for exploration will be the path length from the initial cell to goal. The fitness F_g is the average number of time step taken to reach the goal from any empty cell. There may be one or more cells from which the agent cannot reach the goal when a strategy is applied. In this case we say the strategy cannot *solve* the problem.

To prove that a given Woods Problem cannot be solved by specialized control systems, for example, purely reactive systems, we need to test all possible strategies which belong to that class. This may be a very large exhaustive search. In the experiments here, a control structure is given to solve the problem and evolutionary search will be applied with the control structure to try to find the best performance. If no trial succeeds in finding a solution, then we assume that the control structure is not suitable for the problem. The proof of solvability of problems using particular control structures cannot be validated theoretically in general, but it is based on experimental sampling data.

In this paper, FSM structure was used for memory analysis. Crossover rate was 0.6 with population size 100 and mutation rate was set to 2 divided by the chromosome length. For significance statistics, 25 trials were repeated with the EMO approach over behavior performance and memory size. For each number of internal states, 95% confidence intervals of fitness are estimated by assuming t-distributed results.

3.1 Woods101 Problem

Figure 2(a) is a simple maze example of a Woods Problem. It was tested to show hidden states by McCallum (McCallum, 1996) and it is called *McCallum's maze* or *woods101*. In the woods101 environment, an agent is placed at any empty cell and it can take an action towards one of four directions (up, down, left and right) each time step. Normally Woods Problems allow eight directional moves by considering diagonal movements, but here we restrict the agent's motor actions, making

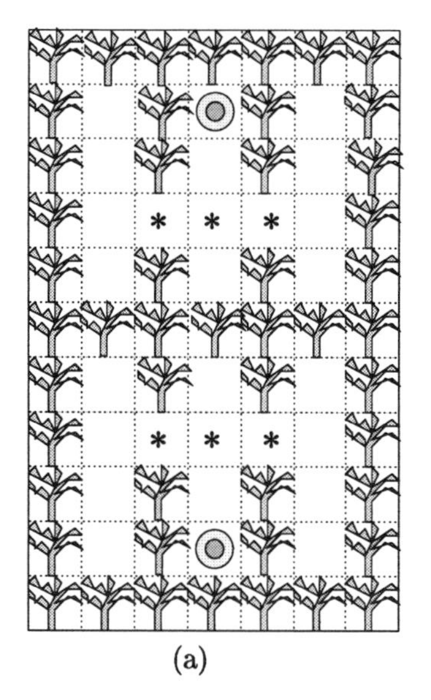

(a)

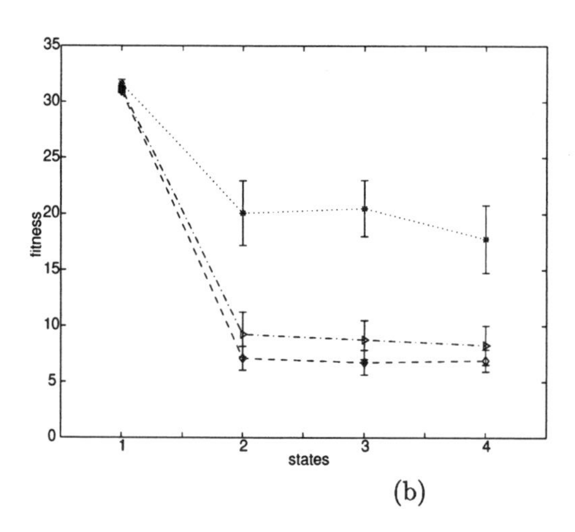

(b)

Figure 3: Woods102 environment (a) grid world (b) memory analysis (* : 10 generations, ▷ : 30 generations, ◇ : 50 generations)

the problem more difficult. The task is to find one food cell at the middle bottom by the shortest path. If a food cell is seen as the goal, this navigation problem becomes a goal search problem.

This problem cannot be solved by purely reactive agents. At some cells with '*' marks in Figure 2(a), agents have the same sensory patterns but need different actions. In the picture, an agent should move left at one cell and move right at the other. It has been shown that temporary memory can help solve such perceptual aliases (Cliff and Ross, 1995, McCallum, 1996).

The Evolutionary multi-objective optimization (EMO) approach was applied to analyze how many internal states are required to solve the perceptual aliasing problem in this Woods environments. The result says two internal states are sufficient to solve woods101 problem. Even if only four sensors are used for an agent instead of eight[3], the result remains that two states are sufficient for this problem and more than two states are luxurious. The environment has six aliased positions for four-sensor agents. Sensor limitation increases the apparent complexity of the problem by generating more sensor aliases but doesn't change the required memory for solution.

Figure 2(b) is the result of the EMO approach with four sensors. It shows for each quantity of internal memory, from one to four internal states, the distribution of behavior fitness obtained with 95% confidence intervals. Performance with two internal states differs significantly from that with one internal state but not from that with 3 or 4 states; hence we conclude that the problem requires two internal states for solution. Purely reactive agents cannot solve the problem and thus the average fitness for one internal state (purely reactive) has a high penalty value. The agent with four sensors needs more time steps to find the goal than eight sensors as it wanders around the environment, looking for the situations which can trigger proper actions.

Using the EMO approach, the best solutions are easily obtained within a small number of generations. An example of solutions is given in Table 1. It is noteworthy to see how to the controller handles sensor aliases with internal memory. Cell 6 and cell 3 have the same sensory pattern. At cell 3, an agent moves right, down and down. At cell 6, it first moves right and sees a corner at right. Then it changes its state and turns left. The agent is now at cell 6. It experiences the same sensory pattern as at cell 3 but it has a different internal state and thus moves left, down and down. Cell 0, 5 and 9 also have the same sensations[4]. By utilizing internal memory states, the agent can reach the goal from any initial cell. Increasing the number of aliases does not necessarily require increasing the number of internal memory states. Agents can efficiently use their internal states by distributing sensor states over memory states. The EMO approach can find compact arrangements of internal states to achieve desirable solutions. In the woods101 problem, one bit of memory is sufficient even for 4-sensor agents. Its performance has total 43 steps; this is the total length of the path from all initial positions to the goal. Its average score is 4.3 time steps, which is the fitness value shown in Figure 2(b).

[3]Four sensors among eight are selected with front, back, left and right directions

[4]Since agents have only binary sensors they cannot distinguish trees and food

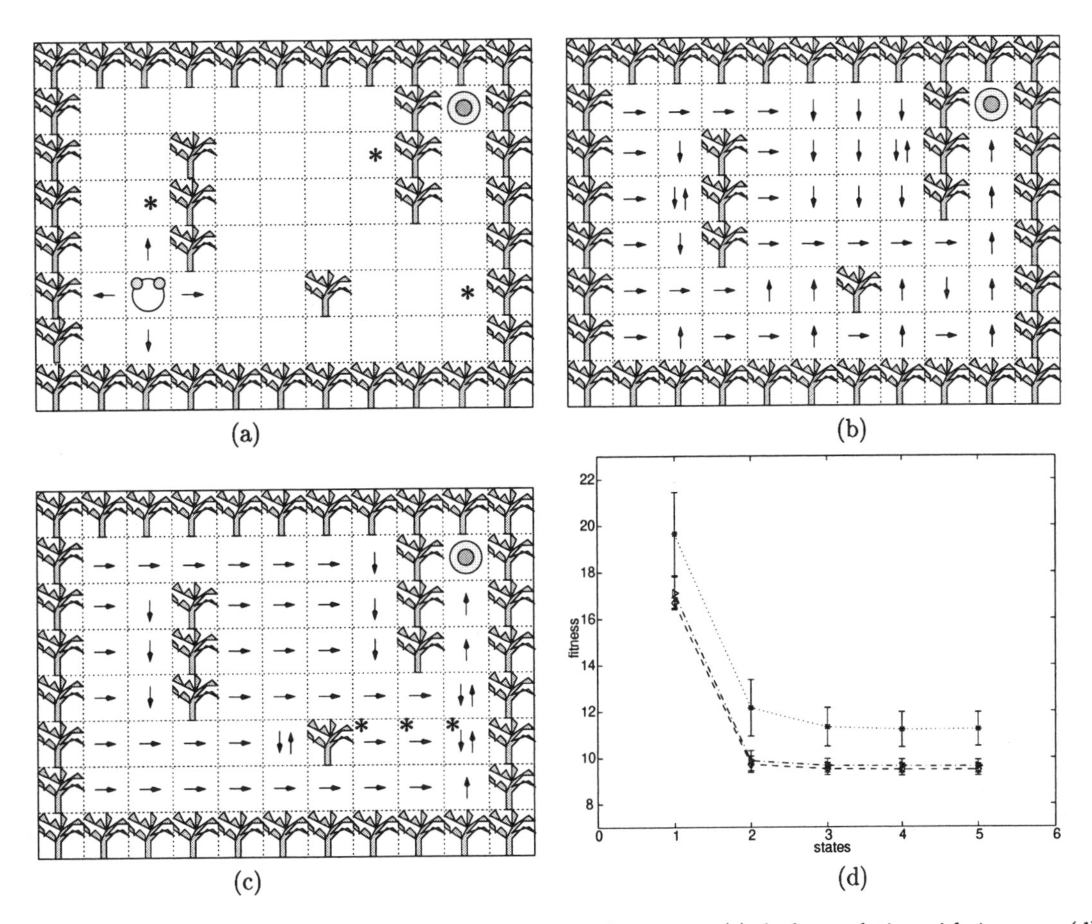

Figure 4: Sutton's grid world (a) grid world (b) the best solution with 8 sensors (c) the best solution with 4 sensors (d) Pareto result (* : 50 generations, ▷ : 200 generations, ◇ : 500 generations)

3.2 *Woods102 Problem*

A new environment called *woods102* was designed to have more difficult situations than the woods101 environment (Cliff and Ross, 1995). This environment is more complex than woods101: there are several cells requiring different actions with sensor aliases marked '*' in Figure 3(a). Cliff and Ross mentioned that two bits of memory are sufficient to disambiguate the sensor aliases (Cliff and Ross, 1995). We restrict the agent to four sensors as in our woods101 example above. An agent will experience more ambiguous situations with this sensor limitation.

When the EMO method with memory states ranging from one to four is applied, the result shows that this problem still requires two states as in Figure 3(b). Thus, a one bit register can solve the problem. However, the memoryless policy — the single state machine — cannot handle perceptual aliasing and the agent cannot reach goal starting from many of the empty cells. It has a high penalty in its fitness and Figure 3(b) shows that it is easily distinguishable from a memory-based policy. Woods102 has more sensor aliases than woods101. In this problem, increasing the number of internal states improves performance. EMO method was again applied to this problem with memory states from two to five, because one state performance is quite distant from the performance of machines with more than one state, to enable the method to concentrate its search on memory-based controllers with best performance. After 1000 generations, results indicate that two states produce the best solution of 155 time steps in total or an average of 5.96 steps to goal. The three state strategy has 146 time steps or an average of 5.62 steps and the four state strategy has 137 time steps or an average of 5.27 steps. More than four states have the same performance as four states. Thus, the best performance can be achieved with 2-bit registers.

3.3 *Sutton's Gridworld*

In Sutton's gridworld (Sutton, 1990), shown in Figure 4(a), an agent can sense eight neighboring cells and takes one of four directional moves. The environment has 46 empty cells, 30 distinct sensor states and one goal position. Littman used branch-and-bound algorithm to

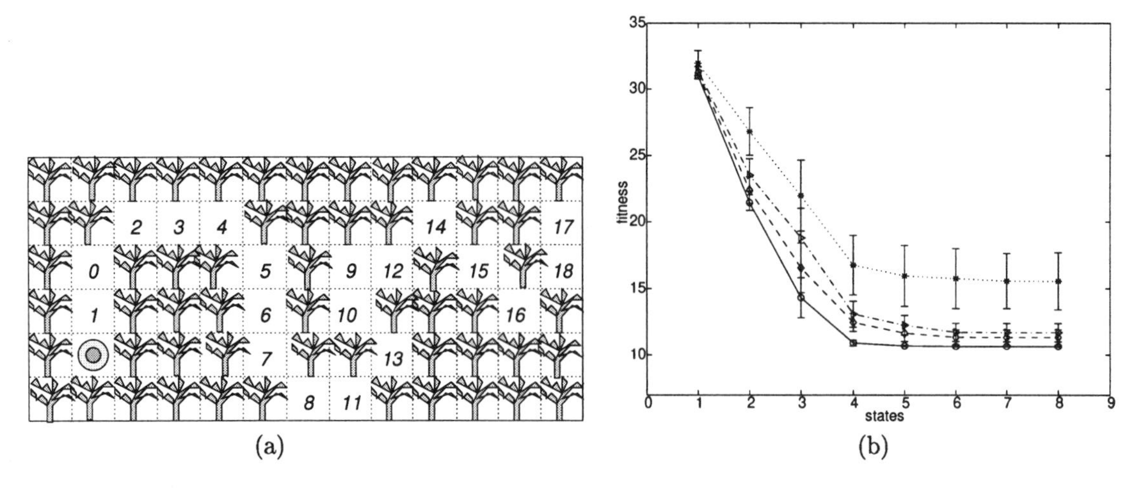

Figure 5: Woods14 environment (a) grid world (b) memory analysis (* : 100 generations, ▹ : 500 generations, ◇ : 1000 generations, ○ : 5000 generations)

find the best memoryless policy (Littman, 1994), which takes a total of 416 steps, an average of 9.04 steps (= 416/46) to reach the goal. Especially at cells with three trees on the right side (the three cells marked with '*') the best memoryless agent experiences ambiguous perceptions and makes inefficient movements, even though it can reach the goal from any initial cell. The agent chooses a roundabout way to escape perceptual aliasing situations.

When internal memory is used, two internal states yield an optimal strategy with performance of 410 total steps to goal, an average of 8.91 steps. Its trace is displayed in Figure 4(b). Two arrows in one cell mean that internal states are involved to give agents appropriate actions. Internal states can be seen as milestones to say which direction is appropriate and they are determined by past history that agents have experienced.

When four out of eight sensors are taken as in the woods101 experiments, its performance is worse than eight sensor experiments. Agents have more sensor aliases for a tree on the right. The memoryless strategy cannot solve this problem, unlike the eight sensor version: there are several cells from which the goal cannot be reached. The best policy with two states for this problem has a performance of total 418 time steps, an average of 9.08 steps to goal. The best strategy as shown in Figure 4(c) is different from the eight sensor strategy in Figure 4(b). An agent chooses to go down when it first sees a tree on the right side. If there are trees on the right and below, it is bounced back to go up with an internal state marker. Otherwise, the agent moves right. The EMO method shows that more than two states give similar performance as displayed in Figure 4(d). One can say that two states are sufficient for this problem with four sensors.

However, 418 time steps is not an optimal solution with four sensors. An optimal strategy with four states was found with a total of 416 time steps. In this strategy, two additional states record the perceptual situation (marked '*' in Figure 4(c)) where a tree is detected on the left side and after moving right two steps, another tree is seen on the right side. This state blocks going down in the rightmost marked cell and the agent can go straight up, while two or three state machines waste time going down and returning to the cell from below.

If an agent can recognize every feature of the entire environment, it will obtain an optimal performance of 404 steps from all empty cells, average of 8.78 steps to goal. However, the agent has only local information about its neighboring cells. Internal memory states help agents overcome the shortcomings of this local information. The performance of the memory-based policy is quite close to the optimal performance of 404 steps. The number of sensors is also an important factor to obtain efficient solutions. In this problem, an optimal memoryless strategy with eight sensors has the same performance as an optimal four-state strategy with four sensors. Reducing sensors or restricting sensor range often requires increasing internal states. In other words, agents with more sensors will have a higher chance of succeeding with purely reactive controllers.

3.4 Woods14 Problem

Woods14 is a Markovian environment designed by Cliff and Ross (Cliff and Ross, 1995) with eight sensors available on the agents. It has a simple linear path of empty cells and agents need to go through a field of trees. At each cell an agent experiences a different sensation and only one appropriate action among eight directional moves should be taken to reach the goal. Cliff and Ross showed that ZCS even with internal memory could not

cell no.	trajectory	length
0	0(2), 0(3)	2
1	4(2), 0(3)	2
2	7(1), 0(2), 0(3)	3
3	3(0), 7(1), 0(2), 0(3)	4
4	3(1), 3(0), 7(1), 0(2), 0(3)	5
5	0(2), 2(2), 6(0), 3(1), 3(0), 7(1), 0(2), 0(3)	8
6	2(2), 6(0), 3(1), 3(0), 7(1), 0(2), 0(3)	7
7	4(2), 6(1), 2(2), 2(2), 6(0), 3(1), 3(0), 7(1), 0(2), 0(3)	10
8	7(1), 6(1), 2(2), 2(2), 6(0), 3(1), 3(0), 7(1), 0(2), 0(3)	10
9	0(0), 4(2), 7(0), 3(1), 6(1), 2(2), 2(2), 6(0), 3(1), 3(0), 7(1), 0(2), 0(3)	13
10	4(2), 7(0), 3(1), 6(1), 2(2), 2(2), 6(0), 3(1), 3(0), 7(1), 0(2), 0(3)	12
11	3(1), 6(1), 2(2), 2(2), 6(0), 3(1), 3(0), 7(1), 0(2), 0(3)	10
12	3(1), 0(0), 4(2), 7(0), 3(1), 6(1), 2(2), 2(2), 6(0), 3(1), 3(0), 7(1), 0(2), 0(3)	14
13	7(1), 7(0), 3(1), 6(1), 2(2), 2(2), 6(0), 3(1), 3(0), 7(1), 0(2), 0(3)	12
14	7(1), 7(0), 4(2), 7(0), 3(1), 6(1), 2(2), 2(2), 6(0), 3(1), 3(0), 7(1), 0(2), 0(3)	14
15	7(1), 6(0), 7(1), 7(0), 4(2), 7(0), 3(1), 6(1), 2(2), 2(2), 6(0), 3(1), 3(0), 7(1), 0(2), 0(3)	16
16	7(1), 6(0), 7(1), 6(0), 7(1), 7(0), 4(2), 7(0), 3(1), 6(1), 2(2), 2(2), 6(0), 3(1), 3(0), 7(1), 0(2), 0(3)	18
17	0(2), 0(3), 7(3), 6(1), 6(0), 7(1), 7(0), 4(2), 7(0), 3(1), 6(1), 2(2), 2(2), 6(0), 3(1), 3(0), 7(1), 0(2), 0(3)	19
18	4(2), 0(3), 7(3), 6(1), 6(0), 7(1), 7(0), 4(2), 7(0), 3(1), 6(1), 2(2), 2(2), 6(0), 3(1), 3(0), 7(1), 0(2), 0(3)	19

Table 2: A solutions of woods14 problem (each pair of numbers in the trajectory represents motor action and state number in FSM, 0 : down, 1 : right, 2 : up, 3 : left, 4 : lower right, 5 : upper right, 6 : upper left, 7 : lower left)

succeed in developing desirable control systems for this problem (Cliff and Ross, 1995). Woods14 shown in Figure 5(a) is a very difficult problem for ZCS to solve because it requires long chains of actions before the reward is given. Also ZCS tends to create a conflicting overgeneral classifier producing inappropriate actions, that is, a classifier that matches multiple sensations and does not cover appropriate actions for each sensation. This happens when the classifier's sensor pattern contains too many don't care ('#') terms.

An FSM with eight sensors, allowing 256 distinct sensor states, was evolved to solve the woods14 problem with a perfect score, average 10 time steps. Without difficulty, evolutionary computation succeeded in finding an appropriate strategy. When only four sensors are used as in the woods101 problem, this woods14 problem generates complex situations. Most empty cells have sensor aliases because of the sensor limitation. The problem becomes a non-Markovian environment.

An EMO analysis for four-sensor agents shows that the memoryless policy, as expected, fails to find solutions to the woods14 problem. Even two or three states were insufficient; there are starting cells from which the agent cannot reach the goal. As shown in Figure 5(b), there is a hierarchy of complexity with memory size. For two or three internal states, performance intermediate between the memoryless strategy and the four state strategy can be achieved. It depends on how many empty cells are successfully built into the path to the goal. An optimal memoryless policy has 5 empty cells that can be connected to the goal. The other 14 cells fail to reach the goal. With two states, the goal is reachable from 11 empty cells. Three states gives 17 successful cells among the total of 19 cells. The diagram in Figure 5(b) shows how closely a given number of states can achieve solutions.

One of the best solutions with four state machines, see Table 2, shows how to handle perceptual aliasing. Cell 13, 14, 15 and 16 have the same sensory pattern. They are surrounded by trees in all four directions. Cells 13 and 14 need the same action of moving to the lower left, while both cell 15 and cell 16 need an action of moving to the upper left. The strategy first tries to move lower-left and if the situation is unchanged, then it tries to move upper-left. The two actions are sequentialized with internal states, depending on sensations. Cell 0, 5 and 17 have sensor aliases. Cells 0 and 17 prefer to take an action of moving down, but cell 5 needs an action of moving to the upper-left. These kinds of sensor aliases require internal states to escape cul-de-sacs. Internal states place marks on each sensor alias state and then block taking the same action. For instance, an agent first moves down at cell 5, changes its internal state, returns to cell 5 and then moves to the upper-left. The agent does not memorize sensation values and put milestones at particular sensations. In this way internal memory can be easily represented by Boolean logical values or finite states.

If the number of sensors is decreased, more sensor aliases may occur, but not necessarily. Instead of four or eight sensors, two sensors are selected among eight sensors to see the influence of sensor limitation. First, FSMs with two sensors, front and right sensors, are evolved and the EMO analysis in Figure 6(a) shows that three internal states are insufficient to solve the problem[5]; there are 9 initial cells which cannot reach the goal for the best three-state machine. (In experiments with one or two sensors, penalty 80 is used instead of 40 in equation 1 to see a clear distinction of solvability for each state machine.) In this case four state machines can solve the problem with a total 301 time steps to reach the goal from all empty cells while four-state agents with four

[5]We assume that the problem cannot be solved with a specified control structure when no experimental run among 25 runs is succeeds in solving the problem.

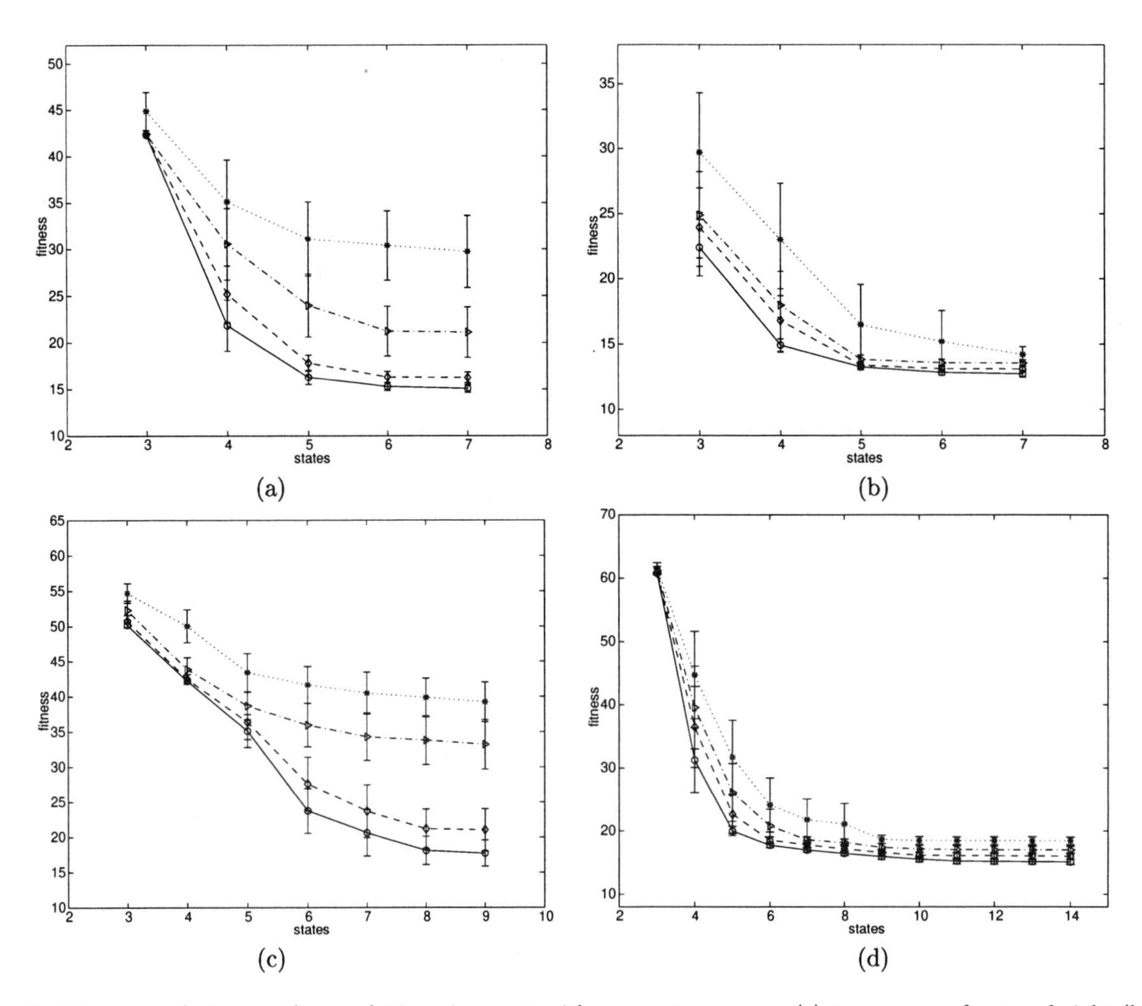

Figure 6: Memory analysis over the woods14 environment with one or two sensors (a) two sensors, front and right (b) two sensors, front and left (c) two sensors, left and right (d) one sensor, front (* : 500 generations, ▷ : 1000 generations, ◇ : 2000 generations, ○ : 5000 generations)

sensors have a score of total 198 time steps[6]. When two sensors in front and left are selected, it still requires four states to solve the problem and the performance is improved with total 245 time steps to the goal as shown in Figure 6(a)-(b). For three state machines, there are two initial cells which cannot reach the goal.

The selection of two sensors on the left and right makes the woods14 problem more difficult. Even with five internal states, there were still three cells unable to reach the goal. Six or more states could solve the problem. Figure 6(c) shows a significant change of performance between five states and six states. If the number of sensors is limited, what sensors are selected will be an important issue to solve problems and produce good performance.

When only one sensor, the front sensor, is used, four states are sufficient to solve the problem with a total 371 time steps to the goal. This agent unexpectedly outperforms the one with two sensors on the left and right. Three state machines had 14 cells which failed to reach the goal. Restricting sensors generally degrades the performance, but does not imply definitely that more internal states are required, as is evident from the above results of two sensors on the left and right versus only one sensor in front. Clearly, sensor morphology is important for agent problems.

It is also notable that the two sensor results including the front sensor are better than the result with only the front sensor, even though all these require four states to solve the woods14 problem. In Figure 6(d), the best performance of the one sensor experiment — even with many states — is significantly worse than the performance of those two sensor experiments including the front sensor.

Thus, we can build a partial ordering relation on the performance for a various number of sensors. For $S_A \subseteq S_B$, where S_A, S_B are a set of sensors, the performance

[6]These score solutions may not be optimal, and they are obtained with 5,000 generations.

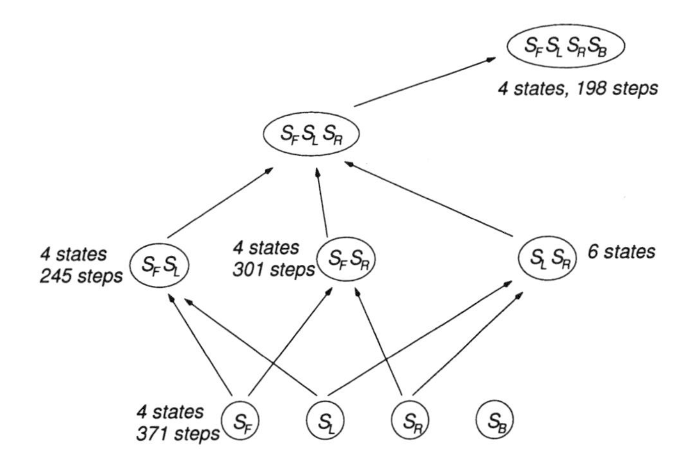

Figure 7: Partial order relation on performance with sensors in the woods14 problem (S_F, S_L, S_R, S_B are sensors in front, left, right and back)

function G keeps the partial order for the same amount of memory, M.

$$G(S_A, M) \leq G(S_B, M)$$

More sensors can see the environment better. Also for $M_1 \leq M_2$, the partial order relation holds with a given set of sensors.

$$G(S_A, M_1) \leq G(S_A, M_2)$$

Using the above property, the following partial order relation may be inferred for the performance of robotic tasks or agent problems.

$$G(S_A, M_1) \leq G(S_B, M_2)$$

where $S_A \subseteq S_B$, and $M_1 \leq M_2$.

When a pair of sensor sets and memory is not ordered with each other — for example, where there is no subset relation between S_A and S_B or $S_A \subseteq S_B$ and $M_1 > M_2$ — the performance order cannot be predictable. It will depend on tasks and environmental situations. The partial order relation in the woods14 environment is displayed in Figure 7. If the front sensor is included, at most four memory states are required to solve the problem. When the number of sensors is increased, the performance becomes better. Thus, it is presumed that agents with front, left and right sensors will solve the problem with four states and have performance in the range ($198 \leq G \leq 245$).

4 Conclusion

Woods Problems are goal search problems with perceptual aliasing. Four Woods Problems, woods101, woods102, Sutton's gridworld and woods14, were solved with a novel Evolutionary Multi-objective Optimization (EMO) analysis. The woods101 environment requires two internal states to solve the problem, and woods102 also needs two states to reach the goal from any initial cell but presumably four states for an optimal solution. In Sutton's gridworld problem, two internal states yield an optimal strategy for eight sensors. Reducing the number of sensors to four sensors requires four states for an optimal solution. Woods14 requires four states for an optimal solution with four sensors, while memoryless agents with eight sensors can solve the problem. The number of sensors influences the necessary memory size, since reducing the number of sensors or sensor range has the potential of creating more sensor aliases. The memory size required to solve a problem is not necessarily proportional to the number of aliases. Instead it is more concerned with what kind of sensor aliases are experienced in the path to goal. Placing the same sensory patterns serially in the path may increase the memory requirement.

Memory plays an important role to solve sensor aliases and can improve an agent's performance. The EMO approach was effective to find the minimum number of memory elements for a variety of tasks and to visualize a hierarchy of performance depending on available memory. All the problems described show that purely reactive systems are inferior to memory-based systems. However, the results also show that a large number of memory elements are not necessarily required. A small number of bits are sufficient to evolve desirable behaviors in several Woods Problems, although many complex control designs such as recurrent neural networks have been suggested for agent problems. The problem set provided in this paper may be simple, but the results indirectly imply that reactive systems with small-sized memory are effective in agent problems. Also the number of sensors is correlated with memory size for desirable performance.

The Woods Problem may be seen as a model of agent navigation tasks. In robotic tasks, robotic agents experience perceptual aliasing in various situations which may cause similar decision problems in which their actions cannot be completely determined by perceptions. The EMO approach to the Woods Problem can be easily applied to robotic tasks requiring a small amount of memory or no memory, in order to measure the necessary memory amount for the tasks as well as to see the memory effect, assuming suitable task simulation environments are available.

References

Bakker, B. and de Jong, M. (2000). The epsilon state count. In *From Animals to Animats 6: Proceedings of the Sixth International Conference on Simulation of Adaptive Behaviour*, pages 51–60. MIT Press.

Brooks, R. (1986). A robust layered control system for a mobile robot. *IEEE Journal of Robotics and Autonomation*, Vol. RA-2, No.1:14–23.

Brooks, R. (1987). Intelligence without representation. In *Workshop on th Foundations of Artificial Intelligence*, Endicott House, Dedham Mass.

Brooks, R. (1991). Intelligence without reason. AI memo 1293, Artificial Intelligence Laboratory, MIT.

Chapman, D. and Kaelbling, L. P. (1991). Learning from delayed reinforcement in a complex domain. In *Proceedings of the 12th Int. Joint Conf. on Artificial Intelligence.*

Cliff, D. and Ross, S. (1995). Adding temporary memory to ZCS. *Adaptive Behavior*, 3(2):101–150.

Colombetti, M. and Dorigo, M. (1994). Training agents to perform sequential behavior. *Adaptive Behavior*, 2(3):305–312.

Fogel, L., Owens, A., and Walsh, M. (1966). *Artificial intelligence through simulated evolution.* Wiley, New York.

Kaelbling, L. P. (1986). An architecture for intelligent reactive systems. In *Reasoning about Actions and Plans: Proceedings of the 1986 Workshop*, Endicott House, Dedham Mass.

Kim, D. and Hallam, J. (2001). Mobile robot control based on boolean logic with internal memory. In *Advances in Artificial Life : European Conference on Artificial Life*, pages 529–538.

Lanzi, P. (2000). Adaptive agents with reinforcement learning and internal memory. In *From Animals to Animats 6: Proceedings of the Sixth International Conference on Simulation of Adaptive Behaviour*, pages 333–342. MIT Press.

Lanzi, P. L. (1998). An analysis of the memory mechanism of XCSM. In *Genetic Programming 98*, pages 643–651. Morgan Kauffman.

Lee, W.-P. (1998). *Applying Genetic Programming to Evolve Behavior Primitives and Arbitrators for Mobile Robots.* Ph. D. dissertation, University of Edinburgh.

Lee, W.-P., Hallam, J., and Lund, H. (1997). Applying genetic programming to evolve behavior primitives and arbitrators for mobile robots. In *Proceedings of IEEE International Conference on Evolutionary Computation*, Indianapolis, USA.

Lin, L. and Mitchell, T. M. (1992). Reinforcement learning with hidden states. In *From Animals to Animats 2: Proceedings of the Second International Conference on Simulation of Adaptive Behaviour*, pages 271–280. MIT Press.

Littman, M. (1994). Memoryless policies: theoretical limitations and practical results. In *From Animals to Animats 3: Proceedings of the Third International Conference on Simulation of Adaptive Behaviour*, pages 238–245. MIT Press.

Maes, P. and Brooks, R. (1990). Learning to coordinate behaviors. In *Proceedings of the Eighth National Conference on Artificial Intelligence(AAAI'90)*, pages 796–802. Boston, MA, August.

Mahadevan, S. and Connell, J. (1991). Scaling reinforcement learning to robotics by exploiting the subsumption architecture. In *Proceedings of the Eighth International Workshop on Machine Learning.*

McCallum, A. K. (1996). *Reinforcemnet Learning with Selective Perception and Hidden State.* Ph. D. dissertation, University of Rochester.

McCallum, R. A. (1993). Overcoming incomplete perception with utile ditinction memory. In *Proceedings of the 10th International Machine Learning Conference.* Morgan Kauffman.

Miller, J. (1996). The coevolution of automata in the repeated prisoner's dilemma. *Journal of Economics Behavior and Organization*, 29(1):87–112.

Nolfi, S. and Floreano, D. (2000). *Evolutionary Robotics : The Biology, Intelligence, and Technology of Self-Organizing Machines.* MIT Press, Cambridge, MA.

Singh, A. P., Jaakkola, T., and Jordan, M. I. (1994). Model-free reinforcement learning for non-markovian decision problems. In *Proceedings of the 11th International Machine Learning Conference.* Morgan Kauffman.

Stanley, E., Ashlock, D., and Smucker, M. (1995). Iterated prisoner's dilemma game with choice and refusal of partners. In *Advances in Artificial Life : European Conference on Artificial Life*, pages 490–502.

Sutton, R. S. (1990). Integrated architectures for learning, planning, and reacting based on approximating dynamic programming. In *Proceedings of the Machine Learning Conference.*

Sutton, R. S. (1991). Planning by incremental dynamic programming. In *Proceedings of the Eighth International Workshop on Machine Learning*, pages 353–357.

Whitehead, S. D. (1992). *Reinforcemnet Learning for the Adaptive Control of Perception and Action.* Ph. D. dissertation, University of Rochester.

Whitehead, S. D. and Ballad, D. H. (1991). Learning to perceive and act by trial and error. *Machine Learning*, 7:45–83.

Wilson, S. W. (1994). ZCS: a zeroth level classifier system. *Evolutionary Computation*, 2(1):1–18.

Evolution of efficient swimming controllers for a simulated lamprey

Jimmy (Hajime) Or, John Hallam, David Willshaw*   Auke Ijspeert**

*Divison of Informatics
University of Edinburgh
5 Forrest Hill, Edinburgh
Scotland EH1 2QL
{hajimeo, john}@dai.ed.ac.uk
david@anc.ed.ac.uk

**Computational Learning and Motor Control Lab
University of Southern California
Hedco Neuroscience Building
3641 Watt Way, Los Angeles
CA 90089-2520, USA
ijspeert@usc.edu

Abstract

This paper investigates the evolutionary design of efficient connectionist swimming controllers for a simulated lamprey. Efficiency is defined as the ratio of forward swimming speed to backward mechanical wave speed.

Using the lamprey model proposed by Ekeberg (1993) and extending the work of Ijspeert et al. (1999) on evolving lamprey swimming central pattern generators (CPGs) through genetic algorithms (GAs), we investigate the space of possible neural configurations which satisfies the property of high swimming efficiency. Techniques are devised to measure efficiency at various swimming speeds. The measurements are incorporated into the fitness function of Ijspeert's original GA and efficient controllers are evolved. Interestingly, the best evolved controller not only is capable of swimming in a similar manner to the real lamprey, but also with the same efficiency (about 0.8). Moreover, it can exhibit a wide range of controllable speeds and efficiencies.

1 Introduction

In recent years, there have been advances in understanding animal motor control due to better physiological measurement techniques, higher density microelectrodes and faster computers for simulations of the neural mechanisms which underlie behavior. However, due to the complexity of the nervous systems, we are still far from being able to understand completely the neural control of higher vertebrates such as humans. The lamprey has been chosen for study by several neurobiologists because it is relatively easy to analyze: firstly, because while it has a brainstem and spinal cord with all the basic vertebrate features, the number of neurons in each category is an order of magnitude fewer than in other vertebrates, and secondly because its swimming gait is simple. Hence, findings on this *prototype vertebrate* can provide a better understanding of vertebrate motor control.

According to Sir James Lighthill, swimming speed and efficiency are the two qualities that fish must maintain in order to survive (Lighthill, 1970). If the swimming efficiency is low, the fish can quickly use up energy derived from food before they can find their next meal. The ability to maintain high swimming efficiency is especially important for lamprey because they do not eat during the long journey up-river from the sea to the breeding grounds (Williams, 1986). Blake (1983) suggested that efficiency is a good criteria to use when comparing the swimming performance of different fish. Its increase with speed is important to the evolutionary ecology of fish. Swimming efficiently is also important from a robotics point of view. An inefficient robotic lamprey can use up its battery power and sink in the ocean easily. Note that in the robotic implementation, it is important to maintain efficiency across a wide range of speeds.

Currently, there are several definitions of swimming efficiency (Sfakiotakis et al., 1999). However, the ratio of forward swimming speed (U) to backward mechanical wave speed (V) has been commonly used (Williams, 1986; Sfakiotakis et al., 1999). Since biological data is available for comparison, we are using the same definition to evolve controllers in this paper.

Over the past 15 years, neurobiologists have achieved a better understanding of the lamprey locomotive networks. However, nobody yet fully understands how the segmental oscillators inside the lamprey CPG are coupled. We believe it is a good idea to consider a few important properties related to the survival of the lamprey (such as swimming efficiency, robustness in speed against changes in body scales and noise in neural connections, etc.) and then use them as a guide towards the discovery of features in its neural organization that are related to such properties. Given that high swimming efficiency is important to both the real and the artificial lamprey, in this paper we extend the idea from Ijspeert's (1998) work on evolutions of lamprey swimming CPGs to evolve efficient swimming controllers for the model lamprey.

Experimental results are encouraging. Most of the evolved controllers are able to swim like the real fish and with high efficiency. Their speed vs. efficiency curves show that they can not only achieve a wide range of speeds but also be able to maintain a fairly constant efficiency (at least for speeds over 0.3 [m/s]).

Most importantly, the best evolved controller has achieved an efficiency of about 0.8, which is close to the one achieved by the real lamprey (Williams, 1986). Thus, through the use of GA, we have found intersegmental couplings which allow the model lamprey to swim at about the same efficiency as the real one. This result could not only provide inspiration to biologists to gain a better understanding of the intersegmental couplings of the real lamprey but could also inspire the development of more efficient swimming controllers for lamprey in both computer simulations and robotic hardware.

2 Background

This section briefly describes the Ekeberg neural and mechanical models. A more detailed description can be found in (Ekeberg, 1993; Or, 2002).

2.1 Neural model

Based on physiological experiments, Ekeberg (1993) hand-crafted a connectionist model for the lamprey swimming CPG. The network is made of 100 copies of interconnected segmental oscillators (Figure 1). Within each segmental oscillator, there are 8 neurons each of which is modeled using a leaky integrator with a saturating transfer function. The output u ($\in [0,1]$) is the mean firing frequency of the population the unit neuron represents. It is calculated using the following set of formulas:

$$\dot{\xi}_+ = \frac{1}{\tau_D}(\sum_{i\in\Psi_+} u_i w_i - \xi_+) \quad (1)$$

$$\dot{\xi}_- = \frac{1}{\tau_D}(\sum_{i\in\Psi_-} u_i w_i - \xi_-) \quad (2)$$

$$\dot{\vartheta} = \frac{1}{\tau_A}(u - \vartheta) \quad (3)$$

$$u = \begin{cases} 1 - \exp\{(\Theta - \xi_+)\Gamma\} - \xi_- - \mu\vartheta & (u > 0) \\ 0 & (u \le 0) \end{cases} \quad (4)$$

where w_i represents the synaptic weights and Ψ_+ and Ψ_- represent the groups of pre-synaptic excitatory and inhibitory neurons respectively. ξ_+ and ξ_- are the delayed 'reactions' to excitatory and inhibitory inputs and ϑ represents the frequency adaptation observed in real neurons. The values of the neural timing (τ_D, τ_A) and gain/threshold (Θ, Γ, μ) parameters and those for the connection weights are set up in such a way that the simulation results from the model agree with physiological observations. For details of the neuron parameters, refer to (Ekeberg, 1993).

As the details of the intersegmental connections of the real lamprey CPG are not yet known, Ekeberg simplified the controller as follows. Except for the CIN neurons which have longer projections in the caudal direction, each neuron has symmetrical connections extending both rostrally and caudally. Since the neurons at both ends of the CPG receive fewer neural connections, synaptic weights are adjusted to account for this by dividing them by the number of segments a neuron receives input from.

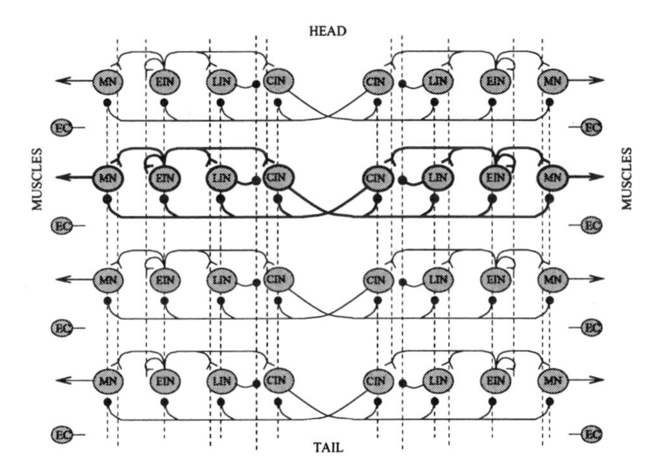

Figure 1: Configuration of the biological swimming controller. The controller is composed of 100 interconnected segmental oscillators (only four segments are shown here). Each segment consists of eight neurons of four types: motoneurons (MN), excitatory interneurons (EIN), lateral inhibitory interneurons (LIN) and contralateral inhibitory interneurons (CIN). A neuron unit represents a population of functionally similar neurons in the real lamprey. Each of them receives excitations from the lamprey brainstem. Connections with a fork ending represent excitatory connections while those with a dot ending represent inhibitory connections. In addition to input signals from the brainstem (not shown here), the controller receives feedback from the stretch sensitive edge cells (EC). Note that the EC cells are not considered in this paper.

The complete swimming CPG functions as follows: Global excitation from the brainstem stimulates all neurons in the CPG; sufficient stimulation results in oscillations in each individual segment at a frequency that depends on the strength of this global excitation signal. Extra excitation is supplied from the brainstem to the five most rostral segments of the CPG. The effect of this, interacting with intersegmental coupling, is to induce a roughly equal relative phase lag between successive segments in the CPG, with the result that caudally traveling waves of neural activity appear. The global excitation controls the amplitude of the motoneurons as well as the frequency of oscillation of the CPG. The extra excitation alters the intersegmental phase lag largely independently from the global excitation.

2.2 Mechanical model

Ekeberg (1993) further proposed a 2D mechanical lamprey model to study how the muscular activity induced by the model CPG affects swimming. The model lamprey is made to approximate the size and shape of the real one. It consists of 10 rigid body links with nine joints of one degree of freedom (Figure 2). Each link is assumed to be a cylinder with an elliptical cross-section. The link is represented by its center of mass coordinate (x_i, y_i) as well as the angle (φ_i) between it and the x-axis.

On each side of the body, muscles connect each link to its immediate neighbors. The muscles are modeled as a combination of springs and dampers. The outputs from the motoneurons control the spring constants of the corresponding muscles. As the neural wave travels along the body from head to tail, the successive contraction of muscles creates a mechanical wave. This in turn generates inertial forces from the surrounding water that propel the lamprey forward.

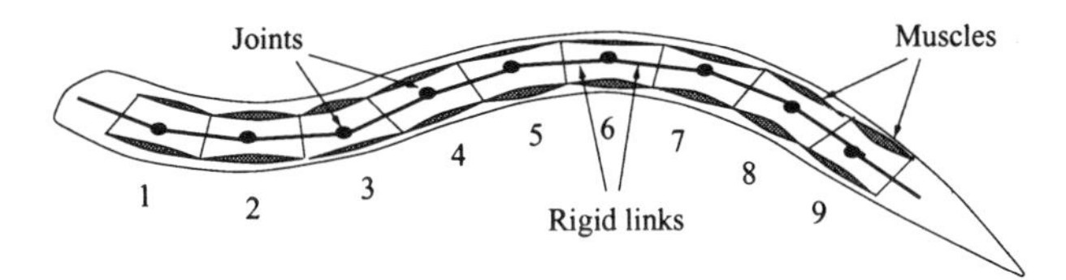

Figure 2: The mechanical lamprey model.

Each mechanical link has an elliptical cross section of constant height (30 [mm]) and variable width. Its length is 30 [mm]. The mass and moment of inertia of each link are calculated by assuming that the density of the lamprey is the same as that of water (Table 1).

3 Methods

3.1 Genetic Algorithms

To evolve intersegmental couplings of 100 copies of any chosen segmental oscillator, the same real number GA with mutation and crossover operators described in Ijspeert et al. (1999) is used here. Two sets of experiments are conducted. The first set is based on evolving controllers with big maximum achievable efficiency while the second set is on evolving controllers with big minimum efficiency (hereby referred to as the bigmax and bigmin approach respectively). The reason for conducting two sets of experiments is to investigate which approach produces better results (i.e. more controllers swimming at higher efficiency). In each set of experiments, the following four prototype controllers are used to seed the otherwise random initial populations of each run.

1. *The Biological Controller*, hand-crafted by Ekeberg (1993) based on physiological data.
2. *Controller 2*, evolved by Ijspeert et al. (1999) using Ekeberg's segmental oscillator and intersegmental couplings evolved by GA.
3. *Controller 3*, evolved by (Ijspeert et al., 1999), with both intra- and intersegmental connections evolved by GA.
4. *Hybrid Robust Controller*, which consists of the hybrid segmental oscillator[1] with the best evolved intersegmental couplings, for that oscillator, from (Or, 2002).

To test whether it was necessary to seed the initial population, with its potential reduction of diversity, we also ran:

5. *Hybrid Random*, in which the initial population was made up of individuals, using the Hybrid Segmental Oscillator, but with random intersegmental couplings, i.e. as in case 4 above but without the pre-evolved seed controller.

For each prototype controller, six evolutions (runs) are performed. This makes a total of 30 runs for each set of experiments. Note that the initial population of each run (for cases 1 to 4) contains a prototype controller together with 39 other randomly generated individuals while for case 5 all 40 are randomly generated.

3.2 Encoding

A chromosome is used to encode the intersegmental couplings among any given segmental oscillator and its neighbors. The couplings are extensions from a neuron in one segment to other post-synaptic neurons in the neighboring segments. For each neuron within a segment, the number of extensions in either the rostral or caudal direction is an integer value between 0 and 12. The reason for choosing this range is that it includes the maximum

[1] In Or (2002), we included two segmental oscillators (one which can exhibit a large range of amplitude and the other which can oscillate at a large range of frequency) along with other randomly generated individuals in the initial population to evolve the hybrid of these two controllers, which we call the Hybrid Segmental Oscillator. Its motoneuron outputs have a larger minimum amplitude and a larger range of oscillation frequency than the motoneuron outputs from the rest of the evolved oscillators. The best complete controller obtained by evolving intersegmental couplings for this segmental oscillator is called the Hybrid Robust Controller.

link	w_i [mm]	m_i [g]	I_i [g mm^2]	$\lambda_\perp$ [Ns^2/m^2]	$\lambda_\parallel$ [Ns^2/m^2]
1	20.0	14.1	1414	0.45	0.3
2	20.0	14.1	1414	0.45	0.2
3	20.0	14.1	1414	0.45	0.1
4	20.0	14.1	1414	0.45	0.0
5	17.2	12.2	1137	0.45	0.0
6	15.0	10.6	944	0.45	0.0
7	11.7	8.3	691	0.45	0.0
8	8.3	5.9	465	0.45	0.0
9	5.0	3.5	271	0.45	0.0
10	1.7	1.2	90	0.45	0.0

Table 1: Parameters for the mechanical simulation. w_i, m_i and I_i are the width, mass and inertia of link i respectively. $\lambda_\perp$ and $\lambda_\parallel$ are the λ factors used to calculate the water forces.

number of extensions (10) in the biological model. The extension is represented by a real-valued gene with range [0, 1]. The gene value is linearly mapped to [0, 12] and rounded to give the extension. Due to left-right symmetry, a chromosome of length 64 can be used to encode the intersegmental connections of the entire CPG.

3.3 Fitness calculation

We defined the fitness function which evolves efficient swimming controllers as a product of five factors so that they can be optimized at the same time. Each fitness factor varies linearly between 0.05 and 1.0 when the corresponding variable varies between the "bad" and "good" boundary. The five factors reward solutions with the abilities to:

1. generate stable and regular oscillations in the 100 segments of the CPG (i.e. *min_fit_oscil* > 0.45),
2. change the wavelength of undulation by changing the global excitation,
3. change the frequency of oscillation by changing the extra excitation,
4. modulate the swimming speed by varying either the frequency of oscillation or the wavelength,
5. swim with high efficiency.

Note that the first two factors are exactly as the ones defined in (Ijspeert, 1998). The third and fourth factors are slightly modified to adapt to the experiments described in this paper.

An evaluation consists of neuromechanical simulations with different control inputs in order to determine how different key characteristics (such as frequency, speed and efficiency, *etc.*) vary. Note that the difference between experiments one and two is that, in the latter case, the fitness function is modified to search for the minimum non-zero efficiency value instead of the maximum efficiency.

3.4 Efficiency calculation

The efficiency and mechanical wave speed are defined as follows:

$$e = \frac{U}{V} \tag{5}$$

$$V = \frac{\lambda}{T} \tag{6}$$

where U, V, λ and T are the swimming speed [m/s], backward mechanical wave speed [m/s], mechanical wavelength [m] and the mechanical period [s] respectively. The forward swimming speed is calculated by fitting a circular arc to the positions and directions of the model lamprey head at two time instants which have the same relative phase in the swimming cycle.

Since both the neural and mechanical simulations are not stable at the beginning, the lamprey can end up swimming straight at any angle. The first step in computing the mechanical wave parameters is to rotate the original coordinate system so that the fish is swimming parallel to the x-axis. Following the transformation, we calculated the mechanical periods of body links 2 to 5. (The mechanical period of each body link can be calculated by using the difference in time instances at which two consecutive wave crests pass through the same link.) If the periods of three or more of the mechanical links are defined, the mechanical wavelength can be calculated using the method described by Videler (1993). Note that due to strange head and tail movements caused by the reduction in neural connections at the ends of the swimming CPG, the mechanical period of the second link, T_2, is used in Equation 6 to calculate the mechanical wave speed. Equation 5 can then be used to compute the swimming efficiency.

4 Results

We monitored the progress of the evolution weekly. Some of the most fit individuals from these evolutions were tested. When the fitness of these individuals stopped increasing significantly, we stopped the evolutions (after 2 months). The best individual of each run is tested over a range of global and extra excitations (under neuromechanical simulations) to determine the ranges of amplitude, frequency, phase lag, speed and efficiency which it can achieve. The corresponding surfaces are plotted for comparison.

Due to space limitations, the results for only 20 of the 60 evolved controllers are presented here. The criteria for choosing these controllers is a balance between high swimming efficiency and high fitness (recall that the fitness function has more factors than efficiency). Based on these criteria, two controllers from each prototype group are chosen for comparison. Since there are five prototypes and two sets of experiments, this makes a total of 20 controllers.

The results for the two sets of experiments are presented in the following subsections. For details of the neural configuration and performance surfaces of each controller, refer to (Or, 2002).

4.1 Results of experiment 1: On evolving controllers with big maximum efficiency

The results for the 10 selected controllers based on the bigmax approach are summarized in Table 2.

The table indicates that except for the run3 controller and those evolved with the hybrid segmental oscillator as the prototype (the bottom four controllers), the rest of the evolved controllers can achieve a higher maximum efficiency than the corresponding prototypes. Among the 10 evolved controllers, the run5 controller has the highest efficiency value of 0.86.

4.2 Results of experiment 2: On evolving controllers with big minimum efficiency

The results for the 10 selected controllers based on the bigmin approach are summarized in Table 3. The table indicates that except for the run7, run9 and run10 controllers (again all evolved with the hybrid segmental oscillator as the prototype), the rest of the evolved controllers are more efficient than their corresponding prototypes. Among the 10 evolved controllers, the run8 controller has the highest efficiency value of 1.03[2]!

5 Discussion of the methods

Rather than starting with random initial populations, we included a prototype controller in each initial population to guide the GAs to search for regions of possible solutions (controllers that can at least swim) in the search space. Although this approach can reduce the amount of time needed to evolve efficient controllers, there is the possibility that all the evolved controllers (under the same prototype) end up similar to each other. Fortunately, this did not pose a serious problem here. Most of the evolved controllers (even evolved with the same prototype) have different neural configurations and performance surfaces. Thus, adding the prototypes helped to accelerate the generation of interesting swimming controllers without significantly biasing the diversity of the evolved controllers. The biological controller, controller 2 and controller 3 prototypes all have a fitness very close to zero under the new fitness function. As a result, they could not dominate the entire population. As for the hybrid robust prototype controller, it has a fitness of 0.11 and 0.06 in experiments one and two respectively, which is relatively low. The mutation and crossover operators of GAs could easily move the search to neighboring regions. Evolutions based on the hybrid random prototype were included just in case this approach failed. In general, if there were plenty of time and computing resources, it might be better to have an initial population with all randomly made individuals. This allows more different types of controllers to be evolved.

In experiment one, we evolved controllers based on big maximum efficiency. The reason for this is that we wanted to obtain controllers which are capable of swimming at high efficiency. Since we only considered positive efficiency to be valid, evolving controllers under this approach implicitly means evolving controllers with a larger efficiency range.

In experiment two, we evolved controllers based on big minimum efficiency. This approach implicitly forces all the measured efficiencies of the controller to be good because the GA is trying to pull up the worst efficiency each controller can achieve. Hence, it can be harder for the evolution system. However, the evolved controllers in this experiment should produce better results than those in experiment 1, and a comparison of the efficiency range achieved by the controllers in Table 2 and Table 3 shows that this is indeed correct. (Also, refer to Table 4 in the discussion which follows.)

To determine pulse regularity, the condition *min_fit_oscil* > 0.45 is used. The threshold value of 0.45 was derived in Ijspeert et al. (1999) based on experience. Generally speaking, this value is good enough to distinguish neural waves which oscillate regularly from those which do not. It seemed to be suitable for the implementation here at the beginning. However, at the end of the evolutions, we realized that the GA had found a way to break this condition to pull up the efficiency (see below). Fortunately, the threshold problem appears in only two of the 60 evolved controllers: run5 (experiment 1) and run8 (experiment

[2]Efficiency greater than one is impossible. This value is caused by the breakdown of the wavelength calculation algorithm. Thus, this value is later considered to be invalid (refer to Section 5 for details).

	Fitness	Amplitude range	Frequency range in [Hz]	Phase lag range in [%]	Speed range in [m/s]	Efficiency range
biological	0.00	[0.0, 0.8]	[1.6, 5.5]	[-0.1, 1.7]	[-0.09, 0.45]	[0.05, 0.58]
run1	0.11	[0.2, 0.8]	[1.6, 7.2]	[-2.9, 2.6]	[-0.09, 0.51]	[0.05, 0.61]
run2	0.10	[0.0, 0.8]	[1.3, 5.7]	[-1.4, 3.2]	[-0.03, 0.53]	[0.05, 0.64]
controller 2	0.00	[0.0, 0.8]	[1.7, 6.0]	[-3.1, 3.2]	[-0.09, 0.49]	[0.02, 0.60]
run3	0.13	[0.0, 0.8]	[1.4, 7.5]	[-2.6, 2.8]	[-0.09, 0.52]	[0.05, 0.59]
run4	0.11	[0.0, 0.8]	[1.6, 5.7]	[-2.2, 3.4]	[-0.03, 0.50]	[0.01, 0.63]
controller 3	0.00	[0.0, 0.6]	[1.3, 5.5]	[-0.2, 1.9]	[-0.08, 0.43]	[0.06, 0.58]
run5	0.10	[0.0, 0.6]	[1.4, 6.4]	[-2.3, 8.6]	[-0.07, 0.49]	[0.02, 0.86]
run6	0.06	[0.0, 0.6]	[1.5, 5.9]	[-0.0, 1.8]	[-0.08, 0.44]	[0.03, 0.64]
hybrid robust	0.11	[0.0, 0.8]	[1.8, 7.1]	[0.0, 3.1]	[-0.02, 0.49]	[0.08, 0.69]
run7	0.15	[0.0, 0.8]	[1.4, 7.1]	[-0.0, 2.8]	[-0.03, 0.48]	[0.18, 0.68]
run8	0.11	[0.0, 0.8]	[1.2, 7.1]	[0.0, 3.3]	[-0.02, 0.48]	[0.30, 0.61]
hybrid random						
run9	0.05	[0.0, 0.8]	[1.5, 7.6]	[-1.3, 2.9]	[-0.05, 0.48]	[0.07, 0.65]
run10	0.09	[0.0, 0.7]	[1.5, 7.0]	[-0.4, 2.2]	[-0.02, 0.38]	[0.07, 0.62]

Table 2: Summary of results for the evolved efficient controllers in experiment one. The table lists the performance of the best individual from each evolution. The evolution is based on the bigmax approach. Note that the hybrid random prototype generates irregular neural waves due to random couplings. As a result, all the parameters are undefined.

	Fitness	Amplitude range	Frequency range in [Hz]	Phase lag range in [%]	Speed range in [m/s]	Efficiency range
biological	0.00	[0.0, 0.8]	[1.6, 5.5]	[-0.1, 1.7]	[-0.09, 0.45]	[0.05, 0.58]
run1	0.11	[0.0, 0.8]	[1.6, 5.6]	[0.0, 6.4]	[-0.05, 0.48]	[0.05, 0.68]
run2	0.06	[0.2, 0.8]	[1.6, 5.5]	[-0.7, 1.8]	[-0.16, 0.47]	[0.05, 0.68]
controller 2	0.00	[0.0, 0.8]	[1.7, 6.0]	[-3.1, 3.2]	[-0.09, 0.49]	[0.02, 0.60]
run3	0.09	[0.0, 0.8]	[1.6, 5.5]	[-0.3, 2.1]	[-0.15, 0.51]	[0.06, 0.76]
run4	0.08	[0.0, 0.8]	[1.6, 5.5]	[-1.7, 3.1]	[-0.03, 0.48]	[0.03, 0.70]
controller 3	0.00	[0.0, 0.6]	[1.3, 5.5]	[-0.2, 1.9]	[-0.08, 0.43]	[0.06, 0.58]
run5	0.06	[0.1, 0.6]	[1.3, 7.9]	[-3.1, 3.3]	[-0.05, 0.48]	[0.03, 0.59]
run6	0.06	[0.0, 0.6]	[1.3, 6.9]	[0.0, 2.4]	[-0.08, 0.48]	[0.03, 0.58]
hybrid robust	0.06	[0.0, 0.8]	[1.8, 7.1]	[0.0, 3.1]	[-0.02, 0.49]	[0.08, 0.69]
run7	0.09	[0.0, 0.8]	[1.2, 7.0]	[-0.1, 3.5]	[-0.02, 0.49]	[0.02, 0.61]
run8	0.09	[0.0, 0.8]	[1.5, 7.1]	[0.0, 3.6]	[-0.05, 0.46]	[0.12, 1.03]
hybrid random						
run9	0.10	[0.0, 0.8]	[1.3, 7.0]	[-1.1, 5.3]	[-0.07, 0.46]	[0.23, 0.68]
run10	0.20	[0.0, 0.7]	[2.0, 7.1]	[0.0, 3.1]	[-0.02, 0.48]	[0.05, 0.63]

Table 3: Summary of results for the evolved efficient controllers in experiment two. The table lists the performance of the best individual from each evolution. The evolution is based on the bigmin approach. Note that the hybrid random prototype generates irregular neural waves due to random couplings. As a result, all the parameters are undefined.

2).

Finally, the methods used to calculate the mechanical wavelength and efficiency have several limitations. According to Videler (1993), the measurement of kinematic parameters such as mechanical wavelength can be achieved accurately only as long as the mechanical wave crests propagating along the body are well pronounced and the amplitude is large, even near the head. This should not pose a problem because these characteristics fit eel-like swimmers such as the lamprey. However, the two controllers with efficiency over 0.8 sometimes swim with a stiff body (due to pulse irregularity) in approximately the first half of the body. This is similar to the sub-carangiform swimming mode described in (Sfakiotakis et al., 1999). It looks like the main problem here is that the wavelength is not constant along the body (i.e. infinite wavelength in the rigid part of the body), as it should be during anguiliform swimming. Under this situation, the measurement algorithm breaks down, underestimating the mechanical wavelength, and efficiency over 1 was obtained.

6 Discussion of results

Using the fitness function presented in this paper, efficient swimming controllers have been evolved successfully. Most of them are more efficient than their corresponding prototypes. The neural configurations of the best individuals from the 20 evolutions are different even with the presence of the same prototype in the initial population. There is not much similarity in the way the segments are coupled.

Generally speaking, controllers based on the bigmin approach can achieve higher efficiency than those based on the bigmax approach. (This is true even when all 60 evolutions are taken into consideration.) Table 4 shows that all the evolved controllers have a maximum efficiency ≥ 0.58. Under the bigmax approach, three of the controllers have efficiency ≥ 0.65, the best of which has efficiency above 0.7. When the bigmin approach is used, six of the evolved controllers have efficiency ≥ 0.65, three of which have efficiency above 0.7. As the evolutions under the bigmin approach can produce more good solutions at the same time as those under the bigmax approach (they all started and terminated at the same time), it is better to evolve efficient controllers based on the bigmin approach.

Approach	$e \geq 0.58$	$e \geq 0.60$	$e \geq 0.65$	$e \geq 0.7$
Bigmax	10	9	3	1
Bigmin	10	8	6	3

Table 4: Comparison of performance of the bigmax and bigmin approaches in terms of the efficiencies of the controllers. Since there is a cheating controller in each experiment (see below), subtract one from each table element if they are considered to be invalid due to the breakdown of the mechanical wavelength calculation algorithm.

6.1 Discussion of the evolved controllers

In order to understand how the evolved controllers achieve high swimming efficiency, we have chosen the best five[3] for further investigation. We looked at the characteristics of their neural waves as well as the corresponding swimming patterns. Based on these investigations, the controllers can be classified into two groups. The first group includes the run5 and run8 controllers (from bigmax and bigmin respectively) while the second group involves the run2, run3 and run9 controllers (all from bigmin). Controllers from the former group are called the "*cheating controllers*" as some of their neural waves contain irregular oscillations.

Figure 3 gives an example of the motoneuron output from a controller (bigmin run8) of this group. Its *min_fit_oscil* value is 0.452 (0.485 for the bigmax run5 controller). The irregular neural waves cause the mechanical wave calculation algorithm to break down and return very short mechanical wavelengths (less than 0.1 m). This results in very high efficiency values. From computer animations, we have found that a lamprey embedded with either of these two controllers swims alternately between sub-carangiform and anguilliform swimming modes. As for the three controllers in the second group, the outputs from the motoneurons are regular.

[3]Based on a balance between high efficiency and smooth performance surfaces.

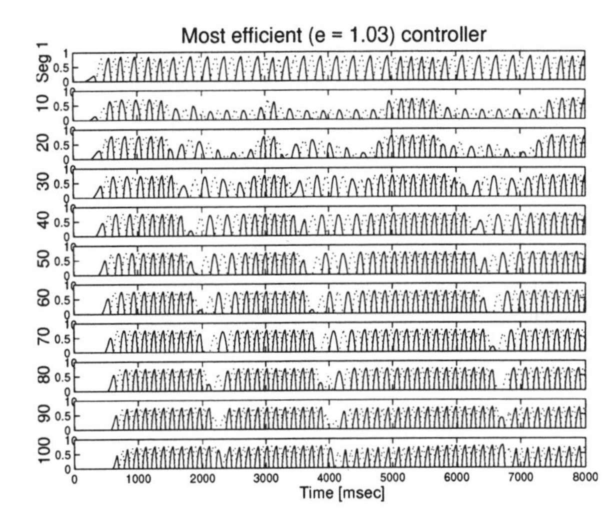

Figure 3: Irregular neural waves generated by the bigmin run8 controller. Solid lines represent outputs from the left motoneurons while dashed lines represent the outputs from the right motoneurons.

Figure 4 shows the neural wave of a typical controller (run3) from this group.

The average amplitude and oscillation frequencies of the three controllers from the second group are about 0.57 and 3.6 [Hz] respectively while the average swimming speed is about 0.34 [m/s] (refer to Table 5). These values are relatively low when compared with those of controllers evolved without taking efficiency into consideration, as reported in (Ijspeert, 1998)[4].

As we require a controller which can both generate regular control signals and swim with high efficiency, the run3 controller evolved under the bigmin approach is considered to be the most efficient one. The connection weight matrix for this controller is shown in Table 6 while its performance surfaces are shown in Figure 5. Note that the maximum efficiency achieved by this controller is 0.76, which is close to 0.8 achieved by the real lamprey (Williams, 1986).

Although the efficiency is very high, the corresponding maximum swimming speed is about 0.32 [m/s], which is lower than typical maximum speeds of about 0.4 [m/s]. This means efficient energy utilization at the cost of speed, as observed in the real lamprey (Williams, 1986).

6.2 Comparison of the best evolved controller with a matched sinusoidal controller

A comparison of the best evolved controller with a matched sinusoidal controller can provide justification

[4]The average frequency and maximum speed for the evolved controllers reported in (Ijspeert, 1998; Ijspeert et al., 1999) are 8.2 [Hz] and 0.54 [m/s] respectively.

Rank	(global, extra)	Amp	Freq [Hz]	Phase lag [%]	Mec λ [m]	Speed [m/s]	Wave speed [m/s]	Efficiency
1	(1.0, 40%)	0.76	5.56	1.22	0.04	0.34	0.33	1.03
2	(0.6, 140%)	0.45	3.60	1.69	0.02	0.27	0.31	0.86
3	(0.4, 100%)	0.54	3.52	1.60	0.12	**0.32**	0.42	**0.76**
4	(0.5, 150%)	0.60	4.03	1.29	0.13	0.37	0.55	0.68
5	(0.6, 30%)	0.58	3.23	1.61	0.15	0.34	0.49	0.68
Sin		0.54	3.52	1.60	0.16	**0.34**	0.58	**0.59**

Table 5: Comparison of neural and mechanical parameters for the five controllers with largest maximum efficiency. The controllers are run8, run5, run3, run2 and run9 (listed in order from top to bottom). Except for the run5 controller, these are evolved based on the bigmin approach. The last line shows the performance of a sinusoidal controller with amplitude, frequency and phase lag matching the rank 3 controller. See text for discussion.

	MNl	EINl	LINl	CINl	CINr	LINr	EINr	MNr	BS
MNl	-	1.0 [4, 9]	-	-	-2.0 [4, 4]	-			5.0
EINl	-	0.4 [3, 3]	-	-	-2.0 [3, 5]	-	-	-	2.0
LINl	-	13.0 [4, 2]	-	-	-1.0 [11, 5]	-	-		5.0
CINl	-	3.0 [11,1]	-1.0 [4, 7]	-	-2.0 [5, 7]	-	-	-	7.0
CINr	-		-	-2.0 [5, 7]	-	-1.0 [4, 7]	3.0 [11, 1]	-	7.0
LINr	-	-	-	-1.0 [11, 5]	-	-	13.0 [4, 2]	-	5.0
EINr	-	-	-	-2.0 [5, 5]	-	-	0.4 [3, 3]	-	2.0
MNr	-	-	-	-2.0 [4, 4]	-	-	1.0 [4, 9]	-	5.0

Table 6: Connection weight matrix for the efficient swimming (bigmin run3) controller. Excitatory and inhibitory connections are represented by positive and negative weights respectively. Left and right neurons are indicated by *l* and *r*. *BS* stands for brainstem. The extensions from a neuron to those in neighboring segments are given in brackets. The first number indicates the number of extensions in the rostral direction while the second number indicates extensions in the caudal direction.

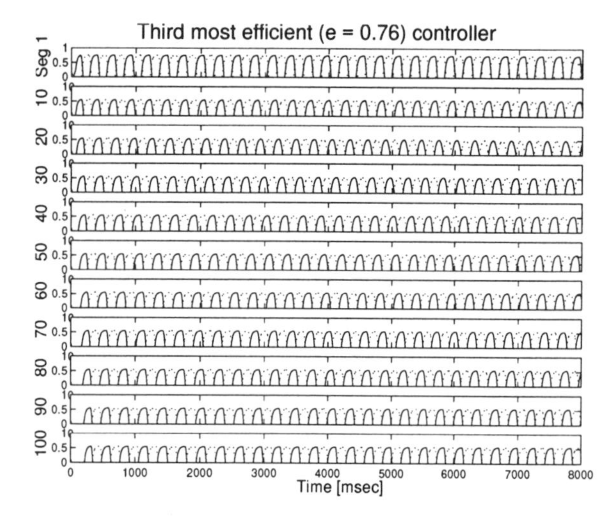

Figure 4: Neural wave of the bigmin run3 controller. Solid lines represent outputs from the left motoneurons while dashed lines represent the outputs from the right motoneurons

for using a neural controller evolved by GA, rather than a simpler analytic controller, to control an artificial lamprey. Table 5 lists the neural parameters and swimming performance of the five controllers with highest efficiency and one matched sinusoidal controller. By matching, we mean a sinusoidal controller whose amplitude, frequency and phase lag are tuned to match the corresponding parameters for the controller under consideration (in this case, the rank 3 one). The table shows that the relatively high efficiency achieved by the first two controllers is caused by the breakdown of the wavelength calculation algorithm. (Based on experience, wavelength much less than 0.1 [m] indicates that the motoneuron outputs are irregular. This in turn corresponds to the breakdown of the wavelength calculation algorithm which often returns a high efficiency value.) As the efficiency of these two controllers is invalid, the sinusoidal controller is compared with the run3 controller which is listed third (but considered to be the best evolved controller) in the table.

The comparison shows that although the analytic sinusoidal controller is able to achieve a slightly higher speed (0.34 [m/s] vs. 0.32 [m/s]), its efficiency is much lower than the run3 controller (0.59 compared with 0.76). Figure 6 shows the joint drive generated by the run3 controller with superimposed joint drive generated by the matched sinusoidal controller.

While the sinusoidal controller and the neural controller are matched in terms of amplitude, frequency and phase lag, they are not matched in terms of pulse shape. The pulse duration of the run3 controller is longer than that of the sinusoidal controller. Since the swimming speed is inversely proportional to burst duration (Grillner and Kashin, 1976; Wallén and Willams, 1984) (later demonstrated for the Ekeberg mechanical model — unpublished results), this agrees with the result that the swimming speed of the run3 controller is lower than that of the matched sinusoidal controller. Given that the sinusoidal controller can achieve a higher swimming speed,

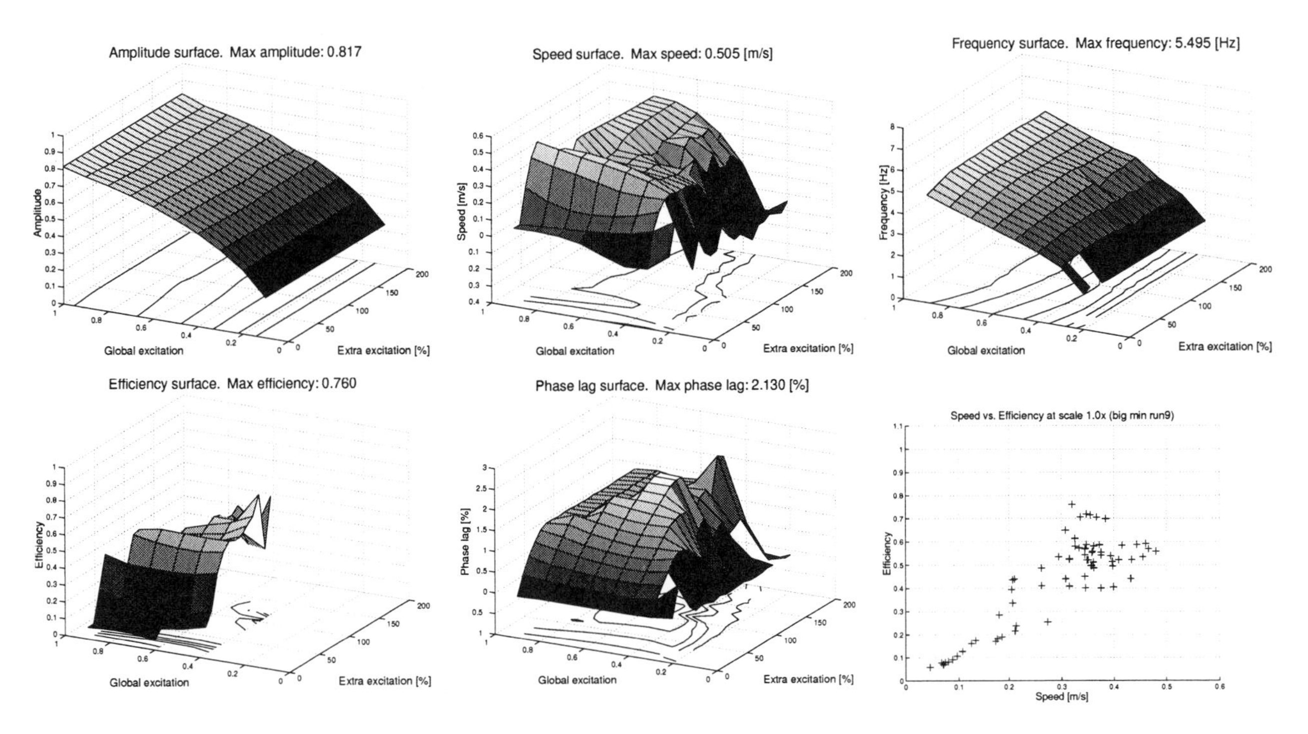

Figure 5: Performance surfaces of run3 controller. Note that efficiencies that cannot be measured (i.e. either *min_fit_oscil* $\leq$ 0.45 or efficiency ≤ 0) are filtered. These filtered values correspond to the empty regions in the efficiency surface.

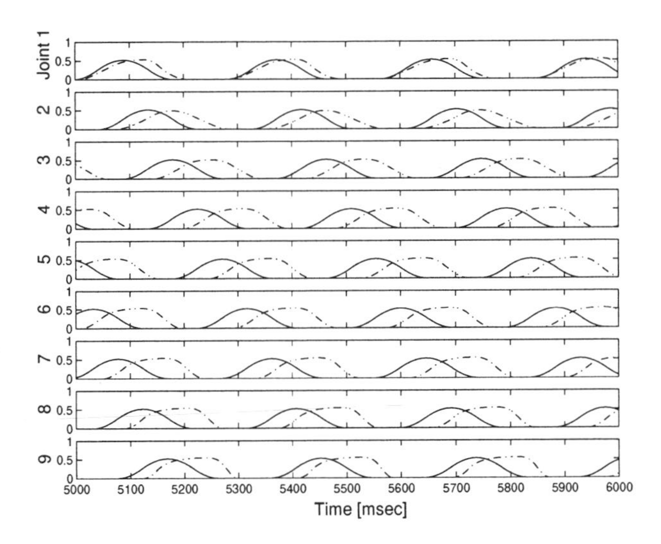

Figure 6: Averaged MNl activity (joint drive) from run3 controller with superimposed averaged MNl activity from a matched sinusoidal controller. The excitation combination used is (0.4, 100%). The amplitude is 0.54 while the frequency is 3.52 Hz. Phase lag is 1.60%. Dashed lines represent outputs from the run3 controller. Solid lines represent outputs from the matched sinusoidal controller

its lower efficiency is caused by the larger wavelength of the induced mechanical wave (0.16 [m] vs. 0.12 [m]).

The reason why a lamprey driven by the run3 controller can have a shorter wavelength is due to the pulse shape. A comparison of the signals that control the first joint of the mechanical lamprey body shows that symmetrical pulses are generated by the sinusoidal controller. On the other hand, the pulses generated by the run3 controller are not symmetrical (Figure 6). Recall that the mechanical wavelength calculation algorithm requires the time instances at which a mechanical wave crest passes through different parts of the body. As the pulse shape can affect the amount of bending at each body link, the mechanical wavelengths of the two controllers can be different. It may be that the rather flat and asymmetrical pulse shape of the run3 controller allows it to achieve a shorter mechanical wavelength. Such a controller is relatively difficult to hand-craft analytically. Note that another difference between the two controllers is that the wavelength in the neural controller varies slightly along the body, while it is perfectly constant in the sinusoidal controller.

6.3 Inherited property: Robustness in swimming efficiency against variations in speed

This section addresses the robustness in swimming efficiency against variation in speed. We want a robotic lamprey which can both achieve a wide range of swimming speed and be able to maintain high efficiency.

In order to determine the relationship between the swimming speed and efficiency of the evolved controllers, we have plotted the efficiency vs. speed curves of these controllers for speed ranges from 0.05 to 0.6 [m/s] in steps of 0.05 (refer to Figure 5 for the plot belonging to the bigmin run3 controller. For the plots of other controllers, refer to (Or, 2002).). Note that the scatter plots are used as the same speed can correspond to more than one efficiency value. For the different controllers, the results suggest that the relation between speed and efficiency can be classified into two types, namely:

- efficiency increases with speed
- efficiency increases with speed initially and then stays fairly constant (for speed over 0.3 [m/s])

Such relationships are favorable to both the real and robotic lampreys as mentioned in the introduction section.

7 Conclusion

In this paper, we successfully used GA to evolve efficient swimming controllers. Most of the evolved controllers exhibit a wide range of controllable speeds and efficiencies. Moreover, some of them are robust in swimming efficiency against speed. Under the same wave characteristics (amplitude, frequency and phase lag), the best evolved controller is able to drive the model lamprey more efficiently than the corresponding analytic sinusoidal controller. The difference appears to be due to subtle differences in the shapes of the signals between the sinusoidal controllers and the neural controllers. The evolved neural controllers can thus potentially be used to efficiently control a robotic lamprey such as the one developed by (McIsaac and Ostrowski, 1999). Based on experimental results, evolutions using the bigmin approach produce more controllers with higher swimming efficiency than those based on the bigmax approach. Most importantly, the GA has found a controller which can achieve the same efficiency as that observed in the real lamprey. Future work involves comparing the evolved neural organization with that of the real lamprey. The results will tell how close they are to each other.

8 Acknowledgments

Jimmy Or is supported by the British ORS Awards, the Canadian Natural Science and Engineering Research Council and the Canadian Space Agency. Facilities for this research were provided by the University of Edinburgh. We thank Catherine Dickie and Elspeth Thomas for proofreading the paper.

References

Blake, R. (1983). *Fish locomotion.* Cambridge University Press.

Ekeberg, Ö. (1993). A combined neuronal and mechanical model of fish swimming. *Biological Cybernetics*, 69:363–374.

Grillner, S. and Kashin, S. (1976). On the generation and performance of swimming in fish. In Cohen, A. H., Rossignol, S., and Grillner, S., (Eds.), *Neural control of locomotion.* Plenum Press.

Ijspeert, A. J. (1998). *Design of artificial neural oscillatory circuits for the control of lamprey- and salamander-like locomotion using evolutionary algorithms.* PhD thesis, University of Edinburgh.

Ijspeert, A. J., Hallam, J., and Willshaw, D. (1999). Evolving swimming controllers for a simulated lamprey with inspiration from neurobiology. *Adaptive Behavior*, 7(2):151–172.

Lighthill, M. J. (1970). Aquatic animal propulsion of high hydrodynamic efficiency. *Journal of Fluid Mechanics*, 44:263–301.

McIsaac, K. A. and Ostrowski, J. P. (1999). A geometric approach to anguilliform locomotion: simulation and experiments with an underwater eel robot. In *Proceedings of IEEE International Conference on Robotics and Automation*, volume 1, pages 2843–2848. IEEE Press.

Or, J. (2002). *An investigation of artificially-evolved robust and efficient connectionist swimming controllers for a simulated lamprey.* PhD thesis, University of Edinburgh. Unpublished PhD thesis being examined.

Sfakiotakis, M., Lane, D., and Davis, J. (1999). Review of fish swimming modes for aquatic locomotions. *IEEE Journal of Oceanic Engineering*, 24(2):237–252.

Videler, J. J. (1993). *Fish swimming.* Chapman and Hall.

Wallén, P. and Willams, T. (1984). Fictive locomotion in the lamprey spinal cord in vitro compared with swimming in the intact and spinal animal. *Journal of Physiology*, 347:225–239.

Williams, T. L. (1986). Mechnical and neural patterns underlying swimming by lateral undulations: Review of studies on fish, amphibia and lamprey. In Grillner, S., Stein, P. S. G., Stuart, D., Forssberg, H., and Herman, R., (Eds.), *Neurobiology of vertebrate locomotion*, pages 141–155. Macmillan.

Evolving Hierarchical Coordination in Simulated Annelid Locomotion

Edgar E. Vallejo*
*Tecnológico de Monterrey
Campus Estado de México
Atizapán de Zaragoza, México
evallejo@campus.cem.itesm.mx

Fernando Ramos**
**Tecnológico de Monterrey
Campus Cuernavaca
Cuernavaca, México
framos@campus.mor.itesm.mx

Abstract

This article presents a model for the evolution of locomotion behavior in a simulated annelid. In our model, segmental muscular coordination is modeled using a two-level hierarchy of cellular automata. We use genetic algorithms to evolve the dynamical law of the controller. We demostrate that this model can be used to evolve patterns of locomotion similar to those observed in natural annelids.

1 Introduction

Annelids are outstandingly successful animals. They possess elegant structures extremely well designed by natural selection. Previously, Maler argued that worms are animals worth modeling (Maler, 1993). Nevertheless, the simulation of worms in animat research has been marginal.

Research in evolving animat controllers has focused for many years on the evolution of neural controllers (Nolfi and Floreano, 2000). More recently, Vallejo and Ramos have demostrated that simpler, more tractable cellular automata controllers can be evolved to produce rhythmic behaviors (Vallejo and Ramos, 2000).

In order to explain the emergence of complex behaviors, controllers should be hierarchical. Ijspeert and colleagues have demostrated the evolution of hierarchical neural controllers capable of producing complex locomotion behaviors (Ijspeert et al., 1998).

We present a model for evolution of annelid locomotion using a two-level hierarchy of cellular automata. We demostrate that this model can be used to evolve locomotion patterns similar to those observed in natural annelids.

2 Annelid locomotion

An annelid is a worm with metameric segmentation and a body wall with both circular and longitudinal muscle. Metamerism improves the capability of muscular contractions, so that active locomotion can be more effective.

Segmentation allows control of particular portions of muscle, enabling each to contract independently of its neighbour. Each segment is a functionally isolated compartment. Locomotion is the result of the alterning contraction of circular and longitudinal muscles as shown in figure 1.

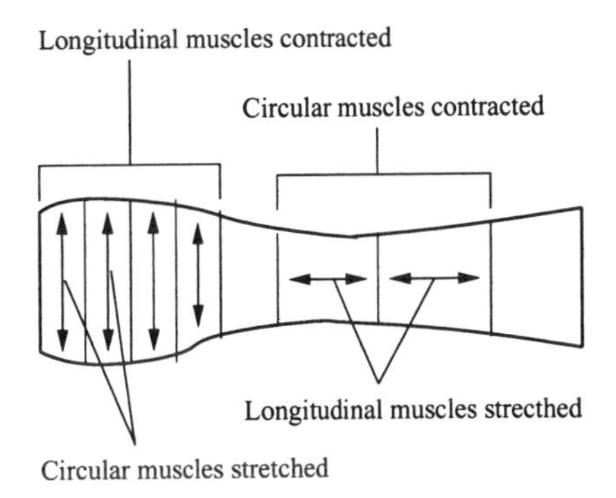

Figure 1: Annelid locomotion (Brusca and Brusca, 1990)

3 The model

The locomotion model consists of two layers of cellular automata. In the first layer, a lattice of longitudinal and circular muscular cells interact to produce coordination of a collection of muscle fibers (Bray, 2001). The second layer consists of a lattice of nerve cells that coordinates the movement of contiguous segments. The architecture of the controller is presented in figure 2.

The state space of cellular automata cells is binary, representing the contracting and relaxing states of muscular fibers, respectively. Nerve cells have four possible states, each representing the possible combinations of circular and longitudinal muscle contraction and relaxation states of each segment.

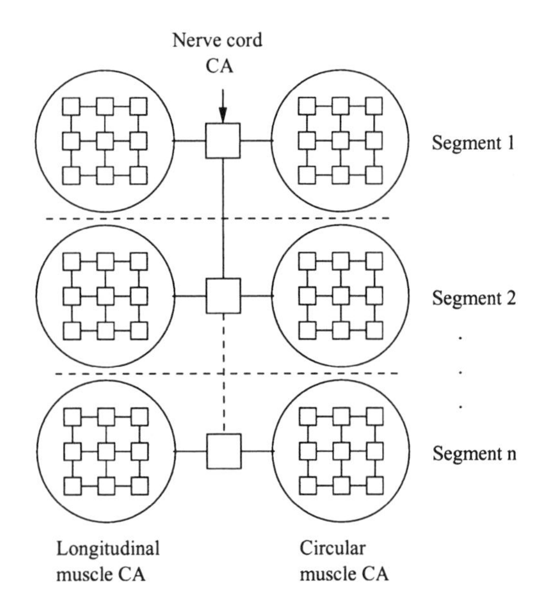

Figure 2: Annelid locomotion controller

4 Experiments and results

The experiments consist of the evolution of cellular automata rules using genetic algorithms. The genome codifies the state transition function of muscular and nerve cells. Fitness is defined as the distance traveled by the simulated annelid. We assume that it advances one step if there exists at least two pairs of contiguous segments that alternates muscular contraction and relaxation in consecutive time steps. We use a population of 256 individuals, and a maximum of 100 generations. We set the crossover and mutation probabilities to $p_c = 0.6$ and $p_m = 0.001$, respectively. In practice, genetic algorithms typically uses these values for the parameters.

We performed several runs with this model. In every run, the evolutionary process yielded controllers capable of sustained forward movement. The patterns produced were similar to those observed in annelids (Brusca and Brusca, 1990). Figure 3 shows a typical controller produced by the evolutionary process.

The controllers thus obtained behave as periodic attractors. From arbitrary initial conditions, the controller converges to a particular locomotion pattern. These results show the existence of periodic attractors that separates the state space into different basins of attraction sets. Attractor behaviors are robust with respect to perturbations in the environment.

5 Conclusions and future work

In this work, we demostrate the evolution of hierarchical cellular automata controllers for the synthesis of locomotion behavior in a simulated annelid.

The focus of this study has been on the evolution of locomotion behavior in a simulated annelid. An immediate extension of this work is the evolution of other annelid behaviors using hierarchical cellular automata controllers. Futhermore, we believe that this model can be used succesfully to evolve more general animat behaviors.

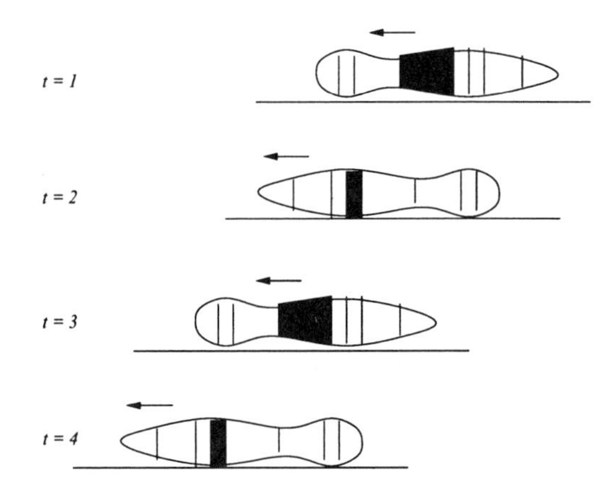

Figure 3: Results. The fourth segment is darkened for reference.

References

Bray, D. (2001). *Cell Movements. From Molecules to Motility.* Garland Publishing.

Brusca, R. and Brusca, G. (1990). *Invertebrates.* Sinauer Assoc.

Ijspeert, A. J., Hallam, J., and Willshaw, D. (1998). From lampreys to salamanders: evolving neural controllers for swimming and walking. In Pfeifer, R., Blumberg, B., Meyer, J. A., and Wilson, S., (Eds.), *From Animals to Animats 5. Proceedings of the Fifth International Conference on Simulation of Adaptive Behavior*, pages 390–399. The MIT Press.

Maler, O. (1993). Why should we build artificial worms and how? In Meyer, J., Roitblat, H. L., and Wilson, S., (Eds.), *From Animals to Animats 2. Proceedings of the Second International Conference on Simulation of Adaptive Behavior*, page 519. The MIT Press.

Nolfi, S. and Floreano, D. (2000). *Evolutionary Robotics. The Biology and Tecnology of Sel-Organizing Machines.* The MIT Press.

Vallejo, E. and Ramos, F. (2000). Evolving insect locomotion using non-uniform cellular automata. In Bedau, M. A., McCaskill, J. S., Packard, N. H., and Rasmussen, S., (Eds.), *Artificial Life VII: Proceedings of the Seventh International Conference*, pages 288–292. The MIT Press.

Evolution of a Circuit of Spiking Neurons for Phototaxis in a Braitenberg Vehicle

R. L. B. French **R. I. Damper**
Image, Speech and Intelligent Systems (ISIS) Research Group
Department of Electronics and Computer Science
University of Southampton
Southampton SO17 1BJ, UK.
{rlbf98r|rid}@ecs.soton.ac.uk

Abstract

Animal nervous systems have evolved to use spiking neurons but the 'artificial nervous systems' of animats typically are designed, not evolved, and use networks of formal, artificial neurons. We describe the evolution of circuits of spiking neurons for a robot, motivated by the desire to study links between neurophysiology and behaviour in artificial and (ultimately) natural animals. Spiking neurons have computational capabilities additional to those of artificial neurons based on activation functions. In particular, they should be better suited to processing temporal sequences. Thus, we describe early work aimed at evolution of neural circuitry which, when implanted in a Braitenberg type 2b vehicle, promotes phototaxis behaviour in the form of movement towards flashing lights of a particular frequency. The longer-term aim is to evolve natural taxis behaviours such as that observed in the cricket.

1 Introduction

Animal nervous systems use spiking neurons but the 'artificial nervous systems' (ANSs) of animats usually do not. The latter typically use a gross simplification in the form of artificial, or 'PDP-type', neurons based on activation functions, so that significant characteristics of real neurons are being ignored. It is likely that spikes evolved mainly as a means of regenerative signaling along the relatively long propagation paths of animal nervous systems (Levitan and Kaczmarek 1997). Because of active regenerative processes, the spike is propagated without loss of amplitude; hence, no information is carried by signal amplitude. Accordingly, detailed timing information for individual spikes, and relative timing between spikes, offers an extra dimension to the neural code (as does the stochastic aspect). Hence, spiking neurons ought to have additional computational capabilities relative to PDP-type neurons (Bugmann 1997; Maass and Ruf 1999).

In previous work (e.g., Damper, French, and Scutt 2000), we have used an autonomous robot to study the links between the 'neurophysiology' of the robot and its external behaviour in the belief that this can inform us about the links between neurophysiology and behaviour in real animals. This dictates that we use spiking neurons in the robot controller. Since real nervous systems are evolved rather than designed, we have more recently studied the evolution of spiking neuron circuitry (or ANSs) for our robot (French and Damper, submitted). A motivation for this work was to see if evolution could profit from the additional computational abilities of spiking neurons. However, the simple obstacle-avoidance task that we set the robot meant that there was no real scope for exploiting temporal-sequence processing capabilities. In this new work, we study a task specifically designed to promote sequence-discrimination behaviour. Specifically, we require the robot to move towards a flashing light of one frequency whilst ignoring a light of another frequency. This sort of discrimination is often exhibited in nature. For instance, female crickets move towards the 'song' of a male conspecific but ignore signals of similar (but different) acoustic structure. This is the phenomenon of *phonotaxis*, and it has been modelled in robot studies by Webb and Scutt (2000). Since our robot is not currently equipped with acoustic sensors, we have switched modalities to study the evolution of *phototaxis* behaviour.

The remainder of this paper proceeds as follows. In Section 2, we briefly review relevant work on spiking neuron models including the Hi-NOON simulator used here. Section 3 then describes the general scheme for evolution of an artificial nervous system (ANS) by genetic programming. Next, Section 4 describes the particular case of the evolution of a flashing-light frequency discriminator, which is then analysed in Section 5. Section 6 details the use of the evolved circuitry to implement phototaxis in a Braitenberg type 2b vehicle.

Section 7 describes plans for extending this work to the simulation of cricket phonotaxis and Section 8 concludes. A general description of the mathematical models used in Hi-NOON is included as Appendix A.

2 Spiking Neuron Models

Spiking neuron models have been implemented both as software (Raymond, Baxter, Buonomano, and Byrne 1992; Maass and Natschlager 1998) and hardware (Maris and Mahowald 1998; Rasche, Douglas, and Mahowald 1998). Generally speaking, such models operate at two different levels of detail: that of individual neurons and 'pools' of many neurons. For example, the model of Raymond et al. was designed to explore the contribution of individual neurons to the behaviour of a simulated nervous system, and the model of Maass and Natschlager was intended to look at the use of pools of neurons. In terms of application, Raymond et al. have investigated a possible circuit for reinforcement learning of head-waving behaviour in *Aplysia*, while Maass and Natschlager have constructed an associative memory of spiking neurons based upon a Hopfield network.

Modelling neurons with software contrasts with the hardware approach adopted by Maris and Mahowald and by Rasche et al. However, instead of building digital circuitry dedicated to modelling spiking neurons, these authors work with low-power analogue circuits. Maris and Mahowald have built a line-following robot, that uses a single analogue spiking neuron to control the steering. The work of Rasche et al., on the other hand, uses complex analogue spiking neurons (each have 25 parameters) to model neocortical pyramidal cells.

Although hardware implementations give the benefit of high-speed execution of the model, committing a neural model to silicon does not allow the model's algorithm to be modified afterwards (the analogue circuitry has to be changed to do this). Hence, the advantage of computer simulation is threefold: (1) Neural algorithms are not fixed and so can be modified as needed during development; (2) Data can be conveniently logged to mass storage from any point (any neural object) in the ANS, and (3) Keeping the neural model as a software construct enables rapid prototyping (even evolution) of circuitry.

In addition to our own work (French and Damper, submitted), Floreano and Mattiussi (2001) have also successfully evolved spiking neuron controllers for a mobile robot. The latter authors claim to have shown that evolved circuits of spiking neurons are superior to evolved circuits of sigmoidal activation function neurons, but this may be because the latter have fixed connection weights (all equal to 1).

Here, as previously, we have used our Hi-NOON software (Scutt and Damper 1992; Damper, French, and Scutt 2001) to simulate the evolved ANSs. The reader is referred to Appendix A for details of the mathematical models used. A brief overview of Hi-NOON's data structures is given as background here, to help understand the evolutionary process. Hi-NOON is a simulator for a Hierarchical Network of Object-Oriented Neurons. For computational efficiency, it attempts to model neuronal function at a relatively high level, using a simple state system to describe trans-membrane potential and its changes. As the name suggests, synapses, neurons and the spiking neuron network itself are all represented as dynamically-created objects (equipped with their own data members and member functions) within an object-oriented hierarchy (Coad and Yourdon 1991; Eliëns 1994). Critically, Hi-NOON was chosen for this work over other simulators because we need intimate knowledge of the simulator and full access to its source code to be able to manipulate ANSs during evolutionary computation. The neural network is held as a list of objects, where each such object corresponds to a single neuron and holds all the information about its state and references synapse objects.

The information held in a neuron object comprises:

- a set of parameters which defines the neuron;
- a set of data structures which defines the 'axon terminals' for the neuron, each of which is itself an object and has its own parameters;
- a set of methods (pointers to functions) which access and alter parameter values and so determine exactly how the neuron functions.

Neurons (together with their synapses) are in turn connected to a linked list of `node` objects as shown in Figure 1. The latter objects form the skeleton of a neural network and are used to manage the network during its dynamic creation, modification, and deletion. (In conventional neural network terminology, *node* and *neuron* are synonyms: Here, they are distinct objects so as to separate structure and function and allow each to be flexibly modified.) The head of the node list is referenced by a `list` object. This is used to maintain the nodes and to allow random access to the neurons via a table, which is used to dereference a postsynaptic target held as a data member of a synapse.

3 Evolving a Nervous System

Traditional genetic programming (GP) does not make a distinction between genotype and phenotype (Koza 1992), unlike genetic algorithms (Goldberg 1989). Even with GP, however, the distinction between genotype and phenotype can usefully be made. Such work is pursued by Banzhaf (1994) where the genotype is a bit-string and the phenotype is source code for languages such as C and FORTRAN (which is then compiled and

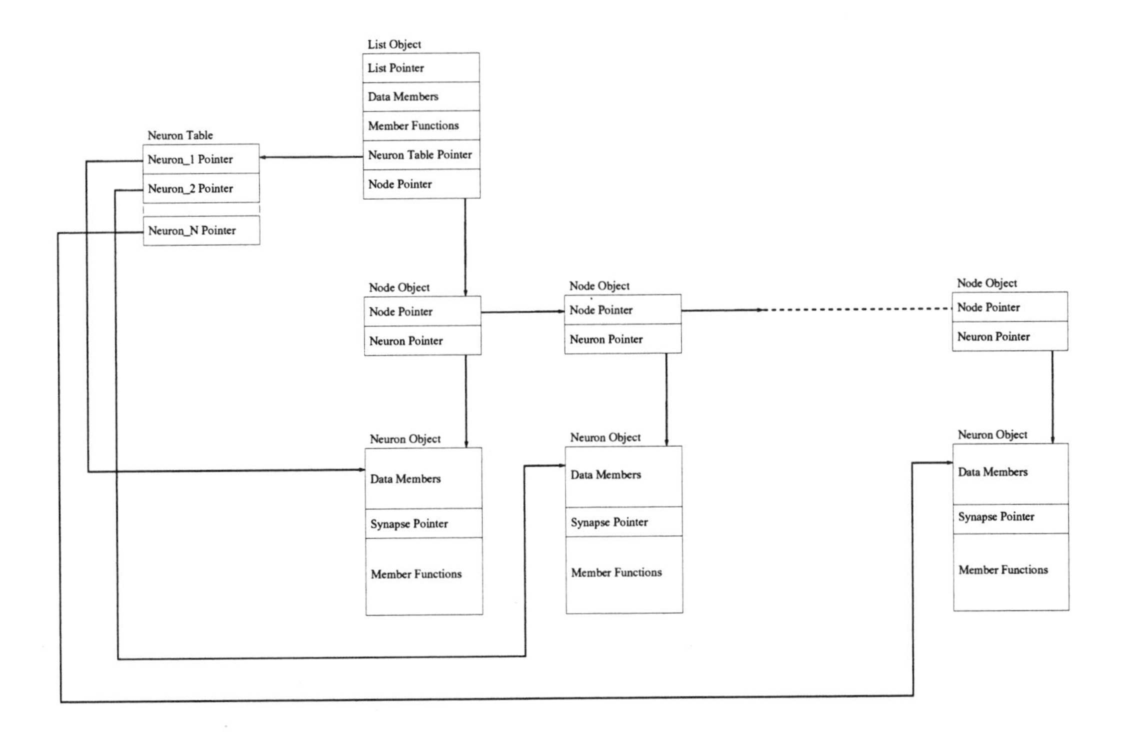

Figure 1: The data structure describing an ANS is a linked list of `node` objects. Each `node` object references a corresponding `neuron` object, which includes a synapse pointer referencing an array of `synapse` objects. The `node` objects are used to manage the ANS during its dynamic creation, modification, and deletion. The head of the linked list is referenced by a `list` object, corresponding to the ANS itself.

evaluated as native code). This distinction between genotype and phenotype allows neutral variants (i.e., different genotypes mapping to the same phenotype) which can improve the search for fit individuals, but has the disadvantage that it may lead to syntactically-invalid offspring (a point taken up in Section 31.2). However, such errors from Banzahf's bit-string to source code translation can be dealt with using a correction mechanism. Citing Banzhaf (1994, p. 325):

> "The main ingredients to this mapping are, first a coding of pieces of the bit-string into the nodes of a program tree ... and, second, a correction mechanism that is able to check statements and to transform them into the nearest correct statement if an error was detected."

Thus the correction mechanism ensures syntactically-valid source code.

So, at the outset, there would appear to be two rather different options available for evolving a nervous system: We can either make a strong distinction between genotype and phenotype, or not. In the first case, a population of genotypes could be 'grown' into a phenotype (i.e., an ANS) by a relatively complex process of neurogenesis (e.g., Michel 1997; Mautner and Belew 1999). This is the approach taken by Michel (1997). In the second case, which is the approach adopted here, genotypes are sought which are as close as possible to the required phenotypes. A solution is to use hierarchical objects as the genotype, which are directly manipulated during evolution. Since these are executable in Hi-NOON, the genotype is identical to the phenotype, remaining close to traditional GP.

There are a few reasons for adopting this alternative. First, to save unnecessary effort, we wish to (re)use as much of the existing Hi-NOON code as possible. Second, because the genotype/phenotype distinction is weak (i.e., there is no complex and error-prone translation from genotype to phenotype that needs subsequent correction), the problem of non-viable offspring can be avoided. Finally, direct manipulation of objects avoids a (possibly) protracted growth phase, and so simplifies fitness evaluation during evolution.

31 *Mutation, crossover and selection*

In what follows, let us assume the prior existence of an initial population of candidate artificial nervous systems (ANSs), each of which is held as a list of objects.

31.1 Mutation

Mutation is used to add components (i.e., neural sub-circuits) to an ANS and/or to delete them. Mutation operators simply create and destroy neurons and synapses. Created objects are given values for their data members which are randomly distributed within a predefined range. The operators are:

- `add_neuron`: May call `add_synapse` for connection to post-synaptic targets.
- `delete_neuron`: Calls delete synapse for a particular neuron as well as any pre-synaptic cell.
- `add_synapse` Called by `add_neuron` for connection to post-synaptic targets. Synapses may be normal, noisy or habituating types.
- `delete_synapse` Also starts a recursive call for deletion of any synapse-on-synapse connections and schedules isolated neurons for deletion.

Mutation operators are randomly selected.

Because the nervous system representation is manipulated directly during the evolutionary process, there is no limit to the size of genotype in this application. Genetic algorithms usually employ a fixed-length genotype (Koza 1992, p. 18) although this is by no means always so (e.g., Goldberg and Deb 1991). This is an additional reason why this work may be viewed as being closer to the paradigm of genetic programming than genetic algorithms, with Hi-NOON viewed as a virtual machine.

31.2 Crossover

Defining an appropriate crossover operator confronts a problem common to a range of applications of GP and which was mentioned above. To quote Koza: "Crossover must be performed in a structure-preserving way so as to preserve the syntactic validity of all offspring" (Koza 1994, p. 85). By a careful choice of a genotype as a hierarchical data structure defining a candidate ANS, a way has been found to ensure that any child resulting from crossover applied to its parents is always syntactically valid for execution on Hi-NOON.

Crossover is single-point. Two parents are chosen as follows. Let the first parent have neurons labeled $1, \ldots, N_1$ and let the second parent have neurons labeled $1, \ldots, N_2$. The crossover point is then randomly chosen after position n (in the linked list in Fig. 1), where $n \in \{2, \ldots, (\min[N_1, N_2] - 1)\}$. Two children are then produced, as demonstrated in Figure 2, and any of their synapses that no longer have valid postsynaptic targets are pruned. In the interests of computational efficiency, this method has been chosen in preference to any attempt at analysing the two parents looking for an 'ideal' crossover point. This is similar in spirit to

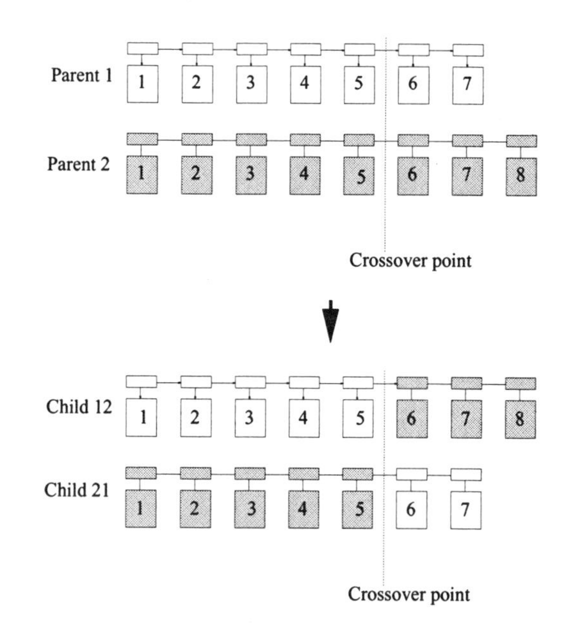

Figure 2: An example of crossover after neuron position 5 applied to the data structures of two parent ANSs.

the crossover operation performed on animat nervous systems by Sims (1994a) in his artificial life research. Here, however, only one crossover point is used whereas Sims uses one or more. Also, he reassigns synapses that no longer have a postsynaptic target and we do not.

31.3 Selection

The selection operator used here is the stochastic universal sampling (SUS) method (Baker 1987; Whitley 1993) with elitism. SUS was chosen because it is computationally efficient and maintains diversity among the population.

32 Initial population

Initially, the evolutionary process must be seeded with a population of randomly-generated individuals of appropriate size, P. To balance the need for a reasonably large population (to maintain diversity) against the practical necessity for reasonably short run times, a value of $P = 100$ has been chosen for this work. Each member of the initial population is created by five applications of the mutation operators described in Section 31.1.

4 Evolving a Frequency Discriminator

In our work to date, circuits of model spiking neurons have not been evolved to process a sensory signal with a temporal component to it. In this section, such work is addressed by evolving a frequency discriminator. In the longer term, it is intended to evolve circuitry which can discriminate between signals with different

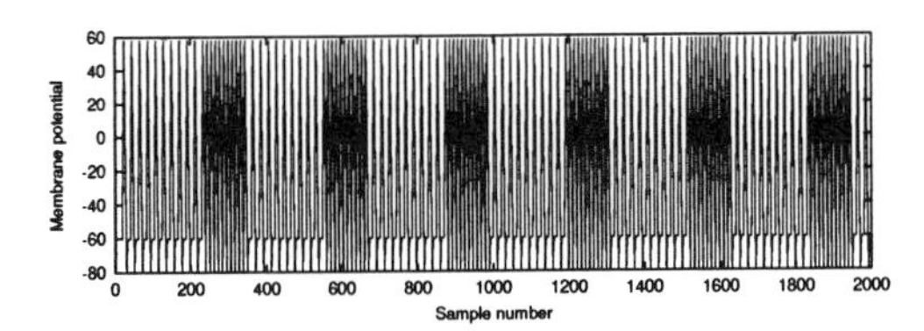

Figure 3: Neuron activity alternating between two firing rates.

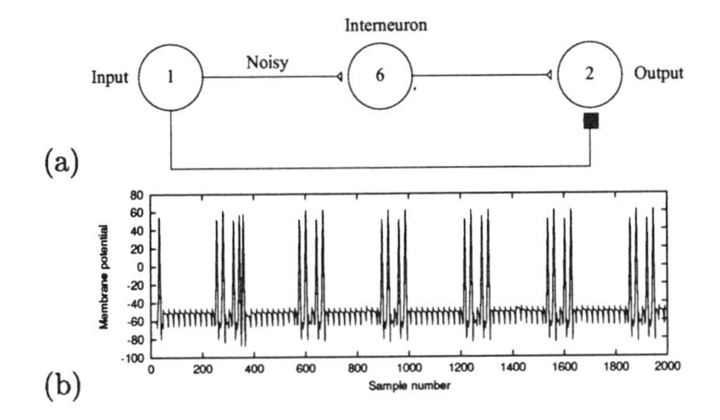

Figure 4: (a) A nervous system that was evolved to select the high firing rate of an input signal (Fig. 3) and (b) its output. Redundant components have been removed for clarity.

temporal structures.

Figure 3 shows the activity of a neuron that alternates between high and low firing rates. This represents the output of a sensory neuron that is being depolarised by a stimulus from an environment and forms the input to our frequency discriminator.

The fitness function of a candidate ANS for this evolutionary task simply increments a counter every time a particular input firing rate (e.g., 'low') coincides with firing of a designated output neuron. The fitness counter is decremented for an incorrect output response. Hence, this fitness function attempts to evolve a circuit of spiking neurons that can recognise a particular (and very simple) signal that has a temporal component.

A population of 1000 individuals was used running for 200 generations. The evolutionary process was repeated 10 times for the low-frequency discrimination task and 10 times for the high-frequency discrimination task. This resulted in only 3 of the high-frequency discrimination and 5 of the low-frequency discrimination candidates being viable. This may be an indicator of a very large search space for this problem.

Figure 4 shows a circuit evolved with a fitness function configured to select the high-frequency component of Fig. 3. In the same way, Figure 5 was evolved with a fitness function configured to select the low-frequency component of Fig. 3.

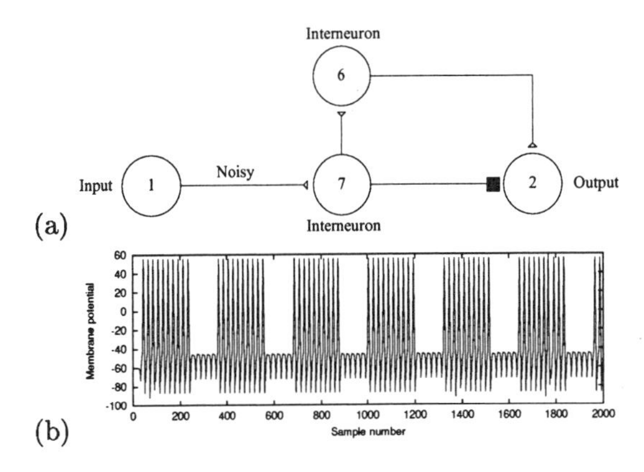

Figure 5: (a) A nervous system that was evolved to select the low firing rate of an input signal (Fig. 3) and (b) its output. Redundant components have been removed for clarity.

5 Analysis of the Evolved Circuits

In his artificial life research, Sims (1994b, p. 34) states "... it can be difficult to analyze exactly how a control system such as this works ... a primary benefit of using artificial evolution is that understanding these representations is not necessary". We believe this view is unnecessarily pessimistic; in our opinion, it is well worth trying to understand the evolved circuits, as we now demonstrate.

In the high-frequency discriminator of Fig. 4, the spikes of neurons 1 and 6 are almost in-phase during the low-frequency input. This is because the delay line effect of neuron 6 has very little influence during low-frequency input. Hence, their excitatory and inhibitory post-synaptic potentials cancel in the membrane of neuron 2. However, during high-frequency input this synchronisation is lost as the delay introduced by neuron 2 effectively stores a spike whose excitatory post-synaptic potentials (EPSPs) in neuron 2 cannot by cancelled by inhibitory post-synaptic potentials (IPSPs) from neuron 1. Hence, these excitatory potentials fire neuron 2 during high-frequency input.

The circuit of Fig. 5 suppresses a high-frequency output by cancellation of EPSPs and IPSPs in neuron 2. However, the mechanism used to detect the high-frequency appears to be based upon the noisy nature of the excitatory synapse between neurons 1 and 7: Editing the evolved circuit to remove this noise destroys the ability to ignore the high-frequency signal. This noisy synapse forms a frequency inverter—the inter-spike interval of neuron 7 increases by a small amount during the high-frequency activity of neuron 1. This is counter-intuitive because we expect an increase in parent neuron firing rate to result in an increase of synapse firing also. However, no bugs have been found in the program that could be exploited by the evolutionary process in this way. As in Fig. 4, a delay line formed from neuron 6

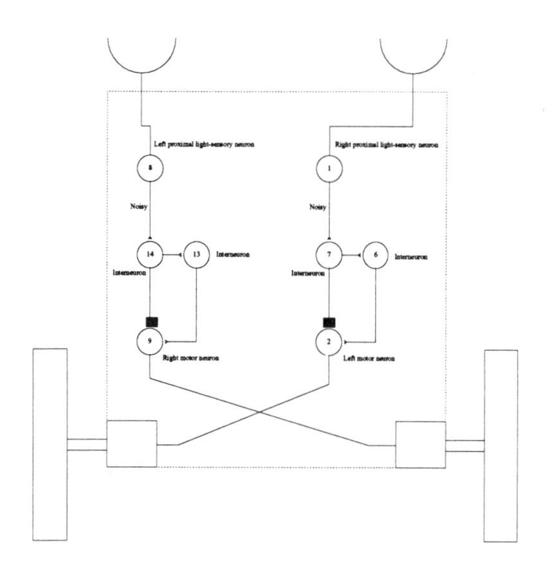

Figure 6: A Braitenberg type 2b vehicle architecture with spiking neurons that seeks low-frequency signals.

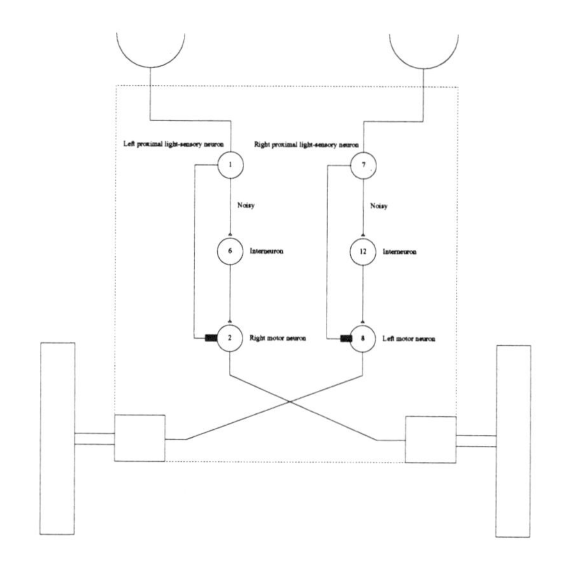

Figure 7: A Braitenberg type 2b vehicle architecture with spiking neurons that seeks high-frequency signals.

(with respect to neuron 7) results in EPSPs and IPSPs cancelling in the membrane of neuron 2 during low-frequency firing of neuron 7 (during high-frequency firing of neuron 1). During low-frequency input (causing a higher firing rate in neuron 7 with respect to its rate during high-frequency input), potentials do not cancel in the membrane of neuron 2. Thus, its membrane reaches threshold and fires, giving an output.

6 Phototaxis in a Braitenberg Vehicle

Now that we have evolved neural circuits that can discriminate between flashing lights of different frequency, they were incorporated in a Braitenberg type 2b vehicle (Braitenberg 1984, p. 7). This should result in a vehicle that can seek out the source of a stimulus, e.g., a light source flashing at a particular rate, so exhibiting phototaxis.

An ANS describing a type 2b vehicle architecture was created using two of the networks in Fig. 5 to connect contralateral *proximal* sensory-motor pairs (i.e., the sensors respond to changes in light level). This is shown in Figure 6. Proximal light sensory neurons were selected to reduce the risk of spurious sensory input. This is because ordinary light-sensor neuron firing is proportional to light intensity.

No changes to the evolved nervous system circuitry were made, with the exception of adding sensory and motor neurons in place of input and output neurons. Experiments were carried out to test the expected phototaxis behaviour.

The experimental setup was as follows. A Khepera robot (Figure 8(a)) was placed in a test environment facing the mid-point between two 2.5 V filament lamps,

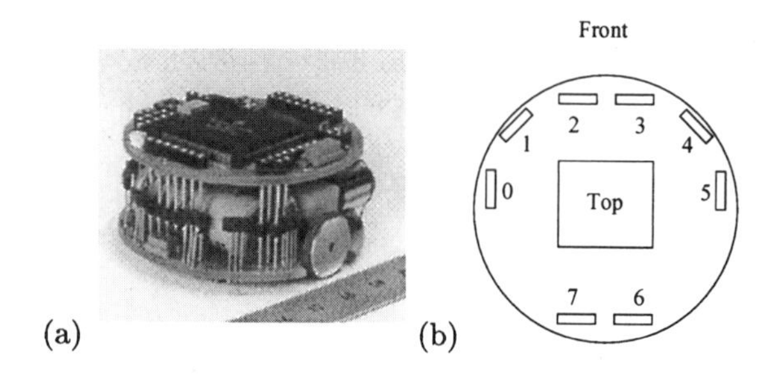

Figure 8: (a) The Khepera robot. (b) Sensor positions on Khepera.

15 cm away. The lamps were positioned 15 cm apart. An electronic circuit switched the lamps at approximately 1 Hz and 10 Hz respectively. (These frequencies were selected to match the spiking input used to evolve the circuits.) A toggle-switch allowed the frequency selection for the lamps to be swapped over (i.e., the low-frequency signal can appear on the left lamp, with the high frequency signal on the right or vice versa). The nervous system models executed on a 80486 Linux PC.

During tests aimed at setting up the intensity of the lamps, it became clear that Khepera's left-hand light sensor (Khepera position 1, see Figure 8(b)) was not as sensitive as that on the right (Khepera position 4). Also, ambient light influenced the robot's phototaxis performance: Bright background light disrupted the ability to discriminate between high- and low-frequency signals, causing the robot to come to rest between the two lamps. Hence, final runs took place in low levels of lighting—low enough not to disrupt phototaxis but sufficient to film the vehicle's activity with a digital

Figure 9: Khepera in the test environment used for phototaxis experiments. The two lamps have different flash frequencies and are shown here at different stages of illumination.

camcorder. The inability to cope robustly with ambient light is probably because the discrimination circuits of Figs. 4 and 5 were evolved in isolation, without the input noise generated by the Khepera simulator sensor models.

The experiments started with a Braitenberg type 2b vehicle that favoured the low-frequency signal. Twenty trials were carried out to observe the robot's behaviour with both settings of the toggle switch mentioned above. For the first 10 trials the low-frequency signal was on the robot's left; For the second group of 10 trials, the low-frequency signal was moved to the robot's right. This was then repeated with a vehicle that favoured the high-frequency signal (shown in Figure 7).

Figure 9 shows the vehicle (Khepera) in the test environment. It was noticed that the response of the high-frequency circuit was not as vigorous as its low-frequency counterpart, resulting in slower movement. In all trials, the vehicles performed their tasks without error, halting in front of the appropriate lamp. For the low-frequency case, the vehicle took approximately 80-90 seconds to reach the lamp and halt. For the high-frequency case, the corresponding time was 25-30 seconds.

7 Towards the Evolution of Phonotaxis

In the Introduction, we mentioned phonotaxis in the cricket, whereby a female insect seeks the calling song of male conspecifics in a noisy environment. This has been extensively studied in the animat research of Webb and colleagues (Lund, Webb, and Hallam 1997; Webb and Scutt 2000; Reeve and Webb 2001). However, throughout their work, the circuitry underlying the behaviour of the animats has been manually designed. This prompts the question: "Can we evolve an ANS that allows a robot to perform phonotaxis behaviour?" This is interesting because an ANS that is generated through evolutionary computation may operate using principles that had previously not been considered for manual designs.

Perhaps the most basic of song characteristics is carrier frequency. For example, the cricket *Gryllus bimaculatus* has a calling song with a carrier frequency of some 4.7 kHz. The song is generated by stridulation—a very common mechanism for sound production in arthropods. The complete song consists of chirps with three syllables, each of about 20 ms in duration and separated by 20 ms silence. Each chirp is separated with a 200 ms break (Lund, Webb, and Hallam 1997).

The early work presented here on phototaxis (which relies on the temporal processing capabilities of evolved circuits of spiking neurons) is intended as an early stage in the evolution of circuitry to drive cricket phonotaxis. Clearly, the cricket song has more complex temporal structure than mere switching between low and high frequencies. Nonetheless, a useful start has been made.

8 Conclusions and Discussion

We have described early work to evolve spiking neural circuits that can discriminate between high- and low-frequency signals. Light-seeking Braitenberg vehicles were constructed that used these circuits as the basis of phototaxis. Although the Braitenberg vehicles were unable to deal with strong ambient light, this is probably because their circuitry was evolved in isolation (for computational convenience) and not onboard a vehicle situated in a real or simulated environment. We intend to extend this work to simulate phonotaxis in the cricket, where the temporal patterning of the relevant signals is considerably more complex.

A motivation for this work was to see if evolution could profit from the additional computational abilities of spiking neurons. At this stage, the question remains open. It seems highly likely, for instance, that a robot equipped with non-spiking recurrent neural networks of the kind evolved by Yamauchi and Beer (1994) would be able to behave appropriately in the simple task described here. Perhaps, if there really are significant differences in computational abilities between spiking and non-spiking neurons, they will only be revealed when we are able to achieve much more complex patterns of behaviour.

Appendix A: Neurons and Synapses

In this section, we present more detailed descriptions of neurons and synapses within Hi-NOON. Since Hi-NOON is intended for (amongst other things) applications in situated robotics studies, there is provision for sensory and motor neurons which connect to the environment, as well as for more prosaic 'basic' (information processing) neurons.

A1 Neurons

The 'basic' neuron type has the state system functionality which is subsequently embedded in all derivatives, such as the sensory and motor cells.

A1.1 Basic neurons

Updating equations for the membrane potential (MP – in millivolts) for this neuron type are:

state A: $MP(t+1) = MP(t) - \tau + S(t)$
state B: $MP(t+1) = MP(t) - \alpha + S(t)$
state C: $MP(t+1) = h + S(t)$
state D: $MP(t+1) = MP(t) - \mu + S(t)$
state E: $MP(t+1) = l + S(t)$
state F: $MP(t+1) = MP(t) + \frac{BaseMP - MP(t)}{\eta} + S(t)$

where:

$$S(t) = \sum_i w_i \kappa \left(MP_i(t) - BaseMP_i\right)$$

is the synaptic potential (SynPot), i is a counter which counts over active pre-synaptic cells, w_i is the synaptic weight from a pre-synaptic neuron, τ is the neuron time constant, $\eta = 1.5$ is the post-undershoot increment rate, $\mu = 25$ is the post-action potential peak-MP decrement, $\kappa = 1/450$ is a heuristically-set constant, $\alpha = 20$ is the post-threshold attack increment, $h = 45$ is the post-threshold maximum MP, and $l = -69$ is the pre-undershoot minimum MP.

Certain of the above parameters (e.g. τ, η) are time-dependent and have been set empirically to suit a range of processor speeds and implementations. However, they may be inappropriate in some circumstances (as when implementing a real-time robotic system using a fast processor).

A1.2 Sensory and motor neurons

These neuron types are important in the specific case of a robotic system which needs input and output from/to its environment. Since, in this paper, we are principally concerned with more general principles, we omit details of these neuron types here.

A2 Synapses

The basic synapse (which is noise free) has functionality which is subsequently embedded in all derivatives such as the habituating, sensitising and conditioning types used in our robot work (see below). These allow us to implement a simple, biologically-based form of learning.

$$w(t) = \begin{cases} w(t) - \beta & \text{if } w(t) > w_{\text{base}} \\ w(t) + \beta & \text{if } w(t) \leq w_{\text{base}} \\ w_{\max} & \text{if } w(t) > w_{\max} \\ w_{\min} & \text{if } w(t) < w_{\min} \\ w_{\min} & \text{otherwise} \end{cases}$$

where β is the MP recovery parameter and w_{base} is the base weight (typically 0). These are individually set (together with $w_{\min}$ and $w_{\max}$, typically ±16) for each neuron.

A2.1 Noise-free synapse

$$\texttt{fired}(t) = \begin{cases} \text{TRUE} & \text{if state B, C, D} \\ \text{FALSE} & \text{otherwise} \end{cases}$$

A2.2 Noisy synapse

$$\texttt{fired}(t) = \begin{cases} \text{TRUE} & \text{if cond1} \\ \text{FALSE} & \text{otherwise} \end{cases}$$

where cond1 is state B, C, D, as for the noise-free synapse, ANDed with:

$$\frac{MP_p - \theta_p}{h - \theta_p} \times 100 \geq \text{rand mod100}$$

and p denotes a parent (pre-synaptic) neuron.

A2.3 Habituating type

$$w(t+1) = \begin{cases} w(t) - d & \text{if state C} \\ w(t) & \text{otherwise} \end{cases}$$

where d is a constant decrement (typically ~ 1).

A2.4 Sensitising type

$$w(t+1)_{\text{targ}} = \begin{cases} w(t)_{\text{targ}} + w(t)_{\text{sos}} & \text{if cond2} \\ w(t)_{\text{targ}} & \text{otherwise} \end{cases}$$

where cond2 is $\texttt{fired}_{\text{targ}} \wedge \texttt{fired}_{\text{sos}}$, 'targ' denotes the target synapse (to be sensitised) and 'sos' denotes the synapse-on-synapse influence.

A2.5 Conditioning type

$$w(t+1)_{\text{targ}} = \begin{cases} w(t)_{\text{targ}} + kw(t)_{\text{sos}} & \text{if cond2} \\ w(t)_{\text{targ}} & \text{otherwise} \end{cases}$$

where:

$$k = \frac{nT}{\psi} \exp\left(\frac{-nT}{\varsigma}\right)$$

and nT is a count of sample periods initiated by encountering state C for the target neuron, ψ (= 250) is an empirically-set scaling factor and ς (= 500) is a constant chosen to maximise the effect of conditioning when the conditioning stimulus precedes the unconditioned stimulus by 0.5 s.

A2.6 Autonomous decay of synapses

To prevent synaptic weights growing without limit, `Weight` is bounded during simulation. This models the finite size of stores of neuro-transmitter in the synaptic terminals of real biological neurons. Also, according to Sutton and Barto (1981, p. 161):

> "If it is assumed that synaptic strength slowly decays in the absence of a reinforcement signal, then a bound on weight size is imposed that is a function of reinforcement level and the decay rate. ... In system theoretic terms, the adaptive element has *definite memory*: it cannot remember anything that occurred arbitrarily far in the past."

Sutton and Barto call this *autonomous decay*. This decay mechanism is also implemented here (via `Recovery`).

References

Baker, J. E. (1987). Reducing bias and inefficiency in the selection algorithm. In J. J. Grenfenstett (Ed.), *Genetic Algorithms and their Applications: Proceedings of the 2nd International Conference on Genetic Algorithms*, pp. 14–21. Hillsdale, NJ: Lawrence Erlbaum Associates.

Banzhaf, W. (1994). Genotype-phenotype mapping and neutral variation – a case study in genetic programming. In Y. Davidor, H.-P. Schwefel, and R. Männer (Eds.), *Proceedings of Parallel Problem Solving from Nature, PPSN III*, pp. 322–332. Berlin, Germany: Springer.

Braitenberg, V. (1984). *Vehicles: Experiments in Synthetic Psychology*. Cambridge, MA: MIT Press.

Bugmann, G. (1997). Biologically plausible neural computation. *BioSystems 40*(1–2), 11–19.

Coad, P. and E. Yourdon (1991). *Object Oriented Analysis* (Second ed.). Englewood Cliffs, NJ: Prentice-Hall.

Damper, R. I., R. L. B. French, and T. W. Scutt (2000). ARBIB: an autonomous robot based on inspirations from biology. *Robotics and Autonomous Systems 31*(4), 247–274.

Damper, R. I., R. L. B. French, and T. W. Scutt (2001). The Hi-NOON neural simulator and its applications. *Microelectronics Reliability 41*(12), 2051–2065.

Eliëns, A. (1994). *Principles of Object-Oriented Software Development*. Wokingham, UK: Addison-Wesley.

Floreano, D. and C. Mattiussi (2001). Evolution of spiking neural controllers for autonomous vision-based robots. In T. Gomi (Ed.), *Evolutionary Robotics IV*. Berlin: Springer-Verlag.

French, R. L. B. and R. I. Damper (2001). Evolving a nervous system of spiking neurons for a behaving robot. In *Proceedings of Genetic and Evolutionary Computation Conference, GECCO 2001*, San Francisco, CA, pp. 1099–1106. Morgan Kaufmann.

French, R. L. B. and R. I. Damper (submitted). Evolution of biologically-inspired nervous systems for a mobile robot. *IEEE Transactions on Evolutionary Computation*.

Goldberg, D. E. (1989). *Genetic Algorithms in Search, Optimization and Machine Learning*. Reading, MA: Addison-Wesley.

Goldberg, D. E. and K. Deb (1991). Don't worry, be messy. In R. K. Belew and L. B. Booker (Eds.), *Proceedings of the 4th International Conference on Genetic Algorithms*, pp. 24–30. La Jolla, CA: Morgan Kaufmann.

Koza, J. R. (1992). *Genetic Programming: On the Programming of Computers by Means of Natural Selection*. Cambridge, MA: MIT Press/Bradford Books.

Koza, J. R. (1994). *Genetic Programming II: Automatic Discovery of Reusable Programs*. Cambridge, MA: MIT Press/Bradford Books. (Pagination refers to 1998 second printing).

Levitan, I. B. and L. K. Kaczmarek (1997). *The Neuron: Cell and Molecular Biology*. New York, NY: Oxford University Press.

Lund, H. H., B. Webb, and J. Hallam (1997). A robot attracted to the cricket species *Gryllus bimaculatus*. In P. Husbands and I. Harvey (Eds.), *Proceedings of Fourth European Conference on Artificial Life, ECAL'97, Brighton, UK*, pp. 246–255. Cambridge, MA: MIT Press/Bradford Books.

Maass, W. and T. Natschlager (1998). Associative memory with networks of spiking neurons in temporal coding. In L. S. Smith and A. Hamilton (Eds.), *Neuromorphic Systems, Engineering Silicon from Neurobiology*, Volume 10, pp. 21–32. Singapore: World Scientific.

Maass, W. and B. Ruf (1999). On computation with pulses. *Information and Computation 148*(2), 202–218.

Maris, M. and M. Mahowald (1998). Neuromorphic sensory-motor mobile robot controller with pre-attention mechanism. In L. S. Smith and A. Hamilton (Eds.), *Neuromorphic Systems, Engineering Silicon from Neurobiology*, Volume 10, pp. 126–137. Singapore: World Scientific.

Mautner, C. and R. K. Belew (1999). Coupling morphology and control in a simulated robot. In *Proceedings of the Genetic and Evolutionary Computation Conference (GECCO-99)*, pp. 1350–1357.

Michel, O. (1997). Dynamical genomic network applied to artificial neurogenesis. *Control and Cybernetics 26*(3), 511–531.

Rasche, C., R. J. Douglas, and M. Mahowald (1998). Characterization of a silicon pyramidal neuron. In L. S. Smith and A. Hamilton (Eds.), *Neuromorphic Systems, Engineering Silicon from Neurobiology*, Volume 10, pp. 169–177. Singapore: World Scientific.

Raymond, J. L., D. A. Baxter, D. V. Buonomano, and J. H. Byrne (1992). A learning based on empirically-derived activity-dependent neuromodulation supports operant conditioning in a small neural network. *Neural Networks 5*, 789–803.

Reeve, R. and B. Webb (2001). Neural network control of a cricket robot. In U. Nehmzow and C. Melhuish (Eds.), *Third British Conference on Autonomous Mobile Robotics and Autonomous Systems: TIMR 01—Towards Intelligent Mobile Robots*, Manchester, UK. No pagination.

Scutt, T. W. and R. I. Damper (1992). Object-oriented modelling of small neuronal systems. *Artificial Intelligence and Simulation of Behaviour Quarterly 80*, 24–33.

Sims, K. (1994a). Evolving virtual creatures. In A. Glassner (Ed.), *SIGGRAPH 94 Conference Proceedings*, pp. 15–22. New York, NY: ACM Press.

Sims, K. (1994b). Evolving 3D morphology and behavior by competition. In R. A. Brooks and P. Maes (Eds.), *Artificial Life IV*, pp. 28–39. Cambridge, MA: MIT Press.

Sutton, R. S. and A. G. Barto (1981). Towards a modern theory of adaptive networks: Expectation and prediction. *Psychological Review 88*, 135–170.

Webb, B. and T. Scutt (2000). A simple latency-dependent spiking-neuron model of cricket phonotaxis. *Biological Cybernetics 82*(3), 247–269.

Whitley, D. (1993). A genetic algorithm tutorial. *Statistics and Computing 2*(4), 65–85.

Yamauchi, B. M. and R. D. Beer (1994). Sequential behavior and learning in evolved dynamical neural networks. *Adaptive Behavior 2*(3), 219–246.

Small is Beautiful : Near Minimal Evolutionary Neurocontrollers Obtained With Self-Organizing Compressed Encoding

Shlomy Boshy
School of Computer Science
Tel Aviv University,Tel-Aviv 69978,Israel
shlomy@post.tau.ac.il

Eytan Ruppin
School of Computer Science and
Sackler School of Medicine
Tel Aviv University,Tel-Aviv 69978,Israel
ruppin@post.tau.ac.il

Abstract

This paper presents a novel method for evolution of artificial autonomous agents. It is based on adaptive, self-organizing compressed genotypic encoding (SOCE) of the phenotypic synaptic efficacies of the agent's neurocontroller. The SOCE encoding implements a parallel evolutionary search for neurocontroller solutions in a dynamically varying and reduced subspace of the original synaptic space. It leads to the robust emergence of compact, near minimal successful neurocontrollers starting from arbitrarily large networks. This is important since in practice the network size needed to solve the problem is unknown beforehand. The SOCE method may also serve to estimate the network size needed to solve a given task, and to delineate the relative importance of the neurons composing the agent's controller network.

1. The SOCE approach : Motivation and Highlights

Many Evolutionary Autonomous Agents (EAAs) use neural networks to control their behavior. Such neurocontrollers of EAAs performing complex tasks can be evolved via genetic algorithms from a population of genomes undergoing natural selection and variation. Much of these EAA studies employ direct genotype-to-phenotype encodings, where the magnitude of each synapse is encoded by a specifically designated gene. These direct encodings are problematic since they scale quadratically with network size and are inadequate for solving complex tasks. This problem has led to considerable efforts for developing "indirect" encodings, where the genome includes a developmental program for specifying the controller neural network and its weights (see (X.Yao, 1999) for a review). Addressing the inherent scaling limitations of direct encodings, we adopt a different, new, approach. Instead of utilizing a developmental program to compactly encode *all* the network information, we present an indirect encoding which adaptively maintains only an *approximate* description of the network via compressed synaptic encodings.

The basic contribution of this paper is to present a novel evolutionary process that maintains *a "good" subspace in which the search is performed.* Our indirect encoding method is based on *adaptive compression.* The synaptic efficacies of each neuron are encoded in a lossy, compressed manner by linear interpolation, and an inverse decompression transform is used to transcribe the genome into a functioning phenotype. Individual neurons may utilize different encoding (compression) levels. The level of compression may vary from containing no information to direct specification of all the synaptic connection values. These levels of compression are themselves encoded in the genome and undergo evolution, adaptively varying in a self-organized manner. We term this method *Self-Organizing Compressed Encoding* (SOCE). Note that we do *not* use a fitness function that *explicitly* encourages shorter solutions. The length of the genomic encodings of solutions to the problem in hand is optimized in an *implicit* manner.

The genome of each agent is hence a compressed indirect encoding of the features of the developed phenotype, i.e. the synaptic weights of the controller network. The evolutionary process is governed by a standard genetic algorithm, with a developmental stage occurring at the beginning of every new generation, creating the controller networks from the genomes by a decoding transform. The distinct encoding (compression) level of each neuron is encoded in the genome. This enables the agent to "choose" during the evolutionary process to which neurons it allocates a more detailed (but hence longer) encoding and which receive coarser (possibly zero length) encodings.

Having adaptive, compressed encodings yields agents

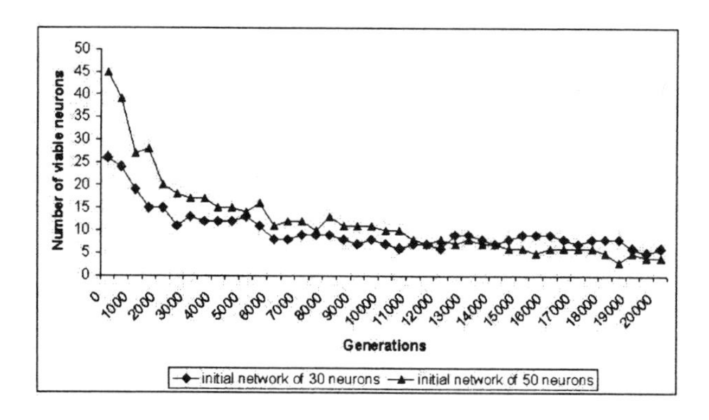

Figure 1: Number of viable neurons (with non zero encoding levels) in the neurocontroller of SOCE agents, starting from initial large networks of 30 and 50 neurons. The SOCE algorithm leads to a near minimal network with only 4-5 effective neurons. In each generation, the agent with minimal network size is shown. At the end of the runs, all agents achieve normalized high performance levels. The results are of the best run out of 20 runs.

with compact genotypes but their phenotypes may remain unnecessary large. Exploiting a specific form of genetic variation in our method leads to the emergence of *compact near-minimal phenotype neurocontrollers.* Thus, our method leads to two important consequences: simpler controller networks that are more amenable for functional analysis, and an estimation of the complexity of the problem in hand. The SOCE algorithm automatically finds compact networks that solve the task, *irrespective of the initial network size.*

Space limitations do not permit us to describe the EAA model in which the SOCE algorithm was studied and the algorithm itself. The reader is referred to *http://www.math.tau.ac.il/~ruppin* for a detailed description.

2. Key Results

The SOCE algorithm was tested on navigation and foraging tasks with both constant and dynamic environments (see (R.Aharonov-Barki et al., 2001)). In dynamic environments the difficulty level of the task is increased in stages, thus changing the solutions and memory needed to solve the task. The SOCE algorithm solves the problems and leads to much smaller networks : the compressed encoded agents were able to solve the navigating problem with a genome length of about 33% of the direct encoding genome. The number of viable neurons (with non zero encoding level) is decreased to the minimum needed in the problem (down to 4 neurons - note that there are 4 motor neurons). Thus the SOCE algorithm is able to find near minimal neurocontrollers starting from an arbitrarily large initial network (Figure 1).

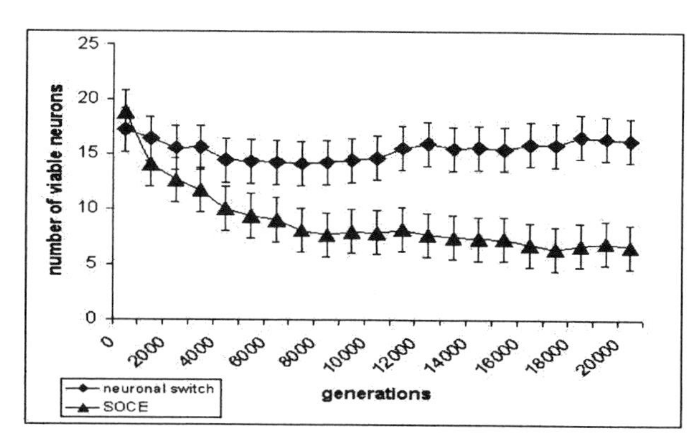

Figure 2: The mean number of viable (with non zero encoding levels) neurons of a maximum fitness agent (averaged over 20 runs, each fitness value tested over 1000 trails) in the population, for SOCE and neuronal switch agents. SOCE agents lead to small, near minimal networks.

The SOCE algorithm is composed of two basic elements: elimination of neurons (which reach a zero encoding level) and synapse values approximation. A simple neuronal switching encoding was used to analyze which of these elements contributes to the emergence of near minimal networks. The neuronal switching encoding uses direct encoding of synapse values and a "switch" that turns each neuron on and off. This encoding was *not* able to achieve near minimal networks (see Figure 2). The gradual decrease in encoding levels of synaptic values of SOCE agents serves to "smooth" the fitness landscape and allows the agents to reach much smaller networks. This relates to [Hinton Nowlan,1987], in which random learning was used to smooth the fitness landscape. The SOCE algorithm accomplishes this smoothing using a different mechanism, on the evolutionary level.

The SOCE algorithm can also be used as an analysis method: The encoding level of each neuron is a good clue for its importance and "input sensitivity". Future research should naturally add learning capabilities into the genome. Learning rules can be encoded using the SOCE method and used in the development stage of the agent with tradeoff in the respective data vs. theory encoding sizes.

References

R.Aharonov-Barki, T.Beker, and E.Ruppin (2001). Emergence of memory-driven command neurons in evolved artificial agents. *Neural Computation,13,pages 691-716.*

X.Yao (1999). Evolving artificial neural networks. *Proceedings of the IEEE,87(9).pages 1423-1447.*

How Useful is Lifelong Evolution for Robotics?

Joanne Walker* **Myra Wilson**
Dept. of Computer Science, UW Aberystwyth, Wales, UK.
*jhw94@aber.ac.uk

Abstract

This paper investigates a system that uses a genetic algorithm to train a robot for a generalised test environment, then an evolutionary strategy to investigate the effect of continuing the evolution as the robot progresses with its task.

The paper concludes that continuing evolution after a training phase has an important role to play in producing truly adaptive robots.

1 Introduction

Researchers in mobile robotics have investigated the use of evolutionary techniques for the development of control methods. Most evolutionary robotics work has concentrated on the use of a robot training phase; few have continued evolution throughout the lifetime of a robot. However the real world is dynamic and a robot working in that world should be able to adapt to changes as they occur. The objective of this work was to investigate a methodology to produce a robot that can work adaptively throughout its task. An evolutionary method was applied in which both a genetic algorithm (GA) and an evolutionary strategy (ES) (Fogel, 1995) were used.

In this work, a clear distinction has been made between the finite training phase in which a robot is adapted to the task and environments it will encounter; and the time after the training phase when the robot is actually carrying out its task. It is hypothesised that the combination of both training phase and lifelong evolution will provide a powerful way for robots to adapt and constantly re-adapt to their world.

2 Training phase

2.1 Method

A simulated world was used as the training ground for a Khepera robot with eight infra-red sensors which were used to detect obstacles and light sources. The robot was controlled by Arkin's schema-based architecture (Arkin, 1989). Three behaviours were used – Move-To-Goal, Avoid-Static-Obstacle and Noise which were controlled by a total of five parameters - goal gain, obstacle gain, obstacle sphere of influence, noise gain and noise persistence. The GA optimised these five parameters, encoded in real number chromosomes.

The aim of the training phase was to produce behaviours for a mobile robot that could generalise to a variety of environments. The method was based on that used in (Ram et al., 1994). A general world was set up which varied the locations of obstacles and the obstacle density from one chromosome to the next. The robot navigated around these obstacles to a light source.

2.2 Results

It was found that over 500 generations there was a clear improvement in robot performance. Further experiments compared the performance of the best evolved chromosome to randomly generated ones. The evolved chromosome performed significantly better then the random ones. It was clear that evolution had successfully adapted solutions to the robot task[1]. The best evolved chromosome from the final generation was tested on the physical robot and was found to transfer well.

3 Lifelong evolution

The aim of lifelong evolution was to improve upon the results of the training phase in new environmental conditions as the robot encountered them.

3.1 Method

The simulated world used for testing the lifelong evolution was similar to that used in the training phase, except that there were distinct phases in which the obstacle density stayed the same and just the locations of the obstacles changed. Two types of world were explored: **Catastrophic** where the obstacle density changed suddenly and massively after a period in which it had remained the same, and **Gradual** where the obstacle density changed slowly, after a period in which it had remained the same.

There was a light source acting as goal on one edge of the robot arena. As soon as the robot got close to the light it was switched off, and a light on an edge adjacent to the robot was switched on. In this way a continuous task was provided.

[1] Full results are at http://users.aber.ac.uk/jhw94/EvolRobotics.htm, see also(Walker, 2002)

A variant of a (1+1)-ES was used, in which one parent gave rise to one offspring by mutation, and the parent in the subsequent generation was chosen by competition between parent and child. A chromosome began its run after a goal was found and completed it when the next goal was found. Fitness was accumulated over one run.

Due to the random positioning of obstacles, the challenge posed could change significantly from one chromosome to the next. This meant that a poor chromosome could get a good run, and perform uncharacteristically well, or vice-versa. An addition was made to the ES – when a child performed better than its parent, it did not immediately replace the parent but was saved, and another offspring of the parent produced (making three chromosomes in total). All three chromosomes were then tested and if the saved chromosome did better than its parent again, it replaced it and became parent of the next generation's offspring.

3.2 Results

In each experiment, the average fitness of each generation of an ES was recorded as well as the time taken to reach each goal. In addition, the results were compared to those when the best chromosome from the training phase was used to control the robot without any lifelong evolution. Fitness results for the two world types can be seen in Figure 1. The times when the robot was navigating in a more cluttered environment are reflected by worse (higher values of) fitness in the graphs. The number of steps taken to reach each goal were also recorded.

In the catastrophic environment, Figure 1.a, there was little improvement to be seen in robot fitness when lifelong evolution was used. However, in terms of the number of steps taken to reach the goal, the robot with lifelong evolution almost consistently reached the goal faster than the robot without.

In the gradually changing world lifelong evolution was more successful. It can be seen from Figure 1.b that the fitness was almost consistently better when using an ES than when not. The time taken to reach the goals was similarly found to be better. The greatest effect of using lifelong evolution could be seen in the more difficult, more densely cluttered environments.

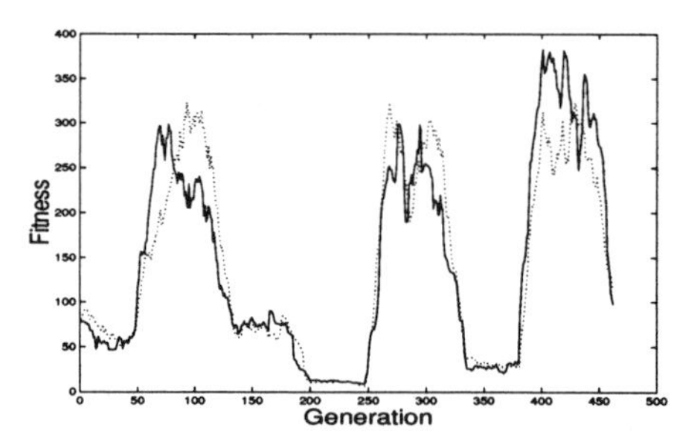

(a) Catastrophic: lifelong evolution and no evolution produced similar results

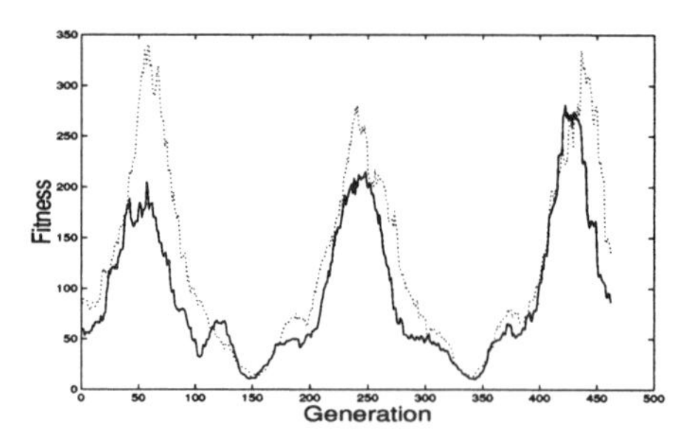

(b) Gradual: lifelong evolution outperformed no evolution

Figure 1: Moving average of fitness in two world types, comparing a robot with lifelong evolution (indicated by solid line —), to a robot without (indicated by dotted line ...). Low fitness values indicate better fitness

4 Conclusions

It has been found that in a gradually changing world the ES was successful in improving the behaviour of the robot. In a catastrophically changing world, however, the ES had less impact.

In this study, lifelong evolution was adapting to the same dynamic conditions that were present in the training phase. In the real world, more factors are likely to be dynamic than the obstacle density alone. For instance the types of obstacles encountered, and the objective of the robot when it finds the goal could change. The real strength of lifelong evolution may lie in evolving for the unpredictable changes that occur in the world and this is what current work is investigating.

References

Arkin, R. C. (1989). Motor schema–based mobile robot navigation. *International Journal of Robotics Research*, 8(4):9–12.

Fogel, D. B. (1995). *Evolutionary Computation - Towards a New Philosophy of Machine Intelligence.* IEEE Press, New York.

Ram, A., Arkin, R., G.Boone, and Pearce, M. (1994). Using genetic algorithms to learn reactive control parameters for autonomous robotic navigation. *Adaptive Behaviour*, 2(3):277–304.

Walker, J. (2002). How useful is lifelong evolution in the physical world? Technical Report UWA-DCS-02-040, Department of Computer Science, University of Wales, Aberystwyth.

Evolvability and analysis of robot control networks*

Tom Smith*[1] Phil Husbands[2] Michael O'Shea[1]
Centre for Computational Neuroscience and Robotics
[1]School of Biological Sciences, [2]School of Cognitive and Computing Sciences
University of Sussex, Brighton, UK. *toms@cogs.susx.ac.uk

Abstract

Identification of control system classes capable of generating adaptive behaviour over time is a black art. Many practitioners rely on systems that have "always worked in the past", others may use trial-and-error until success, but carry out no subsequent analysis of why the particular system actually worked. Here we apply the techniques of dynamical systems theory to the analysis of successfully evolved robot control systems, allowing us to identify potentially useful properties of network classes.

1 Introduction

In previous work, we have investigated the evolvability of a class of artificial neural networks (ANNs) based on a model of diffusing gaseous neuromodulation, the *GasNet* (Husbands et al., 1998). We have shown that we can evolve robot control solutions for a visual discrimination task consistently faster with the gaseous neuromodulatory mechanism, than without (the *NoGas* setup). In a number of papers we have investigated the underlying fitness landscapes in order to identify properties leading to such increased evolvability (Smith et al., 2001b,a, 2002b). In this paper we take a different approach, investigating the operation of successfully evolved controllers. In particular, we frame and test the hypothesis that the GasNet class is more easily tuned to the characteristics of the environment than the NoGas class.

2 The experimental setup

The GasNet is an arbitrarily recurrent ANN augmented with a gas concentration model, in which the instantaneous activation of a node is a function of both the inputs from connected nodes and the current concentration of gas(es) at the node. The basic network model consists of connected sigmoid transfer function nodes overlaid with a model of diffusing gaseous modulators; the gas does not alter the electrical activity in the network directly but rather acts by changing the gain of the transfer function mapping between node input and output.

*A longer version of this paper is published as Smith et al. (2002a).

The evolutionary task at hand is a visual shape discrimination task; starting from an arbitrary position and orientation in a black-walled arena, the robot must navigate under extremely variable lighting conditions to one shape (a white triangle) while ignoring the second shape (a white square). Both the robot control network and the robot sensor input morphology, i.e. the position of the input pixels on the visual array, were under evolutionary control. Fitness over a single trial was taken as the fraction of the starting distance moved towards the triangle by the end of the trial period, and the evaluated fitness was returned as the average over 16 trials of the controller from different initial conditions. For further details see (Husbands et al., 1998; Smith et al., 2002b).

3 Functionally equivalent controllers

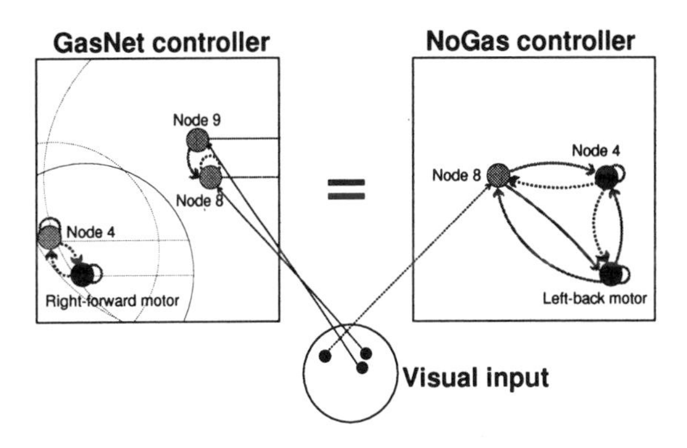

Figure 1: Functionally equivalent GasNet and NoGas timing subnetworks

Two controllers, one evolved using the GasNet class and one evolved using the NoGas class, were investigated using dynamical systems analysis. It was found that both employed the same strategy for the triangle-square discrimination task, based on a method of timing the duration for which bright visual input was received in the upper half of the visual field. Since triangles are narrower at the top than squares, this allows the controllers to successfully discriminate between the two shapes. Figure 1 shows the two functionally equivalent subnetworks; both controllers time the duration over which bright input is received from visual inputs in the

	Double speed		Quarter speed	
	GasNet	NoGas	GasNet	NoGas
Number of cases	20	20	20	20
Mean evaluated fitness (SD)	0.17 (0.074)	0.15 (0.066)	0.36* (0.10)	0.21 (0.016)
Mean re-evolution generations (SD)	10 (5)**	409 (336)	30 (31)**	591 (346)
Median re-evolution generations	10**	360	19**	608

Table 1: Number of generations required to re-evolve functionally equivalent controllers in a modified environment (T-test, Mann-Whitney: $^{*}p < 0.05$, $^{**}p < 0.01$).

	Double speed		Quarter speed	
	GasNet	NoGas	GasNet	NoGas
Number of cases	200	200	200	200
Mean evaluated fitness (SD)	0.27 (0.13)	0.26 (0.18)	0.35 (0.27)	0.29 (0.19)
Mean re-evolution generations (SD)	107 (190)**	240 (363)	108 (229)	116 (252)
Median re-evolution generations	36*	49	13**	21

Table 2: Number of generations required to re-evolve sample of controllers in a modified environment (T-test, Mann-Whitney: $^{*}p < 0.05$, $^{**}p < 0.01$).

upper half of the visual field. Through detailed analysis of the operation of a number of GasNet and NoGas controllers, and the functionally equivalent subnetworks, we hypothesised that the GasNet class is easier to *tune* to the characteristics of the environment.

4 Evolution in modified environments

To test this hypothesis, we seeded successfully evolved controllers back into an evolutionary algorithm, in which fitness was evaluated in environments with properties modified from the initial evolutionary process. Two separate environments were used, where the robot motor speeds are respectively set to double and quarter the usual motor speeds. This has the effect of making the robot move at a different speed in the arena, in particular spinning past the two shapes at very different rates to the speeds encountered during the original evolutionary phase, implying that the timing mechanisms must be altered to complete the task.

Table 1 shows the number of generations required before controllers of 100% fitness are produced, for the re-evolution of functionally equivalent controllers. Table 2 shows the same results for a sample of forty GasNet and NoGas controllers. We see evidence in both cases that the GasNets re-evolve to the modified environments faster than the NoGas solutions.

5 Why GasNets are evolvable

From the analysis of a number of individual networks, and the functionally equivalent subnetworks described above, we propose several hypotheses as to why GasNets are useful for the generation of adaptive behaviour over time.

First, pattern generation is extremely easy to produce using GasNet controllers, which is useful even in tasks such as the visual discrimination task used here. Morevover, tuning these patterns to the characteristics of the environment is also relatively simple using GasNets. Second, the ability to switch between stable states is also extremely useful, e.g. a discontinuous change of behaviour determined by external input, and the ability to mediate such switching. The gas diffusion mechanism allow such a switch to take place over several time-steps, by the build-up of gas concentration levels. Third, filtering out environmental noise through requiring input to be consistent over several time-steps is also straightforward using the GasNets.

References

Husbands, P., Smith, T., Jakobi, N., and O'Shea, M. (1998). Better living through chemistry: Evolving GasNets for robot control. *Connection Science*, 10(3-4):185–210.

Smith, T., Husbands, P., and O'Shea, M. (2001a). Neutral networks and evolvability with complex genotype-phenotype mapping. In Kelemen, J. and Sosík, P., (Eds.), *Advances in Artificial Life, 6th European Conference (ECAL'2001)*, pages 272–281. Springer, Berlin.

Smith, T., Husbands, P., and O'Shea, M. (2001b). Not measuring evolvability: Initial exploration of an evolutionary robotics search space. In *Proceedings of the 2001 Congress on Evolutionary Computation (CEC'2001)*, pages 9–16. IEEE Press, Piscataway, New Jersey.

Smith, T., Husbands, P., and O'Shea, M. (2002a). Adapting to a changing environment: Evolvability and analysis of robot control networks. Cognitive Science Research Paper 549, School of Cognitive and Computing Sciences, University of Sussex, UK.

Smith, T., Husbands, P., and O'Shea, M. (2002b). Local evolvability of statistically neutral GasNet robot controllers. *Biosystems*. In press.

Co-Evolving Robot Soccer Behavior

Esben H. Østergaard and Henrik H. Lund
esben|hhl@mip.sdu.dk
The Maersk McKinney Moller Institute for Production Technology
University of Southern Denmark
Campusvej 55, 5230 Odense M., Denmark

Figure 1: The Khepera Robot, about 5.5 cm in diameter.

1 Introduction

Robot soccer is a complex competitive task. Deciding on which strategy to implement when constructing the robot controller is hard, and no methodology exist for finding the best controller. The task lends itself to co-evolution due to its inherent competitive nature. We present a system that uses competitive co-evolution to develop robot controllers for the task, and show that the evolved controllers are capable competing with hand-coded robots.

In a Khepera robot soccer match two Khepera robots (figure 1) are pitted against each other in an arena, shown in figure 2.

2 The Implemented System

We use an off-line evolutionary system, as shown in figure 3. The evolutionary algorithm works on gene strings, which are translated into a controller. The controller is tested in a simulator, and the performance of the controller is fed back to the evolutionary algorithm. After a pre-specified number of generations, the best performing controller is translated into a binary

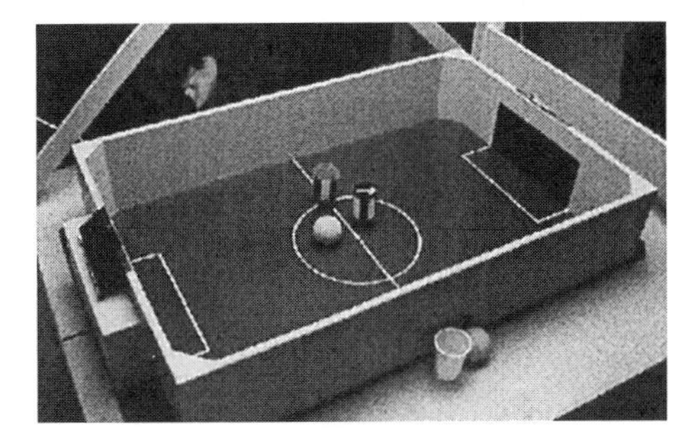

Figure 2: A picture of the football arena.

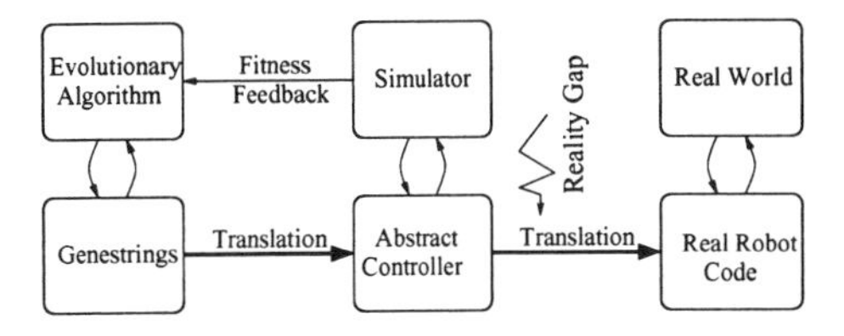

Figure 3: The components of the system.

file that can be executed on a Khepera soccer player. Co-evolutionary is implemented as two separate populations. Individuals from the same population are never pitted against each other, and there is no exchange of gene information between the two populations.

2.1 Behavior Representation

Behavior is represented in a fixed-structure tree of arbitrators and primitives, shown in figure 4. The size

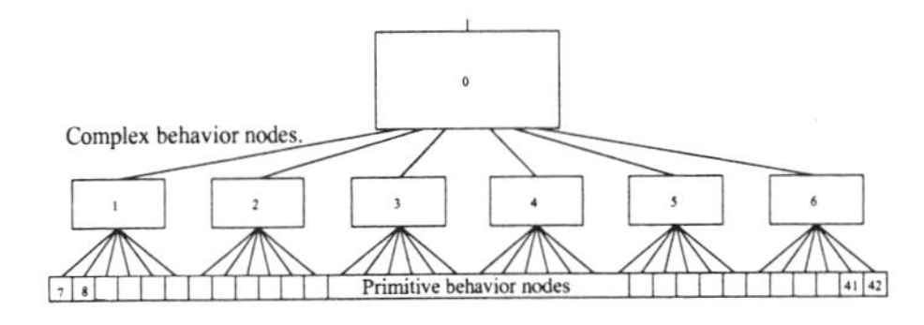

Figure 4: The structure of the behavior tree.

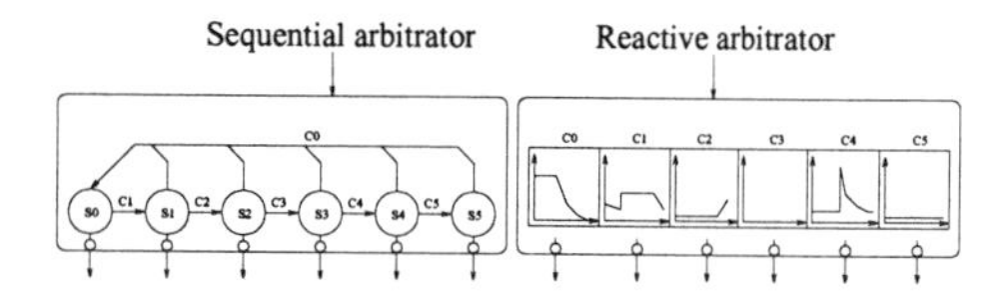

Figure 5: The two types of arbitration nodes. Control is propagated down to a sub module. In the sequential case, control is transferred to the sub module corresponding to the state of a finite state automaton. In the "reactive" arbitrator case, control is transferred to the submodule corresponding to the leaky integrator with highest activation.

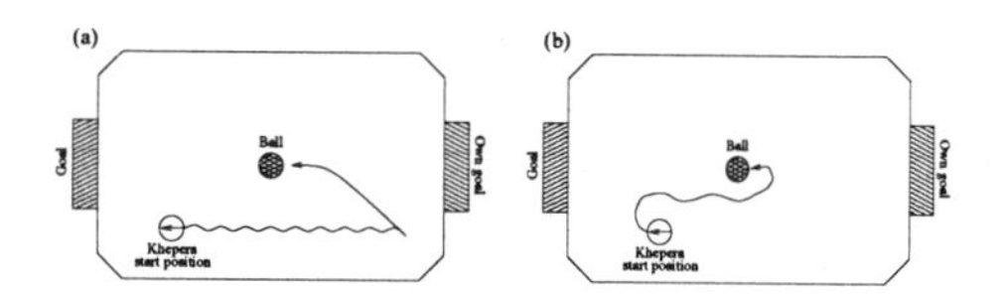

Figure 6: Different evolved football strategies.

and structure of this tree is based on an estimate. Modules 0-6 are arbitrator modules, and modules 7-42 are primitive actions. At each 100 ms time step, control is propagated down through the tree to one of the 36 primitives, which will then have full control of the robot for the duration of the time step. Two types of arbitrators were implemented; sequential and "reactive" arbitration, illustrated in figure 5. This approach is based on the task decomposition approach described by W. Lee, J. Hallam and H. H. Lund in [LHL97].

2.2 The Simulator

A hybrid between the minimal simulation approach, suggested by Nick Jacobi [Jac98], and a geometric mathematical model is used. Basically everything that relatively easily could be modeled geometrical is modeled geometrically, and then the rest is made *unreliable* by adding noise. The world is only modeled in two dimensions, since the third dimension, the height, does not seem relevant for the task at hand.

As stated in [HCH92], producing simulations of visual sensing is a very time-consuming task, both for the programmer when building the simulator, and for the computer during simulation. Instead of simulating the camera pixel by pixel, a filter returning the centroid and with of the ball in the image is simulated. Similar methods were used for the IR sensors and the wheel incremental encoders.

3 Performance

Two distinct strategies for approaching the ball were observed in the evolved robots. The two types are shown in figure 6 (a) and (b). The system was really put to

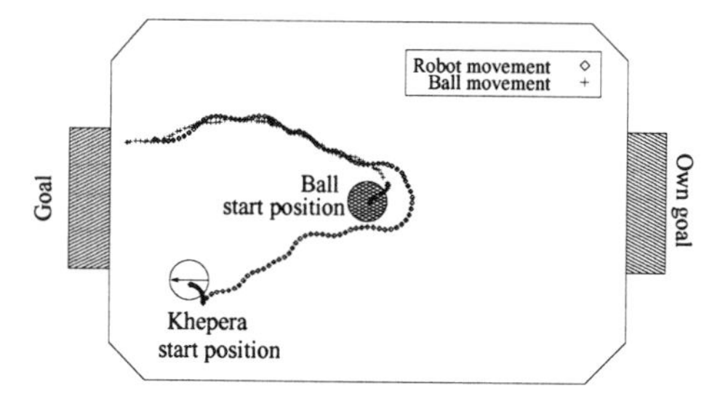

Figure 7: The plotted position of the robot *Brute Force* and the ball in the simulator. The ball starts to move when *Brute Force* pushes the ball.

the test, when an evolved robot controller participated in the Danish Robot Football Competition, Dec. 1999 under the name *Brute Force*. Videos from this event can be found on the home page of this paper[1], and is the main source of documentation for the real robot performance. In the preliminary matches, *Brute Force* won two times and had one draw, which was enough to qualify for the semi finals. In the semi final *Brute Force* won 1-0 without much difficulties, but lost in the final against *KITT*, partly due to a dead battery in the second round. *Brute Force*'s behavior in the simulator is shown in figure 7. By observation, this behavior corresponds to the behavior in the real world.

4 Conclusion

The objective of the work was to explore the co-evolutionary robotics approach and to test whether it could be used to evolve behavior for the Khepera robot soccer task. This seems to be the case. However, existing theories still need further development in several areas to reduce the amount of intuition required to build such a system. Especially the problem with writing complex simulators seems to be difficult to overcome.

References

[HCH92] I. Harvey, D. Cliff, and P. Husbands. Issues in evolutioanry robotics. In Roitblat, H. Meyer, J.-A. and Wilson, S., editors, *Proceedings of SAB92*. MIT Press Bradford Books, Cambridge, MA, jul 1992.

[Jac98] Nick Jacobi. The minimal simulation approach to evolutionary robotics. In Takashi Gomi, editor, *Evolutionary Robotics, Volume II*, 1998.

[LHL97] Wei-Po Lee, John Hallam, and Henrik Hautop Lund. Learning complex robot behaviours by evolutionary approaches. *6th European Workshop on Learning Robots, EWLR-6*, aug 1997.

[1] http://www.mip.sdu.dk/~esben/EvoRobSoc

Conditions for the evolution of mimicry

Daniel W. Franks and Jason Noble
Informatics Research Institute, School of Computing, University of Leeds
Leeds LS2 9JT, UK. Email: {dwfranks;jasonn}@comp.leeds.ac.uk

Abstract

Mimicry is a poorly understood phenomenon; we present a simulation of the evolution of both Batesian (parasitic) and Müllerian (mutualistic) mimicry. In the model, multiple species are preyed upon by a single abstract predator; the appearance of each prey species can evolve but their palatability is fixed. Batesian mimicry evolves regardless of the initial phenotypic distance between the two species, whereas Müllerian mimicry requires an initial resemblance, except when a third (Batesian) species is present.

Mimicry is a textbook example of evolutionary adaptation. However, its evolutionary dynamics are still quite poorly understood. Current models of mimicry tend to focus on the selective pressures on prey brought about by the particular learning abilities of the predator, and employ Monte Carlo or mathematical approaches (see e.g., Turner & Speed, 1996; Huheey, 1988). Although predator learning is an important factor, it is also useful to look directly at the coevolution of mimetic species and their models. To that end, we present a simulation model that explores the conditions for the evolution of various types of mimicry (as far as we know, this is the first such model in the adaptive behaviour literature).

Mimicry is typically classified as either Batesian or Müllerian. Unpalatable species — many insects and snakes are good examples — often evolve conspicuous warning colouration (aposematism) which exploits predators' sensory systems such that an association is rapidly learned between the species' appearance and unpalatability. The evolution of Batesian mimicry takes advantage of aposematism in order to deceive the predator: the tasty (unprotected) species comes to resemble the unpalatable (defended) model species. This is a parasitic relationship between the mimic and the model, and, because the selective pressures on the mimic are greater than those on the model, the mimic is generally expected to keep up in the resulting coevolutionary arms race (Turner, 1995). An example of a Batesian mimic is the hover-fly, for which the bee and the wasp are models.

Müllerian mimicry, on the other hand, is mutualistic. Two defended species converge on each other's appearance to gain extended protection from the predator. Thus, the predator's learning system has been exploited so that it can learn to avoid the warning colouration more quickly. The predators, of course, need to sample some of the co-mimics in order to learn to avoid them. Thus, the main advantage to an individual here is that they have less chance of being sampled by the predator, as they are protected by extra numbers. Bees and wasps are good examples of Müllerian mimicry.

1. The Simulation Framework

Multiple populations of prey species were used in each experiment. Different species of prey were each assigned a fixed palatability level. Each individual had a single gene: a value representing their external appearance or phenotype. The phenotypes were constrained to a 'ring' of values from 1–20 (where 20 and 1 are neighbours). The distance of one phenotype from another represents their level of similarity.

A single, abstract predator was modelled with a simple reinforcement learning system. The predator's experience of each phenotype was represented by a score; after eating prey of a particular external phenotype, the predator would update the relevant score according to the palatability of the individual consumed. The predator generalised on the basis of experience and thus would also, to a lesser extent, update its scores for the four closest neighbour phenotypes. Each predator memory score had a fixed upper and lower bound, and gradually degraded back towards its starting level (usually zero, representing ambivalence).

In each generation the predator was presented with 30 binary forced-choice situations. Two individuals were randomly selected from across all prey populations present and the predator would make a probabilistic choice based on its experience of each phenotype. Random asexual reproduction then took place amongst the surviving prey. Mutation was implemented as a uniform change of ± 1 in the phenotype, and the mutation rate was 0.03. All of the experiments were run over 5000 generations, and prey species populations kept constant at 100. The main variable manipulated was the starting distance between prey species' phenotypes, in order to determine whether an initial chance resemblance is required for the evolution of mimicry.

2. Results

Figure 1 shows the results of two experiments in terms of the initial and final distances between prey species' phenotypes. Experiment 1 was conducted using one palatable and one unpalatable species. Regardless of the starting distance between the two, Batesian mimicry evolved: the palatable species came to have the same or very similar phenotype to the unpalatable one. This was also true when the predator's learning did not involve generalisation.

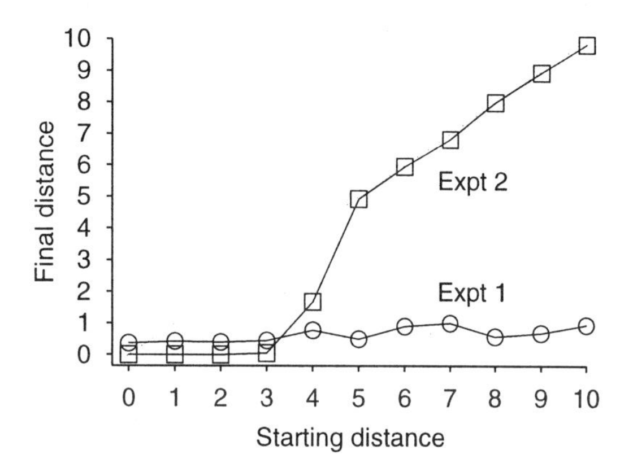

Figure 1: Final distance by initial distance between two prey species' phenotypes; each point is averaged over 40 runs. Dashed line shows zero change in phenotype.

Experiment 2 was conducted in the same way with two unpalatable species. Figure 1 shows that Müllerian mimicry only evolves if the two prey species have some initial resemblance. If they are initially more than about 4 units away on the phenotypic ring, they typically remain distinct in appearance. Selective pressure dictates that unpalatable species, given an initial resemblance, will almost always converge, whereas a palatable species randomly drifts until it 'finds' the model by chance. It follows that, given enough time to drift, all palatable species would eventually become Batesian mimics.

Further experiments were carried out in which additional, neutrally palatable species were added to the contexts of experiments 1 and 2, in order to simulate a more diverse ecosystem. This had no qualitative effect on the results.

In experiment 3 (no graph shown) simultaneous Müllerian and Batesian mimicry was investigated by including two unpalatable and one palatable species. The results showed that the phenotype of the palatable species moved towards that of one of the unpalatable species (i.e., Batesian mimicry). Interestingly, this in turn drove the model species around the phenotypic ring and resulted in Müllerian mimicry with respect to the second unpalatable species, regardless of their initial phenotypic distances.

3. Conclusions and Further Work

Experiment 1 shows that Batesian mimics will close in on the model regardless of how different their initial phenotypes are, and regardless of the predator's ability to generalise. This shows that there is more selective pressure for the palatable species to resemble the model than there is for the model to diverge. However, there is an additional reason why the mimic is successful in "catching" the model: before mimicry has evolved, palatable individuals gain an inherent fitness benefit for mutating away from the modal phenotype of their species. This is because, when presented with a choice, the predator would be more likely to select a well-known phenotype than a newer one.

The results of experiment 2 show that Müllerian mimicry relies on an initial resemblance between the species involved. Such initial resemblance might be caused by any number of factors, such as sexual selection, random drift, etc. (It should be noted that the *particular* resemblance threshold found in the experiment — four units on the phenotypic ring — is of course determined by the nature of the predator's generalisation.) Experiment 2 has a somewhat paradoxical outcome in that the co-mimics, once resemblance is achieved, do not gain any immediate fitness benefit from their resemblance, at least in the absence of alternative prey species. This is because the predator must always eat one of the two choices presented to it. However, the paradox is resolved when we note that a mutant individual would represent a novel and thus appealing meal for the predator.

Experiment 3 demonstrates that pressure due to Batesian mimicry can be a force that drives Müllerian mimics together despite a lack of initial resemblance.

These results have implications for the dynamics of mimicry rings (shared phenotype between many mimetic species) in nature, and could constitute an explination for why they do not all converge into one large ring.

Clearly, one weakness of the current model is that phenotypes are expressed in a single dimension. Future work will incorporate a richer, multi-dimensional phenotypic space. Previous work in biology suggests that predator diet balancing, prey frequency, prey density, predator diversity and the evolution of aposematism itself are factors also worth including in a more complete model.

References

Huheey, J. (1988). Mathematical models of mimicry. *American Naturalist*, 131. Supplement: S22-41.

Turner, J. (1995). Mimicry as a model for coevolution. In Arai, R., Kato, M., & Doi, Y. (Eds.), *Biodiversity and Evolution*, pp. 131–150. Tokyo: National Science Museum Foundation.

Turner, J., & Speed, M. (1996). Learning and memory in mimicry. I. Simulations of laboratory experiments. *Philosophical Transactions of the Royal Society of London: Biological Sciences*, *351*(B), 1157–1170.

Ecological disturbance maintains and promotes biodiversity in an artificial plant ecology

Ben Clark[†] **Seth Bullock**[‡]

[†]Department of Biology
Imperial College at Silwood Park, London University
benjamin.clark@ic.ac.uk

[‡]Informatics Research Institute
University of Leeds
seth@comp.leeds.ac.uk

Abstract

A model of plant growth, competition and reproduction in three dimensions was constructed using L-systems to simulate plant growth, ray tracing to simulate sunlight and shading, and a steady-state genetic algorithm to simulate evolution by natural selection. Simulated plant growth conformed to expected trade-offs between, for instance, growing up and growing out. Simulated cohorts exhibited conventional population-level phenomena such as obeying the self-thinning law. Competition between species was simulated under various disturbance regimes. Undisturbed, a K-selected type of plant species dominated at equilibrium. However, under certain disturbance regimes, diverse life-history strategies were able to coexist at equilibrium, and even speciate.

Competition for light is probably the most important environmental influence on plants. Successful plant life-history strategies maximize light capture (amongst other things) despite the presence of competing plants. As it is often too difficult to measure local interactions between individual plants, most ecological studies take a population-level perspective. The local interactions that give rise to population-level phenomena are acknowledged, but their role in competition is often ignored.

Understanding how plant-plant interactions affect the evolution of plant life-history strategies is made more difficult by the fact that plant structure is far more developmentally plastic than that of most animals. Plant morphological plasticity is largely in response to heterogeneity in the local light environment. Of course, just as light availability influences a plant's structure, so a plant's structure affects its ability to capture light. The intimate, reciprocal relationships between a plant's current morphology, the structure of its local environment (including competing plants), its resultant ability to capture light, and any subsequent changes to its structure are highly complex. However, it is exactly these relationships that drive plant growth and determine the fitness of competing life-history strategies.

Understanding plant competition for light in terms of processes taking place at and below the level of individual plants could provide a unifying account for well-characterized population-level phenomena. This type of individual-level understanding could improve our insight into the role of natural selection in shaping plant life-history strategies. Unfortunately, analytic models at the high-resolution required to achieve this are currently intractable. Here we resort to numerical methods, in particular the use of individual-based models.

1. Method

The life-history strategy of each plant was represented by a set of values used to parameterize a simple L-system (a rule re-writing system used to model plant development; Lindenmayer 1968). These values included branching angles and probabilities, branch strength, and age of onset of reproduction. By varying these parameters we explored a wide range of different plant life-history strategies—from what could be thought of as K-selected plant species (e.g., oak trees) to r-selected plant species (e.g., grasses). A plant's growth was simulated in 3-D by applying the parameterized L-system rules to a seed, and repeatedly applying the same set of rules to all elements of the resultant plant structure in parallel.

We have attempted to model the most important processes involved in plant development (shading, photosynthate production, transport and consumption, leaf and branch shedding, reproduction, and death) in the most basic, and wherever possible, principled manner. Sunlight was simulated using a ray tracing approach which calculated the degree of shading experienced by each of a plant's leaves. The photosynthate production of each leaf varied inversely with degree of shading. Photosynthate transport was simulated using a simple diffusion algorithm that slowly propagated photosynthate from leaves (where it is synthesized) throughout the plant. This enabled us to model plants that were developmentally plastic in response to variation in shade such that they "foraged for light". Leaf and branch shedding was simulated as resulting from over-shading and over-loading, respectively. Once a plant reached it's reproductive age, it began to produce seeds at each branch apex. Seed production was modelled as proportional to the amount of photosynthate present at the apex. Plants with no leaves were considered dead.

In many cases, we were interested in modelling a single plant or a single-species cohort. However, in some cases, we were interested in modelling life-history strategy coevolution directly. For these purposes we considered a plant's set of life-history strategy parameters to constitute its genome. Each plant seed contained a copy of its parent's genome subject to a small chance of mutation (reproduction was thus asexual). By applying a steady-state genetic algorithm to a population of plants, we could simulate the heritable variation and competition for a scarce resource (light) necessary to implement a process analogous to evolution by natural selection.

2. Results

First, we explored the parameter space, varying each parameter in order to assess its impact on maximum attained height and effective leaf area (each taken to be rough indicators of plant fitness). We were, of course unable to explore the entire parameter space. However, these preliminary simulations allowed us to ascertain that various expected developmental trade-offs were indeed present. Higher branching angles can increase effective leaf area (by reducing self-shading) but at the expense of reduced plant height and reduced stability. Laying down stronger tissue can reduce branch shedding and thus increase effective leaf area, but at the result of slower growth rate. Increasing age of reproductive onset can increase overall life-time seed production, but at the expense of higher risk of death before reproduction.

By simulating cohorts of plants, we were able to confirm that population characteristics such as density, size distribution and skewness conformed to results from natural plant populations. In particular, model cohorts obeyed the self-thinning law (Yoda et al., 1963), where plant communities evidence a log-log relationship between biomass and density with an exponent of $-3/2$. Our simulated cohorts (when they achieved canopy closure) all achieved exponents between -1.6 and -1.3. It is interesting that these population-level phenomena resulted solely from simulating the competition for light.

By simulating communities comprising multiple species, we were able to explore the capacity for a plant ecosystem to maintain multiple life-history strategies. Here we explored competition between a single tree-like species (K) that could be considered to represent K-selected species, and a single grass-like species (r) that could be considered to represent r-selected species. In an undisturbed environment, K dominated, eventually driving down numbers of r. However, when random ecological disturbance was simulated by sporadically removing any plant within a certain radius of a randomly chosen location, the balance between the two species altered. High or medium rates (0.5 to 5 units of area per iteration) of small disturbances (0.5 units of area per disturbance) led to increased numbers of both r and K. This type of disturbance weeded plants at random, preventing the formation of an unbroken canopy and promoting the growth of young plants. Increasing the size of the disturbances (to 50 units of area per disturbance) led to increased numbers of r, as this species is more able to quickly exploit newly cleared areas, and sometimes drove K to extinction.

Implementing genotypic mutation allowed us to explore the extent to which the life-history strategies of r and K would change over coevolutionary time as a result of their competition for light. The prohibitive length of coevolutionary simulation runs ensured that these explorations were, by necessity, preliminary. We ran 25 replicate simulations, comparing the genotypes of the initial seed population with that of their descendants 6000 iterations later (the equivalent of 100 generations for K), using K-means clustering to detect speciation. There was evidence of selection having acted on the genotypes, as variation within a genotypic cluster was smaller after 6000 iterations. Results also suggested a tendency for r to speciate, forming a third grass-like species with increased reproductive allocation (although the action of evolutionary drift may have been enough to account for the evidence of a third genotypic cluster).

3. Conclusions

Using an individual-based model at a high resolution, we were able to demonstrate that ecological disturbance has an effect on the evolution of plant life-history strategies. Although the model neglects many aspects of plant growth and development, the system evidenced realistic phenomena at both the individual and population level, and demonstrated the competitive coexistence of at least two species. No doubt many aspects of plant biology so far excluded from the model have implications for diversity. However, the current simulation suggests that the impact of competition for light on plant growth and development alone is capable of maintaining and promoting the diversity of plant life-history strategies.

References

Lindenmayer, A. (1968). Mathematical models for cellular interaction in development, parts I and II. *Journal of Theoretical Biology*, 18:280–315.

Yoda, K., Kira, T., Ogawa, H., and Hozumi, K. (1963). Self-thinning in overcrowded pure stands under cultivated and natural conditions. *Journal of Theoretical Biology*, 14:107–129.

COLLECTIVE AND SOCIAL BEHAVIOR

Competitive Foraging, Decision Making, and the Ecological Rationality of the Matching Law

Anil K Seth [1,2]
[1]The Neurosciences Institute, 10640 John Jay Hopkins Drive, San Diego, CA 92121, USA
[2]CCNR/COGS, University of Sussex, Brighton, BN1 9QH, UK
seth@nsi.edu

Abstract

The matching law describes how *individual* foragers often allocate their choices, occasionally suboptimally, in experimental situations. The 'ideal free distribution' predicts how *groups* of foraging agents should distribute themselves, optimally, over patchy environments. This paper explores the possibility that a single behavioural heuristic can account for both phenomena, allowing the potential suboptimality of matching to be understood in terms of adaptation to a group context. Two simple heuristics are compared, ϵ-sampling and ω-sampling: the latter is successful in both cases, but contrary to prior claims in the literature the former is successful in neither. These results emphasise the importance of multiple environmental value estimates in effective decision making.

1 Introduction

The principle of 'ecological rationality' holds that cognitive mechanisms are best understood as fitting the demands and structure of particular environmental niches, as opposed to the classical view of cognition approximating a 'Laplacean superintelligence' for achieving "general purpose, optimal performance in any situation" (Bullock & Todd, 1999, p.3). An implication of ecological rationality is that cognitive mechanisms operating outside of their proper niche may deliver *ir*rational behaviour, in much the same way that a fish out of water is disadvantaged with respect to breathing.

This paper focuses on Herrnstein's (1961) 'matching law', which describes how *individual* foragers allocate their choices in many experimental situations. Matching behaviour is very widespread, characterising much decision making activity in both humans and animals, and importantly, whilst often optimal, it is not always so. Here, we explore the idea that matching can be accounted for by an ecologically rational decision making heuristic well adapted for decision making in a *group* context, from this perspective the occasional suboptimality (or irrationality) of matching would be a consequence of shifting from a group to an isolated individual environment. A broader motivation is to provide an alternative, empirically grounded perspective on the problem of 'action selection' (see Tyrrell, 1993, for example), a term which is often used synonymously with decision making in the field of adaptive behaviour.

The group context we consider is the 'ideal free distribution' (IFD), which describes the equilibrium (and optimal) distribution of foragers over an environment with patchy resource distribution (Fretwell, 1972). Even though the IFD and the matching law derive from different disciplines (the former from behavioural ecology, the latter from psychology), they present many similarities: (1) the matching law is to do with individual choice and the IFD is to do with its collective consequences; (2) they display striking mathematical congruency (Gray, 1994; Baum & Kraft, 1998, see also sections 2 and 3); and (3) perhaps most importantly, the laboratory environments employed by psychologists are often interpreted as abstractions of natural foraging environments (Dallery & Baum, 1991; Shettleworth, 1988). The suggestion that a single behavioural dynamic may underlie both phenomena is therefore not new (Houston, 1986; Gallistel et al., 1991; Thuisjman et al., 1995; Seth, 1999). The present contribution is primarily to explicitly assess candidate heuristics, to demonstrate success in one case, and to repudiate equivalent claims of success in another (Thuisjman et al., 1995).

Two simple heuristics are compared, ϵ-sampling and ω-sampling (fully defined later), differing primarily in that the former maintains a single estimate of environment 'value' and the latter maintains multiple estimates. Contrary to the claims of Thuisjman et al. it is shown that ϵ-sampling fails to account for either matching or the IFD, but that the novel ω-sampling heuristic is successful in both cases.

2 The ideal free distribution

Given a patchy distribution of resources, the IFD describes the equilibrium distribution of foragers such that no forager can profit by moving elsewhere, regardless of the local resource quality (Fretwell, 1972); in this state all foragers will obtain equal resources.[1] In order to make specific predictions, IFD models require a way of relating the per forager intake rate W_i (s^{-1}) to both the forager density N_i and the resource availability F_i on each patch i. The following equation is adapted from Sutherland (1983) and Milinski and Parker (1991):

$$W_i = \frac{QF_iF^*}{N_i{}^m}, \tag{1}$$

in which Q (ms^{-1}) is a measure of patch-independent forager efficiency, $F_i \in [0.0, 1.0]$ (dimensionless) represents the resource fraction in patch i, F^* represents the total resources available, and m (dimensionless) is the interference constant, which is usually taken to vary between 0.0 (no interference) and 1.0 (high interference), where interference is defined as the (more-or-less immediately reversible) decline in intake due to the presence of conspecifics (Goss-Custard, 1980; Sutherland, 1983).[2] Across two patches A and B, assuming $W_A = W_B$ (the IFD condition):

$$log\frac{N_A}{N_B} = \frac{1}{m}log\frac{F_A}{F_B}, \tag{2}$$

this being the 'generalised habitat matching rule' of Fagen (1987). Also, taking the total forager number to be N_T $(= N_A + N_B)$, it is possible to predict both N_A and N_B directly (Tregenza, Parker, & Thompson, 1996):

$$N_A = \frac{N_T}{(10^{-c} + 1)}, \qquad c = \frac{log\frac{F_A}{F_B}}{m}. \tag{3}$$

In what follows, we will use equation 3 to assess the ability of ϵ-sampling and ω-sampling to lead populations to the IFD under different levels of interference.

2.1 *Resource allocation*

Both ϵ-sampling and ω-sampling operate over discrete time intervals, and as such it is possible to interpret the resource level F_i in at least two ways. The first is simply as specifying a resource level that contributes to the forager intake at every time-step. This process of 'continuous allocation' (C-allocation) is the usual interpretation in the literature (see, for example, Bernstein, Kacelnik, & Krebs, 1991), and in this case equation 1 can be used at every time-step, exactly as it is written. The second approach is to understand F_i as as specifying a probability that patch i will yield the fixed resource quantity F^* to each forager at each time-step. Under this process of 'probabilistic allocation' (P-allocation), W_i becomes a random variable:

$$W_i = \begin{cases} \frac{QF^*}{N_i{}^m}, & p(F_i) \\ 0, & p(1 - F_i) \end{cases} \tag{4}$$

The IFD condition of equal intake rates across all patches in this case must apply to *expected* intake rates over many time-steps. We can write:

$$E(W_i) = \left(\frac{QF^*}{N_i{}^m}\right)F_i.$$

from which the condition $E(W_A) = E(W_B)$ leads to the same generalised habitat matching law described above (equation 2).

No claims are made for the biological relevance of the distinction between C-allocation and P-allocation, it is motivated by analogous resource allocation methods often employed in 'matching law' experiments, described below. One possible intuition, however, is that it may reflect a difference between relatively accessible and widespread types of resource (grass, for example), and relatively inaccessible yet potent types of resource (truffles, for example).

3 The matching law

Moving on to the matching law, Krebs and Kacelnik (1991) offer the following definition: "the matching law states that the animal allocates its behaviour to two alternatives in proportion to the rewards it has obtained from them" (p.131). If the proportionality is direct, this is known as 'strict' matching (Davison & McCarthy, 1988):

$$\frac{B_A}{B_B} = \frac{R_A}{R_B}, \tag{5}$$

where B_A and B_B represent the rate of response to options A and B, and R_A and R_B represent the resources obtained in each case. The 'generalised' matching law (Baum, 1974) includes parameters for bias (b) and sensitivity (s) to account for the departures from strict matching often observed in empirical data:

$$log\frac{B_A}{B_B} = s.log\frac{R_A}{R_B} + log(b). \tag{6}$$

Of the similarities between the IFD and matching noted earlier, their mathematical congruence should now be particularly evident (compare equations 2 and 6), but there are also profound differences: whereas habitat matching predictions are normative, the individual

[1] This last clause requires the assumption that foragers are able to move, without cost, to the patch in which their rewards are maximised.

[2] Values of m in excess of 1.0 are possible, and can be expected in cases in which prey items can be *lost* (for example, by fleeing) as a result of interference. Note also that the model described here is known in the literature as a 'standing stock' or 'interference' model for the reason that it assumes relatively constant resources in each patch. Another popular choice is the 'immediate consumption' model, which assumes a steady input of immediately consumed resources. With m = 1.0 in equation 1, but not otherwise, the two models are equivalent; see Van der Meer and Ens (1997) for further discussion.

matching law is an observed relation, and whereas habitat matching is expressed in terms of *available* resources (F_i), the individual matching law is expressed in terms of *obtained* resources (R_i). (Notice that F_i can still be used in the context of individual matching even if it is not represented in the matching equations themselves, and indeed it is necessary to do so in order to describe the various 'schedules of reinforcement' by which resources are allocated in matching experiments.[3])

Psychologists have investigated matching under many different reinforcement schedules, four of which are considered here, each with two options A, B with associated resource availabilities F_A, F_B:

- *Basic:* Each response is rewarded with an amount determined by the relative values of F_A and F_B. Responses are rewarded at every time-step. This is analogous to C-allocation.
- *Concurrent (conc) VR VR:* A variable ratio (VR) schedule indicates that an option must receive a certain number of responses before a reward is given. This number can vary around a mean value, and can therefore be implemented by associating a probability of reward with each option. F_A and F_B are here interpreted as the mean values, so that conc VR VR is analogous to P-allocation.
- *Concurrent (conc) VI VI:* A variable interval (VI) schedule requires that a certain delay elapse after a reward on a given option until that option can be rewarded again. This delay time can vary around a mean, and these means can differ between response options (F_A and F_B are interpreted as the delays).
- *Concurrent (conc) VI VR:* This is a 'mixed' schedule in which one choice option is rewarded under a VI schedule, and the other under a VR schedule.

Under both basic and conc VR VR schedules, the general consensus in the literature is that exclusive choice for the most profitable option is observed. There is nothing counterintuitive about this; if repeatedly offered a choice between 80p and 40p, any sensible subject would presumably choose the former 100% of the time, and the same would apply to repeated choices between odds of 3:1 and odds of 5:1. Observations of exclusive choice, although consistent with the matching law, are only trivial instances of its applicability, as such these schedules present relatively undemanding assessments of matching behaviour.

The conc VI VI schedule is more interesting. Unlike basic and conc VR VR, the reward rate can be largely independent of the response rate, such that matching to obtained resources can be achieved with a variety of response distributions, including - but not limited to - exclusive choice. Furthermore, under conc VI VI, exclusive choice is no longer optimal (Herrnstein, 1970). Matching to obtained resources under conc VI VI has been observed for both non-humans animals (Davison & McCarthy, 1988) and human subjects (Conger & Killeen, 1974), in all cases *without* exclusive choice.

The final schedule, conc VI VR, also leads to observations of matching to obtained resources, in some cases in the trivial form of exclusive choice, and in other cases non-trivially, depending on the relative productivities of the two component schedules (Herrnstein & Heyman, 1979; Herrnstein & Vaughan, Jr., 1980). The most important feature of this schedule is its relation to the reward maximisation. Unlike all previous schedules, matching to obtained resources (whether trivial or not) is *not* optimal. Conc VI VR therefore enables exploration of the potential suboptimality associated with matching. The consensus in the literature is that matching to obtained resources - not maximisation - is observed under conc VI VR (Herrnstein & Heyman, 1979; Herrnstein, 1997).

4 A Description of the model

With this background in place, we can now turn to the heuristics themselves.

4.1 ϵ-sampling

The idea behind ϵ-sampling is simply that agents stay on a 'current' patch, and occasionally 'sample' other patches, switching if and only if the 'sampled' patch is better.

More formally, given two alternatives A and B, an ϵ-sampler initially selects A or B at random. At each subsequent time interval, it abides by its choice with probability $(1-\epsilon)$, and samples with probability ϵ, remaining with the sampled option (with probability $1-\epsilon$) if the reward from this option exceeds a 'critical level' (E), which is a dynamic estimate of the 'value' of the environment in which more recent rewards are more strongly represented to a degree specified by an *adaptation rate* γ. The operational definition of ϵ-sampling given below is from Thuisjman et al. (1995):

Definition 1 *Let $\gamma, \epsilon \in (0,1)$, let $M(t) \in A, B$ represent the option selected and let $r(t)$ be the resources obtained at time $t \in \{1,2,3\ldots\}$. Define $E(1) = 0$ and*

$$E(t+1) = \gamma E(t) + (1-\gamma) r(t)$$

for $(t \geq 1)$. Then $E(t)$ is called the critical level at time t. Let A_ϵ denote the behaviour of choosing A with probability $(1-\epsilon)$ and B otherwise. Let B_ϵ be defined similarly. The ϵ-sampling strategy is then defined by playing:

[3] The terms 'reinforcement' and 'reward' are used interchangeably; 'reinforcement' is employed only when it helps to maintain consistency with the psychological literature.

at $(t = 1)$ *use* $A_{0.5}$,
at $(t = 2)$ *use* $M(1)_\epsilon$,
at $(t > 2)$ *use* $M(t-1)_\epsilon$ *in case* $M(t-1) \neq M(t-2)$ *and* $r(t-1) > E(t-1)$, *otherwise use* $M(t-2)_\epsilon$.

4.2 ω-sampling

The novel ω-sampling heuristic extends ϵ-sampling by allowing patch switching to be driven directly by estimates of patch value as well as by sampling excursions; ω-samplers also maintain concurrent estimates of *each* (visited) patch, rather than (as for ϵ-sampling) a single estimate of environmental quality as a whole. (The implications of relaxing this strong assumption in relatively complex environments are discussed in section 5.2.)

For a two patch environment, a ω-sampler initially selects A or B at random. At each subsequent time interval, the other option is sampled with probability ϵ, otherwise (with probability $1-\epsilon$) the estimate of the current selection is compared with that of the unselected option, and switching occurs if the former is the lower of the two. Operationally:

Definition 2 *Let* $\gamma, \epsilon, M(t), r(t)$ *be as in Definition 1, let* $E_A(t)$ *and* $E_B(t)$ *represent the estimated values of options* A, B, *and let* $N(t)$ *represent the unselected option at time* $t \in \{1, 2, 3 \ldots\}$. *Define* $E_A(1) = E_B(1) = 0$. *For* $(t \geq 1)$ *then if* $M(t) = A$:

$$E_A(t+1) = \gamma E_A(t) + (1-\gamma) r(t), \qquad E_B(t+1) = E_B(t),$$

otherwise (if $M(t) = B$*):*

$$E_A(t+1) = E_A(t), \qquad E_B(t+1) = \gamma E_B(t) + (1-\gamma) r(t).$$

Let $\mathcal{R} \in (0,1)$ *be a random number. Let* A_ϵ *and* B_ϵ *be as in Definition 1. The* $\omega-$*sampling strategy is then defined by playing:*

at $(t = 1)$ *use* $A_{0.5}$,
at $(t = 2)$ *use* $M(1)_\epsilon$,
at $(t > 2)$ *if* $(\mathcal{R} < \epsilon)$ *use* $N(t-1)$, *else if* $(E_{M(t-1)} < E_{N(t-1)})$ *use* $N(t-1)$, *otherwise use* $M(t-1)$.

It would not do to overstate the novelty of ω-sampling. Many similar strategies are described in the theoretical biology literature, and it is certainly comparatively trivial in relation to the many reinforcement learning algorithms described in the computer science literature. What is novel here is application to matching and the IFD, and simplicity in this context can be considered a bonus.[4]

[4]There are particularly evident similarities between ω-sampling and decision rules based on the 'marginal value theorem' (Charnov, 1976) which have long been associated with the IFD (but not with matching; see, for example, Bernstein et al., 1991), and which specify switching whenever the gain rate in a given patch is lower than the expected gain rate for the environment as a whole. ω-sampling may be considered a marginal-value rule augmented by (1) patch-specific value estimates and (2) sampling-driven switching.

4.3 Model structure

Both heuristics were explored using individual-based models to assess their performance in three conditions: (1) ability to lead groups of agents to the IFD, (2) matching performance of agents when isolated, and (3) matching performance when embedded in a group (this latter condition included because the few empirical biology papers that consider both matching and the IFD generally consider only embedded individuals).

The first condition involved recording the equilibrium distribution (after 1000 time-steps) of populations of 100 ϵ- and ω-samplers, for each of 9 different resource distributions across two patches A and B. Four separate populations were analysed for each strategy, one for each combination of interference level (1.0 or 0.3) and allocation method (C-allocation or P-allocation). In each case, agents were initially randomly allocated to either A or B. Then, each time-step, the resource obtained by each agent was calculated (equation 1 for C-allocation and equation 4 for P-allocation), the appropriate heuristic applied, and the new agent distributions determined. The final equilibrium distributions were compared with the predictions of the IFD (equation 3).

Isolated individual behaviour was analysed under various reinforcement schedules. For the basic and conc VR VR schedules, single ϵ- and ω-samplers foraged in isolation, under C-allocation or P-allocation respectively, for 1000 time-steps under each of 9 different resource distributions. Conc VI VI was implemented by using F_i to set delay intervals (D_i) such that $D_i = 20(1.0 - F_i) + r$, with $r \in [-2, 2]$ an integer random number. The first response to option i on each evaluation procured the full reward F^* and initialised D_i. Subsequent responses to i went unrewarded until D_i time-steps had elapsed, after which a response would again procure F^* and re-initialise D_i, with the incorporation of r ensuring that the schedule was indeed 'variable interval'. Conc VI VR was implemented by applying VI to one option (A), and P-allocation to the other (B). As before, isolated agents were allowed to forage for 1000 time-steps under each of 9 different resource distributions.

The final condition involved recording the behaviour of individuals embedded within their respective groups from each of the 4 original populations over the full 1000 time-steps, under each of the 9 resource distributions, comparing their behaviour with the predictions of the individual matching law.

All three conditions were repeated 30 times each, enabling means and standard deviations to be calculated.

4.4 Parameter values

Both heuristics require values to be chosen for ϵ and γ. Rather than relying on arbitrary choice and holding this choice constant across all conditions, as is usually the case, in this study a genetic algorithm (GA) was used to evolve near-optimal values for each condition (see Appendix A for details). The reasoning behind this is as follows. The objective of comparing heuristic performance over a range of conditions requires some equivalence criteria to be drawn in terms of the parameters ϵ and γ. If a fixed parameter set is chosen, it could be argued that because the parameters themselves are identical in all conditions, any performance differences must be due to inherent strategy properties, however an alternative interpretation is that the arbitrary set may be more appropriate for some conditions than others, and so performance differences may, to some extent, reflect imbalances in parameter suitability rather than inherent strategy properties. An alternative equivalence criteria is that of optimality. Optimal (or near-optimal) parameter values may well vary across conditions, but it can now be asserted that, in each condition, each strategy is performing as well as it possibly can, therefore any performance differences really must reflect inherent strategy properties, and cannot be explained away in terms of parameter (un)suitability. This is the intuition followed in the present study.

Importantly, whilst this methodological point is worth making and has been largely overlooked in the literature, the following results do not depend on it. Identical results (not shown here, see Seth, 2000) were obtained from a control study performed using a fixed parameter set derived from near-optimal values averaged over all conditions ($\bar{\epsilon} = 0.052$ and $\bar{\gamma} = 0.427$). These control results also proved robust to small variations in this mean near-optimal set (Seth, 2000).

5 Results

5.1 *ϵ-sampling*

Figure 1 compares observed distributions of ϵ-sampling agents to the predictions of the IFD (equation 3). Although in most cases there is a good match, ϵ-sampling agents are unable to find the IFD under P-allocation with $m = 0.3$.

With regard to matching (figure 3), ϵ-sampling agents exhibit exclusive choice (trivial matching) under basic reinforcement, in line with the psychological data (3a,e). Under conc VR VR, however, although they continue to match to available resources, they no longer match to obtained resources, and certainly do not exhibit exclusive choice (3b,f). Performance is no better under conc VI VI or conc VI VR; in both cases there are clear departures from strict matching to obtained resources (3g,h).

Embedded ϵ-sampling agents in most cases match closely to obtained resources (figure 4), although there is some divergence from strict matching when $m = 0.3$ under P-allocation (4h).

Contrary to the claims of Thuisjman et al. (1995), these results demonstrate that ϵ-sampling can neither reliably lead populations of agents the IFD, nor reliably lead individual agents to match to obtained resources. It is worth asking why Thuisjman et al. reached such different conclusions, and one likely reason is that they considered only a small set of analytically tractable special cases. With respect to the IFD they explored only C-allocation with $m = 1.0$, a condition in which ϵ-sampling does indeed lead populations to the IFD, but there is at least one other condition (representative of many others) in which it does not. They also analysed isolated ϵ-samplers only under the equivalent of conc VR VR, which not a useful way to explore matching since only trivial adherence (exclusive choice) is to be expected. Moreover, ϵ-sampling does not even deliver this, instead leading agents to match to available resources. Unfortunately, this result, which Thuisjman et al. (1995) *did* obtain, was wrongly asserted by them to be consistent with the individual matching law; it appears they simply misunderstood the matching law as pertaining to available resources. Here we have seen that ϵ-sampling matches to obtained resources only under basic reinforcement. It should be stressed that the present results are in agreement with those of Thuisjman et al. (1995) *in those special cases considered by them.* The problem is that these cases are not sufficient for substantiating their claims.

5.2 *ω-sampling*

Turning to ω-sampling, it is immediately clear that it outperforms ϵ-sampling at least with respect to the IFD. In all 4 conditions ($m = 1.0$ or 0.3, C-allocation or P-allocation) ω-sampling populations closely fit equation 3 (fig 2).

Matching performance is also improved by ω-sampling. ω-samplers agents exhibit exclusive choice (trivial matching to obtained resources) under basic and conc VR VR (figure 5).[5] Furthermore, matching to obtained resources is also observed under both conc VI VI and conc VI VR (figure 5g,h). (The slight deviations from strict matching entailed by ω-sampling under conc VI VR are in the *opposite* direction to that expected if agents were maximising reward; see Herrnstein & Heyman, 1979.)

Embedded ω-samplers also reliably match to obtained resources under all 4 test conditions (figure 6). Notice, however, that these observations are not reflected in the relatively accessible (in the field) statistic of matching to *available resources.* The significance of this is that if embedded agents are *not* observed to match to available

[5]Careful inspection of fig 5(a,b) reveals that the exclusive choice of ω-sampling under conc VR VR is not quite as exclusive as it is under basic. Although this deviation is slight, it is interesting to note that similar deviations have also been observed empirically (see, for example, Sutherland & Mackintosh, 1971).

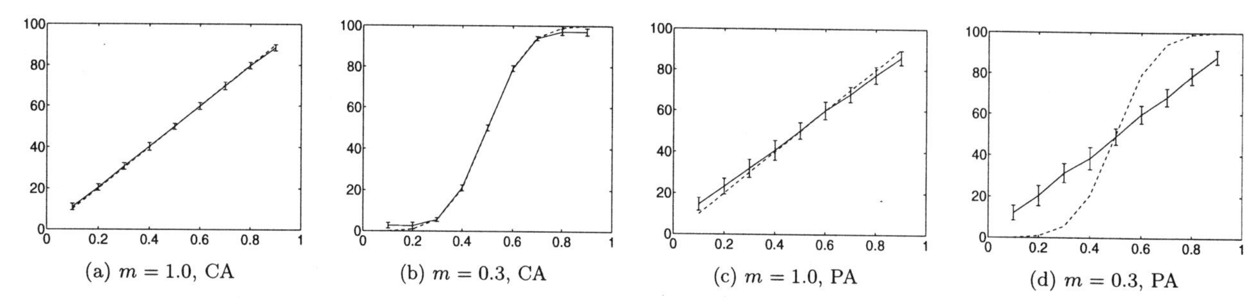
(a) $m = 1.0$, CA (b) $m = 0.3$, CA (c) $m = 1.0$, PA (d) $m = 0.3$, PA

Figure 1: Observed (solid) and predicted (dashed) ϵ-sampling population distributions under 9 different resource distributions (IFD predictions obtained using equation 3). Each observation derives from the mean of 30 distributions, standard deviations are shown. Each abscissa represents F_A and each ordinate represents the percentage of agents on patch A. Four conditions are shown, defined by all combinations of interference level (1.0 or 0.3), and C-allocation or P-allocation (CA or PA).

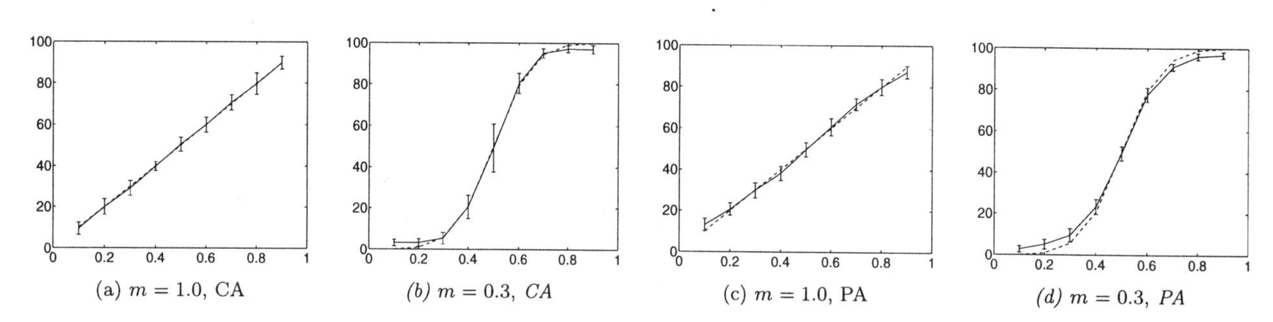
(a) $m = 1.0$, CA (b) $m = 0.3$, CA (c) $m = 1.0$, PA (d) $m = 0.3$, PA

Figure 2: Observed (solid) and predicted (dashed) ω-sampling population distributions under ω-sampling, to be interpreted as figure 1.

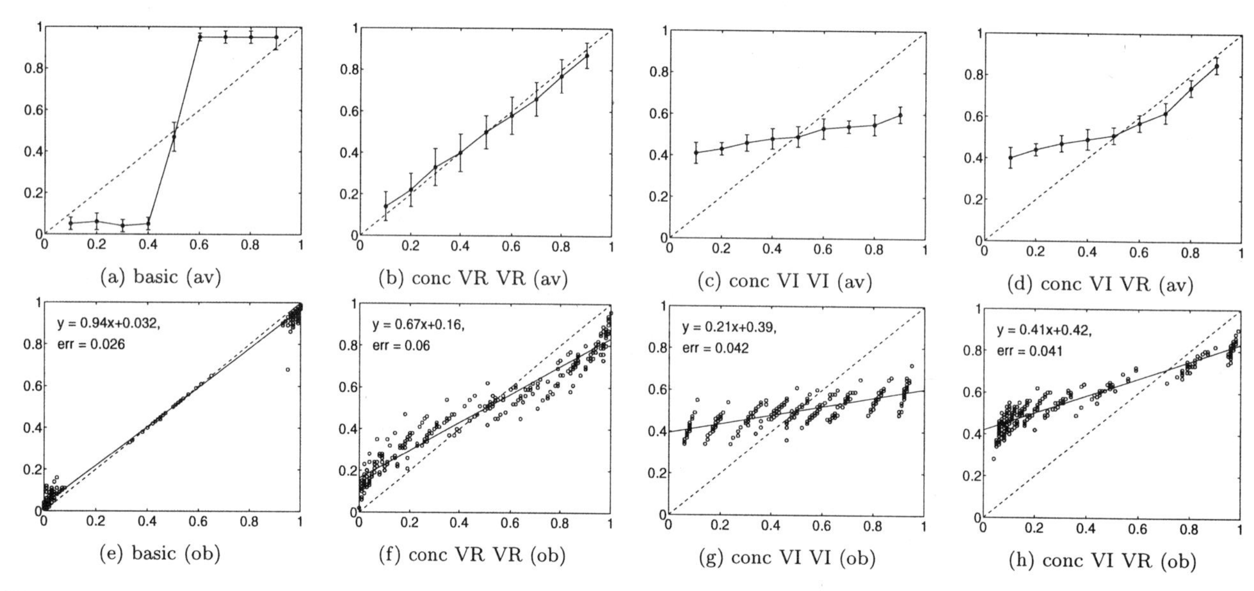

(a) basic (av) (b) conc VR VR (av) (c) conc VI VI (av) (d) conc VI VR (av)

(e) basic (ob) (f) conc VR VR (ob) (g) conc VI VI (ob) (h) conc VI VR (ob)

Figure 3: Matching behaviour of isolated ϵ-sampling agents. Data for each plot is collected from 30 analyses at each of 9 values of F_A, with dashed lines indicating strict matching. Plots labelled (av) concern matching to *available* resources; solid lines show mean proportion of time spent on A (ordinate) as a function of F_A (abscissa), standard deviations are shown. Plots marked (ob) concern matching to *obtained* resources; mean proportion of time spent on A (ordinate) is scatter-plotted as a function of proportion of resources obtained from A (abscissa), with best-fit lines superimposed. The equation of each best-fit line is given together with a measure of goodness-of-fit (this 'error' measure specifies the range around any point on the line that contains at least 50% of the predictions).

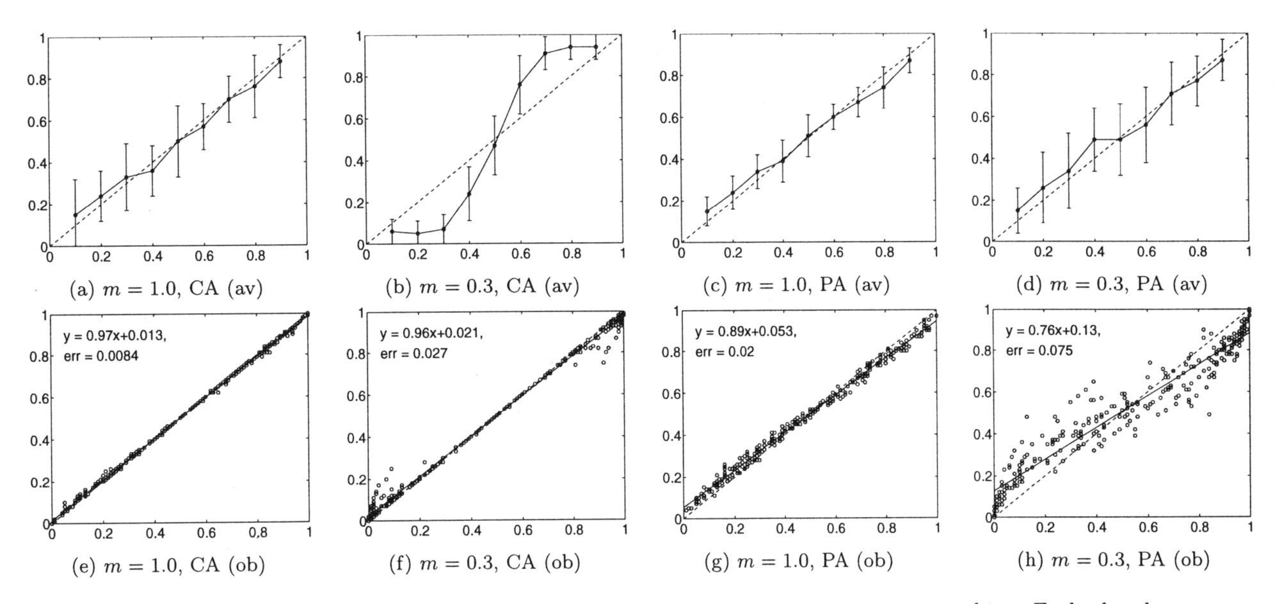

Figure 4: Matching behaviour of embedded ϵ-sampling agents. Dashed lines represent strict matching. Each plot shows mean proportion of time spent on A (ordinate) as a function of resources available (av) or obtained (ob) from A (abscissa), with data collected from 30 analyses at each of 9 values of F_A. Best fit lines are superimposed and equations (with goodness-of-fit) are given as in figure 3.

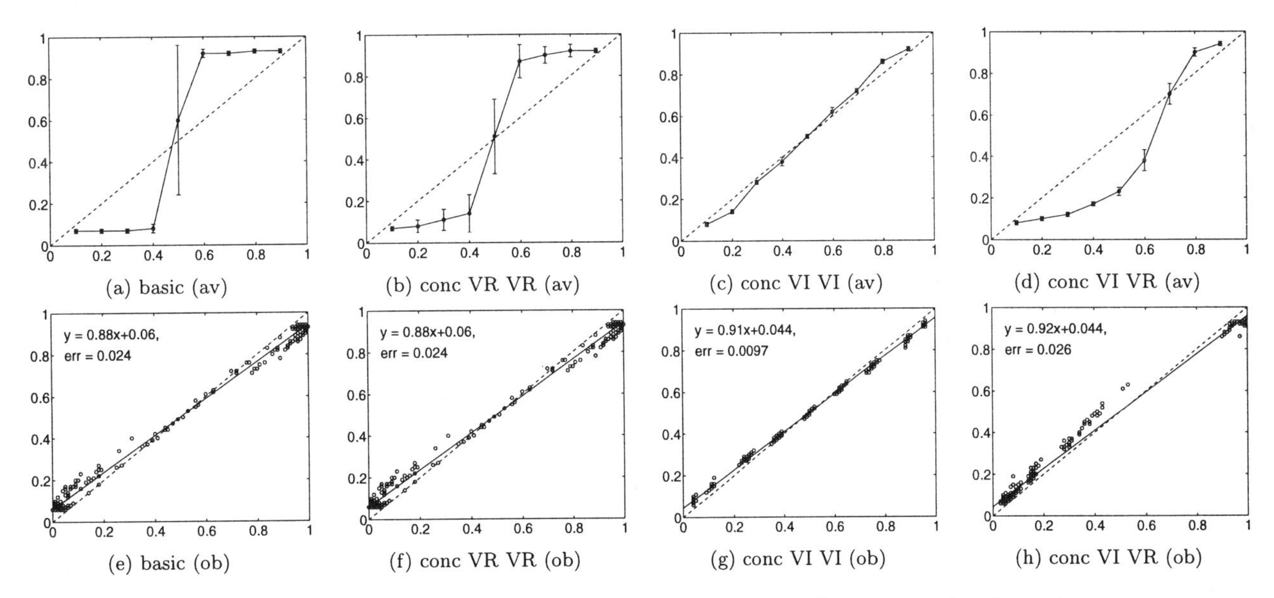

Figure 5: Matching behaviour of isolated ω-sampling agents, to be interpreted as figure 3.

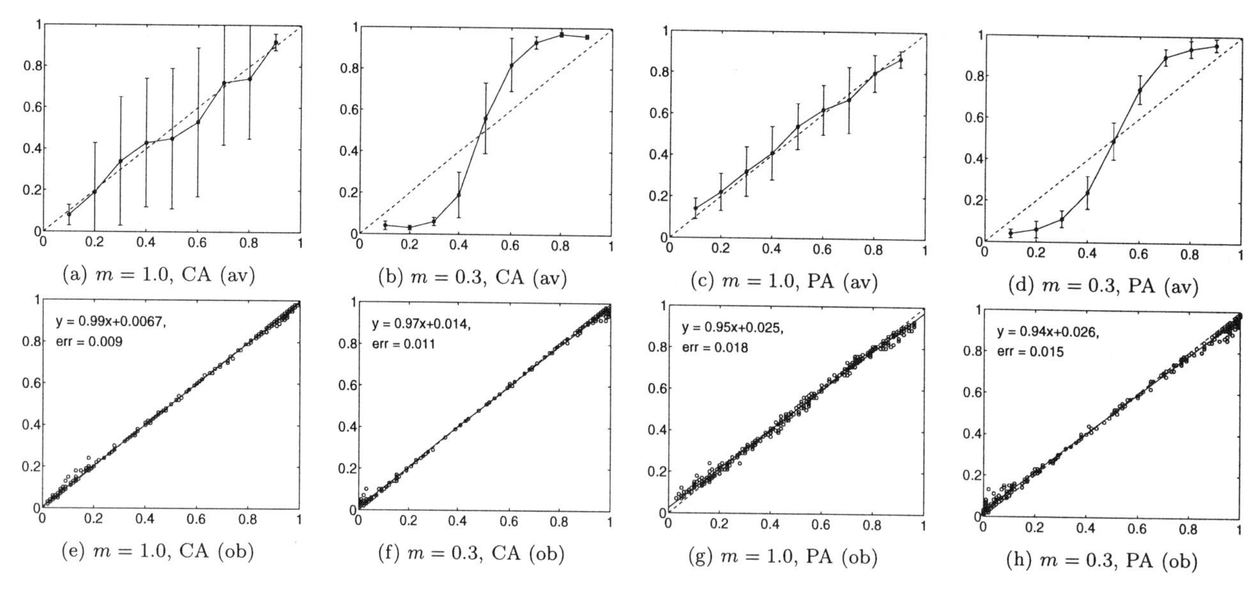

Figure 6: Matching behaviour of embedded ω-sampling agents, to be interpreted as figure 4.

resources, it *cannot* be concluded that isolated agents using the same strategy would fail to match to obtained resources. In other words, matching behaviour in the field may not be a reliable indicator of the performance of isolated individuals with regard to the individual matching law.

As a candidate mechanism underlying both the IFD and individual matching, then, ω-sampling is clearly more successful than ϵ-sampling. This is perhaps unsurprising; by maintaining multiple estimates and employing a more flexible switching rule, ω-sampling is much less likely than ϵ-sampling to be adversely affected by the indeterminacy of P-allocation (with respect to the IFD) or of the conc VR VR, conc VI VI, and conc VI VR schedules (with respect to the matching law), and it is of course in these very conditions that the inadequacies of ϵ-sampling are revealed. One may nonetheless conclude that there do exist simple heuristics capable of underlying both the IFD and individual matching, in several non-trivial situations: ω-sampling is such a strategy, ϵ-sampling is not.

A significant cost of ω-sampling, however, is that it requires concurrent maintenance of more than one value estimate, a cost which would seem rise with environment complexity. Certainly, many 'mechanisms of matching' proposed in the psychology literature use multiple estimates (for example, 'melioration', Herrnstein, 1982, or 'momentary maximisation', Hinson & Staddon, 1983), but is it necessary to assume that agents are able to maintain estimates for *every* patch in their environment? A possible answer is perhaps only to the extent that agents can specify which patch (or option) to choose, which may seem trite, but in fact many patch-switching strategies specify only when to *depart* from a current patch (or option), without specifying where to go afterwards (Bernstein et al., 1991, for example). Consider the IFD. If only random movement is possible, and all areas of the environment can be accessed with equal ease, then it would only be necessary to maintain a single estimate (to prompt departure). If directed (non-random) movement is possible to any part of the environment (again with equal ease) then concurrent estimates of every patch might be valuable. However, if it is assumed that movement is somewhat restricted, but non-random, then some intermediate solution is likely to be best, at which the agent only maintains a few functionally relevant estimates.

6 Discussion and summary

Perhaps the single most important result of those presented above is that ω-sampling entails matching in at least one situation in which such behaviour is suboptimal (conc VI VR). This supports the idea that some instances of individual suboptimal behaviour can be understood in terms of the operation of mechanisms adapted to a group context; ω-sampling can be considered *ecologically rational.* Of course, this does not imply that *all* irrational behaviour must be explained this way, and nor are alternative explanations of matching necessarily excluded (Niv, Joel, Meilijson, & Ruppin, 2001, for example), we have simply demonstrated that the present account is at least plausible, and perhaps parsimonious.

In a previous paper this same idea was explored in the context of foraging within a single patch containing both rich and poor resources (Seth, 1999, see also Seth, in press), and a similar result was uncovered: agents evolved in a group context displayed matching behaviour when assessed in isolation, but those evolved in isolation did not. Compared to the present model, this study was minimal in some ways - no distinct patches, no complex reinforcement schedules, no learning or lifetime adaptation - but rich in others, particularly in the extent to which it captured sensorimotor interactions between agents and their immediate environment, a level of description the present model abstracts away from. Because of these differences, the character of the matching behaviour observed in the two models is also different. In the single patch case, matching was accounted for purely in terms of sensorimotor interaction patterns, to do with inter-agent interference, that acted as historical constraints (Di Paolo, 2001), and was identified only qualitatively. Here, matching has been accounted for in terms of the IFD, and has been identified very closely with its description in the literature. Taken together, these studies light the way to a range of matching phenomena, some of which attach to distinct, separable, and spatially arbitrary choices, others to single patches and basic sensorimotor interaction patterns, and still others to intermediate levels of description, in which sensorimotor interactions may engender patch-switching in the generation of choice behaviour. It remains to be seen what further insights into matching behaviour, and decision making in general, can be attained by modelling at these levels.

A final comment concerns the familiar problem of 'action selection', as it is understood in the adaptive behaviour literature (Tyrrell, 1993). An obvious difference between action selection and matching is that the former normally analyses choices that satisfy *distinct* requirements (for example feeding and sleeping), and the latter concerns different ways of satisfying the *same* requirement. Yet the concepts are clearly very close, and no satisfying and general account of decision making can afford to ignore either. To risk belabouring a commonly made point, the psychological literature contains a wealth of conceptual and empirical resources that relatively novel methodologies directed towards understanding adaptive behaviour, for example agent-based modelling, would do well to consider.

In summary, this paper has assessed two simple heuristics, ϵ-sampling and ω-sampling, on their ability to underlie both matching behaviour and the IFD in a number of non-trivial conditions. Contrary to the claims of Thuisjman et al. (1995), ϵ-sampling proved inadequate at both; ω-sampling, by contrast, was successful, even to the extent of entailing matching under conditions in which such behaviour is suboptimal. ω-sampling can therefore be considered an ecologically rational candidate for a 'mechanism of matching', and from this vantage matching behaviour - and its potential suboptimality - can be interpreted in terms of decision making adapted to a group environment. The extent to which other forms of irrational behaviour can be understood in this way, or in similar ways, remains an interesting and open question. Finally, a novel method of parameter setting has been described which may find useful application in other models of this kind.

	1,CA	.3,CA	1,PA	.3,PA
ϵ-samp.,ϵ	.038	.037	.039	.038
ϵ-samp.,γ	.071	.069	.038	.030
ω-samp.,ϵ	.061	.040	.105	.040
ω-samp.,γ	.021	.065	.022	0.030

Table 1: Evolved parameters. Columns labelled by interference level m and allocation method.

Acknowledgements

I am grateful to the CCNR, the Neurosciences Research Foundation, and to my anonymous reviewers, for a combination of financial support and constructive discussion. The material in this paper is drawn in part from my D.Phil. thesis (Seth, 2000).

Appendix A

For both ϵ- and ω-sampling, initially random populations (size 100) were evolved in each of 4 cases ($m = 1.0$ or 0.3, C-allocation 'CA' or P-allocation 'PA'): 8 conditions in total. Each agent (in each condition) possessed a genome of 2 real numbers (range [0.0,1.0]) specifying ϵ and γ. Each condition applied a tournament GA for 100 generations (mutation rate 0.01, each mutation drawn from Gaussian distribution radius 0.13; range transgressions were truncated). Fitness was averaged over 10 separate evaluations. Each evaluation randomly assigned values for F_A and F_B ($F_A + F_B = 1.0$, total resource $F^* = 200.0$ in all conditions), and randomly allocated agents between A and B. Fitness of each agent was determined by total accumulated resources after 1000 cycles. The entire GA process was repeated 10 times in each condition, from which average (condition-specific) near-optimal parameter values were recovered (table 1). Analysis of each condition used these values. Note that analysis of isolated individual matching utilised the average near-optimal parameters across *all* conditions ($\bar{\epsilon} = 0.052, \bar{\gamma} = 0.427$), since no corresponding populations were evolved in these conditions.

References

Baum, W. (1974). On two types of deviation from the matching law: Bias and undermatching. *Journal of the Experimental Analysis of Behavior, 22*, 231–242.

Baum, W., & Kraft, J. (1998). Group choice: Competition, travel, and the ideal free distribution. *Journal of the Experimental Analysis of Behavior, 69*(3), 227–245.

Bernstein, C., Kacelnik, A., & Krebs, J. (1991). Individual decisions and the distribution of predators in a patchy environment II: The influence of travel costs and structure of the environment. *Journal of Animal Ecology, 60*, 205–225.

Bullock, S., & Todd, P. (1999). Made to measure: Ecological rationality in structured environments. *Minds and Machines, 9*(4), 497–541.

Charnov, E. (1976). Optimal foraging: The marginal value theorem. *Theoretical Population Biology, 9*, 129–136.

Conger, R., & Killeen, P. (1974). Use of concurrent operants in small group research. *Pacific Sociological Review, 17*, 399–416.

Dallery, J., & Baum, W. (1991). The functional equivalence of operant behavior and foraging. *Animal Learning and Behavior, 19*(2), 146–152.

Davison, M., & McCarthy, D. (1988). *The matching law.* Erlbaum, Hillsdale, NJ.

Di Paolo, E. (2001). Artificial life and historical processes. In Kelemen, J., & Sosik, P. (Eds.), *Proceedings of the Sixth European Conference on Artificial Life*, pp. 649–658. Springer-Verlag.

Fagen, R. (1987). A generalized habitat matching law. *Evolutionary Ecology, 1*, 5–10.

Fretwell, S. (1972). *Populations in seasonal environments.* Princeton University Press, Princeton.

Gallistel, C., Brown, A., Carey, S., Gelman, R., & Keil, F. (1991). Lessons from animal learning for the study of cognitive development. In Carey, S., & Gelman, R. (Eds.), *The epigenesis of mind*, pp. 3–37. Lawrence Erlbaum, Hillsdale, NJ.

Goss-Custard, J. (1980). Competition for food and interference amongst waders. *Ardea, 68*, 31–52.

Gray, R. (1994). Sparrows, matching, and the ideal free distribution: Can biological and psychological approaches be synthesised?. *Animal Behaviour, 48*, 411–423.

Herrnstein, R. (1961). Relative and absolute strength of responses as a function of frequency of reinforcement. *Journal of the Experimental Analysis of Behavior, 4*, 267–272.

Herrnstein, R. (1970). On the law of effect. *Journal of the Experimental Analysis of Behavior, 13*(2), 243–266.

Herrnstein, R. (1982). Melioration as behavioural dynamism. In Commons, M., Herrnstein, R., & Rachlin, H. (Eds.), *Quantitative analyses of behavior, vol II: Matching and maximizing accounts*, pp. 433–458. Ballinger Publishing Co., Cambridge, MA.

Herrnstein, R. (1997). *The matching law.* Harvard University Press, Cambridge, MA. A posthumous collection of the papers of R.J. Herrnstein, edited by H. Rachlin and D.I. Laibson.

Herrnstein, R., & Heyman, G. (1979). Is matching compatible with reinforcement maximization on concurrent variable interval, variable ratio?. *Journal of the Experimental Analysis of Behavior, 31*, 209–223.

Herrnstein, R., & Vaughan, Jr., W. (1980). Melioration and behavioral allocation. In Staddon, J. (Ed.), *Limits to action: The allocation of individual behavior.* Academic Press, New York.

Hinson, J., & Staddon, J. (1983). Hill-climbing by pigeons. *Journal of the Experimental Analysis of Behavior, 39*, 25–47.

Houston, A. (1986). The matching law applies to wagtails' foraging in the wild. *Journal of the Experimental Analysis of Behavior, 45*, 15–18.

Krebs, J., & Kacelnik, A. (1991). Decision making. In Krebs, J., & Davies, N. (Eds.), *Behavioural ecology*, pp. 105–137. Blackwell Scientific Publishers, Oxford. 3rd edition.

Milinski, M., & Parker, G. (1991). Competition for resources. In Krebs, J., & Davies, N. (Eds.), *Behavioural ecology*, pp. 137–168. Blackwell Scientific Publishers, Oxford. 3rd edition.

Myers, J. (1976). Probability learning and sequence learning. In Estes, W. (Ed.), *Handbook of learning and cognitive processes*, Vol. 1, pp. 171–205. Erlbaum, Hillsdale, NJ.

Niv, Y., Joel, D., Meilijson, I., & Ruppin, E. (2001). Evolution of reinforcement learning in uncertain environments: emergence of risk aversion and matching. In Kelemen, J., & Sosik, P. (Eds.), *Proceedings of the Sixth European Conference on Artificial Life*, pp. 252–261. Springer-Verlag.

Seth, A. (1999). Evolving behavioural choice: An investigation of Herrnstein's matching law. In Floreano, D., Nicoud, J.-D., & Mondada, F. (Eds.), *Proceedings of the Fifth European Conference on Artificial Life*, pp. 225–236. Springer-Verlag.

Seth, A. (2000). *On the relations between behaviour, mechanism, and environment: Explorations in artificial evolution.* Ph.D. thesis, University of Sussex.

Seth, A. (in press). Modelling group foraging: individual suboptimality, interference, and matching. *Adaptive Behavior.*

Shettleworth, S. (1988). Foraging as operant behavior and operant behavior as foraging: What have we learned?. In Bower, G. (Ed.), *The psychology of learning and motivation: Advances in research and theory*, Vol. 22, pp. 1–49. Academic Press, New York.

Sutherland, N., & Mackintosh, N. (1971). *Mechanisms of animal discrimination learning.* Academic Press, New York.

Sutherland, W. (1983). Aggregation and the 'ideal free' distribution. *Journal of Animal Ecology, 52*, 821–828.

Thuisjman, F., Peleg, B., Amitai, M., & Shmida, A. (1995). Automata, matching, and foraging behaviour of bees. *Journal of Theoretical Biology, 175*, 305–316.

Tregenza, T., Parker, G., & Thompson, D. (1996). Interference and the ideal free distribution: Models and tests. *Behavioral Ecology, 7*(4), 379–386.

Tyrrell, T. (1993). *Computational mechanisms for action selection.* Ph.D. thesis, University of Edinburgh.

Van der Meer, J., & Ens, B. (1997). Models of interference and their consequences for the spatial distribution of ideal and free predators. *Journal of Animal Ecology, 66*, 846–858.

Multi-object Segregation: Ant-like Brood Sorting Using Minimalist Robots

Matt Wilson*
* Intelligent Autonomous Systems Laboratory, Faculty of Computing, Engineering and Mathematical Sciences, University of the West of England
matthew.wilson@uwe.ac.uk

Chris Melhuish*†
†chris.melhuish@uwe.ac.uk

Ana Sendova-Franks*‡
‡Ana.sendova-franks@uwe.ac.uk

Abstract

This study shows that a task as complicated as multi-object annular segregation can be accomplished by robots using a 'minimalist' solution employing a simple adaptive mechanism and minimal hardware. The success of the mechanism, which employs a combined leaky integrator, is demonstrated both in simulation and using real robots.

1. Introduction

The study is inspired by the brood sorting behaviour of *Leptothorax unifaciatus* ants; a species that have evolved to live in between the cracks of rocks. The ants sort their brood so that *"different brood stages are arranged in concentric rings in a single cluster around the eggs and micro-larvae"* (Franks and Sendova-Franks, 1992).

The mechanism described within this paper enables the sorting of any number of objects into an annular pattern. Such annular sorting is defined by Melhuish *et al.* (1998) as *"forming a central cluster of one class of objects, and surrounding it with annular bands of the other classes, each band containing objects of only one type."*

2. Method

This paper presents the results of experiments conducted in a simulation and in a real multi-robot arena. The objects used for sorting are frisbees and different coloured frisbees represent different types of object. Photographs and full descriptions of the real environment, robots and frisbees can be found in Melhuish et al. (1998). In brief, the robots operate in an octagonal arena containing shaded frisbees. They are 23 cm in diameter and are easily portable. Infrared sensors underneath the robots enable them to distinguish between the different shades of frisbee contained within the arena, while sensors on the front and back report collisions that occur with other robots and the sides of the arena.

3. Results

The experiments conducted for this study develop and explore the success 'ant-like brood sorting' algorithm. Six robots and fifteen of each type of frisbee were used throughout the experiments.

Our mechanism involves robots moving in straight lines within an arena. Whenever their scoop is depressed, the robots reverse and turn though a random angle. The robots are able to push a single object (frisbee) in the direction of their motion. If a robot is carrying a frisbee and it collides with another frisbee, the frisbee currently being carried is pulled back a variable distance and deposited. The robot then reverses a little further, turns through a random angle before moving forward again.

Each individual robot holds a separate counter for each type of object. These counters can be seen as analogous to containers (leaky integrators) that contain liquid. Initially the containers are half full. Every time a robot drops an object (unless the container is full) an amount of liquid is added to the container that corresponds to the object type being dropped. Over time the amount of liquid in each container drains away.

The pull back distances are adaptive with respect to the structure within the frisbees and based on the amount of liquid in each of the integrators: the black frisbees are not pulled back at all; the grey frisbees are pulled back a distance based on the sum of the liquid in the grey and the black integrators; and the white frisbees are pulled back a

distance based on the sum of the liquid in the black, grey and white integrators.
The combined leaky integrator mechanism works in the following way:

At the start of a run, the robots are likely to see many frisbees of each colour because all the colours are randomly dispersed. Therefore, there will be a tendency for all of the three integrators to contain a large amount of liquid causing the pullback distances for the grey and white frisbees to be large. This allows a central cluster of black frisbees to form easily. As the black cluster forms, the frequency with which the robots carry black frisbees decreases and therefore there will tend to be less liquid in

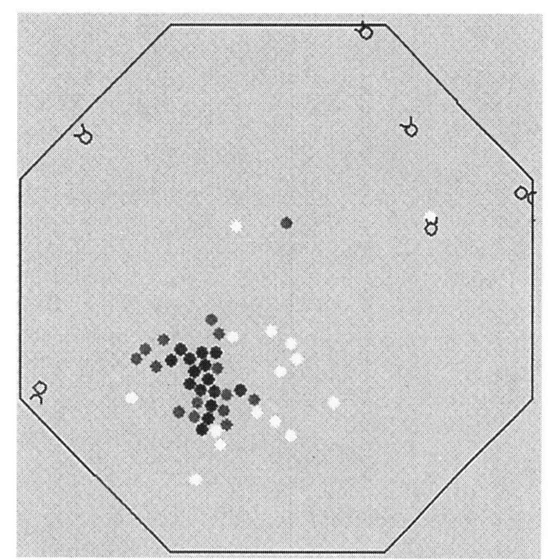

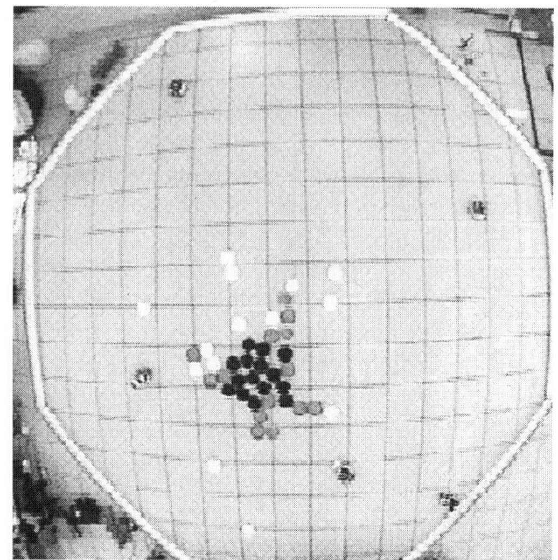

Figure 1: a) Top - simulated arena after 800,000 iterations
b) Bottom – real arena after 6 hours sorting.

the black integrator, reducing the pullback distance for grey and white frisbees. This reduction in pullback distance draws the grey frisbees closer to the central black cluster, which in turn has a tendency to reduce the amount of liquid in the grey integrator. With the black and grey frisbees drawn close to the structure, the pullback distance for white frisbees is reduced and these frisbees are drawn closer to the structure, creating the annular band effect.

The above mechanism was first tested using 3 different coloured frisbees. One hundred runs were carried out in simulation and the results were validated by comparison with 3 real robot runs. Figure 1a shows a simulated run after 800, 000 iterations, which is equivalent to the 6 hours point of a real robot run shown in figure 1b. The separation of object types is clearly visible in both figure 1a and figure 1b. Our mechanism was able a produce separation of types within an annular structures that, while not perfect, is at least as good as the separation of types observed in the structures within real ants nest.

Having shown the mechanism to be successful with 3 types of object, the scalability of the mechanism as the number of different types of objects was increased to four and five different object types was explored. The mechanism was shown to be able to cope well, without a need for an increase in area size.

4. Discussion

We are currently conducting further research into the brood sorting behaviour of *Leptothorax* ants. Whether the mechanism the ants use to sort their brood is not known and at present we have no evidence to suggest this mechanism is based on a combined leaky integrator. It is hoped that a comparison between the results of this novel sorting mechanism and the behaviour of real ants will inspire further biological and robotic research.

In conclusion, a task as complicated as multi-object annular segregation has been accomplished with a 'minimalist' solution employing a simple adaptive mechanism and minimal hardware.

References

Beckers R., Holland O., Deneubourg J-L. (1994). From Local Actions to Global Tasks: Stigmergy & Collective Robots, *Proc. of the 4th International Conference on Artificial Life*

Franks N.R., Sendova-Franks A.B. (1992), Brood sorting by ants: distributing the workload over the work-surface, *Behav. Ecol. Sociobiol. 30 109-123*

Melhuish C., Holland O., Hoddell S (1998). Collective Sorting and Segregation in robots with Minimal Sensing, *5th International Conference on the Simulation of Adaptive Behaviour. Zurich. From Animals to Animats. MIT Press.*

Melhuish C., Wilson M., Sendova-Franks A. (2001). Patch Sorting: Multi-object Clustering using Minimalist Robots, *European Conference on Artificial Life*

Building adaptive structure formations with decentralised control and coordination

Jan Wessnitzer **Chris Melhuish**
Intelligent Autonomous Systems Laboratory
University of the West of England
Coldharbour Lane
Bristol BS16 1QY
United Kingdom

Abstract

We investigate collective decentralised decision-making in simple mobile agents engaged in the task of structure formation where the radius of the formed structure is much larger than the communication radius of an individual agent. We are searching for minimal and local rules allowing adaptive behaviour in collectives of mobile agents. In previous experiments, we presented rules that govern the evolution of the inter-agent communication network toward some task-specific topologies that allow a decentralised collective of mobile agents to dynamically organise into simple geometric structures. In this paper, we study the problem of coordinating and controlling the global behaviour of the group of agents from their local interactions. The agents determine the successful termination of a structure formation and switch to another formation. In simulation we demonstrate global state changes and global state persistance in a decentralised collective of mobile agents.

1 Introduction

The task of formation control is inherent to multi-robot systems and is also useful for larger tasks, such as moving a large object by a group of robots or distributed sensing (Cao et al., 1995). The literature on formation control presents a variety of approaches to achieve coherent global behaviour, for example (Balch and Arkin, 1999) (Fredslund and Mataric, 2001). None of the reviewed papers on robot formations considers the scenario where the formation radius is much larger than the communication radius of an individual robot. What are the salient ingredients required in the design of collective and cooperative behaviour? We assume that a few internal states, primitive sensors and local communication could form a necessary base for collective robot behaviour.

2 Agent-based model

Our simulation is an implementation of a mobile automata network. We simulate a homogeneous, decentralised multi-robot system capable of explicit communication between peer elements (Wessnitzer et al., 2001). A mobile automata network is a set of elementary processing units, where every unit interacts with neighbouring units and moves in a two-dimensional Euclidean space. The number of neighbours of any processor is several orders less than the total number of processors in the network. Each unit makes a decision based on its internal states, the internal states of its immediate neighbours and its local sensor readings. A configuration of the internal states and local sensor readings is then translated to a propulsive motion of the agents. The units self-organise and dynamically assign themselves the positions in the structure to be formed. On recruitment the agents set an internal state which is used by the agents to determine what formation relative to their neighbours should be taken up. This knowledge is hard coded in the form of a look-up table.

3 Experimental results

At the start of each experiment all agents are randomly placed into an enclosed environment. No agent has a predefined position in any of the formations but instead its role is determined dynamically at run-time. Using the model described in the previous section and simple, local rules a number of structures have been achieved. Examples include square, triangle and hexagon formations, etc. The study has also looked into symmetrical structures, global state changes and global state persistance.

Figure 1 shows the last stages in the formation of a hexagon structure and the switch from this hexagon structure to a square structure. The lines between agents represent the communication links. In the third frame of figure 1 the hexagon formation is almost complete and in the following frame the agents start to move into position in order to form a square structure.

No agent can by itself know whether the structure has been successfully formed. Due to the locality of the sensor information and the fact that the agents only actively communicate with their nearest neighbours, the agents can only determine whether they are in formation relative to their neighbours. Rules have been designed that diffuse the information of how many agents are in formation relative to their neighbours. If this number equals the number of agents recruited into building the formation, then a signal is diffused through the network causing the robots to form a different structure.

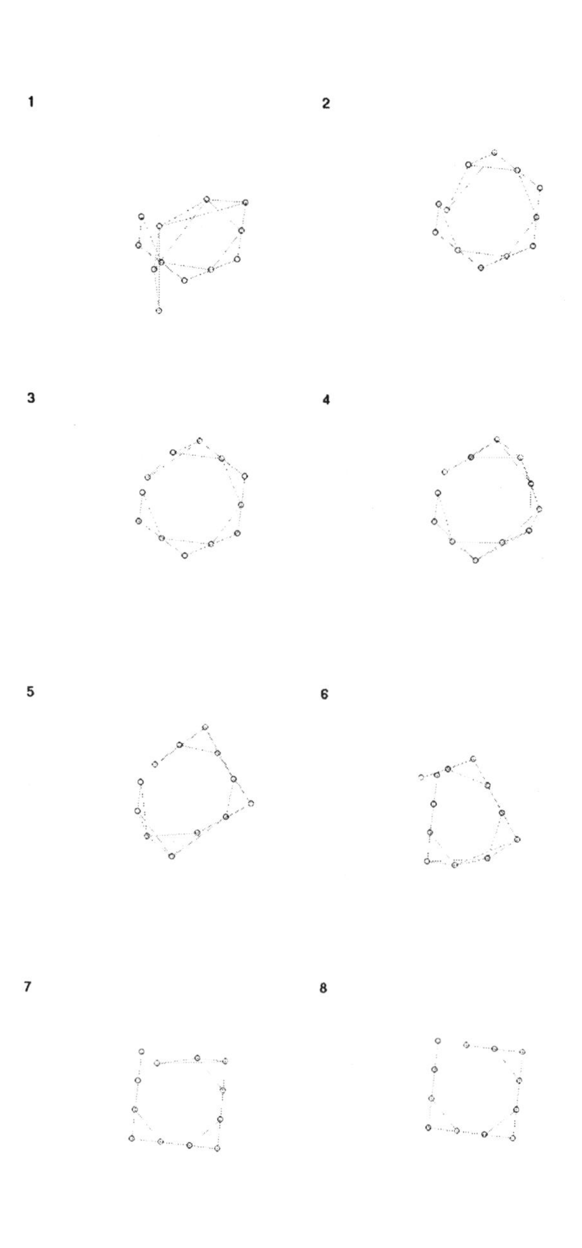

Figure 1: Global state change : from hexagon to square formation

4 Discussion

Adaptive behaviour in a decentralised collective of mobile agents was demonstrated as a formation switch. In all our experiments we assumed the agents and the inter-agent communication to be reliable. Robustness to failing agents and broken communication links must be achieved by designing rules capable of dealing with such events. The ultimate goal of this study is to implement the developed algorithms on real robots. The LinuxBots are equipped with a PC-compatible CPU and wireless local area network technology. The TCP/IP protocols are an integral part of the Linux operating system and provide a robust and reliable data connection. Using these protocols each robot becomes a node on a wireless local area network. Preliminary real-robot experiments validated some of the results presented in (Wessnitzer et al., 2001).

Acknowledgements

We acknowledge the support of the EPSRC grant GR/R79319/01.

References

Balch, T. and Arkin, C. (1999). Behavior-based formation control for multi-robot teams. *IEEE Transactions on Robotics and Automation.*

Cao, Fukunaga, and Kahng (1995). Cooperative mobile robotics: Antecedents and directions. Technical Report 950049.

Fredslund, J. and Mataric, M. (2001). Robot formations using only local sensing and control. *Proceedings of IEEE International Symposium on computational intelligence for Robotics and Automation (CIRA).*

Wessnitzer, Adamatzky, and Melhuish (2001). Towards self-organising structure formations: a decentralised approach. *Proceedings of ECAL 2001.*

Minimalist Coherent Swarming of Wireless Networked Autonomous Mobile Robots

Julien Nembrini **Alan Winfield** **Chris Melhuish**
Intelligent Autonomous Systems Laboratory
University of the West of England
Coldharbour Lane
Bristol, United Kingdom

Abstract

This paper presents the results of simulations investigating a decentralised control algorithm able to maintain the coherence of a swarm of radio-connected robots. In this work the radius of communication is considerably less than the global diameter of the swarm. The study explores coherent movement towards a beacon while avoiding obstacles and maintaining global shape. All behaviours are emergent as the study constrains itself to using only a restricted range omnidirectional radio, a beacon sensor and avoiding sensors. In spite of such restrictions we achieve the proposed aims. Moreover, the fact that the algorithm relies only on local information makes it highly scalable with an increase in the number of robots. The proposed approach also shows graceful degradation in the presence of noise.

1. Introduction

Bacteria and Amoebae are impressively efficient in surviving in their respective niche and such an accomplishment has never been reached by robotic artifacts. Symbolic A.I. has shown its limitations in real robot experiments (Pfeifer and Scheier, 1999), and in contrast behaviour-based robotics has put the emphasis on the relationships of an agent with its environment. It is refered to as *situatedness*. Moreover some researchers argue that the physical implementation of the robot, its *embodiment*, is essential not to bypass assumptions about the real world (Brooks, 1991). This study is conducted within this framework.

Situatedness is now a widely used concept in the robotics field. It is commonly agreed that an intelligent agent cannot be considered apart from the environment with which it is supposed to interact. This has focused research on the morphology of the agent as crucial for its interaction with the environment. It has been shown that a specific morphology adapted to a specific environment can greatly enhance the robot's abilities without an increase in the agent's computational complexity (Pfeifer and Scheier, 1999, Melhuish et al., 1998). Artificial evolution of morphology has also been investigated, mainly in simulation, and demonstrated some impressive results (Lipson and Pollack, 2000, Sims, 1994).

In these examples the change in morphology, either man-made or automatic, made the robot more efficient in a defined task and hence more adapted to its environment. This success raised the idea of a robot able to modify its own shape while executing a task. Such a skill would increase the adaptability of the robot to changing environments.

For a robot to be able to conduct this metamorphosis, it could be composed of smaller components that can rearrange into different shapes. A good deal of research is now investigating reconfigurable robots that consist of basic modules able to perform simple tasks such as attaching with, detaching from or moving relative to a neighbour. Tasks such as locomotion of the whole group is the result of a combination of these basic actions (Kotay and Rus, 1999).

This research extends the work of Winfield (Winfield, 2000) related to sensor networks, as well as the work of Melhuish (Melhuish, 1999) that focused on secondary swarming. This study looks at the employment of mobile robots to combine sensing, locomotion and morphological adaptivity. Following this preliminary direction of research, we propose here to study a swarm of autonomous mobile robots communicating with limited-range radios. Instead of a physically connected system, this approach investigates virtually (wireless) connected robots.We believe that this approach brings greater versatility in morphology and robustness to failures and noise. Also the hardware involved is readily available, allowing us to concentrate on the development of the algorithms. Of course the drawback lies in the fact that much effort has to be made to keep the robots together, *i.e.* to maintain the *coherence* of the swarm.

Working initially in simulation, we show that using only an omni-directional radio and a collision avoidance device is enough to achieve this coherence in an unbounded environment. The swarm then forms a one-connected component communicating network. Adding a sensing ability, we achieve emergent directed swarming towards a light beacon, and present the potential to control the swarm global shape .

In this research we are primarily concerned with developing algorithms that are:

scalable: An increase in the number of agents in the swarm should not lead to complete failure. Indeed the efficiency should show graceful degradation to such an increase. The study is therefore relying on distributed solutions that make use of strictly local information. Our robots are all identical and exchange information only about their neighbourhoods.

robust: The algorithm should provide the swarm with the ability to cope with the unreliability of the real world and show graceful degradation when confronted with noise and component failures. The last requirement is again a reason to seek distributed algorithms controlling homogeneous robots.

2. Related Work

As stated above, the main area of research concerned with online morphology change is the field of reconfigurable robots. The motivation is to achieve function from shape, allowing individual modules to connect and reconnect in various ways in order to achieve the required function.

Because of the technical challenge it represents, recent work has emphasized the design of physical robots. But the control algorithms have as yet mainly been investigated in simulation and the focus appears to be on state representation and planning (Kotay and Rus, 1999, Yoshida et al., 2000) instead of distributed algorithms. An interesting example of distributed control is the recent work of Støy that achieves caterpillar locomotion and multi-legged walking with the use of a fully distributed algorithm that is able to deal with online reconfiguration (Støy et al., 2002).

Our work is concerned with groups of robots that are more loosely connected. While in the field of reconfigurable robots, individuals are normally physically linked together, our approach considers limited-range radio connections that build a dynamic wireless network. Connections are unreliable and often lost. An interesting example of group pattern formation with the use of wireless connection can be found in (Wessnitzer et al., 2001)

Another field related to the work presented here is the study of formation control. The aim is to move a group of robots while maintaining relative position, hence forming a global shape such as, for instance, migrating birds. The work of Balch and Hybinette (Balch and Hybinette,) for instance relies on distributed solutions and shows desirable properties such as scalability and locality. But trying to maintain accurate relative positions involves high-level sensing abilities. On the other hand the control algorithm described in this paper allows some freedom of movement for an individual while maintaining the global shape of the swarm.

Although biomimetics was not the aim at the beginning of our project, it appears that the moving swarm shows comparable properties with amoeboid plasmodium such as *Physarum* or *Dictyostelium*. Interesting work in the field of robotics that follows this idea can be found in (Takahashi et al., 2001).

In the work most closely related to this research, Støy (Støy, 2001) studies real robots trying to stay together by sending messages to others. The approach uses the limited range as a physical advantage: it enables a robot to count how many others are in range and hence get an idea of its neighbourhood density. The control algorithm then makes the robot turn and come back if the number of neighbours is decreasing. The aim is to find whether it can help robots keep together within a bounded arena. The result is positive, but the robots sometimes lose the group and are able to rejoin only by chance thanks to the bounded arena.

Our aim was to remove the constraint of a bounded arena. We have therefore looked for algorithms that conserve the integriy of the group, since the lack of absolute or relative position information in case of deconnection would make the loss of a robot critical. However, our concern for the scalability of the process leads us to use a similar idea for sharing neighbour information.

3. Methods

The research follows a method of investigation that consist of, firstly, designing and implementing the algorithm in simulation to gain insights into the dynamics involved and the problems that may arise. It must be borne in mind that the results depend highly on the degree of accuracy of the simulation. This is the reason that the resulting algorithms must, secondly, be implemented on real robots, while taking full advantage of the simulation work. This paper presents the results of the first stage only, as we are yet to verify our algorithm on real robots.

Following the focus on minimalist design of Melhuish (Melhuish et al., 1998) the aim is to keep the robots as simple as possible. It is believed that coherence is achievable only with a radio device with limited range for communication and proximity sensors for avoidance. The key idea is that the limited range is giving sufficient information on relative position. Such severely constrained conditions oblige us to tie together the act of communicating with other behaviours of the robot. It is referred in (Støy, 2001) as *situated communication.*

These assumptions on hardware make the requirement on coherence even more critical, as a disconnected robot will not be able to return to the swarm, as it lacks any information about relative or absolute position.

Due to the underlying motivation on sensory network applications we also seek to minimise the communication overhead that is necessary to achieve coherence. This is to retain maximum bandwidth for data gathering.

At the same time, our requirement on robustness in the presence of noise is the reason for the solution presented to the problems of directed swarming and global shape control, that both emerge from spatial differentiation.

3.1 Simulation Details

As developing a simulation that tries to take every interaction that contributes to the dynamics into account would be impractical, a simple approach is implemented, bearing in mind the aim of physical realisation to avoid unfeasible solutions. As a consequence noise is not simulated following real models, but instead introduced as false sensor readings or loss of messages in a random manner; the purpose being to test the robustness of the algorithm.

We assume the robots are able to move forward and turn on-the-spot with reasonable precision, that they have infra-red avoidance sensors, are equipped with limited-range radio devices and that they carry an omnidirectional light sensor able to detect whether a robot is illuminated or not.

Communication is implemented as follow: two robots in range are considered to communicate perfectly. Noise is introduced using a constant probability to lose the entire message. This probability can range from zero to 0.1 and we do not attempt to model buffer overflow or any other real phenomena occuring such as signal decay. The motivation for such assumptions can be found in (Winfield, 2000).

In addition to wireless communication an avoidance behaviour is implemented that makes use of three infrared sensors, one on each side of the front of the robot, and the third at the back. All sensors have the same range, smaller than the range of the communication device. The avoidance behaviour causes the robot to turn in the opposite direction of a single activated front sensor, move backwards if both front sensors are activated, or stop if all three sensors are activated. There is also an option to introduce random noise as a probability of false sensor readings to this feature.

In order to study taxis and adaptable morphology, a beacon is introduced as well as its decay and occlusion through line-of-sight obstruction. Again the design was dictated by its possible realisation, the metaphor being a bright light beacon. Noise is also introduced with probabilistic false sensor readings. The simulation also includes the option to introduce into the arena some occlusive obstacles in the path of the swarm.

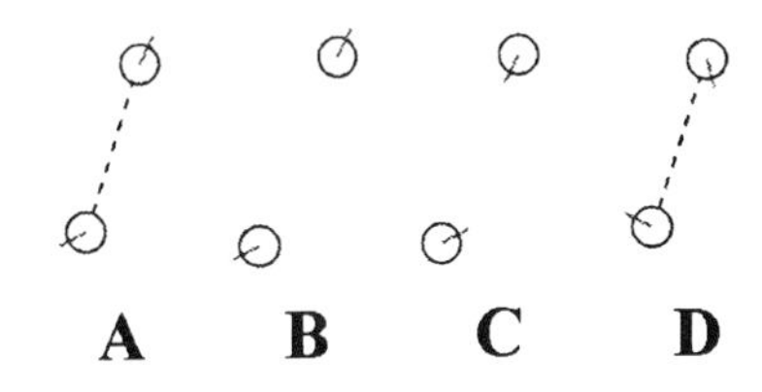

Figure 1: Basic algorithm

3.2 Algorithms

Basic Algorithm

This study restricts itself to use only the information on connections between robots. In other words whether a particular robot is receiving a signal from another or not. The omnidirectionality of the radio device implies that there is no positional indication about where to go in case of deconnection. To transform this limited information into good use we exploit the ability of the robot to turn on the spot with good precision: in the default state the robot moves forward. As soon as the control algorithm detects a deconnection, the robot assumes it is going in the wrong direction and turns back.

For simplicity let us restrict ourselves to the case of two robots (see figure 1):

Assume that the robots are initially in communication range, moving forward with random headings (A). Unless they have parallel or crossing trajectories[1], they will eventually lose contact(B). In order to check whether this is the case or not, the algorithm uses a call-answer mechanism: with a certain periodicity each robot sends a message to the other ("are you there") and waits for a reply ("yes I'm here"). If the reply is not received within a certain time the robot assumes it is out of range (B) and reacts immediately by turning 180 degrees in order to reconnect (C). Then as soon as it receives a reply to its calling messages it chooses a new random heading (D).

As no global time is implemented the robots should react asynchronously. However each robot has the same range of communication so both reactions should occur within a short time, depending on the periodicity of the calling messages.

This behaviour leads the two robots to maintain themselves in range as if they were attached with an elastic band. The choice of a random heading when reconnection occurs makes the pair as a whole follow a random walk. It is important to observe that the reciprocity of reaction even though not simultaneous is crucial to retaining the connection. Homogeneous robots

[1]the former situation wouldn't harm the connection and the latter is dealt by the avoidance behaviour.

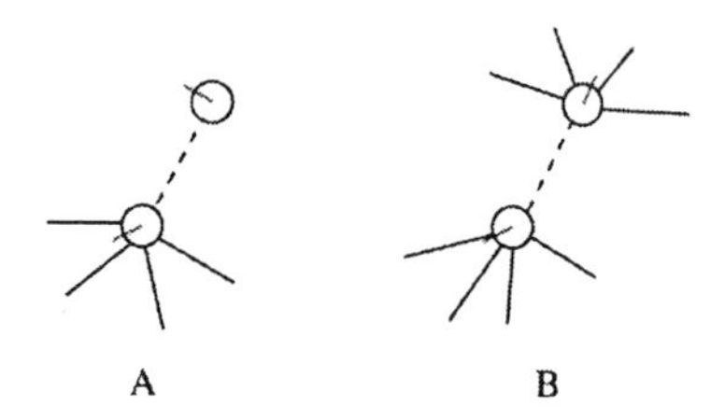

Figure 2: Extreme states

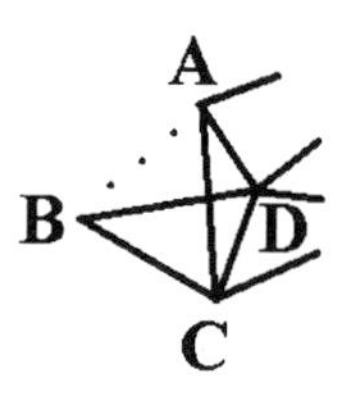

Figure 3: Shared neighbour

have equal velocities and the reaction of only one robot could lead to an endless pursuit.

Following Støy, it is worth emphasizing here the fundamental characteristic of this paradigm to achieve swarming. It is not the semantic content of the message alone that matters but it being tied with an environmental cue, that is the presence -or absence- of its response. This algorithm stands in the framework of situated communication where the semantic content of a message is closely linked to an environmental meaning.

Applying this basic algorithm to a greater number of robots by making a robot react to every loss of connection leads to an over-reactive swarm which clumps together. To react to every connection is equivalent to aiming towards a complete graph where each vertex is connected to every other. This is not our aim.

Trying to make the robots less reactive has demonstrated extreme situations that must be avoided in order to assure the coherence of the swarm (see figure 2). When a robot (or a group) is linked to the rest of the swarm by a single communication link, the danger lies in the possibility of a robot not reacting to the loss of such a connection essential to global connectivity. In graph theory a vertex representing such an important connection is known as a *bridge*.

Coherence

To avoid these situations we make use of the graph theory concept of *clustering*. Instead of considering only its own degree of connection to trigger a reaction, the robot will receive from its neighbours their adjacency table - their neighbours' list - in order to check whether a particular neighbour is *shared* by other ones, that is whether a particular neighbour is the neighbour of other robots' neighbours.

The algorithm works as follows: for each lost connection the robot checks how many of its remaining neighbours still have the lost neighbour in their neighbourhood. If this number is less than or equal to the fixed threshold β the robot turns back. In parallel if its degree of connections is rising the robot chooses a random heading.

For instance in the situation on figure 3, robot A, when losing the connection with robot B, will check its other neighbours and find that robots C and D share B as neighbour. Hence A will react and turn back only if β is equal or greater than two. Our algorithm makes the robot try to maintain the triangulation observable on the picture, therfore avoiding critical states.

The pseudo-code of the algorithm for one robot is set out below.

```
Create list of neighbours for robot, Nlist
k = number of neighbours in Nlist

loop forever {
  Save copy of Nlist in Oldlist
  Save copy of k in LastK
  Set reaction indicator Back to FALSE

  Send radio 'ping' to neighbourhood
       every 100 time steps
  Listen for return calls from robots
       in range that received the 'ping'
  Create Nlist from all returns
  k = number of neighbours in Nlist

  Create LostList, list of robots which have
       lost contact since previous 'ping'

  for (each robot in LostList) {
    Find nShared, number of shared neighbours
    if nShared <= beta (threshold value) {
      Set reaction indicator Back to TRUE
    }
  }

  if Back=TRUE {
    turn robot through 180 degrees
  }
  else if k > LastK {
    make random turn
  }
}
```

It is interesting to note that the robot tries to maintain β shared connections with each neighbour and one might think that this would lead to over-connectivity, as described earlier. But in fact each connection can contribute to different sharings and such a condensed clustering is never reached. On the other hand if the robot for instance establishes a new link with a robot that does not have other connections with the robot's surrounding neighbourhood, it will react to the loss of this neighbour until the shared connections are also established. This is precisely the behaviour we are aiming for.

Running the simulation confirms that such an algorithm increases swarm coherence, as the triangulation is perfectly observable and therefore critical states avoided. A value of β equal to one is enough to achieve coherent spread.

Of course the communication bandwidth of the whole process is somewhat increased compared to the basic algorithm, as well as the processing power needed for the robot. More sensitivity to the message content (semantics) is also introduced. However, the communication is still situated, and hence message loss or misinterpretation only leads to over-reactivity and an increase in connectivity without loss of robots, as introduction of noise in the simulation confirms. Also this increase in bandwidth does not affect the scalability of the algorithm as it concerns only exchanges between neighbouring robots and will therefore not be propagated more than a single hop in the network.

Directed Swarming Algorithm

When following a beacon gradient (chemical, sonic or light...) a possible solution consists of placing two differents sensors on both sides of the robot and then making the robot turn towards the sensor indicating the highest value. This implementation is highly dependent on signal-to-noise ratio as the robot's sensors are not situated far from each other which makes the two different sensing values very similar.

We raise the hypothesis that it could be possible to use different robots to take a sample, share it with their neighbours to generate an approximation of the position of the beacon and then make the whole swarm move towards it.

Although the problem of localization of the beacon might seem the most difficult to answer, the movement also represents a real challenge in our case. Indeed, how is it possible to make a group of robots that do not have any precise idea of relative directions or positions follow a particular path ?

The answer is in binding these two problems together using the light occlusion implementation described earlier. The idea is to make an illuminated robot become special to its neighbours by entering a special state - let us call it 'red'. Then a layer is added on top of the shared neighbour behaviour that always triggers a robot's reaction if the lost connection involved a red robot. This holds whatever the number of shared neighbours.

In figure 4, depicted in grey are the robots illuminated by a beacon situated north of the picture.

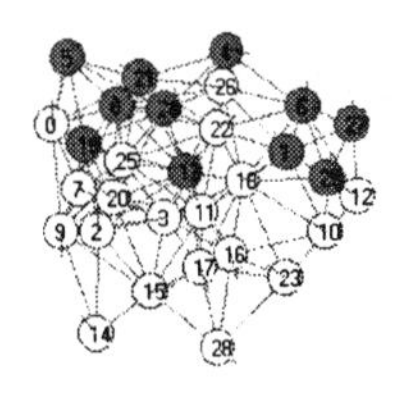

Figure 4: Illuminated robots

The following line is added in the *for* loop of the pseudo-code:

```
if (color of lost robot is red) then
        set Back to TRUE
```

This is in effect a spatial differentiation in the β threshold value, setting it to infinity for the red robots. As a result, these robots try to build complete graphs among themselves, reacting to each loss and therefore not moving much while the others always stick to the red ones. But as the red ones build their complete graph they occlude the previous red ones that stand inside and a slow translation of the swarm starts towards the beacon.

It is important to note that the basic sensing of 'being illuminated' or not does not make a single robot able to reach the beacon. The taxis behaviour only results from the interaction of the red robots and the rest. This is a truly emergent behaviour and as such it is highly dependent on the various parameters in action: communication range, rate of occlusion, avoiding range, etc. These dependencies remain to be investigated in real robot experiments. But it is much less sensitive to noise than a classic two-sensor beacon taxis as the sensing does not need high sensitivity: the robot is either illuminated or not.

With occluding obstacles in the path between the swarm and the beacon, the swarm is able to find its way through. After an initial random spread, the beacon is sensed by a few robots that attract the swarm through the obstacles. If the space between the obstacles is restricted, the swarm will adapt its shape to be able to pass through the gap (see figure 5).

To control the shape of the swarm more differentiation is introduced: either the red robots are made much less rapid compared to the others or the contrary. The result is a swarm that forms a line oriented perpendicularly to the direction of the beacon or towards it, respectively.

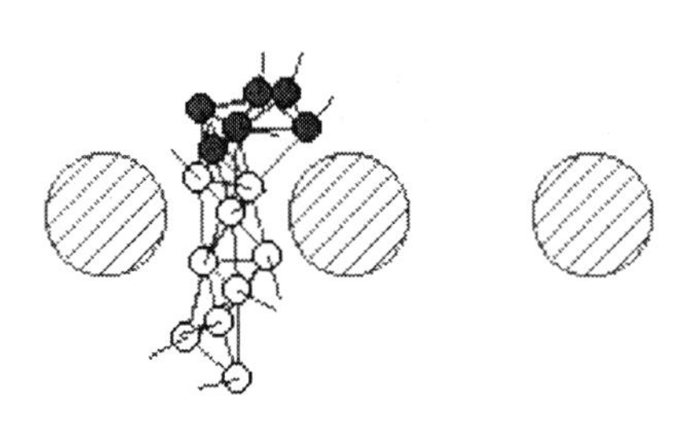

Figure 5: A swarm moving through obstacles

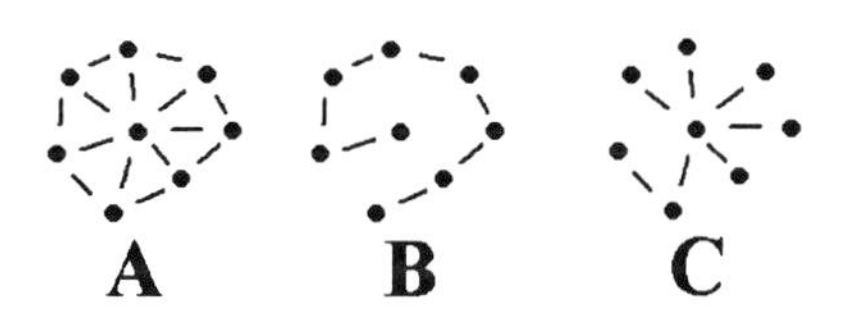

Figure 6: A graph(A) & two disjoint spanning trees(B,C)

3.3 *Measures*

In this study, the first feature that has to be measured is *swarm coherence*. But as the loss of a connection is a discrete event, the quantity that needs to be measured is not obvious. Should we restrict ourselves to detect when the swarm becomes disconnected, or should we look for finer criteria?

To answer this question we introduce here the concept of *disjoint spanning trees* (see figure 6). The aim being to have a gradation in levels of coherence. We stated earlier the importance of bridges as extreme states to be avoided. This raised the idea that in the presence of a bridge there is no possibility to find a spanning tree disjoint to the one that goes through the bridge. Extending this idea, we measure the number of edge-disjoint spanning trees that can be constructed on the graph of the robots' network.

This is an integer value that gives information about the global connectivity of the graph. It is related to the n-connectivity (Hobbs, 1989). Also unlike the n-connectivity which is computationally NP-hard, an approximation algorithm to compute such a value is readily available. As heuristic to search the graph, we use the degree of connections and follow the highest value downwards.

Following our underlying concern for sensor networks, the relevance of such a quantity to the ability of the network to propagate messages is another motivation for measuring it.

Another quantity that is important to measure is the *area coverage* of the whole swarm. This can be done by triangulation of the bounding polygon of the graph. It can also be approximated through the calculation of the mean degree of connections over the whole graph. This number depends on the radius of communication and is therefore highly correlated with area coverage.

To measure success of taxis with or without obstacles, we compute the remaining distance of the centroid of the swarm to the beacon. This is to be able to study the scalability of the algorithm as we increase the number of robots, as well as the influence of noise on the speed of the swarm.

In the case of reactive morphology change, we measure the ratio between the added square distance of each robot to the line starting from the beacon passing through the centroid and the added square distance to its perpendicular. This gives an idea of the shape of the swarm as would the ratio of the two radii of an ellipse.

$$ratio = \frac{\sum_{i=1}^{N} (d_{beacon}(R_{x_i}) - R_{y_i})^2}{\sum_{i=1}^{N} (d_{perpendicular}(R_{x_i}) - R_{y_i})^2}$$

4. Results

In this section we present the results of simulation runs to test coherence, scalability and robustness to noise of our algorithms in different situations.

4.1 *Swarm Coherence*

In this series of runs, the environment contains no beacon and is free of obstacles. The aim is to measure the influence of the parameter β on the area coverage and the coherence of the swarm as defined in section 3.3. As such an influence is dependant on the number of robots involved in the swarm, we plot these results as surfaces. The results are averaged over sampling time on each run and over ten different runs.

We can see in figure 7 that the single parameter β appears to be sufficient to control the area coverage of the swarm. Taking a close look at the plot of the area coverage reveals that this ability is best observed with a higher number of robots. Indeed the possibility for expansion is multiplied by the number of robots.

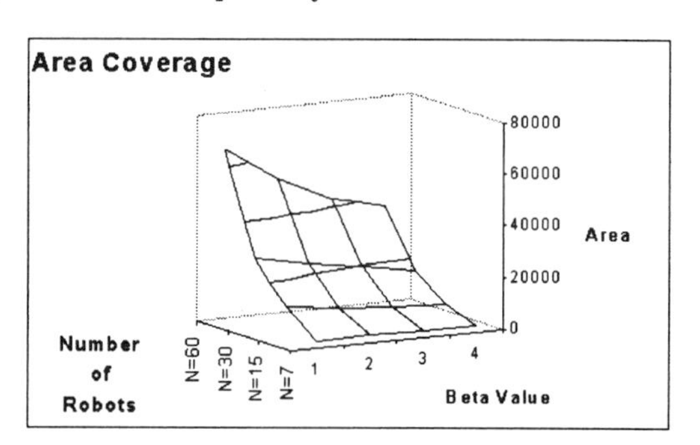

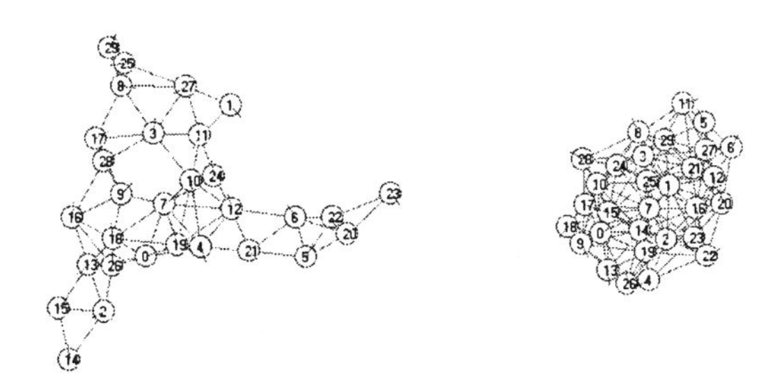

Figure 7: Typical disposition for $\beta = 1$ and $\beta = 4$

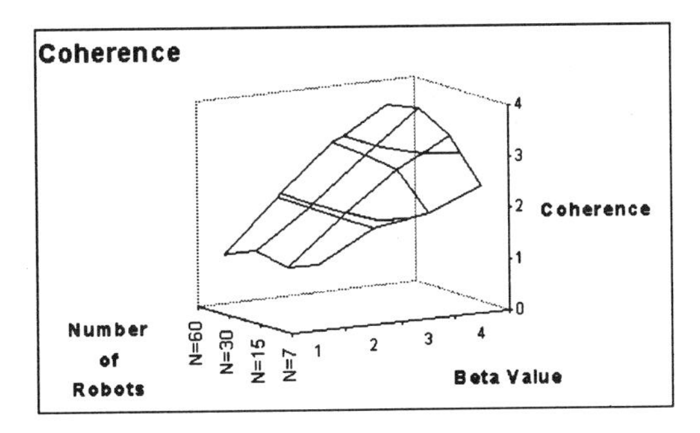

We observe that the coherence increases almost linearly with the parameter β for larger swarms. We expect of course a levelling for larger values as the avoiding behaviour starts to play its role, but in this range only the smaller swarm shows such a feature.

When we take a look at the occurrence of breaking the swarm, we note that the value of $\beta = 1$ is not a safe choice, especially for a small number of robots. In fact small swarms experience high fluctuations that are highly unstable. For the following experiments we will therefore set the value of β to two.

Below is plotted the influence of noise on the area coverage and coherence for 30 robots with $\beta = 4$. We can see that the area coverage is not particularly affected by the introduction of noise, but of course the coherence shows some degradation, as expected.

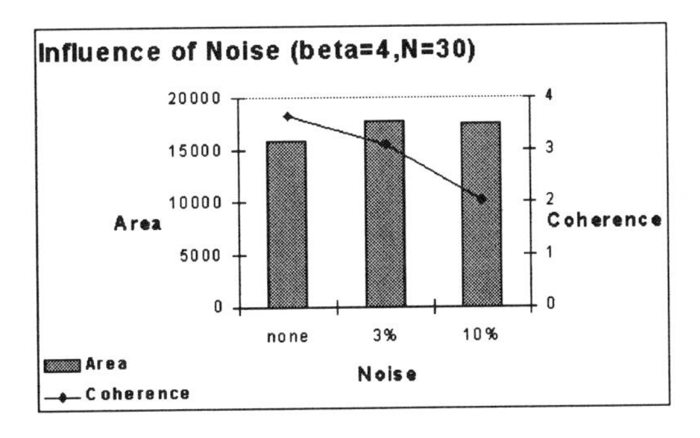

4.2 Taxis

In the experiments of this section we introduce a beacon in the environment to attract the swarm. We measure the progression of the centroid of the swarm towards the beacon every 20,000 steps.

The graph below shows such a progression for a single run of two swarms, one composed of 7 robots and the other of 30 robots. We observe higher fluctuations in the progression of the smaller swarm while the larger one moves almost linearly towards the beacon. This is due to inertia resulting from individual moves to maintain coherence in the swarm.

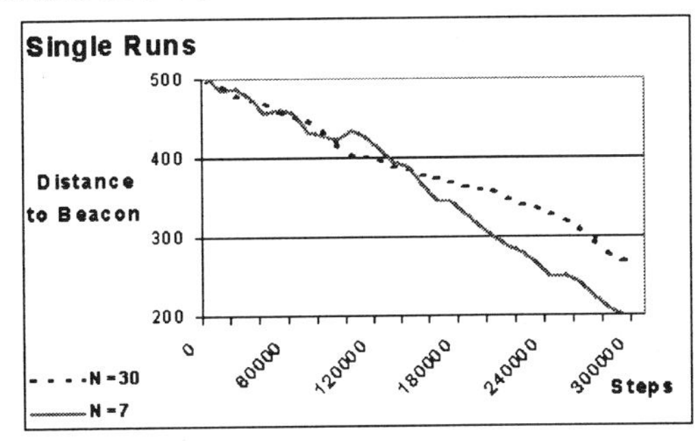

Averaging over ten different runs gives the graph below, where the slope of the lines represents different average progression speed for different sizes of swarms. The loss of speed due to an increase in the number is noticeable. Of interest is the case N=7 that shows that the fluctuations mentioned earlier actually lead to an average progression speed smaller than for a 15 robot swarm.

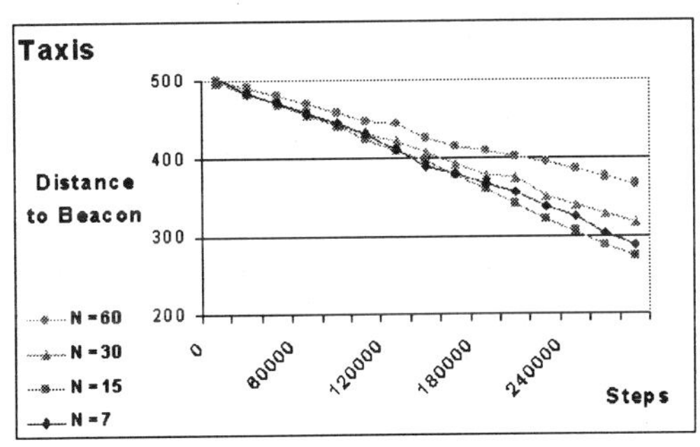

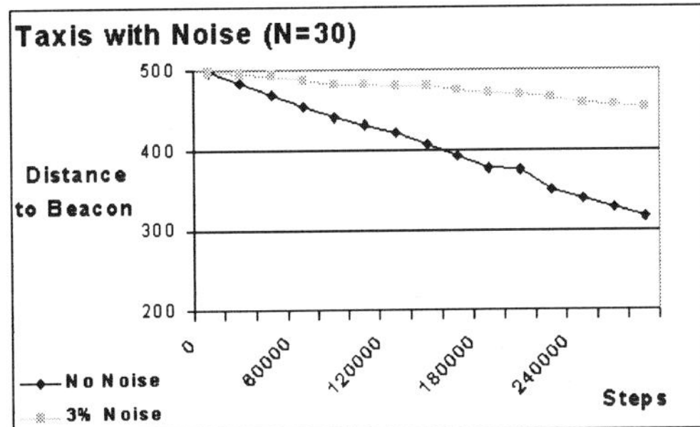

As shown in the graph above, introducing noise dramatically reduces the speed of the swarm. In fact introducing 10% of probability to lose a message does not allow the swarm to move at all. This is due to the fact that a single loss of message triggers a turn back reaction. With such a level of noise, the robots find themselves reacting too often. As a consequence the benefit of differentiation between the illuminated robots and others tends to be cancelled, and the resulting movement no longer occurs. This shows the influence of the accuracy of the radio device (it has to be noted that the illumination sensor does not show such a sensitivity as the swarm moves towards the swarm, even with 10% of noise). A solution to deal with such a problem, following (Støy, 2001), is to make the robot wait two or three listening loops before acting on the loss of the missing robot. This would circumvent message loss due to noise and reduce the sensitivity of the whole algorithm.

Obstacle Avoidance

Adding occlusive obstacles renders the progression towards the beacon more difficult for the swarm. We observe a degradation in speed as the swarm, firstly, has to detect the beacon hidden behind an obstacle and, secondly, has to find its way through.

In figure 8 is a sequence of screenshots showing a swarm of 30 robots moving through the obstacles while aiming towards the beacon due north. Smaller swarms would go through only one aperture as in figure 5.

For larger swarms ($N = 60$ in this case), the attraction tears the swarm apart, as can be seen in figure 9. This is because the robots that have already crossed the obstacle mask the beacon to the remaining ones that are therefore no longer attracted. The remaining block is too big to be attracted through the obstacles only by the coherence requirement and the tearing occurs. Increasing the β value could be a solution, but will also decrease the differentiation, with the already described undesirable effects.

It has to be noted that the critical size of 60 robots for a swarm not to be able to cross the obstacles depends on the width of the gap between them, as well as on the level of noise.

Encapsulation

An interesting side-effect of our algorithm is the behaviour of the swarm when it reaches the beacon. The interaction of the avoidance behaviour, the taxis and the reconnection strategy makes the swarm wrap itself around the beacon and encapsulate it as shown in figure 10. It is an emergent enclosure behaviour.

In a three-dimensional version of the algorithm, such a behaviour could lead to isolation of rogue material from an environment. This is somewhat analogous to Amoeba or blood macrophage phagocyte.

4.3 Reactive Change in Morphology

In this section, we illustrate the potential of our algorithm to make the swarm adopt a specific overall shape reacting to the beacon.

beacon due north

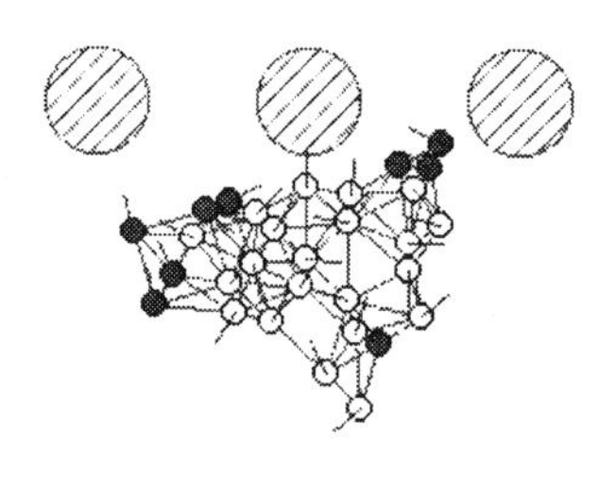

A

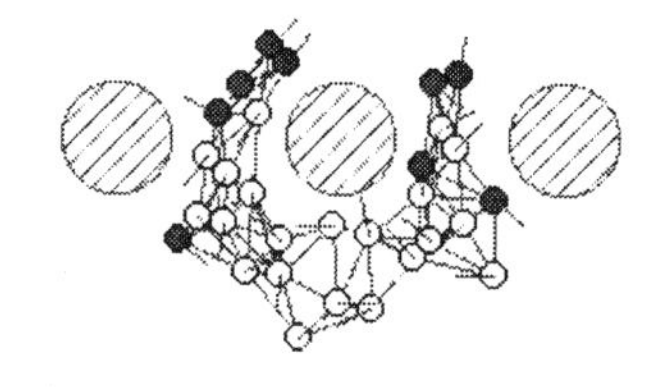

B

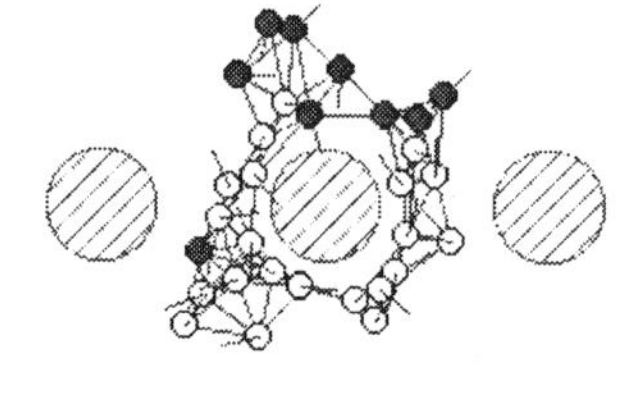

C

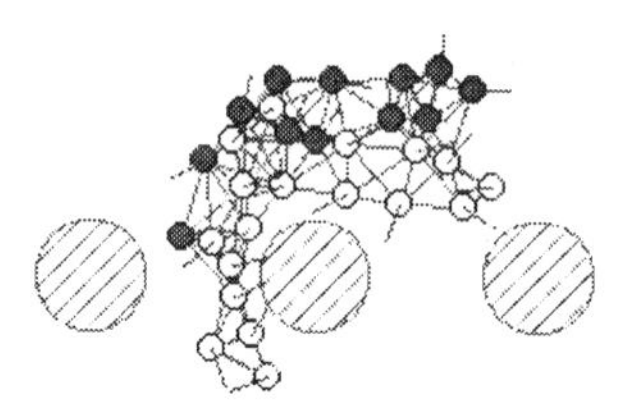

D

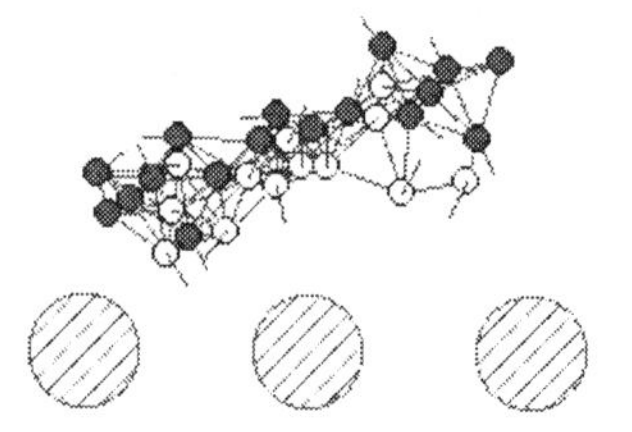

E

Figure 8: Typical run of a swarm of 30 robots

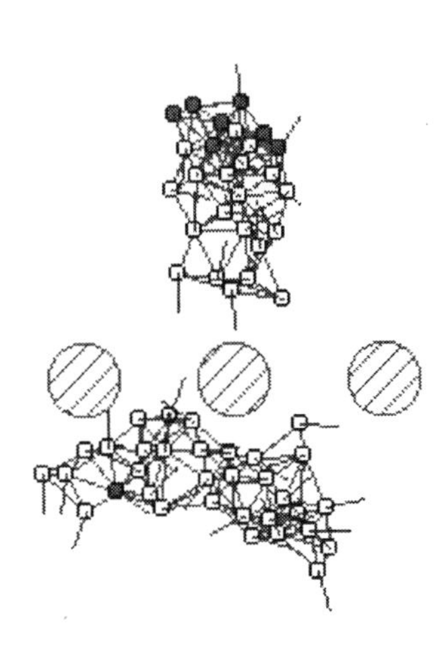

Figure 9: A broken swarm of 60 robots

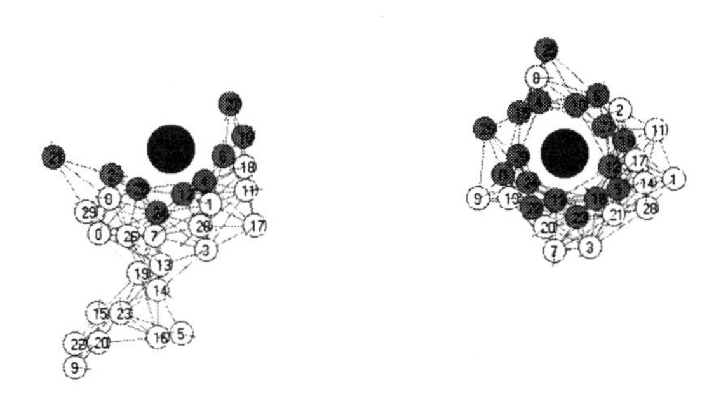

Figure 10: Encapsulation of the beacon

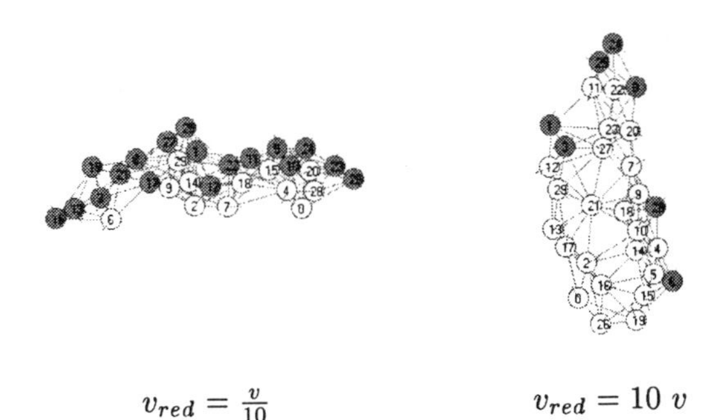

$v_{red} = \frac{v}{10}$ $\qquad$ $v_{red} = 10\ v$

Figure 11: Change in Global Shape

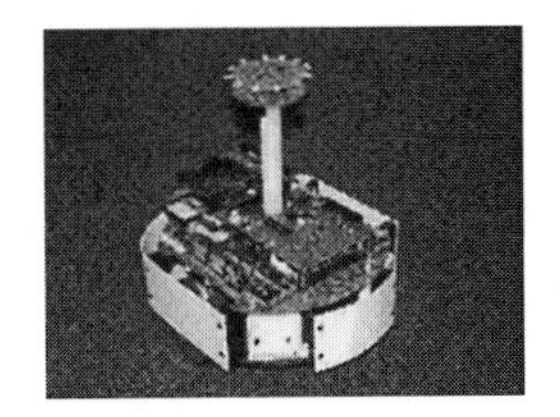

Figure 12: Linuxbot

We introduce a difference between the speed of the red robots and the remaining ones. Again this single parameter is able to control a dramatic change in the shape. When the red ones moves ten times faster, the swarm forms a line perpendicular to the direction of the beacon due north. On the other hand, if the red robots move 10 time slower, the swarm forms a line towards the beacon. Figure 11 shows typical examples of the two cases.

This behaviour allows us to control the exposure of the swarm to the beacon. The perpendicular shape being the most exposed. This relates to column or line formations in rigid formation studies, only tuning a single parameter and without involving high-level sensing capabilities.

In the context of sensor networks, such a behaviour could make use of a natural beacon such as the sun, in order to dispose itself along isoclines to undertake measurements. For instance it could be of interest to dispose underwater sensor robots either at the same depth or vertically. Moreover, the use of judicious switching between speeds according to environmental cues could make the morphology of the swarm adaptive.

The graph presented below shows that the presence of noise slows down the process but does not affect the qualitative result.

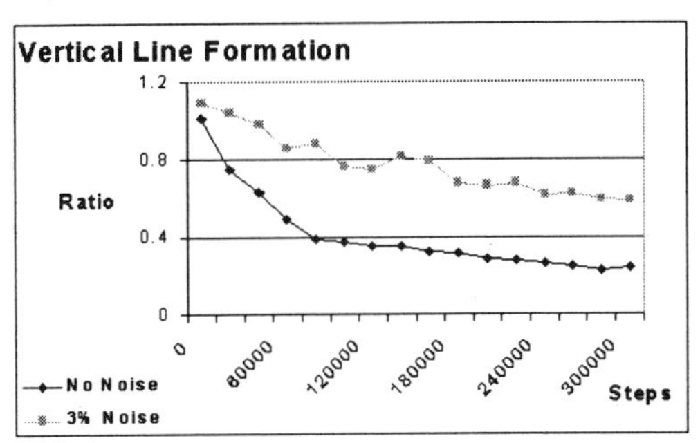

5. Future Work

The next step in our research is to implement these algorithms on real robots. For this purpose, we shall use a fleet of Linuxbots (see figure 12) developed in the IASLab. A description of this platform is to be found in (Winfield and Holland, 2000). We will equip them with an infra-red sensing tower to simulate the locality of the radio-connection (the restricted dimensions of the arena preventing us from using the normal range of our WLAN).

To implement the behaviours presented in this paper we need an omnidirectional light sensor that is sufficiently occluded by the body of other robots to show the same behaviours. A solution could be the integration of different sensors disposed around the body of the linuxbot.

While conducting this hardware research, we shall continue to study the potential of the chosen direction to develop truly adaptive distributed algorithms. We are specially keen to be able to control the morphology without the help of a beacon.

Simultaneously we will investigate the relationship of our work with exemples in biology, principally amoeba and bacteria (Shapiro, 1988, Takahashi et al., 2001)

6. Conclusion

In this paper we have presented a fully distributed algorithm that is able to ensure the global coherence of a swarm of mobile robots, despite their limited-range communication abilities. We achieve also emergent coherent movement of the swarm towards a beacon, making it able to avoid obstacles and adapt its shape to move through them. Moreover the approach allows us to control the swarm global shape by tuning a single parameter (the β threshold in the case of area coverage, and speed in 'linear' formation). The algorithm has been implemented in simulation and real experiments are currently in progress.

As a consequence of the algorithm, global connectivity of the communication network formed by the swarm of robots is also achieved. This could lead to applications in large mobile sensor arrays which require them to adapt their shape to provide appropriate sensing. Indeed we believe this approach has so far revealed only little of its potential for adaptive behaviours and related applications.

References

Balch, T. and Hybinette, M. Social potentials for scalable multi-robot formations. In *Proc.Int.Conf. on Robotics and Automation ICRA'00*, volume 1.

Brooks, R. (1991). Intelligence without reason. In *Proc. Int. Joint Conf.on Artificial Intelligence*, pages 569–595.

Hobbs, A. (1989). Computing edge-toughness and fractional arboricity. In *Contemporary Mathematics*, volume 4, pages 89–106.

Kotay, K. and Rus, D. (1999). Locomotion versatility through self-reconfiguration. *Jour. of Robotics & Autonomous Systems*, 26:217–232.

Lipson, H. and Pollack, J. (2000). Automatic design and manufacture of robotic lifeforms. *Nature*, 406:974–978.

Melhuish, C. (1999). Employing secondary swarming with small scale robots: a biologically inspired collective approach. In *Proc. of the 2nd Int.Conf. on Climbing & Walking Robots CLAWAR.*

Melhuish, C., Holland, O., and Hoddell, S. (1998). Collective sorting and segregation in robots with minimal sensing. In *From Animals to Animat*, volume 5, pages 465–470. MIT Press.

Pfeifer, R. and Scheier, C. (1999). *Understanding Intelligence.* MIT Press.

Shapiro, J. (1988). Bacteria as a multicellular organism. *Scientific American*, June 1988:62–69.

Sims, K. (1994). Evolving 3d morphology and behavior by competition. *Artificial Life VI*, pages 28–39.

Støy, K. (2001). Using situated communication in distributed autonomous mobile robotics. In *7th Scandinavian Conf. on AI.*

Støy, K., Shen, W.-M., and Will, P. (2002). Global locomotion from local interaction in self-reconfigurable robots. In *7th Int.Conf. on Intelligent & Autonomous Systems (IAS7).*

Takahashi, N., Nagai, T., Yokoi, H., and Kakazu, Y. (2001). Control system of flexible structure multicell robot using amoeboid self-organisation mode. In *Proc.6th Europ.Conf. on Artificial Life, ECAL 2001*, pages 563–572. Springer.

Wessnitzer, J., Adamatzky, A., and Melhuish, C. (2001). Towards self-organising structure formations: A decentralised approach. In *Proceedings of ECAL 2001*, pages 573–581. Springer.

Winfield, A. (2000). Distributed sensing and data collection via broken ad hoc wireless connected networks of mobile robots. In *Distributed Autonomous Robotic Systems*, volume IV, pages 273–282.

Winfield, A. and Holland, O. (2000). The application of wireless local area network technology to the control of mobile robots. *Microprocessors and Microsystems*, 23:597–607.

Yoshida, E., Murata, S., Kokaji, S., Tomita, K., and Kurokawa, H. (2000). Micro self-reconfigurable robotic systems using shape memory alloy. In *Distributed Autonomous Robotic System*, volume IV, pages 145–154.

Sharing a Charging Station without Explicit Communication in Collective Robotics

Angélica Muñoz-Meléndez[1*] **François Sempé**[1,2] **Alexis Drogoul**[1]
[1]LIP6 - UPMC
Case 169 - 4, Place Jussieu
75252 Paris Cedex 05. France
[2]France Télécom R&D
38/40 Rue Général Leclerc. 92794
Issy les Moulineaux Cedex 9. France
{Angelica.Munoz, Francois.Sempe, Alexis.Drogoul}@lip6.fr

Abstract

This research focuses on the design of teams of mobile robots that are able to operate in real life situations and non-controlled environments. In order to perform any service, robots have firstly to display two properties: autonomy and self-sufficiency. This paper explores briefly a simple mechanism to enable a team of robots to behave in an autonomous and self-sufficient manner for as long as possible while taking into account the needs of partners.

1 Introduction

This research focuses on the design of teams of mobile robots that are able to operate in real life situations and non-controlled environments (Drogoul and Picault, 1999). The corridors of our laboratory, for instance, are one of the scenarios with which our robots should be able to cope, in order to perform services such as distributing mail, guiding visitors, security, and so on. In order to perform these services, robots have firstly to display two properties: autonomy and self-sufficiency.

Autonomy is concerned with the decision-making process and means self-government (MacFarland, 1994). An autonomous robot is not assumed to be independent of external programming, but of external control when it is running. Thus, it must be able to perform the actions for which it has been designed using its own capabilities, *i.e.* its sensors and its actuators. In contrast, **self-sufficiency** denotes the ability of a system to maintain itself in a viable state for long periods of time (MacFarland, 1994). A self-sufficient robot has to ensure at least its energy supply by itself. For that, it has at its disposal certain recharging facilities, *e.g.* rechargeable batteries and a self-recharge device, and relies on several mechanisms so as to be able to examine its power supply constantly and to locate and use a charging station.

*Supported by the National Council for Science and Technology of Mexico, CONACYT, and the French Ministry of Foreign Affairs.

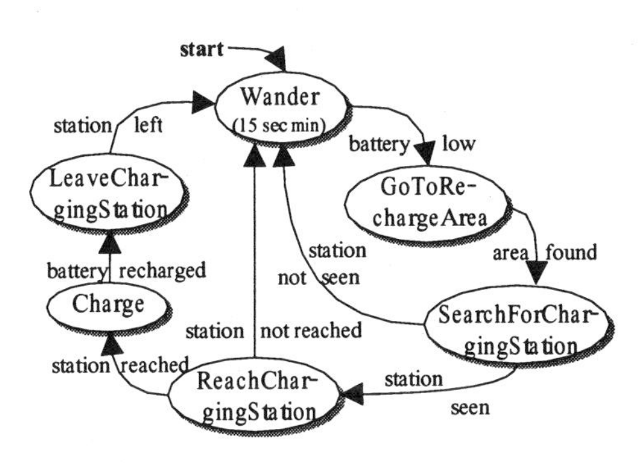

Figure 1: Automaton that summarizes the robot behavior.

2 The Problem of Sharing

The design of a team of robots that require a common resource raises a number of questions: is this resource limited? how should robots manage this resource? how much information should robots exchange? do they need to communicate? if so, how and how much information?.

In our case there is just one resource shared by robots, the charging station. Charging stations are limited resources if there are more robots than charging stations in the environment, or if, even though several charging stations are available in the environment, more than one robot is trying to use the same charging station simultaneously. Several strategies may be used by robots in order to manage a charging station:

1. **Competing for the resource**. In this strategy robots compete for the access to charging stations, and the first robot to reach a station is the first to be served.

2. **Establishing hierarchies to access the resource.** The order of access to charging stations has previously been defined by the designer. Robots must be able to recognize the hierarchy of others, in order to decide whether or not to use a charging station.

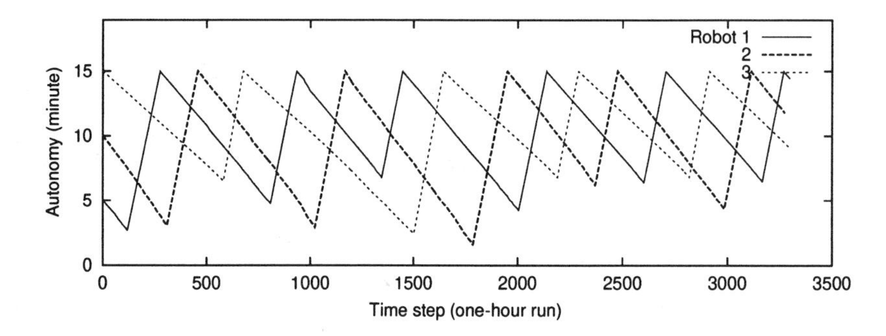

Figure 2: Comparison of the energy autonomy of three robots that share a charging station.

3. **Conflict resolution**. In this way robots solve the conflict created by the sharing of charging stations through negotiation or a similar mechanism. Robots may give reasons such as their power supply, or the tasks that they are doing, to argue their case for a charging station.

The first strategy is the simplest and the cheapest. Robots do not need to communicate directly, but to compete for the resource and to identify the winner. The second strategy may be implemented in a team where hierarchies have been established for some reason, in which case, the sharing of resources benefits from this *social* system. The third strategy seems to be the most efficient, in the sense of providing good functionality, because robots consider their real needs when deciding what to do. However, the existing negotiation models are beyond the possibilities of real robots.

3 Behavior

The research described in this paper is part of the MICRobES[1] project (Drogoul and Picault, 1999). The experiments are conducted using six Pioneer 2-DX mobile robots of ActivMedia©. We also have of a charging station made by our laboratory and France Telecom R&D to enable robots to recharge their batteries by themselves (Sempé et al., 2002).

Robots have a repertoire of basic behaviors that are activated by the external stimuli they perceive. Basic behaviors such as avoiding obstacles, wandering, navigation and localization have been implemented in and tested on our robots. The control of a robot whose main task is to wander is illustrated in figure 1.

4 Results

We have implemented the competition strategy in a team of three robots that share a charging station. The maximum energy autonomy has been set to 15 minutes. The robots start at the same time. Each run lasts one hour at most or until a robot *dies*. In our experiments, the charger gives approximately 5 minutes of energy autonomy for 1 minute of recharge. As three robots work together, the station will be free more or less half of the time.

The size of the environment is about 20 square meters. The three robots start with 5, 10 and 15 minutes of energy autonomy respectively. The threshold that triggers the recharge behavior is 7.5 minutes. Figure 2 shows the energy autonomy of each robot during one hour. In this graph we can observe that robots manage to alternate their use of the charging station.

[1]MICRobES is an acronym in French for Implementation of Robot Collectivities in a Social Environment.

5 Perspectives

The experiments described are in progress and future work will focus on the implementation of alternative strategies. The use of direct communication between robots has not been ruled out, if this kind of communication allows a better use of the charging station. Finally, we are also working on the rescue a robot with a flat battery (Muñoz and Drogoul, 2000).

References

Drogoul A., Picault S. (1999). MICRobES : vers des collectivités des robots socialement situés. In: *Actes des 7èmes Journées Francophones sur l'Intelligence Artificielle Distribuée et les Systèmes Multi-Agents*, Gleizes M.-P., Marcenac (eds.), pages 265-277, Éditions Hermès, Paris.

McFarland D. (1994). Autonomy and Self-Sufficiency in Robots. In: The Artificial Life Route To Artificial Intelligence. Building Embodied, Situated Agents. Steels L.(ed.), pages 187-213, Lawrence Erlbaum Ass. Pub. USA.

Muñoz A., Drogoul A. (2000). Towards the Design of Self-Vigilant Robots using a Principle of Situated Cooperation. In: *Proceedings of the RoboCup Rescue Workshop. ICMAS 2000.* Boston, Mass.

Sempé F., Muñoz A., Drogoul A. (2002). Autonomous Robots Sharing a Charging Station with no Communication: a Case Study. In: *Proceedings of the 6th International Symposium on Distributed Autonomous Robotic Systems (DARS'02).* In press.

Towards a Quantitative Analysis of Individuality and its Maintenance

Alexandra Penn
CCNR, University of Sussex, Brighton. BN1 9QG. alexp@cogs.susx.ac.uk

Abstract

In the context of the study of the evolution of new levels of biological organisation, we present initial work on constructing a metric to detect individuality, or cooperative systems of agents. This is based on the production of new sets of higher-level variables, and their homeostatic maintainence with increasing external perturbation. Results of applying the measure to collectives of simulated Kheperas are presented.

1 Introduction

A prominent new area of study in biology is that of Major Transitions in Evolution (Maynard Smith and Szathmary, 1995). Many of these are the formation of new levels of organisation, or individuality; a process whereby new functional "individuals" are formed from collectives of entities which originally lived separately. Real biological individuals are dynamic physical processes in a constant state of flux, whose extent and nature arise from the physical and chemical interactions of their components with what surrounds them. Simulation models could be a useful tool to explore their formation, but only with realistic and rigourous characterisation of what constitutes an individual.

Both organic and inorganic "individuals" possess a reduced, distinct, set of higher-level system variables, such as temperature or metabolic set points, with respect to the number of degrees of freedom of their components. In addition, all living systems are homeostatic, actively maintaining these characteristic variables in the face of both internal and external perturbation. Hence we can use methods of phase space reconstruction from nonlinear dynamics to estimate the dimension of the attractor in which our system resides; then test the robustness of this property to perturbation to try to determine whether or not our system is homeostatic.

2 Measuring Individuality

A standard method of dimensionality reduction, Principal Components Analysis(PCA), is used to transform the data set, the values of each of the N degrees of freedom of each of the A agents at every time step of the simulation, to a new set of orthogonal variables, the Principal Components(PCs). These are uncorrelated, linear combinations of the original variables, ordered so most of the variation in all the original variables is expressed by the first few. The vector of Principle Values contains the proportion of the variance each component accounts for. To encapsulate the information on the relative sizes of the PCs in a single parameter, we calculate the entropy, S, of the normalised principal values (Wright et al., 2001).

$$S = -\Sigma_{i=1}^{AN} p_i \ln p_i \tag{1}$$

S is maximal when all PCs are the same size and the system has AN effective degrees of freedom, and zero when one normalised principal value is one and the others zero. At intermediate values, it acts as an approximate log count of the number of significant (normalised) PCs weighted by their respective sizes. Hence the effective number of degrees of freedom, $D = e^S$.

3 Application to an Agent System

The dynamical system chosen as an example was a group of 3 simulated Khepera robots, evolved by Quinn (Quinn, 2001). Each robot is $5.8cm$ in diameter, with a sensor range of $5cm$. The group, all members having an identical neural network, was evolved to perform a formation movement task, see fig.1. In our experiment, perturbation in the range $0 : 2cm$ was applied to each robot's position at each time step. Robot orientation, and x and y positions were recorded. Thus each Khepera has 3 degrees of freedom, and the system 9.

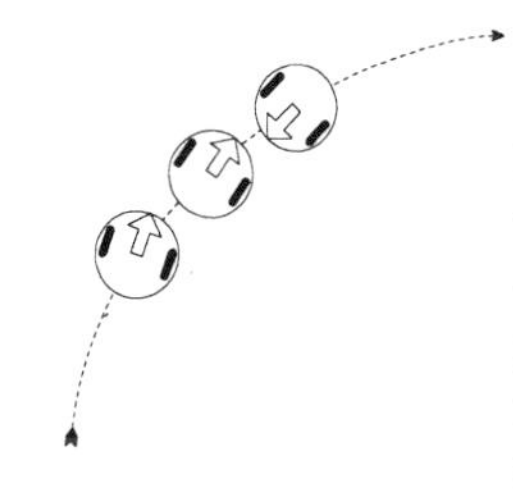

Figure 1: Characteristic configuration of coordinated Kheperas, for low perturbation values, showing orientations and common velocity vector

4 Results and Analysis

A large number of data sets were collected with perturbation values between 0 and 2 cm, and S calculated over the entire run for each. This method obscures some of the within-run variation, but a clear discontinuity in D can be seen as the perturbation is increased, fig.2. For perturbation values up to $0.4cm$, the system has 3 to 4 degrees of freedom. This corresponds visually to the 3 robots moving in the coordinated fashion previously described, fig.1. At $0.4cm$ the total entropy of the system jumps, and D increases to $5:6$, corresponding to the team splitting, with 2 robots remaining correlated, and a third left unattached. This dimensionality is maintained until the system breaks down entirely past $2cm$ perturbation(not shown in graph), and all 3 Kheperas move independently.

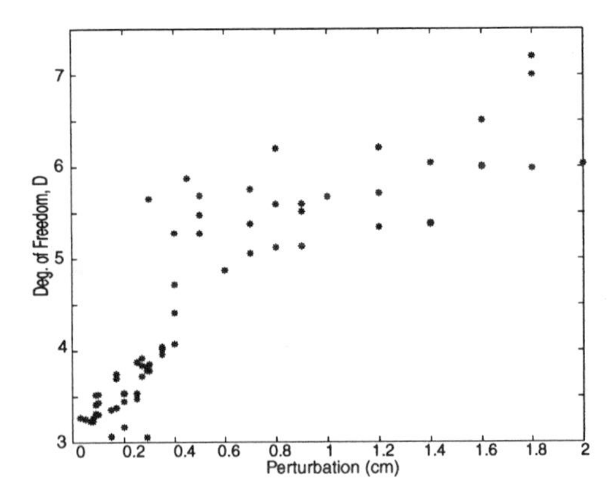

Figure 2: Effective Degrees of Freedom, D, of 3 robot systems against perturbation in cm.

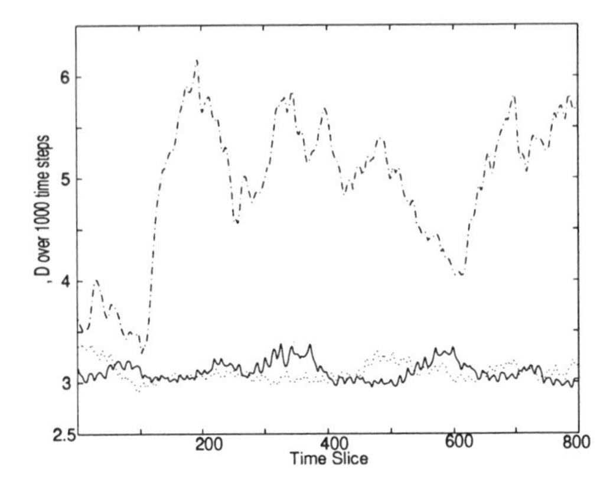

Figure 3: Typical time variation of D for systems subject to perturbations either side of the discontinuity at $0.4cm$. Solid line=$0.15cm$, dotted line=0.2, and dash-dot=$0.4cm$

A time series of entropies was constructed for each run using a "sliding window", PCA was performed for the data from the first t(window size) time steps, where $t > AN$, the window was moved on by dt time steps and the process repeated. The variation in D over the run for perturbation values either side of the transition, shows a qualitative as well as a quantitative change, fig.3. Below $0.4cm$ perturbation, plots show low variance, the system remaining in the same attractor for the duration of the run. After this point the mean number of degrees of freedom of the system increases and becomes highly variable, possibly due to the system jumping between attractors of different dimension. Robot teams often break up and reform at intermediate perturbation values.

5 Conclusions and Further Work

On these initial tests the measure performs well at detecting coordinated groups, or "individuals", in a data sample. Although here we have analysed a low-dimensional system, our approach can deal with data sets of extremely high dimensionality. Correlations difficult to visualise, between large sets of variables, can easily be detected. With further work this measure could potentially be extremely useful in automatic detection, and exploration of such systems. This is a necessary requirement if we intend to produce transition dynamics in an artificial system by evolution, and provide a rigourous definition of individuality.

The next stage is to compare the behaviour of this, and other cooperative systems' under increasing perturbation, with definitively homeostatic systems. Graphs of homeostatic systems are expected to show a characteristic profile, a long plateau at a low entropy, followed by a sharp transition to a high entropy state as the system reaches the boundaries of its homeostatic abilities. For further discussion of the technique and ideas summarised in this paper see (Penn, 2002)

Acknowledgements

Thanks to Matt Quinn, I. Harvey, R. Tateson and J.Allen. Sponsored by a British Telecom BBSRC CASE studentship.

References

Maynard Smith, J. and Szathmary, E. (1995). *Major Transitions in Evolution.* Spektrum.

Penn, A. (2002). Steps towards a quantitative analysis of individuality and its maintenance: A case study with multi-agent systems. In Kim, J., (Ed.), *Proc. GWAL 5.*

Quinn, M. (2001). A comparison of approaches to the evolution of homogeneous multi-robot teams. In *Proc. 2001 Congress on Evolutionary Computation: CEC2001.* IEEE.

Wright, W., Smith, R., Danek, M., and Greenway, P. (2001). A generalisable measure of self-organisation and emergence. In Dorffner, Bischoff, and Hornik, (Eds.), *Artificial Neural Networks-ICANN.*

Learning Social Behaviors without Sensing

Anand Panangadan Michael G. Dyer *
Computer Science Department
University of California, Los Angeles
Los Angeles, California 90095
{anand, dyer}@cs.ucla.edu

Abstract

A learning algorithm is presented that enables agents that do not have the ability to sense other agents to adapt its behaviors (that were learned in a single agent environment) to novel situations (deadlocks arising from existing in an autonomous multi-agent system). This adaptation takes place as the agent continues to perform its construction task. When the agents are confined to narrow spaces, this learned behavior leads to a "bucket brigade". The algorithm also learns the pattern of activations on its spatial map that is associated with deadlocks and the new behaviors are exhibited when this pattern is later observed.

1 Introduction

(Chao et al., 2000) described a behavior-based architecture with connectionist action selection that enabled an agent to rearrange objects in its continuous two-dimensional simulated world into a pre-specified pattern. In this work, a learning mechanism is introduced that enables this architecture to be used in the multi-agent scenario *without* extending the sensory capabilities of the agents. Agents cannot detect other agents and deadlocks can arise between two agents with interfering paths. The learning algorithm enables the agents to learn to drop any "brick" being carried in case of a deadlock so that the other agent can pick it up and replan its path. The learning is unsupervised as an agent uses the progress it has made since trying out an action as reinforcement.

2 Environment and Architecture

The simulated environment is 2-dimensional and continuous. Distance sensors detect bricks but not other agents. If an agent tries to move to a location occupied by another, it will not succeed. Agents can pick up a brick (carrying it as it moves) and drop the brick. Construction in this world thus involves moving toward a brick, grabbing it, moving to one of the specified drop sites and then dropping the brick at that location.

*This work supported in part by an Intel University Research Program grant to the second author.

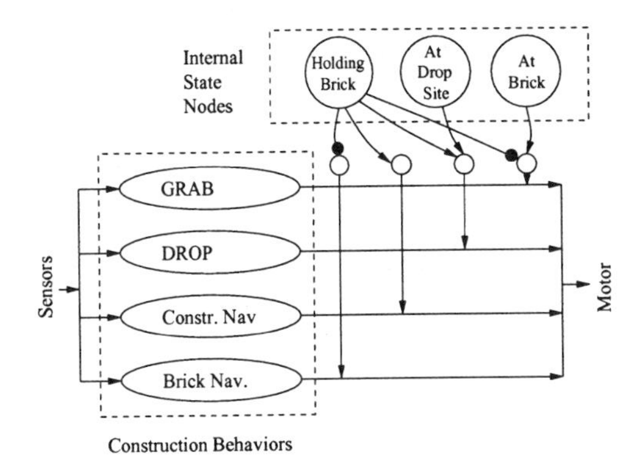

Figure 1: Construction behaviors and internal state nodes

Egocentric Spatial Maps (ESMs) are used to represent the location of bricks around an agent. An ESM is a grid of neurons that divides the area around the agent into small uniform squares such that the central neuron represents the square on which the agent is currently present. Behaviors plan paths to bricks and drop sites by spreading activation on the ESMs. The output of the behaviors is sent to the connectionist action selection module (figure 1) which chooses one of these behaviors depending on the current stage of the construction task (encoded by *Internal State* nodes).

3 Learning Social Behaviors

To recognize deadlocks involving other agents, a new internal state node, f, that measures the "frustration" of the agent is added along with weighted links to behaviors. If the agent is unable to move in a timestep, the activation on this node increases. When this activation exceeds a threshold, θ, then the agent tries to perform a random behavior (grab, drop, move toward brick or drop-site). If the agent is able to get out of the deadlock, then the weights are changed to reinforce this behavior. If the agent remains deadlocked, then that behavior is penalized. Since the action selection network is a single layer network, the Perceptron learning rule is used:

```
1: At time t, if (a_f > θ) then
2:   with probability p_o perform an action o
```

```
3: At time (t + δ), if |p(t) − p(t + δ)| > d
4:     then Δ w_io = sign(0.5 − a_o)η a_i e^(−k w_io^2)
5:     else Δ w_io = sign(a_o − 0.5)η a_i e^(−k w_io^2)
```

where a_i denotes the activation of internal state node i. a_o denotes the activation of output node o (representing some behavior), w_{io} the weight on the link between i and o, and η is the learning rate. p_o in step 2, reduces the probability that both agents are taking random actions at the same time. The test whether the random behavior was successful is carried out at time $t + \delta$ by checking if the change in position $\vec{p}$ is over some distance d. The exponential term is present to bound the weights w_{ij} and to reduce the rate at which large weights (learned previously) change.

These learned behaviors are triggered only when the agent is caught in a deadlock. However, if deadlocks tend to occur in narrow passageways, then the same set of ESM neurons will be activated during deadlocks (since ESMs represent space egocentrically). A simple update rule is used to learn these map activations, activated each time the agent is in a deadlock:

$$l_i \leftarrow l_i + \eta_m(s_i - l_i)e^{-kl_i{}^2}$$

where s_i is the activation of neuron i, l_i is the corresponding learned value, η_m is the learning rate and the exponential term is used to retain values learned from previous time steps. The agent now activates its learned behaviors whenever its ESM activations match the learned weights.

4 Results

Two learning agents: The agents retain the normal construction sequence when f is inactive. When f is active and an agent is holding a brick, it drops it even if it is not at a drop site (if the agent is not holding a brick, no behavior is selected). Learning occurred faster with increasing p_o but behaviors that did not contribute to breaking the deadlock were also reinforced.

Five learning agents: The weights learned were similar to that of the two agent case in environments with open spaces. If there are narrow passageways, more than two agents are often involved in deadlocks and since the time to break such deadlocks will depend on the number of agents, learning occurs slower than in open spaces.

A learner and a previously trained agent: The learner learns to drop its brick when frustrated and holding a brick, but continues to attempt the brick navigation behavior when not holding a brick. This is because the trained agent immediately performs the drop behavior when it is holding a brick, and does not give the learning agent an opportunity to explore its choice of behaviors.

Bucket Brigade: Five agents (trained in pairs), were placed in an environment where bricks and drop sites were separated by a long corridor. When all agents are within this corridor, a bucket brigade is formed through a sequence of pair-wise interactions between agents.

Map association: Two learners were placed in an environment with a narrow corridor. Most deadlocks occurred within this corridor and the agents associated activations on neurons close to the center of the ESM with deadlocks. Agents then drops discs when these neurons are activated (in any corridor) even with no deadlock.

5 Conclusions and Related Work

The adaptation took place while the agents continued to perform their construction task. The mechanism does not require external supervision as it utilizes the feedback provided by the environment. The bucket brigade behavior was learned from purely local interactions.

Learning was faster when there were two agents learning simultaneously compared to the case when one agent was already trained. Also, in the experiments with the learned ESM activations, only one of the agents could use the learned activations (as otherwise both agents would try to drop discs within corridors). The issue of heterogeneity was not studied here and such conditions were explicitly satisified by the experimenters.

(Ostergaard et al., 2001) studies the performance of the "bucket brigade" behavior in different environments. The use of "frustration" to trigger learning is similar in spirit to the impasse driven learning in Soar (Laird et al., 1987) and to the use of progress estimators to speed up reinforcement learning (Mataric, 1994). The architecture described here is novel in its use of a connectionist action selection mechanism that enables simple learning rules to adapt its behaviors, goals and spatial representation. This connectionist approach can be compared to the action selection mechanism in (Maes, 1991) that connects behaviors with activation spreading links.

References

Chao, G., Panangadan, A., and Dyer, M. G. (2000). Learning to integrate reactive and planning behaviors for construction. In *Proc. of the 6th Intl. Conf. on the Simulation Of Adaptive Behavior.*

Laird, J. E., Newell, A., and Rosenbloom, P. S. (1987). Soar: An architecture for general intelligence. *Artificial Intelligence*, 33:1–64.

Maes, P. (1991). A bottom-up mechanism for action selection in an artificial creature. In *Proc. of the First Intl. Conf. on Simulation of Adaptive Behavior.*

Mataric, M. J. (1994). Reward functions for accelerated learning. In *Machine Learning: Proc. of the Eleventh Intl. Conf.*

Ostergaard, E., Sukhatme, G., and Mataric, M. (2001). Emergent bucket brigading. In *Autonomous Agents.*

Spatial Coordination through Social Potential Fields and Genetic Algorithms

Fabien Flacher[*,**] **Olivier Sigaud**[**]
[*]Dassault Aviation-DGTD/DPR/ESA, 78, Quai Marcel Dassault, 92552 ST-CLOUD Cedex
[**]AnimatLab-LIP6, 8, rue du capitaine Scott, 75015 PARIS
fabien.flacher@lip6.fr olivier.sigaud@lip6.fr

1 Introduction

In the context of this paper, we focus on the design of controllers applied to collective spatial coordination problems. Our agents have a local perception of their environment and are able to modify their nominal behavior to manage unpredictable situations. They must also be coordinated, *i.e.* the global task must be achieved through the interdependent behavior of several agents.

Faced with this complex problem, a traditional functional decomposition methodology performs poorly, because it is not adapted to deal with interdependencies: the elementary interactions between agents cannot be isolated from one another.

In this paper, we show that a great improvement can be achieved on these coordination tasks by giving up the functional decomposition approach. We present another methodology consisting in tuning the elementary interactions between the agents at the micro level so that the required global behavior emerges at the macro level. Relying on that methodology based on emergence gives us the ability to solve complex coordination problems with as few design effort as possible.

In order to design the local interactions between agents, we use Potential Fields Methods (PFMs). But, whereas in most frameworks the potential fields have to be designed by hand, we use a Genetic Algorithm (GA) to automate the tuning of the potential fields so that the agents optimize a global criterion.

2 Our framework

PFMs are generally used for robotic navigation. This class of algorithms acts on the agent as if it were a particle attracted or repelled from features in its environment. Here these features are a set of relevant point L_k. The movement $\overrightarrow{D}$ of an agent A is given by the sum (1) of N *forces* $\overrightarrow{F_i}$:

$$\overrightarrow{D} = \Sigma_{i=0}^{N} \overrightarrow{F_i}, \ \overrightarrow{F_i} = G_i \times fs_i(\| \overrightarrow{AP_i} \|) \times \overrightarrow{AP_i} \quad (1)$$

where each P_i is the source of the force $\overrightarrow{F_i}$ and is obtained from the barycentric combination $\overrightarrow{AP_i} = \Sigma_{k=0}^{m_i} \beta_{ik}.\overrightarrow{AL_k}$; each fs_i is a piecewise linear *function of magnitude* modulating the intensity of the force $\overrightarrow{F_i}$ according to the distance $\| \overrightarrow{AP_i} \|$; G_i is a coefficient representing the relative gain of the force $\overrightarrow{F_i}$. This formalism is very similar to Balch and Arkin's one (Balch and Arkin, 1998).

The behavior of each agent can be expressed by the set of all parameters involved in equation (1). Thus, the GA we use to generate an agent encodes it as a genome of N chromosomes representing the combination of parameters (G_i, P_i, fs_i) for each force $\overrightarrow{F_i}$ influencing the agent. A chromosome is encoded with one real number for the coefficient G_i, ($3 \times m_i$) real numbers for the ranges and types (r_{ik}, t_{ik}) together with the coefficient β_{ik} of all relevant points defining the barycentric point P_i, and ($2 \times n_i$) real numbers for the n_i coordinates (x_{il}, y_{il}) of the segment extremities defining the piecewise linear function of magnitude fs_i.

The genetic operators are adapted to our formalism to prevent the generation of meaningless controllers. For mutation, a random value depending on a normal law is added with a probability P_M to each parameter of the genome. The crossover is realized by copying, with a probability P_C, N_3 chromosomes from the set of ($N_1 + N_2$) chromosomes from the parent individuals.

3 A Case Study

Our environment is inspired from the Robot Sheepdog Project, which involves a robot driving a flock of ducks towards a goal position in a circular arena (Vaughan et al., 1998). We extend this experimental setup to the case where the same task has to be solved by the coordinated effort of several simulated agents.

Our simulator includes a flock of six ducks and some shepherds which must drive the flock towards a goal area. The ducks and the shepherds have the same maximum velocity. The ducks tend to keep away from the walls of the arena, to join their mates and to flee from the shepherds when they are too close.

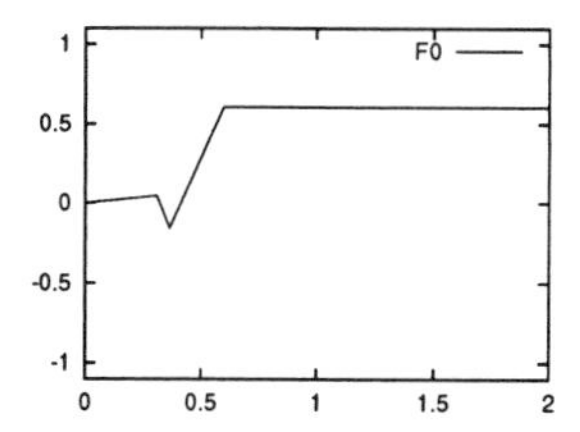

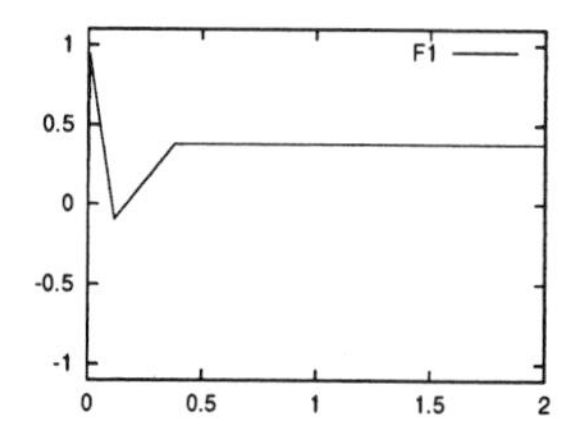

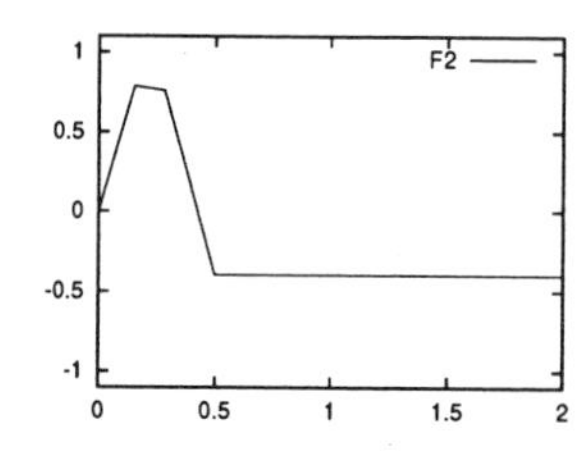

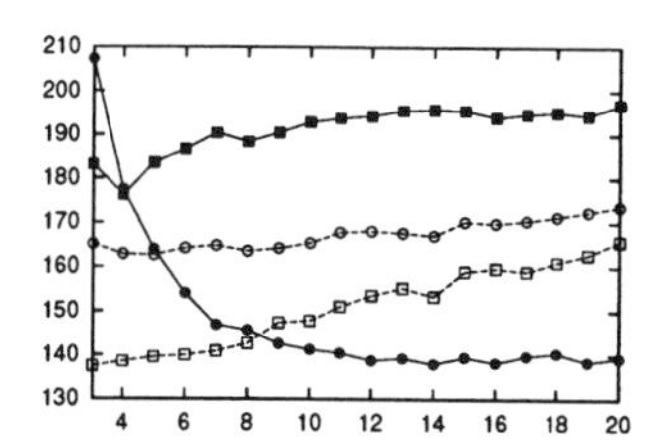

Figure 1: a,b,c) Evoldved functions of magnitude modulating the intensity of the 3 forces F_i influencing the agent d) Performance of the controller C_{150} when the speed of the shepherds is normal (dashed lines with empty shapes) or doubled (plain lines with filled shapes) and when the magnitude functions are nòrmal (circle dots) or constant (square dots). The results are the duration of the run (in time steps) per number of shepherds, averaged over 1000 runs, each run starting with random initial positions

4 Results

In our experiments, all the shepherds share the same controller. Their fitness function is defined by the remaining time after completing the task with respect to the maximum time allowed to fulfill it [1]. The population size N_P is 100 individuals, the maximum time is 500 time steps, and each evaluation of an individual involves 25 trials, with a random number of shepherds ranging from 3 to 20. The probability of mutation P_M is 5% and the probability of crossover P_C is 80%.

In 150 generations we obtain a particularly efficient controller: C_{150}. It reaches its goal in an average of 167.08 time steps (1.51 times better than our best hand-crafted controller obtained through a functional decomposition detailed in (Sigaud and Gérard, 2001)) with a success rate of 99.94% on 1000 different trials. It is scalable since its performance only decreases of 5.18% when we increase the number of shepherds from 3 to 20. C_{150} is also simple as it exhibits only 3 forces $\overrightarrow{F_i}$. Their sources are respectively P_0 associated to the first [2] duck (with a coefficient $\beta_{D1} = 0.85$), P_1 associated to the third ($\beta_{D3} = 9.13$) and fifth ducks ($\beta_{D5} = 2.61$), and P_2 associated to the goal ($\beta_G = 9.08$), to the fourth shepherd ($\beta_S 4 = 2.42$), and to the closest wall landmark ($\beta_{W0} = 0.77$). Their functions of magnitude are shown on figure 1 (a, b, c). The corresponding G coefficients are: $G_0 = 1271.62$, $G_1 = 1758.97$ and $G_2 = 2918.08$.

A detailed study of these parameters allows us to explain how C_{150} drives the shepherds to get aligned with the goal and the flock from behind and, thereafter, to push it towards the goal, with a strategy similar to the one hand-crafted by Vaughan in (Vaughan et al., 1998).

In order to check the robustness of C_{150} and to understand the behavioral interest of its functions of magnitude, we have realized a set of tests, varying the number of shepherds, their speed, and replacing the functions shown in figure 1 (a, b, c) by constant functions. From figure 1 (d), we have shown that there are two different mechanisms involved in the spatial coordination of the behavior of the agents, preventing them from rushing into the flock and scattering the ducks. The first one is due to the shape of the functions of magnitude, the second one to the divergence of attractions when the ducks tend to scatter.

[1] Hence, the higher, the better

[2] *i.e.* closest at that particular time step, since the landmarks are sorted according to their distance to the agent at each time step

5 Future Work and Conclusion

As a general conclusion, we have shown that combining PFMs with GAs is a very powerful methodology which requires a very small design effort and performs better than functional decomposition. Our framework extends Arkin and Balch's approach in a promising direction, which extensively relies on the self-organization mechanisms provided by GA, resulting in a lesser involvement of the designer. As a consequence, however, this methodology requires additional research on analysis tools that would help understanding more accurately how the corresponding controllers work.

References

Balch, T. and Arkin, R. (1998). Behavior-based formation control for multiagent robot teams. *IEEE Transactions on Robotics and Automation.*

Sigaud, O. and Gérard, P. (2001). Being Reactive by Exchanging Roles: an Empirical Study. In Hannebauer, M., Wendler, J., and Pagello, E., (Eds.), *LNAI 2103 : Balancing reactivity and Social Deliberation in Multiagent Systems.* Springer-Verlag.

Vaughan, R., Stumpter, N., Frost, A., and Cameron, S. (1998). Robot Sheepdog Project achieves Automatic Flock Control. In Pfeifer, R., Blumberg, B., Meyer, J.-A., and Wilson, S. W., (Eds.), *From Animals to Animats 5*, pages 489–493, Cambridge, MA. MIT Press.

Learning in Multi-Robot Scenarios through Physically Embedded Genetic Algorithms

Ulrich Nehmzow
Department of Computer Science
University of Essex
udfn@essex.ac.uk, http://cswww.essex.ac.uk/staff/udfn

1 Introduction and Motivation

Motivation. For certain applications of autonomous mobile robots — surveillance, cleaning or exploration come immediately to mind — it is attractive to employ several robots simultaneously. Tasks such as the ones mentioned above are easily divisible between independent robots, and using several robots simultaneously promises a speedup of task execution, as well as more reliable and robust performance.

For any robot operating in the real world, the question of how control is to be achieved is of prime importance. While fixed behavioural strategies, defined by the user, can indeed be used to control robots, they tend to be brittle in practice, due to the noisy and partly unpredictable nature of the real world. Therefore, instead of using *fixed and pre-defined* control procedures, *learning* is an attractive alternative.

To determine a suitable control strategy for a mobile robot operating in noisy and possibly dynamic environments through learning requires a search through a very large state space. By parallelising this process through the use of several robots and collaborative learning, this learning process can be accelerated.

A physically embedded GA (PEGA). In this paper, we present experiments conducted with two communicating mobile robots. Each robot's control policy was encoded through a genetic string. By communicating genetic strings and fitnesses to one another at regular intervals, robots modified their individual control policy, using a genetic algorithm (GA). Contrary to common GA approaches, we did not use a simulate-and-transfer method, but implemented the GA directly on the robots.

We were able to show that the following competences can all be acquired using the PEGA approach:

1. Phototaxis
2. Obstacle avoidance
3. Robot seeking (for communication purposes)
4. Phototaxis *and* obstacle avoidance
5. Phototaxis, obstacle avoidance *and* robot seeking.

2 Experimental Setup

2.1 Mobile Robot Hardware

All experiments were conducted using two small mobile robots, equipped with sonar, infrared, ambient light and tactile sensors (see figure 1). Most importantly, the robots were also able to communicate with each other by means of infrared sensors.

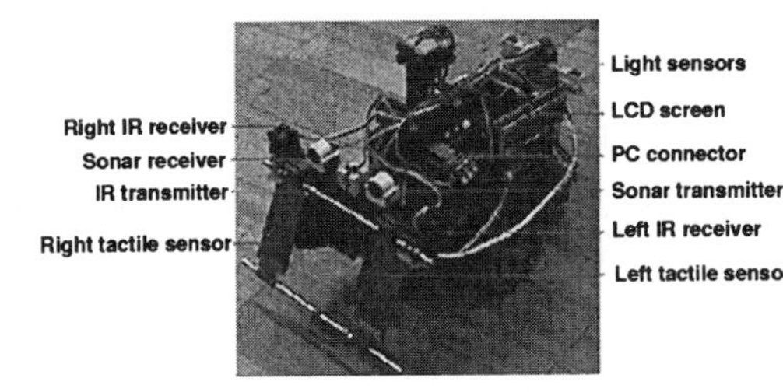

Figure 1: One of the mobile robots used in the experiments.

2.2 PEGA Software

For the first three experiments mentioned in section 1, the robots were equipped with the following pre-installed competences: Obstacle avoidance, using IR and tactile sensors, seeking the other robot, and communication via IR with the other robot.

In the 4th and 5th experiments, more competences were acquired through the PEGA, and fewer were implemented by the designer.

In addition to these competences, each robot had an on-board implementation of a genetic algorithm, including fitness evaluation, mutation, crossover and execution of the control strategy encoded by the string.

In our PEGA, each robot carried two strings (i.e. behavioural strategies): the "active" one that was being evaluated by the GA, and the "best" one found so far (the string population is therefore two). The "best string so far" was used only as a fallback option if the GA did not produce a "better" solution than the "best" one so far. "Better" was defined here by a higher fitness value of the evaluated string.

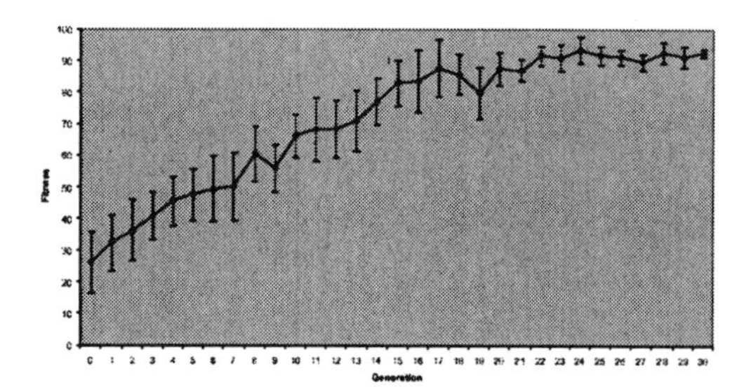

Figure 2: Learning phototaxis: Mean fitness and standard deviation of ten independent runs, against learning time (generations).

Experimental procedure. The two robots were left to execute the behavioural strategy encoded by the current string for a certain amount of time, evaluating the fitness while they were doing this. After the alloted time had expired, the robots initiated a search behaviour (based on infrared signal emissions) to locate the other robot. Once found, robots faced each other and exchanged their current strings and corresponding fitnesses through infrared communication.

2.2.1 Implementation of the PEGA

Crossover. After the exchange of strings and fitnesses, crossover and mutation were applied in the following manner. If the received remote string was fitter than the local one, one half of the local string (determined by a random process), was replaced by the corresponding part of the remote string.

If the local string had a higher fitness than the remote one, no crossover between local and remote string happened. However, if in this case the locally stored "best" string had a higher fitness than the currently used local string, there was a 30% probability that a randomly selected bit of the current string was replaced by the corresponding bit of the "best" string. In the remaining 70% of cases, mutation was applied to the current string.

Mutation. If invoked, the likelihood of mutation of a string was dependent upon a string's fitness. This probability p_m of changing one randomly selected bit is given by $p_m = R\frac{100-F}{100}$, where F stands for the fitness of the string, and $R = 0.3$ is a constant.

3 Experimental Results

The robots acquired all five competences listed in section 1 completely, within a few tens of generations. Average fitness versus the number of generations are shown in figures 2 to 5; full details are given in [Nehmzow 2002].

4 References, background and details

Full details, including related work, detailed description of the algorithm, experimental procedure, results in detail and

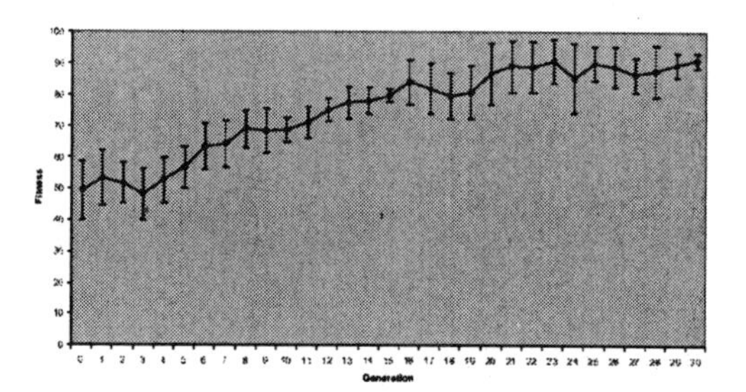

Figure 3: Learning obstacle avoidance: Mean fitness and standard deviation of ten independent runs, against learning time (generations).

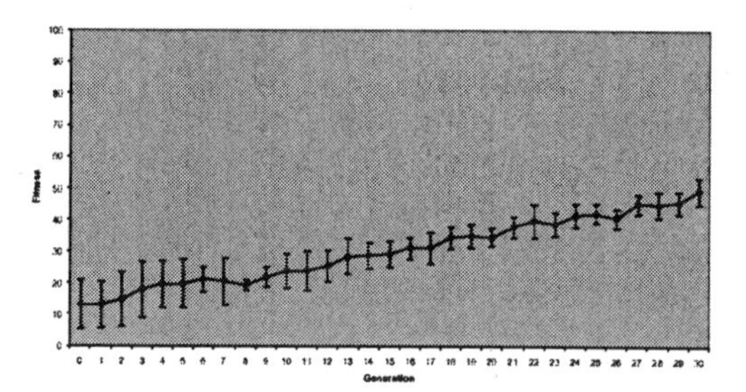

Figure 4: Learning to find other robots: Mean fitnesses and standard deviation of ten independent runs, against learning time (generations).

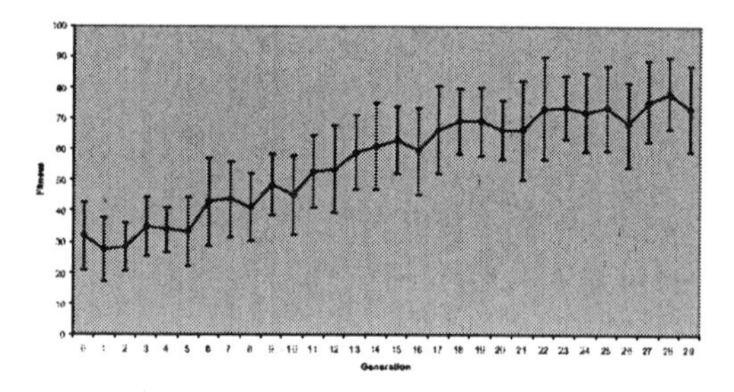

Figure 5: Learning phototaxis *and* obstacle avoidance: Mean fitnesses and standard deviation of ten independent runs, against learning time (generations).

a discussion can be found in

[Nehmzow 2002] Ulrich Nehmzow, *Physically Embedded GA Learning in Multi-Robot Scenarios: The PEGA algorithm*, Technical Report Nr. CSM-367, Department of Computer Science, University of Essex. Available at `http://cswww.essex.ac.uk/technical-reports/2002.htm`.

Acknowledgements This project was carried out in collaboration with my colleague Jonathan Shapiro of Manchester University — I acknowledge his contribution with thanks.

The experimental work reported in this paper was carried out by our project student Mark Johnson, and I acknowledge his contribution gratefully.

Designing Social Force: Control for Collective Behavior of Learning Individuals

Keiki Takadama*
*ATR Human Information Science Labs.
2-2-2 Hikaridai, Seika-cho, Soraku-gun
Kyoto 619-0288 Japan
keiki@atr.co.jp

Katsunori Shimohara**
**ATR Human Information Science Labs.
2-2-2 Hikaridai, Seika-cho, Soraku-gun
Kyoto 619-0288 Japan
katsu@atr.co.jp

Abstract

This paper explores methods for controlling the collective behavior of learning individuals (*e.g.*, robots, agents) to achieve goals at the collective level while these individuals pursue their own goals at the individual level. Toward this enormous goal, we start by categorizing the four levels for controlling individuals in terms of both direct and indirect control, and then investigate the results at each control level to clarify the extent to which degree we should force the learning individuals to achieve goals at the collective level.
Keywords: collective behavior, social force, direct and indirect control, learning individuals

1 Introduction

In complex and dynamic environments where individuals interact with each other to achieve their own goals at the individual level, how can we control the collective behavior of individuals (*e.g.*, robots, agents) to achieve goals at the collective level? This is an important issue requiring clarification for both practical and engineering purposes. Such clarification could contribute to the design of methods that would enable us to control collective behavior for obtaining desired results. Some examples include a communication design in multiagent environments (Balch and Arkin, 1995), a framework for team behaviors (Collinot et al., 1996), primitive behaviors for adaptive group behaviors (Mataric, 1995), and swarm intelligence (Kennedy et al., 2001). However, when controlling *learning* individuals, it is generally difficult to design such control methods at the collective level, because individuals independently learn according to their own goals, evaluations, and local information.

To address this issue, we start by investigating the collective behavior of learning individuals from the control level viewpoint. This is because (1) it remains an open question to determine to what degree we should force learning individuals at the control level to achieve goals at the collective level; and (2) such an investigation has the potential of exploring *social force*, which means here to have (a) the power to change the collective behavior of individuals without excessively restricting them and (b) the capability of encouraging individuals to achieve goals at the collective level. From these reasons, this paper aims to find implications related to the control of the collective behavior of learning individuals.

2 Control Level for Collective Behavior

The several types of control for the collective behavior of learning individuals can roughly be divided into *direct* control and *indirect* control. To understand these two types of control, let's focus on multiple rovers controlled by a centralized system, where all rovers have their own goals as well as their own evaluations for achieving their goals. When a centralized system can *directly* control multiple rovers via a command, the system has the power to force multiple rovers to determine their behaviors. When a centralized system can *indirectly* control multiple rovers by evaluating their behaviors, on the other hand, the system has the power to encourage multiple rovers to cooperate with each other.

Employing the above two fundamental types of control, *i.e.*, direct and indirect control, this paper categorizes the following four levels of control over the collective behavior of learning individuals. This categorization is based on the notions that (1) the level without control is lowest in comparison with the other control levels; (2) the level of indirect control is lower than that of direct control; and (3) the level of an integration of direct and indirect control is the highest.

- **Level 0 (No control):** Individuals (*e.g.*, rovers) pursue their individual-level goals and their behaviors are evaluated according to their individual-level results.
- **Level 1 (Evaluation control):** Individuals pursue their individual-level goals, while their behaviors are evaluated according to their collective-level results by a centralized system through indirect control.
- **Level 2 (Behavior control):** Individuals are forced

to pursue their collective-level goals by a centralized system through direct control, while their behaviors are evaluated according to their individual-level results.

- **Level 3 (Both evaluation and behavior control):** Individuals are forced to pursue their collective-level goals and their behaviors are evaluated according to their collective-level results by a centralized system through both direct and indirect control.

3 Simulation

3.1 Experimental design

A simulation employed the aggregation task of rovers, which aims to control rovers by bringing them together at a certain location. We investigated collective behaviors of rovers from the viewpoint of the four control levels described in Section 2. To simulate this task, each rover is implemented by a Learning Classifier System (Goldberg, 1989), and both goal and evaluation in this task are designed as follows.

- **Goal:** An *individual-level (rover) goal* is to be located near the center of the aggregation area, while a *collective-level (centralized system) goal* is to place all rovers in an area that is as small as possible.
- **Evaluation:** An *individual (rover) evaluation* is made by the distance from the center of the aggregation area, while a *collective (centralized system) evaluation* is made by the size of the area enclosing all of the rovers.

These goal and evaluation designs seem to work toward the same direction, but not all of the rovers can achieve their individual-level goals because some of the rovers have to be placed outside of them. Due to this characteristic of the task, many rovers have to compete with each other to obtain a good location, which makes it difficult to achieve the collective-level goals by pursuing the individual-level goals.

3.2 Experimental results

Through intensive simulations on the aggregation task of rovers, we found the following implications: (1) neither the solutions (*i.e.*, the smallest area) nor the computational costs for finding good solutions become better or smaller, respectively, as the degree of control becomes stronger from Level 0 to Level 3; (2) the behavior control does not offer any improvement of the solutions and merely increases the computational costs, while the evaluation control contributes to reducing the computational costs with keeping the solutions at basically the same level.

3.3 Discussion

From our results, the collective behaviors of learning individuals cannot be controlled by simply increasing the control level from weak to strong but by introducing the evaluation control in comparison with the behavior control. This indicates that the evaluation control can be considered a good *social force* that has (1) the power to change the collective behavior of individuals without excessively restricting them and (2) the capability of encouraging individuals to achieve goals at the collective level. This type of social force is indispensable for the case of many rovers that are limited in control as the number of rovers increases and for the case of individuals such as human beings that are difficult to directly control.

4 Conclusions

This paper explored methods for controlling the collective behavior of learning individuals to achieve goals at the collective level, while these individuals pursue their own goals at the individual level. The results obtained in this paper do not cover all of the available methods for controlling learning individuals and we will need careful qualifications and justifications to strengthen our claims. However, the results contribute to clarifying the extent to which degree we should force learning individuals to achieve goals at the collective level. Specifically, the evaluation control has great potential for achieving collective goals by controlling learning individuals.

Acknowledgements

The research reported here was supported in part by a contract with the Telecommunications Advancement Organization of Japan entitled, "Research on Human Communication."

References

Balch, T. R. and Arkin, R. C. (1995). "Communication in Reactive Multiagent Robotic Systems," *Autonomous Robots*, Vol. 1, No. 1, pp. 27–52.

Collinot, A. et al. (1996). "Agent Oriented Design of a Soccer Robot Team," *The 2nd International Conference on Multi-Agent Systems (ICMAS'96)*, pp. 41–47.

Goldberg, D. E. (1989). *Genetic Algorithms in Search, Optimization, and Machine Learning*, Addison-Wesley.

Kennedy, J. et al. (2001). *Swarm Intelligence*, Morgan Kaufmann Publishers

Mataric, M. J. (1995). "Designing and Understanding Adaptive Group Behavior," *Adaptive Behavior*, Vol. 4, No. 1, pp. 51–80.

Sequential Task Execution in a Minimalist Distributed Robotic System

Chris Jones **Maja J. Matarić**
Computer Science Department
University of Southern California
941 West 37th Place, Mailcode 0781
Los Angeles, CA 90089-0781
{cvjones|maja}@robotics.usc.edu

1 Introduction and Related Work

A Minimalist Distributed Robotic System (MDRS) is a society of simple robots each with limited sensing, communication, and intelligence capabilities. We study ways of providing such MDRS with the capability of *sequential task execution, which requires a set of tasks to be executed in a specified order, with the initiation of a task occurring only after the termination of a required prior task.*. Sequential task execution capabilities greatly increases MDRS functionality.

The accomplishment of a set of sequential tasks requires sufficient information about the progress on the task in order to determine the appropriate action to take at any given time, and in particular at key steps of transitioning between tasks. However, in a MDRS, because of the robots' very limited sensing, intelligence, and communication capabilities, there are many domains in which gathering information on the current state of task progress, part of the global state of the environment, may not be possible for the individuals in the system. Formally, to the individuals in a MDRS, the world is partially-observable and highly non-stationary, yet they must collectively achieve a global goal whose changing state they cannot perceive.

2 Sequential Foraging Task

In the domain of MDRS, the standard, non-sequential foraging task has been studied extensively. We are using a sequential variation of foraging, in order to investigate the capabilities of a MDRS on sequential task execution. Sequential foraging requires a collection of objects (pucks) to be collected in a specified order, based on their color.

Toward proper evaluation of sequential foraging algorithm performance, we developed a cumulative metric that reflects the sequential requirements of the task. The metric, initialized to 0 at the start of every experiment, is updated at every simulation time-step. At each update, for all pucks $Puck_{New}$ deposited in the home region at time t, the utility value, Util(t), is updated according to the procedure:

```
Util(t) = Util(t-1)
for all puck in Puck_New
   if (puck == Puck_Green) then
      Util(t) = Util(t) + Prop_Red
   else if (puck == Puck_Blue) then
      Util(t) = Util(t) + Prop_Red * Prop_Green
```

At the end of an experimental trial, terminated at time t_{Final}, the sequential foraging algorithm is given a final utility value where TP_{Green} and TP_{Blue} are the total number of $Puck_{Green}$ and $Puck_{Blue}$ in the environment, respectively.

$$Util_{Final} = 100 * (Util(t_{Final})/(TP_{Green} + TP_{Blue}) \quad (1)$$

3 Experimental Environments

The experimental environment consists of an arena with an initial collection of randomly distributed pucks and a home region on one side to which the pucks are to be transported. Whenever a puck is deposited in the home region, it is removed from the arena. We used a group size of four robots and fixed initial locations.

Our experimental design involved the use of four different environment variations on the above arena with characteristics shown in Table 1. The environments were designed to evaluate the adaptability of sequential foraging algorithms along two dimensions: 1) the relative puck type proportions and 2) the arena size.

Env	Arena Size(m)	Red/Green/Blue Pucks
1	8.75x8.75	8/8/8
2	8.75x8.75	14/8/2
3	8.75x8.75	2/8/14
4	17.5x17.5	8/8/8

Table 1: Experimental Environments

4 Sequential Foraging Algorithms

Two foraging algorithms were considered: Timer-Based Foraging and Probabilistic Foraging. These were investigated and analyzed to assess their effectiveness in the sequential foraging task and their adaptability to different environmental characteristics.

In the Timer-Based Foraging algorithm, each robot uses an internal timer to dictate which puck type should be foraged. Each robot has its own independent timer; timers across robots are not explicitly synchronized.

Each robot's timer, Timer_{Robot}, is initialized to 0 at the beginning of an experiment and incremented by 1 at each simulation time-step. A set of timer alarms are used to control which puck types can be foraged at a given Timer_{Robot} value. There is a timer alarm for each puck type: Alarm_{Red}, Alarm_{Green}, Alarm_{Blue}, respectively. When a puck is detected, if the Timer_{Robot} value is greater than the timer alarm for the detected puck type, the robot's Timer_{Robot} value will be reset back to the alarm value of the detected puck type and the robot will begin visually servoing toward the detected puck. A full description of the behavior network used to implement the Timer-Based Foraging algorithm can be found in (Jones and Matarić, 2001).

The Probabilistic Foraging algorithm uses two probabilistic behavior activation conditions in each robot's behavior network. A full description of the behavior network for the Probabilistic Foraging algorithm can be found in (Jones and Matarić, 2001).

The first probabilistic activation condition introduced is whether a robot should visually servo toward a detected puck or ignore the detected puck and perform a random walk. Each robot has an assigned probability of ignoring a detected puck of each type. For the three puck types, these probabilities are: PIgnore_{Red}, PIgnore_{Green}, and PIgnore_{Blue}, respectively.

The second probabilistic activation condition is whether a grasped puck should be dropped before reaching the home region or whether it should continue to be transported toward the home region. Each robot has an assigned probability of dropping a grasped puck of each type while not in the home region. For the three puck types, these probabilities are: PDrop_{Red}, PDrop_{Green}, and PDrop_{Blue}, respectively.

5 Experimental Results

All experiments were performed in simulation using Player (Gerkey et al., 2001) and Stage (Vaughan, 2000) with realistic simulations of the Pioneer 2DX mobile robot.

In the Timer-Based Foraging algorithm, the Alarm_{Red}, Alarm_{Green}, and Alarm_{Blue} values used in all experiments were 0, 750, and 1500, respectively. In the Probabilistic Foraging algorithm, the PIgnore_{Red}, PIgnore_{Green}, PIgnore_{Blue} values used were 0, 0.065, and 0.12. The values used for PDrop_{Red}, PDrop_{Green}, PDrop_{Blue} were 0, 0.65, and 0.12. For each experimental environment and sequential foraging algorithm pair a total of five trials were run. The average Util_{Final} and Standard Deviation values are shown in Table 2.

Alg	Env 1	Env 2	Env 3	Env 4
Timer-Based	99.7/0.5	97.7/0.8	97.0/2.4	79.8/2.3
Probabilistic	96.1/1.6	97.7/2.4	86.7/6.4	98.6/1.4

Table 2: Experimental Results (Util_{Final}/S.D.)

As the experimental results show, the Timer-Based Foraging algorithm adapts well along the dimension varying relative puck type proportions, while the Probabilistic Foraging algorithm adapts well along the dimension of varying arena size. A more detailed analysis of the experimental results can be found in (Jones and Matarić, 2001).

6 Conclusions

The robots in our MDRS maintained little or no state information, extract a limited amount of information from available sensors, and cannot explicitly communicate with other robots in the system. The aim of this work is to provide such a MDRS with the capability of sequential task execution. We presented two experimentally validated sequential task execution algorithms, Timer-Based behavior activation and Probabilistic behavior activation, and experimentally verified and compared them in a sequential foraging task.

References

Gerkey, B., Vaughan, R., Stoey, K., Howard, A., Sukhatme, G., and Matarić, M. (2001). Most valuable player: A robot device server for distributed control. In *IEEE/RSJ International Conference on Intelligent Robots and Systems, (IROS-01)*, pages 1226–1231.

Jones, C. and Matarić, M. (2001). Sequential control in a minimalist distributed robotic system. *Institute for Robotics and Intelligent Systems Technical Report IRIS-02-414, Univ. of Southern California.*

Vaughan, R. (2000). Stage: A multiple robot simulator. *Institute for Robotics and Intelligent Systems Technical Report IRIS-00-393, Univ. of Southern California.*

Fleet dynamics and information exchange simulation modeling with artificial neural networks

Michel Dreyfus-Leon [1,2]
[1]Instituto Nacional de la Pesca-PNAAPD,
[2]Facultad de Ciencias Marinas, UABC,
México
PMB-070, PO BOX 189003
Coronado, CA 92178-9003 USA
dreyfus@cicese.mx

Daniel Gaertner[3]
[3]Institut de la Recherche pour le Développement
Avenue Jean Monnet
BP 171, 34203 Sete Cedex FRANCE
gaertner@ird.fr

Abstract

A fishery is simulated in which 20 artificial vessels learn through an artificial neural network to make decisions to search for fish among the fishing grounds available. The simulation model keeps several characteristics of a purse seine tuna fishery. Vessels strategies are chosen by the artificial neural network, where the elements considered for decision are: time searching and performance in an area, knowledge of the quality of several fishing grounds, presence of other vessels and trip duration. In particular an analysis of the effects of sharing information between vessels is done. Results show an effect in performance as well as a bias in the fish abundance index estimates calculated from data from the artificial vessels fishing as a group.

1. Introduction

In fisheries research an important aspect to understand is fleet dynamics since the fishing effort magnitude and distribution is related to fish mortality. It is important for fishery management to know how the fleet will adjust to changing circumstances. Fishermen adapt to regulations in ways managers can not predict. Nevertheless most fishery studies include human behavior, only as an aggregate variable for modeling the displacement of fishing effort to nearby areas in relation to local abundance. Several studies intend to understand fishermen behavior at sea (Gillis et al., 1993; Gaertner et al., 1999) and learning has being considered in modeling by Dreyfus-Leon, (1999).

Another important aspect in fisheries, such as tuna purse-seine fisheries, is communication and information sharing between groups of fishermen (vessels). In this work a simulated tuna fishery is developed. The effects of sharing information between fishermen are analyzed by the use of an artificial neural network (ANN). A comparison of the performance between group and independent artificial fishermen is done as well as an analysis of catch per unit effort (CPUE) as a fish abundance index under such conditions.

2. Methodology

Seven potential fishing grounds were simulated over a square world with 25 areas of 50 x 50 pixels. Starting each year of the simulation, fish recruitment might change in abundance and in location between the fishing grounds. Schools of fish move at random within a fishing ground without the possibility to move elsewhere. Since each pixel represents 6 x 6 nautical miles, according to reported tuna speed, fish schools move every 6 hours in the simulations. Fish natural mortality is also considered in order to keep the fish population relatively stable.

Twenty artificial vessels are considered and depart from the same port, after a time assigned at random (between 10 and 20 days). Movement is performed in an hourly base and trip duration is fixed at 45 days. When a vessel is searching (only during daytime), a tuna school is detected and caught if both have the same position. To account for fishing operations, the vessel stays in a fixed position for 3 hours. Those characteristics are similar to the real purse-seine fishery. When an artificial vessel is searching in a particular area, movement is at random.

The decision to keep searching in the present area or to move to a different one is done at night only if the vessel has been searching for at least 12 hours. This decision is done by an ANN which has been previously trained. If a vessel decides to keep searching in the same area, during the night period it stays at a fixed position, otherwise it keeps moving day and night until it enters the selected area.

The ANN at the input layer receives information of presence of other vessels fishing in the area, trip duration, time spent searching in the area, knowledge of present fishing performance and knowledge of the area quality for fishing based on self-experience or group knowledge and of other areas close or far away. At the output layer 4 neurons predict benefits or costs of each possible decision: stay or move to the best nearby, mid-distant or far away area.

The ANN is trained with standard backpropagation methodology and reinforcement learning. Reinforcement learning is tantalizing because learning occurs through trial and error experimentation within the environment. Feedback is a scalar payoff, hence no explicit teacher is required, and little or no prior knowledge is needed (Whitehead and Lin, 1995). Rewards and punishments are related to benefits and costs of fishing operations of the purse seine tuna fishery in the Eastern Pacific Ocean.

3. Results

Vessels belonging to a group take independent search decisions. Five runs with group size composed with two to twelve artificial vessels each, where performed. The remaining vessels, fishing on their own, were considered as the control group. The difference in performance (number of tuna schools caught per trip) between group and independent vessels increases, reaching an optimum at the 10-vessels group size. Due to competitive interactions, the efficiency of groups larger than 10 vessels starts to decline. The time needed to sample all fishing grounds by artificial vessels belonging to a group sharing knowledge decreases exponentially as the group size increases.
An account of vessels in a high fish density area with an optimum size of 10 vessels for the group sharing information is presented in figure 1. As recruitment occurs,

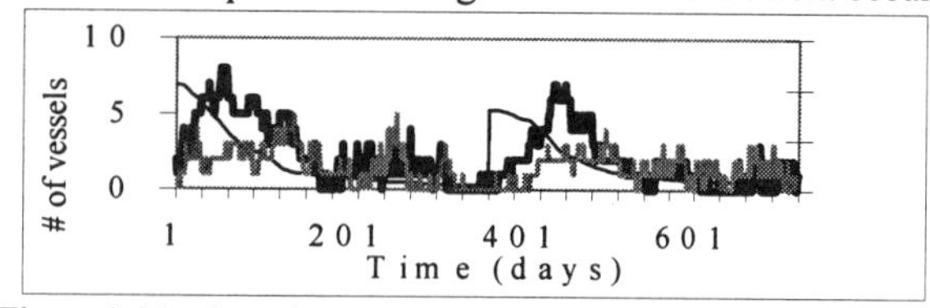

Figure 1. Number of vessels in a high fish density fishing ground. Fish school abundance (thin line), vessels sharing information (bold line), independent vessels (gray line).

some group-vessels start moving in, while as density decreases the opposite occurs. With independent artificial vessels there is no particular trend in the number of vessels in the area.
CPUE, traditionally used as an index of fish abundance, was calculated both for individual and group vessels using days searching as an effort measure. The correlation between CPUE and simulated abundance was calculated each two years of the simulation, and plotted over the simulated time (figure 2).

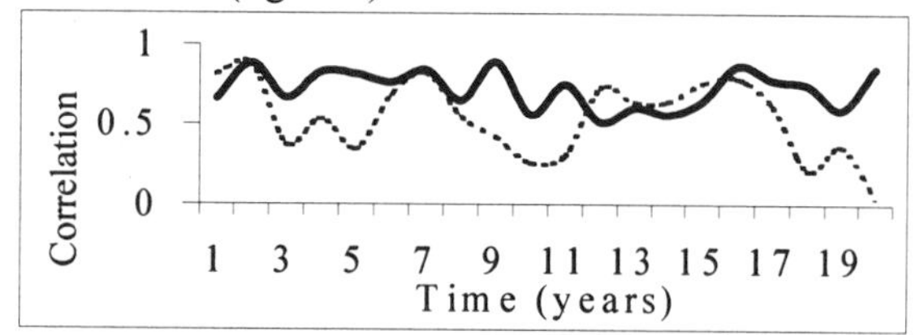

Figure 2. Correlation between real abundance and CPUE for vessels fishing independently (black line) and vessels sharing information (dotted line).

Independent artificial vessels maintain approximately a 0.8 correlation while vessels belonging to a group show a more variable correlation.

4. Discussion

The advantages of belonging to a sharing information group in the real world would be greater specially under some degree of uncertainty related to the distribution and density of fishing resources at sea. Sharing information reduces the time to sample all potential fishing grounds and the acquisition of some knowledge of actual conditions of the resources. In the simulations many of the group vessels exploited high fish density areas close to a peak in fish abundance and several of those vessels started to move elsewhere when fish density decreased. Individual fishing vessels never get concentrated in those high density areas.
The optimum group size in real fisheries seems to depend on uncertainty levels, fish density and the extent of fishing grounds. The fact that group sharing vessels have at some time-intervals low correlation between fish abundance and CPUE is related to a better performance of the former. Vessels belonging to a group tend to concentrate in higher fish density areas although total fish abundance can be low, thus creating a bias in CPUE estimates. Nonrandom fishing strategies generates a nonlinear relationship between abundance and CPUE (Hilborn and Walters, 1987).
In fisheries where information sharing groups occur, a biased CPUE can cause bad management decisions and harm exploited fish populations.

References

Dreyfus-Leon, M.J. 1999. Individual-based modelling of fishermen search behaviour with neural networks and reinforcement learning. *Ecol. Model.* 129 (1-3), 287-297.

Gaertner D., M. Pagavino, J. Marcano, 1999. Influence of fishers behaviour on the catchability of surface tuna schools in the Venezuelan purse-seiner fishery in the Carribbean Sea. *Can. J. Fish. Aquat. Sci.* 56:394-406.

Gillis, M.G., R.M. Peterman, A.V. Tyler, 1993. Movement dynamics in a fishery: application of the ideal free distribution to spatial allocation of effort. *Can. J. Fish. Aquat. Sci.* 50:323-333.

Hilborn, R., C.J. Walters, 1987. A general model for simulation of stock and fleet dynamics in spatially heterogeneous fisheries. *Can. J. Fish Aquat. Sci.* 44:1366-1369.

Whitehead, S.D. and L.J. Lin., 1995. Reinforcement learning of non-Markov decision processes. *Artificial Intelligence*, 73,271-306

What is Important for Realizing an Autonomous Interactive Robot?

Takanori Komatsu, Kentaro Suzuki Natsuki Oka**
Kazuhiro Ueda, Kazuo Hiraki*

*The University of Tokyo
Komaba 3-8-1, Meguro-ku
153-8902 Tokyo, JAPAN
komatsu@cs.c.u-tokyo.ac.jp

**Matsushita Electric Industrial, Co., Ltd.
3-4, Hikaridai, Seika, Souraku
619-0237 Kyoto, JAPAN

Abstract

The requirements for developing an autonomous interactive robot for smooth man-machine interaction were determined by means of experiment: The game "Pong" was played by a pair of subjects (one is an operator and the other is a teacher) who did not share the same language. The results of the experiment revealed two phenomena. First, the process of understanding the instructions can be regarded as a reinforcement learning processes based on different rewards (i.e., one reward for successful game action or the other one from the teacher's high-pitched voice for the wrong action). The second was mutual adaptation in which subjects learned to respond appropriately to one another's behavior. It is expected that these two phenomena can be applied to an autonomous interactive robot that can establish intimate communication with humans.

1. Introduction

The purpose of this study is to clarify the requirements for developing an autonomous interactive robot that could smoothly interact with humans. Several researchers have recently tried to develop robots that can coexist with humans and to realize man-machine communication based on a human language system (Roy, 1999; Kaplan, 2000). However, so far, these attempts have had limited success. With this background in mind, the authors have chosen a different approach; namely, to view speech not as linguistic information, but as a simple sound signal. By applying this idea to an interactive robot, the authors expect to realize the robot that can communicate simply with humans like pet animals do; that is, they cannot understand what its owner actually said linguistically but they can communicate intimately. To imitate the relationship like that between a pet and its owner, an autonomous interactive robot would have to meet certain requirements for its realization. It is these requirements that are clarified in the current study.

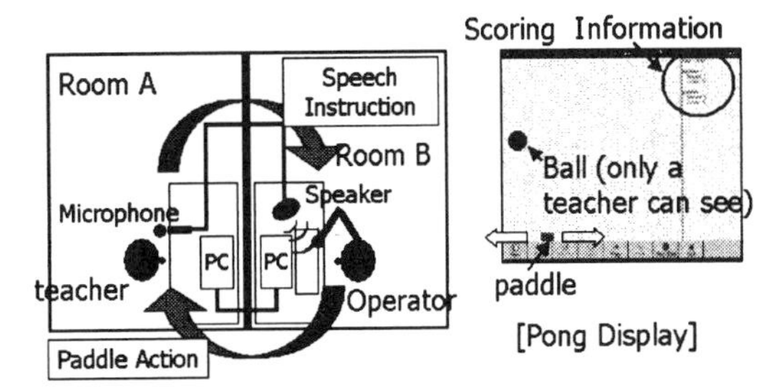

Figure 1: Experimental Setup

2. Experiment

Experiment to observe the interaction processes, like that between a pet and its owner, was carried out in the form of a simple "Pong" video game. In this game, one of the two participants (teacher) gave verbal instructions to the other participant (operator), who then attempted to play the game according to the given instructions (Figure 1). The teacher could see the ball on the display but the operator could not, while both can see the scoring information and the paddle. Moreover the pair of participants did not share the same language, so that the operator was required to understand the teacher's instructions even though s/he could not figure out linguistically what the teacher actually said. If the subjects could get a high score in spite of such a poor communication environment, it can be said that they were able to establish simple communication similar to that between a pet and its owner. Thus, if certain phenomena, which affected the establishing process of successful communication, were observed, these would be the requirements for realizing an autonomous interactive robot.

Seven pairs of subjects participated in this experi-

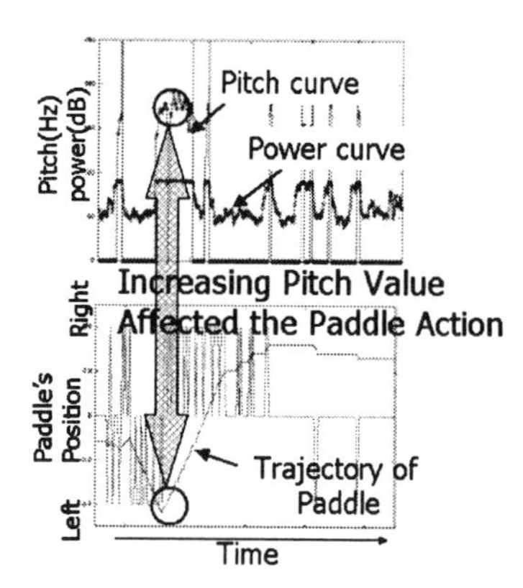

Figure 2: High-pitched Voice and Paddle Action

ment[1]. Each pair played the Pong game for ten minutes. During the game, the roles of the teacher and the operator were fixed, and they did not have the opportunity to face each other. Since they did not share any linguistic elements, the authors focused on the prosodic elements of the teacher's verbal instruction

3. Results and discussion

Out of the seven pairs of subjects, two failed to understand any instructions and among the five remaining pairs succeeded in understanding the instructions. The behaviors of these five successful pairs can be summed up as two phenomena.

First, after the operators understood the instruction types according to different sounds, they acquired the meaning of the instruction as the correspondence between the teacher's speech and their action. To induce the correct correspondence, game points were used as a positive rewards and the teachers' high-pitched voice was used as a negative one. Figure 2 shows that, when the operator heard the teacher's high-pitched voice, s/he changed the direction of the paddle action. Thus, the process of understanding the teacher's instructions can be regarded as a **reinforcement learning processes** based on the two kinds of rewards. Second, **mutual adaptation processes**, in which subjects learned to respond appropriately to one another's behaviors, were observed. Namely, not only the operator adapted to the teacher's instruction, but also the teacher changed the teaching strategy according to the operator's comprehension. Thus, the authors could consider that these two phenomena would be the requirements for realizing an autonomous interactive robot.

In applying these two requirements for an actual robot, the authors consider a concrete application like a pet robot that can obey the owner's instructions such as "sit down' or "come here". With the traditional method, the designer needs to define a mapping list between certain speech instructions and the robot's function (e.g., voiced "sit down" means [kneeling]). However, by applying the two phenomena described above, the designer does not need to define the mapping list *a priori*, but simply need to define "[owner's high-pitched voice = negative reward] and [successful action = positive reward]". Thus, the robot can learn the mapping between unknown speech instructions and its action repertoires by using a reinforcement learning based on the two types of rewards.

This proposed method would have at least three advantages over the traditional one. First, the mapping between the owner's speech and the robot's function does not have to be defined *a priori*. Therefore, the owner can give instructions in her/his own natural way and does not go to the trouble of learning how to give the effective commands. Second, the robot's learning process does not depend on a specific language, so the robot could function effectively regardless of the language spoken by owner. Third, the way of interaction between a robot and its owner is dependent on their own relationship, so that the-pair-specific-communication, which others cannot figure out, could be emerged. Thus, it is expected that this communication would realize an intimate relationship like that between a pet and its owner.

4. Conclusions

To clarify the requirements for an autonomous interactive robot, the authors carried out the experiment to observe the process of establishing smooth communication between subjects who do not share the same language. This experiment revealed two phenomena: First, reinforcement learning based on different rewards (a positive one for successful actions and a negative one from the teacher's high-pitched voice); second, a mutual adaptation process in which subjects learned to respond appropriately to one another's behavior. It is expected that these two phenomena can be applied to an autonomous interactive robot that can establish intimate communication with humans.

References

Kaplan, F.(2000). Talking AIBO : First experimentation of verbal interactions with an autonomous four-legged robot, In A. Nijholt, D. Heylen. & K. Jokinen(Eds.), *Learning to Behave: Interacting agents CELE-TWENTE Workshop on Language Technology*, 57–63.

Roy, D.(1999). *Learning from sights and sounds: A computational model.* Doctoral dissertation, MA: MIT Media Laboratory.

[1] Seven pairs are three Chinese (teacher) and Japanese (operator), one Indonesian and American, one Spanish and Philippine, one Portuguese and Spanish and one Korean and Chinese pair.

Synthetic Social Relationships in Animated Virtual Characters

Bill Tomlinson & Bruce Blumberg
Synthetic Characters Group, MIT Media Lab
NE18-5FL, 77 Massachusetts Ave., Cambridge, MA 02139
+1.617.452.5611
{badger | bruce}@media.mit.edu

ABSTRACT

We describe a multi-agent system based on the social behavior of the gray wolf (*Canis lupus*). This system, shown as an interactive installation at SIGGRAPH 2001, allows several participants to direct semi-autonomous wolf pups in a virtual pack. The heart of the system is a simple, biologically-inspired mechanism by which synthetic entities form social relationships with each other. This mechanism enables the virtual wolves to form relationships with each other that are both biologically plausible and engaging to participants in the installation. Systems like the one described in this paper could be of use in a variety of domains, for example, as platforms for simulation, as educational aids, and as entertainment media.

1. INTRODUCTION

Animals provide an excellent example of autonomous interacting entities. In particular, mammalian social behavior enables multiple complex individuals to interact repeatedly over relatively long time scales. Context preservation [Cohen 1999] is the essence of social behavior – behaving differently toward individual social partners. A social relationship is a remembered construct by which an individual keeps track of its interaction history with another individual, and allows that history to affect its current and future interactions with that individual.

The gray wolf (*Canis lupus*) is one species in which individuals form clear, long-term social relationships with one another [Mech 1998]. In an effort to create a group of virtual creatures that can form relationships with one another, the Synthetic Characters Group at the MIT Media Lab, headed by Professor Bruce Blumberg, has created an interactive multi-agent system based on the behavior of packs of gray wolves. In this paper, we present the system and the mechanism by which the virtual wolves form their relationships.

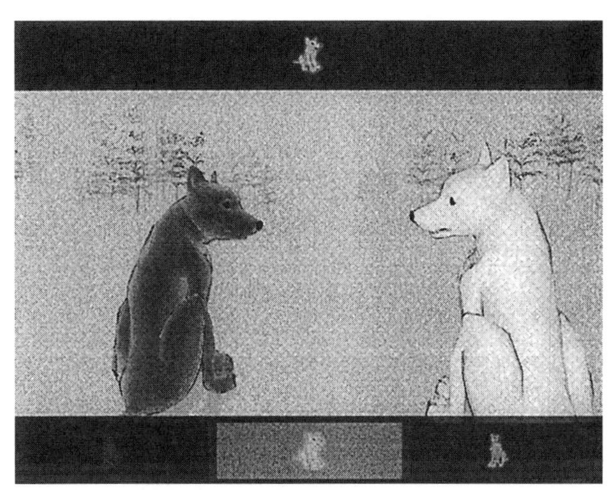

Fig. 1: Two social computational systems exchange a glance.

2. INSTALLATION

An interactive installation featuring the model of social relationship formation presented in this paper was shown in the Emerging Technologies section of SIGGRAPH 2001 (see Figure 1). In the installation, entitled *AlphaWolf*, three participants help direct the actions of virtual wolf pups in a simulated litter. By howling, growling, whining or barking into a microphone, each participant can tell his pup to howl, dominate, submit or play. The actions that the pup takes affect emotional relationships that the pup forms with its littermates and with the adults of the pack. The pups autonomously maintain these relationships and display them by means of the emotional style in which they take the actions suggested by the participants. The relationships are also displayed to the participant through dynamic buttons at the top and bottom of each screen, which show each of the social partners of that wolf in a dominant or submissive pose that reflects how the participant's pup views that partner.

In addition to the three pups, there are three fully autonomous adults who also inhabit the virtual world. They are similar to the pups in their emotional relationship formation, and use these relationships to decide how to interact with the other members of the pack.

Over the course of the five-minute interaction, each puppy grows up from pup to adult size. By the end of the five-minute interaction, the pups, guided by the users, have worked out their social relationships with the other members of the pack. For a short video describing the *AlphaWolf* installation, please visit:

http://www.media.mit.edu/~badger/alphaWolf/alphaWolf.mov

3. RELATED WORK

Various researchers have studied synthetic social systems from natural models (e.g., [Reynolds 1987]). Hemelrijk [Hemelrijk 1996], for example, did experiments using a similar social relationship mechanism to the one we use in the AlphaWolves. We believe that our 3D animated visualization of the social computational entities makes the project significantly different from her work.

For the *AlphaWolf* project, we used a dimensional approach to emotion (e.g., [Mehrabian 1974]), which maps a range of emotional phenomena onto explicitly dimensioned space. Various researchers have implemented models of emotional learning or memory in non-social domains (e.g., [Velasquez 1998]).

4. MECHANISM

The social relationship mechanism in the virtual wolves is similar to Damasio's Somatic Marker Hypothesis [Damasio 1994]. The essence of the mechanism involves emotion, perception, and learning. Each virtual wolf maintains a simple emotional state that is affected by its interactions with the world. A wolf is able to recognize specific pack mates across multiple encounters. After its first interaction with a certain individual, the wolf forms an "emotional memory" of that social partner. When it again encounters that individual, the emotional memory influences its current emotional state, so that it can "pick up where it left off" with regard to its emotional relationship. At the end of each interaction, the wolf revises its emotional relationship with that social partner. This mechanism allows individual wolves to interact differently with specific social partners, based on the history of interactions between them. As such, we believe that it captures an element of mammalian social behavior in a simple way.

5. CONCLUSION

We have presented a multi-agent system featuring an emotional-memory-based mechanism by which the agents form social relationships with each other. This mechanism incorporates models of emotion, perception, and learning. The system was presented as the *AlphaWolf* installation, shown in the Emerging Technologies section of SIGGRAPH 2001.

The social relationship mechanism is derived from biological studies of animal behavior, simulations of animal behavior and models of emotion. It allows the virtual wolves in *AlphaWolf* to form relationships with each other that resemble the relationships of wild gray wolves.

We believe that the novelty of displaying this social relationship mechanism in a 3D-animated virtual world represents a significant step with regard to the explicability of synthetic social relationships. The interactive experience of *AlphaWolf* proved to be quite engaging for participants, and helped them understand the social relationships of the virtual wolves. We believe that systems like *AlphaWolf* could serve a significant role as platforms for simulation, education and entertainment.

6. ACKNOWLEDGMENTS

We would like to thank all the members and friends of the Synthetic Characters Group who have helped build the wolves over the last several years.

REFERENCES

Cohen, M. D., Riolo, R.L., and Axelrod, R. (1999). The Emergence of Social Organization in the Prisoner's Dilemma, Santa Fe Institute Working Paper 99-01-002.

Damasio, A. (1994). Descartes' Error: Emotion, Reason, and the Human Brain. New York, G. P. Putnam's Sons.

Hemelrijk, C. K. (1996). Dominance interactions, spatial dynamics and emergent reciprocity in a virtual world. From Animals to Animats 4, SAB '96.

Mech, L. D., Adams, L. G., Meier, T. J., Burch, J. W., and Dale, B. W. (1998). The Wolves of Denali. Minneapolis, University of Minnesota Press.

Mehrabian, A., and Russell, J. (1974). An Approach to Environmental Psychology. Cambridge, MA, MIT Press.

Reynolds, C. (1987). Flocks, Herds and Schools: A Distributed Behavioral Model. Proceedings of the ACM Computer Graphics, SIGGRAPH 87.

Velasquez, J. (1998). "When Robots Weep: Emotional Memories and Decision-Making." Proceedings of the Fifteenth National Conference on Artificial Intelligence.

Can a dog tell the difference? Dogs encounter AIBO, an animal-like robot in two social situations

Enikö Kubinyi[a], Ádám Miklósi[a], Frédéric Kaplan[b]
Márta Gácsi[c], József Topál[c] and Vilmos Csányi[c]

[a] Department of Ethology; Eötvös Loránd University,
[b] Sony Computer Science Laboratory, Paris
[c] Comparative Ethology Group, Hungarian Academy of Sciences

Corresponding author: Frédéric Kaplan – Sony CSL, Paris – 6 rue Amyot, 75005, Paris, France, e-mail: kaplan@csl.sony.fr

Abstract

The use of autonomous animal-like robots can offer new possibilities for studying social recognition ability in animals. We report on an experiment where 24 adult dogs and 16 puppies were studied in a spontaneous and in a competitive situation with AIBO, a four-legged autonomous robot. A puppy and a remote controlled car were used as controls.

1. Introduction

Several studies have already shown that robots can be used in an interesting fashion for biological research (Holland and McFarland, 2001). We argue for a novel use of autonomous robots in this area: carrying out behavioural experiments on animals. There are only a few reports of the application of robotics in animal behaviour tests. Honeybees have been shown to adopt the dance of a robotic 'bee', when collecting food out of their home (Michelsen et al., 1992). A new experimental system was developed to test rat and rat-robot interaction in a social learning set-up by Takanishi et al. (1998). With this kind of tests the experimenters can learn about the social recognition ability of the studied animal.

Social recognition is based on measuring the occurrence of behavioural interactions in terms of species specific behaviours between two individuals or an individual and a dummy (Colgan, 1983). By designing different series of dummies one can find out the relative importance of different types of perception clues for the recognition process (visual, acoustic, olfactory). Based on these results, it is possible to extend the resemblance between the dummy and the real animal.

In this paper, we investigate whether the AIBO, an animal-like autonomous robot – although it has not been designed to perfectly mimick a real dog - could be an appropriate scale model for the study of social recognition in dogs, because some aspects of its shape and its movement could be compared to the ones of a dog-like animal. To our knowledge, the only experiment that tested species recognition ability in dogs and puppies used a two-dimensional painting of a dog (Fox, 1971). We assume that AIBO, as a three-dimensional, independently moving creature, can evoke more investigative behaviours.

Naturally, in order to obtain real interactive situations, the robot must be able to detect and react to some elements of the environment it is sharing with the tested animal. In our case, a red plate for food is a salient object both for the animal (it contains food) and the robot (it is red, a color the robot is spontaneously attracted to).

Based on behavioural observations, it is clear that in many species, younger conspecifics can be "fooled" by scale models for a longer time than adult. It is also known that social recognition depends often on the context In the case of a competion for food, animals might be less sensitive for species specific appearance, since they defend food against all kind of different species. Therefore we compare young and adult animals in a spontaneous encounter and in a competitive situation.

2. Method

24 adult dogs and 16 18-22-week-old puppies were tested in a room after having spent a 1.5 min period for habituation. The owner was sitting in a chair, faced to the door. In front of him/her there was a red bowl on the floor. The owner was asked not to talk to the dog. Each individual dog encountered four test-partners in two test situations. Each test lasted one minute.

The four test-partners were : a real puppy, two different versions of AIBO (one "naked" and one with fur) and a remote controlled car. They can easily be ranked on a scale of dog-similarity. The remote control car is moving, but does not produce any other animal-like behaviour. AIBO's general appearance and movement can be considered to be dog-like in some aspects. But an AIBO covered with a dog-smelled fur is supposed to stimulate both the olfactory and the visual perception. Finally, a real puppy was used as a control. One testing situation involved simple encounter between the adult dog or the puppy and the test-partner. In the competitive situation, a bowl fool of food is placed near the dog and the test partner moved towards it as a potential competitor. All tests were recorded on tape.

Figure 1. Dog and AIBO ERS-210 covered with artificial fur in the competitive situation

3. Results and discuss ion

In general, dogs sniffed the puppy longer, approached and growled it sooner than the non-living test-partners, although the furry robot was found to be similar to the puppy in some behavioural measures. For example, adult dogs spent the same amount of time to identify the furry robot and the puppy. (Fig 2.). Contact seeking was not stable with the non-living partners, dogs preferred to avoid them after the identification (Fig. 3).

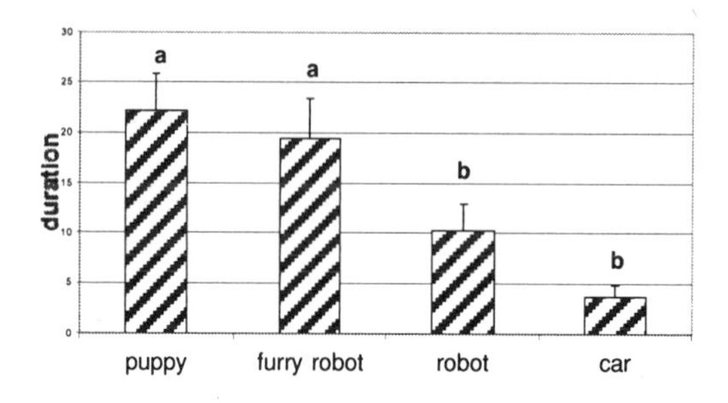

Figure 2. Orientation towards the partners in adult dogs during the encounters. Duration (mean+SE). Different letters indicate significant differences. One-way ANOVA with S-N-K post hoc test; F=9.59, p <0.05.

As expected, AIBO induced a lot more interactions than a two-dimensional painting (Fox, 1971). Statistics could not reveal that two young dogs out of 16 attacked directly and knocked over the furry robot during competitions, two adults growled, and one showed the intention of attacking the furry robot. In the latter case, we intervened to stop the test. But in most cases, if the similar outward appearance may have deceived the dogs for a few seconds, this effect did not last long. In the competitional situation, adult dogs behaved as if only the puppy was a real competitor.

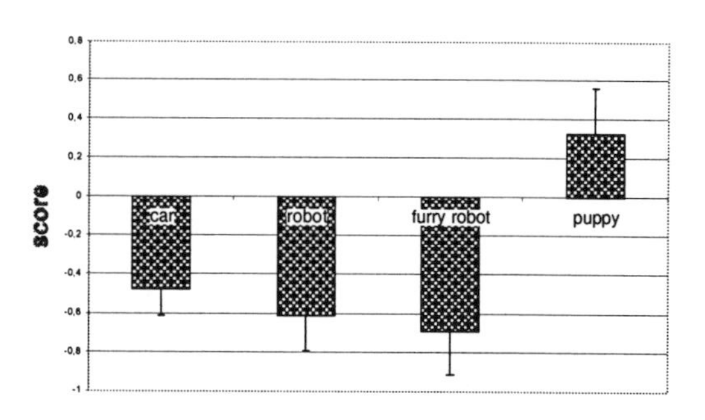

Figure 3. Contact seeking with the partners. Score (mean +SE). Repeated measures of analysis of variance; F = 17.71; p<0.001.

Though, according to the results of this first study, dogs cannot be fooled for more than a few seconds by AIBO, it is important to realise that autonomous robots go far beyond simple dummies or systems that produce a mirror-image of the animal behaviour. Today, for this kind of experiments, the main limitation of such a robot as AIBO is speed. Dogs react much faster than the robot. A lot of interactions, like playing, will not be possible until the robot's reactions are faster. But, as this is also a crucial factor for interactions with humans, we can reasonably hope that faster autonomous robots will soon appear on the market.

4. Acknowledgements

This work has been supported by OTKA (T 029705), with a grant (F 226/98) allowed by the Hungarian Academy of Sciences and by a scholarship of the Hungarian Republic to E. K. (MÖB 39/2001).

5. References

P. W. Colgan. Comparative Social Recognition. Wiley-Interscience Publication, 1983.

Fox. Integrative Development of brain and behavior in the dog. The University of Chicago Press, pp.; 55-60, 1971.

O. Holland and D. Mc Farland. Artificial Ethology. Oxford University Press, 2001.

A. Michelsen, B. B. Andersen, J. Storm, W. H. Kirchner and M. Lindauer. How honeybees perceive communication dances, studied by means of a mechanical model. Behavioral Ecology and Sociobiology, 30É143-150, 1992

A. Takanishi, T. Aoki, M. Ho, Y. Ohkawa and J. Yamaguchi. Interaction between creature and robot: development of an experiment system for rat and rat-robot interaction. In: IEEE/RSJ International Conference on Intelligent Robotics and Systems, pp. 1975-1980. Los Alamitos, California: IEEE Computer Society Press, 1998.

Exploring the impact of contextual input on the evolution of word-meaning

Paul Vogt and Hans Coumans
IKAT - Universiteit Maastricht
P. O. Box 616 - 6200 MD Maastricht - The Netherlands
p.vogt@cs.unimaas.nl

1 Introduction

It is widely acknowledged in the field of language evolution that language users learn word-meanings, for which the following question is relevant: what kind of input is required to learn word-meanings? The most obvious form of input is speech, but in this paper we are concerned with the pragmatic of contextual cues that indicate the meaning of spoken words. Such cues may, for instance, be provided by establishing joint attention or by evaluating corrective feedback, but there is some evidence that children do not need such directed cues to learn the meaning of their first words (Lieven, 1994).

Recent computational studies on the evolution of language show how agents can learn word-meanings successfully. In most of these studies it is assumed that either joint attention was established or that agents receive corrective feedback. Both conditions have also been studied successfully with mobile robots (Vogt, 2000). Only few studies have investigated whether cues such as joint attention and corrective feedback are really necessary (Smith, 2001, Vogt, 2000). Although Smith's simulations indicate that neither type of input is required, this is not confirmed by Vogt's robotic experiments. Both studies used a minimal experimental setup - only 2 agents and, in Vogt's case, a very limited number of objects to communicate about. It is therefore interesting to study the impact of the three conditions on lexicon formation in a scaled experiment, which is the focus of this paper. In addition, we look how the lexicon evolves under the three conditions and a population dynamics as modeled by the iterated learning model (Kirby and Hurford, 2002).

2 Three language games

In the simulations of this paper, populations of agents play adaptive language games to bootstrap a shared lexicon (Steels and Kaplan, 1999). In a language game two agents - a speaker and a hearer - are selected from a population and observe a context that contains a number of predefined meanings. The speaker selects a topic from the context and produces an utterance by searching in its lexicon for a matching word-meaning association, which the hearer tries to interpret. Selection of an association depends both on its applicability in the context and the strength of its association given by a score. At the end of the game both agents adapt their lexicons: new associations may be constructed and existing ones may be strengthened or weakened. By playing a large number of language game, a lexicon emerges.

The simulations of this paper involve three variants of a language game: the observational game, the guessing game and the selfish game, which all differ in their use or disuse of joint attention and corrective feedback. The *observational game* (OG) is based on (Oliphant, 1997) and uses joint attention to indicate the topic of a game and it uses Hebbian learning to regulate the strength of associations. The *guessing game* (GG) is based on (Steels and Kaplan, 1999) and uses corrective feedback to allow reinforcement learning. In the *selfish game* (SG), based on (Smith, 2001, Vogt, 2000), neither source of input is used and the agents learn the associations with Bayesian learning techniques.

The iterated learning model (ILM) of (Kirby and Hurford, 2002) applies to all types of language games and models a population dynamics. In the ILM a population consists of a group of adults speakers and an, in our case, equal sized group of learners who only act as hearers and start with empty lexicons like the adult speakers of the first iteration. After a fixed number of language games the adults are replaced by the learners and new learners enter the population. This process is iterated over and over again.

3 The results

With the three models, a number of simulations were done in which the world consisted of 100 meanings and the context size in each game was fixed at 5. Figure 1 shows the results of these experiments.

The upper left figure shows the communicative success of simulations with the three types of language games us ing a population size of 2 without iterated learning. As shown both the OG and the GG converge to 1 within 1,000 games and are indistinguishable from each other,

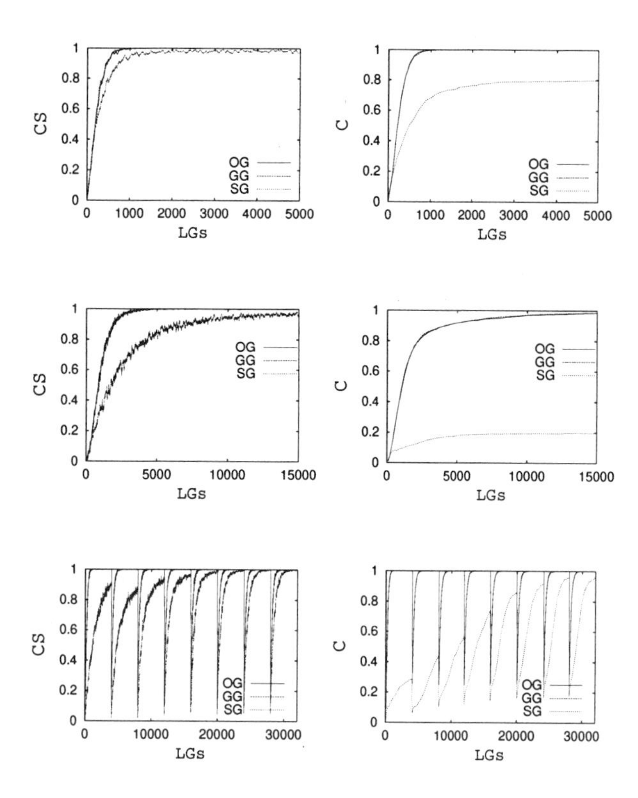

Figure 1: The results of the simulations with population sizes of 2 (top row), 5 (second row) and 8 (bottom row) where the latter also applies the ILM. The figures show the evolution of the communicative success CS (left column) and the coherence C of the lexicon (right column).

which holds for all following experiments. The SG approaches 1, but does not reach this after 5,000 games. The upper left figure shows the evolution of the coherence in the same simulations. The coherence measures to what extent the agents use the same vocabulary as a speaker and converges to 1 almost equally fast as the communicative success for the OG and the GG, but it does not exceed 0.8 for the SG. This means that the lexicons for the OG and GG show no ambiguities, while the lexicon of the SG is still ambiguous.These results confirm previous results as reported by, e.g., (Steels and Kaplan, 1999, Oliphant, 1997, Smith, 2001).

The graphs in the middle row show the results of simulating the language games with a population size of 5 and no iterated learning. Although the communicative success shows a similar but slower evolution for all games, the coherence of the SG does no exceed 0.2, which is a drastic decrease. We have done simulations with population sizes up to 20 agents, and the results get worse with increasing population sizes, although no clear dependency between the results and population size has been observed.

The bottom figures show the communicative success and coherence of applying the ILM to the language games for a population size of 8. The simulation was done over 8 iterations of 4,000 games each. Clearly the communicative success and coherence of the OG and GG converge to 1 pretty fast in each iteration. The results of the SG reveal that the communicative success converges to 1 from the sixth iteration and the coherence reaches an increasingly higher end-value in each iteration.

4 Conclusion

The selfish game shows that agents can learn word-meanings without using directed information concerning a word's meaning, but this only works for larger populations when agents can learn the lexicon from adult speakers This is probably because learners in the ILM receive more consistent speech that they can pass on to a next generation of learners even more consistently. So although the selfish game may explain some phenomena of lexicon acquisition, it is not a likely scenario for explaining the origins of language as it appears too difficult to bootstrap a coherent lexicon to be advantageous for a population. The observational or guessing games provide more likely strategies.

References

Kirby, S. and Hurford, J. R. (2002). The emergence of linguistic structure: An overview of the iterated learning model. In Cangelosi, A. and Parisi, D., (Eds.), *Simulating the Evolution of Language*, pages 121–148, London. Springer.

Lieven, E. V. M. (1994). Crosslinguistic and cross-cultural aspects of language addressed to children. In Gallaway, C. and Richards, B. J., (Eds.), *Input and interaction in language acquisition*, Cambridge. Cambridge University Press.

Oliphant, M. (1997). *Formal Approaches to Innate and Learned Communication: Laying the Foundation for Language.* PhD thesis, University of California, San Diego.

Smith, A. D. M. (2001). Establishing communication systems without explicit meaning transmission. In Kelemen, J. and Sosík, P., (Eds.), *Proceedings of the 6th European Conference on Artificial Life, ECAL 2001*, LNAI 2159, pages 381–390, Berlin Heidelberg. Springer-Verlag.

Steels, L. and Kaplan, F. (1999). Situated grounded word semantics. In *Proceedings of IJCAI 99.* Morgan Kaufmann.

Vogt, P. (2000). Bootstrapping grounded symbols by minimal autonomous robots. *Evolution of Communication*, 4(1):89–118.

Phonemic Coding Might Result From Sensory-Motor Coupling Dynamics

Pierre-yves Oudeyer
Sony Computer Science Lab, Paris
e-mail : py@csl.sony.fr

Abstract

Human sound systems are invariably phonemically coded. Furthermore, phoneme inventories follow very particular tendancies. To explain these phenomena, there existed so far three kinds of approaches : "Chomskyan"/cognitive innatism, morpho-perceptual innatism and the more recent approach of "language as a complex cultural system which adapts under the pressure of efficient communication". The two first approaches are clearly not satisfying, while the third, even if much more convincing, makes a lot of speculative assumptions and did not really bring answers to the question of phonemic coding. We propose here a new hypothesis based on a low-level model of sensory-motor interactions. We show that certain very simple and non language-specific neural devices allow a population of agents to build signalling systems without any functional pressure. Moreover, these systems are phonemically coded. Using a realistic vowel articulatory synthesizer, we show that the inventories of vowels have striking similarities with human vowel systems.

1. The origins of phonemic coding and other related puzzling questions

Human sound systems have very particular properties. First of all, they are phonemically coded. This means that syllables, defined as oscillations of the jaw (MacNeilage, 1998), are composed of re-usable parts. These are called phonemes. Thus, syllables of a language may look rather like la, li, na, ni, bla, bli, etc ... than like la, ze, fri, won, etc .... This might seem unavoidable for us who have a phonetic writing alphabet, but in fact our vocal tract allows to produce syllable systems in which each syllable is holistically coded and has no parts which is also used in another syllable. Yet, as opposed to writing systems for which there exists both "phonetic" coding and holistic/pictographic coding (for e.g. Chinese), all human languages are invariably phonemically coded.

Secondly, the set of re-usable parts of syllable systems, as well as the way they are combined, follows precise and surprising tendancies. For example, our vocal tract allows us to produce hundreds of different vowels. Yet, each particular vowel system uses most often only 5 or 6 vowels, and extremely rarely more than 12 (Maddieson and Ladefoged, 1996). Moreover, there are vowels that appear in these sets much more often than others. For example, most of languages contain the vowels [a], [i] and [u] (87 percent of languages) while some others are very rare, like [y], [oe] and [ui] (5 percent of languages). Also, there are structural regularities that caracterize these sets : for example, if a language contains a back rounded vowel of a certain height, for example an [o], it will usually also contain the front, unrounded vowel of the same height.

The questions are then : Why are there these regularities ? How did they appear ? What are the genetic, glosso-genetic/cultural, and ontogenetic components of this formation process ? Several approaches have already been proposed in the litterature.

The first one, known as the "post-structuralist" Chomskian view, defends the idea that our genome contains some sort of program which is supposed to grow a language specific neural device (the so-calles Language Acquisition Device) which knows a priori all the algebric structures of language. This concerns all aspects of language, ranging from syntax (Chomsky, 1958; Archangeli and Langendoen, 1997) to phonetics (Chomsky and Halle, 1968). For example this neural device is supposed to know that syllables are composed of phonemes which are made up by the combination of a few binary features like the nasality or the roundedness. Learning a particular language only amounts to the tuning of a few parameters like the on or off state of these features. It is important to note that in this approach, the innate knowledge is completely cognitive, and no reference to morpho-perceptual properties of the human articulatory and perceptual apparatuses appears. This view is becoming more and more incompatible with neuro-biological findings (which have basically failed to find a LAD), and genetics/embryology which tend to show that the genome can not contain specific and detailed information for the growth of so complex neural devices. Finally, even if it revealed to be true, it is not really an answer to the questions we asked earlier : it is only a displacement of the problem. How do the concerned genes get there in the course of evolution ? Why were they selected ? No answer has been proposed by post-structuralist linguistics.

Another approach is that of "morpho-perceptual" innatists. They argue (Stevens 1972) that the properties of human articulatory and perceptual systems explain totally the properties of sound systems. More precisely, their theory relies on the fact that the mapping between the articulatory space and the acoustic and then perceptual spaces is highly non-linear : there are a number of "plateaus" separated by sharp boundaries. Each plateau is supposed to naturally define a category. Hence in this view, phonemic coding and phoneme inventories are direct consequences of the physical properties of the body. Convincing experiments have been conducted concerning certain stop concsonants (Damper 2000) with physical models of the vocal tract and the cochlea. Yet, there are flaws to this view : first of all, it gives a poor account of the great diversity that caracterize human languages. All humans have approximately the same articulatory/perceptual mapping, and yet different language communities use different systems of categories. One could imagine that it is because some "plateaus"/natural categories are just left unused in certain languages, but perceptual experiments (Kuhl 2000) have shown that very often there are sharp perceptual non-linearities in some part of the sound space for people speaking language L1, corresponding to boundaries in their category system, which are not perceived at all by people speaking another language L2. This means for instance that japanese speakers cannot hear the difference between the "l" in "lead" and the "r" in "read". As a consequence, it seems that there are no natural categories, and most probably the results concerning certain stop consonants are anecdotal. Moreover, the physical models of the vocal tract and of our perceptual system that have been developped in the litterature (Boersma 1998) show clearly that there are important parts of the mapping which is not at all looking like plateaus separated by sharp boundaries. Clearly, considering only physical properties of the human vocal tract and cochlea is not sufficient to explain both phonemic coding and structural regularities of sound systems.

A more recent approach proposes that the phenomena we are interested in come from self-organisation processes occuring mainly at the cultural and ontogenetic scale. The basic idea is that sound systems are good solutions to the problem of finding an efficient communicative system given articulatory, perceptual and cognitive constraints. And good solutions are caracterized by the regularities that we try to explain. This approach was initially defended by (Lindblom 1992) who showed for example that if one optimizes the energy of vowel systems as defined by a compromise between articulatory cost and perceptual distinctiveness, one finds systems which follow the structural and frequency regularities of human languages. (Schwartz et al. 1997) reproduced and extended the results to CV syllables regularities. As far as phonemic cogding is concerned, Lindblom made only simple and abstract experiments in which he showed that the optimal systems in terms of compromise between articulatory cost and acoustic distinctiveness are those in which some targets composing syllables are re-used (note that Lindblom presuposes that syllables are sequences of targets, which we will do also in this paper). Yet, these results were obtained with very low-dimensional and discrete spaces, and it remains to be seen if they remain valid when one deals with realistic spaces. Lindblom proposed another possible explanation for phonemic coding, which is the storage cost argument. It states that re-using parts requires less biological material to store the system, and thus is more advantageous. This argument seems weak for two reasons : first the additional cost of storing un-related parts is not so important, and there are many examples of cultural systems which are extremely memory unefficient (for example the pictogram based writing systems) ; secondly, it does suppose that the possibility of re-using is already there, but what "re-using" means and how it is performed by our neural systems is a fundamental question (this is similar to models of the origins of compositionality (Kirby, 1998) which in fact pre-suppose that the ability to compose basic units is already there, and in fact only show in which conditions it is used or not).

These experiments were a breakthrough as comparared to innatists theories, but provide unsatisfaying explanations : indeed, they rely on explicit optimization procedures, which never occur as such in nature. There are no little scientists in the head of humans which make calculations to find out which vowel system is cheaper. Rather, natural processes adapt and self-organise. Thus, one has to find the processes which formed these sound systems, and can be viewed only a posteriori as optimizations. It has been proposed by (de Boer 2001) that these are imitation behaviors among humans/agents. He built a computational model which consisted of a society of agents playing culturally the so-called "imitation game". Agents were given a physical model of the vocal tract, a model of the cochlea, and a simple prototype based cognitive memory. Their memory of prototypes was initially empty and grew through invention and learning from others, and scores were used to assess them and possibly prune the unefficient ones. One round of the game consisted in picking up two agents, the speaker and the hearer. The speaker utters one sound of its repertoire, and the hearer tries to imitate it. Then the speaker evaluates the imitation by checking if he categorizes the imitation as the item he initially uttered. Finally, he gives feedback to the hearer about the result of this evaluation (good or not). de Boer showed that after a while, a society of agents forms a shared vowel system, and that the formed vowel systems follow the structural regularities of human languages. They are somewhat optimal, but this is a side effect to adaptation for efficient communication under the articulatory, perceptual and cognitive pressures and biases. These results were extended by (Oudeyer 2001b) for the case of syllable systems, where phonological rules were shown to emerge within the same process. As far as phonemic coding is concerned, (Oudeyer 2002) has made experiments which tend to indicate that the conclusions drawn from the simple experiments of Lindblom can hardly be extended to realistic settings. It seems that with realistic articulatory and perceptual spaces, non phonemically coded syllable systems that are perfectly sufficient for efficient communication emerge easily. Thus it seems that new hypothesis are needed.

This paper will present a model that follows a similar approach, yet with a crucial difference : no functional pressure will be used here. Another difference is that the cognitive architecture of the agents that we use is modeled at a lower level, which is the neural level. We will show that phonemic coding and shared vowel systems following the right regularities emerge as a consequence of basic sensory-motor coupling on the one hand, and of unsupervised interactions among agents on the other hand. In particular, we will show that phonemic coding can be explained without any reference to the articulatory/perceptual mapping, and yet how this mapping explains some of the structural regularities. The emergent vowel systems will be shown to have great efficiency if they were to be recruited for communication, and yet were not formed under any communicative pressure. This is a possible example of what has been sometimes termed "exaptation". An important aspect to keep in mind is that the neural devices of our agents are very generic and could be used to learn for example hand-eye coordination. Thus they are not at all language specific and at odds with neural devices like the LAD.

2. A low-level model of agents that interact acoustically

The model is a generalization of the one described in (Oudeyer 2001a), which was used to model a particular phenomenon of acoustic illusion, called the perceptual magnet effect. (Oudeyer 2001a) also described a first simple experiment which coupled agent and neural maps, but it involved only static sounds/articulations and abstract articulatory models. In particular, the question of phonemic coding was not studied. The present paper extends it to dynamic articulations, hence complex sounds, and will use both abstract and realistic articulatory models. We also describe in details the resulting dynamics by introducing entropy-based measures which allow to follow precisely what happens.

The model is based on topological neural maps. This type of neural network has been widely used for many models of cortical maps (Morasso et al., 1998), which are the neural devices that humans have to represent parts of the outside world (acoustic, visual, touch etc...). There are two neuroscientific findings on which our model relies, and that were initially made popular with the experiments of Georgopoulos (1988) : on the one hand, for each neuron/receptive field in the map there exist a stimulus vector to which it responds maximally (and the response decreases when stimuli get further from this vector) ; on the other hand, from the set of activities of all neurons at a given moment one can predict the perceived stimulus or the motor output, by computing what is termed the population vector (see Georgopoulos 1988) : it is the sum of all prefered vectors of the neurons ponderated by their activity (normalized like here since we are interested in both direction and amplitude of the stimulus vector). When there are many neurons and the preferred vectors are uniformly spread across the space, the population vector corresponds accurately to the stimulus that gave rise to the activities of neurons, while when the distribution is inhomogeneous, some imprecisions appear. This imprecision has been the subjects of rich research, and many people proposed more precise variants (see Abbot and Salinas, 1996) to the formula of Georgopoulos because they assumed the sensory system coded exactly stimuli (and hence that the formula of Georgopoulos must be somewhat false). On the contrary we have shown in (Oudeyer 2001a) that this imprecision allows the interpretation of "magnet effect" like psychological phenomena, i.e. sensory illusions, and so may be a fundamental characteristic of neural maps. Moreover, the neural maps are recurrent, and their relaxation consists in iterating the coding/decoding with the population vector : the imprecision coupled with positive feedback loop forming neuron clusters will provide well-define non-trivial attractors which can be interpreted as (phonemic) categories.

A neural map consists of a set of neurons n_i whose "preferred" stimulus vector is noted v_i. The activity of neuron n_i when presented stimulus v is computed with a gaussian function :

$$act(n_i) = e^{-dist(v_i,v)^2/\sigma^2} \quad (1)$$

with sigma being a parameter of the simulation (to which it is very robust). The population vector is then :

$$pop(v) = \frac{\sum_i act(n_i) * v_i}{\sum_i act(n_i)}$$

The normalizing term is necessary here since we are not only interested in the direction of vectors. There are arguments for this being biologically acceptable (see Reggia 1992). Stimuli are here 2 dimensional, corresponding to the first two formants of sounds. Each neural map is fully recurrent : all the neurons n_i of the map are connected with symmetric synapses of weight

$$w_{i,j} = e^{-dist(v_i,v_j)^2/\sigma^2}$$

, which represent the correlation of activity between 2 neurons (and so could be learnt with a hebbian rule for instance, but are computed directly here for sake of efficiency of the simulation). When presented an input stimulus, two computations take place with a neural map : the population vector is calculated with the initial activation of neurons, and gives what is often interpreted as what the agent senses ; then the network is relaxed using local dynamics : the activity of each neuron is updated as :

$$act(n_{i,t+1}) = \frac{\sum_j act(n_{i,t}) * w_{i,j}}{\sum_i act(n_{i,t})}$$

This is the mechanism of competitive distribution of activation as described in (Reggia et al. 1992, Morasso et al. 1998), together with its associated dynamical properties. The fact that weights are symmetric makes that this dynamic system has point attractors. As a consequence, the iterated update of neurons activity soon arrive at a fixed point, which can be interpreted as a categorizing behavior. So the population vector allows to model how we perceive for example a particular vowel, say [e], in a particular sentence and spoken by a particular person, and the attractor in which the map falls models the behavior of categorizing this sound as being of class [e]. Moreover, it is easy to show that this local dynamics is equivalent to the global process of iteratively coding and decoding with the population vector, and each time feeding back to the input the decoded vector.

There are two neural maps : one articulatory which represents the motor space (neurons n_i), and one acoustic which represent the perceptual space (neurons l_i). Both spaces are typically isomorphic to $[0,1]^n$. The two maps are fully connected to each other : there are symmetric connections of weights $w'_{i,j}$ between every neuron of one map and the other. These weights are supposed to represent the correlation of activity between neurons, and will allow to perform the double direction acoustic/articulatory mapping. They are learnt with a hebbian learning rule (Sejnowsky 1977)

$$\delta w'_{i,j} = c_2(act_i - < act_i >)(act_j - < act_j >)$$

where act_i is the activation of neuron i and $< act_i >$ the mean activation of neuron i over a certain time interval (correlation rule). The propagation of signals in this paper always happen from the acoustic map to the articulatory map (the articulatory map can only give information to the acoustic map through the environment, as in the babbling phase described below where activity in the motor map moves articulators, which then produces sound and activate the cochlea which finnaly activate the acoustic map). Figure 1 gives an overview of the architecture.

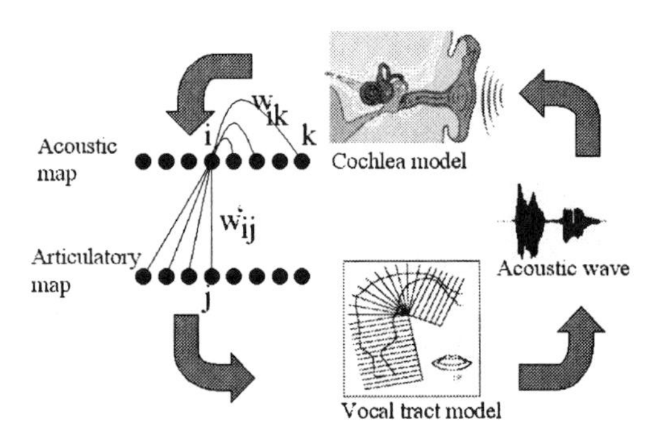

Figure 1: Overview of the architecture

The network is initially made by initializing the preferred vectors of neurons (i.e. their weights in a biological implementations) to random vectors following a uniform distribution, while the $w'_{i,j}$ are set to 0. Part of the initial state can be visualized by plotting all the v_i as in one of the upper squares of figure 2 which represents the acoutic maps of two agents (the perceptual space is 2-dimensional, and points represents the preferred vectors of neurons). Also, one can visualize how agents initially perceive the sound world by plotting all the *pop*(v) corresponding to a set of stimulus whose vectors values are the intersections of a regular grid covering the whole space. This gives graphs similar to those used in (Kuhl 2000) who discovered the perceptual magnet effect. The lower squares of figure 2 are examples of such an initial perceptual state : we see that the grid is nearly not deformed, which means as predicted by theoretical results, that the population vector is a rather accurate coding of the input space. It means that initially, agents are not subject to auditory illusions as will be the case later on. One can also visualize the initial attractors of the acoustic neural maps : figure 3 shows one example, in which each arrow has its ending point being the population coded vector after one iteration of the relaxation rule with initial activation of neurons corresponding to the population vector represented as the beginning of the arrow. In brief, if one views the relaxation rule as a function which maps one set of neuronal activities, caracterized by a unique population vector, to another one, caracterized by another population vector, then the ending point of arrows are the image of the starting points through this function. What one can notice is that initially, attractors are few, trivial and random (most often there is only one).

Before having agents interact, there is a first phase of babbling during which appropriate initial values of the $w'_{i,j}$'s are learnt with the correlation rule : random articulations are performed, which on the one hand provides activations for the neurons in the articulatory map, and on the other hand produces sounds which are perceived with the cochlea which then provides activations in the acoustic map. Also, acoustic neurons who get very low activity or equivalently whose arriving $w'_{i,j}$ are very low are simply pruned. This is consistent with the well know phenomena of activity dependant growth, and in particular allows a better visualization of the neurons in the acoustic map. Once this initial phase is over, the $w'_{i,j}$'s still continue to evolve with the associated hebbian rule. This is indeed necessary since neurons in each map change their "preferred vector" during the simulation.

Then there is a learning mechanism used to update the weights/preferred vectors in the two neural maps when one agent hears a sound stimulus

$$v = (v_{t_0}, v_{t_1}, v_{t_2}, v_{t_3} ...),$$

which is represented by a temporal sequence of feature vectors here in $[0, 1]^n$, typically corresponding to the formants of the sound at a moment t (formants are the frequencies for which there is a peak in the power spectrum). The delay between two feature vectors is typically a few milliseconds, corresponding to the time resolution of the cochlea. For each of these feature vectors, the activation of the neurons (just after perception) in the acoustic map is computed, and propagates to the motor map. Then, the population vector of both maps is computed, giving two vectors v_{acoust} and v_{motor} corresponding to what the networks perceived of the input. Then, each neuron of each map is updated so as to be a little bit more responsive to the perceived input next time it will occur (which means that their preferred vectors are shifted towards the perceived vectors). The actual formula is

$$\delta v_i = c_1 * e^{-\frac{dist(v_{acoustormotor}, v_i)^2}{\sigma^2}} * (v_{acoustormotor} - v_i).$$

The agents in this model produce dynamic articulations. These are generated by choosing N articulatory targets (which are configurations of the vocal tract), and then using a control mechanism which drives the articulators successively to these targets. Articulators are the parts of the vocal tract that control its shape (for example the jaw). In the experiments presented here, N=3 for sake of simplicity, but experiments have been conducted for N=2,...,5 and showed this does not change the results. The choice of the articulatory targets is made by activating successively and randomly 3 neurons of the articulatory map. Their preferred vectors code for the articulatory configuration of the target. The control mechanism that moves the articulators which we used here was very simple : it is simply a linear interpolation between the successive targets. We did not use realistic mechanisms like the propagation techniques of population codes proposed in (Morasso et al., 1998), because these would have been rather computationally unefficient for this kind of experiment, and does not alter the results. Finally, gaussian noise is introduced just before sending the target values to the control system. This noise is fixed in the present paper : the standard deviation of the gaussian is equal to 5 percent of the articulatory range (similar to experiments of de Boer).

When an articulation is performed, a model of the vocal tract is used to compute the corresponding acoustic trajectory. There are two models. The first one is abstract and serves as a test model to see which properties are due to the coupling of neural systems and which are due to the particular shape of the articulatory/acoustic mapping. This is simply a random linear mapping between he articulatory space and the acoustic space. In this paper the articulatory space is always three-dimensional, isomorphic to $[0, 1]^3$, and the perceptual space is always 2-dimensional.

The second model is realistic in the sense that it reproduces the human articulatory to perceptual mapping concerning the production of vowels. We model only vowels here for sake of computational efficiency. The three major vowel articulatory parameters are used : (Ladefoged and Maddieson, 1996) tongue height, tongue position and lip rounding. To produce the acoustic output of an articulatory configuration, a simple model of the vocal tract was used, as described in (de Boer, 1999), which generates the 4 first formant values of the sound. Then, from these four values one extracts the first formant and what is called the second effective formant (de Boer, 2001), which is a highly non-linear combination of the first 4 formants. The first and second effective formant are known to represent well human perception of vowels (de Boer, 2001).

The experiment presented consists in having a population of agents (typically 20 agents) who are going to interact through the production and perception of sounds. They are endowed with the neural system and one of the articulatory synthesizers described previously. Each neural map contains 500 neurons in the simulations. Typically, they interact by pairs of two (following the evolutionary cultural scheme devised in many models of the origins of language, see Steels 1997, Steels and Oudeyer, 2000) : at each round, one agent is chosen randomly and produces a dynamic articulation according to its articulatory neural map as described earlier. This produces a sound. Then another random agent is chosen, perceives the sound, and updates its neural map with the learning rule described earlier. It is crucial to note that as opposed to all simulations on the origins of language that exist in the litterature (Hurford et al., 1998) our agents do not play here a "language game", in the sense that there is no need to suppose an extra-linguitic protocol of interaction such as who should give a feedback to whom and at what particular moment and for what particular purpose. Indeed, there are no "purpose" in our agents heads. Actually, the simulation works exactly in the same way in the following setup :

imagine that agents are in a world in which they have particular locations. Then, the only thing they do is to wander randomly around, produce sounds at random times, and listen to the sounds that they hear in their neighborhood. In particular, they might not make any difference between sounds produced by themselves and sounds produced by other agents. No concept of "self" is needed. They learn also on their own productions. As a consequence, the model presented here for example makes a lot less assumptions about cognitive and social pre-requisites than the model in (de Boer 2001) for the origins of vowel systems.

3. Shared crystalisation with phonemic coding : the case of abstract linear articulatory/acoustic mappings

Let us describe first what we obtain when agents use the abstract articulator. Initially, as the receptive fields of neurons are randomly and uniformly distributed across the space, the different targets that compose the productions of agents are also randomly and uniformly distributed. What is very interesting, is that this initial state situation is not stable : rapidly, agents get in the a situation like on figures 4 (for the unbiased case) or 8 which are respectively the correspondances of figures 2 and 8 after 1000 interactions in a population of 20 agents. These figures show that the distribution of receptive fields is not anymore uniform but clustered. The associated point attractors are now several, very well-defined, and non-trivial. Moreover, the receptive fields distribution and attractors are approximately the same for all agents. This means that now the targets that agents use belong to one of well-defined clusters, and moreover can be classified automatically as such by the relaxation of the network. In brief, agents produce phonemically coded sounds. The code is the same for all agents at the end of a simulation, but different across simulations due to the inherent stochasticity of the process.

Also, what we observe is that the point attractors that appear are relatively well spread across the space. The prototypes that they define are thus perceptually quite distinct. In terms of Lindblom's framework, the energy of these systems is high. Yet, there was no functional pressure to avoid close prototypes. They are distributed in that way thanks to the intrinsic dynamic of the recurrent networks and rather large tuning function of receptive fields : indeed, if two neuron clusters just get too close, then the summation of tuning functions in the iterative process of relaxation smoothes locally their distribution and only one attractor appears.

To show this shared crystalisation phenomenon in a more systematic manner, measures were developped that track on the one hand the evolution of the clusteredness of targets for each agent, and on the other hand the similarity of target distributions between agents. The basic idea is to make an approximation of these distributions. A first possibility would have been standard binning, where the space is discretized into a number of boxes, and one counts how many receptive fields fall in each bin, and then normalize. The drawback is that the choice of the bin size is not very convenient and robust; also, as far as distribution comparison is concerned, this can lead to inadequate measures if for example there are small translations among clusters from one distribution to another. As a consequence, we did decide to make approximations of the local probability density function at a set of particular points using parzen windows (Duda et al., 2001). This can be viewed as a fuzzy binning. For a given point x, the approximation of the probability density function is calculated using a gaussian window :

$$p_n(x) = \frac{1}{n}\sum_{i=1}^{n} \frac{1}{2\pi\sigma} e^{-\frac{||x-x_i||}{\sigma^2}}$$

where the x_i are the set of targets. The width of the windows is determined by σ, and the range of satisfaying values is large and so this is easy to tune. This approximation is repeated for a set of points distributed on the crossings of a refular grid. Typically, for the 2D perceptual map/space, the grid is 10 points wide and gaussian have a variance equal to that of the noise (5 percent of range).

In order to track the evolution of clusteredness, we chose to use the concept of entropy (Duda et al. 2001). The entropy is minimal for a completely uniform distribution, and maximal for a distribution in which all points are the same (1 perfect cluster). It is defined here as :

$$entropy = -\sum_{i=1}^{l} p_n(xgrid_i) ln(p_n(xgrid_i)$$

where $xgrid_i$ are the crossings of the regular grid at which we evaluated the probability density function. As far as the comparison between two target distributions is compared, one used a symmetric measure based on the Kullback-Leibler distance defined as :

$$distance(p(x), q(x)) = \frac{1}{2}\sum_{xgrid} q(x) log(\frac{q(x)}{p(x)}) + p(x) log\frac{p(x)}{q(x)}$$

where $p(x)$ and $q(x)$ are distributions of targets.

Figure 6 shows the evolution of the mean clusteredness for 10 agents during 2000 games. We clearly see the process of crystalisation. Figure 7 shows the evolution of similarity among the distributions of targets. Each point in the curve represents the mean distance among distributions of all pairs of agents. What we expect is that the curve stays relatively stable, and does not increase. Indeed, initially, all distributions are approximately uniform, so approximately identical. What we verify here is that while each distribution becomes peaked and non-trivial, it remains close to the distributions of other agents.

Why does this phenomenon occur ? To understand intuitively, one has to view the neural map that agents

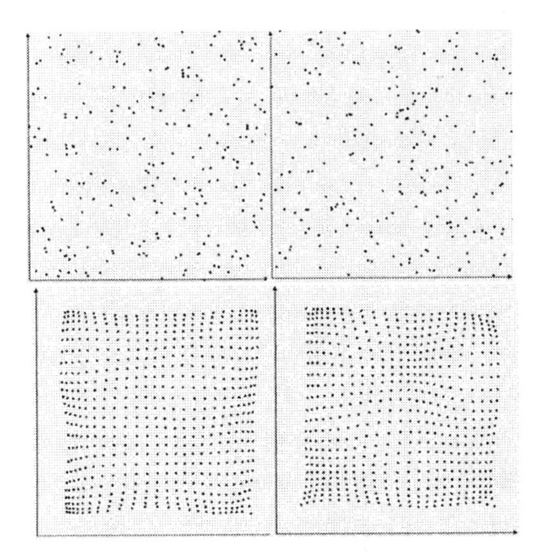

Figure 2: Acoustic neural maps at the beginning (top), and associated initial perceptual warping, i.e. images the points of a regular grid through the population vectir function (bottom). As with all other figures, the horizontal axis represents the first formant (F1), and the vertical axis represents the second effective formant (F2')

use, in particular the perceptual map, as modeling the distribution of sounds that are perceived, and which are produced by members of the society. The crucial point is that the acoustic map is coupled and evolves with the articulatory map so that the distribution of sounds which are produced is very close to the distribution of sounds which is modeled in the acoustic map. As a consequence, agents learn to produce utterances composed of sounds following the same distribution as what they hear around them. All agents initially produce, and so perceive, the same distribution. Logically, one would expect that this state remains unchanged. Yet, this is not what happens : indeed, at some point, symmetry breaks due to chance. The "uniform" state is unstable. And positive feed-back loops make that this symmetry breaking, which might happen simultaneously in several parts of the space or in several agents, gets amplified and converges to a state in which the distribution is multi-peaked, and of course still shared by agents.

These results show a real alternative to earlier described theories to explain phonemic coding as well as the formation of shared sound systems : the neural device is very generic and could have been used to learn the correspondence between other modalities (e.g. hand-eye coordination, see Morasso et al., 1998, who use similar networks), so no LAD is required (Chomskian innatists); the articulatory to perceptual mapping is linear and trivial, so there is no need for innate particularities of this mapping (morpho-perceptual innatists); agents are not playing any sort of particular language game, and there

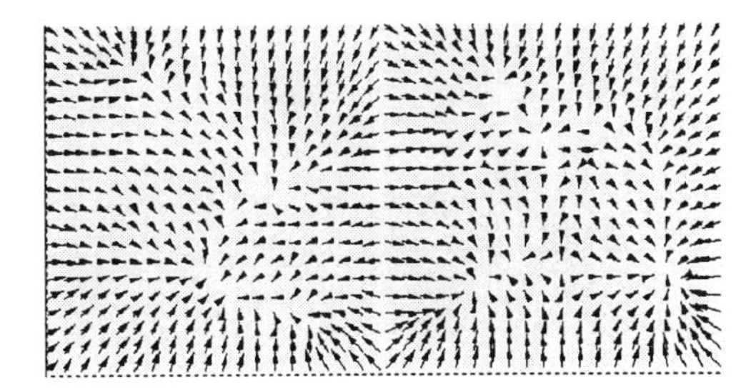

Figure 3: Representation of the population vector function for the initial neural maps of figure 1 : each arrow gives information about in which direction are shifted stimuli in the local area where they are drawn

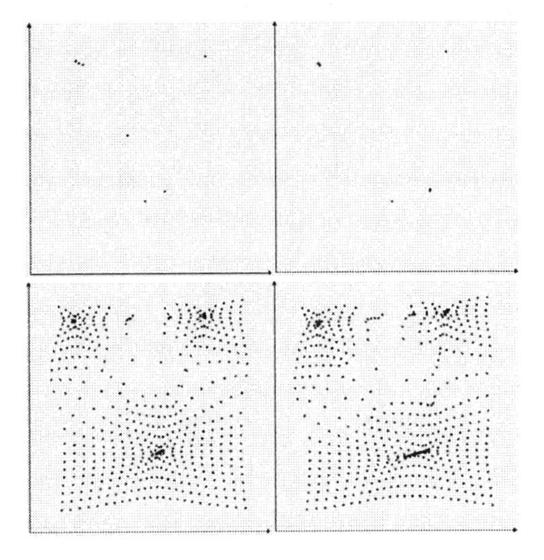

Figure 4: Neural maps and perceptual warping after 1000 interactions, corresponding to the intial states of figure 1)

is no pressure for developing an efficient and shared signalling system (they do develop it, but this is a side effect !), so there are many fewer assumptions needed than in Lindblom's or de Boer's approach, and as a consequence the hypothesis presented in this paper should be preferred for simplicity sake, following Occam's razor law.

4. The use of realistic articulatory/acoustic mapping

Yet, we have so far not been able to reproduce the structural regularities of for example human vowel systems as done by de Boer's model. By "structure" we mean what set of vowels (and how many of them) appear together in a vowel system. Indeed, our vocal tract theoretically allows us to have thousands of different vowel systems, but yet only very few are actually used in human languages (Ladefoged and Maddison, 1996). This is due to the

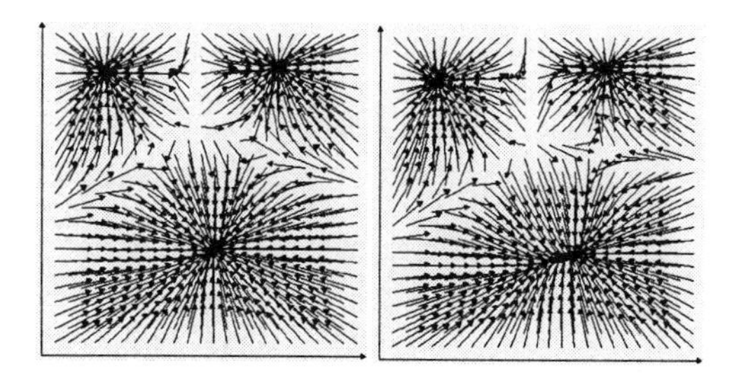

Figure 5: Representation of the population vector function for the final neural maps of figure 3 : each arrow gives information about in which direction are shifted stimuli in the local area where they are drawn

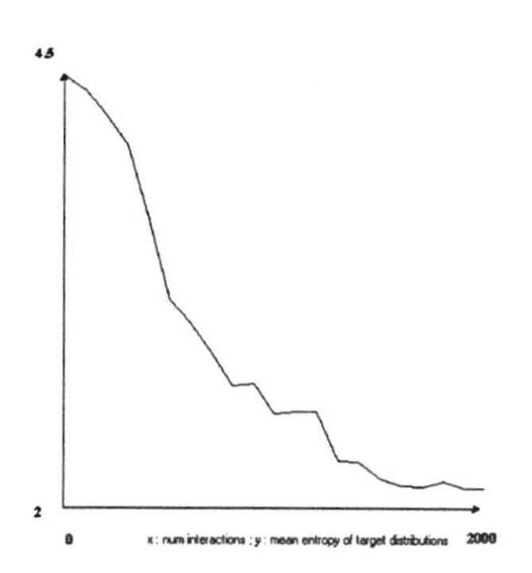

Figure 6: Evolution of entropy of target distributions during 2000 interactions : the emergence of clusteredness

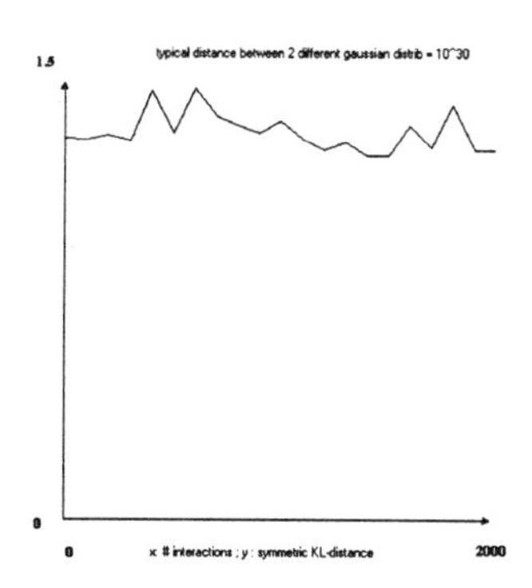

Figure 7: Evolution of target distributions similarity during 2000 interactions : emergent clusters are similar in different agents

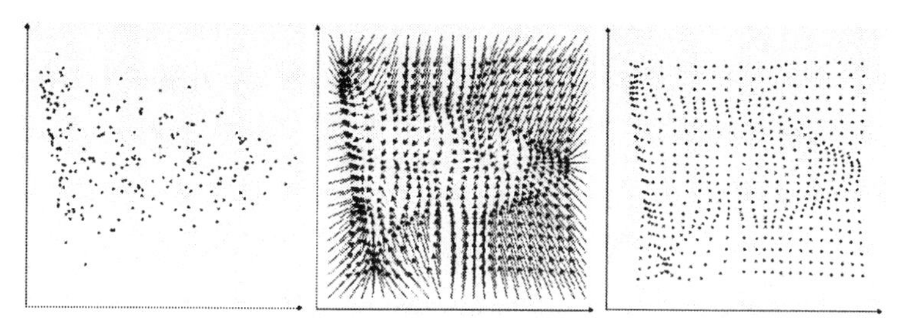

Figure 8: Initial neural map, population vector function representation and perceptual warping of one agent within a population of 20 agents. Here the realistic articulatiry synthesizer is used

fact that we used an abstract articulatory synthesizer. We are now going to use the realistic vowel articulatory synthesizer presented earlier. The mapping that it implements is not any more linear. To get an idea of it, figure 8 shows the state of the acoustic neural maps of agents just after the initial babbling phase which allows to set up initial weights for the connections with the articulatory map, and after the pruning phase which got rid of never used acoustic neurons. We see that the image of the cube $[0, 1]^3$ which is uniformly explored during babbling is a triangle (the so-called vocalic triangle). A series of 500 simulations were ran with the same set of parameters, and each time the number of vowels as well as the structure of the system was checked. The first result shows that the distribution of vowel inventory size is very similar to the one of human vowel systems (Ladefoged and Maddison, 1996) : figure 10 shows the 2 distributions (in plain line the distribution corresponding to the emergent systems of the experiment, in dotted line the distribution in human languages), and in particular the fact that there is a peak at 5 vowels, which is remarkable since 5 is neither the maximum nor the minimum number of vowels found in human languages. Also, among these 5 vowel systems, it appeared that one of them is generated much more frequently (79 percent) than others : figure 9 shows an instance of it. The remaining 5 vowel systems are either with a central vowel together with more front vowels, or with more back vowels. This agrees very well with what has been found in natural languages. (Schwartz et al. 1997) found that 89 percent of the languages had the symmetric system, while the two other types with the central vowel occur in 5 percent of the cases. For different system sizes similarly good matches between predicted systems and human vowel systems are found.

5. Conclusion

Functional and computational models of the origins of language (Hurford et al., 1998) typically make a lot of initial assumptions such as the ability to play language games or in general coordinate. The present paper presented an experiment concerning sound systems which might be a possible example of how to bootstrap these linguistic interactions. Indeed, with very simple and non-specific neural systems, without any need for explicit coordination schemes, without any deus exmachina functional pressure, we obtained here a signalling system which could very well be recruited as a

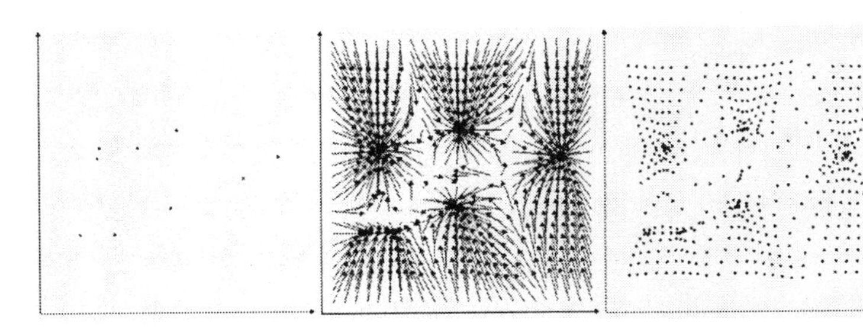

Figure 9: neural map, population vector function representation and perceptual warping of the agent of figure 4 after 2000 interactions with other 20 agents. The corresponding figures of other agents are nearly identical, as in figures 2 and 3. The produced vowel system corrresponds to the most frequent 5 vowel system in human languages.

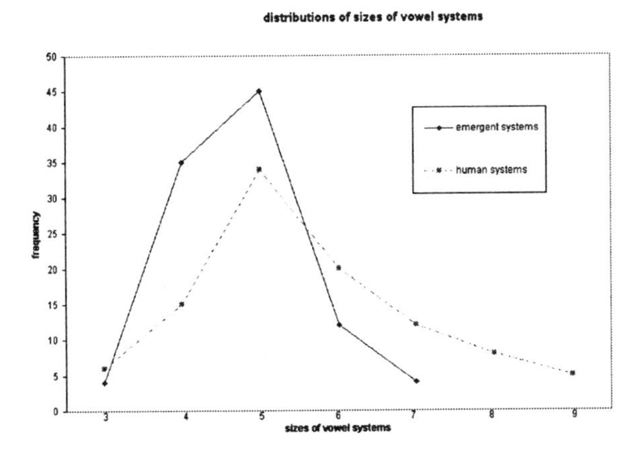

Figure 10: Distribution of vowel inventories sizes in emergent and exsiting human vowel systems

building block for a naming game for example. (Oudeyer 2002) presents a more traditional functional model of higher aspects of sound systems (phonological rules) which is based on the bootstrapping mechanism presented here.

Moreover, it provides a very simple explanation for phonemic coding, which received only poor account in previous research. Yet, understanding truly phonemic coding might be of crucial importance to understand the origins of language : indeed, syntax is thought to be the keystone and one the hottest topics is how compositionality appeared. Interestingly, phonemic coding is a form of primitive compositionality.

Finally, the use of a realistic vowel synthesizer allowed to show that the model also predicts inventories regularities. We showed how the use of a realistic synthesizer is crucial for the prediction of these regularities, but is certainly not the explanation of phonemic coding. As far as phonemic coding and vowel inventories are concerned, the model presented in this paper is more biology-compliant than innatists models, and makes less assumptions than traditional cultural functional models.

6. References

Abbot L., Salinas E. (1994) Vector reconstruction from firing rates, Journal of computational Neuroscience, 1, 89-116.

Anderson J., Silverstein, Ritz, Jons (1977) Distinctive features, categorical perception and probability learning : some applications of a neural model, Psychologycal Review, 84, 413-451.

Archangeli D., Langendoen T. (1997) Optimality theory, an overview, Blackwell Pulishers.

de Boer, B. (1999) Investigating the Emergence of Speech Sounds. In: Dean, T. (ed.) Proceedings of IJCAI 99. Morgan Kauffman, San Francisco. pp. 364-369.

de Boer, B. (2000) The origins of vowel systems, Oxford Linguistics, Oxford University Press.

Boersma, P. (1998) Functional phonology, PhD Thesis, Amsterdam University.

Chomsky, N. and M. Halle (1968) The Sound Pattern of English. Harper Row, New york.

Chomsky, N. (1958) Syntactic Structures.

Duda R., Hart P., Stork D. (2001) Pattern Classification, Wiley-interscience.

Watson G.S. (1964) Smooth regression analysis, Sankhya : The Indian Journal of Statistics. Series A, 26 359-372.

Reggia J.A., D'Autrechy C.L., Sutton G.G., Weinrich M. (1992) A competitive distribution theory of neocortical dynamics, Neural Computation, 4, 287-317.

R.I. Damper and S.R. Harnad (2000) Neural network modeling of categorical perception. Perception and Psychophysics, 62 p.843-867.

R. I. Damper (2000) Ontogenetic versus phylogenetic learning in the emergence of phonetic categories. 3rd International Workshop on the Evolution of Language, Paris, France. p.55-58.

Georgopoulos, Kettner, Schwartz (1988), Primate motor cortex and free arm movement to visual targets in three-dimensional space. II. Coding of the direction of movement by a neuronal population. Journal of Neurosciences, 8, pp. 2928-2937.

Guenther and Gjaja (1996) Magnet effect and neural maps, Journal of the Acoustical Society of America, vol. 100, pp. 1111-1121.

Hurford, J., Studdert-Kennedy M., Knight C. (1998), Approaches to the evolution of language, Cambridge, Cambridge University Press.

Kirby, S. (1998), Syntax without natural selection: how compositionnality emerges from vocabulary in a population of learners, in Hurford, J., Studdert-Kennedy M., Knight C. (eds.), Approaches to the evolution of language, Cambridge, Cambridge University Press.

Kuhl, Williams, Lacerda, Stevens, Lindblom (1992), Linguistic experience alters phonetic perception in infants by 6 months of age. Science, 255, pp. 606-608.

Kuhl (2000) Language, mind and brain : experience alters perception, The New Cognitive Neurosciences, M. Gazzaniga (ed.), The MIT Press.

Ladefoged, P. and I. Maddison (1996) The Sounds of the World's Languages. Blackwell Publishers, Oxford.

Lindblom, B. (1992) Phonological Units as Adaptive Emergents of Lexical Development, in Ferguson, Menn, Stoel-Gammon (eds.) Phonological Development: Models, Research, Implications, York Press, Timonnium, MD, pp. 565-604.

MacNeilage, P.F. (1998) The Frame/Content theory of evolution of speech production. *Behavioral and Brain Sciences*, 21, 499-548.

Maddieson, I., Ladefoged P. (1996) The sounds of the world's languages, Oxford Publishers.

Morasso P., Sanguinetti V., Frisone F., Perico L., (1998) Coordinate-free sensorimotor processing: computing with population codes, Neural Networks 11, 1417-1428.

Oudeyer, P-Y. (2001a) Coupled Neural Maps for the Origins of Vowel Systems. Proceedings of ICANN 2001, International Conference on Artificial Neural Networks, Vienna, Austria, LNCS, springer verlag, Lectures Notes in Computer Science, 2001. Springer Verlag.

Oudeyer P-Y. (2001b) The Origins Of Syllable Systems : an Operational Model. to appear in proceedings of the International Conference on Cognitive science, COGSCI'2001, Edinburgh, Scotland., 2001.

Oudeyer, P-Y. (2002) Emergent syllable systems : why functional pressure is not sufficient to explain phonemic coding, submitted.

Pinker, S., Bloom P., (1990), Natural Language and Natural Selection, The Brain and Behavioral Sciences, 13, pp. 707-784.

Sejnowsky, T. (1977) Storing covariance woth nonlinearly interacting neurons, Journal of mathematical biology, 4:303-312, 1977.

Schwartz J.L., Boe L.J., Valle N., Abry C. (1997) The Dispersion/Focalisation theory of vowel systems , Journal of phonetics, 25:255-286, 1997.

Steels, L. (1997a) The synthetic modeling of language origins. *Evolution of Communication*, 1(1):1-35.

Steels L., Oudeyer P-y. (2000) The cultural evolution of phonological constraints in phonology, in Bedau, McCaskill, Packard and Rasmussen (eds.), Proceedings of the 7th International Conference on Artificial Life, pp. 382-391, MIT Press.

Stevens, K.N. (1972) The quantal nature of speech : evidence from articulatory-acoustic data, in David, Denes (eds.), Human Communication : a unified view, pp. 51-66, New-York:McGraw-Hill.

Vennemann, T. (1988), Preference Laws for Syllable Structure, Berlin: Mouton de Gruyter.

Language adaptation helps language acquisition

Willem Zuidema
AI Lab – Vrije Universiteit Brussel
Pleinlaan 2, B-1050, BRUSSELS, Belgium
jelle@arti.vub.ac.be

Abstract

Language acquisition is a very particular type of learning problem: it is a problem where the target of the learning process is itself the outcome of a learning process. Language can therefore adapt to the learning algorithm. I present a model that shows that due to this effect – and contrary to some claims from the Universal Grammar tradition – "unlearnable" grammars can be successfully acquired, and grammatical coherence in a population can be maintained.

1 Introduction

Human language is one of the most intriguing adaptive behaviors that has emerged in evolution. Language makes it possible to express an unbounded number of different messages, and it serves as the vehicle for transmitting knowledge that is acquired over many generations. Not surprisingly, the origins of language are a central issue in both evolutionary biology and the cognitive sciences.

The dominant explanation for the origins and nature of human language postulates a "Universal Grammar": an innate system of principles and parameters, that is universal, genetically specified and independent from other cognitive abilities. In this paper, I study an argument that lies at the heart of this dominant position: the argument from the poverty of stimulus. This argument states that children have insufficient evidence to learn the language of their parents without innate knowledge about which languages are possible and which are not. This claim is backed-up with a series of mathematical models. Here, we will focus our discussion on two such models: Gold (1967) and Nowak et al. (2001).

Gold (1967) introduced the criterion "identification in the limit" for evaluating the success of a learning algorithm: with an infinite number of training samples all hypotheses of the algorithm should be identical, and equivalent to the target. Gold showed that context-free grammars are in general not learnable by this criterion from positive samples alone. This proof is based on the fact that if one has a grammar G that is consistent with all the training data, one can always construct a grammar G' that is slightly more general: i.e. the language of G, $L(G)$ is a subset of $L(G')$.

Nowak et al. (2001) provide a novel variant of the argument from the poverty of stimulus, that is based on a mathematical model of the evolution of grammars. The first step of their argument is a "coherence threshold". This threshold is the minimum learning accuracy of an individual that is consistent with grammatical coherence in a population, i.e. with a majority of individuals to use the same grammar. The second step relates this coherence threshold to a lower bound (b_0) on the number of sample sentences that a child needs. They derive that b_0 is proportional to the total number of possible grammars N. From this and the fact that the number of sample sentences is finite, Nowak et al. conclude that only if N is relatively small can a stable grammar emerge in a population. I.e. the population dynamics require a restrictive Universal Grammar.

2 Model design

These models have in common that they implicitly assume that every possible grammar is equally likely to become the target grammar for learning. If even the best possible learning algorithm cannot learn such a grammar, the set of allowed grammars must be restricted. There is, however, reason to believe that this assumption is not the most useful for language learning. Language learning is a very particular type of learning problem, because the outcome of the learning process at one generation is the input for the next.

The model study I present here is motivated by this observation. The model consists of an evolving population of language learners, that learn a grammar from their parents and get offspring proportional to the success in communicating with other individuals in their generation. The grammar induction procedure is fixed; it is inspired by Kirby (2000). The details of the grammatical formalism (context-free grammars) and the population structure are deliberately close to Gold (1967) and Nowak et al. (2001) respectively.

I use context-free grammars to represent the linguistic abilities. In particular, the representation is limited to grammars G where all rules are of one of the following forms: $A \mapsto t$, $A \mapsto BC$, or $A \mapsto Bt$. Since every context-free grammar can be transformed to such a grammar, the restrictions on the rule-types above do not limit the scope of languages that can be represented. They are, however, relevant for the language acquisition algorithm that will be discussed below. Note that the class of languages that the formalism can represent is unlearnable by Gold's criterion.

The language acquisition algorithm used in the model consists of three operations: (i) incorporation (extend the language, such that it includes the encountered string), (ii) compression (substitute frequent and long substrings with a nonterminal, such that the grammar becomes smaller and the language remains unchanged), (iii) generalization (equate two nonterminals, such that the grammar becomes smaller and the language larger).

3 Results

The main result is in figure 1, which shows two curves: (i) the average communicative success of agents speaking with their parents which is the measure for the *learnability* of the language (labeled "between generation C"), and (ii) the average communicative success of agents speaking with other agents of the same generation (labeled "within generation C") which gives the fitness of agents and is a measure for the grammatical variation in the population.

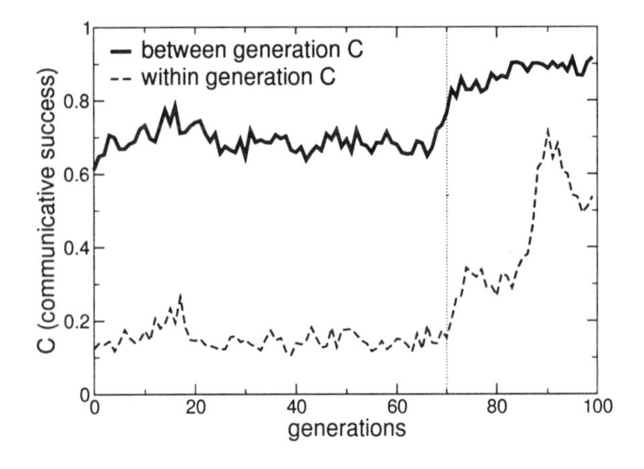

Figure 1: *Parameters are:* $V_t = \{0,1,2,3\}$, $V_{nt} = \{S,a,b,c,d,e,f\}$, *P=20, T=100, M=100,* l_0*=12*

For a long period the learning is not very successful. The between generation C is low (grammars are unlearnable), and consequently the within generation C is also low (the dynamics are below the "coherence threshold" of Nowak et al. 2001). In other words, individuals are so bad at learning that members of the population can not understand each other. Around generation 70 this situation suddenly changes. The between generation C rises, and very quickly also the within generation C rises to non-trivial levels. With always the same number of sample sentences, and with always the same grammar space, there are regions of that space where the dynamics are apparently under the coherence threshold, while there are other regions where the dynamics are above this threshold. The language has adapted to the learning algorithm, and, consequently, the coherence does not satisfy the prediction of Nowak et al. In many runs (not shown here) I have also observed 100% learning accuracy of children. The grammars in this situations are thus learnable by Gold's criterion. In some, but not all cases, these emergent grammars are recursive.

4 Discussion

I believe that these results, simple and preliminary as they may be, have some important consequences for our thinking about language acquisition. In studies like the mathematical models of Gold and Nowak et al., one derives from the properties of the learning procedure (the search procedure), fundamental constraints on the nature of the target grammar (the search space). My results, like those of Kirby (2000) and others, indicate that in *iterated learning* it is not necessary to put the (whole) explanatory burden on constraints on the search space. In my model, the target grammars are learnable, not because the used formalism imposes restrictions on the grammars, but because the targets dynamically change and – in the iteration of learners learning from learners – adapt to the used learning algorithm. In other words, neither the search space nor the search procedure directly determine which grammars "exist"; the set of target grammars at the end of the simulation is the emergent result of iterating a search process over and over again.

Isn't this Universal Grammar in disguise? Learnability is – consistent with the undisputed proof of Gold (1967) – still achieved by constraining the set of targets. However, unlike in usual *interpretations* of this proof, these constraints are not strict (some grammars are better learnable than others, allowing for an infinite "Grammar Universe"), and they are not a-priori: they are the outcome of iterated learning. The poverty of stimulus is here no longer a problem; instead, the ancestors' poverty is the solution for the child's.

References

Gold, E. M. (1967). Language identification in the limit. *Information and Control (now Information and Computation)*, 10:447–474.

Kirby, S. (2000). Syntax without natural selection. In Knight, C., Hurford, J., and Studdert-Kennedy, M., (Eds.), *The Evolutionary Emergence of Language.* Cambridge University Press.

Nowak, M. A., Komarova, N., and Niyogi, P. (2001). Evolution of universal grammar. *Science*, 291:114–118.

AUTHOR INDEX